Energy, Environment, and Sustainability

Series Editor

Avinash Kumar Agarwal, Department of Mechanical Engineering, Indian Institute of Technology Kanpur, Kanpur, Uttar Pradesh, India

This books series publishes cutting edge monographs and professional books focused on all aspects of energy and environmental sustainability, especially as it relates to energy concerns. The Series is published in partnership with the International Society for Energy, Environment, and Sustainability. The books in these series are edited or authored by top researchers and professional across the globe. The series aims at publishing state-of-the-art research and development in areas including, but not limited to:

- Renewable Energy
- Alternative Fuels
- Engines and Locomotives
- Combustion and Propulsion
- Fossil Fuels
- Carbon Capture
- Control and Automation for Energy
- Environmental Pollution
- Waste Management
- Transportation Sustainability

More information about this series at http://www.springer.com/series/15901

Sudhir P. Singh · Santosh Kumar Upadhyay ·
Ashutosh Pandey · Sunil Kumar
Editors

Molecular Approaches in Plant Biology and Environmental Challenges

Springer

Editors
Sudhir P. Singh
Center of Innovative and Applied Bioprocessing (CIAB)
Mohali, India

Santosh Kumar Upadhyay
Department of Botany
Panjab University
Chandigarh, India

Ashutosh Pandey
National Institute of Plant Genome Research (NIPGR)
New Delhi, India

Sunil Kumar
Technology Development Centre
National Environmental Engineering Research Institute
Nagpur, India

ISSN 2522-8366 ISSN 2522-8374 (electronic)
Energy, Environment, and Sustainability
ISBN 978-981-15-0689-5 ISBN 978-981-15-0690-1 (eBook)
https://doi.org/10.1007/978-981-15-0690-1

This Springer imprint is published by the registered company Springer Nature Singapore Pte Ltd.
The registered company address is: 152 Beach Road, #21-01/04 Gateway East, Singapore 189721, Singapore

Preface

Energy demand has been rising remarkably due to increasing population and urbanization. Global economy and society are significantly dependent on the energy availability because it touches every facet of human life and activities. Transportation and power generation are two major examples. Without the transportation by millions of personalized and mass transport vehicles and availability of 24×7 power, human civilization would not have reached contemporary living standards.

The International Society for Energy, Environment and Sustainability (ISEES) was founded at Indian Institute of Technology Kanpur (IIT Kanpur), India in January 2014 with an aim to spread knowledge/awareness and catalyze research activities in the fields of energy, environment, sustainability, and combustion. The Society's goal is to contribute to the development of clean, affordable, and secure energy resources and a sustainable environment for the society and to spread knowledge in the above-mentioned areas and create awareness about the environmental challenges, which the world is facing today. The unique way adopted by the society was to break the conventional silos of specializations (engineering, science, environment, agriculture, biotechnology, materials, fuels, etc.) to tackle the problems related to energy, environment, and sustainability in a holistic manner. This is quite evident by the participation of experts from all fields to resolve these issues. The ISEES is involved in various activities such as conducting workshops, seminars, and conferences in the domains of its interests. The society also recognizes the outstanding works done by the young scientists and engineers for their contributions in these fields by conferring them awards under various categories.

Third International Conference on "Sustainable Energy and Environmental Challenges" (III-SEEC) was organized under the auspices of ISEES from December 18 to 21, 2018, at Indian Institute of Technology Roorkee. This conference provided a platform for discussions between eminent scientists and engineers from various countries including India, USA, Norway, Finland, Sweden, Malaysia, Austria, Hong Kong, Bangladesh, and Australia. In this conference, eminent speakers from all over the world presented their views related to different aspects of energy, combustion, emissions, and alternative energy resource for sustainable development and cleaner environment. The conference presented five high-voltage plenary talks from globally renowned experts on topical themes, namely "The Evolution of Laser Ignition Over more than Four Decades" by Prof. Ernst Wintner, Technical University of Vienna, Austria; "Transition to Low Carbon Energy Mix for India" by Dr. Bharat Bhargava, ONGC Energy Center; "Energy Future of India" by Dr. Vijay Kumar Saraswat, Hon. Member (S&T) NITI Ayog, Government of India; "Air Quality Monitoring and Assessment in India" by Dr. Gurfan Beig, Safar; and "Managing Large Technical Institutions and Assessment Criterion for Talent Recruitment and Retention" by Prof. Ajit Chaturvedi, Director, IIT Roorkee.

The conference included 24 technical sessions on topics related to energy and environmental sustainability including 5 plenary talks, 27 keynote talks, and 15 invited talks from prominent scientists, in addition to 84 contributed talks and 50 poster presentation by students and researchers. The technical sessions in the conference included Advances in IC Engines, Solar Energy, Environmental Biotechnology, Combustion, Environmental Sustainability, Coal and Biomass Combustion/Gasification, Air and Water Pollution, Biomass to Fuels/Chemicals, Combustion/Gas Turbines/Fluid Flow/Sprays, Energy and Environmental Sustainability, Atomization and Sprays, Sustainable Transportation and Environmental Issues, New Concepts in Energy Conservation, and Waste to Wealth. One of the highlights of the conference was the Rapid-Fire Poster Sessions in (i) Engine/Fuels/Emissions, (ii) Renewable and Sustainable Energy, and (iii) Biotechnology, where 50 students participated with great enthusiasm and won many prizes in a fiercely competitive environment. 200+ participants and speakers attended this four-day conference, which also hosted Dr. Vijay Kumar Saraswat, Hon. Member (S&T) NITI Ayog, Government of India, as the chief guest for the book release ceremony, where 14 ISEES books published by Springer, Singapore, under a special dedicated series "Energy, Environment and Sustainability" were released. This was second time in a row that such significant and high-quality outcome has been achieved by any society in India. The conference concluded with a panel discussion on "Challenges, Opportunities and Directions for National Energy Security," where the panelists were Prof. Ernst Wintner, Technical University of Vienna; Prof. Vinod Garg, Central University of Punjab, Bhatinda; Prof. Avinash Kumar Agarwal, IIT Kanpur; and Dr. Michael Sauer, Boku University of Natural Resources, Austria. The panel discussion was moderated by Prof. Ashok Pandey, Chairman, ISEES. This conference laid out the roadmap for technology development, opportunities and challenges in energy, environment, and sustainability domain. All these topics are very relevant to the country and the world in present context. We acknowledge the support received from various funding agencies and organizations for the successful conduct of the Third ISEES Conference (III-SEEC), where these books germinated. We would, therefore, like to acknowledge NIT Srinagar, Uttarakhand (TEQIP) (special thanks to Prof. S. Soni, Director, NIT, UK), SERB, Government of India (special thanks to Dr. Rajeev Sharma, Secretary); UP Bioenergy Development Board, Lucknow (special thanks to Sh. P. S. Ojha), CSIR, and our publishing partner Springer (special thanks to Swati Mehershi).

The increasing global population, urbanization, and industrialization are generating a plethora of unfavorable and adverse environmental factors for the living organisms in the ecosystem. The environmental factors, such as temperature variations, drought, salinity, flood, metal concentration, are the major stresses to the plants which are negatively affecting the plant's physiology, like photosynthetic capacity, plant's growth and biomass, sustainability of the crop yield, and nutrient quantity and quality. These environmental stresses pose higher mutation rates, metabolomic changes, and epigenetic modifications leading to the genotypic and phenotypic alterations in plants. Many transcription factors have been described to play a key role in developing tolerance mechanism against a variety of environmental challenges. Secondary metabolites may not be essential for growth and development of plants; however, they play important role in plant survival under certain sets of environmental conditions, such as biotic and abiotic stresses. Therefore, these compounds may lead to many agronomic traits including quality, yield, resistance, and stress tolerance. Further, secondary metabolites constitute important dietary components as well as phytomedicines. Cellular events, such as autophagy and apoptosis, play an important role in conferring stress tolerance in plants. Numerous defense-related proteins, including receptor-like kinases, facilitate stress signaling for adaptation under various abiotic and biotic stress conditions in plants. Further, the environmental stress increases reactive oxygen species (ROS), which are useful in signaling at low concentration but toxic at a higher level. Several protein families, such as superoxide dismutase, ascorbate peroxidase, glutathione peroxidases, are known to perform the homeostasis of ROS in the plant cell to maintain optimal concentration. This book intends to compile a comprehensive knowledge about the plant molecular approaches to take up the environmental challenges for productivity and sustainability of agricultural crops. The book will also cover the natural tolerance mechanism, which the plants adopt to cope with the adverse environmental factors, as well as the novel molecular strategies to engineer the plants in human interest.

The editors would like to express their sincere gratitude to large number of authors from all over the world for submitting their high-quality work in a timely manner and revising it appropriately at a short notice. We would like to express our special thanks to Dr. Martin Sagasser, Dr. Rakesh K. Upadhyay, Dr. Prashant Mishra, Dr. Smita Kumari, Dr. Rahul Singh, Dr. Jitendra Kumar, Dr. Lokesh Narnoliya, Dr. Jyoti Singh, and all others, who reviewed various chapters of this monograph and provided their valuable suggestions to improve the manuscripts.

Mohali, India | Sudhir P. Singh
Chandigarh, India | Santosh Kumar Upadhyay
New Delhi, India | Ashutosh Pandey
Nagpur, India | Sunil Kumar

Contents

Editors and Contributors

About the Editors

Dr. Sudhir P. Singh is currently Scientist at the Center of Innovative and Applied Bioprocessing, Mohali, India. He has been working in the area of molecular biology and biotechnology for more than a decade. He developed novel stringently regulated gene expression systems for inducing male sterility and fertility restoration in plants. He revealed tissue-specific distribution pattern of mineral nutrients in grains and its probable impacts on mineral bioavailability. He provided the first molecular insights into fruit and seed development in custard apple and litchi by analyzing de novo transcriptome of developing fruits and ovules of contrasting genotypes. His de novo transcriptomic study provided the first molecular details of the specialized metabolic pathways in rose-scented geranium. Currently, his main focus of research is gene mining and biocatalyst engineering for development of approaches for transformation of agro-industrial residues and under- or un-utilized side-stream biomass into value-added bioproducts. His group has discovered novel genes for D-allulose 3-epimerase, amylosucrase, xylanase, and cellulase enzymes from the extreme habitat metagenomes. Dr. Singh has over 50 scientific publications and 11 patents (10 filed, and 01 granted) to his credit. He has been conferred International Bioprocessing Association-Young Scientist Award-2017, School of

Biosciences-Madurai Kamraj University (SBS-MKU) Genomics Award-2017, and Professor Hira Lal Chakravarty Award, Indian Science Congress Association (ISCA, DST)- 2018. He has also been the recipient of Early Career Research Award, SERB, 2016.

Dr. Santosh Kumar Upadhyay is currently Assistant Professor in the Department of Botany, Panjab University, Chandigarh, India. He has been working in the field of Plant Biotechnology for more than 12 years. His present research focuses in the area of functional genomics. He is involved in the characterization of various insect toxic proteins from plant biodiversity, and defense and stress signaling genes in bread wheat. He has played significant role in identification of whitefly toxic proteins from lower plants and involved in the development of whitefly resistant cotton plants. His research group at PU has characterized numerous important gene families related to the abiotic and biotic stress tolerance and signaling in bread wheat. He has also established the method for genome editing in bread wheat using CRISPR-Cas system. His research contribution led publication of more than 40 research papers in leading journals of international repute. Further, there are more than five national and international patents and five book chapters in his credit. In recognition of his strong research record he has been awarded NAAS Young scientist award 2017-18, NAAS-Associate (2018), INSA Medal for Young Scientist (2013), NASI- Young Scientist Platinum Jubilee Award (2012) and Altech Young Scientist Award (2011). He has also been the recipient of the prestigious DST-INSPIRE Faculty Fellowship (2012), and Early Career Research Award, SERB (2016) from Ministry of Science and Technology, Government of India.

Dr. Ashutosh Pandey is currently Staff Scientist at National Institute of Plant Genome Research, New Delhi, India. His research area is metabolic engineering of flavonoid biosynthesis in plants. He has overexpressed AtMYB12 and AtMYB11 in tomato, which led large-scale differential modulation in transcriptome and flavonoid content in leaf and fruit tissues, and it conferred insect attack tolerance. Further, he coexpressed Arabidopsis transcription factor, AtMYB12, and soybean isoflavone synthase, GmIFS1, genes in tobacco, which enhanced biosynthesis of isoflavones and flavonols resulting in osteoprotective activity. He has characterized isoflavone biosynthesis in Psoralea corylifolia. He developed callus culture for large scale production of rutin with biopesticidal potential. He has performed genome-wide analysis of carotenoid biosynthesis pathway in banana which facilitates metabolic engineering for enhance provitamin A biosynthesis and bioavailability in biofortified banana cell lines. Further, he achieved several folds increase in β-carotene content by overexpressing phytoene synthase1 of Nendran (NEN-PSY1) in banana cell lines. He has been conferred membership of Indian National Young Academy of Sciences (INYAS) in 2018 (Plant Sciences), INSA Medal for Young Scientist (2017), and the Alexander von Humboldt Post Doctoral Research Fellowship (2016) from Alexander von Humboldt Foundation, Germany.

Dr. Sunil Kumar is currently Principal Scientist and Head, Technology Development Centre, Council of Scientific and Industrial Research-National Environmental Engineering Research Institute (CSIR-NEERI), Nagpur, India. Dr. Kumar has carried out several national and internationally sponsored projects. Most of the projects are related to solid and hazardous waste management and environmental impact and risk assessment related aspects. He has authored 117 research articles, 3 books, edited 8 book volumes, 25 book chapters, 35 technical reports, and 25 conference presentations/ publications. He has been conferred CSIR-NEERI

Foundation Day Award (2009), CSIR-NEERI Outstanding Scientist Award (2015), and Alexander von Humboldt-Stiftung Jean-Paul-Str.12 D-53173 Bonn, Germany (2018).

Contributors

Aditya Vikram Agarwal DST—Centre for Policy Research, Babasaheb Bhimrao Ambedkar University, Lucknow, India

Parul Agarwal Council of Scientific and Industrial Research—National Botanical Research Institute (CSIR-NBRI), Lucknow, Uttar Pradesh, India

Anshu Alok Department of Biotechnology, Panjab University, Chandigarh, India

S. Anil Kumar Department of Biotechnology, Vignan's Foundation for Science, Technology and Research, Vadlamudi, Guntur, Andhra Pradesh, India

Surjeet Kumar Arya Council of Scientific and Industrial Research—National Botanical Research Institute (CSIR-NBRI), Lucknow, Uttar Pradesh, India;
Academy of Scientific and Innovative Research (AcSIR), Ghaziabad, India

Monika Bhuria Department of Biotechnology, CSIR—Institute of Himalayan Bioresource Technology, Palampur, India

Subhankar Biswas Department of Botany, Centre of Advanced Study in Botany, Institute of Science, Banaras Hindu University, Varanasi, India

Amit Choudhary Council of Scientific and Industrial Research—National Botanical Research Institute (CSIR-NBRI), Lucknow, Uttar Pradesh, India;
Academy of Scientific and Innovative Research (AcSIR), Ghaziabad, Uttar Pradesh, India

Aksar Ali Chowdhary Department of Botany, Central University of Jammu, Samba, India

Sameer Dixit Council of Scientific and Industrial Research—National Botanical Research Institute (CSIR-NBRI), Lucknow, Uttar Pradesh, India;
Department of Biology, University of Western Ontario, London, ON, Canada

Umar Gani Plant Biotechnology Division, CSIR—Indian Institute of Integrative Medicine, Jammu, India;
Academy of Scientific and Innovative Research (AcSIR), New Delhi, India

Deepak Ganjewala Amity Institute of Biotechnology, Amity University Uttar Pradesh, Noida, Uttar Pradesh, India

Devina Ghai Department of Botany, Panjab University, Chandigarh, India

Parul Goel Department of Biotechnology, CSIR—Institute of Himalayan Bioresource Technology, Palampur, India

Parul Gupta Department of Botany and Plant Pathology, Oregon State University, Corvallis, USA

P. Hima Kumari Department of Genetics, Osmania University, Hyderabad, India

Shweta Jha Plant Functional Genomics Lab, Biotechnology Unit, Department of Botany (UGC-CAS), J.N.V. University, Jodhpur, Rajasthan, India

Shourabh Joshi Department of Plant Biotechnology, Agriculture University, Jodhpur, India

Gurminder Kaur Faculty of Bioscience, Institute of Bioscience and Technology, Shri Ramswaroop Memorial University, Barabanki, Uttar Pradesh, India

S. Krupanidhi Department of Biotechnology, Vignan's Foundation for Science, Technology and Research, Vadlamudi, Guntur, Andhra Pradesh, India

Sunil Kumar CSIR—National Environmental Engineering Research Institute, Nagpur, India

Vinod Kumar Biochemistry Laboratory, Forage Section, Department of Genetics and Plant Breeding, CCS Haryana Agricultural University, Hisar, India

Maridul Kundan Plant Biotechnology Division, CSIR—Indian Institute of Integrative Medicine, Jammu, India;
Academy of Scientific and Innovative Research (AcSIR), New Delhi, India

Charu Lata Council of Scientific and Industrial Research—National Botanical Research Institute (CSIR-NBRI), Lucknow, Uttar Pradesh, India;
CSIR—National Institute of Science Communication and Information Resources, New Delhi, India

Shakti Mehrotra Department of Biotechnology, Institute of Engineering and Technology, Lucknow, India

Sonal Mishra School of Biotechnology, University of Jammu, Jammu, India

Yogesh Mishra Department of Botany, Centre of Advanced Study in Botany, Institute of Science, Banaras Hindu University, Varanasi, India

Prashant Misra Plant Biotechnology Division, CSIR—Indian Institute of Integrative Medicine, Jammu, India;
Academy of Scientific and Innovative Research (AcSIR), New Delhi, India

Akshay Nag Department of Biotechnology, Panjab University, Chandigarh, India

Abhishek Kumar Nautiyal Plant Biotechnology Division, CSIR—Indian Institute of Integrative Medicine, Jammu, India;
Academy of Scientific and Innovative Research (AcSIR), New Delhi, India

Ashutosh Pandey National Institute of Plant Genome Research, New Delhi, India

Naveen C. Pant Department of Plant Biochemistry, SKN Agriculture University, Jobner, India

Sumya Pathak DST—Centre for Policy Research, Babasaheb Bhimrao Ambedkar University, Lucknow, India

Boas Pucker Genetics and Genomics of Plants, Center for Biotechnology (CeBiTec), Bielefeld University, Bielefeld, Germany

Punesh Sangwan Department of Biochemistry, Eternal University, Baru Sahib, India

Abhishake Saxena Department of Biotechnology, Eternal University, Baru Sahib, India

Gauri Saxena Department of Botany, University of Lucknow, Lucknow, Uttar Pradesh, India

Hanna Marie Schilbert Genetics and Genomics of Plants, Center for Biotechnology (CeBiTec), Bielefeld University, Bielefeld, Germany

Jaspreet K. Sembi Department of Botany, Panjab University, Chandigarh, India

Tilak Raj Sharma ICAR—Indian Institute of Agricultural Biotechnology, Ranchi, India

Akanchha Shukla Department of Biology, University of Western Ontario, London, ON, Canada

Shumayla Department of Botany and Department of Biotechnology, Panjab University, Chandigarh, India

Anil Kumar Singh Department of Biotechnology, CSIR—Institute of Himalayan Bioresource Technology, Palampur, India;
ICAR-Indian Institute of Agricultural Biotechnology, Ranchi, India

Dharmendra Singh Department of Biotechnology, Eternal University, Baru Sahib, India

Kashmir Singh Department of Biotechnology, Panjab University, Chandigarh, India

Sanchita Singh Department of Botany, University of Lucknow, Lucknow, Uttar Pradesh, India;
Council of Scientific and Industrial Research—National Botanical Research Institute (CSIR-NBRI), Lucknow, Uttar Pradesh, India

Sudhir P. Singh Center of Innovative and Applied Bioprocessing, Mohali, India

Sunil Kumar Singh Department of Biology, University of North Dakota, Grand Forks, USA

Ragini Sinha ICAR—Indian Institute of Agricultural Biotechnology, Ranchi, India

Akanksha Srivastava Department of Botany, Centre of Advanced Study in Botany, Institute of Science, Banaras Hindu University, Varanasi, India

Nidhi Srivastava Amity Institute of Biotechnology, Amity University Uttar Pradesh, Noida, Uttar Pradesh, India

Vikas Srivastava Department of Botany, Central University of Jammu, Samba, India

Sushmita Council of Scientific and Industrial Research—National BotanicalResearch Institute (CSIR-NBRI), Lucknow, Uttar Pradesh, India;
Academy of Scientific and Innovative Research (AcSIR), Ghaziabad, India

Manish Tiwari Council of Scientific and Industrial Research—National Botanical Research Institute (CSIR-NBRI), Lucknow, Uttar Pradesh, India

Shalini Tiwari Council of Scientific and Industrial Research—National Botanical Research Institute (CSIR-NBRI), Lucknow, Uttar Pradesh, India

Prabodh Kumar Trivedi Council of Scientific and Industrial Research—National Botanical Research Institute (CSIR-NBRI), Lucknow, Uttar Pradesh, India

Shivi Tyagi Department of Botany, Panjab University, Chandigarh, India

Santosh Kumar Upadhyay Department of Botany, Panjab University, Chandigarh, India

K. Venkatesh Department of Genetics, Osmania University, Hyderabad, India

Jagdeep Verma Department of Botany, Government College, Rajgarh, Sirmaur, Himachal Pradesh, India

Praveen Chandra Verma Council of Scientific and Industrial Research—National Botanical Research Institute (CSIR-NBRI), Lucknow, Uttar Pradesh, India;
Academy of Scientific and Innovative Research (AcSIR), Ghaziabad, Uttar Pradesh, India

Ajar Nath Yadav Department of Biotechnology, Eternal University, Baru Sahib, India

Sandhya Yadav Department of Botany, Centre of Advanced Study in Botany, Institute of Science, Banaras Hindu University, Varanasi, India

Chapter 1
Molecular Approaches in Plant Biology and Environmental Challenges

Sudhir P. Singh, Santosh Kumar Upadhyay, Ashutosh Pandey and Sunil Kumar

Abstract Global development has generated a plethora of changes in the ecosystem, which directly or indirectly affect living organisms in the biosphere. Plants are crucial to sustaining life on this planet. Therefore, it is desirable to understand manifold biological processes in plants at molecular levels in response to the environmental challenges. The development in molecular technologies worldwide provides a great hope to evolve approaches for augmenting stress tolerance in plants, keeping in mind the extreme environmental conditions in different parts of the world. Research strategies need to be designed in executing the favorable genetic manipulations in the crucial molecular components of the plants such as transcription factors, superoxide dismutases, receptor kinases, histone acetyltransferases and histone deacetylases, signaling pathways, secondary metabolic pathways, etc.

Keywords Environmental challenges · Transcription factors · Stress tolerance · Histone acetyltransferases · Superoxide dismutases · CRISPR-Cas9

Plants are sessile organisms, and they face enormous environmental challenges, related to abiotic and biotic stresses, which can induce numerous molecular and physiological changes in the biology of a plant. Abiotic stresses such as heat, drought, salinity, UV, metal stress, etc., and biotic stresses like fungal, bacterial and viral pathogens, and insect pests are the major challenges to the various grain and horticultural crops, negatively influencing the yield and quality of agricultural produce

S. P. Singh (✉)
Center of Innovative and Applied Bioprocessing, Mohali, India
e-mail: sudhirsingh@ciab.res.in

S. K. Upadhyay
Department of Botany, Panjab University, Chandigarh, India

A. Pandey
National Institute of Plant Genome Research, New Delhi, India

S. Kumar
CSIR—National Environmental Engineering Research Institute, Nagpur, India

S. P. Singh et al. (eds.), *Molecular Approaches in Plant Biology and Environmental Challenges*, Energy, Environment, and Sustainability,
https://doi.org/10.1007/978-981-15-0690-1_1

(Mittler and Blumwald 2010). However, the environmental challenges in the biosphere should be investigated, keeping in mind the evolutionary processes like adaptation, dispersal and genetic drift, which are crucial in shaping the ecological habitat for organisms and the species.

Plants have evolved various molecular mechanisms to resist these environmental challenges. They have a system of communication through numerous signaling molecules and molecular processes that function in response to various abiotic and biotic inducers (Saijo and Loo 2019). The emitter plants produce signals in the form of common or specific molecules that are perceived by the plants in the vicinity called as receiver plants. The receiver plants then enable themselves for a better adaptation to the changing conditions by producing the relevant proteins. The communication among plants can occur either with the help of a biological or chemical mediator or sometimes without any mediator. Usually, the emitters could not receive benefits in this communication process, but it provides significant benefits to the other plant species. This type of communication among plants is beyond the boundary of species, genus, and family (Heil and Karban 2010). This communication phenomenon in plants can be very useful in future crop designing.

Activation of a cell signaling pathway is one of the most efficient and primary responses of plants and the abiotic and biotic stresses that ultimately produce cellular responses like the expression of stress proteins and/or gathering of compatible solutes to cope up with the changing conditions. Receptor-like kinases (RLKs) are important and stress-specific defense-related proteins that play vital functions stress conditions by activating the signaling cascade. They are not only involved in stress signaling but also play significant roles in plants growth and developments (Liang and Zhou 2018). Structurally, plants have an extracellular domain to perceive the signal, a transmembrane domain to anchor the protein into the membrane, and a cytoplasmic serine/threonine kinase domain to stimulate the signaling for immunity development in plants. Till date, 15 different classes of RLKs are identified on the basis of variation in the extracellular domain in plants. These extracellular domains are diverse in their structural organization as well as signal perception. The detail about structural and functional divergence is explained inside the chapter.

Further, variation in environmental conditions is responsible for overproduction of reactive oxygen species (ROS) such as superoxide ion ($O_2^{\cdot-}$), hydrogen peroxide (H_2O_2), singlet oxygen ($^1O_2$), etc. in plants. The limited production of ROS is useful to the plants as some of them act as signaling molecules, but a higher amount of ROS produces a negative impact on plants growth and development. To overcome the effect of ROS, various enzymatic and non-enzymatic antioxidant systems have been involved in plants. The superoxide dismutases (SODs) are a major class of antioxidant proteins that provide the first line of defense. SODs are metalloenzymes that perform dismutation of superoxide radicals into molecular oxygen and hydrogen peroxide. There are groups of SODs have been identified in plants based on metal cofactor, i.e., Cu/ZnSODs, FeSODs, and MnSODs. They are known to play vital roles in plant development and combating oxidative stress (Alscher et al. 2002).

Researches on transcription factors (TFs) have revealed the involvement of several TFs belonging to the families such as NAC, MYB, WRKY, AP2/EREBP, bHLH, and bZIP, etc., in conferring multiple stress tolerance to plants (Lan Thi Hoang et al. 2017). The gene regulatory networks is a complex molecular mechanism to manage the cellular physiology in response to biotic and abiotic stresses. Furthermore, the potential of specific transcription factors in developing stress tolerance in plants has been discussed. Heat shock transcription factors (Hsfs), and histone acetyltransferases (HATs) and histone deacetylases (HDACs) also perform vital functions for adaptation in the presence of abiotic stresses by different mechanisms (Priya et al. 2019; Luo et al. 2017). HsFs exhibited stress response by regulating the expression of heat shock proteins (Hsps). However, HATs and HDACs are involved in epigenetic modifications by controlling the level of histone acetylation. There are four different HATs families, i.e., CBP/p300, GNAT, MYST, and $TAF_{II}250$, in which GNAT/MYST is associated with UV-B induced DNA damage repair and various other stress responses. Apart from the staple food crops like rice and wheat, this book has also covered the molecular interventions in the flowering plant like orchid, and lower plants like bryophytes. These plants can be used as marker plants to address various environmental challenges, keeping in mind the global climate dynamics. The adaptation mechanism of the lower plants can be a role model for molecular engineering of the environmental robustness in economically important crops.

Recent progress in sequencing technologies and bioinformatics aspects has generated enormous genomic resource, facilitating the engineering and editing of plant genetic material. Numerous genetic engineering strategies have been used for the development of superior varieties concerning stress tolerance and yield. Now a days, targeted genome editing using artificial nucleases such as Zinc-finger nucleases (ZFNs), transcription activator-like effector nucleases (TALENs), and clustered regularly interspaced short palindromic repeat (CRISPR)/CRISPR-associated protein (Cas) have been used in various crop plants like rice, wheat, maize, tomato, etc. to successfully edit or modulate the target genes as crop improvement strategies. These methods are useful in developing improved plant lines by adding important traits or by removing undesirable traits (Jaganathan et al. 2018). Herein, the recent development and application of the CRISPR/Cas9 system and its advancement in comparison to the ZFNs and TALENs, and its application in the agricultural crops like rice and tomato have been discussed. These technologies have paved ways to engineer biosynthesis of secondary metabolites, which play essential roles in adaptation to changing the environment, management of abiotic stresses, including drought, temperature, salinity, etc. Several specialized secondary metabolites such as isoprenoids, carotenoids, and flavonoids have been established as important stress protectant biomolecules, biosynthesis of which should be taken into consideration for the development of designer crops to face the environmental challenges.

This monograph presents the dynamics of molecular responses, and molecular approaches to combat the environmental challenges in plants. Specific topics covered in the monograph include:

- Molecular Approaches in Plant Biology and Environmental Challenges
- Promising Transcription Factors for Salt and Drought Tolerance in Plants
- Role of Superoxide Dismutases (SODs) in Stress Tolerance in Plants

- Receptor-like Kinases and Environmental Stress in Plants
- Role of Histone Acetyltransferases in Plant Abiotic Stress
- Function of Plant Heat Shock Transcription Factors in Abiotic Stress
- Mode of Communication Among Plants During Environmental Stress
- Molecular Approaches for Combating Multiple Abiotic Stresses in Crops of Arid and Semi-arid Region
- Applications of Landscape Genetics to Study the Effect of Varying Landscapes and Environmental Challenges in Plant Populations
- Arsenic in Rice Grain: Role of Transporters in Arsenic Accumulation
- Metabolic Engineering of Stress Protectant Secondary Metabolites to Confer Abiotic Stress Tolerance in Plants
- An update on molecular strategies of transgenic rice tolerance to abiotic stresses
- An Update on the Applications of CRISPR/Cas9 Technology in Tomato
- Transgenic Approaches for Enhancement of Salinity Stress Tolerance in Plants
- Genome Engineering in Rice: Applications, Advancements and Future Perspectives
- Secondary Metabolite Pathways in Medicinal plants: Approaches in Reconstruction and Analysis
- Molecular Biology of Glandular Trichomes and Their Functions in Environmental Stresses
- Gene Regulatory Networks: Current Updates and Applications in Plant Biology
- Genomics and Transcriptomics Advance in Plant Sciences
- Molecular Interventions to Ameliorate Environmental Stresses in Orchids
- Development of Bryophytes as A New Model System to Understand the Phenomenon of Terrestrialization with Environmental Changes
- Role of Endosymbionts in Nutritional Uptake of Sap Sucking Insects.

References

Alscher RG, Erturk N, Heath LS (2002) Role of superoxide dismutases (SODs) in controlling oxidative stress in plants. J Exp Bot 53(372):1331–1341

Heil M, Karban R (2010) Explaining evolution of plant communication by airborne signals. Trends Ecol Evol 25(3):137–144

Jaganathan D, Ramasamy K, Sellamuthu G, Jayabalan S, Venkataraman G (2018) CRISPR for crop improvement: an update review. Front Plant Sci 9

Lan Thi Hoang X, Du Nhi NH, Binh Anh Thu N, Phuong Thao N, Phan Tran LS (2017) Transcription factors and their roles in signal transduction in plants under abiotic stresses. Curr Genomics 18(6):483–497

Liang X, Zhou JM (2018) Receptor-like cytoplasmic kinases: central players in plant receptor kinase–mediated signaling. Annu Rev Plant Biol 69:267–299

Luo M, Cheng K, Xu Y, Yang S, Wu K (2017) Plant responses to abiotic stress regulated by histone deacetylases. Front Plant Sci 8:2147

Mittler R, Blumwald E (2010) Genetic engineering for modern agriculture: challenges and perspectives. Annu Rev Plant Biol 61:443–462. https://doi.org/10.1146/annurev-arplant-042809-112116

Priya M, Dhanker OP, Siddique KH, HanumanthaRao B, Nair RM et al (2019) Drought and heat stress-related proteins: an update about their functional relevance in imparting stress tolerance in agricultural crops. Theor Appl Genet 1–32

Saijo Y, Loo EP (2019) Plant immunity in signal integration between biotic and abiotic stress responses. New Phytol. https://doi.org/10.1111/nph.15989

Chapter 2
Promising Transcription Factors for Salt and Drought Tolerance in Plants

Parul Goel, Monika Bhuria, Ragini Sinha, Tilak Raj Sharma and Anil Kumar Singh

Abstract Drought and salinity are the most common environmental stresses that cause major loss to crop productivity. To combat such stresses and to satisfy the world food demand, it is desired to develop stress tolerant crop plants. To date, transgenic approach, mainly by overexpressing stress-responsive transcription factors (TFs) is highly appreciated. TFs can activate large set of genes that participate towards multiple stress response. TFs are thus considered as potential candidates for developing stress tolerant plants. Most of the TFs, which play important role in abiotic stress tolerance in plants, fall into major TF families namely AP2/EREBP, NAC, MYB/MYC, WRKY, bZIP, bHLH and ZFP etc. In the current chapter, we have discussed the role of major TF families in enhancing salt and/or drought tolerance in plants. Additionally, we have also discussed the importance of multiple-stress responsive promoters in engineering plants with sustainable abiotic stress tolerance.

Keywords Abiotic stress · Drought stress · Salinity stress · Stress signaling · Stress tolerant plants · Transcription factor · Transgenic plants

2.1 Introduction

Being sessile, plants have to continuously face several biotic and abiotic stresses throughout their life cycle. Abiotic stresses such as extreme temperature (high or low), salinity, drought, UV-radiation, flooding and submergence are detrimental to plants as they negatively influence their survival and yield. In field conditions, plants may encounter a combination of these stresses together, that leads to extensive losses in agriculture productivity (Mittler and Blumwald 2010). It has been estimated that abiotic stresses cause 50–70% loss to the world crop productivity (Wang et al. 2003).

P. Goel · M. Bhuria · A. K. Singh
Department of Biotechnology, CSIR—Institute of Himalayan Bioresource Technology, Palampur 176061, India

R. Sinha · T. R. Sharma · A. K. Singh (✉)
ICAR—Indian Institute of Agricultural Biotechnology, Ranchi 834010, India
e-mail: anils13@gmail.com

S. P. Singh et al. (eds.), *Molecular Approaches in Plant Biology and Environmental Challenges*, Energy, Environment, and Sustainability,
https://doi.org/10.1007/978-981-15-0690-1_2

It is predicted that, the world population will reach 9 billion by 2050; and to satisfy their food demand, agriculture productivity should be enhanced approximately by 70% (Bruinsma 2009).

Among abiotic stresses, salinity and drought are considered as major abiotic stresses, which commonly trigger osmotic and oxidative stresses in plants (Reynolds and Tuberosa 2008; Landi et al. 2017). In addition, salinity stress also leads to ionic stress due to high proportion of Na^+, Cl^-, Ca^{2+} and Mg^{2+} ions (Hasegawa et al. 2000). Salinity stress is one of the major abiotic stresses affecting more than 45 million hectare of irrigated land, worldwide (Munns and Tester 2008). In comparison to the salt stress, the consequence of drought stress is more widespread and devastating (Boyer 1982). Drought stress is the condition of low water availability in the soil. In general, plants experiences drought stress, when soil water potential ranges −0.5 to − 1.5 megapascal (MPa) (Onaga and Wydra 2016). In 21st century, approximately 30% of the lands will face extreme drought condition due to climate changes (Burke et al. 2006). Several factors such as low rainfall; high evaporation rate and low water retention capacity of soil contribute to drought stress in crops plants (Wery et al. 1994). However, plant response towards salinity and drought is quite similar as both these stresses lead to the osmotic stress in plants and also trigger ABA accumulation that induces several adaptive responses.

The abiotic stress tolerance is a highly complex trait in plants, and governed by various mechanisms at physiological, biochemical and molecular levels. However, the advancements in various omics (transcriptomics, proteomics and metabolomics) approaches has made feasible to understand these complex mechanisms, comprehensibly. In order to cope up with a stress or combination of stresses, plants have evolved various strategies. Better understanding of these strategies and availability of natural genetic variability within crop species have provided opportunities to introgress desirable traits in important crops through breeding approaches. Traditional breeding approaches have been extremely successful in developing various crop varieties with desirable traits. However, development of abiotic stress tolerant crops through breeding has achieved low success, due to low selection efficiency using agronomic characters, complexity of phenotyping methods and limited availability of natural genetic resource with inherent stress tolerance trait (Richards 1996; Singh et al. 2008). Transgenic approach has emerged as an alternative to breeding with several advantages, such as, transgenic approach is faster than conventional breeding; it can transfer desirable gene isolated from any organism; the expression of transgene can be controlled spatio-temporally. During recent past, availability of whole genome sequence of several organisms has made it possible to identify novel genes associated with stress response and deploy them in economically important plants. However, abiotic stress tolerance in plants is a multigenic trait, thus, targeting a single gene may not be very efficient in imparting desired level of stress tolerance. Therefore, targeting a transcription factor (TF) gene that can regulate expression of genes involved in multiple stress responsive pathways seems an interesting alternative. Functional characterization of several members of TF families such as NAC, MYB, WRKY, AP2/EREBP, bHLH and bZIP etc., has revealed their role in multiple stress tolerance. In the present chapter, we firstly discussed the effect of salinity and drought on plant growth and development and how plants respond to these stresses. Secondly, we discussed the potential of transcription factor genes in developing salt and drought tolerant transgenic plants.

2.2 Plant Growth During Drought and/or Salinity Stress

The intensity and the duration of stresses are the two major factors that determine the degree of damage on plant growth and development under unfavorable conditions (Dolferus 2014). Plants can utilize escape, tolerance and avoidance strategies in order to counter the abiotic stress conditions (Touchette et al. 2009). On the basis of tolerance or sensitivity towards salt stress, plants are commonly categorized as glycophytes and halophytes (Flowers et al. 1977). Glycophytes are salt sensitive plants that cannot tolerate high concentration of salts. Unfortunately, most of the agricultural crops are glycophytes. The halophytes can grow and complete their life cycle under relatively high concentration of salt. Halophytes can be further classified as euhalophytes (true halophytes), pseudohalophytes (salt-avoider) and crinohalophytes (salt excretors). Consequences of drought and salinity stress on plant growth are depicted in Fig. 2.1. Inhibition of plant growth during salt stress is a consequence of osmotic stress, ionic stress and nutritional imbalance. Two-phase effect of salinity is observed on plant growth (Munns and Termaat 1986). In first phase (osmotic phase), reduction in soil water potential leads to stomatal closure and inhibition of leaf expansion. The second phase (ionic phase) leads to premature senescence resulting in cell death due to increases cytotoxic ion level. In general, plants face drought stress, when transpiration rate exceeds the root water absorption rate. In case of drought stress, reduced plant growth is mainly due to loss of turgor pressure that impairs cell elongation and

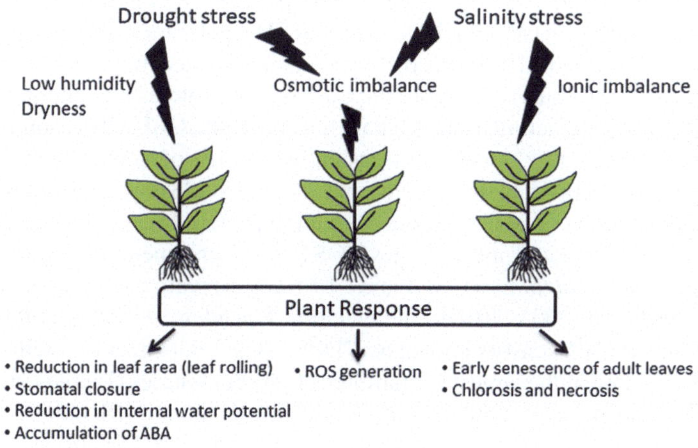

Fig. 2.1 Consequence of drought and salinity stress on plant growth and development

expansion (Farooq et al. 2009). Shoot growth is affected more under both the stresses, while, root elongation in search of groundwater could be observed in plants facing drought stress (Brunner et al. 2015). Whereas, reduced leaf size, early senescence and abscission of older leaves are observed in above ground parts (Forni et al. 2017), similarly salinity stress reduces shoot growth more severely as compared to the root growth (De Oliveira et al. 2013).

2.3 Plant Response Towards Stress

Plant response towards abiotic stresses is highly complex that involves modulation in expression of several genes and also diverse changes at physiological and biochemical levels. The following sub-section describes the mechanism of signal perception and also the changes that take place in plants upon stress exposure.

2.3.1 Stress Perception and Signal Transduction

Plant has to first perceive the outside stress signal by means of stress sensors or receptors localized on the cell surface. These sensors include histidine kinases, receptor like kinases and G-protein coupled receptors (Fig. 2.2). The signal perception is followed by signal transduction that leads to the generation of second messenger molecules mainly Ca^{2+}, ROS, cyclic nucleotides and inositol phosphate (Jain et al. 2018). These messenger molecules activate various signal transduction pathways that ultimately phosphorylate and activate transcription factors regulating the expression of several downstream genes involved in plant stress adaptation. Three most common signal transduction pathways are (i) calcium dependent SOS pathway; (ii) calcium dependent signaling; (iii) MAP Kinase pathway. The calcium dependent SOS pathway is the first identified CBL-CIPK pathway that is activated during ionic stress. The calcineurin B-like proteins (CBLs) are the Ca^{2+} sensor proteins that interact and activate CIPKs (CBL-interacting protein kinase), forming the CBL-CIPK complex. As a consequence of stress, Ca^{2+} level rises transiently, which is perceived by SOS3/CBL4 (Kolukisaoglu et al. 2004). In the presence of Ca^{2+}, the SOS3 physically interacts with SOS2/CIPK24, which encodes a serine/threonine protein kinase (Gao et al. 2008). The SOS2 phosphorylates and activates SOS1 in plasma membrane that acts as a Na^+/H^+ antiporter. *Arabidopsis* CBL1 and CBL5 act as positive regulators of drought stress (Cheong et al. 2003, 2010). The calcium dependent signaling pathway activates specialized Ca^{2+} sensors known as CDPKs (Ca^{2+} dependent protein kinases). The CDPKs are one of the largest subfamilies of protein kinases (Ludwig et al. 2004) and considered as positive regulators of abiotic stress tolerance (Schulz et al. 2013). The CDPKs have calmodulin Ca^{2+} binding domain and a Ser/Thr kinase domain for transmitting calcium signals. CDPKs are activated by Ca^{2+} binding and then transfer the signal via phosphorylation of downstream target genes. The downstream targets

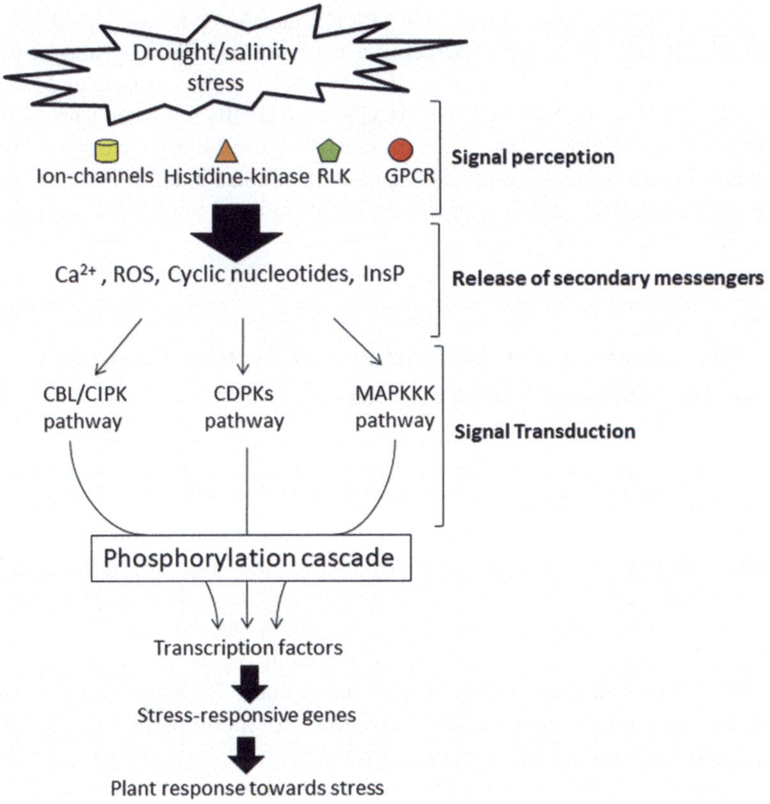

Fig. 2.2 Mechanism of abiotic stress signaling in plants Perceiving stress signal from the outside environment is the very first step initiated by receptors, such as ion channels, histidine kinases, receptor like kinase (RLK) and G-protein coupled receptors (GPCR). Signal perception is then followed by release of secondary messengers, like Ca^{2+}, ROS, cyclic nucleotides and inositol phosphate (InsP). These secondary messengers in turn activate major signal transduction pathways namely CBL-CIPK, CDPK, and MAPK, which lead to phosphorylation and activation of TFs These TFs bind to the promoter regions of stress-responsive genes and regulate their expression to bring about morphological, physiological, biochemical and molecular response

of CDPKs are ABA response factors, ion channels or transporters. The TFs activated via Ca^{2+} mediated signaling pathways are reported to contain ABA-responsive element (ABRE; CACGTG[T/C/G]) and its coupling element ([C/A]ACGCG[T/C/G]) in their upstream promoter regions (Kaplan et al. 2006). Transcriptomic study of Arabidopsis seedlings, subjected to specific $[Ca^{2+}]$cyt transients has categorized TFs families mainly MYB, bZIP, bHLH and zinc finger protein as early Ca^{2+}-responsive genes (Kaplan et al. 2006). Another signaling pathway activated mainly in response to osmotic stress is Mitogen-activated protein kinase (MAPK) pathway (Sinha et al. 2011). ROS are considered as central players in activating this pathway. MAPK pathway is comprised of three-components, namely a MAPK kinase kinase kinase

(MAPKKK), a MAPK kinase kinase (MAPKK), and MAP kinase (MAPK). Activation of MAPKKK upon signal perception activates MAPKK by activating two serine or threonine residues. The activated MAPKK then phosphorylates MAPK on its threonine and tyrosine residues. The MAPK then finally activates various downstream genes. Two MAPK pathways, namely MPK4 and MPK6 have been found to be involved in salinity stress response in plants (Ichimura et al. 2000). TFs mainly MYB/MYC and WRKY are known to be phosphorylated via MAPK pathway (Li et al. 2017).

2.3.2 Physiological, Biochemical and Molecular Responses to Drought and Salinity Stress

In plants, drought stress leads to water deficit condition that trigger reduction in cell volume and turgor pressure, which ultimately reduce the cell wall extensibility (De Oliveira et al. 2013). Drought stress also reduces leaf water potential and stomatal conductance. Plant photosynthetic efficiency also declines due to decrease in Rubisco activity (Bota et al. 2004). Decline in net photosynthetic rate, chlorophyll content and reduction in water potential are also observed in plants facing salinity stress (Forni et al. 2017). During salt and drought stress, plants also produce wide spectrum of metabolites that include amino acids, sugars and amines. The most important amino acid, accumulated under such stresses is proline. As a compatible solute, proline protects plants via osmotic adjustment and ROS detoxification (Hayat et al. 2012). Another compatible solute is non-reducing disaccharide named as trehalose. Transgenic plants overexpressing trehalose biosynthetic genes confer tolerance towards drought and salinity (Iordachescu and Imai 2008). Sugar alcohols mainly mannitol and sorbitol also serve as compatible solutes during osmotic stress. Some plants also produce group of aliphatic amines mainly polyamines and glycine-betaine, which act as osmo-protectants. The major polyamines imparting stress tolerance are, putrescine, spermidine and spermine. Beside these metabolites, various classes of defense proteins mainly osmotins, dehydrins, and late embryogenesis abundant proteins (LEA) are also produced by plants in response to drought and salinity. Osmotins and dehydrins are mainly involved in cellular protection against osmotic shocks (Liu et al. 2017). LEAs are group of low-molecular weight hydrophilic proteins that contribute in stabilizing structural integrity of cell (Hand et al. 2011). Reactive oxygen species (ROS), such as hydrogen peroxide (H_2O_2), superoxide radical ($O_2^{\cdot -}$) and hydroxyl radical ($OH^{\cdot}$) etc. generated under abiotic stress leads to the oxidative damage to cell membrane, DNA, RNA and proteins (You and Chan 2015). To protect cell from such damage, plants produce antioxidant enzymes such as superoxide dismutase (SOD), catalase (CAT), ascorbate peroxidase (APX) and glutathione reductase (GR) (Kusvuran 2015). SOD converts superoxide radicals to H_2O_2, whereas APX, GPX and CAT detoxify H_2O_2 (Choudhury et al. 2013). Molecular responses include transcriptional changes in hundreds of stress responsive genes that ultimately help plant to tolerate such adversities (Fraire-Velazquez S Balderas-Hernandez 2013). Transcriptional regulators mainly transcription factors are the major component of the plant molecular response towards drought and salinity. As a regulator, the TF protein binds to the cis-acting element in the promoter region of genes that are responsive to various stresses. The downstream targets of these TFs are discussed in detail in sub Sect. 4.2. Majority of transcription factors have positive regulatory effect on stress responsive genes. For instance, enhanced expression of antioxidant genes (SOD, CAT etc.) under drought stress has been observed in transgenic cotton overexpressing potato DREB2 TF (El-Esawi and Alayafi 2019).

2.4 Transcription Factors: Potential Candidates for Enhancing Multiple Stress Tolerance

Transcription factors (TFs) are the regulatory proteins that can turn specific genes on (activate) or off (repress) by binding to the *cis*-acting sequence in the promoter region of their target genes. So far, a total of 58 TF families have been identified in higher plants (Jin et al. 2017). Moreover, whole genome sequencing of majority of plants revealed that approximately 7% of the coding sequences, code for TFs (Udvardi et al. 2007). Stress tolerance mechanism in plants is controlled by complex transcriptional network (Umezawa et al. 2006) and TFs are the major player in this network. In general, the TFs involved in plant abiotic or biotic stress response are mainly referred as stress-responsive transcription factors. These TFs are induced early in response to various stresses and in turn regulate the expression of several downstream target genes. The induction of these TFs is regulated by the presence of conserved cis-elements present in their promoter regions. For e.g. the induction of TFs like AP2/EREBP, MYB/MYC whose promoter contains ABA-responsive element (ABRE) is mediated by binding of ABRE-binding proteins (Sah et al. 2016). Additionally, histone modification such as phosphorylation mediated by protein kinases can also activate TFs expression (Pfluger and Wagner 2007). Developing stress tolerant plants by overexpressing TFs is much promising approach due to the reason that a single TF can regulate expression of vast array of stress-responsive genes (Goel and Singh 2018). The following sub-section covers major TF families which have been targeted to develop salt and/or drought tolerant transgenic plants and their downstream targets.

2.4.1 Transcriptional Regulatory Network Under Drought and Salinity Stress

The phytohormone ABA plays a central role in abiotic stress response especially under drought and salinity (Zhang et al. 2006). Water deficit resulting from drought and salinity, leads to ABA accumulation, which in turn triggers the complex regulatory network enabling plants to withstand such conditions (Kim et al. 2010; Hubbard et al. 2010). Under salt and drought stresses, ABA-biosynthesis is induced due to transcriptional activation of genes encoding enzymes of ABA biosynthesis pathway. Moreover, ABA degradation is suppressed under salt and drought stresses (Xiong and Zhu 2003). Thus, de novo synthesis of ABA and suppression of its degradation lead to accumulation of ABA that bind to the ABA receptors to initiate signal transduction that in turn activates set of stress-responsive genes (Tuteja 2007). Transcriptional regulation of stress-responsive genes via transcription factors takes place either by ABA-dependent or ABA-independent pathway (Fig. 2.3). In general, both pathways participate during drought and salt stress, whereas in response to cold stress only ABA-independent pathway operates (Shinozaki and Yamaguchi-Shinozaki 2007). Majority of ABA-responsive genes have ABA-responsive element (ABRE, PyACGTGG/TC) in their promoter regions (Busk and Pages 1997). Transcription factors mainly AREB (ABA-responsive element binding protein)/ABF

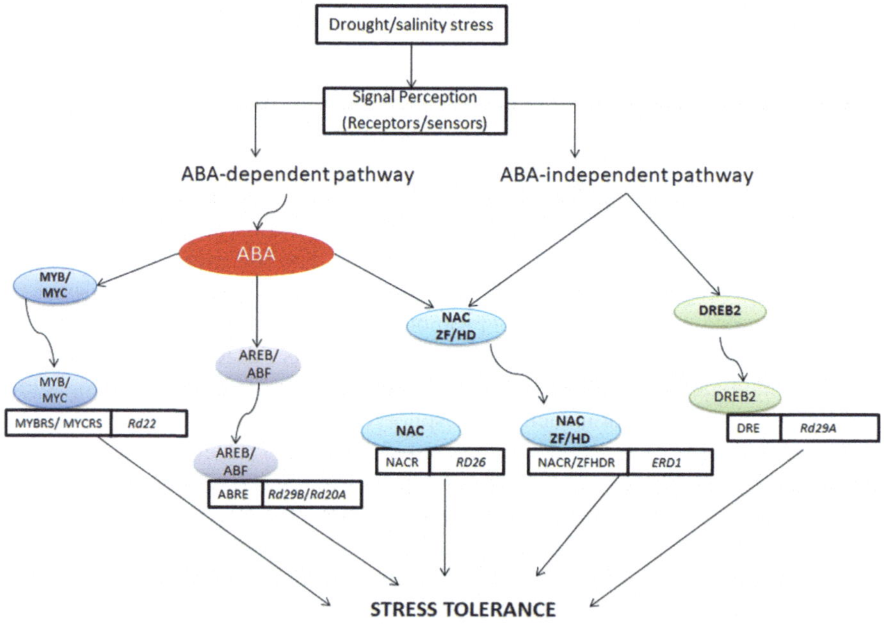

Fig. 2.3 Transcriptional regulatory network via ABA-dependent and ABA-Independent pathway In ABA-dependent pathway, ABA responsive *cis*-acting elements (ABRE) are present in the promoters of certain genes, like RD29B and Rd22. Transcription factors namely AREB/ABFs (ABA-responsive elementary binding proteins/factors), MYB/MYC and NAC bind to ABRE, MYBRS/MYCRS (MYB/MYC recognition sequence) and NACRS (NAC recognition sequence) motifs, respectively, to activate the expression of these genes in response to stress. Transcription factors mainly HD-ZIP and DREB2 are involved in ABA-independent manner during drought and salinity stress

(ABRE binding factors), MYB/MYC, RD26, NAC and ERF are involved in ABA-dependent signaling pathway (Sah et al. 2016). The ABA-dependent stress signaling first activates the transcription factors, which in turn induce the stress responsive gene for e.g. ABA-inducible basic leucine zipper (bZIP) transcription factor named as AREB induces *RD29B* and *RD20A* genes by binding to the ABRE motif of these genes (Abe et al. 1997). Similarly, MYB/MYC TFs activate *RD22* gene by binding to MYBRS (C/TAACNA/G) and MYCRS (CANNTG) elements in its promoter region (Abe et al. 1997). The transcription factors like, HD-ZIP and DREB2 are involved in ABA-independent manner during drought and salinity stress (Jia et al. 2012). The ABA-independent pathway is mainly regulated by DREB proteins that gets activated rapidly (within 20 min) in response to salinity and drought stresses. These proteins then bind to the DRE (dehydration-responsive element) *cis*-elements of several stress responsive genes encoding LEA proteins, heat shock proteins and proteins involved in detoxification etc. (Agarwal and Jha 2010). However, some TFs like NAC are found to be involved in both ABA-dependent and independent pathways (Sah et al. 2016).

2.4.2 Downstream Targets of Stress-Responsive Transcription Factors

As mentioned previously that the transcription factors regulate (activate or repress) the expression of downstream target genes by binding to the *cis*-acting element in the promoter region. One of the downstream targets of TFs is *RD* (responsive to dehydration) genes, which are induced under drought and high salinity. These genes possess stress responsive TF binding sites in their promoter regions. For e.g. *RD22* promoter has binding sites for MYB and MYC TFs, whereas, *RD29A* promoter has binding site for DREB1A TF (Jia et al. 2012). Stress-responsive TFs can also regulate the expression of genes encoding for antioxidant enzymes. Drought and salt stress escalate the level of ROS production that leads to oxidative damage and ultimately plant death (You and Chan 2015). To overcome this, plants generate antioxidant enzymes (SOD, catalases, peroxidases etc.) that participate in ROS detoxification. Another downstream targets of TFs are osmo-regulatory genes, such as those which are involved in biosynthesis of proline, glycine betaine and mannitol that accumulate during osmotic imbalance caused by drought and salinity. The promoter of *P5CS1* (Pyrroline-5-carboxylate synthetase1) that is involved in proline biosynthesis contains binding site for transcription factors MYB/MYC and AP2/EREBP (Szabados 2017). In addition, one of the genes encoding for proline dehydrogenase enzyme involved in proline degradation is also under the control of bZIP TFs (Weltmeier et al. 2006). Group of hydrophilic proteins named as LEA that protect plants from damage caused by drought stress are also the target of stress-responsive TFs (Hong-Bo et al. 2005). Stress related TF binding sites are present in the promoter region of LEA genes. For instance, cotton LEA genes contain several stress responsive TF binding elements such as ABA-responsive element (ABRE), dehydration-responsive element/C-repeat (DRE/CRT) and MYB in their promoter region (Magwanga et al. 2018). Polyphenolic secondary metabolites known as flavonoids impart stress tolerance to plants by scavenging ROS (Fini et al. 2011). The expression of flavonoid biosynthetic genes (FBPs) are found to be regulated by TFs mainly MYB, WRKY and bZIP (Ramsay and Glover 2005).

2.4.3 Major Transcription Factor Families that Participate in Drought and Salinity Tolerance

Transcriptional regulatory network in response to various abiotic stresses is highly complex and being considered as a master regulator of such network. TFs are considered as potential candidates for enhancing plant tolerance towards abiotic stresses (Goel and Singh 2018). Genome-wide expression profiling of various TF families in response to abiotic stresses provides an excellent opportunity to select candidate stress-responsive TF genes that could be functionally characterized for assessing their role in combating plant stresses. Past few years have witnessed a surge in interest in targeting TF genes for raising drought and salt tolerant transgenic plants. In the following sub-section, we focus on major TF families along with their role in enhancing drought and salinity tolerance in plants.

2.4.3.1 AP2/EREBP Transcription Factors

AP2/EREBP (APETALA2/ethylene-responsive element binding proteins) family is a large group of plant specific TFs implicated in multiple plant processes, such as plant growth and development, hormonal response and diverse abiotic and biotic stresses (Nakano et al. 2006; Licausi et al. 2010; Gutterson and Reuber 2004). The members of AP2/EREBP superfamily have been identified in several plant species including *Arabidopsis*, poplar, rice, potato, tomato, cotton and pepper (Riechmann and Meyerowitz 1998; Zhuang et al. 2008; Sharoni et al. 2011; Charfeddine et al. 2015; Sharma et al. 2010; Liu and Zhang 2017; Jin et al. 2018). AP2/EREBPs are characterized by the presence of 50–70 amino acid long conserved AP2-DNA-binding domain. This domain was first identified in the homeotic gene of *Arabidopsis AP2* (*APETALA2*), which is mainly involved in flower and seed development (Jofuku et al. 1994). The AP2/EREBP family was subdivided into four subfamilies: AP2 (Apetala 2), RAV (related to ABI3/VP1), DREB (dehydration-responsive element binding protein) and ERF (ethylene-response factor) on the basis of number and similarity of AP2/ERF domain (Dietz et al. 2010). Several other members of DREB and ERF subfamilies have also been overexpressed in various plant species, which clearly demonstrated potential of these genes in imparting salinity and/or drought tolerance in transgenic plants (Table 2.1) (El-Esawi and Alayafi 2019; Li et al. 2017,

Table 2.1 Selective examples of major transcription factors families overexpressed in plants to enhance drought and/or salinity tolerance

TF family	Gene name	Host plant	Transgenic	Abiotic stress tolerance	References
AP2/EREBP	*StDREB2*	*Solanum tuberosum*	Cotton	Drought	El-Esawi and Alayafi (2019)
	ScDREB10	*Syntrichia caninervis*	Arabidopsis	Salinity	Li et al. (2019)
	AhDREB1	*Arachis hypogaea*	Arabidopsis	Drought	Zhang et al. (2018)
	OsERF48	*Oryza sativa*	Rice	Drought	Jung et al. (2017)
	BjABR1	*Brassica juncea*	Arabidopsis	Salinity	Xiang et al. (2018)
	MdDREB76	*Malus domestica*	Tobacco	Salinity and drought	Sharma et al. (2019)
	CsDREB	*Camellia sinensis*	Arabidopsis	Salinity and drought	Wang et al. (2017)
	SmDREB1	*Sophora moorcroftiana*	Arabidopsis	Drought	Li et al. (2017)
	MsDREB6.2	*Malus sieversii*	Apple	Drought	Liao et al. (2017)
	MrCBF2	*Muscadinia rotundifolia*	Arabidopsis	Drought	Wu et al. (2017)
	ThDREB	*Tamarix hispida*	Tobacco, *T. hispida*	Salinity and drought	Yang et al. (2017)
	LoERF017	*Larix olgensis*	Arabidopsis	Salinity	Hu et al. (2018)
	NnDREB1	*Nelumbo nucifera*	Arabidopsis	Drought	Cheng et al. (2017)
	IbRAP2-12	*Ipomoea batatas*	Arabidopsis	Salinity and drought	Li et al. (2019)
	OsLG3	*Oryza sativa*	Rice	Drought	Xiong et al. (2018)
	PsnERF75	*Populus simonii* × *P. nigra*	Arabidopsis	Salinity	Wang et al. (2018)
NAC	*MlNAC10*	*Miscanthus lutarioriparius*	Arabidopsis	Salinity and drought	He et al. (2019)
	MlNAC12	*Miscanthus lutarioriparius*	Arabidopsis	Salinity and drought	Yang et al. (2018)

(continued)

Table 2.1 (continued)

TF family	Gene name	Host plant	Transgenic	Abiotic stress tolerance	References
	TsNAC1	*Thellungiella halophila*	Arabidopsis, *T. halophila*	Salinity and drought	Liu et al. (2018)
	CmNAC1	*Cucurbita moschata*	Arabidopsis	Salinity and drought	Cao et al. (2017)
	OsNAC14	*Oryza sativa*	Rice	Drought	Shim et al. (2018)
	NAC57	*Populus alba x Populus glandulosa*	Arabidopsis	Salinity	Yao et al. (2018)
	EsNAC1	*Eutrema salsugineum*	Arabidopsis	Salinity	Liu et al. (2018)
	GmNAC085	*Glycine max*	Arabidopsis	Drought	Nguyen et al. (2018)
	PvNAC1	*Panicum virgatum*	Switchgrass	Salinity	Wang et al. (2018)
	NtNAC2	*Nicotiana tabacum*	Tobacco	Drought	Xu et al. (2018)
	DgNAC1	*Dendranthema grandiflorum*	*Chrysanthemum*	Drought	Zhao et al. (2018)
	SlNAC2	*Solanum lycopersicum*	Arabidopsis	Salinity and drought	Borgohain et al. (2019)
	ThNAC13	*Tamarix hispida*	Arabidopsis, *T. hispida*	Salinity	Wang et al. (2017)
	PgNAC21	*Pennisetum glaucum*	Arabidopsis	Salinity	Shinde et al. (2019)
	PbeNAC1	*Pyrus betulifolia*	*P. betulifolia*	Drought	Jin et al. (2017)
	SlNAC8	*Suaeda liaotungensis*	Arabidopsis	Salinity and drought	Wu et al. (2018)
	NAC13	*Populus. alba × P. glandulos*	Poplar	Salinity	Zhang et al. (2019)
MYB	*TaSIM*	*Triticum aestivum*	Arabidopsis	Salinity and drought	Yu et al. (2017, 2019)
	ZmMYB3R	*Zea mays*	Arabidopsis	Salinity and drought	Wu et al. (2019)
	ScMYBAS1-3	*Saccharum officinarum*	Rice	Drought	Favero Peixoto-Junior et al. (2018)

(continued)

Table 2.1 (continued)

TF family	Gene name	Host plant	Transgenic	Abiotic stress tolerance	References
	TaMYB31	*Triticum aestivum*	Arabidopsis	Drought	Zhao et al. (2018)
	OsMYB6	*Oryza sativa*	Rice	Salinity and drought	Tang et al. (2019)
	GsMYB15	*Glycine soja*	Arabidopsis	Salinity	Shen et al. (2018)
	RhMYB96	*Rosa hybrida*	Arabidopsis	Salinity	Jiang et al. (2018)
	BplMYB46	*Betula platyphylla*	Birch	Salinity	Guo et al. (2017)
	PbrMYB21	*Pyrus betulaefolia*	Tobacco	Drought	Li et al. (2017)
	GmMYB84	*Glycine max*	Soybean	Drought	Wang et al. (2017)
	GaMYB85	*Gossypium arboretum*	Arabidopsis	Drought	Butt et al. (2017)
	MYB49	*Solanum lycopersicum*	Tomato	Salinity and drought	Cui et al. (2018)
	GmMYB118	*Glycine max*	Arabidopsis, soybean	Salinity and drought	Du et al. (2018)
	ZmMYB48	*Zea mays*	Arabidopsis	Drought	Wang et al. (2017)
	FtMYB13	*Fagopyrum tataricum*	Arabidopsis	Salinity and drought	Huang et al. (2018)
WRKY	*TaWRKY2*	*Triticum aestivum*	Wheat	Drought	Gao et al. (2018)
	OsWRKY42	*Oryza sativa*	Arabidopsis	Salinity	Pillai et al. (2018)
	RtWRKY1	*Reaumuria trigyna*	Arabidopsis	Salinity	Du et al. (2017)
	DgWRKY4	*Dendranthema grandiflorum*	Chrysanthemum	Salinity	Wang et al. (2017)
	TaWRKY146	*Triticum aestivum*	Arabidopsis	Drought	Ma et al. (2017)
	GmWRKY16	*Glycine max*	Arabidopsis	Salinity and drought	Ma et al. (2019)
	AtWRKY30	*Arabidopsis thaliana*	Wheat	Drought	El-Esawi et al. (2019)
	DgWRKY2	*D. grandiflorum*	Chrysanthemum	Salinity	He et al. (2018)

(continued)

Table 2.1 (continued)

TF family	Gene name	Host plant	Transgenic	Abiotic stress tolerance	References
	GmWRKY12	*Glycine max*	Soybean	Salinity and drought	Shi et al. (2018)
	VvWRKY2	*Vitis vinifera*	Tobacco	Salinity	Mzid et al. (2018)
	MsWRKY11	*Medicago sativa*	Soybean	Salinity	Wang et al. (2018)
	GmWRKY49	*Glycine max*	Arabidopsis	Salinity	Xu et al. (2018)
	VlWRKY48	*Vitis labrusca × V. vinifera*	Arabidopsis	Drought	Zhao et al. (2018)
	MbWRKY2	*Malus baccata*	Tobacco	Drought	Han et al. (2018)
	MbWRKY5	*M. baccata*	Tobacco	Salinity and drought	Han et al. (2018)
	VaWRKY14	*Vitis amurensis*	Arabidopsis	Drought	Zhang et al. (2018)
bZIP	*GmFDL19*	*Glycine max*	Soybean	Salinity and drought	Li et al. (2017)
	ABP9	*Zea mays*	Cotton	Salinity and drought	Wang et al. (2017)
	ABP2	*Zea mays*	Arabidopsis	Salinity and drought	Zong et al. (2018)
	VlbZIP30	*Vitis labrusca × V. vinifera*	Arabidopsis	Drought	Tu et al. (2018)
	VvABF2	*Vitis vinifera*	Arabidopsis	Drought	Liu et al. (2019)
	OsABF1	*Oryza sativa*	Rice	Drought	Zhang et al. (2017)
	TabZIP14-B	*Triticum aestivum*	Arabidopsis	Salinity	Zhang et al. (2017)
	CaDILZ1	*Capsicum annuum*	Arabidopsis	Drought	Lim et al. (2018)
	OsbZIP16	*Oryza sativa*	Arabidopsis	Salinity and drought	Pandey et al. (2018)
	SlbZIP1	*Solanum lycopersicum*	Tomato	Salinity and drought	Zhu et al. (2018)
	EcbZIP17	*Eleusine coracana*	Tobacco	Salinity and drought	Ramakrishna et al. (2018)

(continued)

Table 2.1 (continued)

TF family	Gene name	Host plant	Transgenic	Abiotic stress tolerance	References
	TabZIP	*Triticum aestivum*	Arabidopsis	Salinity and drought	Agarwal et al. (2019)
	VqbZIP39	*Vitis quinquangularis*	Arabidopsis	Salinity and drought	Tu et al. (2016)
	VlbZIP36	*Vitis labrusca* × *V. vinifera*	Arabidopsis	Drought	Tu et al. (2016)
	OsbZIP66	*Oryza sativa*	Rice	Drought	Yoon et al. (2017)
	OsbZIP42	*Oryza sativa*	Rice	Drought	Joo et al. (2019)
bHLH	*PebHLH35*	*Populus euphratica*	Arabidopsis	Drought	Dong et al. (2014)
	EcbHLH57	*Eleusine coracana*	Tobacco	Salinity and drought	Babitha et al. (2015)
	AtbHLH112	*Arabidopsis*	Arabidopsis	Salinity and drought	Liu et al. (2015)
	OrbHLH2	*Oryza rufipogon*	Arabidopsis	Salinity	Zhou et al. (2009)
	OsbHLH035	*Oryza sativa*	Rice	Salinity	Chen et al. (2018)
	FtbHLH3	*Fagopyrum tataricum*	Arabidopsis	Drought	Yao et al. (2017)
	AtbHLH68	*Arabidopsis thaliana*	Arabidopsis	Drought	Le Hir et al. (2017)
	OsbHLH068	*Oryza sativa*	Arabidopsis	Salinity	Chen et al. (2017)
	OrbHLH001	*Oryza rufipogon*	Rice	Salinity	Chen et al. (2013)
	TabHLH39	*Triticum aestivum*	Arabidopsis	Salinity and drought	Zhai et al. (2016)
	ThbHLH1	*Tamarix hispida*	Arabidopsis	Salinity and drought	Ji et al. (2016)
	VvbHLH1	*Vitis vinifera*	Arabidopsis	Salinity and drought	Wang et al. (2016)
ZFP	*ZAT18*	*Arabidopsis thaliana*	Arabidopsis	Drought	Yin et al. (2017)
	AtRZFP	*Arabidopsis thaliana*	Arabidopsis	Salinity	Zang et al. (2016)

(continued)

Table 2.1 (continued)

TF family	Gene name	Host plant	Transgenic	Abiotic stress tolerance	References
	ZAT6	*Arabidopsis thaliana*	Rice, cotton, *Pinus melliottii*	Salinity	Tang and Luo (2018)
	ZjZFN1	*Zoysia japonica*	Arabidopsis	Salinity	Teng et al. (2018)
	IbZFP1	*Ipomoea batatas*	Arabidopsis	Salinity and drought	Wang et al. (2016)
	PtrZPT2-1	*Poncirus trifoliate*	Tobacco	Salinity and drought	Liu et al. (2017)
	GaZnF	*Gossypium arboretum*	Cotton	Salinity and drought	Iqbal et al. (2017)
	OsMSR15	*Oryza sativa*	Arabidopsis	Drought	Zhang et al. (2016)
	OsDRZ1	*Oryza sativa*	Rice	Drought	Yuan et al. (2018)
	TaZNF	*Triticum aestivum*	Arabidopsis	Salinity	Ma et al. (2016)
	SL-ZH13	*Solanum lycopersicum*	Tomato	Salinity	Zhao et al. (2018)

2019a, b; Zhang et al. 2018; Jung et al. 2017; Xiang et al. 2018; Sharma et al. 2019; Wang et al. 2017, 2018; Liao et al. 2017; Wu et al. 2017; Yang et al. 2017; Hu et al. 2018; Cheng et al. 2017; Xiong et al. 2018).

DREBs are classified into two subgroups, DREB1/CBF (C-repeat binding factor) and DREB2. It has been observed that expression of DREB1 was mainly induced under cold stress, whereas expression of DREB2A and 2B was induced by drought and high salinity (Liu et al. 1998; Nakashima et al. 2000). However, some exceptions are also there, such as *Ca-DREBLP1* from hot pepper was found to be induced in response to drought and salinity but not by cold stress (Hong and Kim 2005). DREB proteins bind to A/GCCGAC motif (DRE/CRT *cis*-element) in the promoter region of several COR (*cor6.6*, *cor15a*) and RD genes (*RD29A*, *RD17*, *ERD10*) (Wu et al. 2017; Yang et al. 2017; Hu et al. 2018; Cheng et al. 2017; Li et al. 2019; Xiong et al. 2018; Wang et al. 2018; Liu et al. 1998; Nakashima et al. 2000; Hong and Kim 2005; Nordin et al. 1991; Agarwal et al. 2006; Lucas et al. 2011). DREB TFs are known to activate the expression of multiple genes and therefore help plants to combat several abiotic stresses. The overexpression of an apple gene *MdDREB76* enhanced salinity and drought tolerance in transgenic tobacco plants by elevating antioxidant enzymes and also the expression of some stress-responsive genes such as LEA5, ERD10B, ERD10D, HSP26 etc. (Sharma et al. 2019). A *DREB* gene from *Camellia sinensis* enhances drought and salt tolerance by increasing the expression of both ABA-dependent (*AtRD29B*, *AtRAB18*, *AtABI1*, and *AtABI2*) and ABA-independent genes (*AtCOR15a* and *AtRD29A*) in transgenic *Arabidopsis* (Jung et al. 2017). Salinity and drought inducible *StDREB1* gene increased the content of osmo-protectant and expression of stress-related genes in transgenic potato in response to salt stress (Bouaziz et al. 2013). Overexpression studies pertaining to DREB2 are limited. The very first report of *DREB2A* transgenic *Arabidopsis* plants did not show any stress tolerance (Liu et al. 1998), but thereafter Sakuma (Sakuma et al. 2006) overexpressed the active form of *AtDREB2A* that imparted drought tolerance in transgenic Arabidopsis. Overexpressing *OsDREB2A* gene under control of *RD29A* promoter showed salinity and drought tolerance in transgenic rice (Mallikarjuna et al. 2011). The overexpression of *DREB2* gene from *Pennisetum glaucum* (*PgDREB2A*) in tobacco resulted in tolerance towards hyperionic and hyperosmotic stresses (Agarwal et al. 2010). In a recent report, *StDREB2* overexpressing transgenic cotton plants displayed drought tolerance by exhibiting high proline content, increased antioxidant activities and upregulation of some stress-responsive genes (*ERF2*, *Rd22*, *DREB1B*, *DREB1C*) (El-Esawi and Alayafi 2019). Another AP2/EREBP sub-family that imparts stress tolerance in plants is ERF (Ethylene responsive factor). They bind to the AGCCGCC, known as the GCC box of their target genes (Hao et al. 1998) and constitute the largest sub-family of AP2/EREBP. A tomato ERF gene, *TSRF1* enhances drought tolerance in transgenic rice by activating set of genes involved in ABA, proline synthesis and photosynthesis-related genes (Quan et al. 2010). The *TaERF3* gene also increases drought and salinity tolerance in transgenic wheat by activating some stress-responsive genes, such as *LEA3*, *RAB*, *POX* etc. (Rong et al. 2014). The root-specific overexpression of *OsERF71* in rice mediates drought tolerance in transgenic rice by modulating root architecture (Lee et al. 2016). The transgenic *OsERF71* plants showed enhanced expression of several cell wall-associated genes and lignin biosynthesis genes.

2.4.3.2 NAC Transcription Factors

NAC TF family represents one of the largest families of plant specific transcription factors. The NAC TF derived its name from the first three reported members that contain a particular domain (NAC domain): Petunia NO APICAL MERISTEM (NAM), Arabidopsis ACTIVATION FACTOR (ATAF), and CUP SHAPED COTYLEDON (CUC). The Arabidopsis gene *RD26* (RESPONSIVE TO DEHYDRATION 26) was the first gene reported to encode NAC protein (Yamaguchi-Shinozaki et al. 1992). Till date, large number of NAC genes have been identified in several plants through the availability of their complete genome sequences, such as 117 in Arabidopsis, 151 in rice, 110 in potato, 104 in tomato, 204 in Chinese cabbage, 104 in pepper, 152 in maize, soybean and tobacco (Nuruzzaman et al. 2013; Singh et al. 2013; Su et al. 2014, 2015; Liu et al. 2014; Diao et al. 2018; Shiriga et al. 2014; Le et al. 2011). A typical NAC protein contains a highly conserved N-terminal DNA-binding NAC domain and a highly diverged transcriptional regulatory C-terminal region (Fang et al. 2008; Puranik et al. 2012). The NAC domain comprises of approximately 150-160 amino acids and is divided into five sub-domains (A–E) (Ooka et al. 2003) (Fig. 2.4a). The NAC domain is found to be rich in basic amino acids; however, distribution of positive and negative amino acids varies in each sub-domain. The N-terminal region is involved in the formation of homodimers or heterodimers with other NAC protein. The diverged C-terminal region is involved in transcriptional regulation by activating or repressing the transcription process. This region is rich in simple amino acids, like serine, threonine, proline and glutamine (Puranik et al. 2012). The C-terminal region in some NAC TFs contains alpha-helical transmembrane domain motifs that are responsible for anchoring to plasma membrane or ER and are named as membrane bound NAC TFs. NAC TFs play important roles in plant developmental processes and response to biotic and abiotic stresses (Ooka et al. 2003; Olsen et al. 2005; Nakashima et al. 2012; Tran et al. 2004; Jiang and Deyholos 2006). NAC TFs also play widespread roles in plant abiotic and biotic stress response (Singh et al. 2013; Su et al. 2015; Nakashima et al. 2012). The expression of NAC genes under these stresses is mainly regulated by the presence of stress responsive *cis*-acting elements, like ABRE, DREs, jasmonic acid and salicylic acid responsive elements in their promoter regions. NAC TFs are involved in both ABA-dependent (e.g. *RD26*) and independent pathways (e.g. *ERD1*, *TaNAC8*). Moreover, post-transcriptional regulations, like microRNA mediated cleavage, alternative splicing and several post-translational changes that include dimerization, phosphorylation and ubiquitination also modulate their expressions (Puranik et al. 2012). The NAC genes in turn can regulate the expression of several downstream target genes by binding to their promoter regions (Tran et al. 2004). Majority of stress responsive genes have two core DNA-binding motif CGTA and CGTC in their consensus NAC recognition site (NACRS). The expression analysis of *NAC* genes through genome-wide transcriptomic studies has

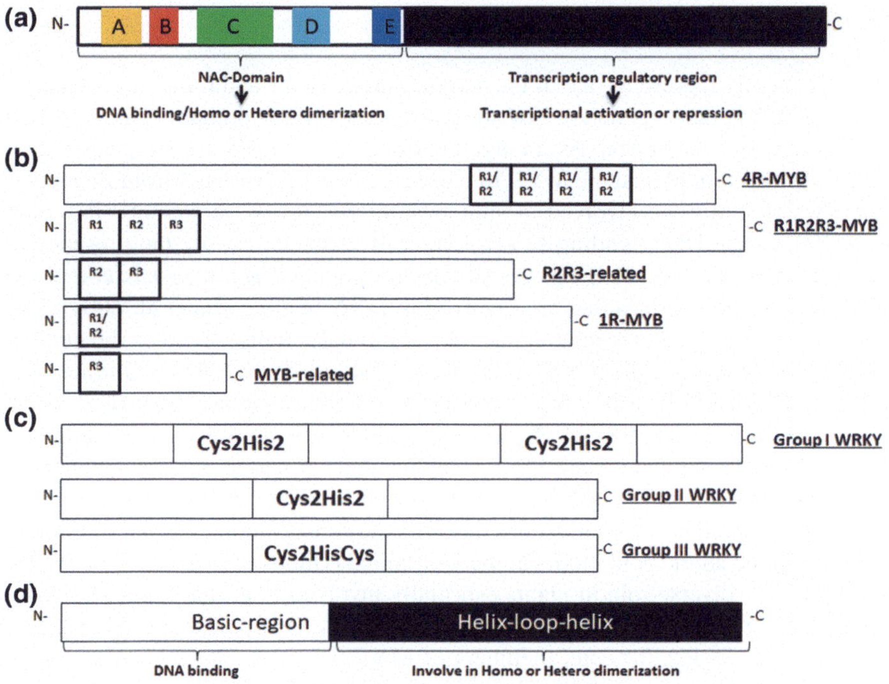

Fig. 2.4 Pictorial representation of structure of major TF proteins involved in drought and/or salinity tolerance

revealed their involvement in majority of abiotic stresses in plants. In Arabidopsis, the expression of 33 *NAC* genes was significantly affected under salt stress (Jiang and Deyholos 2006), whereas in rice, 40 *NAC* genes were found to respond towards salt and/ or drought stress (Fang et al. 2008). Several members of NAC family have been overexpressed to develop salt and drought tolerant plants (Table 2.1) (He et al. 2019; Yang et al. 2018; Liu et al. 2018a, b; Cao et al. 2017; Shim et al. 2018; Yao et al. 2018; Nguyen et al. 2018; Wang et al. 2018; Xu et al. 2018; Zhao et al. 2018; Borgohain et al. 2019; Wang et al. 2017; Shinde et al. 2019; Jin et al. 2017; Wu et al. 2018; Zhang et al. 2019). Majority of rice *NAC* genes impart enhanced salt and drought tolerance without any negative impact on transgenic plants for e.g. overexpression of *OsNAC063* showed enhanced salt and drought tolerance in transgenic Arabidopsis (Yokotani et al. 2009). Similarly, overexpression of *OsNAC6*in rice also conferred tolerance to high salt and dehydration stress by up-regulating the expression of *LEA3* (Nakashima et al. 2007). In addition, stress-responsive NAC from sorghum (*SbSNAC1*), maize (*ZmNAC55*) and wheat (*TaNAC29*) positively improved dehydration and/or salt tolerance in transgenic plants (Lu et al. 2013; Mao et al. 2014; Xu et al. 2015).

2.4.3.3 MYB/MYC Transcription Factors

MYB TF family represents one of the diverse classes of TF and found in all eukaryotes. The MYB name is derived by the presence of MYB domain. The very first *MYB* gene was isolated from Avian myeloblastosis virus (AMV) (Klempnauer et al. 1982). However, in plants, the MYB like gene *C1* was first isolated from *Zea mays*, which was found to be involved in anthocyanin biosynthesis in kernels (Paz-Ares et al. 1987). The DNA-binding MYB-domain is characterized by the presence of one to four conserved MYB repeats (R) at N-terminus. Each repeat is composed of 52 amino acid residues that are involved in DNA-binding (Jia et al. 2004). The C-terminus is engaged in regulating the transcriptional activity. Based on the number and position of repeats, the MYB TF family is subdivided into four subfamilies: 4R-MYB, R1R2R3-MYB, R2R3-MYB and 1R-MYB (Dubos et al. 2010) (Fig. 2.4b). The 4R-MYB is the smallest MYB-subfamily and R2R3-MYB is the most common MYB-subfamily in plants. Plant MYBs having single or partial repeat are generally known as MYB-related or MYB-like proteins. The plant R2R3-MYB proteins are further categorized into 23 sub-groups based on the conservation of the DNA binding domain and the amino acid motifs at the C-terminal (Dubos et al. 2010). The R2R3-MYB TFs has diverse role in plants especially involved in primary and secondary metabolism, plant development, cell fate and response to abiotic and biotic stresses (Baldoni et al. 2015). To date, members of MYB TF family have been identified in several plants, such as 197 in Arabidopsis, 183 in rice, 97 in pineapple, 85 in bamboo and 244 in soybean (Katiyar et al. 2012; Yanhui et al. 2006; Liu et al. 2017; Yang et al. 2019; Du et al. 2012; Cao et al. 2013). AtMYB2 was the first MYB TF found to be induced in response to drought and high salinity (Urao et al. 1993). Transcriptomic and microarray analysis of several plants have shown the involvement of MYB-TF in response to drought (Cao et al. 2013; Yan et al. 2012). The drought induced *AtMYB96*, genes imparts drought tolerance in transgenic plants by activating *RD22* gene (Yan et al. 2012). In rice, approximately 65% of MYB genes were found to be differentially regulated under drought stress (Katiyar et al. 2012). Several stress tolerant transgenic plants have been developed by overexpressing MYB TFs (Table 2.1) (Yu et al. 2017, 2019; Wu et al. 2019; Favero Peixoto-Junior et al. 2018; Zhao et al. 2018; Tang et al. 2019; Shen et al. 2018; Jiang et al. 2018; Guo et al. 2017; Li et al. 2017; Wang et al. 2017; Butt et al. 2017; Cui et al. 2018; Du et al. 2018; Wang et al. 2017; Huang et al. 2018). Overexpression of *OsMYB48-1* improved drought and salinity tolerance in rice by upregulating the expression ABA biosynthetic genes and several stress-responsive genes, mainly *LEA3*, *RAB16C* and *RAB21* etc. (Xiong et al. 2014). Overexpression of *TaMYB30-B*, *TaMYB33* and *TaMYB2A* from wheat also displayed drought tolerance in Arabidopsis (Zhang et al. 2012; Mao et al. 2011; Qin et al. 2012). In case of tobacco, overexpression of *TaODORANT1* (R2-R3 MYB) leads to high relative water content and lower Na^+ accumulation under drought and salt stress (Wei et al. 2017). The overexpression of wheat MYB TF *TaSIM* improved salt tolerance in transgenic Arabidopsis by activating stress-responsive *RD22* and *RD29A* genes (Yu et al. 2017). Recently, overexpression of *OsMYB6* in transgenic rice showed increased drought and salt tolerance by possessing higher proline content and higher CAT and SOD activities (Tang et al. 2019). MYB TFs are also the major regulator of genes involved in cell wall development for e.g. overexpression of *MYB46* or *MYB83* enhanced the biosynthetic pathway of lignin, cellulose and xylan (Zhong and Ye 2012), thus making this TF a good candidate for enhancing plant biomass production especially under stress condition (Joshi et al. 2018).

2.4.3.4 WRKY Transcription Factors

WRKY TF family is one of the largest family of transcription factors that are ubiquitously present from green algae to land plants (Rinerson et al. 2015). The first WRKY gene named as *SPF1* (SWEET POTATO FACTOR 1) was identified from sweet potato (Ishiguro and Nakamura 1994). Thereafter, two WRKY proteins (ABF1, ABF2) were also isolated from wild oat (Rushton et al. 1995). Large number of WRKY genes have been identified in plants, such as 74 in Arabidopsis (Ulker and Somssich 2004), 103 in rice (Ramamoorthy et al. 2008), 55 in cucumber (Ling et al. 2011), 107 in desert poplar (Ma et al. 2015) and 54 in pineapple (Xie et al. 2018). WRKY TFs contain a 60-amino acid long DNA binding domain with highly conserved WRKYGQK motif at N-terminus and a novel zinc-finger like motif at C-terminus (Babu et al. 2006). The WRKY TFs mainly bind to the consensus *cis*-acting element TTGACT/C (W-box) in the promoter regions of target genes. WRKY proteins are categorized into three groups based on the number of WRKY domain and type of zinc finger-like motif. The group-I WRKY proteins have two WRKY domains and group-II WRKY proteins have one WRKY domain (Fig. 2.4c). Both the groups have similar Cys2-His2 zinc-finger motif. Group II is further subdivided into five subgroups (IIa-IIe) on the basis of primary amino acid sequences. Group III WRKY proteins have one WRKY domain containing the different Cys2-His/Cys Cys2-His2 zinc-finger motif (Fig. 2.4c). WRKY proteins are involved in various developmental processes, including embryogenesis, seed coat and trichome development (Johnson et al. 2002; Lagace and Matton 2004). WRKY TFs are of prime importance for imparting multiple stress tolerance as a single WRKY protein has potential to respond against both abiotic and biotic stresses. Moreover, WRKY TFs (like AtWRKY18, AtWRY40) are involved in multiple phytohormones (ABA, Jasmonic and salicylic acid)-mediated signaling pathways, which are activated during stress response (Shang et al. 2010; Chen et al. 2010; Li et al. 2010). WRKY proteins also act as important regulator in ABA signaling, for e.g. OsWRKY24 and OsWRKY45 act as repressor while OsWRKY72 and OsWRKY77 act as activator of the ABA-inducible promoter (Xie et al. 2005). Arabidopsis roots subjected to salt treatment showed early induction of 8 *WRKY* genes (Jiang and Deyholos 2006). The expression of 24 *CiWRKYs* from *Caragana* was found to be differentially regulated by salt and drought treatment (Wan et al. 2018). Several WRKY proteins exhibited similar responses under drought and salinity stress. For e.g., overexpression of salt and drought inducible *TaWRKY2* and *TaWRKY19* in transgenic tobacco and wheat, respectively showed enhanced tolerance to these stresses by activating multiple stress-responsive genes and exhibiting high levels of proline and soluble sugar (Gao et al. 2018; Niu et al. 2012). The *GmWRKY54* showed enhanced salt and drought tolerance in transgenic Arabidopsis by regulating the expression of salt tolerance zinc finger STZ/Zat10 and DREB2A transcription factors (Zhou et al. 2008). The overexpression of *DnWRKY11* improved the drought and salt tolerance in transgenic tobacco by exhibiting higher activity of antioxidant enzymes, like CAT, POD (Peroxidase) and SOD (Xu et al. 2014). Few examples of transgenic plants showing drought or salinity tolerance by overexpressing WRKY TF are shown in Table 2.1 (Pillai et al. 2018; Du et al. 2017; Wang et al. 2017, 2018; Ma et al. 2017, 2019; El-Esawi et al. 2019; He et al. 2018; Shi et al. 2018; Mzid et al. 2018; Xu et al. 2018; Zhao et al. 2018; Han et al. 2018; Han et al. 2018; Zhang et al. 2018).

2.4.3.5 bZIP Transcription Factors

The bZIP TF derived its name due to the presence of 40–80 amino acid long basic region/leucin zipper domain that consists of a highly uninterrupted α-helix region for nuclear localization and DNA binding at N-terminus followed by leucine rich motif at C-terminus responsible for dimerization (Schumacher et al. 2000; Miller et al. 2003). Several bZIP proteins have been reported in dicots and monocots (Jakoby et al. 2002; Nijhawan et al. 2008; Wei et al. 2012). The bZIP TFs are considered as family of dimeric TFs and can undergo both homo or hetero-dimerization (Llorca et al. 2014). This dimerization is highly important in determining the function of bZIPs. Like other TFs, bZIP protein also respond to drought and salinity stress. The bZIP TFs are mainly involved in stress response via ABA-dependent pathway for e.g. overexpression of *OsbZIP23* and *OsbZIP71* enhanced salt and drought tolerance by upregulating ABA-inducible genes in rice (Xiang et al. 2007; Liu et al. 2014). In Arabidopsis, overexpression of *bZIP* gene (*AtTGA4*) conferred drought tolerance by improving nitrate transport and assimilation (Zhong et al. 2015). A bZIP protein, GmFDL19 also enhanced the salt and drought tolerance in transgenic soybean by up-regulating the expression of several salt-responsive genes (like *SOS*, *NHX*) and also some TF genes like *bZIP1*, *NAC11*, *NAC29* and *MYB174* (Li et al. 2017). Maize ABP9 (bZIP TF) gene, when overexpressed in transgenic cotton conferred enhanced tolerance to drought and salinity by exhibiting higher proline, cholorophyll and soluble sugar content. Moreover, the expression of several stress related genes like *ZFP1*, *ERF1*, *NCED2* and *SAP11* was also found to be upregulated (Wang et al. 2017). To date, several transgenic plants overexpressing bZIP TFs has been developed which exhibited enhanced tolerance towards salinity and drought (Zong et al. 2018; Tu et al. 2018; Liu et al. 2019; Zhang et al. 2017a, b; Lim et al. 2018; Pandey et al. 2018; Zhu et al. 2018; Ramakrishna et al. 2018; Agarwal et al. 2019; Tu et al. 2016a, 2016; Yoon et al. 2017; Joo et al. 2019).

2.4.3.6 bHLH Transcription Factors

The basic-helix-loop-helix TFs constitute a large superfamily, ubiquitously distributed among both plants and animals. In plants, bHLH TFs regulate many biological processes, like photosynthesis, flowering and flavonoid biosynthesis (Dong et al. 2014; Ito et al. 2012; Ohno et al. 2011). The bHLH proteins can undergo both homo and hetero dimerization (Pireyre and Burow 2015). The typical bHLH proteins are characterized by the presence of highly conserved bHLH domain having two distinct functional regions: 10–15 amino acids long basic region at N-terminus of the domain and 40 amino acids long helix-loop-helix (HLH) region at its C-terminus (Feller et al. 2011) (Fig. 2.4d). The bHLH proteins regulate expression of target genes by recognizing and binding to the E-box (CANNTG) and G-box (CACGTG) DNA motifs in their promoter regions (Toledo-Ortiz et al. 2003). The members of bHLH family have been identified in Arabidopsis (Bailey et al. 2003), rice (Li et al. 2006), tomato (Wang et al. 2015), grapes (Wang et al. 2018) and in peach (Zhang et al. 2018). Involvement of bHLH proteins in abiotic stress response, especially under drought and salt is well known. Transcriptome-wide expression analysis of bHLH genes from tea (*C. sinensis*) has revealed the upregulation of total 39 *CsbHLH* genes under drought stress (Cui et al. 2018). Arabidopsis roots showed upregulation in the expression of at least 15 *bHLH* genes upon salt exposure (Jiang and Deyholos 2006). Recent reports demonstrating transgenic plants development by overexpressing bHLH TFs, which exhibited stress tolerance, are shown in Table 2.1 (Babitha et al. 2015; Liu et al. 2015; Zhou et al. 2009; Chen et al. 2013, 2017, 2018; Yao et al. 2017; Le Hir et al. 2017; Zhai et al. 2016; Ji et al. 2016; Wang et al. 2016). The *OrbHLH2* gene isolated from wild rice positively regulates the salt tolerance in transgenic Arabidopsis in ABA-independent manner (Zhou et al. 2009). Transgenic Arabidopsis overexpressing drought inducible *FtbHLH3* gene showed enhanced drought tolerance by preventing ion leakage and maintaining higher osmolyte content (Yao et al. 2017). In another report, the *VvbHLH1* gene from grapes, when overexpressed in Arabidopsis increased drought and salt tolerance by elevating flavonoid and ABA content (Wang et al. 2016). Ectopic expression of *PebHLH35* from *Populus euphratica* conferred drought tolerance in transgenic plants through changes in stomatal aperture, stomatal density and transpiration rates (Dong et al. 2014). The *AtbHLH112* also displayed salt tolerance in transgenic plants by enhancing proline content and reducing ROS generation (Lim et al. 2018). Similar results were observed for transgenic Arabidopsis overexpressing *ThbHLH1* gene of *Tamarix hispida* (Ji et al. 2016).

2.4.3.7 Zinc Fingers Protein Transcription Factors

Zinc finger protein (ZFP) family is a large family of eukaryotic TFs (Sakamoto et al. 2000), first identified as TFIIIA from *Xenopus oocytes* (Miller et al. 1985). ZFPs contain a conserved zinc finger motif, which binds to DNA with zinc ion through its cysteine (Cys) and histidine (His) residues. The ZFP gene families have been identified in Arabidopsis (Englbrecht et al. 2004), rice (Agarwal et al. 2007), foxtail millet (Muthamilarasan et al. 2014), tobacco (Yang et al. 2016) and soybean (Yuan et al. 2018). ZFPs are categorized into different families, like C2H2, C3HC4, C4, CCCH, C4HC3 and C2H5 on the basis of Cys and His combination for coordination with zinc ion. Among these, C2H2 ZFPs have major role in plant abiotic stress tolerance (Singh et al. 2010). For instance, the C2H2 ZFP from Arabidopsis (*AtZAT7*) and rice (*OsZFP179*) showed enhanced expression in response to drought and salt stress (Ciftci-Yilmaz et al. 2007; Sun et al. 2010). A drought inducible *ZAT18* acts as a positive regulator of drought tolerance by enhancing the expression of several downstream target genes, including *COR47*, *ERD7*, *LEA6*, and *RAS1* (Yin et al. 2017). Gain and loss-of-function studies of AtR*ZFP* showed enhanced and reduced tolerance, respectively, towards salt and osmotic stress in transgenic Arabidopsis (Zang et al. 2016). The overexpression of *AtZAT6* in cells of rice, cotton and slash pine enhanced salt tolerance by increasing antioxidant activities and phytohormone (ABA, GA_8) levels. In addition, expression of Ca^{2+} dependent protein kinases (*OsCPK9*, *OsCPK25*) also increased tolerance in rice cells towards drought stress (Tang and Luo 2018). The recent reports of transgenic plant overexpressing member of ZFP are shown in Table 2.1 (Yin et al. 2017; Zang et al. 2016; Tang and Luo 2018; Teng et al. 2018; Wang et al. 2016; Liu et al. 2017; Iqbal et al. 2017; Zhang et al. 2016; Yuan et al. 2018; Ma et al. 2016; Zhao et al. 2018).

2.5 Stress-Inducible Promoters: A Potential Tool to Develop Plants with Sustainable Stress Tolerance

For transgenic studies, regulation of transgenes is very important step. Both the strength of the expression and tissue specificity of transgenes are the important parameters that need to be regulated. These regulations are achieved by the use of specific promoters (Bajaj et al. 1999). Promoters are the regulatory regions upstream to the gene, to which RNA polymerase binds to initiate transcription. The most extensively used promoters for developing stress tolerant transgenic crops (monocots or dicots) is constitutive CaMV (Cauliflower mosaic virus) 35S promoter from plant virus (Odell et al. 1985). Another set of constitutive promoters attained from the plant genes are mainly actin and ubiquitin (Dhankher et al. 2002). Rice actin (*Act1*) and maize ubiquitin (*Ubi1* and *Ubi2*) promoters have been used to enhance stress tolerance, especially in monocot crops (Chen et al. 2016; Ito et al. 2006). No doubt, constitutive promoters are valuable for achieving higher gene expression in all tissues and organs constitutively, but in several cases it has been observed that constitutive expression of some stress responsive TFs may sometimes exert detrimental effects on plant growth and productivity. For instance, constitutive overexpression of *TaDREB1* in transgenic rice showed dwarf phenotype under normal conditions (Shen et al. 2003). In another example, constitutive expression of *OsNAC6* gene also resulted in reduced plant growth (Nakashima et al. 2007). Therefore, to minimize these deleterious effects of constitutive promoters, use of stress inducible promoters is advisable. These promoters get activated in response to various stresses such high temperature, cold and desiccation. The stress induced promoters have been isolated from several stress responsive genes such as *rd, erd* and *cor* (Singhal et al. 2016). There are several success stories of TFs conferring drought and salt tolerance under the control of stress-inducible promoters. The *AtDREB1A* gene under the control of RD29A promoter enhanced tolerance towards drought and salinity without showing any adverse effects on plant growth (Kasuga et al. 1999). Similar results were also obtained for transgenic alfalfa, tobacco, potato and peanut harboring RD29A promoter (Jin et al. 2010; Kasuga et al. 2004; Behnam et al. 2007; Bhatnagar-Mathur et al. 2008). The other advantage of using stress-inducible promoters is the avoidance of any recombination event with its host species DNA that could be possible in case of CaMV35S (Daniell 2002). In field conditions, plants often encounter a combination of two or more stresses simultaneously, therefore, engineering plants with gene(s) of interest, under the control of multiple stress inducible promoters is a desirable strategy (Mittler 2006). Several TFs and proteins that are common to different stress signaling and response pathways can serve as potential candidates for the isolation of multiple stress inducible promoters (Matsui et al. 2008). The promoters of such genes contain a subset of *cis*-elements that are responsive to multiple stresses viz. ABRE, DRE and MYC. The sweet potato oxidative stress inducible promoter, SWPA2 has been successfully employed to develop multiple stress tolerance in alfalfa plants (Wang et al. 2014; Li et al. 2014; Wang et al. 2016). In past few years, multiple-stress responsive promoters have been isolated and characterized in plants like *Arabidopsis*, rice, soybean, common buckwheat, maize and wheat (Wang et al. 2016; Nakashima et al. 2014; Fang et al. 2015; Bhuria et al. 2016; Conforte et al. 2017; Niu et al. 2018). The detailed knowledge of *cis*-regulatory elements and their effects can be helpful in designing synthetic stress-inducible promoters. The strength and spatio-temporal control of such synthetic promoters can be modulated by varying the type, spacing and copy number of *cis*-elements (Wang et al. 2018; Liu and Stewart 2016).

2.6 Conclusions

Exposure of plants to abiotic stresses, such as salinity and drought leads to severe reduction in plant growth and productivity. Presently, overexpressing regulatory genes i.e. TFs to develop stress tolerant plants is the most accepted transgenic approach. These regulatory genes include transcription factors controlling the simultaneous expression of multiple genes involved in plant stress tolerance. TFs are thus considered as potential candidates, which have provided the greater opportunity to enhance tolerance towards multiple stresses, such as drought and salinity. Although, extensive studies have been done to identify and characterize stress-responsive TFs, there are certain issues, such as functional redundancy between different members of TFs and constitutive overexpression of TFs that may exert negative effect on growth, need to be addressed. Therefore, to develop stress tolerant plants via manipulating

TF genes, firstly it is necessary to identify multiple stress-responsive TFs by examining their expression under various stresses. Secondly, the authenticity of these stress-responsive TFs via overexpression studies should be confirmed in both, model plants and crop plants. Additionally, the growth of stress tolerant transgenic plants should also be evaluated under laboratory and field conditions. Finally, engineering TFs is a promising approach to develop multiple-stress tolerant crop plants.

Acknowledgements AKS acknowledges funding support by Indian Council of Agricultural Research, New Delhi in the form of projects IXX12585 and IXX12644; Science and Engineering Research Board (SERB), Department of Science and Technology (DST), Government of India for the project under "Scheme for Young Scientists" (SB/YS/LS-255/2013). RS acknowledges DST-SERB for the National-Postdoctoral Fellowship (PDF/2016/000924), and DST, Govt. of India for funding under the Women Scientists Scheme-A (SR/WOS-A/LS-160/2018).

References

Abe H, Yamaguchi-Shinozaki K, Urao T, Iwasaki T, Hosokawa D, Shinozaki K (1997) Role of Arabidopsis MYC and MYB homologs in drought- and abscisic acid-regulated gene expression. Plant Cell 9:1859–1868. https://doi.org/10.1105/tpc.9.10.1859

Agarwal PK, Jha B (2010) Transcription factors in plants and ABA dependent and independent abiotic stress signalling. Biol Plant 54:201–212. https://doi.org/10.1007/s10535-010-0038-7

Agarwal PK, Agarwal P, Reddy MK, Sopory SK (2006) Role of DREB transcription factors in abiotic and biotic stress tolerance in plants. Plant Cell Rep 25:1263–1274. https://doi.org/10.1007/s00299-006-0204-8

Agarwal P, Arora R, Ray S, Singh AK, Singh VP, Takatsuji H, Kapoor S, Tyagi AK (2007) Genome-wide identification of C2H2 zinc-finger gene family in rice and their phylogeny and expression analysis. Plant Mol Biol 65:467–485. https://doi.org/10.1007/s11103-007-9199-y

Agarwal P, Agarwal PK, Joshi AJ, Sopory SK, Reddy MK (2010) Overexpression of PgDREB2A transcription factor enhances abiotic stress tolerance and activates downstream stress-responsive genes. Mol Biol Rep 37:1125–1135. https://doi.org/10.1007/s11033-009-9885-8

Agarwal P, Baranwal VK, Khurana P (2019) Genome-wide analysis of bZIP transcription factors in wheat and functional characterization of a *TabZIP* under abiotic stress. Sci Rep 9:4608. https://doi.org/10.1038/s41598-019-40659-7

Aleksza D, Horvath GV, Sandor G, Szabados, L (2017) Proline accumulation is regulated by transcription factors associated with phosphate starvation. Plant Physiol 175:555–567. https://doi.org/10.1104/pp.17.00791

Babitha KC, Vemanna RS, Nataraja KN, Udayakumar M (2015) Overexpression of EcbHLH57 transcription factor from *Eleusine coracana* L. in tobacco confers tolerance to salt, oxidative and drought stress. PLoS ONE 10:e0137098. https://doi.org/10.1371/journal.pone.0137098

Babu MM, Iyer LM, Balaji S, Aravind L (2006) The natural history of the WRKY-GCM1 zinc fingers and the relationship between transcription factors and transposons. Nucleic Acids Res 34:6505–6520. https://doi.org/10.1093/nar/gkl888

Bailey PC, Martin C, Toledo-Ortiz G, Quail PH, Huq E, Heim MA, Jakoby M, Werber M, Weisshaar B (2003) Update on the basic helix-loop-helix transcription factor gene family in *Arabidopsis thaliana*. Plant Cell 15:2497–2502. https://doi.org/10.1105/tpc.151140

Bajaj S, Targolli J, Liu LF, Ho THD, Wu R (1999) Transgenic approaches to increase dehydration-stress tolerance in plants. Mol Breed 5:493–503. https://doi.org/10.1023/A:1009660413133

Baldoni E, Genga A, Cominelli E (2015) Plant MYB transcription factors: their role in drought response mechanisms. Int J Mol Sci 16:15811–15851. https://doi.org/10.3390/ijms160715811

Behnam B, Kikuchi A, Celebi-Toprak F, Kasuga M, Yamaguchi-Shinozaki K, Watanabe KN (2007) Arabidopsis rd29A: DREB1A enhances freezing tolerance in transgenic potato. Plant Cell Rep 26:1275–1282. https://doi.org/10.1007/s00299-007-0360-5

Bhatnagar-Mathur P, Devi MJ, Reddy DS, Lavanya M, Vadez V, Serraj R, Yamaguchi-Shinozaki K, Sharma KK (2008) Stress-inducible expression of At DREB1A in transgenic peanut (*Arachis hypogaea* L.) increases transpiration efficiency under water-limiting conditions. Plant Cell Rep 26:2071–2082. https://doi.org/10.1007/s00299-007-0406-8

Bhuria M, Goel P, Kumar S, Singh AK (2016) The promoter of *AtUSP* is co-regulated by phytohormones and abiotic stresses in *Arabidopsis thaliana*. Front Plant Sci 7:1957. https://doi.org/10.3389/fpls.2016.01957

Borgohain P, Saha B, Agrahari R, Chowardhara B, Sahoo S, van der Vyver C, Panda SK (2019) *SlNAC2* overexpression in *Arabidopsis* results in enhanced abiotic stress tolerance with alteration in glutathione metabolism Protoplasma 1–13. https://doi.org/10.1007/s00709-019-01368-0

Bota J, Flexas J, Medrano H (2004) Is photosynthesis limited by decreased Rubisco activity and RuBP content under progressive water stress? New Phytol 162:671–681. https://doi.org/10.1111/j.1469-8137.2004.01056.x

Bouaziz D, Pirrello J, Charfeddine M, Hammami A, Jbir R, Dhieb A, Bouzayen M, Gargouri-Bouzid R (2013) Overexpression of StDREB1 transcription factor increases tolerance to salt in transgenic potato plants. Mol Biotechnol 54:803–817. https://doi.org/10.1007/s12033-012-9628-2

Boyer JS (1982) Plant productivity and environment. Science 218:443–448. https://doi.org/10.1126/science.218.4571.443

Bruinsma J (2009) The resource outlook to: by how much do land, water and crops yields need to increase by 2050? In: Prepared for the FAO Expert Meeting on 'How to Feed the World in 2050; Food and Agriculture Organization (FAO), Rome, Italy pp 1–33

Brunner I, Herzog C, Dawes MA, Arend M, Sperisen C (2015) How tree roots respond to drought. Front Plant Sci 6:547. https://doi.org/10.3389/fpls.2015.00547

Burke EJ, Brown SJ, Christidis N (2006) Modeling the recent evolution of global drought and projections for the twenty-first century with the Hadley Centre climate model. J Hydrometeorol 7:1113–1125. https://doi.org/10.1175/JHM544.1

Busk PK, Pages M (1997) Protein binding to the abscisic acid-responsive element is independent of VIVIPAROUS 1 in vivo. Plant Cell 9:2261–2270. https://doi.org/10.1105/tpc.9.12.2261

Butt HI, Yang Z, Gong Q, Chen E, Wang X, Zhao G, Ge X, Zhang X, Li F (2017) GaMYB85, an R2R3 MYB gene, in transgenic Arabidopsis plays an important role in drought tolerance. BMC Plant Biol 17:142. https://doi.org/10.1186/s12870-017-1078-3

Cao ZH, Zhang SZ, Wang RK, Zhang RF, Hao YJ (2013) Genome wide analysis of the apple MYB transcription factor family allows the identification of MdoMYB121 gene conferring abiotic stress tolerance in plants. PLoS ONE 8:e69955. https://doi.org/10.1371/journal.pone.0069955

Cao H, Wang L, Nawaz MA, Niu M, Sun J, Xie J, Kong Q, Huang Y, Cheng F, Bie Z (2017) Ectopic expression of pumpkin NAC transcription factor CmNAC1 improves multiple abiotic stress tolerance in *Arabidopsis*. Front Plant Sci 8:2052. https://doi.org/10.3389/fpls.2017.02052

Charfeddine M, Saidi MN, Charfeddine S, Hammami A, Gargouri Bouzid R (2015) Genome-wide analysis and expression profiling of the ERF transcription factor family in potato (*Solanum tuberosum* L.). Mol Biotechnol 57:348–358. https://doi.org/10.1007/s12033-014-9828-z

Chen H, Lai Z, Shi J, Xiao Y, Chen Z, Xu X (2010) Roles of *Arabidopsis* WRKY18, WRKY40 and WRKY60 transcription factors in plant responses to abscisic acid and abiotic stress. BMC Plant Biol 10:281. https://doi.org/10.1186/1471-2229-10-281

Chen Y, Li F, Ma Y, Chong K, Xu Y (2013) Overexpression of OrbHLH001, a putative helix-loop-helix transcription factor, causes increased expression of AKT1 and maintains ionic balance under salt stress in rice. J Plant Physiol 170:93–100. https://doi.org/10.1016/j.jplph.2012.08.019

Chen D, Richardson T, Chai S, Lynne McIntyre C, Rae A, Xue GP (2016) Drought-Up-Regulated TaNAC69-1 is a transcriptional repressor of TaSHY2 and TaIAA7, and enhances root length and biomass in wheat. Plant Cell Physiol 10:2076–2090. https://doi.org/10.1093/pcp/pcw126

Chen HC, Hsieh-Feng V, Liao PC, Cheng WH, Liu LY, Yang YW, Lai MH, Chang MC (2017) The function of OsbHLH068 is partially redundant with its homolog, AtbHLH112, in the regulation of the salt stress response but has opposite functions to control flowering in Arabidopsis. Plant Mol Biol 94:531–548. https://doi.org/10.1007/s11103-017-0624-6

Chen HC, Cheng WH, Hong CY, Chang YS, Chang MC (2018) The transcription factor OsbHLH035 mediates seed germination and enables seedling recovery from salt stress through ABA-dependent and ABA-independent pathways, respectively. Rice 11:50. https://doi.org/10.1186/s12284-018-0244-z

Cheng LB, Yang JJ, Yin L, Hui LC, Qian HM, Li SY, Li LJ (2017) Transcription factor NnDREB1 from lotus improved drought tolerance in transgenic *Arabidopsis thaliana*. Biol Plant 61:651–658. https://doi.org/10.1007/s10535-017-0718-7

Cheong YH, Kim KN, Pandey GK, Gupta R, Grant JJ, Luan S (2003) CBL1, a calcium sensor that differentially regulates salt, drought, and cold responses in Arabidopsis. Plant Cell 15:1833–1845. https://doi.org/10.1105/tpc.012393

Cheong YH, Sung SJ, Kim BG, Pandey GK, Cho JS, Kim KN, Luan S (2010) Constitutive overexpression of the calcium sensor CBL5 confers osmotic or drought stress tolerance in Arabidopsis. Mol Cells 29:159–165. https://doi.org/10.1007/s10059-010-0025-z

Choudhury S, Panda P, Sahoo L, Panda SK (2013) Reactive oxygen species signaling in plants under abiotic stress. Plant Signal Behav 8:e23681. https://doi.org/10.4161/psb.23681

Ciftci-Yilmaz S, Morsy MR, Song L, Coutu A, Krizek BA, Lewis MW, Warren D, Cushman J, Connolly EL, Mittler R (2007) The EAR-motif of the Cys2/His2-type zinc finger protein Zat7 plays a key role in the defense response of Arabidopsis to salinity stress. J Biol Chem 282:9260–9268. https://doi.org/10.1074/jbc.M611093200

Conforte AJ, Guimaraes-Dias F, Neves-Borges AC, Bencke-Malato M, Felix-Whipps D, Alves-Ferreira M (2017) Isolation and characterization of a promoter responsive to salt, osmotic and dehydration stresses in soybean. Genet Mol Biol 40:226–237. https://doi.org/10.1590/1678-4685-GMB-2016-0052

Cui J, Jiang N, Zhou XX, Hou XX, Yang GL, Meng J, Luan YS (2018a) Tomato MYB49 enhances resistance to *Phytophthora infestans* and tolerance to water deficit and salt stress. Planta 248:1487–1503. https://doi.org/10.1007/s00425-018-2987-6

Cui X, Wang YX, Liu ZW, Wang WL, Li H, Zhuang J (2018b) Transcriptome-wide identification and expression profile analysis of the bHLH family genes in *Camellia sinensis*. Funct Integr Genomics 5:489–503. https://doi.org/10.1007/s10142-018-0608-x

Daniell H (2002) Molecular strategies for gene containment in transgenic crops. Nat Biotech 20:581–586. https://doi.org/10.1038/nbt0602-581

De Oliveira VP, Marques EC, De Lacerda CF, Prisco JT, Gomes-Filho E (2013) Physiological and biochemical characteristics of *Sorghum bicolor* and *Sorghum sudanense* subjected to salt stress in two stages of development. Afr J Agric Res 8:660–670. https://doi.org/10.5897/AJAR12.861

Dhankher OP, Li Y, Rosen BP, Shi J, Salt D, Senecoff JF, Sashti NA, Meagher RB (2002) Engineering tolerance and hyperaccumulation of arsenic in plants by combining arsenate reductase and [gamma]-glutamylcysteine synthetase expression. Nat Biotech 20:1140–1145. https://doi.org/10.1038/nbt747

Diao W, Snyder JC, Wang S, Liu J, Pan B, Guo G, Ge W, Dawood MHSA (2018) Genome-Wide analyses of the NAC transcription factor gene family in pepper (*Capsicum annuum* L.): chromosome location, phylogeny, structure, expression patterns, cis-elements in the promoter, and interaction network. Int J Mol Sci 19:1028. https://doi.org/10.3390/ijms19041028

Dietz KJ, Vogel MO, Viehhauser A (2010) AP2/EREBP transcription factors are part of gene regulatory networks and integrate metabolic, hormonal and environmental signals in stress acclimation and retrograde signaling. Protoplasma 245:3–14. https://doi.org/10.1007/s00709-010-0142-8

Dolferus R (2014) To grow or not to grow: a stressful decision for plants. Plant Sci 229:247–261. https://doi.org/10.1016/j.plantsci.2014.10.002

Dong Y, Wang C, Han X, Tang S, Liu S, Xia X, Yin W (2014) A novel bHLH transcription factor PebHLH35 from *Populus euphratica* confers drought tolerance through regulating stomatal development, photosynthesis and growth in Arabidopsis. Biochem Biophys Res Commun 450:453–458. https://doi.org/10.1016/j.bbrc.2014.05.139

Du H, Yang SS, Liang Z, Feng BR, Liu L, Huang YB, Tang YX (2012) Genome-wide analysis of the MYB transcription factor superfamily in soybean. BMC Plant Biol 12:106. https://doi.org/10.1186/1471-2229-12-106

Du C, Zhao P, Zhang H, Li N, Zheng L, Wang Y (2017) The *Reaumuria trigyna* transcription factor RtWRKY1 confers tolerance to salt stress in transgenic Arabidopsis. J Plant Physiol 215:48–58. https://doi.org/10.1016/j.jplph.2017.05.002

Du YT, Zhao MJ, Wang CT, Gao Y, Wang YX, Liu YW, Chen M, Chen J, Zhou YB, Xu ZS (2018) Identification and characterization of *GmMYB118* responses to drought and salt stress. BMC Plant Biol 18:320. https://doi.org/10.1186/s12870-018-1551-7

Dubos C, Stracke R, Grotewold E, Weisshaar B, Martin C, Lepiniec L (2010) MYB transcription factors in Arabidopsis. Trends Plant Sci 15:573–581. https://doi.org/10.1016/j.tplants.2010.06.005

El-Esawi MA, Alayafi AA (2019) Overexpression of StDREB2 Transcription Factor enhances drought stress tolerance in cotton (*Gossypium barbadense* L.). Genes (Basel) 14:2. https://doi.org/10.3390/genes10020142

El-Esawi MA, Al-Ghamdi AA, Ali HM, Ahmad M (2019) Overexpression of AtWRKY30 transcription factor enhances heat and drought stress tolerance in wheat (*Triticum aestivum* L.). Genes 10:163. https://doi.org/10.3390/genes10020163

Englbrecht CC, Schoof H, Bohm S (2004) Conservation, diversification and expansion of C2H2 zinc finger proteins in the *Arabidopsis thaliana* genome. BMC Genom 5:39. https://doi.org/10.1186/1471-2164-5-39

Fang Y, You J, Xie K, Xie W, Xiong L (2008) Systematic sequence analysis and identification of tissue-specific or stress-responsive genes of NAC transcription factor family in rice. Mol Genet Genomics 280:547–563. https://doi.org/10.1007/s00438-008-0386-6

Fang ZW, Xu XY, Gao JF, Wang PK, Liu ZX, Feng BL (2015) Characterization of FeDREB1 promoter involved in cold- and drought-inducible expression from common buckwheat (Fagopyrum esculentum). Genet Mol Res 14:7990–8000. https://doi.org/10.4238/2015

Farooq M, Wahid A, Kobayashi N, Fujita D, Basra SMA (2009) Plant drought stress: effects, mechanisms and management. Agron Sustain Dev 29:185–212. https://doi.org/10.1051/agro:2008021

Favero Peixoto-Junior R, Mara de Andrade L, Dos Santos Brito M, Macedo Nobile P, Palma Boer Martins A, Domingues Carlin S, Vasconcelos Ribeiro R, de Souza Goldman MH, Carlos Nebo, de Oliveira JF, de Oliveira Vargas, Figueira A, Creste S (2018) Overexpression of ScMYBAS1 alternative splicing transcripts differentially impacts biomass accumulation and drought tolerance in rice transgenic plants. PLoS ONE 13:e0207534. https://doi.org/10.1371/journal.pone.0207534

Feller A, Machemer K, Braun EL, Grotewold E (2011) Evolutionary and comparative analysis of MYB and bHLH plant transcription factors. Plant J 66:94–116. https://doi.org/10.1111/j.1365-313X.2010.04459.x

Fini A, Brunetti C, Di Ferdinando M, Ferrini F, Tattini M (2011) Stress-induced flavonoid biosynthesis and the antioxidant machinery of plants. Plant Signal Behav 6:709–711. https://doi.org/10.4161/psb.6.5.15069

Flowers TJ, Troke PF, Yeo AR (1977) The mechanism of salt tolerance in halophytes. Annu Rev Plant Physiol 28:89–121. https://doi.org/10.1146/annurev.pp.28.060177.000513

Forni C, Duca D, Glick BR (2017) Mechanisms of plant response to salt and drought stress and their alteration by rhizobacteria. Plant Soil 410:335–356. https://doi.org/10.1007/s11104-016-3007-x

Fraire-Velazquez S Balderas-Hernandez VE (2013) Abiotic stress in plants and metabolic responses. In: Vahdati K, Leslie C (ed) Abiotic stress—plant responses and applications in agriculture. InTech. https://doi.org/10.5772/45842

Gao JP, Chao DY, Lin HX (2008) Toward understanding molecular mechanisms of abiotic stress responses in rice. Rice 1:90006. https://doi.org/10.1007/s12284-008-9006-7

Gao H, Wang Y, Xu P, Zhang Z (2018) Overexpression of a WRKY transcription factor TaWRKY2 enhances drought stress tolerance in transgenic wheat. Front Plant Sci 9:997. https://doi.org/10.3389/fpls.2018.00997

Goel P, Singh AK (2018) Single-versus multigene transfer approaches for crop abiotic stress tolerance. In: Wani SH (ed) Biochemical, physiological and molecular avenues for combating abiotic stress tolerance in plants. Academic Press

Guo HY, Wang YC, Wang LQ, Hu P, Wang YM, Jia YY, Zhang CR, Zhang Y, Zhang YM, Wang C (2017) Expression of the MYB transcription factor gene BplMYB46 affects abiotic stress tolerance and secondary cell wall deposition in *Betula platyphylla*. Plant Biotechnol J 15:107–121. https://doi.org/10.1111/pbi.12595

Gutterson N, Reuber TL (2004) Regulation of disease resistance pathways by AP2/ERF transcription factors. Curr Opin Plant Biol 7:465–471. https://doi.org/10.1016/j.pbi.2004.04.007

Han DG, Zhang ZY, Ding HB, Chai LJ, Liu W, Li HX, Yang GH (2018a) Isolation and characterization of MbWRKY2 gene involved in enhanced drought tolerance in transgenic tobacco. J Plant Interact 13:163–172. https://doi.org/10.1080/17429145.2018.1447698

Han DG, Hou Y, Wang Y, Ni B, Li Z, Yang G (2018b) Overexpression of a *Malus baccata* WRKY transcription factor gene (MbWRKY5) increases drought and salt tolerance in transgenic tobacco. Can J Plant Sci 99:173–183. https://doi.org/10.1139/cjps-2018-0053

Hand SC, Menze MA, Toner M, Boswell L, Moore D (2011) LEA proteins during water stress: not just for plants anymore. Annu Rev Physiol 73:115–134. https://doi.org/10.1146/annurev-physiol-012110-142203

Hao D, Ohme-Takagi M, Sarai A (1998) Unique mode of GCC box recognition by the DNA-binding domain of ethylene-responsive element-binding factor (ERF domain) in plants. J Biol Chem 273:26857–26861. https://doi.org/10.1074/jbc.273.41.26857

Hasegawa PM, Bressan RA, Zhu JK, Bohnert HJ (2000) Plant cellular and molecular responses to high salinity. Annu Rev Plant Physiol Plant Mol Biol 51:463–499. https://doi.org/10.1146/annurev.arplant.51.1.463

Hayat S, Hayat Q, Alyemeni MN, Wani AS, Pichtel J, Ahmad A (2012) Role of proline under changing environments. A Rev Plant Signal Behav 11:1456–1466. https://doi.org/10.4161/psb.21949

He L, Wu Y, Zhao Q, Wang B, Liu Q, Zhang L (2018) Chrysanthemum DgWRKY2 gene enhances tolerance to salt stress in transgenic chrysanthemum. Int J Mol Sci 19:2062. https://doi.org/10.3390/ijms19072062

He K, Zhao X, Chi X, Wang Y, Jia C, Zhang H, Zhou G, Hu R (2019) A novel *Miscanthus* NAC transcription factor MlNAC10 enhances drought and salinity tolerance in transgenic *Arabidopsis*. J Plant Physiol 233:84–93. https://doi.org/10.1016/j.jplph.2019.01.001

Hong JP, Kim WT (2005) Isolation and functional characterization of the Ca-DREBLP1 gene encoding a dehydration-responsive element binding-factor-like protein 1 in hot pepper (*Capsicum annuum* L. cv Pukang). Planta 220:875–888. https://doi.org/10.1007/s00425-004-1412-5

Hong-Bo S, Zong-Suo L, Ming S (2005) LEA proteins in higher plants: structure, function, gene expression and regulation. Colloids Surf B Biointerfaces 45:131–135. https://doi.org/10.1016/j.colsurfb.2005.07.017

Hu XQ, Xu XM, Li CH (2018) Ectopic expression of the LoERF017 transcription factor from *Larix olgensis* Henry enhances salt and osmotic-stress tolerance in *Arabidopsis thaliana*. Plant Biotechnol Rep 12:93–104. https://doi.org/10.1007/s11816-018-0473-z

Huang Y, Zhao H, Gao F, Yao P, Deng R, Li C, Chen H, Wu Q (2018) A R2R3-MYB transcription factor gene, FtMYB13, from Tartary buckwheat improves salt/drought tolerance in Arabidopsis. Plant Physiol Biochem 132:238–248. https://doi.org/10.1016/j.plaphy.2018.09.012

Hubbard KE, Nishimura N, Hitomi K, Getzoff ED, Schroeder JI (2010) Early abscisic acid signal transduction mechanisms: newly discovered components and newly emerging questions. Genes Dev 24:1695–1708. https://doi.org/10.1101/gad.1953910

Ichimura K, Mizoguchi T, Yoshida R, Yuasa T, Shinozaki K (2000) Various abiotic stresses rapidly activate Arabidopsis MAP kinases ATMPK4 and ATMPK6. Plant J 24:655–665. https://doi.org/10.1046/j.1365-313x.2000.00913.x

Iordachescu M, Imai R (2008) Trehalose biosynthesis in response to abiotic stresses. J Integr Plant Biol 50:1223–1229. https://doi.org/10.1111/j.1744-7909.2008.00736.x

Iqbal F, Sadique S, Batool F, Sarwar MB, Rashid B, Shahid MN, Shahid AA, Hussain T (2017) Zinc finger transcription factor induces the drought, salt and cold stress tolerance in transgenic cotton. Indian J Biotech 16:333–340. http://nopr.niscair.res.in/handle/123456789/43338

Ishiguro S, Nakamura K (1994) Characterization of a cDNA encoding a novel DNA-binding protein, SPF1, that recognizes SP8 sequences in the 5′ upstream regions of genes coding for sporamin and β-amylase from sweet potato. Mol Gen Genet 244:563–571. https://doi.org/10.1007/BF00282746

Ito Y, Katsura K, Maruyama K, Taji T, Kobayashi M, Seki M, Shinozaki K, Yamaguchi-Shinozaki K (2006) Functional analysis of rice DREB1/CBF-type transcription factors involved in cold-responsive gene expression in transgenic rice. Plant Cell Physiol 47:141–153. https://doi.org/10.1093/pcp/pci230

Ito S, Song YH, Josephson-Day AR, Miller RJ, Breton G, Olmstead RG, Imaizumi T (2012) FLOWERING BHLH transcriptional activators control expression of the photoperiodic flowering regulator CONSTANS in Arabidopsis. Proc Natl Acad Sci U S A 109:3582–3587. https://doi.org/10.1073/pnas.1118876109

Jain M, Nagar P, Goel P, Singh AK, Kumari S, Mustafiz A (2018) Second messengers: central regulators in plant abiotic stress response. In: Zargar SM, Zargar MY (eds) Abiotic stress-mediated sensing and signaling in plants: an omics perspective. https://doi.org/10.1007/978-981-10-7479-0_2

Jakoby M, Weisshaar B, Droge-Laser W, Vicente-Carbajosa J, Tiedemann J, Kroj T, Parcy F (2002) bZIP transcription factors in Arabidopsis. Trends Plant Sci 7:106–111. https://doi.org/10.1016/S1360-1385(01)02223-3

Ji X, Nie X, Liu Y, Zheng L, Zhao H, Zhang B, Huo L, Wang Y (2016) A bHLH gene from *Tamarix hispida* improves abiotic stress tolerance by enhancing osmotic potential and decreasing reactive oxygen species accumulation. Tree Physiol 6:193–207. https://doi.org/10.1093/treephys/tpv139

Jia L, Clegg MT, Jiang T (2004) Evolutionary dynamics of the DNA-binding domains in putative R2R3-MYB genes identified from rice subspecies *indica* and *japonica* genomes. Plant Physiol 134:575–585. https://doi.org/10.1104/pp.103.027201

Jia H, Zhang S, Ruan M, Wang Y, Wang C (2012) Analysis and application of RD29 genes in abiotic stress response. Acta Physiol Plant 34:1239–1250. https://doi.org/10.1007/s11738-012-0969-z

Jiang Y, Deyholos MK (2006) Comprehensive transcriptional profiling of NaCl-stressed Arabidopsis roots reveals novel classes of responsive genes. BMC Plant Biol 6:2510. https://doi.org/10.1186/1471-2229-6-25

Jiang X, Li S, Ding A, Zhang Z, Hao Q, Wang K, Liu K, Liu Q (2018) The novel rose MYB transcription factor RhMYB96 enhances salt tolerance in transgenic Arabidopsis. Plant Mol Biol Rep 36:406–417. https://doi.org/10.1007/s11105-018-1094-y

Jin T, Chang Q, Li W, Yin D, Li Z, Wang D, Liu B, Liu L (2010) Stress-inducible expression of GmDREB1 conferred salt tolerance in transgenic alfalfa. Plant Cell, Tissue Organ Cult 100:219–227. https://doi.org/10.1007/s11240-009-9628-5

Jin JP, Tian F, Yang DC, Meng YQ, Kong L, Luo JC, Gao G (2017a) PlantTFDB 4.0: toward a central hub for transcription factors and regulatory interactions in plants. Nucleic Acids Res 45:D1040–D1045. https://doi.org/10.1093/nar/gkw982

Jin C, Li KQ, Xu XY, Zhang HP, Chen HX, Chen YH, Hao J, Wang Y, Huang XS, Zhang SL (2017b) A novel NAC Transcription factor, PbeNAC1, of *Pyrus betulifolia* confers cold and drought tolerance via interacting with *PbeDREBs* and activating the expression of stress-responsive genes. Front Plant Sci 8:1049. https://doi.org/10.3389/fpls.2017.01049

Jin JH, Wang M, Zhang HX, Khan A, Wei AM, Luo DX, Gong ZH (2018) Genome-wide identification of the AP2/ERF transcription factor family in pepper (*Capsicum annuum* L.). Genome 61:663–674. https://doi.org/10.1139/gen-2018-0036

Jofuku KD, den Boer BGW, Van Montagu M, Okamuro JK (1994) Control of Arabidopsis flower and seed development by the homeotic gene APETALA2. Plant Cell 6:1211–1225. https://doi.org/10.1105/tpc.6.9.1211

Johnson CS, Kolevski B, Smyth DR (2002) TRANSPARENT TESTA GLABRA2, a trichome and seed coat development gene of Arabidopsis, encodes a WRKY transcription factor. Plant Cell 14:1359–1375. https://doi.org/10.1105/tpc.001404

Joo J, Lee YH, Song SI (2019) OsbZIP42 is a positive regulator of ABA signaling and confers drought tolerance to rice. Planta 249:1521–1533. https://doi.org/10.1007/s00425-019-03104-7

Joshi R, Singla-Pareek SL, Pareek A (2018) Engineering abiotic stress response in plants for biomass production. J Biol Chem 293:5035–5043. https://doi.org/10.1074/jbc.TM117.000232

Jung H, Chung PJ, Park SH, Redillas MCFR, Kim YS, Suh JW, Kim JK (2017) Overexpression of OsERF48 causes regulation of OsCML16, a calmodulin-like protein gene that enhances root growth and drought tolerance. Plant Biotechnol J 15:1295–1308. https://doi.org/10.1111/pbi.12716

Kaplan B, Davydov O, Knight H, Galon Y, Knight MR, Fluhr R, Fromm H (2006) Rapid transcriptome changes induced by cytosolic Ca^{2+} transients reveal ABRE-related sequences as Ca^{2+}-responsive cis elements in Arabidopsis. Plant Cell 18:2733–2748. https://doi.org/10.1105/tpc.106.042713

Kasuga M, Liu Q, Miura S, Yamaguchi-Shinozaki K, Shinozaki K (1999) Improving plant drought, salt, and freezing tolerance by gene transfer of a single stress-inducible transcription factor. Nat Biotechnol 17:287–291. https://doi.org/10.1038/7036

Kasuga M, Miura S, Shinozaki K, Yamaguchi-Shinozaki K (2004) A combination of the Arabidopsis DREB1A gene and stress-inducible rd29A promoter improved drought- and low-temperature stress tolerance in tobacco by gene transfer. Plant Cell Physiol 45:346–350. https://doi.org/10.1093/pcp/pch037

Katiyar A, Smita S, Lenka SK, Rajwanshi R, Chinnusamy V, Bansal KC (2012) Genome-wide classification and expression analysis of MYB transcription factor families in rice and Arabidopsis. BMC Genom 13:544. https://doi.org/10.1186/1471-2164-13-544

Kim TH, Bohmer M, Hu H, Nishimura N, Schroeder JI (2010) Guard cell signal transduction network: advances in understanding abscisic acid, CO_2, and Ca^{2+} Signaling. Annu Rev Plant Biol 61:561–591. https://doi.org/10.1146/annurev-arplant-042809-112226

Klempnauer KH, Gonda TJ, Bishop JM (1982) Nucleotide sequence of the retroviral leukemia gene v-myb and its cellular progenitor c-myb: the architecture of a transduced oncogene. Cell 31:453–463. https://doi.org/10.1016/0092-8674(82)90138-6

Kolukisaoglu U, Weinl S, Blazevic D, Batistic O, Kudla J (2004) Calcium sensors and their interacting protein kinases: genomics of the Arabidopsis and rice CBL-CIPK signaling networks. Plant Physiol 134:43–58. https://doi.org/10.1104/pp.103.033068

Kusvuran A (2015) The effects of salt stress on the germination and antioxidative enzyme activity of Hungarian vetch (*Vicia pannonica* Crantz.) varieties. Legume Res 38:51–59. https://doi.org/10.5958/0976-0571.2015.00009.0

Lagace M, Matton DP (2004) Characterization of a WRKY transcription factor expressed in late torpedo-stage embryos of *Solanum chacoense*. Planta 219:185–189. https://doi.org/10.1007/s00425-004-1253-2

Landi S, Hausman JF, Guerriero G, Esposito S (2017) Poaceae vs. abiotic stress: focus on drought and salt stress, recent insights and perspectives. Front Plant Sci 8:1214. https://doi.org/10.3389/fpls.2017.01214

Le Hir R, Castelain M, Chakraborti D, Moritz T, Dinant S, Bellini C (2017) AtbHLH68 transcription factor contributes to the regulation of ABA homeostasis and drought stress tolerance in Arabidopsis thaliana. Physiol Plant 160:312–327. https://doi.org/10.1111/ppl.12549

Le DT, Nishiyama R, Watanabe Y, Mochida K, Yamaguchi-Shinozaki K, Shinozaki K, Tran LS (2011) Genome-wide survey and expression analysis of the plant-specific NAC transcription factor family in soybean during development and dehydration stress. DNA Res 18:263–276. https://doi.org/10.1093/dnares/dsr015

Lee DK, Jung H, Jang G, Jeong JS, Kim YS, Ha SH, Do Choi Y, Kim JK (2016) Overexpression of the OsERF71 transcription factor alters rice root structure and drought resistance. Plant Physiol 172:575–588. https://doi.org/10.1104/pp.16.00379

Li X, Duan X, Jiang H, Sun Y, Tang Y, Yuan Z, Guo J, Liang W, Chen L, Yin J, Ma H, Wang J, Zhang D (2006) Genome-wide analysis of basic/helix-loop-helix transcription factor family in rice and Arabidopsis. Plant Physiol 141:1167–1184. https://doi.org/10.1104/pp.106.080580

Li S, Zhou X, Chen L, Huang W, Yu D (2010) Functional characterization of Arabidopsis thaliana WRKY39 in heat stress. Mol Cells 29:475–483. https://doi.org/10.1007/s10059-010-0059-2

Li H, Wang Z, Ke Q, Ji CY, Jeong JC, Lee HS, Lim YP, Xu B, Deng XP, Kwak SS (2014) Overexpression of *codA* gene confers enhanced tolerance to abiotic stresses in alfalfa. Plant Physiol Biochem 85:31–40. https://doi.org/10.1016/j.plaphy.2014.10.010

Li S, Yu X, Cheng Z, Yu X, Ruan M, Li W, Peng M (2017a) Global gene expression analysis reveals crosstalk between response mechanisms to cold and drought stresses in cassava seedlings. Front Plant Sci 8:1259. https://doi.org/10.3389/fpls.2017.01259

Li H, Zhang Y, Guo Q, Yao W (2017b) Molecular characterization of a DREB gene from *Sophora moorcroftiana*, an endemic species of plateau. Protoplasma 254:1735–1741. https://doi.org/10.1007/s00709-016-1065-9

Li KQ, Xing CH, Yao ZH, Huang XS (2017c) PbrMYB21, a novel MYB protein of Pyrus betulaefolia, functions in drought tolerance and modulates polyamine levels by regulating arginine decarboxylase gene. Plant Biotechnol J 15:1186–1203. https://doi.org/10.1111/pbi.12708

Li Y, Chen Q, Nan H, Li X, Lu S, Zhao X, Liu B, Guo C, Kong F, Cao D (2017d) Overexpression of GmFDL19 enhances tolerance to drought and salt stresses in soybean. PLoS ONE 12:e0179554. https://doi.org/10.1371/journal.pone.0179554

Li X, Liang Y, Gao B, Mijiti M, Bozorov TA, Yang H, Zhang D, Wood AJ (2019a) ScDREB10, an A-5c type of DREB gene of the desert moss *Syntrichia caninervis*, confers osmotic and salt tolerances to Arabidopsis. Genes 10:146. https://doi.org/10.3390/genes10020146

Li Y, Zhang H, Zhang Q, Liu Q, Zhai H, Zhao N, He S (2019b) An AP2/ERF gene, *IbRAP2-12*, from sweet potato is involved in salt and drought tolerance in transgenic *Arabidopsis*. Plant Sci 281:19–30. https://doi.org/10.1016/j.plantsci.2019.01.009

Liao X, Guo X, Wang Q, Wang Y, Zhao D, Yao L, Wang S, Liu G, Li T (2017) Overexpression of MsDREB6.2 results in cytokinin-deficient developmental phenotypes and enhances drought tolerance in transgenic apple plants. Plant J 89:510–526. https://doi.org/10.1111/tpj.13401

Licausi F, Giorgi F, Zenoni S, Osti F, Pezzotti M, Perata P (2010) Genomic and transcriptomic analysis of the AP2/ERF superfamily in *Vitis vinifera*. BMC Genom 11:719. https://doi.org/10.1186/1471-2164-11-719

Lim CW, Baek W, Lee SC (2018) Roles of pepper bZIP protein CaDILZ1 and its interacting partner RING-type E3 ligase CaDSR1 in modulation of drought tolerance. Plant J 96:452–467. https://doi.org/10.1111/tpj.14046

Ling J, Jiang W, Zhang Y, Yu H, Mao Z, Gu X, Huang S, Xie B (2011) Genome-wide analysis of WRKY gene family in *Cucumis sativus*. BMC Genom 12:471. https://doi.org/10.1186/1471-2164-12-471

Liu W, Stewart CN (2016) Plant synthetic promoters and transcription factors. Curr Opin Biotechnol 37:36–44. https://doi.org/10.1016/j.copbio.2015.10.001

Liu C, Zhang T (2017) Expansion and stress responses of the AP2/EREBP superfamily in cotton. BMC Genom 18:118. https://doi.org/10.1186/s12864-017-3517-9

Liu Q, Kasuga M, Sakuma Y, Abe H, Miura S, Goda H, Shimada Y, Yoshida S, Shinozaki K, Yamaguchi-Shinozaki K (1998a) Two transcription factors, DREB1 and DREB2, with an EREBP/AP2 DNA binding domain separate two cellular signal transduction pathways in drought- and low-temperature-responsive gene expression, respectively, in Arabidopsis. Plant Cell 10:391–406. https://doi.org/10.1105/tpc.10.8.1391

Liu Q, Kasuga M, Sakuma Y, Abe H, Miura S, Yamaguchi-Shinozaki K, Shinozaki K (1998b) Two transcription factors, DREB1 and DREB2, with an EREBP/AP2 DNA binding domain separate two cellular signal transduction pathways in drought- and low-temperature-responsive gene expression, respectively, in Arabidopsis. Plant Cell 10:1391–1406. https://doi.org/10.1105/tpc.10.8.1391

Liu T, Song X, Duan W, Huang Z, Liu G, Li Y, Hou X (2014a) Genome-Wide analysis and expression patterns of NAC transcription factor family under different developmental stages and abiotic stresses in Chinese cabbage. Plant Mol Biol Rep 32:1041–1056. https://doi.org/10.1007/s11105-014-0712-6

Liu C, Mao B, Ou S, Wang W, Liu L, Wu Y, Chu C, Wang X (2014b) OsbZIP71, a bZIP transcription factor, confers salinity and drought tolerance in rice. Plant Mol Biol 84:19–36. https://doi.org/10.1007/s11103-013-0115-3

Liu Y, Ji X, Nie X, Qu M, Zheng L, Tan Z, Zhao H, Huo L, Liu S, Zhang B, Wang Y (2015) Arabidopsis AtbHLH112 regulates the expression of genes involved in abiotic stress tolerance by binding to their E-box and G Cao CG-box motifs. New Phytol 207:692–709. https://doi.org/10.1111/nph.13387

Liu Y, Song Q, Li D, Yang X, Li D (2017a) Multifunctional roles of plant dehydrins in response to environmental stresses. Front Plant Sci 8:1018. https://doi.org/10.3389/fpls.2017.01018

Liu C, Xie T, Chen C, Luan A, Long J, Li C, Ding Y, He Y (2017b) Genome-wide organization and expression profiling of the R2R3-MYB transcription factor family in pineapple (*Ananas comosus*). BMC Genom 18:503. https://doi.org/10.1186/s12864-017-3896

Liu D, Yang L, Luo M, Wu Q, Liu S, Liu Y (2017c) Molecular cloning and characterization of PtrZPT2-1, a ZPT2 family gene encoding a Cys2/His2-type zinc finger protein from trifoliate orange (*Poncirus trifoliata* (L.) Raf.) that enhances plant tolerance to multiple abiotic stresses. Plant Sci 263:66–78. https://doi.org/10.1016/j.plantsci.2017.07.012

Liu C, Wang B, Li Z, Peng Z, Zhang J (2018a) TsNAC1 is a key transcription factor in abiotic stress resistance and growth. Plant Physiol 176:742–756. https://doi.org/10.1104/pp.17.01089

Liu C, Sun Q, Zhao L, Li Z, Peng Z, Zhang J (2018b) Heterologous expression of the transcription factor EsNAC1 in Arabidopsis enhances abiotic stress resistance and retards growth by regulating the expression of different target genes. Front Plant Sci 9:1495. https://doi.org/10.3389/fpls.2018.01495

Liu J, Chu J, Ma C, Jiang Y, Ma Y, Xiong J, Cheng ZM (2019) Overexpression of an ABA-dependent grapevine bZIP transcription factor, VvABF2, enhances osmotic stress in Arabidopsis. Plant Cell Rep 38:587–596. https://doi.org/10.1007/s00299-019-02389-y

Llorca CM, Potschin M, Zentgra U (2014) bZIPs and WRKYs: two large transcription factor families executing two different functional strategies. Front Plant Sci 5:169. https://doi.org/10.3389/fpls.2014.00169

Lu M, Zhang D, Shi Y, Song Y, Wang T, Li Y (2013) Expression of *SbSNAC1*, a NAC transcription factor from sorghum, confers drought tolerance to transgenic Arabidopsis. Plant Cell, Tissue Organ Cult 115:443–455. https://doi.org/10.1007/s11240-013-0375-2

Lucas S, Durmaz E, Akpinar BA, Budak H (2011) The drought response displayed by a DRE-binding protein from *Triticum dicoccoides*. Plant Physiol Biochem 49:346–351. https://doi.org/10.1016/j.plaphy.2011.01.016

Ludwig AA, Romeis T, Jones JDG (2004) CDPK-mediated signalling pathways: specificity and cross-talk. J Exp Bot 55:181–188. https://doi.org/10.1093/jxb/erh008

Ma JC, Lu J, Xu J, Duan B, He X, Liu J (2015) Genome-wide identification of WRKY genes in the desert poplar *Populus euphratica* and adaptive evolution of the genes in response to salt stress. Evol Bioinform 11:47–55. https://doi.org/10.4137/EBO.S22067

Ma X, Liang W, Gu P, Huang Z (2016) Salt tolerance function of the novel C2H2-type zinc finger protein TaZNF in wheat. Plant Physiol Biochem 106:129–140. https://doi.org/10.1016/j.plaphy.2016.04.033

Ma J, Gao X, Liu Q, Shao Y, Zhang D, Jiang L, Li C (2017) Overexpression of TaWRKY146 increases drought tolerance through inducing stomatal closure in *Arabidopsis thaliana*. Front Plant Sci 8:2036. https://doi.org/10.3389/fpls.2017.02036

Ma Q, Xia Z, Cai Z, Li L, Cheng Y, Liu J, Nian H (2019) GmWRKY16 enhances drought and salt tolerance through an ABA-mediated pathway in *Arabidopsis thaliana*. Front Plant Sci 9:1979. https://doi.org/10.3389/fpls.2018.01979

Magwanga RO, Lu P, Kirungu JN, Lu H, Wang X, Cai X, Zhou Z, Zhang Z, Salih H, Wang K, Liu F (2018) Characterization of the late embryogenesis abundant (LEA) proteins family and their role in drought stress tolerance in upland cotton. BMC Genet 19:6. https://doi.org/10.1186/s12863-017-0596-1

Mallikarjuna G, Mallikarjuna K, Reddy M, Kaul T (2011) Expression of OsDREB2A transcription factor confers enhanced dehydration and salt stress tolerance in rice (*Oryza sativa* L.). Biotechnol Lett 33:1689–1697. https://doi.org/10.1007/s10529-011-0620-x

Mao X, Jia D, Li A, Zhang H, Tian S, Zhang X, Jia J, Jing R (2011) Transgenic expression of TaMYB2A confers enhanced tolerance to multiple abiotic stresses in Arabidopsis. Funct Integr Genomics 11:445–465. https://doi.org/10.1007/s10142-011-0218-3

Mao X, Chen S, Li A, Zhai C, Jing R (2014) Novel NAC transcription factor TaNAC67 confers enhanced multi-abiotic stress tolerances in Arabidopsis. PLoS ONE 9:e84359. https://doi.org/10.1371/journal.pone.0084359

Matsui A, Ishida J, Morosawa T, Mochizuki Y, Kaminuma E, Endo TA, Okamoto M, Nambara E, Nakajima M, Kawashima M, Satou M, Kim JM, Kobayashi N, Toyoda T, Shinozaki K, Seki M (2008) *Arabidopsis* transcriptome analysis under drought, cold, high-salinity and ABA treatment conditions using a tiling array. Plant Cell Physiol 49:1135–1149. https://doi.org/10.1093/pcp/pcn101

Miller J, McLachlan AD, Klug A (1985) Repetitive zinc-binding domains in the protein transcription factor IIIA from Xenopus oocytes. EMBO J 4:1609–1614. https://doi.org/10.1002/j.1460-2075.1985.tb03825.x

Miller M, Shuman JD, Sebastian T, Dauter Z, Johnson PF (2003) Structural basis for DNA recognition by the basic region leucine zipper transcription factor CCAAT/enhancer-binding protein alpha. J Biol Chem 278:15178–15184. https://doi.org/10.1074/jbc.M300417200

Mittler R (2006) Abiotic stress, the field environment and stress combination. Trends Plant Sci 11:15–19. https://doi.org/10.1016/j.tplants.2005.11.002

Mittler R, Blumwald E (2010) Genetic engineering for modern agriculture: challenges and perspectives. Annu Rev Plant Biol 61:443–462. https://doi.org/10.1146/annurev-arplant-042809-112116

Munns R, Termaat A (1986) Whole-plant responses to salinity. Funct Plant Biol 13:143–160. https://doi.org/10.1071/PP986014

Munns R, Tester M (2008) Mechanisms of salinity tolerance. Annu Rev Plant Biol 59:651–681. https://doi.org/10.1146/annurev.arplant.59.032607.092911

Muthamilarasan M, Bonthala VS, Mishra AK, Khandelwal R, Khan Y, Roy R, Prasad M (2014) C2H2 type of zinc finger transcription factors in foxtail millet define response to abiotic stresses. Funct Integr Genomics 14:531–543. https://doi.org/10.1007/s10142-014-0383-2

Mzid R, Zorrig W, Ayed RB, Hamed KB, Ayadi M, Damak Y, Lauvergeat V, Hanana M (2018) The grapevine VvWRKY2 gene enhances salt and osmotic stress tolerance in transgenic *Nicotiana tabacum*. Biotech 8:277. https://doi.org/10.1007/s13205-018-1301-4

Nakano T, Suzuki K, Fujimura T, Shinshi H (2006) Genome-wide analysis of the ERF gene family in Arabidopsis and rice. Plant Physiol 140:411–432. https://doi.org/10.1104/pp.105.073783

Nakashima K, Shinwar ZK, Sakuma Y, Seki M, Miura S, Shinozaki K, Yamaguchi-Shinozaki K (2000) Organization and expression of two Arabidopsis DREB2 genes encoding DRE-binding proteins involved in dehydration- and high salinity-responsive gene expression. Plant Mol Biol 42:657–665. https://doi.org/10.1023/A:1006321900483

Nakashima K, Tran LS, Van Nguyen D, Fujita M, Maruyama K, Todaka D, Ito Y, Hayashi N, Shinozaki K, Yamaguchi-Shinozaki K (2007) Functional analysis of a NAC-type transcription factor OsNAC6 involved in abiotic and biotic stress-responsive gene expression in rice. Plant J 51:617–630. https://doi.org/10.1111/j.1365-313X.2007.03168.x

Nakashima K, Takasaki H, Mizoi J, Shinozaki K, Yamaguchi-Shinozaki K (2012) NAC transcription factors in plant abiotic stress responses. Biochim Biophys Acta 1819:97–103. https://doi.org/10.1016/j.bbagrm.2011.10.005

Nakashima K, Jan A, Todaka D, Maruyama K, Goto S, Shinozaki K, Yamaguchi-Shinozaki K (2014) Comparative functional analysis of six drought-responsive promoters in transgenic rice. Planta 239:47–60. https://doi.org/10.1007/s00425-013-1960-7

Nguyen KH, Mostofa MG, Li W, Van Ha C, Watanabe Y, Le DT, Thao NP, Tran LSP (2018) The soybean transcription factor GmNAC085 enhances drought tolerance in Arabidopsis. Environ Exp Bot 151:12–20. https://doi.org/10.1016/j.envexpbot.2018.03.017

Nijhawan A, Jain M, Tyagi AK, Khurana JP (2008) Genomic survey and gene expression analysis of the basic leucine zipper transcription factor family in rice. Plant Physiol 146:333–350. https://doi.org/10.1104/pp.107.112821

Niu CF, Wei W, Zhou QY, Tian AG, Hao YJ, Zhang WK, Ma B, Lin Q, Zhang ZB, Zhang JS, Chen SY (2012) Wheat WRKY genes TaWRKY2 and TaWRKY19 regulate abiotic stress tolerance in transgenic Arabidopsis plants. Plant, Cell Environ 35:1156–1170. https://doi.org/10.1111/j.1365-3040.2012.02480.x

Niu GL, Gou W, Han XL, Qin C, Zhang LX, Abomohra AE, Ashraf M (2018) Cloning and functional analysis of Phosphoethanolamine Methyltransferase promoter from Maize (*Zea mays* L.). Int J Mol Sci 19: E191. https://doi.org/10.3390/ijms19010191

Nordin K, Heino P, Palva ET (1991) Separate signal pathways regulate the expression of low temperature-induced gene in *Arabidopsis thaliana* (L.) Heyn. Plant Mol Biol 16:1061–1071. https://doi.org/10.1007/BF00016077

Nuruzzaman M, Sharoni AM, Kikuchi S (2013) Roles of NAC transcription factors in the regulation of biotic and abiotic stress responses in plants. Front Microbiol 4:248. https://doi.org/10.3389/fmicb.2013.00248

Odell JT, Nagy F, Chua NH (1985) Identification of DNA sequences required for activity of the cauliflower mosaic virus 35S promoter. Nature 313:810–812. https://doi.org/10.1038/313810a0

Ohno S, Hosokawa M, Hoshino A, Kitamura Y, Morita Y, Park KI, Nakashima A, Deguchi A, Tatsuzawa F, Doi M, Iida S, Yazawa S (2011) A bHLH transcription factor, DvIVS, is involved in regulation of anthocyanin synthesis in dahlia (*Dahlia variabilis*). J Exp Bot 62:5105–5116. https://doi.org/10.1093/jxb/err216

Olsen AN, Ernst HA, Leggio LL, Skriver K (2005) NAC transcription factors: structurally distinct, functionally diverse. Trends Plant Sci 10:79–87. https://doi.org/10.1016/j.tplants.2004.12.010

Onaga G, Wydra K (2016) Advances in plant tolerance to abiotic stresses. In: Abdurakhmonov IY (ed) Plant genomics. (Rijeka: InTech). https://doi.org/10.5772/64350

Ooka H, Satoh K, Doi K, Nagata T, Otomo Y, Murakami K, Matsubara K, Osato N, Kawai J, Carninci P, Hayashizaki Y, Suzuki K, Kojima K, Takahara Y, Yamamoto K, Kikuchi S (2003) Comprehensive analysis of NAC family genes in Oryza sativa and Arabidopsis thaliana. DNA Res. 20:239–247. https://doi.org/10.1093/dnares/10.6.239

Pandey AS, Sharma E, Jain N, Singh B, Burman N, Khurana JP (2018) A rice bZIP transcription factor, OsbZIP16, regulates abiotic stress tolerance when over-expressed in Arabidopsis. J Plant Biochem Biotech 27:393–400. https://doi.org/10.1007/s13562-018-0448-8

Paz-Ares J, Ghosal D, Wienand U, Peterson P, Saedler H (1987) The regulatory c1 locus of Zea mays encodes a protein with homology to myb proto-oncogene products and with structural similarities to transcriptional activators. EMBO J 6:3553. https://doi.org/10.1002/j.1460-2075.1987.tb02684.x

Pfluger J, Wagner D (2007) Histone modifications and dynamic regulation of genome accessibility in plants. Curr Opin Plant Biol 10:645–652. https://doi.org/10.1016/j.pbi.2007.07.013

Pillai SE, Kumar C, Patel HK, Sonti RV (2018) Overexpression of a cell wall damage induced transcription factor, OsWRKY42, leads to enhanced callose deposition and tolerance to salt stress but does not enhance tolerance to bacterial infection. BMC Plant Biol 18:177. https://doi.org/10.1186/s12870-018-1391-5

Pireyre M, Burow M (2015) Regulation of MYB and bHLH transcription factors: a glance at the protein level. Mol Plant 3:378–388. https://doi.org/10.1016/j.molp.2014.11.022
Puranik S, Sahu PP, Srivastava PS, Prasad M (2012) NAC proteins: regulation and role in stress tolerance. Trends Plant Sci 17:369–381. https://doi.org/10.1016/j.tplants.2012.02.004
Qin Y, Wang M, Tian Y, He W, Han L, Xia G (2012) Over-expression of TaMYB33 encoding a novel wheat MYB transcription factor increases salt and drought tolerance in Arabidopsis. Mol Biol Rep 39:7183–7192. https://doi.org/10.1007/s11033-012-1550-y
Quan R, Hu S, Zhang Z, Zhang H, Zhang Z, Huang R (2010) Overexpression of an ERF transcription factor TSRF1 improves rice drought tolerance. Plant Biotechnol J 8:476–488. https://doi.org/10.1111/j.1467-7652.2009.00492.x
Ramakrishna C, Singh S, Raghavendrarao S, Padaria JC, Mohanty S, Sharma TR, Solanke AU (2018) The membrane tethered transcription factor EcbZIP17 from finger millet promotes plant growth and enhances tolerance to abiotic stresses. Sci Rep 8:2148. https://doi.org/10.1038/s41598-018-19766-4
Ramamoorthy R, Jiang SY, Kumar N, Venkatesh PN, Ramachandran S (2008) A comprehensive transcriptional profiling of the WRKY gene family in rice under various abiotic and phytohormone treatments. Plant Cell Physiol 49:865–879. https://doi.org/10.1093/pcp/pcn061
Ramsay NA, Glover BJ (2005) MYB-bHLH-WD40 protein complex and the evolution of cellular diversity. Trends Plant Sci 10:63–70. https://doi.org/10.1016/j.tplants.2004.12.011
Reynolds M, Tuberosa R (2008) Translational research impacting on crop productivity in drought-prone environments. Curr Opin Plant Biol 11:171–179. https://doi.org/10.1016/j.pbi.2008.02.005
Richards RA (1996) Defining selection criteria to improve yield under drought. Plant Grow Reg 20:157–166. https://doi.org/10.1007/BF00024012
Riechmann JL, Meyerowitz EM (1998) The AP2/EREBP family of plant transcription factors. Biol Chem 379:633–646. https://doi.org/10.1515/bchm.1998.379.6.633
Rinerson CI, Rabara RC, Tripathi P, Shen QJ, Rushton PJ (2015) The evolution of WRKY transcription factors. BMC Plant Biol 15:66. https://doi.org/10.1186/s12870-015-0456-y
Rong W, Qi L, Wang A, Ye X, Du L, Liang H, Xin Z, Zhang Z (2014) The ERF transcription factor TaERF3 promotes tolerance to salt and drought stresses in wheat. Plant Biotechnol J 12:468–479. https://doi.org/10.1111/pbi.12153
Rushton PJ, Macdonald H, Huttly AK, Lazarus CM, Hooley R (1995) Members of a new family of DNA-binding proteins bind to a conserved cis-element in the promoters of alpha-Amy2 genes. Plant Mol Biol 29:691–702. https://doi.org/10.1007/BF00041160
Sah SK, Reddy KR, Li J (2016) Abscisic acid and abiotic stress tolerance in crop plants. Front Plant Sci 7:571. https://doi.org/10.3389/fpls.2016.00571
Sakamoto H, Araki T, Meshi T, Iwabuchi M (2000) Expression of a subset of the Arabidopsis Cys(2)/His(2)-type zinc-finger protein gene family under water stress. Gene 248:23–32. https://doi.org/10.1016/S0378-1119(00)00133-5
Sakuma Y, Maruyama K, Osakabe Y, Qin F, Seki M, Shinozaki K, Yamaguchi-Shinozaki K (2006) Functional analysis of an Arabidopsis transcription factor, DREB2A, involved in drought-responsive gene expression. Plant Cell 18:1292–1309. https://doi.org/10.1105/tpc.105.035881
Schulz P, Herde M, Romeis T (2013) Calcium-dependent protein kinases: hubs in plant stress signaling and development. Plant Physiol 163:523–530. https://doi.org/10.1104/pp.113.222539
Schumacher MA, Goodman RH, Brennan RG (2000) The structure of a CREB bZIP middle dot Somatostatin CRE complex reveals the basis for selective dimerization and divalent cation-enhanced DNA binding. J Biol Chem 275:35242–35247. https://doi.org/10.1074/jbc.M007293200
Shang Y, Yan L, Liu ZQ, Cao Z, Mei C, Xin Q, Wu FQ, Wang XF, Du SY, Jiang T, Zhang XF, Zhao R, Sun HL, Liu R, Yu YT, Zhang DP (2010) The Mg-chelatase H subunit of *Arabidopsis* antagonizes a group of WRKY transcription repressors to relieve ABA-responsive genes of inhibition. Plant Cell 22:1909–1935. https://doi.org/10.1105/tpc.110.073874
Sharma MK, Kumar R, Solanke AU, Sharma R, Tyagi AK, Sharma AK (2010) Identification, phylogeny, and transcript profiling of ERF family genes during development and abiotic stress

treatments in tomato. Mol Genet Genomics 284:455–475. https://doi.org/10.1007/s00438-010-0580-1

Sharma V, Goel P, Kumar S, Singh AK (2019) An apple transcription factor, MdDREB76, confers salt and drought tolerance in transgenic tobacco by activating the expression of stress-responsive genes. Plant Cell Rep 38:221–241. https://doi.org/10.1007/s00299-018-2364-8

Sharoni AM, Nuruzzaman M, Satoh K, Shimizu T, Kondoh H, Sasaya T, Choi IR, Omura T, Kikuchi S (2011) Gene structures, classification and expression models of the AP2/EREBP transcription factor family in rice. Plant Cell Physiol 52:344–360. https://doi.org/10.1093/pcp/pcq196

Shen YG, Zhang WK, Yan DQ, Du BX, Zhang JS, Liu Q, Chen SY (2003) Characterization of a DRE-binding transcription factor from a halophyte *Atriplex hortensis*. Theor Appl Genet 107:155–161. https://doi.org/10.1007/s00122-003-1226-z

Shen XJ, Wang YY, Zhang YX, Guo W, Jiao YQ, Zhou XA (2018) Overexpression of the wild soybean R2R3-MYB transcription factor GsMYB15 enhances resistance to salt stress and *Helicoverpa armigera* in transgenic Arabidopsis. Int J Mol Sci 19:3958. https://doi.org/10.3390/ijms19123958

Shi WY, Du YT, Ma J, Min DH, Jin LG, Chen J, Chen M, Zhou YB, Ma YZ, Xu ZS, Zhang XH (2018) The WRKY transcription factor GmWRKY12 confers drought and salt tolerance in soybean. Int J Mol Sci 19:4087. https://doi.org/10.3390/ijms19124087

Shim JS, Oh N, Chung PJ, Kim YS, Choi YD, Kim JK (2018) Overexpression of *OsNAC14* improves drought tolerance in rice. Front Plant Sci 9:310. https://doi.org/10.3389/fpls.2018.00310

Shinde H, Dudhate A, Tsugama D, Gupta SK, Liu S, Takano T (2019) Pearl millet stress-responsive NAC transcription factor PgNAC21 enhances salinity stress tolerance in Arabidopsis. Plant Physiol Biochem 135:546–553. https://doi.org/10.1016/j.plaphy.2018.11.004

Shinozaki K, Yamaguchi-Shinozaki K (2007) Gene networks involved in drought stress response and tolerance. J Exp Bot 58:221–227. https://doi.org/10.1093/jxb/erl164

Shiriga K, Sharma R, Kumar K, Yadav SK, Hossain F, Thirunavukkarasu N (2014) Genome-wide identification and expression pattern of drought-responsive members of the NAC family in maize. MetaGene 2:407–417. https://doi.org/10.1016/j.mgene.2014.05.001

Singh AK, Ansari MW, Pareek A, Singla-Pareek SL (2008) Raising salinity tolerant rice: recent progress and future perspectives. Physiol Mol Biol Plants 14:137–154

Singh AK, Sopory SK, Wu R, Singla-Pareek (2010) Transgenic approaches. In: Pareek A, Sopory SK, Bohnert HJ, Govindjee (eds), Abiotic stress adaptation in plants: physiological, molecular and genomic foundation, pp 417–450. https://doi.org/10.1007/978-90-481-3112-9_19

Singh AK, Sharma V, Pal AK, Acharya V, Ahuja PS (2013) Genome-wide organization and expression profiling of the NAC transcription factor family in potato (*Solanum tuberosum* L.). DNA Res 20:403–423. https://doi.org/10.1093/dnares/dst019

Singhal P, Jan AT, Azam M, Haq QMR (2016) Plant abiotic stress: a prospective strategy of exploiting promoters as alternative to overcome the escalating burden. Front Life Sci 9:52–63. https://doi.org/10.1080/21553769.2015.1077478

Sinha AK, Jaggi M, Raghuram B, Tuteja N (2011) Mitogen-activated protein kinase signaling in plants under abiotic stress. Plant Signal Behav 2:196–203. https://doi.org/10.4161/psb.6.2.14701

Su LT, Li JW, Liu DQ, Zhai Y, Zhang HJ, Li XW, Zhang QL, Wang Y, Wang QY (2014) A novel MYB transcription factor, GmMYBJ1, from soybean confers drought and cold tolerance in Arabidopsis thaliana. Gene 538:46–55. https://doi.org/10.1016/j.gene.2014.01.024

Su H, Zhang S, Yin Y, Zhu D, Han L (2015) Genome-wide analysis of NAM-ATAF1,2-CUC2 transcription factor family in *Solanum lycopersicum*. J Plant Biochem Biotechnol 24:176–183. https://doi.org/10.1007/s13562-014-0255-9

Sun SJ, Guo SQ, Yang X, Bao YM, Tang HJ, Sun H, Huang J, Zhang HS (2010) Functional analysis of a novel Cys2/His2-type zinc finger protein involved in salt tolerance in rice. J Exp Bot 61:2807–2818. https://doi.org/10.1093/jxb/erq120

Tang W, Luo C (2018) Overexpression of zinc finger transcription factor ZAT6 enhances salt tolerance. Open Life Sci 13:431–445. https://doi.org/10.1515/biol-2018-0052

Tang Y, Bao X, Zhi Y, Wu Q, Guo Y, Yin X, Zeng L, Li J, Zhang J, He W, Liu W, Wang Q, Jia C, Li Z, Liu K (2019) Overexpression of a MYB family gene, OsMYB6, increases drought and salinity stress tolerance in transgenic rice. Front Plant Sci 10:168. https://doi.org/10.3389/fpls.2019.00168

Teng K, Tan P, Guo W, Yue Y, Fan X, Wu J (2018) Heterologous expression of a novel *Zoysia japonica* C2H2 zinc finger gene, ZjZFN1, improved salt tolerance in Arabidopsis. Front Plant Sci 9:1159. https://doi.org/10.3389/fpls.2018.01159

Toledo-Ortiz G1, Huq E, Quail PH (2003) The Arabidopsis basic/helix-loop-helix transcription factor family. Plant Cell 15:1749–1770. https://doi.org/10.1105/tpc.013839

Touchette BW, Smith GA, Rhodes KL, Poole M (2009) Tolerance and avoidance: two contrasting physiological responses to salt stress in mature marsh halophytes *Juncus roemerianus* Scheele and *Spartina alterniflora* Loisel. J Exp Mar Bio Ecol 380:106–112. https://doi.org/10.1016/j.jembe.2009.08.015

Tran LS, Nakashima K, Sakuma Y, Simpson SD, Fujita Y, Maruyama K, Fujita M, Seki M, Shinozaki K, Yamaguchi-Shinozaki K (2004) Isolation and functional analysis of Arabidopsis stress-inducible NAC transcription factors that bind to a drought-responsive cis-element in the early responsive to dehydration stress 1 promoter. Plant Cell 16:2481–2498. https://doi.org/10.1105/tpc.104.022699

Tu M, Wang X, Huang L, Guo R, Zhang H, Cai J, Wang X (2016a) Expression of a grape bZIP transcription factor, *VqbZIP39*, in transgenic *Arabidopsis thaliana* confers tolerance of multiple abiotic stresses. Plant Cell, Tissue Organ Cult 125:537–551. https://doi.org/10.1007/s11240-016-0969-6

Tu M, Wang X, Feng T, Sun X, Wang Y, Huang L, Gao M, Wang Y, Wang X (2016b) Expression of a grape (*Vitis vinifera*) bZIP transcription factor, VlbZIP36, in *Arabidopsis thaliana* confers tolerance of drought stress during seed germination and seedling establishment. Plant Sci 252:311–323. https://doi.org/10.1016/j.plantsci.2016.08.011

Tu M, Wang X, Zhu Y, Wang D, Zhang X, Cui Y, Li Y, Gao M, Li Z, Wang Y, Wang X (2018) VlbZIP30 of grapevine functions in dehydration tolerance via the abscisic acid core signaling pathway. Hortic Res 5:49. https://doi.org/10.1101/231811

Tuteja N (2007) Abscisic acid and abiotic stress signaling. Plant Signal Behav 2:135–138. https://doi.org/10.4161/psb2.3.4156

Udvardi MK, Kakar K, Wandrey M, Montanari O, Murray J, Andriankaja A, Zhang J-Y, Benedito V, Hofer JMI, Chueng F, Town CD (2007) Legume transcription factors: global regulators of plant development and response to the environment. Plant Physiol 144:538–549. https://doi.org/10.1104/pp.107.098061

Ulker B, Somssich IE (2004) WRKY transcription factors: from DNA binding towards biological function. Curr Opin Plant Biol 7:491–498. https://doi.org/10.1016/j.pbi.2004.07.012

Umezawa T, Fujita M, Fujita Y, Yamaguchi-Shinozaki K, Shinozaki K (2006) Engineering drought tolerance in plants: discovering and tailoring genes to unlock the future. Curr Opin Biotechnol 17:113–122. https://doi.org/10.1016/j.copbio.2006.02.002

Urao T, Yamaguchi-Shinozaki K, Urao S, Shinozaki K (1993) An Arabidopsis myb homolog is induced by dehydration stress and its gene product binds to the conserved MYB recognition sequence. Plant Cell 5:1529–1539. https://doi.org/10.1105/tpc.5.11.1529

Wan Y, Mao M, Wan D, Yang Q, Yang F, Mandlaa Li G, Wang R (2018) Identification of the WRKY gene family and functional analysis of two genes in *Caragana intermedia*. BMC Plant Biol 18:31. https://doi.org/10.1186/s12870-018-1235-3

Wang W, Vinocur B, Altman A (2003) Plant responses to drought, salinity and extreme temperatures: towards genetic engineering for stress tolerance. Planta 218:1–14. https://doi.org/10.1007/s00425-003-1105-5

Wang Z, Li H, Ke Q, Jeong JC, Lee HS, Xu B, Deng XP, Lim YP, Kwak SS (2014) Transgenic alfalfa plants expressing AtNDPK2 exhibit increased growth and tolerance to abiotic stresses Plant Physiol Biochem 84:67–77. https://doi.org/10.1016/j.plaphy.2014.08.025

Wang J, Hu Z, Zhao T, Yang Y, Chen T, Yang M, Yu W, Zhang B (2015) Genome-wide analysis of bHLH transcription factor and involvement in the infection by yellow leaf curl virus in tomato (*Solanum lycopersicum*). BMC Genom 16:39. https://doi.org/10.1186/s12864-015-1249-2

Wang F, Zhu H, Chen DH, Li ZJ, Peng R, Yao QH (2016a) A grape bHLH transcription factor gene, VvbHLH1, increases the accumulation of flavonoids and enhances salt and drought tolerance in transgenic Arabidopsis thaliana. Plant Cell, Tissue Organ Cult 125:387–398. https://doi.org/10.1007/s11240-016-0953-1

Wang F, Tong W, Zhu H, Kong WL, Peng R, Liu Q, Yao Q (2016b) A novel Cys2/His2 zinc finger protein gene from sweet potato, IbZFP1, is involved in salt and drought tolerance in transgenic Arabidopsis. Planta 243:783–797. https://doi.org/10.1007/s00425-015-2443-9

Wang Z, Su G, Li M, Ke Q, Kim SY, Li H, Huang J, Xu B, Deng XP, Kwak SS (2016c) Overexpressing Arabidopsis ABF3 increases tolerance to multiple abiotic stresses and reduces leaf size in alfalfa. Plant Physiol Biochem 109:199–208. https://doi.org/10.1016/j.plaphy.2016.09.020

Wang M, Zhuang J, Zou Z, Li Q, Xin H, Li X (2017a) Overexpression of a *Camellia sinensis* DREB transcription factor gene (*CsDREB*) increases salt and drought tolerance in transgenic *Arabidopsis thaliana*. J Plant Biol 60:452–461. https://doi.org/10.1007/s12374-016-0547-9

Wang L, Li Z, Lu M, Wang Y (2017b) ThNAC13, a NAC transcription factor from *Tamarix hispida*, confers salt and osmotic stress tolerance to transgenic *Tamarix* and *Arabidopsis*. Front Plant Sci 8:635. https://doi.org/10.3389/fpls.2017.00635

Wang N, Zhang W, Qin M, Li S, Qiao M, Liu Z, Xiang F (2017c) Drought tolerance conferred in soybean (*Glycine max* L.) by GmMYB84, a novel R2R3-MYB transcription factor. Plant Cell Physiol 58:1764–1776. https://doi.org/10.1093/pcp/pcx111

Wang Y, Wang Q, Liu ML, Bo C, Wang X, Ma Q, Cheng B, Cai R (2017d) Overexpression of a maize *MYB48* gene confers drought tolerance in transgenic Arabidopsis plants. J Plant Biol 60:612–621. https://doi.org/10.1007/s12374-017-0273-y

Wang K, Wu YH, Tian XQ, Bai ZY, Liang QY, Liu QL, Pan YZ, Zhang L, Jiang BB (2017e) Overexpression of *DgWRKY4* enhances salt tolerance in chrysanthemum seedlings. Front Plant Sci 8:1592. https://doi.org/10.3389/fpls.2017.01592

Wang C, Lu G, Hao Y, Guo H, Guo Y, Zhao J, Cheng H (2017f) ABP9, a maize bZIP transcription factor, enhances tolerance to salt and drought in transgenic cotton. Planta 246:453–469. https://doi.org/10.1007/s00425-017-2704-x

Wang S, Zhou B, Yao W, Jiang T (2018a) PsnERF75 transcription factor from *Populus simonii* × *P. nigra* confers salt tolerance in transgenic *Arabidopsis*. J Plant Biol 61:61–71. https://doi.org/10.1007/s12374-017-0450-z

Wang J, Zhang L, Wang X, Liu L, Lin X, Wang W, Qi C, Cao Y, Li S, Ren S, Zhang Y, Zhang W, Guo YD (2018b) *PvNAC1* increases biomass and enhances salt tolerance by decreasing Na^+ accumulation and promoting ROS scavenging in switchgrass (*Panicum virgatum* L.). Plant Sci 280:66–76. https://doi.org/10.1016/j.plantsci.2018.11.007

Wang Y, Jiang L, Chen J, Rao L, An Y, Cai H, Guo C (2018c) Overexpression of the alfalfa *WRKY11* gene enhances salt tolerance in soybean. PLoS ONE 13:e0192382. https://doi.org/10.1371/journal.pone.0192382

Wang P, Su L, Gao H, Jiang X, Wu X, Li Y, Zhang Q, Wang Y, Ren F (2018d) Genome-Wide characterization of bHLH genes in grape and analysis of their potential relevance to abiotic stress tolerance and secondary metabolite biosynthesis. Front Plant Sci 9:64. https://doi.org/10.3389/fpls.2018.00064

Wang J, Song Z, Jia H, Yang S, Zhang H (2018e) Characterization of wheat TaSnRK2.7 promoter in *Arabidopsis*. Planta 248:1393–1401. https://doi.org/10.1007/s00425-018-2984-9

Wei K, Chen J, Wang Y, Chen Y, Chen S, Lin Y, Pan S, Zhong X, Xie D (2012) Genome-wide analysis of bZIP-encoding genes in maize. DNA Res 19:463–476. https://doi.org/10.1093/dnares/dss026

Wei Q, Luo Q, Wang R, Zhang F, He Y, Zhang Y (2017) A wheat R2R3-type MYB transcription factor TaODORANT1 positively regulates drought and salt stress responses in transgenic tobacco plants. Front Plant Sci 8:1374. https://doi.org/10.3389/fpls.2017.01374

Weltmeier F, Ehlert A, Mayer CS, Dietrich K, Wang X, Schutze K, Harter K, Vicente-Carbajosa J, Droge-Laser W (2006) Combinatorial control of Arabidopsis proline dehydrogenase transcription by specific heterodimerisation bZIP transcription factors. EMBO J 25:3133–3143. https://doi.org/10.1038/sj.emboj.7601206

Wery J, Silim SN, Knights EJ, Malhotra RS, Cousin R (1994) Screening techniques and sources and tolerance to extremes of moisture and air temperature in cool season food legumes. Euphytica 73:73–83. https://doi.org/10.1007/BF00027184

Wu J, Folta KM, Xie Y, Jiang W, Lu J, Zhang Y (2017) Overexpression of *Muscadinia rotundifolia* CBF2 gene enhances biotic and abiotic stress tolerance in Arabidopsis. Protoplasma 254:239–251. https://doi.org/10.1007/s00709-015-0939-6

Wu D, Sun Y, Wang H, Shi H, Su M, Shan H, Li T, Li Q (2018) The *Sl*NAC8 gene of the halophyte *Suaeda liaotungensis* enhances drought and salt stress tolerance in transgenic *Arabidopsis thaliana*. Gene 662:10–20. https://doi.org/10.1016/j.gene.2018.04.012

Wu J, Jiang Y, Liang Y, Chen L, Chen W, Cheng B (2019) Expression of the maize MYB transcription factor ZmMYB3R enhances drought and salt stress tolerance in transgenic plants. Plant Physiol Biochem 137:179–188. https://doi.org/10.1016/j.plaphy.2019.02.010

Xiang Y, Tang N, Du H, Ye H, Xiong L (2007) Characterization of OsbZIP23 as a key player of the basic leucine zipper transcription factor family for conferring abscisic acid sensitivity and salinity and drought tolerance in rice. Plant Physiol 148:1938–1952. https://doi.org/10.1104/pp.108.128199

Xiang L, Liu C, Luo J, He L, Deng Y, Yuan J, Wu C, Cai Y (2018) A tuber mustard AP2/ERF transcription factor gene, *BjABR1*, functioning in abscisic acid and abiotic stress responses, and evolutionary trajectory of the ABR1 homologous genes in Brassica species. PeerJ 6:e6071. https://doi.org/10.7717/peerj.6071

Xie Z, Zhang ZL, Zou X, Huang J, Ruas P, Thompson D, Shen QJ (2005) Annotations and functional analyses of the rice WRKY gene superfamily reveal positive and negative regulators of abscisic acid signaling in aleurone cells. Plant Physiol 137:176–189. https://doi.org/10.1104/pp.104.054312

Xie T, Chen C, Li C, Liu J, Liu C, He Y (2018) Genome-wide investigation of WRKY gene family in pineapple: evolution and expression profiles during development and stress. BMC Genom 19:490. https://doi.org/10.1186/s12864-018-4880-x

Xiong L, Zhu JK (2003) Regulation of abscisic acid biosynthesis. Plant Physiol 133:29–36

Xiong H, Li J, Liu P, Duan J, Zhao Y, Guo X, Li Y, Zhang H, Ali J, Li Z (2014) Overexpression of OsMYB48-1, a novel MYB-related transcription factor, enhances drought and salinity tolerance in rice. PLoS ONE 9:e92913. https://doi.org/10.1371/journal.pone.0092913

Xiong H, Yu J, Miao J, Li J, Zhang H, Wang X, Liu P, Zhao Y, Jiang C, Yin Z, Li Y, Guo Y, Fu B, Wang W, Li Z, Ali J, Li Z (2018) Natural variation in OsLG3 increases drought tolerance in rice by inducing ROS scavenging. Plant Physiol 178:451–467. https://doi.org/10.1104/pp.17.01492

Xu XB, Pan YY, Wang CL, Ying QC, Song HM, Wang HZ (2014) Overexpression of DnWRKY11 enhanced salt and drought stress tolerance of transgenic tobacco. Biologia 69:99–1000. https://doi.org/10.2478/s11756-014-0398-0

Xu Z, Gongbuzhaxi WC, Xue F, Zhang H, Ji W (2015) Wheat NAC transcription factor TaNAC29 is involved in response to salt stress. Plant Physiol Biochem 96:356–363. https://doi.org/10.1016/j.plaphy.2015.08.013

Xu X, Yao X, Lu L, Zhao D (2018a) Overexpression of the transcription factor NtNAC2 confers drought tolerance in tobacco. Plant Mol Biol Report 4:543–552. https://doi.org/10.1007/s11105-018-1096-9

Xu Z, Raza Q, Xu L, He X, Huang Y, Yi J, Zhang D, Shao HB, Ma H, Ali Z (2018b) GmWRKY49, a salt-responsive nuclear protein, improved root length and governed better salinity tolerance in transgenic Arabidopsis. Front Plant Sci 9:809. https://doi.org/10.3389/fpls.2018.00809

Yamaguchi-Shinozaki K, Koizumi M, Urao S, Shinozaki K (1992) Molecular cloning and characterization of 9 cDNAs for genes that are responsive to desiccation in *Arabidopsis thaliana*:

sequence analysis of one cDNA clone that encodes a putative transmembrane channel protein. Plant Cell Physiol 33:217–224. https://doi.org/10.1093/oxfordjournals.pcp.a078243

Yan L, Liu S, Zhao S, Kang Y, Wang D, Gu T, Xin Z, Xia G, Huang Y (2012) Identification of differentially expressed genes in sorghum (*Sorghum bicolor*) brown midrib mutants. Physiol Plant 146:375–387. https://doi.org/10.1111/j.1399-3054.2012.01646.x

Yang ML, Chao JT, Wang DW, Hu JH, Wu H, Gong D, Liu GS (2016) Genome-wide identification and expression profiling of the C2H2-type zinc finger protein transcription factor family in tobacco. Yi Chuan 38:337–349. https://doi.org/10.16288/j.yczz.15-440

Yang G, Yu L, Zhang K, Zhao Y, Guo Y, Gao C (2017) A ThDREB gene from *Tamarix hispida* improved the salt and drought tolerance of transgenic tobacco and *T. hispida*. Plant Physiol Biochem 113:187–197. https://doi.org/10.1016/j.plaphy.2017.02.007

Yang X, He K, Chi X, Chai G, Wang Y, Jia C, Zhang H, Zhou G, Hu R (2018) Miscanthus NAC transcription factor MlNAC12 positively mediates abiotic stress tolerance in transgenic Arabidopsis. Plant Sci 277:229–241. https://doi.org/10.1016/j.plantsci.2018.09.013

Yang K, Li Y, Wang S, Xu X, Sun H, Zhao H, Li X, Gao Z (2019) Genome-wide identification and expression analysis of the MYB transcription factor in moso bamboo (*Phyllostachys edulis*). Peer J 6:e6242. https://doi.org/10.7717/peerj.6242

Yanhui C, Xiaoyuan Y, Kun H, Meihua L, Jigang L, Zhaofeng G, Zhiqiang L, Yunfei Z, Xiaoxiao W, Xiaoming Q, Yunping S, Li Z, Xiaohui D, Jingchu L, Xing-Wang D, Zhangliang C, Hongya G, Li-Jia Q (2006) The MYB transcription factor superfamily of Arabidopsis: expression analysis and phylogenetic comparison with the rice MYB family. Plant Mol Biol 60:107–124. https://doi.org/10.1007/s11103-005-2910-y

Yao PF, Li CL, Zhao XR, Li MF, Zhao HX, Guo JY, Cai Y, Chen H, Wu Q (2017) Overexpression of a tartary buckwheat gene, FtbHLH3, enhances drought/oxidative stress tolerance in transgenic Arabidopsis. Front Plant Sci 8:625. https://doi.org/10.3389/fpls.2017.00625

Yao W, Zhao K, Cheng Z, Li X, Zhou B, Jiang T (2018) Transcriptome analysis of poplar under salt stress and over-expression of transcription factor NAC57 gene confers salt tolerance in transgenic Arabidopsis. Front Plant Sci 9:1121. https://doi.org/10.3389/fpls.2018.01121

Yin M, Wang Y, Zhang L, Li J1, Quan W, Yang L, Wang Q, Chan Z (2017) The Arabidopsis Cys2/His2 zinc finger transcription factor ZAT18 is a positive regulator of plant tolerance to drought stress. J Exp Bot 68:2991–3005. https://doi.org/10.1093/jxb/erx157

Yokotani N, Ichikawa T, Kondou Y, Matsui M, Hirochika H, Iwabuchi M, Oda K (2009) Tolerance to various environmental stresses conferred by the salt-responsive rice gene ONAC063 in transgenic Arabidopsis. Planta 229:1065–1075. https://doi.org/10.1007/s00425-009-0895-5

Yoon S, Lee DK, Yu IJ, Kim YS, Yang DC, Kim JK (2017) Overexpression of the OsbZIP66 transcription factor enhances drought tolerance of rice plants. Plant Biotechnol Rep 11:53–62. https://doi.org/10.1007/s11816-017-0430-2

You J, Chan J (2015) ROS regulation during abiotic stress responses in crop plants. Front Plant Sci 6:1092. https://doi.org/10.3389/fpls.2015.01092

Yu Y, Ni Z, Chen Q, Qu Y (2017) The wheat salinity-induced R2R3-MYB transcription factor TaSIM confers salt stress tolerance in *Arabidopsis thaliana*. Biochem Biophys Res Commun 491:642–648. https://doi.org/10.1016/j.bbrc.2017.07.150

Yu Y, Bi C, Wang Q, Ni Z (2019) Overexpression of TaSIM provides increased drought stress tolerance in transgenic Arabidopsis. Biochem Biophys Res Commun 512:66–71. https://doi.org/10.1016/j.bbrc.2019.03.007

Yuan S, Li X, Li R, Wang L, Zhang C, Chen L, Hao Q, Zhang X, Chen H, Shan Z, Yang Z1, Chen S, Qiu D, Ke D, Zhou X (2018) Genome-Wide Identification and classification of soybean C2H2 zinc finger proteins and their expression analysis in legume-rhizobium symbiosis. Front Microbiol 9:126. https://doi.org/10.3389/fmicb.2018.00126

Yuan X, Huang P, Wang R, Li H, Lv X, Duan M, Tang H, Zhang H, Huang J (2018b) A zinc finger transcriptional repressor confers pleiotropic effects on rice growth and drought tolerance by down-regulating stress-responsive genes. Plant Cell Physiol 59:2129–2142. https://doi.org/10.1093/pcp/pcy133

Zang D, Li H, Xu H, Zhang W, Zhang Y, Shi X, Wang Y (2016) An arabidopsis zinc finger protein increases abiotic stress tolerance by regulating sodium and potassium homeostasis, reactive oxygen species scavenging and osmotic potential. Front Plant Sci 7:1272. https://doi.org/10.3389/fpls.2016.01272

Zhai Y, Zhang L, Xia C, Fu S, Zhao G, Jia J, Kong X (2016) The wheat transcription factor, TabHLH39, improves tolerance to multiple abiotic stressors in transgenic plants. Biochem Biophys Res Commun 473:1321–1327. https://doi.org/10.1016/j.bbrc.2016.04.071

Zhang J, Jia W, Yang J, Isamail AM (2006) Role of ABA in integrating plant responses to drought and salt stresses. Field Crop Res 97:111–119. https://doi.org/10.1016/j.fcr.2005.08.018

Zhang L, Zhao G, Xia C, Jia J, Liu X, Kong X (2012) A wheat R2R3-MYB gene, TaMYB30-B, improves drought stress tolerance in transgenic Arabidopsis. J Exp Bot 63:5873–5885. https://doi.org/10.1093/jxb/ers237

Zhang X, Zhang B, Li MJ, Yin XM, Huang LF, Cui YC, Wang ML, Xia X (2016) OsMSR15 encoding a rice C2H2-type zinc finger protein confers enhanced drought tolerance in transgenic Arabidopsis. J Plant Biol 59:271–281. https://doi.org/10.1007/s12374-016-0539-9

Zhang C, Li C, Liu J, Lv Y, Yu C, Li H, Zhao T, Liu B (2017a) The OsABF1 transcription factor improves drought tolerance by activating the transcription of COR413-TM1 in rice. J Exp Bot 68:4695–4707. https://doi.org/10.1093/jxb/erx260

Zhang L, Zhang L, Xia C, Gao L, Hao C, Zhao G, Jia J, Kong X (2017b) A novel wheat C-bZIP gene, TabZIP14-B, participates in salt and freezing tolerance in transgenic plants. Front Plant Sci 8:710. https://doi.org/10.3389/fpls.2017.00710

Zhang B, Su L, Hu B, Li L (2018a) Expression of *AhDREB1*, an AP2/ERF transcription factor gene from peanut, is affected by histone acetylation and increases abscisic acid sensitivity and tolerance to osmotic stress in *Arabidopsis*. Int J Mol Sci 19:1441. https://doi.org/10.3390/ijms19051441

Zhang L, Cheng J, Sun X, Zhao T, Li M, Wang Q, Li S, Xin H (2018b) Overexpression of VaWRKY14 increases drought tolerance in Arabidopsis by modulating the expression of stress-related genes. Plant Cell Rep 37:1159–1172. https://doi.org/10.1007/s00299-018-2302-9

Zhang C, Feng R, Ma R, Shen Z, Cai Z, Song Z, Peng B, Yu M (2018c) Genome-wide analysis of basic helix-loop-helix superfamily members in peach. PLoS ONE 13:e0195974. https://doi.org/10.1371/journal.pone.0195974

Zhang X, Cheng Z, Zhao K, Yao W, Sun X, Jiang T, Zhou B (2019) Functional characterization of poplar *NAC13* gene in salt tolerance. Plant Sci 281:1–8. https://doi.org/10.1016/j.plantsci.2019.01.003

Zhao Q, Zhong M, He L, Wang B, Liu Q, Pan Y, Jiang B, Zhang L (2018a) Overexpression of a chrysanthemum transcription factor gene *DgNAC1* improves drought tolerance in chrysanthemum. Plant Cell Tiss Org 135:119–132. https://doi.org/10.1007/s11240-018-1449-y

Zhao Y, Cheng X, Liu X, Wu H, Bi H, Xu H (2018b) The wheat MYB transcription factor TaMYB31 is involved in drought stress responses in Arabidopsis. Front Plant Sci 9:1426. https://doi.org/10.3389/fpls.2018.01426

Zhao J, Zhang XM, Guo RR, Wang YQ, Guo CL, Li Z, Chen ZP, Guo H, Wang XP (2018c) Over-expression of a grape WRKY transcription factor gene, VlWRKY48, in *Arabidopsis thaliana* increases disease resistance and drought stress tolerance. Plant Cell, Tissue Organ Cult 132:359–370. https://doi.org/10.1007/s11240-017-1335-z

Zhao T, Hu J, Gao Y, Wang Z, Bao Y, Zhang X, Yang H, Zhang D, Jiang J, Zhang H, Li J (2018d) Silencing of the SL-ZH13 transcription factor gene decreases the salt stress tolerance of tomato. J Amer Soc Hort Sci 143:391–396. https://doi.org/10.21273/JASHS04477-18

Zhong R, Ye ZH (2012) MYB46 and MYB83 bind to the SMRE sites and directly activate a suite of transcription factors and secondary wall biosynthetic genes. Plant Cell Physiol 53:368–380. https://doi.org/10.1093/pcp/pcr185

Zhong L, Chen D, Min D, Li W, Xu Z, Zhou Y, Li L, Chen M, Ma Y (2015) AtTGA4, a bZIP transcription factor, confers drought resistance by enhancing nitrate transport and assimilation in Arabidopsis thaliana. Biochem Biophys Res Commun 457:433–439. https://doi.org/10.1016/j.bbrc.2015.01.009

Zhou QY, Tian AG, Zou HF, Xie ZM, Lei G, Huang J, Wang CM, Wang HW, Zhang JS, Chen SY (2008) Soybean WRKY-type transcription factor genes, GmWRKY13, GmWRKY21, and GmWRKY54, confer differential tolerance to abiotic stresses in transgenic Arabidopsis plants. Plant Biotechnol J 6:486–503. https://doi.org/10.1111/j.1467-7652.2008.00336.x

Zhou J, Li F, Wang JL, Ma Y, Chong K, Xu YY (2009) Basic helix-loop-helix transcription factor from wild rice (OrbHLH2) improves tolerance to salt- and osmotic stress in Arabidopsis. J Plant Physiol 16:1296–1306. https://doi.org/10.1016/j.jplph.2009.02.007

Zhu M, Meng X, Cai J, Li G, Dong T, Li Z (2018) Basic leucine zipper transcription factor SlbZIP1 mediates salt and drought stress tolerance in tomato. BMC Plant Biol 18:83. https://doi.org/10.1186/s12870-018-1299-0

Zhuang J, Cai B, Peng RH, Zhu B, Jin XF, Xue Y, Gao F, Fu XY, Tian YS, Zhao W, Qiao YS, Zhang Z, Xiong AS, Yao QH (2008) Genome-wide analysis of the AP2/ERF gene family in *Populus trichocarpa*. Biochem Biophys Res Commun 371:468–474. https://doi.org/10.1016/j.bbrc.2008.04.087

Zong N, Li XJ, Wang L, Wang Y, Wen HT, Li L, Zhang X, Fan YL, Zhao J (2018) Maize ABP2 enhances tolerance to drought and salt stress in transgenic Arabidopsis. J Integr Agric 17:2379–2393. https://doi.org/10.1016/S2095-3119(18)61947-1

Chapter 3
Role of Superoxide Dismutases (SODs) in Stress Tolerance in Plants

Shivi Tyagi, Shumayla, Sudhir P. Singh and Santosh Kumar Upadhyay

Abstract The plants get exposed to different abiotic stresses and pathogen attack due to their sessile nature. These stresses result in the overproduction of reactive oxygen species (ROS) such as superoxide radical ($O_2^{\cdot -}$), hydrogen peroxide (H_2O_2), singlet oxygen ($^1O_2$) etc. To overcome the effect of these ROS, different classes of antioxidants are involved in providing tolerance to plants. The superoxide dismutase (SOD) consists of one such major class of antioxidant proteins, which provide primary defense against different stress conditions. These are ubiquitous metalloenzymes, which carry out the dismutation of superoxide radicals ($O_2^{\cdot -}$) into molecular oxygen and hydrogen peroxide (H_2O_2). In plants, a total of three classes of SODs are reported i.e., Cu-ZnSODs, FeSODs, and MnSODs, which have cytoplasmic or apoplastic or nuclear, chloroplastic and mitochondrial subcellular localization, respectively. SODs are well known for their role in plant growth and development and in providing tolerance against biotic and abiotic stress conditions by combating oxidative stress. These enzymes are stable and active over a broad range of pH and temperature. Due to their astonishing enzymatic properties, they are widely used in industries for various purposes. In this chapter, we have focused on the myriad functions of SODs in response to biotic and abiotic stresses and their utilization in enhancing the stress tolerance in plants.

Keywords Growth · Reactive oxygen species · Stress · Superoxide dismutase

S. Tyagi · Shumayla · S. K. Upadhyay (✉)
Department of Botany, Panjab University, Chandigarh 160014, India
e-mail: skupadhyay@pu.ac.in

Shumayla
Deparment of Biotechnology, Panjab University, Chandigarh 160014, India

S. P. Singh
Center of Innovative and Applied Bioprocessing, Sector-81, Mohali 140306, India

S. P. Singh et al. (eds.), *Molecular Approaches in Plant Biology and Environmental Challenges*, Energy, Environment, and Sustainability,
https://doi.org/10.1007/978-981-15-0690-1_3

3.1 Introduction

The reactive oxygen species (ROS) are detrimental derivatives of molecular oxygen (O_2) and produced during both normal and stressed growth conditions (Mittler et al. 2004; Scandalios 2005). Normally, a trivial amount of ROS is generated as by-products of regular metabolic processes due to incomplete reduction or excitation of O_2 within the cell (Halliwell 2006). These ROS comprise free radicals like alkoxyradical ($RO^{\cdot}$), hydroxyl radical ($OH^{\cdot}$), superoxide radical ($O_2^{-\cdot}$), etc. and non-radical derivatives such as hydrogen peroxide (H_2O_2), singlet oxygen ($^1O_2$) and so forth (Apel and Hirt 2004; Karuppanapandian et al. 2011). They act as a signalling molecule at low concentrations and play an important role in various plant physiological processes (Apel and Hirt 2004; Foyer and Noctor 2009). However, the changing environmental conditions and exposure of plants to various biotic and abiotic stresses lead to elevated production and accumulation of ROS (Mittler et al. 2004). These stress conditions disrupt the balance between the production and eradication of ROS within the cell (Suzuki and Mittler 2006). It results in serious damage to photosynthetic apparatus and biological macromolecules within the cell, leading to metabolic dysfunction and eventually cell death (Mittler 2002; Leonowicz et al. 2018). The plants have also evolved an array of ROS scavenging systems, which diminishes the cytotoxic effects of oxidative stress. It comprises superoxide dismutase (SOD), catalase (CAT), ascorbate peroxidase (APX), glutathione peroxidase (GPX), glutathione S-transferase (GST), etc. Apart from this, non-enzymatic system comprises ascorbic acid, carotenoids, flavonoids, α-tocopherol etc. (Halliwell 2006; ASADA 1987; Tyagi et al. 2017) (Fig. 3.1). These systems work together within a cell in combating the excessive amount of ROS produced during stress conditions. Among all the antioxidant systems, SOD contributes a major part in the ROS detoxification as it is highly efficient in the elimination of superoxide radicals.

SODs are imperious metalloenzymes, which catalyse the dismutation of $O_2^{-\cdot}$ into H_2O_2 and O_2, and avert the cascade of noxious ROS formation such as $HO^{\cdot}$, peroxynitrate (ONO_2^-), hypochlorite (OCl^-) etc. (Miller 2012). These are ubiquitous enzymes, which provide the first line of defence to the plants against oxidative stress by directly modulating the amount of two Haber-Weiss reaction substrates i.e. $O_2^{-\cdot}$ and H_2O_2 (Bowler et al. 1992). In plants, SODs can be classified into copper and zinc (Cu-ZnSOD), iron (FeSOD) and manganese (MnSOD), on the basis of metal co-factor at their active centre (Miller 2012). The catalytic function of SOD was first discovered in 1969 by McCord and Fridovich (McCord and Fridovich 1969). In plants, the presence of SOD isozymes and their enzymatic role was initially demonstrated in maize (Scandalios 1993). Moreover, the first SOD gene in the plant kingdom was cloned from maize (Cannon et al. 1987) and later on, different SOD isozymes and the entire SOD gene family was identified and characterized in different plant species (Table 3.1). For instance, six *SOD* genes (3 *Cu-ZnSODs*, 1 *MnSOD* and 2 *FeSODs*) in *Hordeum vulgare*, seven *SODs* (4 *Cu-ZnSODs*, 1 *MnSOD* and 2 *FeSODs*) in *Oryza sativa*, 12 *SODs* (7 *Cu-ZnSODs*, 3 *FeSODs* and 2 *MnSODs*) in *Populus trichocarpa*, nine *SODs* (5 *Cu-ZnSODs*, 1 *MnSOD* and 3 *FeSODs*) in *Phalaenopsis*

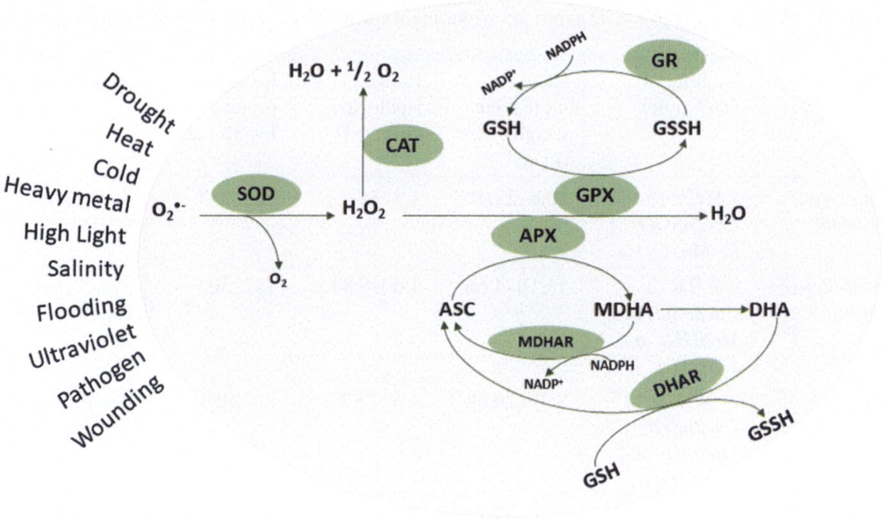

Fig. 3.1 Different antioxidant enzymes involved in ROS scavenging pathways under various stresses. The figure has been adapted and modified from Raja et al. (2017) with permission from "Elsevier" with licence number 4671290733046

equestris and seven *SODs* (4 *Cu-ZnSODs*, 1 *MnSOD* and 2 *FeSODs*) in *Setaria italic* etc. were identified (Feng et al. 2015; Molina-Rueda et al. 2013). The role of SODs in plant development and their protection against numerous stresses has been well documented (Alscher et al. 2002). Availability of high-throughput RNAseq data of a specific developmental stage or stress condition also enabled the study on the gene expression of entire SOD gene family within a plant (Tyagi et al. 2017; Feng et al. 2016). It has become the subject of tremendous research in the field of redox biology among all the genera.

Unlike plants, a nickel ion dependent SOD or NiSOD is known in *Streptomyces coelicolor*, a bacteria (Kim et al. 1996). Animals and fungi consist of only Cu-ZnSOD and MnSOD isoforms (Bowler et al. 1992). Additionally, a copper and zinc containing isoform of SOD localized in extracellular space or EcSOD is well known in animals and contain a classical signal peptide, which lacked in plants. Apart from this, an enzymatically active hybrid of FeSOD and MnSOD is also known in some lower organisms, which can use either of the metal ions depending upon the availability and termed as cambialistic Fe–Mn SOD (Alscher et al. 2002). Recently, a new class of SOD enzyme designated as copper-only SOD, which functions without Zn ion and dependent only on Cu ion for the catalysis has been reported. These are extra-cellular and present in fungi and some bacteria like Mycobacteria and play role in signalling processes and protection against the host oxidative stress (Robinett et al. 2017). Normally, the production of $O_2^{-\bullet}$ is higher at the site of electron transport chain due to the intermittent transfer of an electron to O_2 (Elstner 1991). Thus, SODs are mainly present in chloroplasts, mitochondria and peroxisomes in addition to the cytosol,

Table 3.1 The occurrence of *SOD* genes in various plant species and the physicochemical properties of their proteins

Plant species	Number of *SOD* genes	Range of molecular weight (kDa)	Range of isoelectric point (pI)	Range of protein length (aa)	Reference
Aegilops tauschii	4 *SODs* (3 *Cu-ZnSODs*, 1 *Fe-MnSODs*)	10.8–27.4	4.5–7.6	108–243	Tyagi et al. (2017)
Arabidopsis thaliana	8 *SODs* (3 *Cu-ZnSODs*, 2 *MnSODs and 3 FeSODs*)	15.10–4.66	4.60–8.84	152–305	Wang et al. (2016)
Brassica juncea	29 *SODs* (12 *Cu-ZnSODs*, 7 *MnSODs* and 10 *FeSODs*)	8.79–70.88	4.6–9.57	84–646	Verma et al. (2019)
B. rapa	18 *SODs* (8 *Cu-ZnSODs*, 4 *MnSODs* and 6 *FeSODs*)	6.20–72.17	3.76–8.97	54–657	Verma et al. (2019)
Cucumis sativus	9 *SODs* (5 *Cu-ZnSODs*, 3 *FeSODs*, and 1 MnSOD)	15.3–42.5	4.97–8.77	152–377	Zhou et al. (2017)
Dimocarpus longan	9 *SODs* (3 *Cu-ZnSODs*, 3 *FeSODs*, and 1 MnSOD)	15.3–35.6	5.08–6.86	152–310	Lin and Lai (2013)
Gossypium hirsutum	18 *SODs* (6 *Cu-ZnSODs*, 4 *MnSODs* and 4 *FeSODs*)	15.11–49.64	4.84–8.50	152–467	Wang et al. (2017)
G. raimondii	9 *SODs* (5 *Cu-ZnSODs*, 2 MnSOD, 2 *FeSODs*)	15.11–49.57	4.85–8.81	152–467	Wang et al. (2016)
G. arboreum	9 *SODs* (5 *Cu-ZnSODs*, 2 MnSOD, 2 *FeSODs*)	15.34–35.65	4.84–8.50	151–300	Wang et al. (2016)
Musa acuminata	12 *SODs* (6 *Cu-ZnSODs* and 6 *Fe-MnSODs*)	15.0–34.2	4.87–7.90	152–301	Feng et al. (2015)
Sorghum bicolor	8 *SODs* (5 *Cu-ZnSODs* and 3 *Fe-MnSODs*)	15.1–43.5	5.13–7.14	151–392	Filiz and Tombuloğlu (2015)

(continued)

Table 3.1 (continued)

Plant species	Number of *SOD* genes	Range of molecular weight (kDa)	Range of isoelectric point (pI)	Range of protein length (aa)	Reference
Solanum lycopersicum	9 *SODs* (4 *Cu-ZnSODs* and 5 *Fe-MnSODs*)	15.3–34.6	5.38–7.13	152–311	Feng et al. (2016)
Triticum aestivum	23 *SODs* (14 *Cu-ZnSODs*, 9 *Fe-MnSODs*)	15–41.4	5.1–9.8	151–391	Tyagi et al. (2017)
Triticum urartu	7 *SODs* (3 *Cu-ZnSODs*, 4 *Fe-MnSODs*)	10.9–45.3	4.6–9.6	108–408	Tyagi et al. (2017)
Vitis vinifera	10 *SODs* (6 *Cu-ZnSODs*, 2 *FeSODs*, and 2 *MnSODs*)	8.6–37.8	4.93–9.29	79–329	Hu et al. (2019)

apoplasts, glyoxysomes, and microsomes (Alscher et al. 2002). The diverse sub-cellular localization of SODs prevent cellular organelles from oxidative damage. In plants, the study on SOD is necessary for boosting up the immunity against different biotic and abiotic stresses and in understanding their role in signalling pathways.

3.2 Different Classes of SODs

3.2.1 Cu-ZnSODs

They are present typically in eukaryotes but a few bacteria, for instance, *Caulobacter crescentus*, and *Photobacterium leiognathi* also consist of *Cu-ZnSODs*. Primarily, a blue copper protein known as "hemocuprein" (presently known as Cu-ZnSOD) was isolated from bovine erythrocytes (Mann and Keilin 1938). However, the enzymatic property of this protein was unknown. On the basis of sub-cellular localization, they are divided into two groups in plants. The first group comprises cytoplasm and periplasm localized Cu-ZnSODs, which form homodimeric structure. The second group consists of chloroplast and extracellular localized Cu-ZnSODs, which are homotetramers (Bordo et al. 1994). Furthermore, the active-site functions autonomously in each subunit and the functional interaction among these subunits are not necessary for the catalysis (Fridovich 1986). The tertiary structure is composed of usually eight antiparallel β-strands, which are arranged in a Greek-key pattern, forming the β-barrels (Khanna-Chopra and Sabarinath 2004). They form a highly stable structure and consist of various disulfide bridges amid cysteine residues (Ogawa et al.

1996; Perry et al. 2010). The structure also has one variable loop (species-specific) between β2 and β3, and an electrostatic loop between β7 and β8 strands (Perry et al. 2010). Amid these loops and β barrels, the active site of an enzyme is formed and an active site channel is created by β4-β5 and β7-β8 barrels (Ogawa et al. 1996; Perry et al. 2010). A local electrostatic force produced by electrostatic loop directs the metal ion to the active site for the activity (Getzoff et al. 1992). However, the activity of these enzymes is sensitive to H_2O_2 and cyanide treatment (Sheng et al. 2014).

The chloroplastic and cytosolic Cu-ZnSODs share approximately 68% identity with each other, while approximately 90% and 80–90% identity with themselves, respectively. In *Spinacia oleracea*, one chloroplastic and two cytosolic isoforms have been reported (Ogawa et al. 1996; Kanematsu and Asada 1990). The chloroplastic Cu-ZnSOD is a soluble enzyme and localized mainly on the thylakoid membranes at the stromal side (Ogawa et al. 1995). Though, the cytosolic Cu-ZnSODs were localized in the apoplast and the nucleus (Ogawa et al. 1996). It suggested the unique role of each SOD isoform at the site of their action, for instance, chloroplastic Cu-ZnSODs may scavenge the $O_2^{-\cdot}$ produced during the process of photosynthesis, cytosolic Cu-ZnSODs in nucleus protect the genetic material from mutations and those in apoplast have a role in lignification process (Ogawa et al. 1996). In plants, the extracellular SODs (Ec-SODs) are usually absent, however, a few evidence supports their presence in plants (Ogawa et al. 1996; Streller and Wingsle 1994).

3.2.2 FeSODs

FeSODs are found in both lower organisms and higher plants, though in latter they are limited to the chloroplast. The chloroplastic localization has been confirmed in various plants such as *Nuphar luteum* (Salin and Bridges 1981). Additionally, FeSOD of soybean consisted of a chloroplast target peptide in the sequence indicating their chloroplastic localization (Crowell and Amasino 1991). FeSODs are resistant to cyanide inhibition thought gets strongly inactivated by H_2O_2 treatment (Sheng et al. 2014). FeSODs exist in homo-dimer and homo-tetramer forms, which are made up of 20 kDa subunit proteins. The former type enzymes consist of 1–2 g of iron at the active site and has been reported in lower organisms such as *Photobacterium sepoa*, *P. leiognathi*, *Escherichia coli*, *Thiobacillus denitrificans*, *Chromatium vinosum* etc. and plants like *Brassica campestris*, *Ginkgo biloba* and *Nuphar luteum* (Salin and Bridges 1981; Puget and Michelson 1974; Yost and Fridovich 1973; Baldensperger 1978; Kanematsu and Asada 1978). While, the latter comprises 2–4 g of iron and has been mostly described in higher plants and few prokaryotes, for instance, *Methanobacterium bryantii*, *Mycobacterium tuberculosis* and *Thermoplasma acisophilum* (Kirby et al. 1981; Kusunose et al. 1976; Searcy and Searcy 1981).

3.2.3 MnSODs

These are present both in prokaryotes and eukaryotes, however, unlike FeSODs, mammals also consist of MnSODs along with the plants. They showed mitochondrial and peroxisomal sub-cellular localization and exist as homo-dimer or tetramer with only one metal ion for each subunit. Though they share high similarity with FeSODs, but they cannot function in the absence of Mn ion at their active site, unlike cambialistic Fe-MnSOD (Fridovich 1986). The positively charged amino acids at the active site attract the negatively charged $O_2^{-\cdot}$, and then Mn^{3+} ion directly donates an electron to $O_2^{-\cdot}$ molecule and catalyze the disproportionation reaction (Bowler et al. 1991). In plants, MnSODs share around 65% sequence identity with each other and are insensitive to KCN or H_2O_2 inhibition like other isoforms. The mitochondrial localized MnSODs have been reported in various plants such as *Dianthus caryophyllus, Nicotiana tabacum, Pisum sativum, Spinacia oleracea, Vigna mungo,* watermelon, *Zea mays* etc. (Alscher et al. 2002; Zhu and Scandalios 1993; Sandalio and Del Río 1987). However, the peroxisomal localization of MnSOD has been recognized in watermelon and pea (Sandalio and Del Río 1987; del Río et al. 1992).

3.3 Evolution of *SOD* Isoforms

Around 2.4 billion years ago, the transition of environment became oxidizing from reducing due to the evolution of oxygenic photosynthesis (Blankenship 2010). Therefore, the SODs enabled an organism to survive in the oxidizing environment on the earth and confronted high evolutionary pressure and evolved into different forms at separate junctures (Miller 2012). The higher availability of Fe ion in the reducing environment and the presence of *FeSODs* in both aerobic and anaerobic bacteria suggested them as the most ancient form of *SODs*. They are apparently lacking in animals and fungi, and present in the plants where they are localized only to the chloroplast. With the transition in the environment, a switch in the Fe-based activity could happen to utilize the Mn ion which fits in the chemistry of superoxide anion dismutation reaction. It probably have occurred with the reduction in Fe bioavailability and enhanced iron toxicity in the oxygenic environment (Miller 2012). Moreover, the cambialistic *SODs* in bacteria and archaeal shows both Fe and Mn-based activity which behave as a potential precursor for the endeavor of modern MnSODs evolution. Since these cambialistic SOD are present in primitive anaerobic organisms, they could have evolved into the *FeSODs* and MnSODs found in higher plants. Further, the evolution of MnSODs was also concurrent with eukaryotes (Miller 2012).

To understand the dependency on different metal ions the FeSOD and MnSOD of *Escherichia coli* were compared. They show high structural resemblance in terms of conserved metal ion binding and active site residues and shares similar coordination geometries with both the metal ions with corresponding electronic structures (Miller 2012; Fink and Scandalios 2002). Moreover, both Fe and Mn have akin ionic radii and ligand fondness and oscillate between 2^+ and 3^+ oxidation states. However, the similar oxidation states of Fe and Mn resulted in different d-shell electron configuration, due to which the reduction of Fe^{3+} to Fe^{2+} is less favored in comparison of Mn^{3+} to Mn^{2+}. It might suggest a significant divergence between modern FeSOD and MnSOD. Further, the clustering of both the types in different phylogenetic clades and occurrence in different organelles provides an insight into the divergence in both the forms during evolution (Fink and Scandalios 2002).

The origin of Cu-ZnSOD is not much clear, although the utilization of Cu^{2+} as a metal cofactor started because of the reduction in the availability of Fe^{2+} and the conversion of insoluble Cu^+ to soluble Cu^{2+} during the oxidizing environment. The variations in structural and chemical properties of Cu-ZnSOD from others suggested the occurrence of high divergence after Cu became a metal cofactor (Bannister et al. 1991).

3.4 Role of SODs Under Abiotic Stress Conditions

Abiotic stress factors influence the growth and yield of plants while *SODs* facilitate their survival during these stresses. The extent of abiotic stresses for instance heat, drought, salt, heavy metal, high light stress, etc. or combination of any of these stresses is increasing due to varying climatic conditions (Mittler 2006). Moreover, certain developmental stages of a plant are highly prone to these kind of stresses which affect the vigor of plants. Herein, the roles of SODs have been discussed in different plant species during major abiotic stresses.

3.4.1 Heat Stress

The rise in temperatures may lead to the heat stress which is often combined with drought or salt stress resulting in the severe devastation in plant growth and development (Mittler 2006). Extremely high temperatures cause various biochemical, physiological and morphological changes in plants and affect the permeability and fluidity of the cell membrane. The plants have developed a class of thermotolerant proteins known as heat shock proteins which provide tolerance against the heat stress. Secondly, the high temperatures also trigger the oxidative stress, where, the SODs and other antioxidants come into the role for plant protection (Kotak et al. 2007) (Table 3.2). For instance, in the case of *Jatropha curcas* and mulberry exposed to heat stress, the total SOD activity increased remarkably in the leaves which in turn caused the accumulation of H_2O_2 (Silva et al. 2013; Chaitanya et al. 2002). Within heat tolerant and susceptible genotypes of *B. juncea,* the SOD activities increased in

Table 3.2 SOD activities during abiotic stress in plants

Abiotic stress	Plant	Tissue analysed	Affected SOD	Alteration in SOD activity[a]	References
Heat stress	*Cicer arietinum*	Roots	Cu/Zn-SOD	↑↑ and later ↓↓	Ceylan et al. (2013)
			Mn-SOD	↑↑ and later ↑	
	Festuca arundinacea	Leaves	Total SOD	No change	Jiang and Huang (2001)
	Jatropha curcas	Leaves	Total SOD	↑↑	Silva et al. (2013)
	Morus alba	Leaves	Total SOD	↑↑	Chaitanya et al. (2002)
	Pennisetum glaucum	Leaves	Total SOD	↑↑	Mahanty et al. (2012)
	Poa pratensis	Leaves	Total SOD	No change	Jiang and Huang (2001)
Drought	*Beta vulgaris*	Shoot tip	Total SOD	↑↑	Sen and Alikamanoglu (2013)
	Cicer arietinum	Roots	Cu/Zn-SOD	↑↑ later ↓↓	Ceylan et al. (2013)
			FeSOD	↓	
			MnSOD	↓	
	Gossypium herbaceum	Leaves	Total SOD	↑↑ later ↓↓	Deeba et al. (2012)
	Oryza sativa	Roots	Total SOD	↑↑	Sharma and Dubey (2005)
	Olea europaea	Leaves	Total SOD	↑↑	Doupis et al. (2013)
	Triticum aestivum	Leaves	Total SOD	↓↓	Alexieva et al. (2001)
Cold	*Cucumis sativus*	Leaves	Total SOD	↑↑	Lee (2000)
	Hordeum vulgare	Leaves	Total SOD	↓↓	Mutlu et al. (2013)
	Musa acuminata	Leaves	Total SOD	↓↓	Kang et al. (2003)
	Nicotiana tabacum	Shoot	Total SOD	↑↑	Xu et al. (2010)

(continued)

Table 3.2 (continued)

Abiotic stress	Plant	Tissue analysed	Affected SOD	Alteration in SOD activity[a]	References
	Oryza sativa	Shoot	Total SOD	No change	Huang and Guo (2005)
	Pennisetum glaucum	Leaves	Total SOD	↓↓	Mahanty et al. (2012)
Salinity (NaCl)	*Arabidopsis thaliana*	Seedlings	Total SOD	↑↑	Zsigmond et al. (2012)
	Beta vulgaris	Leaves	Total SOD	↑↑	Bor et al. (2003)
	Brassica juncea	Leaves	Total SOD	↑	Kumar et al. (2013)
	Cicer arietinum	Leaves	Total SOD	↑↑	Rasool et al. (2013)
	Gossypium hirsutum	Leaves	Total SOD	↑↑	Meloni et al. (2003)
	Nicotiana tabacum	Seedlings	Cu-ZnSOD	↑↑	Lee et al. (2013)
	Phaseolus vulgaris	Roots	Total SOD	↓	Jebara et al. (2005)
			Mn-SOD	No change	
			Cu-ZnSOD	↓↓	
			Fe-SOD	↑↑	
	Vigna unguiculata	Leaves	Total SOD	↓↓	Hernández et al. (1994)
			Mn-SOD	↓↓	
			Cu-ZnSOD	↓↓ and no change	
		Leaf protoplast	Mn-SOD	↓↓	
			Cu/Zn-SOD	↓↓	

[a]↑↑ and ↑ shows significant and insignificant increase, whereas ↓↓ and ↓ shows significant and insignificant decrease in SOD activity

both but it was higher in tolerant genotypes. During the recovery period, the activity remarkably reduced but remained higher from the control plants (Rani 2009). Similarly, the increased SOD activities were also observed in wheat genotypes under heat stress (Almeselmani et al. 2006). In contrast, two cool season turfgrasses namely *Poa pratensis* and *Festuca arundinacea* had no effect on SOD activity due to heat stress (Jiang and Huang 2001). The accumulation of H_2O_2 has also been reported in tomato exposed to the heat stress along with the increase in total phenol content (Rivero et al. 2001). The study in *Lotus japonicus* cv. Gifu showed the degradation of chloroplastic Cu-ZnSOD due to heat and the intensification in superoxide levels due to combined heat and drought stress. Further, the sternness of the combined heat and drought stress on Photosystem II (PSII) was higher in comparison to a particular stress. They vitiate the antioxidant defense system of chloroplast and reduced the activity of PSII (Sainz et al. 2010).

3.4.2 Drought Stress

Drought stress is also becoming more severe with the changing climatic conditions and the water deficit in the ground levels. It is mostly accompanied by other abiotic stresses such as extreme heat or cold and has major effects on the growth and yield of the plants. It targets various physiological processes such as reduction in growth, photosynthetic rate, CO_2 fixation due to stomata closure to avoid transpiration, etc. and leads to the ROS burst within a plant (Alexieva et al. 2001; Gill and Tuteja 2010). During drought stress, the total SOD activity remarkably increased in case of *Olea europae* leaves, *O. sativa* shoots and in various organs of other plant species (Doupis et al. 2013; Sharma and Dubey 2005) (Table 3.2). The effect of drought, cadmium and their combination was observed in drought tolerant (*Lycopersicon peruvianum* (L.) Mill.) and sensitive (*Lycopersicon esculentum* Mill. cv. Lukullus) species of tomato. The SOD activity enhanced in both the species but more in the case of *L. esculentum*, suggesting the over-production of ROS in drought-sensitive tomato (Cekic et al. 2006). The increase in total SOD activity followed by reduction was observed in *Gossypium herbaceum*, *Festuca arundinacea* and *Poa pratensis* leaves under drought stress (Jiang and Huang 2001; Deeba et al. 2012). In contrast, the reduction in total SOD activity was also observed in the case of wheat and pea (Alexieva et al. 2001).

3.4.3 Cold Stress

The chilling or cold stress is another temperature-based abiotic stress, which affects the growth and productivity of non-acclimatized plants. It can cause various physiological and morphological damage to the plant resulting in the ROS over-production (Mittler 2006; Rivero et al. 2001). Under cold stress, the enhanced SOD and other antioxidant activities provide increased tolerance to the plants. For instance, the tolerant cultivars of tobacco and rice showed an increase in total SOD activity in comparison to the sensitive cultivars (Xu et al. 2010; Huang and Guo 2005). Similarly, in the case of citrus callus and other plants, the higher activity of SOD conferred enhanced resistance to cold stress (Gueta-Dahan et al. 1997) (Table 3.2). The SOD activity shows an abrupt rise in initial days of cold stress given at both 1 °C and −10 °C, though remain unaffected in later days in comparison to the control as observed in *Avena nuda* (Liu et al. 2013). In *Cynodon dactylon* (Bermudagrass), the effect of cold stress was studied along with melatonin hormone common in both plants and animals. The results indicated high cold stress tolerance at −5 °C for 8 h in Bermudagrass pre-treated with melatonin (100 μM) in comparison to the non-melatonin treated plants. Further, the elevated SOD activities in pre-treated plants suggested the role of melatonin in inducing the SOD for ROS scavenging during stress (Fan et al. 2015).

3.4.4 Salinity Stress

Salinity stress is a serious abiotic stress for non-halophytes as they are sensitive to high salt concentration and exhibit various morphological disorders (Meloni et al. 2003; Sheokand et al. 2008). During salinity stress, the activity of SOD gets affected in response to the ROS over-production. The experimental evidence in various plants such as maize, pea, tomato, tea, and many more depicted the role of *SODs* during salt stress (Tuna et al. 2008; Ahmad et al. 2008; Alharby et al. 2016; Upadhyaya et al. 2008). In some cases, the total activity of SODs markedly increased during salt stress such as in *Arabidopsis*, *Beta vulgaris*, *B. juncea, B. napus, Cicer arietinum, Morus alba* and *O. sativa* (Sheokand et al. 2008; Zsigmond et al. 2012; Bor et al. 2003; Kumar et al. 2013; Ashraf and Ali 2008; Harinasut et al. 2003; Khan and Panda 2002) (Table 3.2). However, in some plants the activity showed reduction or remain unaffected depending upon the plant tissue used for the measurement and the concentration of salt used for the experimentation. In tomato and *Plantago,* the reduced and enhanced SOD activities were observed in salt sensitive and tolerant cultivar, respectively, during salt stress (Shalata et al. 2001; Hediye Sekmen et al. 2007). Further, the enhancement in activities of chloroplast SOD was observed in *Suaeda salsa* under salt stress (Qiu-Fang et al. 2005). The SODs have high tendencies to combat the oxidative stress formed under certain concentrations of salt stress but both the amount and activities declines as the concentration is increased above the threshold tolerated by the plant as observed in the leaves of *Ulmus pumila* (Song et al. 2006).

In pea plant, the intensified SOD activities were observed particularly for the FeSOD in comparison of Cu-ZnSOD under severe salt stress. Whereas, the salt stress of low intensity enhanced the activities of Cu-ZnSOD (Gomez et al. 2003; Hernandez et al. 1995). Likewise, *FeSODs* and MnSODs also show differences in their response to a particular stress condition. These variances proposed the involvement of the *SODs* in the signaling pathways and the distinct subcellular localization of their isoforms within the cell. For instance, the stresses which affect the chloroplast might not affect the MnSOD as discussed in case of Arabidopsis treated with methyl viologen altered the activity of both FeSOD and MnSOD. While DCMU treatment caused a change in the activity of only FeSOD and MnSOD remained unaffected (Kliebenstein et al. 1998).

3.4.5 Heavy Metal Stress

The upsurge in anthropogenic activities such as industrialization, mining, use of the extensive amount of chemicals in agriculture, etc. have developed the problem of heavy metals in soils which affect the wellbeing of both plants and animals. The accumulation of heavy metals such as cobalt (Co), nickel (Ni), cadmium (Cd), mercury (Hg), etc. in the plants results in the production of ROS. These heavy metals cause morphological and physiological disorders in plants and influence growth, photosynthesis, metabolic pathways, etc. The elevated activity of SOD and H_2O_2 formation was observed in plants under Co toxicity (Sinha et al. 2012). The phytotoxicity of Hg results in the disruption in water flow by binding to water-channel proteins of the root cell, hinder the nutrient uptake and translocation, enzymes malfunction, lipid peroxidation, etc. (Israr et al. 2006; Lomonte et al. 2010). A significant elevation in SOD activities has been observed in the entire plant due to Hg toxicity (Lomonte et al. 2010). Similarly, Cr and Pb toxicity also resulted in increased SOD activities in various plants (Table 3.3). In case of Cd toxicity, the plant show chlorosis in leaves, reduction in growth, inhibition in seed germination, reduced photosynthesis rate (Benavides et al. 2005). The activity of SOD showed an elevation in the leaves of *B. juncea, Lepidium sativum, P. sativum,* O. sativa, and some others, in contrast, reduced in Arabidopsis and durum wheat (Kumar et al. 2013; Dixit et al. 2001) (Table 3.3). The concentration of heavy metals and their exposure time strongly affects the activity of SODs along with the other antioxidants within the plants.

3.5 Biotic Stress

In spite of abiotic stress, plants encounter elevated oxidative stress at the time of the pathogen attack. Though, SODs are well known to have an important role in providing tolerance against biotic stress. For instance, the resistance against virus-induced hypersensitive necrosis has been studied in *Nicotiana tabacum* transgenic line over-expressing the chloroplast superoxide dismutase (*SlChSOD*) of tomato (Viczián et al. 2014). Another instance revealed the role of cytosolic *Cu-ZnSOD* alone and in combination with an H_2O_2 reducing cytosolic ascorbate peroxidase enzyme against *Pseudomonas syringae* pv.tabaci and *Agrobacterium tumefaciens* which cause bacterial wildfire and crown gall diseases, respectively , in tobacco. The varying levels of disease tolerance were detected in all the transgenic tobacco line against both the pathogens. However, in double transformants, more resistance was observed against *P. syringae* pv. tabaci (Faize et al. 2012). The increased SOD activity was observed in TMV infected leaves of Xanthi tobacco with systemic acquired resistance (SAR) induced with salicylic acid or through previous TMV infection. However, leaves deprived of SAR showed no increase in SOD activity (Barna et al. 2003). The role of extracellular SOD (EcSOD) against

Table 3.3 SOD activities in response to heavy metal stress in plants

Heavy metal stress	Plant	Tissue analysed	Affected SOD	Alteration in SOD activity[a]	References
Cd toxicity	*Arabidopsis thaliana*	Leaves	Total SOD	↓	Cho and Seo (2005)
	Brassica juncea	Leaves	Total SOD	↑	Kumar et al. (2013)
	Oryza sativa	Leaves	Total SOD	↑↑	Hsu and Kao (2004)
	Triticum durum	Leaves	Total SOD	↓↓	Milone et al. (2003)
	Vigna radiata	Leaves	Total SOD	↑↑	Anjum et al. (2011)
Co toxicity	*Brassica campestris*	leaves	Total SOD	↑↑	Sinha et al. (2012)
		Root	Total SOD	↑↑	
	Brassica juncea	leaves	Total SOD	↑↑ later ↓↓	Arora et al. (2012)
Cr toxicity	*Brassica juncea*	Shoot	Total SOD	↑↑	Diwan et al. (2010)
	Ocimum tenuiflorum	Leaves	Total SOD	↑↑, later ↓↓	Rai et al. (2004)
	Oryza sativa	Root	Total SOD	↑↑	Panda (2007)
	Triticum aestivum	Shoot	Total SOD	↑	Subrahmanyam (2008)
	Zea mays	Leaves	Total SOD	↑↑	(Maiti et al. 2012)
Hg toxicity	*Atriplex codonocarpa*	Shoot	Total SOD	↑↑	Lomonte et al. (2010)
		Root	Total SOD	↑↑, later ↓↓	
	Lycopersicon esculentum	Leaves	Total SOD	↑↑	Cho (2000)
		Roots	Total SOD	↑↑	
Pb toxicity	*Cassia angustifolia*	Leaves	Total SOD	↑↑, later ↓↓	Qureshi et al. (2007)
	Hordeum vulgare	Root tip	Total SOD	↑↑	Tamás et al. (2010)
			MnSOD	↑↑	
	Lupinus luteus	Root tip	Total SOD	↑↑	Rucińska et al. (1999)
	Oryza sativa	Shoot	Total SOD	↑↑	Verma and Dubey (2003)

[a]↑↑ and ↑ shows significant and insignificant increase, whereas ↓↓ and ↓ shows significant and insignificant decrease in SOD activity

pathogen attack has been studied in pea plant, where the elicitor and supprescins from *Mycosphaerella pinodes* (pea pathogen) increase and inhibit the activity of EcSOD, respectively (Kasai et al. 2006).

Plants susceptible to necrotrophic fungus show the ROS production to a large extent such as in Arabidopsis, barley and wheat infected with *Botrytis cinerea, Pyrenophora teres* and *Zymoseptoria tritici*, respectively (Govrin and Levine 2000; Liu et al. 2015; Shetty et al. 2007). These pathogens may induce cell death by exploiting ROS production by host plant or itself contributing to ROS production (Govrin and Levine 2000). However, an *in planta* H_2O_2 infiltration assay in wheat leaves during necrotrophic stage of *Z. tritici* using catalase resulted in reduced growth of pathogen but it still grows *in planta* (Shetty et al. 2007). Moreover, in barley leaves *in planta* infiltration of H_2O_2 partially reduces the symptoms but did not affect *P. teres* growth (Lightfoot et al. 2017). It suggests the complex interaction between plant and pathogen in terms of ROS overproduction, fungal growth rate, virulence, etc. which require more detailed analysis in understanding these associations. The function of *HvCSD1* (*Cu-ZnSOD*) of barley was studied in RNA interference (RNAi) knockdown lines in response to *Magnaporthe oryzae* and *Blumeria graminis.* Unexpectedly, it has no role in susceptibility and disease expansion caused by these pathogens, however, a slower growth rate and enhanced resistance to *Pyrenophora teres* f. *teres* (Lightfoot et al. 2017). It signifies the differential regulations mediated by H_2O_2 in response to different plant-pathogen and their signaling behaviors.

3.6 Industrial Uses of SODs

Nature has developed SODs as most potent antioxidant enzyme known for its astonishing properties. Benefits of SODs in the field of agriculture, horticulture, medical, dermatology, cosmetics, nutraceuticals, food preservatives, etc. include them to the pool of industrial enzymes (Table 3.4).

3.6.1 Agriculture and Horticulture Uses

Various stress tolerant plants have been developed with the *SOD* genes in past years using transgenic approaches. For instance, the enhanced tolerance against drought stress and methyl viologen (MV) was observed in *O. sativa* transformed with *MnSOD* gene from *P. sativum* (Wang et al. 2005). The overexpression of *MnSOD* in transgenic *Arabidopsis*, showed tolerance against salinity stress (150 mM NaCl) and reduced lipid peroxidation product (MDA), though, the wild type plants showed gradual withering (Wang et al. 2004). The improved resistance to oxidative stress was observed in *Nicotania tabacum* overexpressing the *Cu-ZnSOD* gene (Gupta et al. 1993). Further, the SOD gene of *Lycopersicon esculentum* provide tolerance to pure cercosporin and MV in transformed sugar beet (Tertivanidis et al. 2004). In *O. sativa*,

Table 3.4 List of some plant *SODs* having astonishing physicochemical properties

Plant name	SOD type	Extraordinary properties	References
Camellia sinensis	MnSOD	Low temperature responsive (0 °C optimum), active over wide temperature range	Vyas and Kumar (2005)
Citrus limon	Cu-ZnSOD	Stable in 2.3–11 pH, resistant to chymotrypsin and trypsin digestion, 99 min half-life at 90 °C	Lin et al. (2002)
Chenopodium murale	Cu-ZnSOD	Stable after 10 min boiling	Khanna-Chopra and Sabarinath (2004)
Curcuma longa	Cu-ZnSOD	Stable after autoclaving (6–20 bars, 10 min), 20 min boiling, microwaving (2450 MHz, 1–3 min), and wide pH range, SDS and alcohol and concentrations	Kochhar and Kochhar (2008)
Curcuma aromatica	Cu-ZnSOD	Wide pH (3—10) and temperature range (−10 to +80 °C); retained 50% activity after autoclaving; resistant to inactivation at 80 °C for 180 min, reductants, denaturating agent etc.	Kumar et al. (2014)
Nicotiana sp.	MnSOD	Stable after prolonged incubations at 90 °C	Carter and Thornburg (2000)
Pisum sativum	MnSOD	>95% and >80% stability after 15 min and 75 min at 85 °C, respectively	Gucciardo et al. (2007)
Potentilla atrosanguinea	Cu-ZnSOD	Retains activity after 1 h of boiling and autoclaving; wide temperature range (−10 to +80 °C)	Yogavel et al. (2008)

the tolerance against salinity stress attained after the transformation with *NaMnSOD* gene of *Natrinema altunense* (a halophilic archaeon). The increased expression of *NaMnSOD* gene, total SOD activity, and photosynthesis in the transformants suggested the efficient detoxification of ROS produced due to the salt stress (Chen et al. 2013). Numerous reports suggest the role of SODs in providing tolerance to the plants against various environmental challenges. Conversely, in some cases, the extragenetic SODs failed in providing enhanced tolerance to the transgenic plants (Tepperman and Dunsmuir 1990). It might be attributed to the differences in SOD isozymes, their subcellular localization and the complexity of ROS detoxification system. Further, the finding indicating the overexpression of *Cu-ZnSOD* or *FeSOD* rendered the stress tolerance in some transgenic plants, might be due to their inhibition by H_2O_2, unlike *MnSOD*. In such cases, other approaches like antisense technology should be used instead of overexpression studies.

3.6.2 Human Health Benefits

ROS induces atherosclerosis, neurological disorders, infertility, diabetes, aging, cancer, etc. in human beings (McCord 1993). Supplements of SOD have shown the prevention and reduction in the adverse effects of these diseases. Mostly bovine, recombinant human and human Cu-ZnSOD is used as the intervention agent (Manzanas García et al. 2008). However, some studies include catalase along with SOD in their formulation because H_2O_2 is a potent oxidant and is also a product of SOD and substrate for catalase catalyzed reaction. The consumption of cigarette and alcohol results in the reduction in SOD levels and cause oxidative stress (Russo et al. 2011). The formulations of SOD can prevent the alcohol-induced hangover and may be incorporated in tobacco products and cigarettes to reduce the ROS induced damage in the respiratory tract and oropharynx (Colin and N'Guyen 2007; Hersh and Hersh 2002).

3.6.3 Uses in Cosmetics

The commercial use of SOD is well established in cosmetic formulations, such as sunscreens, moisturizers, anti-hair fall sprays, eye creams, skin-lightening creams, nail polish marketed by various prestigious brands (Bafana et al. 2011). SOD can be very useful in averting the harmful effects of some cosmetic products itself. For instance, facial creams and lipsticks have been known to produce ROS themselves after exposure to sunlight and cause lipid peroxidation and hemolysis in human erythrocytes (Hans et al. 2008). In topical creams, the fusion with lysine-rich peptide or HIV-1 tat protein transduction domain improved the penetration of SOD into the skin (Park et al. 2002). Further, coalescing chaperone and SOD proteins prevent them from H_2O_2 inhibition and stabilized to at least 45 °C temperatures (Bresson-Rival et al. 1997). The importance and use of SOD in cosmetics was first recognized by L'Oreal (France) and got patent in 1973 for SOD of marine source. Afterward, several manufacturing processes and formulations consisting of SOD were patented by L'Oreal (Colin and N'Guyen 1999). The marine bacteria such as *Photobacterium leiognathi, P. sepia,* and *P. phosphoreum* are used for the extraction. The products also showed positive results in erythema caused by UV-radiations. Further, a yeast-derived SOD having high stability at 45 °C was developed by Brooks Industries in 1987 (Lods et al. 2000). The overexpression of *Cu-ZnSOD* gene in recombinant *Saccharomyces cerevisiae* enhanced the production as well as the stability of the enzyme. Hence, the stability of products increased for at least 2 years at 4–8 °C.

3.7 Conclusions

In conclusion, SODs can be selected as a potential source for targeting the stress-tolerance in the economically important plant. Since the rise in ROS production is an initiation of every stress, so the SODs can minimize the oxidative damage and prevent physiological impairments in plants. The scenario of changing climate conditions and increasing stress-inducing factors like pollutants and harmful chemicals can be detrimental to the plant vigor. Enormous properties of SODs including the stress tolerance and stability in a wide range of temperature and pH suggest them as an important enzyme for agriculture and industrial benefits. The overexpression of SODs along with any H_2O_2 detoxifying enzyme like catalase will provide more promising results in stress tolerance. Besides this, the commercial use of SODs can be explored in other dimensions like as an additive in food products, preservatives, medicines, biosensors for $O_2^{\bullet -}$, anti-depressants, for increasing the yield of agricultural and horticultural plants, etc. The detailed study on each SOD isoform would further reveal their roles in different plant species.

Acknowledgements Authors are grateful to Panjab University, Chandigarh, India for research facilities. ST is grateful Science and Engineering Research Board (SERB), Government of India for financial support. We are also grateful to the Department of Science and Technology, Government of India for partial financial support under the Promotion of University Research and Scientific Excellence (PURSE) grant scheme.

Conflict of Interest

All the authors declare that there are no conflicts of interest

References

Ahmad P, John R, Sarwat D, Umar S (2008) Responses of proline, lipid peroxidation and antioxidative enzymes in two varieties of *Pisum sativum* L. under salt stress. Int J Plant Prod 2(4)

Alexieva V, Sergiev I, Mapelli S, Karanov E (2001) The effect of drought and ultraviolet radiation on growth and stress markers in pea and wheat. Plant, Cell Environ 24:1337–1344. https://doi.org/10.1046/j.1365-3040.2001.00778.x

Alharby HF, Metwali EMR, Fuller MP, Aldhebiani AY (2016) The alteration of mRNA expression of SOD and GPX genes, and proteins in tomato (Lycopersicon esculentum Mill) under stress of NaCl and/or ZnO nanoparticles. Saudi J Biol Sci 23:773–781. https://doi.org/10.1016/j.sjbs.2016.04.012

Almeselmani M, Deshmukh PS, Sairam RK, Kushwaha SR, Singh TP (2006) Protective role of antioxidant enzymes under high temperature stress. Plant Sci 171:382–388. https://doi.org/10.1016/J.PLANTSCI.2006.04.009

Alscher RG, Erturk N, Heath LS (2002) Role of superoxide dismutases (SODs) in controlling oxidative stress in plants. J Exp Bot 53:1331–1341. https://doi.org/10.1093/jexbot/53.372.1331
Anjum NA, Umar S, Iqbal M, Khan NA (2011) Cadmium causes oxidative stress in mung bean by affecting the antioxidant enzyme system and ascorbate-glutathione cycle metabolism. Russ J Plant Physiol 58:92–99. https://doi.org/10.1134/S1021443710061019
Apel K, Hirt H (2004) Reactive oxygen species: metabolism, oxidative stress, and signal transduction. Annu Rev Plant Biol 55:373–399. https://doi.org/10.1146/annurev.arplant.55.031903.141701
Arora P, Bhardwaj R, Kanwar MK (2012) Effect of 24-epibrassinolide on growth, protein content and antioxidative defense system of *Brassica juncea* L. subjected to cobalt ion toxicity. Acta Physiol Plant 34:2007–2017. https://doi.org/10.1007/s11738-012-1002-2
ASADA K (1987) Production and scavenging of active oxygen in photosynthesis. Photoinhibition 227–287
Ashraf M, Ali Q (2008) Relative membrane permeability and activities of some antioxidant enzymes as the key determinants of salt tolerance in canola (*Brassica napus* L.). Environ Exp. Bot 63:266–273. https://doi.org/10.1016/j.envexpbot.2007.11.008
Bafana A, Dutt S, Kumar S, Ahuja PS (2011) Superoxide dismutase: an industrial perspective. Crit Rev Biotechnol 31:65–76. https://doi.org/10.3109/07388551.2010.490937
Baldensperger JB (1978) An iron-containing superoxide dismutase from the chemolithotrophic Thiobacillus denitrificans? RT? strain. Arch Microbiol 119:237–244. https://doi.org/10.1007/BF00405401
Bannister WH, Bannister JV, Barra D, Bond J, Bossa F (1991) Evolutionary aspects of superoxide dismutase: the copper/zinc enzyme. Free Radic Res Commun 12–13(Pt 1):349–361
Barna B, Fodor J, Pogány M, Király Z (2003) Role of reactive oxygen species and antioxidants in plant disease resistance. Pest Manag Sci 59:459–464. https://doi.org/10.1002/ps.706
Benavides MP, Gallego SM, Tomaro ML (2005) Cadmium toxicity in plants. Brazilian J Plant Physiol 17:21–34. https://doi.org/10.1590/S1677-04202005000100003
Blankenship RE (2010) Early evolution of photosynthesis. Plant Physiol 154:434–438. https://doi.org/10.1104/pp.110.161687
Bor M, Özdemir F, Türkan I (2003) The effect of salt stress on lipid peroxidation and antioxidants in leaves of sugar beet *Beta vulgaris* L. and wild beet Beta maritima L Plant Sci 164:77–84. https://doi.org/10.1016/s0168-9452(02)00338-2
Bordo D, Djinovic K, Bolognesi M (1994) Conserved patterns in the Cu, Zn superoxide dismutase family. J Mol Biol 238:366–386. https://doi.org/10.1006/jmbi.1994.1298
Bowler C, Slooten L, Vandenbranden S, De Rycke R, Botterman J, Sybesma C, Van Montagu M, Inzé D (1991) Manganese superoxide dismutase can reduce cellular damage mediated by oxygen radicals in transgenic plants. EMBO J 10:1723–1732
Bowler C, Montagu MV, Inze D (1992) Superoxide dismutase and stress tolerance. Annu Rev Plant Physiol Plant Mol Biol 43:83–116 (1992). https://doi.org/10.1146/annurev.pp.43.060192.000503
Bresson-Rival D, Boivin P, Linden G, Perrier E, Humbert G (1997) Stabilized compositions of superoxide dismutase obtained from germinated plant seeds. US Pat. 5904921
Cannon RE, White JA, Scandalios JG (1987) Cloning of cDNA for maize superoxide dismutase 2 (SOD2). Proc Natl Acad Sci 84:179–183. https://doi.org/10.1073/pnas.84.1.179
Carter C, Thornburg RW (2000) Tobacco Nectarin I. J Biol Chem 275:36726–36733. https://doi.org/10.1074/jbc.M006461200
Cekic FO, Ünyayar S, Keleş Y, Çekiç FÖ (2006) The antioxidative response of two tomato species with different drought tolerances as a result of drought and cadmium stress combinations. Plant Soil Environ 51. https://doi.org/10.17221/3556-pse
Ceylan HA, Türkan I, Sekmen AH (2013) Effect of coronatine on antioxidant enzyme response of chickpea roots to combination of PEG-induced osmotic stress and heat stress. J Plant Growth Regul 32:72–82. https://doi.org/10.1007/s00344-012-9277-5

Chaitanya KV, Sundar D, Masilamani S, Ramachandra Reddy A (2002) Variation in heat stress-induced antioxidant enzyme activities among three mulberry cultivars. Plant Growth Regul 36:175–180. https://doi.org/10.1023/A:1015092628374

Chen Z, Pan YH, An LY, Yang WJ, Xu LG, Zhu C (2013) Heterologous expression of a halophilic archaeon manganese superoxide dismutase enhances salt tolerance in transgenic rice. Russ J Plant Physiol 60:359–366. https://doi.org/10.1134/S1021443713030059

Cho Park (2000) Mercury-induced oxidative stress in tomato seedlings. Plant Sci 156:1–9

Cho U-H, Seo N-H (2005) Oxidative stress in Arabidopsis thaliana exposed to cadmium is due to hydrogen peroxide accumulation. Plant Sci 168:113–120. https://doi.org/10.1016/J.PLANTSCI.2004.07.021

Colin C, N'Guyen Q (1999) Cosmetic composition containing, in combination, a superoxide-dismutase and a melanin pigment. US Pat. 5925363

Colin C, N'Guyen QL (2007) Composition including superoxide dismutase and prickly-pear cactus for minimizing and preventing hangovers. US Pat. 5925363 (2007)

Crowell DN, Amasino RM (1991) Induction of Specific mRNAs in cultured soybean cells during cytokinin or auxin starvation. Plant Physiol 95:711–715. https://doi.org/10.1104/pp.95.3.711

Deeba F, Pandey AK, Ranjan S, Mishra A, Singh R, Sharma YK, Shirke PA, Pandey V (2012) Physiological and proteomic responses of cotton (*Gossypium herbaceum* L.) to drought stress. Plant Physiol Biochem 53:6–18 (2012). https://doi.org/10.1016/j.plaphy.2012.01.002

del Río LA, Sandalio LM, Palma JM, Bueno P, Corpas FJ (1992) Metabolism of oxygen radicals in peroxisomes and cellular implications. Free Radic Biol Med 13:557–580

Diwan H, Khan I, Ahmad A, Iqbal M (2010) Induction of phytochelatins and antioxidant defence system in Brassica juncea and Vigna radiata in response to chromium treatments. Plant Growth Regul 61:97–107. https://doi.org/10.1007/s10725-010-9454-0

Dixit V, Pandey V, Shyam R (2001) Differential antioxidative responses to cadmium in roots and leaves of pea (*Pisum sativum* L. cv. Azad)1. J Exp Bot 52:1101–1109 (2001). https://doi.org/10.1093/jexbot/52.358.1101

Doupis G, Bertaki M, Psarras G, Kasapakis I, Chartzoulakis K (2013) Water relations, physiological behavior and antioxidant defence mechanism of olive plants subjected to different irrigation regimes. Sci Hortic (Amsterdam) 153:150–156. https://doi.org/10.1016/J.SCIENTA.2013.02.010

Elstner FE (1991) Metabolisms of oxygen activation in different compartments of plant cells. Act Oxyg Stress Plant Metab 13–25 (1991)

Faize M, Burgos L, Faize L, Petri C, Barba-Espin G, Díaz-Vivancos P, Clemente-Moreno MJ, Alburquerque N, Hernandez JA (2012) Modulation of tobacco bacterial disease resistance using cytosolic ascorbate peroxidase and Cu. Zn-superoxide dismutase Plant Pathol 61:858–866. https://doi.org/10.1111/j.1365-3059.2011.02570.x

Fan J, Hu Z, Xie Y, Chan Z, Chen K, Amombo E, Chen L, Fu J (2015) Alleviation of cold damage to photosystem II and metabolisms by melatonin in Bermudagrass. Front Plant Sci 6:925. https://doi.org/10.3389/fpls.2015.00925

Feng X, Lai Z, Lin Y, Lai G, Lian C (2015) Genome-wide identification and characterization of the superoxide dismutase gene family in Musa acuminata cv. Tianbaojiao (AAA group). BMC Genomics 16:823. https://doi.org/10.1186/s12864-015-2046-7

Feng K, Yu J, Cheng Y, Ruan M, Wang R, Ye Q, Zhou G, Li Z, Yao Z, Yang Y, Zheng Q, Wan H (2016) The SOD gene family in tomato: identification, phylogenetic relationships, and expression patterns. Front Plant Sci 7:1279. https://doi.org/10.3389/fpls.2016.01279

Filiz E, Tombuloğlu H (2015) Genome-wide distribution of superoxide dismutase (SOD) gene families inSorghum bicolor. TURKISH J Biol 39:49–59. https://doi.org/10.3906/biy-1403-9

Fink RC, Scandalios JG (2002) Molecular evolution and structure–function relationships of the superoxide dismutase gene families in angiosperms and their relationship to other eukaryotic and prokaryotic superoxide dismutases. Arch Biochem Biophys 399:19–36. https://doi.org/10.1006/abbi.2001.2739

Foyer CH, Noctor G (2009) Redox regulation in photosynthetic organisms: signaling, acclimation, and practical implications. Antioxid Redox Signal 11:861–905. https://doi.org/10.1089/ars.2008.2177

Fridovich I (1986) Superoxide dismutases. Adv Enzymol Relat Areas Mol Biol 58:61–97

Getzoff ED, Cabelli DE, Fisher CL, Parge HE, Viezzoli MS, Banci L, Hallewell RA (1992) Faster superoxide dismutase mutants designed by enhancing electrostatic guidance. Nature 358:347–351. https://doi.org/10.1038/358347a0

Gill SS, Tuteja N (2010) Reactive oxygen species and antioxidant machinery in abiotic stress tolerance in crop plants. Plant Physiol Biochem 48:909–930. https://doi.org/10.1016/j.plaphy.2010.08.016

Gomez JM, Jimenez A, Olmos E, Sevilla F (2003) Location and effects of long-term NaCl stress on superoxide dismutase and ascorbate peroxidase isoenzymes of pea (Pisum sativum cv. Puget) chloroplasts. J Exp Bot 55:119–130 (2003). https://doi.org/10.1093/jxb/erh013

Govrin EM, Levine A (2000) The hypersensitive response facilitates plant infection by the necrotrophic pathogen Botrytis cinerea. Curr Biol 10:751–757

Gucciardo S, Wisniewski J-P, Brewin NJ, Bornemann S (2007) A germin-like protein with superoxide dismutase activity in pea nodules with high protein sequence identity to a putative rhicadhesin receptor. J Exp Bot 58:1161–1171. https://doi.org/10.1093/jxb/erl282

Gueta-Dahan Y, Yaniv Z, Zilinskas BA, Ben-Hayyim G (1997) Salt and oxidative stress: similar and specific responses and their relation to salt tolerance in Citrus. Planta 203:460–469. https://doi.org/10.1007/s004250050215

Gupta AS, Heinen JL, Holaday AS, Burke JJ, Allen RD (1993) Increased resistance to oxidative stress in transgenic plants that overexpress chloroplastic Cu/Zn superoxide dismutase. Proc Natl Acad Sci U S A 90:1629–1633. https://doi.org/10.1073/pnas.90.4.1629

Halliwell B (2006) Reactive species and antioxidants. Redox biology is a fundamental theme of aerobic life. Plant Physiol 141:312–322. https://doi.org/10.1104/pp.106.077073

Hans RK, Agrawal N, Verma K, Misra RB, Ray RS, Farooq M (2008) Assessment of the phototoxic potential of cosmetic products. Food Chem Toxicol 46:1653–1658. https://doi.org/10.1016/J.FCT.2008.01.005

Harinasut P, Poonsopa D, Roengmongkol K, Charoensataporn R (2003) Salinity effects on antioxidant enzymes in mulberry cultivar. Sci. Asia 29:109–113. https://doi.org/10.2306/scienceasia1513-1874.2003.29.109

Hediye Sekmen A, Türkan İ, Takio S (2007) Differential responses of antioxidative enzymes and lipid peroxidation to salt stress in salt-tolerant Plantago maritima and salt-sensitive Plantago media. Physiol Plant 131:399–411. https://doi.org/10.1111/j.1399-3054.2007.00970.x

Hernandez J, Olmos E, Corpas F, Sevilla F, Del Río L (1995) Salt-induced oxidative stress in chloroplasts of pea plants. Plant Sci 105:151–167. https://doi.org/10.1016/0168-9452(94)04047-8

Hernández JA, Del Río LA, Sevilla F (1994) Salt stress-induced changes in superoxide dismutase isozymes in leaves and mesophyll protoplasts from *Vigna unguiculata* (L.) Walp New Phytol 126:37–44. https://doi.org/10.1111/j.1469-8137.1994.tb07527.x

Hersh T, Hersh R (2002) Antioxidants to neutralize tobacco free radicals. US Pat. 6415798

Hsu YT, Kao CH (2004) Cadmium toxicity is reduced by nitric oxide in rice leaves. Plant Growth Regul 42:227–238. https://doi.org/10.1023/B:GROW.0000026514.98385.5c

Hu X, Hao C, Cheng Z-M, Zhong Y (2019) Genome-wide identification, characterization, and expression analysis of the grapevine superoxide dismutase (SOD) family. Int J Genomics 2019:1–13. https://doi.org/10.1155/2019/7350414

Huang M, Guo Z (2005) Responses of antioxidative system to chilling stress in two rice cultivars differing in sensitivity. Biol Plant 49:81–84. https://doi.org/10.1007/s00000-005-1084-3

Israr M, Sahi S, Datta R, Sarkar D (2006) Bioaccumulation and physiological effects of mercury in Sesbania drummondii. Chemosphere 65:591–598. https://doi.org/10.1016/j.chemosphere.2006.02.016

Jebara S, Jebara M, Limam F, Aouani ME (2005) Changes in ascorbate peroxidase, catalase, guaiacol peroxidase and superoxide dismutase activities in common bean (Phaseolus vulgaris) nodules under salt stress. J Plant Physiol 162:929–936. https://doi.org/10.1016/j.jplph.2004.10.005

Jiang Y, Huang B (2001) Drought and heat stress injury to two cool-season turfgrasses in relation to antioxidant metabolism and lipid peroxidation. Crop Sci—Crop SCI 41. https://doi.org/10.2135/cropsci2001.412436x

Kanematsu S, Asada K (1978) Superoxide dismutase from an anaerobic photosynthetic bacterium Chromatium vinosum. Arch Biochem Biophys 185:473–482. https://doi.org/10.1016/0003-9861(78)90191-1

Kanematsu S, Asada K (1990) Characteristic amino acid sequences of chloroplast and cytosol isozymes of CuZn-superoxide dismutase in Spinach. Rice and Horsetail Plant Cell Physiol 31:99–112. https://doi.org/10.1093/oxfordjournals.pcp.a077887

Kang G, Wang C, Sun G, Wang Z (2003) Salicylic acid changes activities of H2O2-metabolizing enzymes and increases the chilling tolerance of banana seedlings. Environ Exp Bot 50:9–15. https://doi.org/10.1016/S0098-8472(02)00109-0

Karuppanapandian T, Moon JC, Kim C, Manoharan K, Kim W (2011) Reactive oxygen species in plants: their generation, signal transduction, and scavenging mechanisms. Aust J Crop Sci 5:709–725

Kasai T, Suzuki T, Ono K, Ogawa K, Inagaki Y, Ichinose Y, Toyoda K, Shiraishi T (2006) Pea extracellular Cu/Zn-superoxide dismutase responsive to signal molecules from a fungal pathogen. J Gen Plant Pathol 72:265–272. https://doi.org/10.1007/s10327-006-0283-y

Khan MH, Panda SK (2002) Induction of oxidative stress in roots of oryza *sativa* L. Response to Salt Stress Biol Plant 45:625–627. https://doi.org/10.1023/A:1022356112921

Khanna-Chopra R, Sabarinath S (2004) Heat-stable chloroplastic Cu/Zn superoxide dismutase in Chenopodium murale. Biochem Biophys Res Commun 320:1187–1192. https://doi.org/10.1016/j.bbrc.2004.06.071

Kim FJ, Kim HP, Hah YC, Roe JH (1996) Differential expression of superoxide dismutases containing Ni and Fe/Zn in Streptomyces coelicolor. Eur J Biochem 241:178–185

Kirby TW, Lancaster JR, Fridovich I (1981) Isolation and characterization of the iron-containing superoxide dismutase of Methanobacterium bryantii. Arch Biochem Biophys 210:140–148

Kliebenstein DJ, Monde RA, Last RL (1998) Superoxide dismutase in Arabidopsis: an eclectic enzyme family with disparate regulation and protein localization. Plant Physiol 118:637–650

Kochhar S, Kochhar VK (2008) Identification and characterization of a super-stable Cu–Zn SOD from leaves of turmeric (*Curcuma longa* L.). Planta 228:307–318. https://doi.org/10.1007/s00425-008-0738-9

Kotak S, Larkindale J, Lee U, von Koskull-Döring P, Vierling E, Scharf K-D (2007) Complexity of the heat stress response in plants. Curr Opin Plant Biol 10:310–316. https://doi.org/10.1016/j.pbi.2007.04.011

Kumar D, Yusuf MA, Singh P, Sardar M, Sarin NB (2013) Modulation of antioxidant machinery in α-tocopherol-enriched transgenic Brassica juncea plants tolerant to abiotic stress conditions. Protoplasma 250:1079–1089. https://doi.org/10.1007/s00709-013-0484-0

Kumar A, Kaachra A, Bhardwaj S, Kumar S (2014) Copper, zinc superoxide dismutase of Curcuma aromatica is a kinetically stable protein. Process Biochem 49:1288–1296. https://doi.org/10.1016/J.PROCBIO.2014.04.010

Kusunose E, Ichihara K, Noda Y, Kusunose M (1976) Superoxide dismutase from mycobacterium tuberculosis. J Biochem 80:1343–1352. https://doi.org/10.1093/oxfordjournals.jbchem.a131407

Lee Lee (2000) Chilling stress-induced changes of antioxidant enzymes in the leaves of cucumber: in gel enzyme activity assays. Plant Sci 159:75–85

Lee Y-P, Ahmad R, Lee H-S, Soo Swak S, Shafqat M, Kwon S-Y (2013) Improved tolerance of Cu/Zn superoxide dismutase and ascorbate peroxidase expressing transgenic tobacco seeds and seedlings against multiple abiotic stresses. Int J Agric Biol 15:725–730

Leonowicz G, Trzebuniak KF, Zimak-Piekarczyk P, Ślesak I, Mysliwa-Kurdziel B (2018) The activity of superoxide dismutases (SODs) at the early stages of wheat deetiolation. PLoS ONE 13:1–21. https://doi.org/10.1371/journal.pone.0194678

Lightfoot DJ, Mcgrann GRD, Able AJ (2017) The role of a cytosolic superoxide dismutase in barley-pathogen interactions. Mol Plant Pathol 18:323–335. https://doi.org/10.1111/mpp.12399

Lin Y-L, Lai Z-X (2013) Superoxide dismutase multigene family in longan somatic embryos: a comparison of CuZn-SOD, Fe-SOD, and Mn-SOD gene structure, splicing, phylogeny, and expression. Mol Breed 32:595–615. https://doi.org/10.1007/s11032-013-9892-2

Lin M-W, Lin M-T, Lin C-T (2002) Copper/zinc-superoxide dismutase from lemon cDNA and enzyme stability. J Agric Food Chem 50:7264–7270

Liu W, Yu K, He T, Li F, Zhang D, Liu J (2013) The low temperature induced physiological responses of *Avena nuda* L., a cold-tolerant plant species. Sci World J 658793. https://doi.org/10.1155/2013/658793

Liu Z, Holmes DJ, Faris JD, Chao S, Brueggeman RS, Edwards MC, Friesen TL (2015) Necrotrophic effector-triggered susceptibility (NETS) underlies the barley- *Pyrenophora teres* f. *teres* interaction specific to chromosome 6H. Mol Plant Pathol 16:188–200. https://doi.org/10.1111/mpp.12172

Lods LM, Dres C, Johnson C, Scholz DB, Brooks GJ (2000) The future of enzymes in cosmetics. Int J Cosmet Sci 22:85–94. https://doi.org/10.1046/j.1467-2494.2000.00012.x

Lomonte C, Sgherri C, Baker AJM, Kolev SD, Navari-Izzo F (2010) Antioxidative response of Atriplex codonocarpa to mercury. Environ Exp Bot 69:9–16. https://doi.org/10.1016/J.ENVEXPBOT.2010.02.012

Mahanty S, Kaul T, Pandey P, Reddy RA, Mallikarjuna G, Reddy CS, Sopory SK, Reddy MK (2012) Biochemical and molecular analyses of copper–zinc superoxide dismutase from a C4 plant Pennisetum glaucum reveals an adaptive role in response to oxidative stress. Gene 505:309–317. https://doi.org/10.1016/j.gene.2012.06.001

Maiti S, Ghosh N, Mandal C, Das K, Dey N, Adak MK (2012) Responses of the maize plant to chromium stress with reference to antioxidation activity. Brazilian J Plant Physiol 24:203–212. https://doi.org/10.1590/S1677-04202012000300007

Mann T, Keilin D (1938) Haemocuprein and hepatocuprein, copper-protein compounds of blood and liver in mammals. Proc R Soc London B Biol Sci 126

Manzanas García A, Carrizosa C, Vallejo Ocaña C, Samper Ots P, María Delgado Pérez J, Carretero Accame E, Gómez-Serranillos M, de la Morena del Valle L (2008) Superoxidase dismutase (SOD) topical use in oncologic patients: Treatment of acute cutaneous toxicity secondary to radiotherapy. Clin Transl Oncol 10:163–167. https://doi.org/10.1007/s12094-008-0174-0

McCord JM (1993) Human disease, free radicals, and the oxidant/antioxidant balance. Clin Biochem 26:351–357

McCord JM, Fridovich I (1969) Superoxide dismutase. An enzymic function for erythrocuprein (hemocuprein). J Biol Chem 244:6049–6055 (1969)

Meloni DA, Oliva MA, Martinez CA, Cambraia J (2003) Photosynthesis and activity of superoxide dismutase, peroxidase and glutathione reductase in cotton under salt stress. Environ Exp Bot 49:69–76. https://doi.org/10.1016/S0098-8472(02)00058-8

Miller A-F (2012) Superoxide dismutases: Ancient enzymes and new insights. FEBS Lett 586:585–595. https://doi.org/10.1016/j.febslet.2011.10.048

Milone MT, Sgherri C, Clijsters H, Navari-Izzo F (2003) Antioxidative responses of wheat treated with realistic concentration of cadmium. Environ Exp Bot 50:265–276. https://doi.org/10.1016/S0098-8472(03)00037-6

Mittler R (2002) Oxidative stress, antioxidants and stress tolerance. Trends Plant Sci 7:405–410. https://doi.org/10.1016/S1360-1385(02)02312-9

Mittler R (2006) Abiotic stress, the field environment and stress combination. Trends Plant Sci 11:15–19. https://doi.org/10.1016/j.tplants.2005.11.002

Mittler R, Vanderauwera S, Gollery M, Van Breusegem F (2004) Reactive oxygen gene network of plants. Trends Plant Sci 9:490–498. https://doi.org/10.1016/j.tplants.2004.08.009

Molina-Rueda JJ, Tsai CJ, Kirby EG (2013) The Populus superoxide dismutase gene family and its responses to drought stress in transgenic poplar overexpressing a pine cytosolic glutamine synthetase (GS1a). PLoS ONE 8:e56421. https://doi.org/10.1371/journal.pone.0056421

Mutlu S, Karadağoğlu Ö, Atici Ö, Nalbantoğlu B (2013) Protective role of salicylic acid applied before cold stress on antioxidative system and protein patterns in barley apoplast. Biol Plant 57:507–513. https://doi.org/10.1007/s10535-013-0322-4

Ogawa K, Kanematsu S, Takabe K, Asada K (1995) Attachment of CuZn-superoxide dismutase to thylakoid membranes at the site of superoxide generation (PSI) in Spinach chloroplasts: detection by immuno-gold labeling after rapid freezing and substitution method. Plant Cell Physiol 36:565–573. https://doi.org/10.1093/oxfordjournals.pcp.a078795

Ogawa K, Kanematsu S, Asada K (1996) Intra- and extra-cellular localization of & cytosolic & CuZn-superoxide dismutase in spinach leaf and hypocotyl. Plant Cell Physiol 37:790–799. https://doi.org/10.1093/oxfordjournals.pcp.a029014

Panda SK (2007) Chromium-mediated oxidative stress and ultrastructural changes in root cells of developing rice seedlings. J Plant Physiol 164:1419–1428. https://doi.org/10.1016/j.jplph.2007.01.012

Park J, Ryu J, Jin LH, Bahn JH, Kim JA, Yoon CS, Kim DW, Han KH, Eum WS, Kwon HY, Kang T-C, Won MH, Kang JH, Cho S-W, Choi SY (2002) 9-polylysine protein transduction domain: enhanced penetration efficiency of superoxide dismutase into mammalian cells and skin. Mol Cells 13:202–208

Perry JJP, Shin DS, Getzoff ED, Tainer JA (2010) The structural biochemistry of the superoxide dismutases. Biochim Biophys Acta—Proteins Proteomics 1804:245–262. https://doi.org/10.1016/j.bbapap.2009.11.004

Puget K, Michelson AM (1974) Iron containing superoxide dismutases from luminous bacteria. Biochimie 56:1255–1267

Qiu-Fang Z, Yuan-Yuan L, Cai-Hong P, Cong-Ming L, Bao-Shan W (2005) NaCl enhances thylakoid-bound SOD activity in the leaves of C3 halophyte Suaeda salsa L. Plant Sci 168:423–430. https://doi.org/10.1016/J.PLANTSCI.2004.09.002

Qureshi MI, Abdin MZ, Qadir S, Iqbal M (2007) Lead-induced oxidative stress and metabolic alterations in Cassia angustifolia Vahl. Biol Plant 51:121–128. https://doi.org/10.1007/s10535-007-0024-x

Rai V, Vajpayee P, Singh SN, Mehrotra S (2004) Effect of chromium accumulation on photosynthetic pigments, oxidative stress defense system, nitrate reduction, proline level and eugenol content of *Ocimum tenuiflorum* L. Plant Sci 167:1159–1169. https://doi.org/10.1016/J.PLANTSCI.2004.06.016

Raja V, Majeed U, Kang H, Andrabi KI, John R (2017) Abiotic stress: interplay between ROS, hormones and MAPKs. Environ Exp Bot 137:142–157

Rani B (2009) High temperature induced changes in oxidative stress, antioxidant system and polypeptide pattern in Indian mustard [Brassica juncea (L.) Czern & Coss.]

Rasool S, Ahmad A, Siddiqi TO, Ahmad P (2013) Changes in growth, lipid peroxidation and some key antioxidant enzymes in chickpea genotypes under salt stress. Acta Physiol Plant 35:1039–1050. https://doi.org/10.1007/s11738-012-1142-4

Rivero RM, Ruiz JM, García PC, López-Lefebre LR, Sánchez E, Romero L (2001) Resistance to cold and heat stress: accumulation of phenolic compounds in tomato and watermelon plants. Plant Sci 160:315–321. https://doi.org/10.1016/s0168-9452(00)00395-2

Robinett NG, Peterson RL, Culotta VC (2017) Eukaryotic Cu-only superoxide dismutases (SODs): a new class of SOD enzymes and SOD-like protein domains. J Biol Chem. https://doi.org/10.1074/jbc.TM117.000182

Rucińska R, Waplak S, Gwóźdź EA (1999) Free radical formation and activity of antioxidant enzymes in lupin roots exposed to lead. Plant Physiol Biochem 37:187–194. https://doi.org/10.1016/S0981-9428(99)80033-3

Russo M, Cocco S, Secondo A, Adornetto A, Bassi A, Nunziata A, Polichetti G, De Felice B, Damiano S, Serù R, Mondola P, Di Renzo G (2011) Cigarette smoke condensate causes a decrease

of the gene expression of Cu–Zn superoxide dismutase, Mn superoxide dismutase, glutathione peroxidase, catalase, and free radical-induced cell injury in SH-SY5Y human neuroblastoma cells. Neurotox Res 19:49–54. https://doi.org/10.1007/s12640-009-9138-6
Sainz M, Díaz P, Monza J, Borsani O (2010) Heat stress results in loss of chloroplast Cu/Zn superoxide dismutase and increased damage to Photosystem II in combined drought-heat stressed Lotus japonicus. Physiol Plant 140:46–56. https://doi.org/10.1111/j.1399-3054.2010.01383.x
Salin ML, Bridges SM (1981) Absence of the iron-containing superoxide dismutase in mitochondria from mustard (Brassica campestris). Biochem J 195:229–233. https://doi.org/10.1042/bj1950229
Sandalio LM, Del Río LA (1987) Localization of superoxide dismutase in glyoxysomes from citrullus vulgaris. Functional Implications in Cellular Metabolism. J Plant Physiol 127:395–409. https://doi.org/10.1016/s0176-1617(87)80248-1
Scandalios JG (1993) Oxygen stress and superoxide dismutases. Plant Physiol 101:7–12. https://doi.org/10.1104/pp.101.1.7
Scandalios JG (2005) Oxidative stress: molecular perception and transduction of signals triggering antioxidant gene defenses. Brazilian J. Med. Biol. Res. 38:995–1014
Searcy KB, Searcy DG (1981) Superoxide dismutase from the archaebacterium thermoplasma acidophilum. Biochim Biophys Acta 670:39–46
Sen A, Alikamanoglu S (2013) Antioxidant enzyme activities, malondialdehyde, and total phenolic content of PEG-induced hyperhydric leaves in sugar beet tissue culture. Vitr Cell Dev Biol—Plant 49:396–404. https://doi.org/10.1007/s11627-013-9511-2
Shalata A, Mittova V, Volokita M, Guy M, Tal M (2001) Response of the cultivated tomato and its wild salt-tolerant relative Lycopersicon pennellii to salt-dependent oxidative stress: the root antioxidative system. Physiol Plant 112:487–494
Sharma P, Dubey RS (2005) Drought induces oxidative stress and enhances the activities of antioxidant enzymes in growing rice seedlings. Plant Growth Regul 46:209–221. https://doi.org/10.1007/s10725-005-0002-2
Sheng Y, Abreu IA, Cabelli DE, Maroney MJ, Miller A-F, Teixeira M, Valentine JS (2014) Superoxide dismutases and superoxide reductases. Chem Rev 114:3854–3918. https://doi.org/10.1021/cr4005296
Sheokand S, Kumari A, Sawhney V (2008) Effect of nitric oxide and putrescine on antioxidative responses under NaCl stress in chickpea plants. Physiol Mol Biol Plants 14:355–362. https://doi.org/10.1007/s12298-008-0034-y
Shetty NP, Mehrabi R, Lütken H, Haldrup A, Kema GHJ, Collinge DB, Jørgensen HJL (2007) Role of hydrogen peroxide during the interaction between the hemibiotrophic fungal pathogen Septoria tritici and wheat. New Phytol 174:637–647. https://doi.org/10.1111/j.1469-8137.2007.02026.x
Silva EN, Vieira SA, Ribeiro RV, Ponte LFA, Ferreira-Silva SL, Silveira JAG (2013) Contrasting physiological responses of jatropha curcas plants to single and combined stresses of salinity and heat. J Plant Growth Regul 32:159–169. https://doi.org/10.1007/s00344-012-9287-3
Sinha P, Khurana N, Nautiyal N (2012) Induction of oxidative stress and antioxidant enzymes by excess cobalt in mustard. J Plant Nutr 35:952–960. https://doi.org/10.1080/01904167.2012.663636
Song F, Yang C, Liu X, Li G (2006) Effect of salt stress on activity of superoxide dismutase (SOD) in Ulmus pumila L. J For Res 17:13–16. https://doi.org/10.1007/s11676-006-0003-7
Streller S, Wingsle G (1994) *Pinus sylvestris* L. needles contain extracellular CuZn superoxide dismutase. Planta 192:195–201. https://doi.org/10.1007/bf01089035
Subrahmanyam, D.: Effects of chromium toxicity on leaf photosynthetic characteristics and oxidative changes in wheat (*Triticum aestivum* L.). Photosynthetica. 46, 339–345 (2008). https://doi.org/10.1007/s11099-008-0062-4
Suzuki N, Mittler R (2006) Reactive oxygen species and temperature stresses: a delicate balance between signaling and destruction. Physiol Plant 126:45–51. https://doi.org/10.1111/j.0031-9317.2005.00582.x

Tamás L, Mistrík I, Huttová J, Halušková L, Valentovičová K, Zelinová V (2010) Role of reactive oxygen species-generating enzymes and hydrogen peroxide during cadmium, mercury and osmotic stresses in barley root tip. Planta 231:221–231. https://doi.org/10.1007/s00425-009-1042-z

Tepperman JM, Dunsmuir P (1990) Transformed plants with elevated levels of chloroplastic SOD are not more resistant to superoxide toxicity. Plant Mol Biol 14:501–511. https://doi.org/10.1007/BF00027496

Tertivanidis K, Goudoula C, Vasilikiotis C, Hassiotou E, Perl-Treves R, Tsaftaris A (2004) Superoxide dismutase transgenes in sugarbeets confer resistance to oxidative agents and the Fungus C. beticola. Transgenic Res 13:225–233. https://doi.org/10.1023/b:trag.0000034610.35724.04

Tuna AL, Kaya C, Dikilitas M, Higgs D (2008) The combined effects of gibberellic acid and salinity on some antioxidant enzyme activities, plant growth parameters and nutritional status in maize plants. Environ Exp Bot 62:1–9. https://doi.org/10.1016/j.envexpbot.2007.06.007

Tyagi S, Sharma S, Taneja M, Shumayla Kumar R, Sembi JK, Upadhyay SK (2017) Superoxide dismutases in bread wheat (*Triticum aestivum* L.): comprehensive characterization and expression analysis during development and, biotic and abiotic stresses. Agri Gene 6. https://doi.org/10.1016/j.aggene.2017.08.003

Upadhyaya H, Panda SK, Dutta BK (2008) Variation of physiological and antioxidative responses in tea cultivars subjected to elevated water stress followed by rehydration recovery. Acta Physiol Plant 30:457–468. https://doi.org/10.1007/s11738-008-0143-9

Verma S, Dubey RS (2003) Lead toxicity induces lipid peroxidation and alters the activities of antioxidant enzymes in growing rice plants. Plant Sci 164:645–655. https://doi.org/10.1016/S0168-9452(03)00022-0

Verma D, Lakhanpal N, Singh K (2019) Genome-wide identification and characterization of abiotic-stress responsive SOD (superoxide dismutase) gene family in Brassica juncea and B. rapa. BMC Genomics 20:227. https://doi.org/10.1186/s12864-019-5593-5

Viczián O, Künstler A, Hafez Y, Király L (2014) Catalases may play different roles in influencing resistance to virus-induced hypersensitive necrosis. Acta Phytopathol Entomol Hungarica 49:189–200. https://doi.org/10.1556/APhyt.49.2014.2.5

Vyas D, Kumar S (2005) Purification and partial characterization of a low temperature responsive Mn-SOD from tea (*Camellia sinensis* (L.) O. Kuntze). Biochem. Biophys Res Commun 329:831–838. https://doi.org/10.1016/j.bbrc.2005.02.051

Wang Y, Ying Y, Chen J, Wang X (2004) Transgenic Arabidopsis overexpressing Mn-SOD enhanced salt-tolerance. Plant Sci 167:671–677. https://doi.org/10.1016/J.PLANTSCI.2004.03.032

Wang F-Z, Wang Q-B, Kwon S-Y, Kwak S-S, Su W-A (2005) Enhanced drought tolerance of transgenic rice plants expressing a pea manganese superoxide dismutase. J Plant Physiol 162:465–472. https://doi.org/10.1016/j.jplph.2004.09.009

Wang W, Xia M, Chen J, Deng F, Yuan R, Zhang X, Shen F (2016) Genome-wide analysis of superoxide dismutase gene family in Gossypium raimondii and G. arboreum. Plant Gene 6:18–29. https://doi.org/10.1016/j.plgene.2016.02.002

Wang W, Zhang X, Deng F, Yuan R, Shen F (2017) Genome-wide characterization and expression analyses of superoxide dismutase (SOD) genes in Gossypium hirsutum. BMC Genom 18:376. https://doi.org/10.1186/s12864-017-3768-5

Xu S, Li Y, Hu J, Guan Y, Ma W, Zheng Y, Zhu S (2010) Responses of antioxidant enzymes to chilling stress in tobacco seedlings. Agric Sci China 9:1594–1601. https://doi.org/10.1016/S1671-2927(09)60256-X

Yogavel M, Mishra PC, Gill J, Bhardwaj PK, Dutt S, Kumar S, Ahuja PS, Sharma A (2008) Structure of a superoxide dismutase and implications for copper-ion chelation. Acta Crystallogr Sect D: Biol Crystallogr 64:892–901. https://doi.org/10.1107/S0907444908019069

Yost FJ, Fridovich I (1973) An iron-containing superoxide dismutase from Escherichia coli. J Biol Chem 248:4905–4908

Zhou Y, Hu L, Wu H, Jiang L, Liu S (2017) Genome-wide identification and transcriptional expression analysis of cucumber superoxide dismutase (SOD) family in response to various abiotic stresses. Int J Genomics 2017:1–14. https://doi.org/10.1155/2017/7243973

Zhu D, Scandalios JG (1993) Maize mitochondrial manganese superoxide dismutases are encoded by a differentially expressed multigene family. Proc Natl Acad Sci 90:9310–9314. https://doi.org/10.1073/pnas.90.20.9310

Zsigmond L, Szepesi Á, Tari I, Rigó G, Király A, Szabados L (2012) Overexpression of the mitochondrial PPR40 gene improves salt tolerance in Arabidopsis. Plant Sci 182:87–93. https://doi.org/10.1016/j.plantsci.2011.07.008

Chapter 4
Receptor-Like Kinases and Environmental Stress in Plants

Shumayla, Shivi Tyagi and Santosh Kumar Upadhyay

Abstract Various biotic and abiotic environmental stresses negatively affect the diverse aspects of plant growth and development, and crop productivity. Plants being sessile organisms have developed effective strategies to avoid, tolerate, or acclimatized to various kinds of stress conditions. Several stress factors of plants activate cellular responses and signaling pathways such as secretion of stress proteins. Receptor-like kinases (RLKs) is a class of defense-related proteins comprising more than a thousand members. The RLKs mostly consisted of an extracellular domain for signal perception, a transmembrane domain to anchor the protein into membrane and a cytoplasmic serine/threonine kinase domain for stimulating the immunity of plants. They are known to play a diverse range of functions in plants, ranging from growth and development to responses against various environmental stresses. RLK signaling is arbitrated by phosphorylation events which take place amid proteins present in receptor complexes. Several RLKs such as BR1, *CLAVATA1*, S-locus receptor kinase, *Flagellin Insensitive 2,* etc. provide fruitful information on the roles arbitrated by the members of RLK gene family. Plants recognize numerous number of RLKs as pattern recognition receptors (PRRs) which detect host and microbe-derived molecular patterns as the first layer of inducible defense. The studies have revealed the mechanism of PRR activation and signaling and their ligands. In this chapter, the systematic analyses of plant RLKs responses to different stresses have been explained in detail.

Keywords Developmental stages · Environmental stress · Immunity · Pathogen recognition receptors (PRRs) · Receptor-like kinases · Signaling

---

Shumayla · S. Tyagi · S. K. Upadhyay (✉)
Department of Botany, Panjab University, Chandigarh 160014, India
e-mail: skupadhyay@pu.ac.in

Shumayla
Department of Biotechnology, Panjab University, Chandigarh 160014, India

S. P. Singh et al. (eds.), *Molecular Approaches in Plant Biology and Environmental Challenges*, Energy, Environment, and Sustainability,
https://doi.org/10.1007/978-981-15-0690-1_4

4.1 Introduction

Plants compromise their life as the optimal growth condition is rare in nature. Certainly, most plants grow to adapt to survive themselves under sub optimal conditions. The adaptation to the changing environmental conditions results in metabolism and growth compromises and also costs energy, which is why it is considered as stress (Boyer 1982). The most essential criteria for plant life is reproduction. All other criteria such as growth and yield become meaningless if the ability to reproduce is lost. From an economical point of view, the most significant criteria is yield. The environmental stress often decreases growth and yield therefore, the research has been focus on improving the stress tolerance of most important crop plants (Nemali et al. 2015; Prado et al. 2014; Ronald 2014; Tran and Mochida 2010). Nowadays an important global problem in the improvement of agriculture is the major changes in annual crop yield, which is due to alteration in environmental stress conditions such as salinity, drought, temperature variations and various kinds of pathogen attacks. However, plants encounter a variety of environmental stresses and have developed many ways to optimize growth in response to changing conditions. For example, plants can collect water and nutrients in roots and stems and regulate their use; they can also adapt to changes in light intensity and temperature; they also have sophisticated methods to defend against many pathogens attacks (Dodds and Rathjen 2010; Jones and Dangl 2006). Over the past decade, a stress tolerance mechanism underlying both biotic and abiotic stress has been immensely studied with much emphasis on tolerance against individual stress (AbuQamar et al. 2009; Mengiste 2003). The positive and negative influence on the plants opened to abiotic and biotic stresses have been discovered independently (AbuQamar et al. 2009; Fujita et al. 2006; Jakab et al. 2005). To rescue the plants from life-threatening situations, the entry and growth of pathogens is limited by the process called plant programmed cell death (PCD) (Jones and Dangl 2006; Bruggeman et al. 2015; Coll et al. 2011). For example, by leaf abscission, plants can eliminate the infected leaves to save the mother plant (González-Carranza et al. 1998; Taylor and Whitelaw 2001). Additionally, plant hormones also have a vital role in the improvement of stress tolerance, survival and growth (Choudhary et al. 2012; Ha et al. 2012; Lozano-Durán and Zipfel 2015; Peleg and Blumwald 2011). Plants perceive heat stress signal and activate signaling pathways which regulate diverse responses for stress tolerance (Zhu 2016). The response against certain stimuli requires cofactors and signaling molecules interaction in cellular compartment or tissues. These signaling molecules showed involvement in the activation of stress-responsive genes. There are various signal transduction molecules depending upon the plant type, which activate stress-responsive genes along with transcriptional factors (Ahmad et al. 2012). There are other mechanisms studied against environmental stress like antioxidant defense; use of exogenous protectants such as phytohormones, signaling molecules and osmoprotectants (Hasanuzzaman et al. 2012), genetic engineering, and other approaches (Hasanuzzaman et al. 2013). Recently, for the identification of stress-responsive genes, the data of individual gene expression

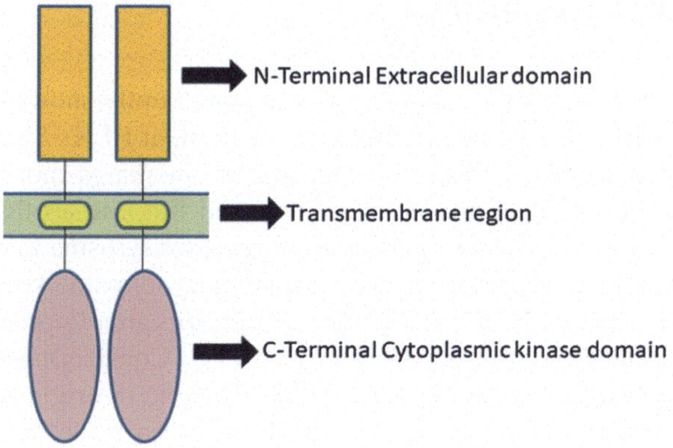

Fig. 4.1 Basic structure of receptor like kinases (RLKs)

under abiotic and biotic stress have been explored (Luo et al. 2005; Narsai et al. 2013; Shaik and Ramakrishna 2013).

Receptor-like kinases (RLKs) belong to a class of defense-related proteins which help plants to protect themselves from different kind of stress elicitors. In this chapter, we will learn about how plants protect themselves via RLKs from these conditions. RLKs form one of the largest subfamily of defense-related proteins, which play a role ranging from growth and development to stress response. The main role of RLKs is to sense the environment and further activates the downstream signalling pathways in order to protect the plants from various environmental challenges. They consist of an N-terminal extracellular domain (ECDs), a single membrane-spanning domain and a C-terminal intracellular kinase domain (Fig. 4.1). Plant RLKs have evolved a diverse balance of extracellular domains (>20 have been identified) including self-incompatibility (S) domains, epidermal growth factor repeats, leucine-rich repeats (LRR), lectin domains and various others Shiu and Bleecker (2001). It has been observed that the variety of the extracellular domains in RLKs emulate their need to develop rapidly in order to protect against an ever-changing population of biotic and abiotic stress elicitors Shiu and Bleecker (2001).

They act as signaling cascade during various stress conditions (Walker 1994). There are 15 sub-classes of RLKs in plants. Some of the RLKs of leucine-rich repeat receptor-like kinases (LRR-RLK) class such as FLS2 and EFR receptors of Arabidopsis and Xa21 protein of rice are identified for being involved in plant defense which are the components of signal transduction. Further, the lysine-motif (LysM) receptor kinase of rice has a role in the recognition of the chitin elicitor of fungus (Kaku et al. 2006), and a lectin receptor kinase (LecRK) involved in disease resistance, cysteine-rich repeat receptor-like kinases against fungal pathogen (Acharya et al. 2007) which further shows the overlapping roles of other RLK subfamilies in plants from growth and development to defense response and pathogen recognition.

4.2 Origin of RLKs

The evolutionary study of kinase gene family shows that animal Pelle kinases and IRAKs show close resemblance with plant RLKs Shiu and Bleecker (2001). Additionally, the RLK/Pelles kinase domain shares more similarity with the tyrosine kinase domain present in animals and Raf kinases than other kinase families Shiu and Bleecker (2001). Therefore, receptor tyrosine kinases and RLK/Pelle possibly have a monophyletic origin. Apparently, tyrosine kinases in animals belong to the largest family of transmembrane receptors after G-protein coupled receptors (Hunter et al. 1992; van der Geer et al. 1994). Considering that animal receptor tyrosine kinases and plant RLKs are monophyletic in origin and have highly similar structural configurations, it is possible that they might have a common ancestor which has a role in extracellular stimuli perception. However, it does not make it clear if these ancestral kinases were a transmembrane receptor protein which interacted with receptor complexes proteins that bind to extracellular signals. Initially, the RLK/Pelle family was recognized from the homologs of plant and metazoan Shiu and Bleecker (2001). Further, studies with extended taxonomic sampling showed the absence of RLK/Pelle homologs in fungi (Shiu and Li 2004). The taxonomic distribution of RLK/Pelle is unexpectedly rare or less among eukaryotes. There are no homologs in basal metazoan such as *Monosiga brevicollis*, although IRAK and Pelle sequences are identified in numerous invertebrates and vertebrates (King et al. 2008). Additionally, RLK/Pelle homologs are found merely in *Plasmodium, Perkinsus* and *Toxoplasma* among non-metazoan and non-plant eukaryotes. In most of the eukaryotic species, the absence of RLK/Pelle genes can be explained due to the loss of genes in various lineages or sequence divergence. There is a difficulty in explaining the gene loss, because to explain the pattern, independent multiple losses had to occur. More than 1–2 billion years ago the eukaryotic divergence occurred (Hedges 2002) and at that time RLK/Pelle divergent evolution had resulted in eukaryotes. One interesting possibility is the horizontal transfer in apicomplexan RLK/Pelles as a result of a secondary symbiotic event involving chlorophytes (Köhler et al. 1997), although this further needs to be validated. We wondered that the conserved ancestral function between animal and plant RLK/Pelle might have limited the level of sequence divergence in metazoan RLK/Pelle. The members of RLK/Pelle family in Viridiplantae are found in all land plants and several green algal species. In green alga *Ostreococcus tauri,* regarded as the smallest eukaryote lacked RLK/ Pelle gene (Courties et al. 1994), however, *Chlamydomonas reinhardtii* consists of two RLCK genes (Lehti-Shiu et al. 2009). In the charophyte division, several RLK/Pelle are identified in *Nitella axillaris* and *Closterium ehrenbergii* (unicellular species), which share their common ancestry with the land plants (Sasaki et al. 2007). The number of RLK/Pelle members rose abruptly in the land plants, with 610, 1070, 1192 and 329 and number of genes in, *Arabidopsis thaliana*, moss, *Oryzae sativa* and poplar, respectively (Lehti-Shiu et al. 2009). But in animals, the number of RLK/Pelle genes is less as compared to land plant lineages, which indicated the dramatic expansion of the gene family in plants. In addition, before the divergence of charophytes, the configuration of receptor arose possibly and their RLKs were found from the sequences of the transcript. Therefore, detail evolutionary study has to be done using an entire range of RLK/Pelle genes in the genome of charophyte to check if the receptor configuration originated from single or multiple sources. Now the question arises that why there is a dramatic expansion of RLK/Pelle family in plant lineages? It is apparent that expansion of receptor kinase families has taken place in different lineages. In a relative analysis, the expansion of tyrosine kinase ranging from *Caenorhabditis elegans* to human occurred significantly over the time of metazoan evolution (Shiu and Li 2004). The study reported the presence of RLK/Pelle family in plants and other different receptor kinase families in oomycetes and brown algae (Cock et al. 2010). Consequently, it suggests the possible interaction has occurred between kinases and ECD in different lineages and different families was possibly interacted with ECDs in different lineages, and the adaptive advantage of signal perception allowed the expansion of different receptor kinases independently.

4.3 Types of RLKs

There are 15 sub-classes of RLKs on the basis diverse variety of extracellular domains. They have been listed in the table with their extracellular domain name, number of sub-families and pfam ids. Some of the RLKs family member has subfamilies which are listed in Table 4.1. The information about some of the families is not known. Out of fifteen subfamilies the information of four subfamilies and their pfam ids are not known yet.

4.4 Evolutionary Study of RLKs in Plant Lineages

The phylogenetic analysis divided the RLK/Pelle family into several subfamilies Shiu and Bleecker (2001). The complexity of domain in animal receptor kinases is somewhat similar to the extracellular region diversity of plant RLKs (Cock et al. 2002). It has been seen that variety of RLKs bind directly to lipid, polysaccharides, protein, and other ligands. Over the course of evolution, RLK/Pelle gene fusions were repeated because of the efficacy of transmembrane signaling. The activation of downstream signaling by chimeric RLKs containing nonnative ECDs (He et al. 2000; Albert et al. 2010; Brutus et al. 2010) illustrates the creation of novel chimeric RLKs in nature. Numerous membrane spanning proteins in plants share similar domains as found in RLKs (Shiu and Li 2004; Fritz-Laylin et al. 2005), so these proteins are named as Receptor-Like Proteins (RLPs). They consist of signal perceiving extracellular region and transmembrane domain but lack intracellular kinase domain. Several RLPs function together with RLKs to regulate defense responses and developmental processes. For instance, to regulate meristem development, *CLAVATA1* of RLKs and *CLAVATA2* of RLP interact with *CORYNE* (Jeong et al. 1999; Müller et al. 2008).

Table 4.1 The list of RLK sub families along with their domain description

S. No	RLK type	Domain type	Number of subfamilies	Pfam ids
1	Leucine rich repeat receptor like kinases	Leucine rich repeat	~15	PF13855
2	Lectin receptor like kinases	legume lectin, bulb lectin and c-type lectin	3	PF00139, PF01453, PF00059
3	Cysteine rich receptor like kinases	Cysteine rich repeat/ DUF26	1	PF01657
4	Thaumatin receptor like kinases	Thaumatin	1	PF00314
5	Lysine Motif receptor like kinases	Lysine motif	1	PF01476
6	Wall associated receptor like kinases	EGF repeat	1	PF08488
7	S domain receptor like kinases	Agglutinin, EGF, PAN	3	PF07468, PF00008
8	PR5 like receptor kinases	PR5 K	1	
9	CR4 like receptor like kinases	TNFR	1	PF00020
10	Extension like	Proline rich	1	PF15240
11	PERK	Proline rich	1	PF15240
12	RKF3 like	Unknown domain	1	
13	URK 1		1	
14	CrRLK1 like		2	
15	LRK10 like		2	

The ERECTA family members form a complex with RLP and regulate stomatal patterning (Shpak et al. 2005), likethe defense response pathway is activated on treatment with chitin oligosaccharides by rice *CEBiP* LysM domain containing RLP and *OsCERK1*RLK (Shimizu et al. 2010). There is a consistency amid RLKs and RLPs similarities and their functional relatedness due to which, their fusion results in the creation of RLKs with novel domain configuration. The membrane spanning proteins of RLPs are important in extracellular signaling networks. Therefore, the association of ancestral RLK/Pelle kinases and RLPs could have introduced the novel signal transduction pathways based on the ligand perception to downstream kinase targets. On the other hand, the RLK/Pelle and RLPs association might have occurred as they are part of the same signaling network. In the animal system, the association between the cytoplasmic kinases and receptor without kinase domain is known, for instance, in *D. melanogaster* the innate immunity and development are mediated by the association of Pelle and toll receptor which are a cytoplasmic kinase and transmembrane receptor without a kinase domain, respectively (Shelton and Wasserman 1993; Hecht and Anderson 1993; Belvin and Anderson 1996). Likewise, mammalian IRAKs and toll-like receptors mediate innate immunity through signaling networks (Flannery and Bowie 2010). In various plants, innate immunity is induced by multiple RLK/Pelle members (Boller and Felix 2009). Therefore it can be stated that innate immunity function of some RLK/Pelle family members is an expected ancestral trait. Hence, the fusion between RLP and RLCK must be responsible for the known receptor configuration and it can also be assumed that the RLKs might have evolved from the fusion of RLPs and RLCKs and provided innate immunity.

4.5 Role of RLKs in Signal Transduction Pathways

Receptor-like kinases (RLKs) play an important role in various aspects of plant life such as growth, development and defense responses. The phosphatases and kinases control the phosphorylation process which is regulated by RLKs. The receptor localization and its abundance are controlled by these RLKs to create a balance between signal perception and downstream gene-expression. A basic model depicting RLK signaling under different stress has been shown in Fig. 4.2.

Also there are various other complexes involved in plant growth and developments. These complexes are explained as follows.

4.5.1 Brassinosteroid Complexes Mediated Signaling

BRI1 (Brassinosteroid insensitive 1) binds directly or form a part of receptor complex to its ligand brassinolide (BL). This binding activates downstream signaling by inducing transphosphorylation of BRI1-BAK1 heterodimers (Li et al. 2002; Nam and Li 2002). BAK1 is a LRR-RLK consisting of five extracellular LRRs that exists as a member of the BRI1 receptor complex in vivo. The downstream component of the BR signalling pathway has been identified by novel mutants which show either BR-insensitive or hypersensitive phenotypes. *BIN2* (allelic to DWF12 and UCU1) encodes a glycogen synthase kinase *(GSK3)/SHAGGY-like* cytoplasmic serine/threonine kinase. Interestingly there is missense mutation within a short segment of four amino acids which has led to an increased kinase activity in sixth semi-dominant *bin2* (Nam and Li 2002; Choe et al. 2002; Pérez-Pérez et al. 2002). BR signaling is negatively regulated by *BIN2* as indicated by *bin2* kinase mutant. The downstream targets of *BIN2* are BZR1 (BRASSINAZOL RESISTANT1) and BES1

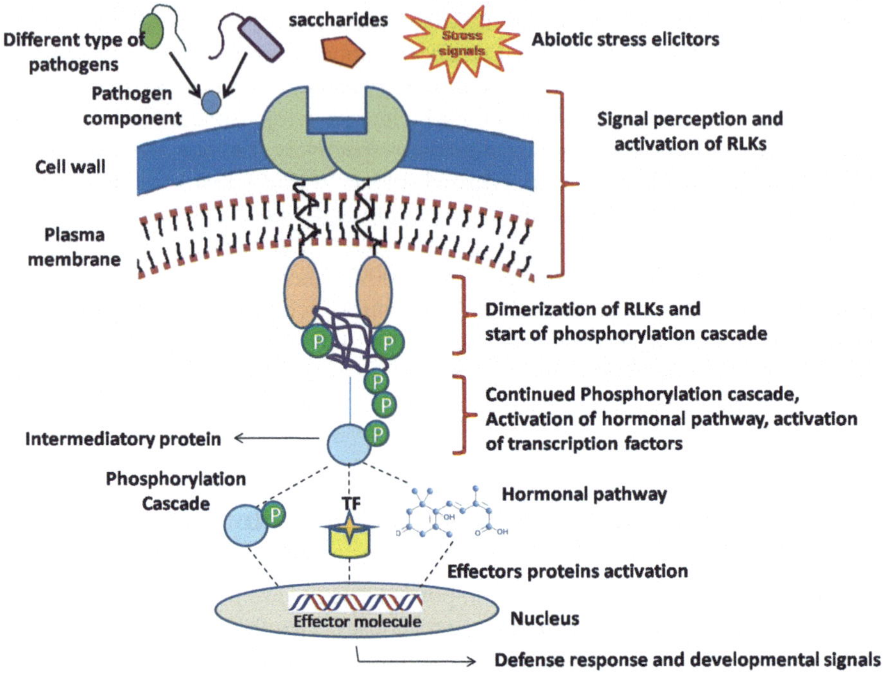

Fig. 4.2 A model depicting the role of RLKs in signalling pathway and their crosstalk under different stress conditions. The figure has been adapted from Vaid et al. (2015) with slight modifications and permission from "Elsevier" Lic. no. 4662911283603

(BRI1-EMS-SUPPRESSOR1), which are novel proteins with 88% sequence similarity. BES1 and BZR1 are positive components for mediating BL-induced gene expression (Wang et al. 2002; Yin et al. 2002). In response to BL, dephosphorylated forms of BES1 and BZR1 get accumulated. The gain-of-function mutations in these dephosphorylated forms create highly stable mutant proteins in the nucleus, resulting in upregulation of BR-inducible gene expression. The unphosphorylated, stable BES1/BZR1 proteins accumulation is dependent on functional BRI1 (Wang et al. 2002; Yin et al. 2002). Furthermore, from the various evidence BES1 and BZR1 are the substrate proteins of *BIN2* (Yin et al. 2002; He et al. 2002; Zhao et al. 2002). *BIN2* directly interacts with BES1 and BZR1 and phosphorylates both in vitro. The stable accumulation of BZR1and BES1 is negatively regulated by *BIN2* in vivo because the phosphorylation of BES1 and BZR1 by *BIN2* triggers their degradation through the proteosome-dependent pathway (He et al. 2002). The function of BES1 and BZR1 is distinct in BR signaling, because BZR1 shows BR biosynthesis regulation by a negative feedback mechanism. The binding of ligands (BRs), at the 70-amino acid island of BRI1, stimulates dimerization and further transphosphorylation of BRI1–BAK1 receptor pairs. The initiation of signaling events inhibits a cytoplasmic glycogen synthase kinase 3 (GSK-3) *BIN2*, which is a negative regulator of BR signal transduction. To prevent phosphorylation and further degradation of BES1 and BZR1,

BIN2 is inhibited. As a result, stable forms of BZR1 and BES1accumulate within the nucleus and hence promote BR-dependent gene-expression. *BIN2* kinase is always active in BR absence and phosphorylates BES1 and BZR1, which leads to proteolysis of BES1 and BZR1, thus preventing BR-responsive gene-expression. Yin et al. in 2002 (Yin et al. 2002) pointed out that the BR signaling pathway resembles the Wnt signaling pathway in animals, which plays vital roles in differentiation, cell proliferation, and cancer development.

4.5.2 CLAVATA *Complexes Mediated Signaling*

The molecular genetic studies showed that the maintenance of the stem cell population is determined by a regulatory loop amid *WUS* and *CLV* at the shoot apical meristem (SAM) (Brand et al. 2000; Schoof et al. 2000). The three *CLV* genes (*CLV1, CLV2,* and *CLV3*) control stem cell numbers, while *WUS* encourage stem cell fate (Clark et al. 1997; Mayer et al. 1998), and their complex regulatory loop defines the *CLV* and *WUS* expression domain. At the shoot apical meristem central zone *WUS* up- regulates *CLV3* expression at the surface layers (L1 and L2) (Schoof et al. 2000). CLV3 is the secretory ligand for *CLV1 LRR-RK* activates the downstream signaling pathway and in turn, restricts the expression of *WUS* at the organizing center. Therefore, it was seen that in the *clv* mutant SAM the expression of *WUS* is up-regulated while overexpression of *CLV3* inhibits the *WUS* expression at the SAM (Brand et al. 2000). This explains the stem cell population maintenance despite being continually displaced by newly forming daughter cells. *CLV3* associates with an active form of *CLV1* receptor complex at the surface layer of the SAM CZ underneath meristem cell layers. Gel-filtration analysis suggested that in the absence of ligand *CLV3*, *CLV1 LRR-RK* and *CLV2 LRR-RP* form a core heterodimeric receptor complex (185 kDa) (Trotochaud et al. 1999). Possibly the binding of *CLV3* activates the *CLV1–CLV2* core complex and forms receptor complex with high molecular weight (450-kDa). KAPP and ROP are the two downstream components which are specifically employed into the active receptor complex of *CLV1* (Trotochaud et al. 1999), out of two *KAPP* negatively regulate signal of *CLV* but *ROP* a small GTPase acts a positive regulator of *CLV* pathway. *KAPP* overexpression study showed that wild type plant resembles *clv1* mutant whereas reduction in *KAPP* expression suppressed the phenotype of weak *clv1* mutant (Stone et al. 1998; Williams et al. 1997). A further signal is transmitted through unknown steps, which involve POLTERGUIST (POL), and ultimately the expression of *WUS* is repressed. POL does not confer any developmental defects to plants but a mutation in the POL locus can cause the suppression of *clv* phenotype (Yu et al. 2000, 2003). The genetic analysis suggests that the repression of *WUS* expression might be inhibited by POL through *CLV* signaling. It would be interesting to check whether POL modulates the phosphorylation status of *WUS* directly and effects *WUS* activity. The molecular mechanisms by which *WUS* promotes *CLV3* expression remain unknown. It has been observed that *CLV3* expression is regulated through its cis-regulatory elements (Brand et al. 2000). However, *CLV3* and *WUS* are expressed at cell layers apart from each other; it is fascinating to speculate that some unknown factors, which might be *WUS* itself, moves from cell to cell or, alternatively, an unknown signal transduction pathway relays signals to up-regulate *CLV3* expression.

4.6 Role of RLKs in Biotic Stress

Plants have the ability to sense the presence of microbes invading through the conserved pathogen-associated molecular patterns (PAMPs) which are found specifically in microbes (Medzhitov and Janeway 1997). PAMP-triggered immunity (PTI) is initiated by the perception of surface-localized transmembrane pattern recognition receptors (PRRs), inhibiting the growth of the pathogen. Though numerous PAMPs are known, but the number is still less in plants. Many PRRs are LRR-RLKs which have established the role of PRR associated proteins for PTI signaling. These different kinds of receptors along with their action mechanism against pathogens have been discussed in this section.

The FLS2 (flagellin sensing 2) is a LRR-RLK receptor which recognizes the well-conserved protein flagellin present in a wide range of bacteria such as *Pseudomonas syringae* pv. *tomato* (*Pto*) DC3000 (Gómez-Gómez and Boller 2000). The flg22 peptide derived from flagellin directly bind to the FLS2 (Chinchilla et al. 2006), but recently its binding to an unsulfonated *Xoo* Ax21 peptide has been reported. Moreover, these two peptides do not share sequence homology and is quite surprising (Danna et al. 2011).

FLS2 independently form homo-dimers of flg22 binding, however, the role of this event for receptor function is unknown (Sun et al. 2012). The heterodimers formation of FLS2 with brassinosteroid receptor-associated kinase 1 (BAK1) in the presence of bound flg22 is well-established (Chinchilla et al. 2007; Schulze et al. 2010). BAK1 was a first identified component in brassinosteroid signaling via the receptor brassinosteroid insensitive 1 (BRI1) and is very common in many RLK signaling complexes (Li et al. 2002). An important function of BAK1 in flg22 sensing in bak1 plants marked the decline of flg22 induced responses (Chinchilla et al. 2007; Heese et al. 2007). Interestingly, the interaction between BAK1-FLS2 does not have competition with other interactors of BAK1 for instance, BRI1 and its interaction does not regulate BR mediated PAMP defense response. (Albrecht et al. 2012). BAK1 belongs to the somatic embryogenesis receptor kinase (SERK) family which comprises of five members, *SERK1*, *SERK2*, *BAK1/SERK3*, *BAK1-like (BKK1)/SERK4*, and *SERK5*. The interaction of FLS2 with these five members has been identified, but the strongest association of FLS2 was with BAK1. BKK1and BAK1 act in a synergistic manner in PAMP signaling (Roux et al. 2011). The function of BIK1is dependent on complicated interactions with immune-response regulators and provide RLK signaling complexes with the ability to differentiate among the various kinds of pathogens (Laluk et al. 2011). For instance, *bik1* mutants are susceptible to *Pto* DC3000, while lower expression of flg22 responsiveness showed resistance to virulent *Pto* DC3000. The BAK1 and FLS2 also show interaction with PBS-like kinase (PBL1) but its mutant *pbl1* showed reduced PTI response. (Zhang et al. 2010).

In *O. sativa,* the *Xa21* gene provides resistance against various strains of the Gram-negative bacterium *Xanthomonas oryzae* pv. oryzae (Xoo) (Song et al. 1995). *Xa21* is a LRR-RLK which acts as a receptor for a molecular pattern of the pathogen. LRR-RLK also recognizes an elf18 peptide of EF-Tu, an abundant bacterial protein. EFR (elongation factor Tu receptor) also interacts with BAK1 in a ligand-dependent manner (Roux et al. 2011). Both EFR and FLS2 can associates with all the members of the SERK family. But the interaction of *BKK1*, *SERK1, SERK2* is stronger with *EFR* than with FLS2 (Roux et al. 2011). Due to this EFR avoids pathogen effectors action on the single SERKs. The novel discovery of bak1 and bak1-5 has enabled the studies of non-lethal double mutants' *bak1-5 bkk1*. This study shows the cooperative action of both BKK1 and BAK1 in PAMP signaling (Roux et al. 2011; Schwessinger et al. 2011).

Chitin elicitor receptor kinase 1 (*CERK1*) is the LysM-RLK (Kaku et al. 2006; Miya et al. 2007), and its binding to chitin has been reported in *A. thaliana* (Iizasa et al. 2010; Petutschnig et al. 2010). *CERK1*'s binding is BAK1-independent, unlike EFR and FLS2. In *O. sativa* the association of *OsCerk1* and *CeBIP* a LysM domain containing protein, functions together as a receptor complex, in a ligand-dependent manner to obtain chitin signaling (Shimizu et al. 2010). *cerk1* shows reduced sensitivity to PGN, and hypersusceptible to *Pto* DC300, although *CERK1* does not bind to PGN. *FERONIA* (FER) another RLK regulates pollen tube reception (Escobar-Restrepo et al. 2007). *FER* perform penetration of powdery mildew (PM) into host cells (Kessler et al. 2010). It is supposed that FER may be involved in regulating the localization of MLO protein family members, which are necessary for PM infection (Consonni et al. 2006). *FERONIA* (FER) functions in signaling pathways by regulating ROS production. FER interact with guanine nucleotide exchange factors (GEFs) which regulate RHO GTPases (RAC/ROPs). RAC/ROP plays crucial roles in stress responses. The association of a RAC/ROP (Rac GTPase) to NADPH oxidases has been reported in rice (Wong et al. 2007). In *Arabidopsis*, Rop2 along with FER show co-immunoprecipitation. In addition, *FER* has also been shown to involve in ROS accumulation during PAMP signaling in leaves.

4.7 Role of RLKs in Abiotic Stress

Heat, cold, drought, salinity, toxic metals and metalloids, UV-B radiation and ozone, are major abiotic factors causing stress in plants. Their adverse effects on plants include retarded growth, early senescence, decreased crop yields, and even death. Plants adapt through various strategies to respond to the changes in environmental conditions (Srivastava et al. 2014; Latif et al. 2016). Various genes have been identified in plants that have essential roles in abiotic stress management, including many protein kinases and transcription factors. Recent years have seen an increase in research on the roles of RLKs that reveals their involvement in multiple molecular mechanisms for resistance against different abiotic stresses and diseases and signal transduction. These findings can have a profound impact on the development of modern agriculture (Wu et al. 2015).

4.7.1 RLKs Against Drought Stress

Drought stress is multi-dimensional stress for plants, causing changes in morphological, physiological, molecular and biochemical traits (Salehi-Lisar and Bakhshayeshan-Agdam 2016). It causes a reduction in leaf size, root proliferation and stems extension, disturbs plant water relation and therefore reduces water use efficiency (Farooq et al. 2009). Plants have evolved to develop a diverse range of strategies to combat drought stress at morphological, physiological, cellular and molecular levels (Bartels and Sunkar 2005; Yu et al. 2013). ABA has a pivotal role in the management of abiotic stress. It regulates the expression of genes involved in osmotic stress response, other adaptive physiological responses and regulation of plant development (Kim et al. 2012). In addition, ABA has been reported as an essential mediator of RLKs' function in drought stress management. Expression of *FON1*, an LRR-RLK gene in rice, was shown to be induced by drought conditions and treatment with ABA (Feng et al. 2014). Transgenic rice plants with FON1 over-expression exhibited higher sensitivity towards ABA and increased drought tolerance. Another LRR-RLK gene, *Leaf Panicle 2* (LP2) showed down-regulation under ABA and drought stress (Wu et al. 2015). The two pathways act in coordination during drought stress, for sustaining the growth and development of the plant. When stress conditions are introduced, ABA levels increase in the plants and cause activation of positive regulatory pathways. This results in closure of stomata due to which less water evaporates and hence growth conditions are maintained. Simultaneously a zinc finger transcription factor C_2H_2 directly binds to *LP2* (a negative regulator of drought response) promoter and inhibits its expression. In addition, few LRRK genes and L-LRK genes showed an enhanced level of transcripts after exposure of plants to drought stress through the quantitative real-time PCR (Shumayla et al. 2016a, b). Further, LRK10 and LRK10L1.2 genes, Arabidopsis orthologs of the wheat were found to positively regulate the closure of stomata in response to drought stress. That may take place through ABA-mediated signaling, directly or indirectly (Lim et al. 2015). *GsRLCK*, an ABA-responsive receptor-like cytoplasmic kinase (RLCK) of *Glycine soja*, caused a decrease in sensitivity towards ABA and altered expression of ABA-responsive genes in countering the drought stress (Sun et al. 2013).

4.7.2 RLKs Against Heat Stress

Heat stress affects the plants in a variety of ways such as a reduction in seed germination, water loss, improper growth, yield reduction, oxidative stress and altered metabolism etc. Various studies have been done in this direction. Several members of RLKs have been reported to play an important role during heat stress conditions. Some of the recent reports showed the involvement of lectin receptor-like kinases in heat stress. In *T. aestivum* Shumayla et al. (2016a) has reported the increase in the level of *TaLRKs* transcript expression after heat treatment to plants. Likewise, few members of leucine-rich repeat receptor-like kinases also showed elevated expression level using quantitative real-time PCR in *T. aesstivum* (Shumayla et al. 2016b). Also, the involvement of cysteine rich receptor-like kinases in heat stress tolerance has been reported. *TaCRK68-A* gene of *T. aestivum* showed enhanced tolerance against heat stress (50 °C) in the cells of transformed *E. coli* and yeast (Shumayla et al. 2019).

4.7.3 RLKs Against Salt Stress

High salinity is a major abiotic factor which can affect large agricultural lands and cause stress in plants, resulting in slow plant growth and decreased yield (Sairam 2004). Multiple RLK genes have been reported to be associated with salt stress management in plants. The studies have been focused on distinguishing and identifying the functions of lectin receptor-like kinases (*LecRLKs)* in response to abiotic stresses, particularly salinity. The plasma membrane localized transcripts of *Pisum sativum* LecRLK (*PsLecRLK*), were upregulated in high salinity conditions (Vaid et al. 2015). Plants with *PsLecRLK* over-expression showed higher tolerance towards salt stress due to ROS-scavenging enzymes, that can reduce the accumulation ROS and prevent damage to the membrane. *PsLecRLK* overexpression also enhanced water uptake by activating Na^+ transporters/ water channel, which further overcomes the osmotic stress caused by high salinity. Various transporters like *HKT1, NHXs, AVP,* and *SOS1* maintain the water uptake by lowering the sodium (Na^+)/K^+ ratio. Another *LecRLK* gene named *SIT1* was found by Li in 2014 (Li et al. 2014), that was expressed predominantly in rice root epidermal cells and was involved in salt sensitivity. *SIT1* produces and accumulates ROS through MPK3/6 phosphorylation, regulates ethylene production and ethylene-mediated signaling in saline conditions. Therefore, *SIT1* leads to inhibition of plant growth and senescence due to salt stress. Shumayla et al. (2016a, b) performed quantitative RT-PCR and identified few *L-RLK* and *B-LRK* genes which showed higher transcripts level after salt stress treatment of 48 h to plants. There are other subfamilies of RLKs too, that have salient functions in salinity conditions. Recent reports showed that *FERONIA* (FER), a gene from RLK subfamily, played a key role in ABA and salt stress responses in *Catharanthus roseus* (Chen et al. 2016). Knockout FER mutant showed hypersensitivity to ABA and salt stress, suggesting that FER may be involved in salt stress management and may act through ABA signaling (Chen et al. 2016). Under salt stress conditions, *PnRLK-1*, a homolog of RLCKs found in the Antarctic moss *Pohlia nutans,* was reported to cause increased ABA sensitivity and induce higher expression of ROS-scavenger genes like *AtZAT10, AtAPX1* and *AtCAT1*, which reduces the accumulation of ROS and ultimately lowers the adverse effects of salt stress (Li et al. 2014). Additionally, expression of *OsRMC*, a cysteine-rich RLK (CRK), was induced by salinity stress conditions (Serra et al. 2013). Further functional characterization of *CRK68-A* of *T. aestivum* showed enhanced tolerance under salinity tolerance in yeast and *E. coli* (Shumayla et al. 2019). *TaPRK2697a*, an NCBI predicted LRR-RLK gene, improved the Na^+ efflux and increased salt stress tolerance in *T. aestivum* (Ma et al. 2016). Some of the *LRRK* genes from *T. aestivum* reported by Shumayla et al. (2016b) showed increased expression level after salt stress treatment at various time intervals.

4.7.4 RLKs Against Cold Stress

Cold stress is also an important abiotic stress which affects plants enzyme kinetics and cell membrane fluidity, which in turn cause metabolic disorder, reduction in photosynthesis, disruption in the transport of materials and consequently damages the plant (Janská et al. 2010). Some of the calcium-regulated genes have been reported to play a crucial part in managing the plant response to cold stress. *CRLK1* is a calcium-regulated receptor-like kinase gene localized on plasma membrane that is crucial for plants in cold stress management (Yang et al. 2010). There is a rapid increase in CRLK1 protein levels after cold treatment and hydrogen peroxide treatment, suggesting that it is involved in signaling pathways related to cold-related oxidative stress. Plants with *CRLK1* knockout showed higher sensitivity to cold and delayed induction of genes involved in cold-response, compared to the wild type. These findings show that *CRLK1* is a positive regulator of cold stress tolerance and acts as a bridge between calcium signaling and cold signaling. CRLK1-interacting MEKK1 protein was isolated by Yang et al in 2010 to understand the CRLK1 mediated signal pathway. Plants possess multiple MAP kinases family members such as *MAPKs*, *MAPKKs,* and *MAPKKKs*, that are activated in response to upstream signals and induce various downstream pathways. Through a MAPK cascade in plants, the expression of genes involved in cold stress response is promoted, which leads to the adaptation in plants to survive under low-temperature conditions (Furuya et al. 2013). LRR-RLK proteins are also presumed to have essential roles in the cold stress management, for instance, an LRR-RLK gene in *A. thaliana*, named Phloem intercalated with Xylem-Like1 (*AtPXL1*) was reported to be highly inducible in cold stress (Jung et al. 2015). Autophosphorylation activity was found in *AtPXL-1* by yeast two-hybrid and complementation assays. *AtPXL1*-knockout lines showed hypersensitivity to cold stress, while plants with *AtPXL1*-overexpression had increased germination rates despite low temperature conditions. *GsLRPK* (an LRR-RLK gene in *G. soja*) overexpression in yeast and *A. thaliana* showed higher tolerance towards cold stress and an increase in expression of numerous genes related to cold tolerance, including

KIN1 and *COR15b* (Yang et al. 2014). *CRK68-A* gene from *T. aestivum* also showed increased tolerance against cold stress in *E. coli* and yeast (Shumayla et al. 2019).

4.7.5 RLKs Against Metal Stress

Metals/metalloids can cause various toxicity problems to the plants if stored in excess and show various symptoms including wilting, chlorosis and even cell death. To avoid these symptoms plants have evolved different protective measures such as metal transportation, chelation, and sequestration (Hall 2002). RLKs are one of the major molecular players involved in detoxification of toxic metals/metalloids in plants. The WAKs (wall-associated kinases) is one of the members of RLKs that were shown to be involved in metal detoxification. In *A. thaliana*, the expression of *WAK1* was induced in roots after 3 h of Al treatment suggesting its role in Al detoxification for plant defense (Sivaguru et al. 2003). Also, T-DNA insertion in the promoter region of *WAK4* significantly altered its expression under different mineral nutrient conditions and as a result, it showed an enhanced nickel (Ni) tolerance (Hou et al. 2005). The expression of *OsWAK11* was upregulated strongly by Copper (Cu) and Aluminium (Al) and its promoter was strongly induced in response to metals (Hu et al. 2014). The cis-element localized in the promoter region of *OsWAK11* was found to be possibly involved in its expression regulation. The promoter region of *PvSR2* indicated heavy metal specific responsive activity (Qi et al. 2007). Cu-responsive elements (CuREs) are known to be involved in Cu tolerance. WAK also showed the presence of CuREs in the promoter. These results lead to the hypothesis that Al or Cu response is due to CuREs localized to the WAK promoter region. Besides WAKs, there are other plants RLKs that are also involved in response to toxic metals/metalloids stress. The comparative research was performed to reveal altered gene expression in response to metals/metalloids stress (Fu et al. 2014). Knockout study of four *LRR-RLK VIII* genes showed decreased as sensitivity, indicating their specific and significant role in response to As. Further to gain knowledge of chromium affected genes quantitative RT-PCR was performed by Trinh et al. (2014) and found an induced expression of genes by Cr, such as CDPK and MAPKs. The microarray data revealed that CRK and *LRK10-L* (PR5 K) were upregulated in rice roots under Cr stress. RLK genes, *DUF26, RLCK* and *LRK10-L* were also found to be involved in transcriptional regulation. This reveals that RLKs of different subfamilies may show a similar regulatory network under toxic metal stress (Trinh et al. 2014).

4.7.6 RLKs Against Other Abiotic Stresses

RLKs are also known to play an important role in other abiotic stress responses, including mechanical wounding and mineral deficiency. Mineral deficiency being the utmost prevalent health problem affects more than half of the world's population especially in developing countries. In plants, Iron (Fe) deficiency results in chlorosis and reduced both the crop yield and quality. The non-bioavailability of iron in food is one of the major causes of anemia in humans. Earlier it was shown that *O. sativa OsRMC* is involved in salt stress management and also one novel finding highlighted that *OsRMC* also plays a role in Fe acquisition under Fe-deficient condition (Serra et al. 2013; Yang et al. 2013). *OsRMC* regulates Fe-procurement by enhancing the biosynthesis of MA, increase Fe-MA transport, IRT1-mediated Fe transport and root development, which finally confers Fe accumulation in mature seeds under adequate iron conditions. During mechanical wounding due to animal feeding and freezing, plants initiate regulation of plant hormones, depolarization of their cell membrane, calcium channels activation and a series of other responses. *lecRK-a1* gene of *A. thaliana* showed an enhanced expression when the plant was subjected to the mechanical wounding (Riou et al. 2002). Plants are known to release oligo galacturonides upon mechanical wounding which cause the disruption of pectic cell wall constituents. It was also observed that *lecRK-a1* was not activated by jasmonic acid, which suggests that lecRKa1 might be a constituent of the wound-inducible jasmonic acid-independent pathway.

4.8 Conclusion

From the above literature, we conclude that the time and extent of RLK/Pelle family expansion suggested their significant role in the land plant evolution. Plants being sessile organism face an innumerable environmental stress. To fight for their survival they have multiple strategies. Plants consist of large RLKs members that play roles ranging from tolerance against abiotic and biotic stress to plant development and growth. The diverse receptors of RLKs fight against different stress by sensing either microbes recognition patterns or various abiotic stress elicitors outside their environment. They are extremely highly specific and selective. Also, various signaling cascade help plants to respond to various environmental stresses through sensing different elicitors. The role of receptor like kinases are induced by the various kinds of abiotic stress elicitors, which leads to the activation of signaling cascade to further activate the downstream target gene for stress response through the process of phosphorylation, activation of hormonal pathway, activation of transcription factors. Although much literature has been developed about the various aspects of plant RLKs but many more questions are yet to be answered. For example on the link between pathogen and RLK member to activate a particular pathway is still missing like on the perception of signal on PRR receptors and production of ROS and activation of MAP kinases is still missing. Therefore, it can be said that RLKs are one of the important protein family through which plants can save themselves from environmental challenges and could become a key protein family in the development of stress resistant crops.

Acknowledgements Authors are grateful to Panjab University, Chandigarh, India for research facilities. S is grateful to UGC for the senior research fellowship. We are also grateful to the Department of Science and Technology, Government of India for partial financial support under the Promotion of University Research and Scientific Excellence (PURSE) grant scheme.
Conflict of interest All the authors declare that there are no conflicts of interest.

References

AbuQamar S, Luo H, Laluk K, Mickelbart MV, Mengiste T (2009) Crosstalk between biotic and abiotic stress responses in tomato is mediated by the *AIM1* transcription factor. Plant J 58:347–360. https://doi.org/10.1111/j.1365-313X.2008.03783.x

Acharya BR, Raina S, Maqbool SB, Jagadeeswaran G, Mosher SL, Appel HM, Schultz JC, Klessig DF, Raina R (2007) Overexpression of CRK13, an Arabidopsis cysteine-rich receptor-like kinase, results in enhanced resistance to Pseudomonas syringae. Plant J 50:488–499. https://doi.org/10.1111/j.1365-313X.2007.03064.x

Ahmad P, Bhardwaj R, Tuteja N (2012) Plant signaling under abiotic stress environment. in: environmental adaptations and stress tolerance of plants in the era of climate change. Springer, New York, NY, pp 297–323

Albert M, Jehle AK, Mueller K, Eisele C, Lipschis M, Felix G (2010) *Arabidopsis thaliana* pattern recognition receptors for bacterial elongation factor Tu and flagellin can be combined to form functional chimeric receptors. J Biol Chem 285:19035–19042. https://doi.org/10.1074/jbc.M110.124800

Albrecht C, Boutrot F, Segonzac C, Schwessinger B, Gimenez-Ibanez S, Chinchilla D, Rathjen JP, de Vries SC, Zipfel C (2012) Brassinosteroids inhibit pathogen-associated molecular pattern-triggered immune signaling independent of the receptor kinase BAK1. Proc Natl Acad Sci 109:303–308. https://doi.org/10.1073/pnas.1109921108

Bartels D, Sunkar R (2005) Drought and salt tolerance in plants. CRC Crit Rev Plant Sci 24:23–58. https://doi.org/10.1080/07352680590910410

Belvin MP, Anderson KV (1996) A conserved signaling pathway: the *Drosophila* toll-dorsal pathway. Annu Rev Cell Dev Biol 12:393–416. https://doi.org/10.1146/annurev.cellbio.12.1.393

Boller T, Felix G (2009) A renaissance of elicitors: perception of microbe-associated molecular patterns and danger signals by pattern-recognition receptors. Annu Rev Plant Biol 60:379–406. https://doi.org/10.1146/annurev.arplant.57.032905.105346

Boyer JS (1982) Plant productivity and environment. Science (80-.). 218:443–448. https://doi.org/10.1126/science.218.4571.443

Brand U, Fletcher JC, Hobe M, Meyerowitz EM, Simon R (2000) Dependence of stem cell fate in Arabidopsis on a feedback loop regulated by CLV3 activity. Science 289:617–619

Bruggeman Q, Raynaud C, Benhamed M, Delarue M (2015) To die or not to die? Lessons from lesion mimic mutants. Front Plant Sci 6:24. https://doi.org/10.3389/fpls.2015.00024

Brutus A, Sicilia F, Macone A, Cervone F, De Lorenzo G (2010) A domain swap approach reveals a role of the plant wall-associated kinase 1 (WAK1) as a receptor of oligogalacturonides. Proc Natl Acad Sci 107:9452–9457. https://doi.org/10.1073/pnas.1000675107

Chen J, Yu F, Liu Y, Du C, Li X, Zhu S, Wang X, Lan W, Rodriguez PL, Liu X, Li D, Chen L, Luan S (2016) FERONIA interacts with ABI2-type phosphatases to facilitate signaling cross-talk between abscisic acid and RALF peptide in *Arabidopsis*. Proc Natl Acad Sci 113:E5519–E5527. https://doi.org/10.1073/pnas.1608449113

Chinchilla D, Bauer Z, Regenass M, Boller T, Felix G (2006) The Arabidopsis receptor kinase FLS2 Binds flg22 and determines the specificity of flagellin perception. Plant Cell Online 18:465–476. https://doi.org/10.1105/tpc.105.036574

Chinchilla D, Zipfel C, Robatzek S, Kemmerling B, Nürnberger T, Jones JDG, Felix G, Boller T (2007) A flagellin-induced complex of the receptor FLS2 and BAK1 initiates plant defence. Nature 448:497–500. https://doi.org/10.1038/nature05999

Choe S, Schmitz RJ, Fujioka S, Takatsuto S, Lee M-O, Yoshida S, Feldmann KA, Tax FE (2002) Arabidopsis brassinosteroid-insensitive dwarf12 mutants are semidominant and defective in a glycogen synthase kinase 3beta-like kinase. Plant Physiol 130:1506–1515. https://doi.org/10.1104/pp.010496

Choudhary SP, Yu J-Q, Yamaguchi-Shinozaki K, Shinozaki K, Tran L-SP (2012) Benefits of brassinosteroid crosstalk. Trends Plant Sci 17:594–605. https://doi.org/10.1016/j.tplants.2012.05.012

Clark SE, Williams RW, Meyerowitz EM (1997) The CLAVATA1 gene encodes a putative receptor kinase that controls shoot and floral meristem size in Arabidopsis. Cell 89:575–585

Cock JM, Vanoosthuyse V, Gaude T (2002) Receptor kinase signalling in plants and animals: distinct molecular systems with mechanistic similarities. Curr Opin Cell Biol 14:230–236

Cock JM, Sterck L, Rouzé P, Scornet D, Allen AE, Amoutzias G, Anthouard V, Artiguenave F, Aury J-M, Badger JH, Beszteri B, Billiau K, Bonnet E, Bothwell JH, Bowler C, Boyen C, Brownlee C, Carrano CJ, Charrier B, Cho GY, Coelho SM, Collén J, Corre E, Da Silva C, Delage L, Delaroque N, Dittami SM, Doulbeau S, Elias M, Farnham G, Gachon CMM, Gschloessl B, Heesch S, Jabbari K, Jubin C, Kawai H, Kimura K, Kloareg B, Küpper FC, Lang D, Le Bail A, Leblanc C, Lerouge P, Lohr M, Lopez PJ, Martens C, Maumus F, Michel G, Miranda-Saavedra D, Morales J, Moreau H, Motomura T, Nagasato C, Napoli CA, Nelson DR, Nyvall-Collén P, Peters AF, Pommier C, Potin P, Poulain J, Quesneville H, Read B, Rensing SA, Ritter A, Rousvoal S, Samanta M, Samson G, Schroeder DC, Ségurens B, Strittmatter M, Tonon T, Tregear JW, Valentin K, von Dassow P, Yamagishi T, Van de Peer Y, Wincker P (2010) The Ectocarpus genome and the independent evolution of multicellularity in brown algae. Nature 465:617–621. https://doi.org/10.1038/nature09016

Coll NS, Epple P, Dangl JL (2011) Programmed cell death in the plant immune system. Cell Death Differ 18:1247–1256. https://doi.org/10.1038/cdd.2011.37

Consonni C, Humphry ME, Hartmann HA, Livaja M, Durner J, Westphal L, Vogel J, Lipka V, Kemmerling B, Schulze-Lefert P, Somerville SC, Panstruga R (2006) Conserved requirement for a plant host cell protein in powdery mildew pathogenesis. Nat Genet 38:716–720. https://doi.org/10.1038/ng1806

Courties C, Vaquer A, Troussellier M, Lautier J, Chrétiennot-Dinet MJ, Neveux J, Machado C, Claustre H (1994) Smallest eukaryotic organism. Nature 370:255–255. https://doi.org/10.1038/370255a0

Danna CH, Millet YA, Koller T, Han S-W, Bent AF, Ronald PC, Ausubel FM (2011) The Arabidopsis flagellin receptor FLS2 mediates the perception of Xanthomonas Ax21 secreted peptides. Proc Natl Acad Sci U S A 108:9286–9291. https://doi.org/10.1073/pnas.1106366108

Dodds PN, Rathjen JP (2010) Plant immunity: towards an integrated view of plant–pathogen interactions. Nat Rev Genet 11:539–548. https://doi.org/10.1038/nrg2812

Escobar-Restrepo J-M, Huck N, Kessler S, Gagliardini V, Gheyselinck J, Yang W-C, Grossniklaus U (2007) The FERONIA receptor-like kinase mediates male-female interactions during pollen tube reception. Science (80-.) 317:656–660. https://doi.org/10.1126/science.1143562

Farooq M, Wahid A, Kobayashi N, Fujita D, Basra SMA (2009) Plant drought stress: effects, mechanisms and management. Agron Sustain Dev 29:185–212. https://doi.org/10.1051/agro:2008021

Feng L, Gao Z, Xiao G, Huang R, Zhang H (2014) Leucine-rich repeat receptor-like kinase FON1 regulates drought stress and seed germination by activating the expression of ABA-responsive genes in rice. Plant Mol Biol Report 32:1158–1168. https://doi.org/10.1007/s11105-014-0718-0

Flannery S, Bowie AG (2010) The interleukin-1 receptor-associated kinases: critical regulators of innate immune signalling. Biochem Pharmacol 80:1981–1991. https://doi.org/10.1016/j.bcp.2010.06.020

Fritz-Laylin LK, Krishnamurthy N, Tör M, Sjölander KV, Jones JDG (2005) Phylogenomic analysis of the receptor-like proteins of rice and arabidopsis. PLANT Physiol 138:611–623. https://doi.org/10.1104/pp.104.054452

Fu S-F, Chen P-Y, Nguyen QT, Huang L-Y, Zeng G-R, Huang T-L, Lin C-Y, Huang H-J (2014) Transcriptome profiling of genes and pathways associated with arsenic toxicity and tolerance in Arabidopsis. BMC Plant Biol 14:94. https://doi.org/10.1186/1471-2229-14-94

Fujita M, Fujita Y, Noutoshi Y, Takahashi F, Narusaka Y, Yamaguchi-Shinozaki K, Shinozaki K (2006) Crosstalk between abiotic and biotic stress responses: a current view from the points of convergence in the stress signaling networks. Curr Opin Plant Biol 9:436–442. https://doi.org/10.1016/j.pbi.2006.05.014

Furuya T, Matsuoka D, Nanmori T (2013) Phosphorylation of Arabidopsis thaliana MEKK1 via Ca2+ signaling as a part of the cold stress response. J Plant Res 126:833–840. https://doi.org/10.1007/s10265-013-0576-0

Gómez-Gómez L, Boller T (2000) FLS2: an LRR receptor-like kinase involved in the perception of the bacterial elicitor flagellin in Arabidopsis. Mol Cell 5:1003–1011

González-Carranza ZH, Lozoya-Gloria E, Roberts JA (1998) Recent developments in abscission: shedding light on the shedding process. Trends Plant Sci 3:10–14. https://doi.org/10.1016/S1360-1385(97)01132-1

Ha S, Vankova R, Yamaguchi-Shinozaki K, Shinozaki K, Tran L-SP (2012) Cytokinins: metabolism and function in plant adaptation to environmental stresses. Trends Plant Sci 17:172–179. https://doi.org/10.1016/j.tplants.2011.12.005

Hall JL (2002) Cellular mechanisms for heavy metal detoxification and tolerance. J Exp Bot 53:1–11. https://doi.org/10.1093/jexbot/53.366.1

Hasanuzzaman M, Hossain MA, da Silva JAT, Fujita M (2012) Plant response and tolerance to abiotic oxidative stress: antioxidant defense is a key factor. In: Crop stress and its management: perspectives and strategies. Springer, Netherlands, Dordrecht, pp 261–315

Hasanuzzaman M, Nahar K, Alam MM, Roychowdhury R, Fujita M (2013) Physiological, biochemical, and molecular mechanisms of heat stress tolerance in plants. Int J Mol Sci 14:9643–9684. https://doi.org/10.3390/ijms14059643

He Z, Wang ZY, Li J, Zhu Q, Lamb C, Ronald P, Chory J (2000) Perception of brassinosteroids by the extracellular domain of the receptor kinase BRI1. Science 288:2360–2363

He J-X, Gendron JM, Yang Y, Li J, Wang Z-Y (2002) The GSK3-like kinase BIN2 phosphorylates and destabilizes BZR1, a positive regulator of the brassinosteroid signaling pathway in Arabidopsis. Proc Natl Acad Sci 99:10185–10190. https://doi.org/10.1073/pnas.152342599

Hecht PM, Anderson K (1993) V: genetic characterization of tube and pelle, genes required for signaling between Toll and dorsal in the specification of the dorsal-ventral pattern of the Drosophila embryo. Genetics 135:405–417

Hedges SB (2002) The origin and evolution of model organisms. Nat Rev Genet 3:838–849. https://doi.org/10.1038/nrg929

Heese A, Hann DR, Gimenez-Ibanez S, Jones AME, He K, Li J, Schroeder JI, Peck SC, Rathjen JP (2007) The receptor-like kinase SERK3/BAK1 is a central regulator of innate immunity in plants. Proc Natl Acad Sci 104:12217–12222. https://doi.org/10.1073/pnas.0705306104

Hou X, Tong H, Selby J, Dewitt J, Peng X, He Z-H (2005) Involvement of a cell wall-associated kinase, WAKL4. Arabidopsis mineral responses. Plant Physiol. 139:1704–1716. https://doi.org/10.1104/pp.105.066910

Hu W, Lv Y, Lei W, Li X, Chen Y, Zheng L, Xia Y, Shen Z (2014) Cloning and characterization of the Oryza sativa wall-associated kinase gene OsWAK11 and its transcriptional response to abiotic stresses. Plant Soil 384:335–346. https://doi.org/10.1007/s11104-014-2204-8

Hunter T, Lindberg RA, Middlemas DS, Tracy S, van der Geer P (1992) Receptor protein tyrosine kinases and phosphatases. Cold Spring Harb Symp Quant Biol 57:25–41

Iizasa E, Mitsutomi M, Nagano Y (2010) Direct binding of a plant LysM receptor-like kinase, LysM RLK1/CERK1, to Chitin in Vitro. J Biol Chem 285:2996–3004. https://doi.org/10.1074/jbc.M109.027540

Jakab G, Ton J, Flors V, Zimmerli L, Métraux J-P, Mauch-Mani B (2005) Enhancing arabidopsis salt and drought stress tolerance by chemical priming for its Abscisic acid responses. Plant Physiol 139:267–274. https://doi.org/10.1104/pp.105.065698

Janská A, Maršík P, Zelenková S, Ovesná J (2010) Cold stress and acclimation—what is important for metabolic adjustment? Plant Biol 12:395–405. https://doi.org/10.1111/j.1438-8677.2009.00299.x

Jeong S, Trotochaud AE, Clark SE (1999) The arabidopsis CLAVATA2 gene encodes a receptor-like protein required for the stability of the CLAVATA1 receptor-like kinase. Plant Cell Online 11:1925–1934. https://doi.org/10.1105/tpc.11.10.1925

Jones JDG, Dangl JL (2006) The plant immune system. Nature 444:323–329. https://doi.org/10.1038/nature05286

Jung CG, Hwang S-G, Park YC, Park HM, Kim DS, Park DH, Jang CS (2015) Molecular characterization of the cold- and heat-induced Arabidopsis PXL1 gene and its potential role in transduction pathways under temperature fluctuations. J Plant Physiol 176:138–146. https://doi.org/10.1016/j.jplph.2015.01.001

Kaku H, Nishizawa Y, Ishii-Minami N, Akimoto-Tomiyama C, Dohmae N, Takio K, Minami E, Shibuya N, Clarence Ryan by A (2006) Sciences of the USA, PNAS, pp 11086–11091

Kaku H, Nishizawa Y, Ishii-Minami N, Akimoto-Tomiyama C, Dohmae N, Takio K, Minami E, Shibuya N (2006b) Plant cells recognize chitin fragments for defense signaling through a plasma membrane receptor. Proc Natl Acad Sci 103:11086–11091. https://doi.org/10.1073/pnas.0508882103

Kessler SA, Shimosato-Asano H, Keinath NF, Wuest SE, Ingram G, Panstruga R, Grossniklaus U (2010) Conserved molecular components for pollen tube reception and fungal invasion. Science (80-.). 330:968–971 (2010). https://doi.org/10.1126/science.1195211

Kim H, Hwang H, Hong J-W, Lee Y-N, Ahn IP, Yoon IS, Yoo S-D, Lee S, Lee SC, Kim B-G (2012) A rice orthologue of the ABA receptor, OsPYL/RCAR5, is a positive regulator of the ABA signal transduction pathway in seed germination and early seedling growth. J Exp Bot 63:1013–1024. https://doi.org/10.1093/jxb/err338

King N, Westbrook MJ, Young SL, Kuo A, Abedin M, Chapman J, Fairclough S, Hellsten U, Isogai Y, Letunic I, Marr M, Pincus D, Putnam N, Rokas A, Wright KJ, Zuzow R, Dirks W, Good M, Goodstein D, Lemons D, Li W, Lyons JB, Morris A, Nichols S, Richter DJ, Salamov A, Sequencing J, Bork P, Lim WA, Manning G, Miller WT, McGinnis W, Shapiro H, Tjian R, Grigoriev IV, Rokhsar D (2008) The genome of the choanoflagellate Monosiga brevicollis and the origin of metazoans. Nature 451:783–788. https://doi.org/10.1038/nature06617

Köhler S, Delwiche CF, Denny PW, Tilney LG, Webster P, Wilson RJ, Palmer JD, Roos DS (1997) A plastid of probable green algal origin in Apicomplexan parasites. Science 275:1485–1489

Laluk K, Luo H, Chai M, Dhawan R, Lai Z, Mengiste T (2011) Biochemical and genetic requirements for function of the immune response regulator BOTRYTIS-INDUCED KINASE1 in plant growth, ethylene signaling, and PAMP-triggered immunity in *Arabidopsis*. Plant Cell 23:2831–2849. https://doi.org/10.1105/tpc.111.087122

Latif F, Ullah F, Mehmood S, Khattak A, Khan AU, Khan S, Husain I (2016) Effects of salicylic acid on growth and accumulation of phenolics in *Zea mays* L. under drought stress. Acta Agric. Scand. Sect. B—Soil Plant Sci 66:325–332. https://doi.org/10.1080/09064710.2015.1117133

Lehti-Shiu MD, Zou C, Hanada K, Shiu S-H (2009) Evolutionary history and stress regulation of plant receptor-like kinase/pelle genes. Plant Physiol 150:12–26. https://doi.org/10.1104/pp.108.134353

Li J, Wen J, Lease KA, Doke JT, Tax FE, Walker JC (2002) BAK1, an Arabidopsis LRR receptor-like protein kinase, interacts with BRI1 and modulates brassinosteroid signaling. Cell 110:213–222

Li C-H, Wang G, Zhao J-L, Zhang L-Q, Ai L-F, Han Y-F, Sun D-Y, Zhang S-W, Sun Y (2014) The receptor-like kinase SIT1 mediates salt sensitivity by activating MAPK3/6 and regulating ethylene homeostasis in rice. Plant Cell 26:2538–2553. https://doi.org/10.1105/tpc.114.125187

Lim CW, Yang SH, Shin KH, Lee SC, Kim SH (2015) The AtLRK10L1.2, Arabidopsis ortholog of wheat LRK10, is involved in ABA-mediated signaling and drought resistance. Plant Cell Rep 34:447–455. https://doi.org/10.1007/s00299-014-1724-2

Lozano-Durán R, Zipfel C (2015) Trade-off between growth and immunity: role of brassinosteroids. Trends Plant Sci 20:12–19. https://doi.org/10.1016/j.tplants.2014.09.003

Luo M, Liang XQ, Dang P, Holbrook CC, Bausher MG, Lee RD, Guo BZ (2005) Microarray-based screening of differentially expressed genes in peanut in response to Aspergillus parasiticus infection and drought stress. Plant Sci 169:695–703. https://doi.org/10.1016/J.PLANTSCI.2005.05.020

Ma X-L, Cui W-N, Zhao Q, Zhao J, Hou X-N, Li D-Y, Chen Z-L, Shen Y-Z, Huang Z-J (2016) Functional study of a salt-inducible *TaSR* gene in *Triticum aestivum*. Physiol Plant 156:40–53. https://doi.org/10.1111/ppl.12337

Mayer KF, Schoof H, Haecker A, Lenhard M, Jürgens G, Laux T (1998) Role of WUSCHEL in regulating stem cell fate in the arabidopsis shoot meristem. Cell 95:805–815. https://doi.org/10.1016/S0092-8674(00)81703-1

Medzhitov R, Janeway CA (1997) Innate immunity: impact on the adaptive immune response. Curr Opin Immunol 9:4–9

Mengiste T (2003) The BOTRYTIS SUSCEPTIBLE1 gene encodes an R2R3MYB transcription factor protein that is required for biotic and abiotic stress responses in arabidopsis. Plant Cell Online 15:2551–2565. https://doi.org/10.1105/tpc.014167

Miya A, Albert P, Shinya T, Desaki Y, Ichimura K, Shirasu K, Narusaka Y, Kawakami N, Kaku H, Shibuya N (2007) CERK1, a LysM receptor kinase, is essential for chitin elicitor signaling in Arabidopsis. Proc Natl Acad Sci 104:19613–19618. https://doi.org/10.1073/pnas.0705147104

Müller R, Bleckmann A, Simon R (2008) The receptor kinase CORYNE of arabidopsis transmits the stem cell-limiting signal CLAVATA3 independently of CLAVATA1. Plant Cell 20:934. https://doi.org/10.1105/TPC.107.057547

Nam KH, Li J (2002) BRI1/BAK1, a receptor kinase pair mediating brassinosteroid signaling. Cell 110:203–212

Narsai R, Wang C, Chen J, Wu J, Shou H, Whelan J (2013) Antagonistic, overlapping and distinct responses to biotic stress in rice (Oryza sativa) and interactions with abiotic stress. BMC Genom 14:93. https://doi.org/10.1186/1471-2164-14-93

Nemali KS, Bonin C, Dohleman FG, Stephens M, Reeves WR, Nelson DE, Castiglioni P, Whitsel JE, Sammons B, Silady RA, Anstrom D, Sharp RE, Patharkar OR, Clay D, Coffin M, Nemeth MA, Leibman ME, Luethy M, Lawson M (2015) Physiological responses related to increased grain yield under drought in the first biotechnology-derived drought-tolerant maize. Plant, Cell Environ 38:1866–1880. https://doi.org/10.1111/pce.12446

Peleg Z, Blumwald E (2011) Hormone balance and abiotic stress tolerance in crop plants. Curr Opin Plant Biol 14:290–295. https://doi.org/10.1016/j.pbi.2011.02.001

Pérez-Pérez JM, Ponce MR, Micol JL (2002) The UCU1 Arabidopsis Gene Encodes a SHAGGY/GSK3-like kinase required for cell expansion along the proximodistal axis. Dev Biol 242:161–173. https://doi.org/10.1006/dbio.2001.0543

Petutschnig EK, Jones AME, Serazetdinova L, Lipka U, Lipka V (2010) The lysin motif receptor-like kinase (LysM-RLK) CERK1 is a major chitin-binding protein in *Arabidopsis thaliana* and subject to chitin-induced phosphorylation. J Biol Chem 285:28902–28911. https://doi.org/10.1074/jbc.M110.116657

Prado JR, Segers G, Voelker T, Carson D, Dobert R, Phillips J, Cook K, Cornejo C, Monken J, Grapes L, Reynolds T, Martino-Catt S (2014) Genetically engineered crops: from idea to product. Annu Rev Plant Biol 65:769–790. https://doi.org/10.1146/annurev-arplant-050213-040039

Qi X, Zhang Y, Chai T (2007) Characterization of a novel plant promoter specifically induced by heavy metal and identification of the promoter regions conferring heavy metal responsiveness. Plant Physiol 143:50–59. https://doi.org/10.1104/PP.106.080283

Riou C, Hervé C, Pacquit V, Dabos P, Lescure B (2002) Expression of an Arabidopsis lectin kinase receptor gene, lecRK-a1, is induced during senescence, wounding and in response

to oligogalacturonic acids. Plant Physiol Biochem 40:431–438. https://doi.org/10.1016/S0981-9428(02)01390-6
Ronald PC (2014) Lab to farm: applying research on plant genetics and genomics to crop improvement. PLoS Biol 12:e1001878. https://doi.org/10.1371/journal.pbio.1001878
Roux M, Schwessinger B, Albrecht C, Chinchilla D, Jones A, Holton N, Malinovsky FG, Tör M, de Vries S, Zipfel C (2011) The arabidopsis leucine-rich repeat receptor-like kinases BAK1/SERK3 and BKK1/SERK4 are required for innate immunity to hemibiotrophic and biotrophic pathogens. Plant Cell 23:2440–2455. https://doi.org/10.1105/tpc.111.084301
Sairam RK (2004) Tyagi a physiology and molecular biology of salinity stress tolerance in plants. https://www.jstor.org/stable/24108735
Salehi-Lisar SY, Bakhshayeshan-Agdam H (2016) Drought stress in plants: causes, consequences, and tolerance. In: Drought stress tolerance in plants, vol 1. Springer International Publishing, Cham, pp 1–16
Sasaki G, Katoh K, Hirose N, Suga H, Kuma K, Miyata T, Su Z-H (2007) Multiple receptor-like kinase cDNAs from liverwort Marchantia polymorpha and two charophycean green algae, Closterium ehrenbergii and Nitella axillaris: extensive gene duplications and gene shufflings in the early evolution of streptophytes. Gene 401:135–144. https://doi.org/10.1016/j.gene.2007.07.009
Schoof H, Lenhard M, Haecker A, Mayer KF, Jürgens G, Laux T (2000) The stem cell population of Arabidopsis shoot meristems in maintained by a regulatory loop between the CLAVATA and WUSCHEL genes. Cell 100:635–644
Schulze B, Mentzel T, Jehle AK, Mueller K, Beeler S, Boller T, Felix G, Chinchilla D (2010) Rapid heteromerization and phosphorylation of ligand-activated plant transmembrane receptors and their associated kinase BAK1. J Biol Chem 285:9444–9451. https://doi.org/10.1074/jbc.M109.096842
Schwessinger B, Roux M, Kadota Y, Ntoukakis V, Sklenar J, Jones A, Zipfel C (2011) Phosphorylation-dependent differential regulation of plant growth, cell death, and innate immunity by the regulatory receptor-like kinase BAK1. PLoS Genet 7:e1002046. https://doi.org/10.1371/journal.pgen.1002046
Serra TS, Figueiredo DD, Cordeiro AM, Almeida DM, Lourenço T, Abreu IA, Sebastián A, Fernandes L, Contreras-Moreira B, Oliveira MM, Saibo NJM (2013) OsRMC, a negative regulator of salt stress response in rice, is regulated by two AP2/ERF transcription factors. Plant Mol Biol 82:439–455. https://doi.org/10.1007/s11103-013-0073-9
Shaik R, Ramakrishna W (2013) Genes and co-expression modules common to drought and bacterial stress responses in arabidopsis and rice. PLoS ONE 8:e77261. https://doi.org/10.1371/journal.pone.0077261
Shelton CA, Wasserman SA (1993) pelle encodes a protein kinase required to establish dorsoventral polarity in the Drosophila embryo. Cell 72:515–525
Shimizu T, Nakano T, Takamizawa D, Desaki Y, Ishii-Minami N, Nishizawa Y, Minami E, Okada K, Yamane H, Kaku H, Shibuya N (2010) No Title Plant J 64:204–214. https://doi.org/10.1111/j.1365-313X.2010.04324.x
Shiu S-H, Bleecker AB (2001) Plant receptor-like kinase gene family: diversity, function, and signaling. Sci Signal 2001:re22–re22. https://doi.org/10.1126/stke.2001.113.re22
Shiu S-H, Bleecker AB (2001b) Receptor-like kinases from Arabidopsis form a monophyletic gene family related to animal receptor kinases. Proc Natl Acad Sci 98:10763–10768. https://doi.org/10.1073/pnas.181141598
Shiu S-H, Li W-H (2004) Origins, lineage-specific expansions, and multiple losses of tyrosine kinases in eukaryotes. Mol Biol Evol 21:828–840. https://doi.org/10.1093/molbev/msh077
Shpak ED, McAbee JM, Pillitteri LJ, Torii KU (2005) Stomatal patterning and differentiation by synergistic interactions of receptor kinases. Science (80-.) 309:290–293. https://doi.org/10.1126/science.1109710

Shumayla, Sharma S, Pandey AK, Singh K, Upadhyay SK (2016a) Molecular characterization and global expression analysis of lectin receptor kinases in bread wheat (Triticum aestivum). PLoS One 11:e0153925. https://doi.org/10.1371/journal.pone.0153925

Shumayla, Sharma S, Kumar R, Mendu V, Singh K, Upadhyay SK (2016b) Genomic dissection and expression profiling revealed functional divergence in triticum aestivum leucine rich repeat receptor like kinases (TaLRRKs). Front Plant Sci 7:1374 (2016). https://doi.org/10.3389/fpls.2016.01374

Shumayla, Tyagi S, Sharma A, Singh K, Upadhyay SK (2019) Genomic dissection and transcriptional profiling of Cysteine-rich receptor-like kinases in five cereals and functional characterization of TaCRK68-A. Int J Biol Macromol 134:316–329. https://doi.org/10.1016/j.ijbiomac.2019.05.016

Sivaguru M, Ezaki B, He Z-H, Tong H, Osawa H, Baluska F, Volkmann D, Matsumoto H (2003) Aluminum-induced gene expression and protein localization of a cell wall-associated receptor kinase in Arabidopsis. Plant Physiol 132:2256–2266. https://doi.org/10.1104/pp.103.022129

Song WY, Wang GL, Chen LL, Kim HS, Pi LY, Holsten T, Gardner J, Wang B, Zhai WX, Zhu LH, Fauquet C, Ronald P (1995) A receptor kinase-like protein encoded by the rice disease resistance gene, Xa21. Science 270:1804–1806

Srivastava AK, Penna S, Nguyen D Van, Tran L-SP (2014) Multifaceted roles of aquaporins as molecular conduits in plant responses to abiotic stresses. Crit Rev Biotechnol 36:1–10. https://doi.org/10.3109/07388551.2014.973367

Stone JM, Trotochaud AE, Walker JC, Clark SE (1998) Control of meristem development by CLAVATA1 receptor kinase and kinase-associated protein phosphatase interactions. Plant Physiol 117:1217–1225. https://doi.org/10.1104/pp.117.4.1217

Sun W, Cao Y, Jansen Labby K, Bittel P, Boller T, Bent AF (2012) Probing the arabidopsis flagellin receptor: FLS2-FLS2 association and the contributions of specific domains to signaling function. Plant Cell 24:1096–1113. https://doi.org/10.1105/tpc.112.095919

Sun X, Sun M, Luo X, Ding X, Ji W, Cai H, Bai X, Liu X, Zhu Y (2013) A glycine soja ABA-responsive receptor-like cytoplasmic kinase, GsRLCK, positively controls plant tolerance to salt and drought stresses. Planta 237:1527–1545. https://doi.org/10.1007/s00425-013-1864-6

Taylor JE, Whitelaw CA (2001) Signals in abscission. New Phytol 151:323–340. https://doi.org/10.1046/j.0028-646x.2001.00194.x

Tran L-SP, Mochida K (2010) Functional genomics of soybean for improvement of productivity in adverse conditions. Funct Integr Genomics 10:447–462. https://doi.org/10.1007/s10142-010-0178-z

Trinh N-N, Huang T-L, Chi W-C, Fu S-F, Chen C-C, Huang H-J (2014) Chromium stress response effect on signal transduction and expression of signaling genes in rice. Physiol Plant 150:205–224. https://doi.org/10.1111/ppl.12088

Trotochaud AE, Hao T, Wu G, Yang Z, Clark SE (1999) The CLAVATA1 Receptor-like kinase requires CLAVATA3 for its assembly into a signaling complex that includes KAPP and a Rho-related protein. Plant Cell Online 11:393–406. https://doi.org/10.1105/tpc.11.3.393

Vaid N, Pandey P, Srivastava VK, Tuteja N (2015) Pea lectin receptor-like kinase functions in salinity adaptation without yield penalty, by alleviating osmotic and ionic stresses and upregulating stress-responsive genes. Plant Mol Biol 88:193–206. https://doi.org/10.1007/s11103-015-0319-9

van der Geer P, Hunter T, Lindberg RA (1994) Receptor protein-tyrosine kinases and their signal transduction pathways. Annu Rev Cell Biol 10:251–337. https://doi.org/10.1146/annurev.cb.10.110194.001343

Walker JC (1994) Structure and function of the receptor-like protein kinases of higher plants. Plant Mol Biol 26:1599–1609

Wang ZY, Nakano T, Gendron J, He J, Chen M, Vafeados D, Yang Y, Fujioka S, Yoshida S, Asami T, Chory J (2002) Nuclear-localized BZR1 mediates brassinosteroid-induced growth and feedback suppression of brassinosteroid biosynthesis. Dev Cell 2:505–513

Williams RW, Wilson JM, Meyerowitz EM (1997) A possible role for kinase-associated protein phosphatase in the Arabidopsis CLAVATA1 signaling pathway. Proc Natl Acad Sci 94:10467–10472. https://doi.org/10.1073/pnas.94.19.10467

Wong HL, Pinontoan R, Hayashi K, Tabata R, Yaeno T, Hasegawa K, Kojima C, Yoshioka H, Iba K, Kawasaki T, Shimamoto K (2007) Regulation of rice NADPH oxidase by binding of Rac GTPase to its N-terminal extension. Plant Cell Online 19:4022–4034. https://doi.org/10.1105/tpc.107.055624

Wu F, Sheng P, Tan J, Chen X, Lu G, Ma W, Heng Y, Lin Q, Zhu S, Wang J, Wang J, Guo X, Zhang X, Lei C, Wan J (2015) Plasma membrane receptor-like kinase leaf panicle 2 acts downstream of the DROUGHT AND SALT TOLERANCE transcription factor to regulate drought sensitivity in rice. J Exp Bot 66:271–281. https://doi.org/10.1093/jxb/eru417

Yang T, Shad Ali G, Yang L, Du L, Reddy ASN, Poovaiah BW (2010) Calcium/calmodulin-regulated receptor-like kinase CRLK1 interacts with MEKK1 in plants. Plant Signal Behav 5:991–994. https://doi.org/10.4161/psb.5.8.12225

Yang A, Li Y, Xu Y, Xu Y, Zhang W-H (2013) A receptor-like protein RMC is involved in regulation of iron acquisition in rice. J Exp Bot 64:5009–5020. https://doi.org/10.1093/jxb/ert290

Yang L, Wu K, Gao P, Liu X, Li G, Wu Z (2014) GsLRPK, a novel cold-activated leucine-rich repeat receptor-like protein kinase from Glycine soja, is a positive regulator to cold stress tolerance. Plant Sci 215–216:19–28. https://doi.org/10.1016/j.plantsci.2013.10.009

Yin Y, Wang ZY, Mora-Garcia S, Li J, Yoshida S, Asami T, Chory J (2002) BES1 accumulates in the nucleus in response to brassinosteroids to regulate gene expression and promote stem elongation. Cell 109:181–191

Yu LP, Simon EJ, Trotochaud AE, Clark SE (2000) POLTERGEIST functions to regulate meristem development downstream of the CLAVATA loci. Development 127:1661–1670

Yu LP, Miller AK, Clark SE (2003) POLTERGEIST encodes a protein phosphatase 2C that regulates CLAVATA pathways controlling stem cell identity at Arabidopsis shoot and flower meristems. Curr Biol 13:179–188

Yu L, Chen X, Wang Z, Wang S, Wang Y, Zhu Q, Li S, Xiang C (2013) Arabidopsis enhanced drought tolerance1/HOMEODOMAIN GLABROUS11 confers drought tolerance in transgenic rice without yield penalty. Plant Physiol 162:1378–1391. https://doi.org/10.1104/pp.113.217596

Zhang J, Li W, Xiang T, Liu Z, Laluk K, Ding X, Zou Y, Gao M, Zhang X, Chen S, Mengiste T, Zhang Y, Zhou J-M (2010) Receptor-like cytoplasmic kinases integrate signaling from multiple plant immune receptors and are targeted by a pseudomonas syringae effector. Cell Host Microbe 7:290–301. https://doi.org/10.1016/j.chom.2010.03.007

Zhao J, Peng P, Schmitz RJ, Decker AD, Tax FE, Li J (2002) Two putative BIN2 substrates are nuclear components of brassinosteroid signaling. Plant Physiol 130:1221–1229. https://doi.org/10.1104/pp.102.010918

Zhu J-K (2016) Abiotic stress signaling and responses in plants. Cell 167:313–324. https://doi.org/10.1016/j.cell.2016.08.029

Chapter 5
Role of Histone Acetyltransferases in Plant Abiotic Stress

Amit Choudhary, Sanchita Singh and Praveen Chandra Verma

Abstract In regulation of chromatin dynamics and gene expression, epigenetic modification plays a fundamental role. Acetylation is a histone modification responsible for making the DNA sequence available for transcription factors thus, leading to gene expression. Two functionally antagonistic enzymes, namely Histone acetyltransferase (HAT) and histone deacetylase (HDAC) control histone acetylation level. On encountering stress, several plant genes involved in stress response are activated by the action of HATs. Here, in the chapter we will discuss the role of acetylation in stress management by plants.

Keywords Histone acetyltransferases · Abiotic stress · Epigenetic modification · Histone acetylation · SAGA complex

5.1 Introduction

The literal meaning of the word **epigenetic** is "in addition to the classical basis of genetic inheritance". Epigenetics is defined as a section of genetics which deals with the alteration in the gene expression without any variation in the underlying DNA sequence. The term **epigenome** comprises the biogenesis of all the biochemical changes in proteins, non-coding RNAs (ncRNAs), nuclear DNA, and histone in a cell. In various types of RNAs approximately 150 modifications have been identified (Cantara et al. 2010). To compress the eukaryotic DNA in nucleus, the DNA is packed with histones into a complex structure called chromatin (Luger et al. 1997). The nucleosome is the basic unit of chromatin. Each nucleosome consists of histone

A. Choudhary · S. Singh · P. C. Verma (✉)
Council of Scientific and Industrial Research—National Botanical Research Institute (CSIR-NBRI), Rana Pratap Marg, Lucknow, Uttar Pradesh 226001, India
e-mail: praveencverma@nbri.res.in

A. Choudhary · P. C. Verma
Academy of Scientific and Innovative Research (AcSIR), Ghaziabad, Uttar Pradesh 201002, India

S. Singh
Department of Botany, University of Lucknow, Lucknow, Uttar Pradesh 226007, India

S. P. Singh et al. (eds.), *Molecular Approaches in Plant Biology and Environmental Challenges*, Energy, Environment, and Sustainability,
https://doi.org/10.1007/978-981-15-0690-1_5

octamer, comprised of around 146 base pairs of DNA wrapped on an octamer having two copies core of histone H2A, H2B, H3 and H4 (Kornberg and Lorch 1999; Richmond and Davey 2003). There is an additional histone protein called as linker histone (H1), this one acts as stabilizer of the octameric core (Luger et al. 1997). Numerous epigenetical processes, such as histone modifications, adenosine triphosphate (ATP) dependent chromatin remodeling, DNA methylation, histone variants placement and non-coding RNA mediated regulation, are operated to regulate the chromatin structure and function (Berger 2007). The unstructured amino-acid tail of each histone offers the sites for number of posttranslational modifications (PTMs) like methylation, ubiquitination, phosphorylation, ADP-ribosylation and acetylation (Berger 2007). One of the most significant and easily reversible PTMs is acetylation of histone, controlled by histone modifying enzymes called histone acetyltransferases (HATs) and histone deacetylases (HDACs) (Kuo and Allis 1998). Both HATs and HDACs have antagonistic activity which is ultimately accountable for the acetylation of histone. HATs adds the acetyl group of acetyl-CoA to the amino group of the lysine residue of the histones, whereas by the action of HDACs acetyl group from histones is eliminated (Hassig and Schreiber 1997; Ma et al. 2013). Histone hyperacetylation relaxes the DNA and thus makes the DNA sequence more accessible to the transcription factors leading to gene transcriptional activation on the other hand, a lesser acetylated DNA is more compact and this leads to repression of gene transcription (Struhl 1998).

5.1.1 Types of HATs

On the basis of the subcellular distribution, HATs have been classified into two classes: HAT-A and HAT-B (Brownell and Allis 1996; Roth et al. 2001). Type A HATs are found inside the nucleus and they add the acetyl group to the core histones of the nucleosome. As they have an important role in the gene regulation therefore they are often addressed as transcriptional co-activators. Type A HATs are further classified into five families that include the General control non-repressed protein 5 (GCN5)-related N terminal acetyltransferases (GNATs), MYST (MOZ, Ybf2/Sas3, Sas2 and Tip60), p300/CREB binding protein (CBP) and transcription initiation factor TAFII-250 [for TATA-binding protein (TBP)- associated factor]. Type B HATs are positioned in the cytoplasm and these acetylate the free histones (those which are not associated with DNA), only the lysine 5 and lysine 12 of histone H4 are acetylated by these (Verreault et al. 1998; Parthun et al. 1996). They are reported in maize (Lusser et al. 1999) and selectively acetylate histone H4 (Kölle et al. 1998).

5.2 Abiotic Stress and Epigenetics

There are plenty of evidences suggesting the importance of histone modification in regulation of plant responses to abiotic stresses (Chinnusamy et al. 2008; Liu et al. 2012). This chapter highlights the role of plant HATs in abiotic stress. The knowledge gained on this topic can further be used for investigation of acetylation and deacetylation of non-histone proteins of plants during abiotic stress response. Unraveling the role of HATs in plant's abiotic stress response will contribute to detailed understanding of plant's adaptability to the changing environment, which could be useful in improving the agricultural productivity.

Here in this segment of the chapter, we will discuss the significance of histone acetyltransferases that can increase the accessibility of the genomic DNA in response to different abiotic stresses.

5.2.1 Cold Stress

At different plant developmental stages, there is ~25% of crop yield is lost due to low temperature (Jeon and Kim 2013; Park et al. 2018). When plants are exposed to moderate and/or short nonfreezing temperatures they become capable of increasing their tolerance to freezing. This phenomenon is defined as Cold acclimation (Chinnusamy et al. 2006; Thomashow 1999). The consequences of cold acclimation are cell membrane stabilization and oxidative stress maintenance (Mahajan and Tuteja 2005). The expression of more than 100 genes are altered by CRT (C-repeat) binding factor (CBF) transcription factor, which are related to freezing tolerance (Fowler and Thomashow 2002; Vogel et al. 2005). Several studies have shown that the transcription stimulation of many cold-regulatory genes of *Arabidopsis* plant is done by binding of CBF transcription factor to the CRT/DRE regulatory elements present in their promoters (Stockinger et al. 2001). In order to do this, CBF1 transcription factor binds to GCN5, ADA2a and ADA2b proteins containing Spt-Ada-Gcn5-Acetyltransferase (SAGA) complex (Mao et al. 2006). In a study on maize, at total protein levels there was a reduction in acetylation at H3K9, H4K5 and H4K4 positions on short low temperatures treatments. A reversion of effects was observed on returning the plants to control temperature (Hu et al. 2011). The information gained from these reports provide us the basics behind the involvement of HAT in cold stress. This understanding can in turn help us to overcome the limiting factor like suitable geographical locations for agricultural and horticultural plants (Table 5.1).

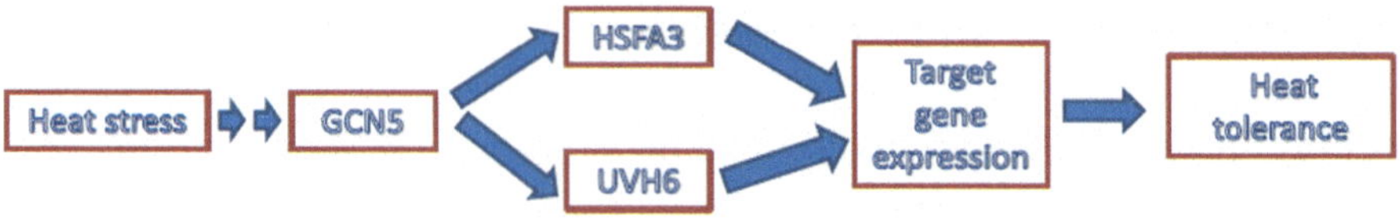

Fig. 5.1 Transcriptional cascade model under heat stress, regulated by GCN5

5.2.2 Heat Stress

Plants respond to heat stress by histone modification (acetylation and/or SUMOylation) and histone variant deposition. In thermal stress responses, H2A.Z deposition takes an important part. Studies have shown histone chaperon AtASF1A/B facilitates the acetylation of the H3K56 resulting in loss of nucleosome which in turn causes RNA pol II accumulation (Weng et al. 2014) as heat stress response. It is known that H2A.Z deposition in yeast requires histone acetylation (Watanabe et al. 2013), it is possible that plant H2A.Z or other histone variants deposition may also require histone modification upon heat stress. So what kind of histone modification and deposition contribute in response to heat stress has to be analyzed to uncover. In an another experiment, *Arabidopsis* leaves were excised and kept at high temperature (37 °C). Further comparative analysis of gene expression was done between SAGA components and respective controls. Thus, this study revealed useful evidences of participation of the SAGA complex in regulation of genes and plant stress responses (Srivastava et al. 2015). In another experiment AtGCN5 mutation led to serious thermotolerance defects and significant impairment in heat responsive gene expression was caused due to down regulation of related transcription factors *HSFA2*, *HSFA3*, *UV-HYPERSENSITIVE 6* (*UVH6*) and *Multiprotein Bridging Factor 1c* (*MBF1c*). In *Arabidopsis* AtGCN5 facilitates H3K9 and H3K14 acetylation at the promoter regions of *UVH6* and *HSFA3* genes under heat stress. This study suggested that GCN5 plays an important role in the thermotolerance preservation via versatile regulation in *Arabidopsis* (Hu et al. 2015; Scharf et al. 2012) (Fig. 5.1 and Table 5.1).

5.2.3 Water Stress

Water is important for generation of turgor pressure, transport system and plant metabolism (Des Marais and Juenger 2010). Apart from this water also has an adverse effect on plant growth aspects, for example, when the legumes are under water stress there is a reduction in the rate of nitrogen fixation by them and their symbionts (Gil-Quintana et al. 2013). The studies involving plant mutants deprived of histone-modifying enzymes gave more concert evidences for the significance of histone modification when the plant is under water stress. A number of reports have shown induction of changes in histone modifications in plants when treated with ABA (the stress hormone) or on sensing drought conditions (Kim et al. 2010; Yuan

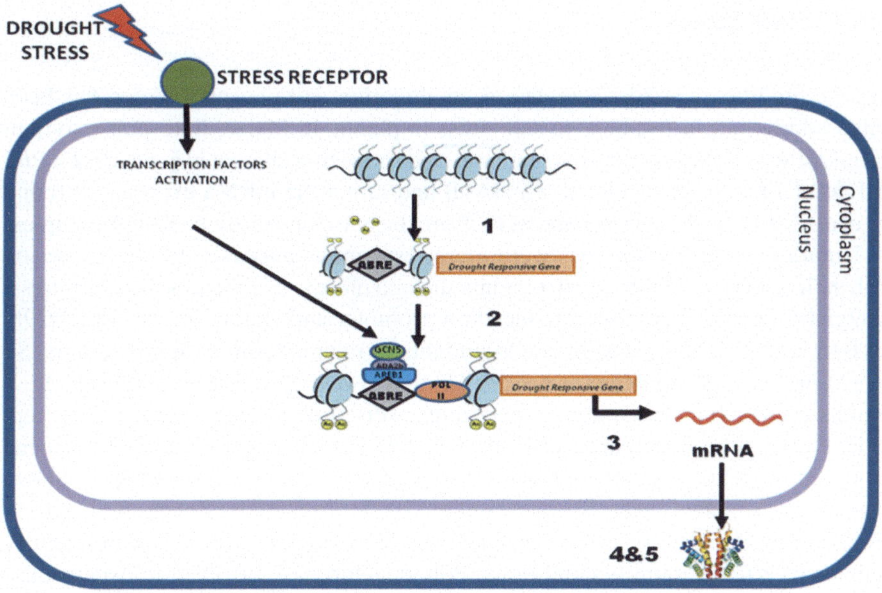

Fig. 5.2 This figure is showing an overview of epigenetic changes and signal transduction in a plant cell under drought stress. Step 1 Is showing DNA unwinding on histone acetylation at position H3K9/23/27, making the promoter of drought responsive genes available for transcription factors, step 2 is showing attachment of TF and assimilation of RNA pol II, step 3 is showing expression of drought responsive genes, step 4 and 5 are showing biochemical and physiological changes. AREB (ABRE binding factor). ABRE (abscisic acid-Responsive Element). Gnc5 (General control non-repressed protein 5)

et al. 2013). In cultured cells of tobacco and *Arabidopsis* there was an induction of acetylation of H4K14 and phosphorylation of H3S10 by salt stress or ABA application (Sokol et al. 2007). On giving a short drought treatment the 15 days old seedlings of *Arabidopsis* displayed H3K27, H3K9 and H3K23 acetylation enrichment at the drought stress-responsive gene's coding regions (Kim et al. 2008). The alterations in the modifications of histones are directed by the transcriptional changes (Zentner and Henikoff 2013). In addition to this, studies are required to clarify whether water stress triggers changes in transcription which in turn is an outcome of changes in histone PTMs. In another experiment, Li et al. (2019) revealed that in *Populus trichocarpa*, formation of a ternary complex comprising of abscisic acid-Responsive Element (ABRE) binding protein PtrAREB1-2, HAT unit ADA2b and GCN5, causes activation of drought responsive genes *PtrNAC006*, *PtrNAC007* and *PtrNAC120*. An increase in drought-sensitivity of *P. trichocarpa* was observed when down-regulation of any of the members of ternary protein complex (AREB1-ADA2b-GCN5) was done (Li et al. 2019) (Fig. 5.2 and Table 5.1).

5.2.4 Salt Stress

Histone modifications like acetylation, methylation and phosphorylation has been found taking part in response to salt stress in plants. In Maize, in response to high salinity stress, H3K9 acetylation increases and root cell wall related genes *ZmEXPB2* and *ZmXET1* were up regulated. ZmHATB and ZmGCN5 mRNA expressions found (Li et al. 2014). GCN5, ADA2 and SGF29 are the components of the SAGA complex. In *Arabidopsis*, sgf29a mutants were more resistant against salinity in comparison with their wild type. While ada2b mutants showed more sensitivity towards salt stress. Acetylation marks were found reduced on promoter and coding regions of RD29b, RAB18, and COR6.6 genes in ada2b mutants in relation with wild type plants. So Ada2b maintains acetylation of histone H3 and H4 on specific loci upon salt stress (Kaldis et al. 2011). In another experiment, excised *Arabidopsis* leaves were kept under high salt concentration (150 mM NaCl) for 24 h and compared the gene expression of SAGA components with their respective genes and found up regulated expression in treated ones (Srivastava et al. 2015). In a study, Zheng et al. (2019) revealed that the histone acetyltransferase GCN5 is required for cell wall integrity maintenance under salinity conditions. Polygalacturonase involved in expansion-3 (PGX3), chitinase-like gene CTL1 and MYB domain protein-54 (MYB54) genes are activated as direct target of GCN5 by acetylation of H3K9 and H3K14 of their respective promoters under salt stress. Taken together, GCN5 plays an important role in the preservation of salt tolerance in plants (Table 5.1).

5.2.5 Nutritional Stress

Zinc, manganese and iron are the components of plant nutrition and recently a report showed that histone acetyltransferase GCN5 takes part in accumulation of these minerals in the roots of *Arabidopsis* (Xing et al. 2015). GCN5 mutant exhibited iron translocation from root to shoot and this impaired translocation was rescued by treatment of histone deacetylase inhibitor, trichostatin-A (Xing et al. 2015). Under iron deficiency, GCN5 acetylates H3K9/14 of the promoters of ferric reductase defective3 (*FRD3*), a gene responsible for iron homeostasis (Table 5.1).

5.2.6 Light Stress

Different wavelengths of light is perceived by different photoreceptors of plants, like cryptochromes and phytochromes. These light signals are associated with downstream gene expression which are expressed through light responsive elements present in the promoters of these genes. HAF2 and GCN5 function together and acetylate light-inducible gene promoters (Bertrand et al. 2005; Benhamed et al. 2006).

Table 5.1 Table shows HATs activating different stress related genes through various acetylation marks

Stress	Hat involved	TF associated	Domains associated	Gene	Acetylation	References
Cold	GCN-5	CBF	CRT/DRE	COR (Cold stress responsive genes)	H3K9 H4K5 H4K4	Fowler and Thomashow (2002), Vogel et al. (2005), Stockinger et al. (2001), Mao et al. 2006, Hu et al. (2011)
Heat	GCN-5	DREB2A	DRE (dehydration-responsive element)	*UVH6* *HSFA3*	H3K9 H3K14	Scharf et al. (2012), Hu et al. (2015)
Water	GCN-5	PtrAREB1-2	ABRE motifs	*PtrNAC006* *PtrNAC007* *PtrNAC120*	H3K27 H3K9 H3K23	Li et al. (Li et al. 2019)
Salt	GCN-5	ND	MYB domain	*AtPGX3* *AtCTL1* *AtMYB54* *ZmEXPB2* *ZmXET1*	H3K9 H3K14	Zheng et al. (2019)
Nutrition	GCN-5	ND	ND	*FRD3*	H3K9 H3K14	Xing et al. (2015)
Light	GCN-5 TAFII-250 (HAF2)	HY5 HYH	G-box element	CAB2 RBCS-1A	H3 H4	Bertrand et al. (2005), Benhamed et al. (2006)

ND not described

Recently, a study has shown that expression of light-inducible genes is significantly reduced in *Arabidopsis* mutants with six SAGA subunits (Srivastava et al. 2015) (Table 5.1).

5.3 Conclusion

Chromatin remodeling is a conserved mechanism involved in stress responses. Alterations in the chromatin organization make the plants capable of dealing with multiple stresses. Gene expression studies and the acetylation levels of HATs have shown their role in stress responses. The continuously increasing human population and the ever-changing climatic conditions are resulting into great crop productivity loss. A detailed knowledge of molecular mechanisms regulating the stress responses and inducing stress tolerance in the plants is highly needed. The knowledge gained from this chapter along with the available high throughput techniques can give us a better insight into the interrelationship of HATs and abiotic stress. Future research can explore and emphasis on, (i) enhancing the heritable and/or primary stress tolerance, (ii) understanding of cellular and biochemical responses causing augmented stress tolerance, (iii) decoding the epigenetic machinery and thus providing us a better management for agricultural problems.

References

Benhamed M, Bertrand C, Servet C, Zhou DX (2006) Arabidopsis GCN5, HD1, and TAF1/HAF2 interact to regulate histone acetylation required for light-responsive gene expression. Plant Cell 18(11):2893–2903

Berger SL (2007) The complex language of chromatin regulation during transcription. Nature 447(7143):407

Bertrand C, Benhamed M, Li YF, Ayadi M, Lemonnier G, Renou JP, Delarue M, Zhou DX (2005) Arabidopsis HAF2 gene encoding TATA-binding protein (TBP)-associated factor TAF1, is required to integrate light signals to regulate gene expression and growth. J Biol Chem 280(2):1465–1473

Brownell JE, Allis CD (1996) Special HATs for special occasions: linking histone acetylation to chromatin assembly and gene activation. Curr Opin Genet Dev 6(2):176–184

Cantara WA, Crain PF, Rozenski J, McCloskey JA, Harris KA, Zhang X, Vendeix FA, Fabris D, Agris PF (2010) The RNA modification database, RNAMDB: 2011 update. Nucleic Acids Res 39(suppl_1):D195–D201

Chinnusamy V, Zhu J, Zhu JK (2006) Gene regulation during cold acclimation in plants. Physiol Plant 126(1):52–61

Chinnusamy V, Gong Z, Zhu JK (2008) Abscisic acid-mediated epigenetic processes in plant development and stress responses. J Integr Plant Biol 50(10):1187–1195

Des Marais DL, Juenger TE (2010) Pleiotropy, plasticity, and the evolution of plant abiotic stress tolerance. Ann NY Acad Sci 1206(1):56–79

Fowler S, Thomashow MF (2002) Arabidopsis transcriptome profiling indicates that multiple regulatory pathways are activated during cold acclimation in addition to the CBF cold response pathway. Plant Cell 14(8):1675–1690

Gil-Quintana E, Larrainzar E, Seminario A, Díaz-Leal JL, Alamillo JM, Pineda M, Arrese-Igor C, Wienkoop S, González EM (2013) Local inhibition of nitrogen fixation and nodule metabolism in drought-stressed soybean. J Exp Bot 64(8):2171–2182

Hassig CA, Schreiber SL (1997) Nuclear histone acetylases and deacetylases and transcriptional regulation: HATs off to HDACs. Curr Opin Chem Biol 1(3):300–308

Hu Y, Zhang L, Zhao L, Li J, He S, Zhou K, Yang F, Huang M, Jiang L, Li L (2011) Trichostatin A selectively suppresses the cold-induced transcription of the ZmDREB1 gene in maize. PLoS ONE 6(7):e22132

Hu Z, Song N, Zheng M, Liu X, Liu Z, Xing J, Ma J, Guo W, Yao Y, Peng H, Xin M (2015) Histone acetyltransferase GCN 5 is essential for heat stress-responsive gene activation and thermotolerance in Arabidopsis. Plant J 84(6):1178–1191

Jeon J, Kim J (2013) Cold stress signaling networks in Arabidopsis. J Plant Biol 56(2):69–76

Kaldis A, Tsementzi D, Tanriverdi O, Vlachonasios KE (2011) Arabidopsis thaliana transcriptional co-activators ADA2b and SGF29a are implicated in salt stress responses. Planta 233(4):749–762

Kim JM, To TK, Ishida J, Morosawa T, Kawashima M, Matsui A, Toyoda T, Kimura H, Shinozaki K, Seki M (2008) Alterations of lysine modifications on the histone H3 N-tail under drought stress conditions in Arabidopsis thaliana. Plant Cell Physiol 49(10):1580–1588

Kim JM, To TK, Nishioka T, Seki M (2010) Chromatin regulation functions in plant abiotic stress responses. Plant Cell Environ 33(4):604–611

Kölle D, Sarg B, Lindner H, Loidl P (1998) Substrate and sequential site specificity of cytoplasmic histone acetyltransferases of maize and rat liver. FEBS Lett 421(2):109–114

Kornberg RD, Lorch Y (1999) Twenty-five years of the nucleosome, fundamental particle of the eukaryote chromosome. Cell 98(3):285–294

Kuo MH, Allis CD (1998) Roles of histone acetyltransferases and deacetylases in gene regulation. BioEssays 20(8):615–626

Li H, Yan S, Zhao L, Tan J, Zhang Q, Gao F, Wang P, Hou H, Li L (2014) Histone acetylation associated up-regulation of the cell wall related genes is involved in salt stress induced maize root swelling. BMC Plant Biol 14(1):105

Li S, Lin YCJ, Wang P, Zhang B, Li M, Chen S, Shi R, Tunlaya-Anukit S, Liu X, Wang Z, Dai X (2019) The AREB1 transcription factor influences histone acetylation to regulate drought responses and tolerance in Populus trichocarpa. Plant Cell 31(3):663–686

Liu X, Luo M, Zhang W, Zhao J, Zhang J, Wu K, Tian L, Duan J (2012) Histone acetyltransferases in rice (*Oryza sativa L.*): phylogenetic analysis, subcellular localization and expression. BMC plant biology 12(1):145

Luger K, Mäder AW, Richmond RK, Sargent DF, Richmond TJ (1997) Crystal structure of the nucleosome core particle at 2.8 Å resolution. Nature 389(6648):251

Lusser A, Eberharter A, Loidl A, Schramel MG, Horngacher M, Haas H, Loidl P (1999) Analysis of the histone acetyltransferase B complex of maize embryos. Nucleic Acids Res 27(22):4427–4435

Ma X, Lv S, Zhang C, Yang C (2013) Histone deacetylases and their functions in plants. Plant Cell Rep 32(4):465–478

Mahajan S, Tuteja N (2005) Cold, salinity and drought stresses: an overview. Arch Biochem Biophys 444(2):139–158

Mao Y, Pavangadkar KA, Thomashow MF, Triezenberg SJ (2006) Physical and functional interactions of Arabidopsis ADA2 transcriptional coactivator proteins with the acetyltransferase GCN5 and with the cold-induced transcription factor CBF1. Biochim Biophys Acta (BBA)-Gene Struct Expr 1759(1–2):69–79

Park J, Lim CJ, Shen M, Park HJ, Cha JY, Iniesto E, Rubio V, Mengiste T, Zhu JK, Bressan RA, Lee SY (2018) Epigenetic switch from repressive to permissive chromatin in response to cold stress. Proc Natl Acad Sci 115(23):E5400–E5409

Parthun MR, Widom J, Gottschling DE (1996) The major cytoplasmic histone acetyltransferase in yeast: links to chromatin replication and histone metabolism. Cell 87(1):85–94
Richmond TJ, Davey CA (2003) The structure of DNA in the nucleosome core. Nature 423(6936):145
Roth SY, Denu JM, Allis CD (2001) Histone acetyltransferases. Annu Rev Biochem 70(1):81–120
Scharf KD, Berberich T, Ebersberger I, Nover L (2012) The plant heat stress transcription factor (Hsf) family: structure, function and evolution. Biochim Biophys Acta (BBA)-Gene Regul Mech 1819(2):104–119
Sokol A, Kwiatkowska A, Jerzmanowski A, Prymakowska-Bosak M (2007) Up-regulation of stress-inducible genes in tobacco and Arabidopsis cells in response to abiotic stresses and ABA treatment correlates with dynamic changes in histone H3 and H4 modifications. Planta 227(1):245–254
Srivastava R, Rai KM, Pandey B, Singh SP, Sawant SV (2015) Spt-Ada-Gcn5-Acetyltransferase (SAGA) complex in plants: genome wide identification, evolutionary conservation and functional determination. PLoS ONE 10(8):e0134709
Stockinger EJ, Mao Y, Regier MK, Triezenberg SJ, Thomashow MF (2001) Transcriptional adaptor and histone acetyltransferase proteins in Arabidopsis and their interactions with CBF1, a transcriptional activator involved in cold-regulated gene expression. Nucleic Acids Res 29(7):1524–1533
Struhl K (1998) Histone acetylation and transcriptional regulatory mechanisms. Genes Dev 12(5):599–606
Thomashow MF (1999) Plant cold acclimation: freezing tolerance genes and regulatory mechanisms. Annu Rev Plant Biol 50(1):571–599
Verreault A, Kaufman PD, Kobayashi R, Stillman B (1998) Nucleosomal DNA regulates the core-histone-binding subunit of the human Hat1 acetyltransferase. Curr Biol 8(2):96–108
Vogel JT, Zarka DG, Van Buskirk HA, Fowler SG, Thomashow MF (2005) Roles of the CBF2 and ZAT12 transcription factors in configuring the low temperature transcriptome of Arabidopsis. Plant J 41(2):195–211
Watanabe S, Radman-Livaja M, Rando OJ, Peterson CL (2013) A histone acetylation switch regulates H2A. Z deposition by the SWR-C remodeling enzyme. Science 340(6129):195–199
Weng M, Yang YUE, Feng H, Pan Z, Shen WH, Zhu YAN, Dong A (2014) Histone chaperone ASF1 is involved in gene transcription activation in response to heat stress in Arabidopsis thaliana. Plant Cell Environ 37(9):2128–2138
Xing J, Wang T, Liu Z, Xu J, Yao Y, Hu Z, Peng H, Xin M, Yu F, Zhou D, Ni Z (2015) GENERAL CONTROL NONREPRESSED PROTEIN5-mediated histone acetylation of FERRIC REDUCTASE DEFECTIVE3 contributes to iron homeostasis in Arabidopsis. Plant Physiol 168(4):1309–1320
Yuan L, Liu X, Luo M, Yang S, Wu K (2013) Involvement of histone modifications in plant abiotic stress responses. J Integr Plant Biol 55(10):892–901
Zentner GE, Henikoff S (2013) Regulation of nucleosome dynamics by histone modifications. Nat Struct Mol Biol 20(3):259
Zheng M, Liu X, Lin J, Liu X, Wang Z, Xin M, Yao Y, Peng H, Zhou DX, Ni Z, Sun Q (2019) Histone acetyltransferase GCN 5 contributes to cell wall integrity and salt stress tolerance by altering the expression of cellulose synthesis genes. Plant J 97(3):587–602

Chapter 6
Function of Plant Heat Shock Transcription Factors in Abiotic Stress

Sonal Mishra, Aksar Ali Chowdhary, Shakti Mehrotra and Vikas Srivastava

Abstract Plants under natural environment necessitate dealing with various abiotic constraints that have significant impact on plant growth and development. These abiotic constraints include low and high temperature, salinity, drought, chemical pollutants etc which adversely affect plant production potential. Plants respond to such discomfort by various phenological, physiological, biochemical and molecular alterations/mechanisms to minimize their negative impact. Many of these responses require high expression of stress-responsive genes mediated by various transcription factors. The heat shock transcription factor (Hsfs) is one such transcription factor that offers a crucial role in abiotic stress response by regulation of heat shock proteins (Hsps). Hsfs in plants are represented by high numbers as compared with other eukaryotes, thus giving more opportunity for Hsfs associated functions mainly in plant stress management. Taking the account of this background, the present chapter has been structured to cover aspects of plant Hsfs along with their contribution in abiotic stress management, which will offer a better understanding for managing adequate crop productivity.

Keywords Abiotic stress · Drought stress · Heat shock transcription factors (Hsfs) · Heat stress · Heavy metal stress · Salt stress

---

Sonal Mishra and Aksar Ali Chowdhary contributed equally to this work.

S. Mishra
School of Biotechnology, University of Jammu, Jammu 180006, India

A. A. Chowdhary · V. Srivastava (✉)
Department of Botany, Central University of Jammu, Samba 181143, India
e-mail: vikas.bot@cujammu.ac.in

S. Mehrotra
Department of Biotechnology, Institute of Engineering and Technology, Lucknow 226021, India

S. P. Singh et al. (eds.), *Molecular Approaches in Plant Biology and Environmental Challenges*, Energy, Environment, and Sustainability,
https://doi.org/10.1007/978-981-15-0690-1_6

6.1 Introduction

Plants' natural environment is the composite of abiotic and biotic stress. These stress lead to significant losses worldwide in agriculture production. Focussing on abiotic stress particularly drought, salinity, heat etc, and their management by plants are the matter of intense research (Mittler 2006). However, with unpredictable rain conditions and also due to environmental pollutants plants also face hypoxia/anoxia and heavy metal stress, respectively. Further, plants respond to these stress and acquire molecular, physiological, biochemical and phenological alterations to better survive under such stressful conditions. One of the effective strategies is to transcriptionally modulate the expression of stress-related genes that may be associated with transcriptional activator/inhibitor activity or any crucial element of the plant signalling network, such as hormonal signalling, MAPK-mediated signalling, ROS/NO/H_2S signalling, etc. Heat shock transcription factors (Hsfs) is one such protein that function as a terminal component of signal transduction chain and is responsible for activation of stress responsive genes under adverse environmental conditions. It recognizes conserved heat stress or heat shock promoter elements (HSEs; 5′-AGAAnnTTCT-3′) found in the promoters of heat stress inducible genes of all eukaryotes (Nover et al. 2001). Nover et al. (2001) reported more than 20 Hsfs in plants which is comparatively very high when compared to others life forms of distant relations for example vertebrates (four), *Drosophila* (one), *Caenorhabditis elegans* (one), and yeast (one Hsf plus three Hsf-related proteins) (Nover et al. 1996, 2001; Nakai 1999). Nevertheless, the recent surveys of many plant genomes demonstrated the range of plant Hsfs from 16 to 56, where least was observed in tea and the most in wheat (Xue et al. 2014; Liu et al. 2016; Guo et al. 2016). The Hsfs have been demonstrated to play significant role in various abiotic stress (Scharf et al. 2012). Most of these studies have been performed under individual stress conditions, however, the condition which prevails under nature that include synergistic effect of various abiotic stress factors such as drought, heat and salinity have not been considered (Sewelam et al. 2014). The Hsfs are one of the prime transcription factors identified against plant heat stress response. Besides, their contributions have also been reported for the regulation of other abiotic and biotic stress (Scharf et al. 2012; Ashraf et al. 2018). In certain cases, it offers contradictory stress tolerance for instance; the transgenics of *SlHsfA3*, *LlHsfA3A* and *LlHsfA3B* revealed high thermotolerance but at the same time they were hypersensitive to salinity (Li et al. 2013; Wu et al. 2018). However, clear evidence of Hsfs in improving salt tolerance in many plants justify them as an effective strategy also for salt tolerance development. Till date, the contribution of Hsfs in offering tolerance to several abiotic stress such as drought (Ma et al. 2016), salt (Ogawa et al. 2007; Pérez-Salamó et al. 2014), cold (Deng et al. 2018), anoxia (Banti et al. 2010), oxidative stress (Song et al. 2016), desiccation (Lang et al. 2017) and heavy metal stress (Cai et al. 2017) have been reported. In the present chapter, the details of the plant Hsfs structure followed by their functions under abiotic stress conditions have been presented in brief.

6.2 Structure of Plant Heat Shock Transcription Factors

Structurally, the fundamental architecture of plant Heat shock transcription factors (Hsfs) comprised of domains (DBD, DNA binding domain; OD, oligomerization domain; and RD, repressor domain), transcriptional activation motifs (AHA motif) and signals (NLS and NES). The distribution and variation of these motifs/domains within the protein sequence constitute the basis of the functionality of plant Hsfs. These plant Hsfs are classified into A, B and C subtype. The N terminal of Hsfs comprised of DBD, which typically have conserved central helix-turn-helix motif responsible for binding to heat stress elements (HSEs) on the promoters of stress-inducible genes and responsible for their regulation (Guo et al. 2016). Analysis of crystal and NMR solution structures of DBD of selected Hsfs from *Drosophila*, and yeast revealed that it is made up of three helical bundles (H1, H2 and H3) and four stranded antiparallel β-sheets (Harrison et al. 1994; Damberger et al. 1994; Vuister et al. 1994; Schultheiss et al. 1996). The DBD is attached to OD (oligomerization domain or HR-A/B region) with the flexible linker of amino acid stretch (15–80 aa). The OD domain is a bipartite heptad pattern of hydrophobic amino acid residues (HR-A/B region) (Baniwal et al. 2004). The Hsfs—flexible linker is of 9–39 amino acid residues for class A, 50–78 amino acid residues for class B, and 14–49 amino acid residues for class C. Further, the HR-A/B of class B Hsfs is comparatively compact and resemble with non-plant Hsfs; however, extended HR-A/B region is present in class A and C with insertion of 21 and 7 amino acids residues in between A and B region, respectively (Nover et al. 2001; Scharf et al. 2012; Ashraf et al. 2018). The activation domain of Hsfs comprises of short peptide motif (AHA motif, comprised of aromatic, hydrophobic and acidic aa) at C-terminal. Here, also the subtypes have variation, in which the Hsf-A motif is formed of aromatic, large hydrophobic, and acidic amino acid residues (Döring et al. 2000; Kotak et al. 2004). Besides, the members of class B (Excluding HsfB5) of Hsfs characteristically exhibit tetrapeptide—LFGV as repressor domain at C-terminal (Ikeda and Ohme-Takagi 2009; Fragkostefanakis et al. 2015). The presence of both nuclear localization signal (NLS) and nuclear export signal (NES) of Hsfs helps in the assembly of nuclear import complex (Baniwal et al. 2004; Görlich and Kutay 1999; Heerklotz et al. 2001).

6.3 Hsfs in Abiotic Stress Response

The contribution of heat shock transcription factors have been demonstrated in many abiotic stresses (Table 6.1; Fig. 6.1). Some of the important interventions in this direction are as follows;

Table 6.1 Recent studies related to heat shock transcription factor and abiotic stress tolerance (2015 onwards)

Plant system	Gene	Response to stress	References
Arabidopsis thaliana	*HsfAIb*	Studies with over-expressed lines revealed its contribution in regulation of multiple genes and stress conditions	Albihlal et al. (2018)
A. thaliana	*HsfA3*	Over expression of this gene increased oxidative stress tolerance in *Arabidopsis*	Song et al. (2016)
Brassica napus	*BnHsfA4a*	Over expression enhanced desiccation tolerance in seeds of *Arabidopsis*	Lang et al. (2017)
Chrysanthemum morifolium	*CmHsfA4*	Over expression enhanced salinity stress tolerance in transgenic *chrysanthemum*	Li et al. (2018)
Cicer arietinum	*CarHsfB2*	Over expression improved tolerance to drought and heat stress in transgenic *Arabidopsis*	Ma et al. (2016)
Cyanodon transvaalensis	*CtHsfA2b*	Expression of this gene confers heat tolerance in *Arabidopsis*	Wang et al. (2016)
Festucabarundinaceae	*FaHsfC1b*	Over expression confer heat tolerance in *Arabidopsis thaliana*	Zhuang et al. (2018)
F.arundinaceae	*FaHsfA2C*	Over expression increased thermo tolerance in *Arabidopsis* and fescue	Wang et al. (2017)
Helianthus annuus	*HaHsfA9* and *HaHsfA4a*	Over expression improved tolerance to drastic dehydration and oxidative stress in transgenic tobacco	Almoguera et al. (2015)
H.annuus	*HaHsfA9* and *HaHsfA4a*	Co-over expression enhanced seed longevity and in synergistic effects on seedling tolerance to serve dehydration and oxidative stress	Personat et al. (2014)
Lilium longiflorum	*LIHsfA2b*	Over expression enhance tolerance to heat and oxidative stress in transgenic *Arabidopsis* seedling	Xin et al. (2017)
L.longiflorum	*LIHsfA1*	Over expression enhance thermo tolerance by interacting with LiHSFA2 of transgenic *A. thaliana*	Gong et al. (2014)
L.longiflorum	*LIHsfA3* and *LIHsfA3B*	Over expression improved thermo tolerance but hypersensitivity to salt stress in *Arabidopsis*	Wu et al. (2018)
Manihotesculanta	*MeHsf3*	Transient expression in leaves of *Nicotiana benthamiana* and virus-induced gene silencing (VIGS) in cassava revealed its role in plant disease resistance	Wei et al. (2018)

(continued)

Table 6.1 (continued)

Plant system	Gene	Response to stress	References
Oryza sativa	*OsHsfA2dI*	Involved in re-establishment of cellular protein homeostasis under heat stress in rice	Cheng et al. (2015)
Hevea brasiliensis	*HbHsfA1* and *HbHsfB1*	Over expression enhance cold stress tolerance in *Saccharomyces cerevisiae*	Deng et al. (2018)
Solanum lycopersicum	*HsfA1a*	Over expression of this gene increased cadmium tolerance in tomato plants	Cai et al. (2017)
Triticum aestivum	*TaHsfC2a-B*	Over expression improved thermo tolerance in transgenic wheat but not dehydration tolerance	Hu et al. (2018)
T.aestivum	*TaHsfA6e*	Involved in the regulation of heat shock proteins under multiple stresses (heat and drought)	Kumar et al. (2018)
T.aestivum	*HsfA4A*	Expression improved growth and tolerance of wheat against salt stress	Iranbakhsh et al. (2018)
Zea mays	*ZmHsf05*	Increased thermotolerance was observed in *A. thaliana* expressing *ZmHsf05*; Rescues thermo tolerance defects of *A. thaliana* athsfa2 mutant	Li et al. (2019)
Z.mays	*ZmHsf04*	Over expression increased salt and thermo tolerance in transgenic *A. thaliana*	Jiang et al. (2018)
Z.mays	*ZmHsf06*	Over expression of this gene enhance the thermo and drought tolerance in transgenic *A. thaliana*	Li et al. (2015)

6.3.1 Heat Stress

Due to continuous increase in average global temperature, plants are bound to deal with high heat or temperature (HT) stress. The high heat affects the plant physiology and metabolism and creates a negative impact on plant growth and development. Usually, under natural conditions, HT stress mostly accompanied with drought and salt stress, thus create a synergistic impact. Plants manage this negative impact through phenological, physiological and biochemical makeshifts, which also involve various adjustments at molecular levels. The Hsfs is one such molecular candidate that assists plant HT management system by regulation of a number of stress responsive genes. In fact, studies on tomato and *Arabidopsis* demonstrated HsfA1 as a master regulator for heat stress response (Liu et al. 2011; Mishra et al. 2002). In tomato, transgenic plants were generated with altered *HsfA1* expression (Mishra et al. 2002). The *HsfA1* co-suppressed plants were similar to wild plants in normal growth parameters, but were extremely sensitive to elevated temperatures. This sensitivity was supposed to be due to reduction or lack of heat stress-induced synthesis of chaperones and Hsfs,

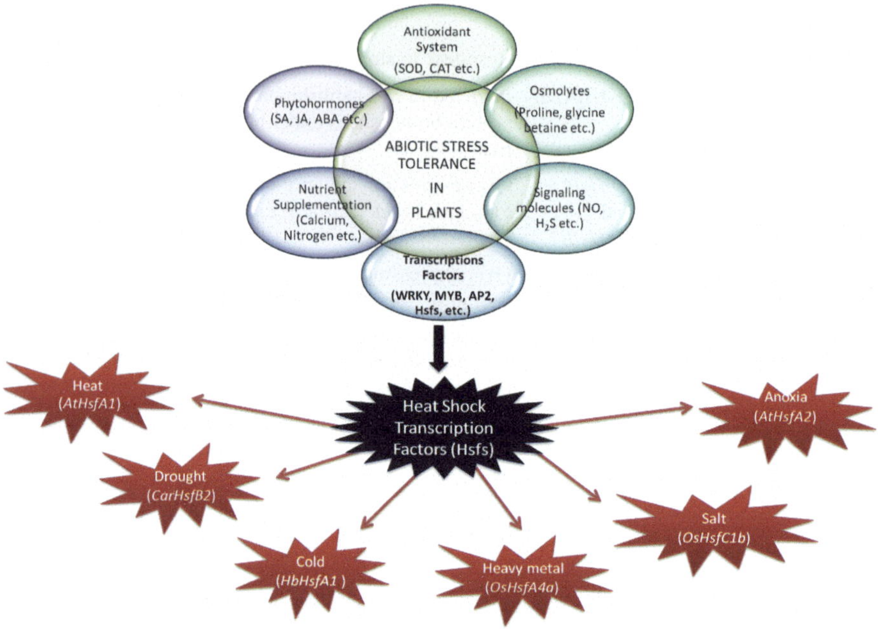

Fig. 6.1 Approaches for abiotic stress tolerance development in plants and contribution of plant Hsfs in abiotic stress management

thus demonstrating master regulation of HsfA1 for induced thermotolerance. Another evidence for HsfA1 role as master regulator in heat stress response was reported in *Arabidopsis* (Liu et al. 2011). *Arabidopsis* has four homologs of class A1 and studies with double mutant could not prove its key property because of functional redundancy (Lohmann et al. 2004). However, the quadruple KO (*QK*) and four triple KO mutants were able to analyse their functions. With these mutants, it was observed that basal and acquired thermotolerance is noticeably reduced in the *QK* mutant but has been varied in triple KO mutants at diverse developmental stages. Further, the transcriptome profile has suggested that more than 65% of heat stress regulated genes are HsfA1 dependent (Liu et al. 2011). Additionally, HsfA1 has also found to be associated with the expression of several heat stress genes induced by H_2O_2, salt and mannitol. Shi et al. (2015) have observed heat stress induced endogenous melatonin (N-acetyl-5-methoxytryptamine) production in *Arabidopsis* leaves (Shi et al. 2015). The exogenous application of melatonin has exhibited improved thermotolerance, and both heat stress and exogenous melatonin treatment in *Arabidopsis* can up-regulate the expression of the master regulator *AtHsfA1*. Additionally, the exogenously applied melatonin-enhanced thermotolerance has been found to be mostly alleviated in *HsfA1s* quadruple knockout (QK) mutants; further, *HsfA1s*-activated transcripts of heat-responsive genes possibly contributed to melatonin-mediated thermotolerance. Similarly, the overexpression of *GmHsfA1* enhances thermotolerance in transgenic soybeans by the activation of downstream Hsps under heat stress (Zhu et al. 2006).

Out of 21 *Arabidopsis Hsfs, HsfA2* has exhibited the strongest expression under heat stress and its accumulation is similar to other heat shock proteins (Hsps) (Schramm et al. 2006). The transcriptome of *HsfA2* knockout lines under heat stress has demonstrated *APX2* as most affected transcript along with Hsps, genes of Hsp70 and Hsp100 family, as well as many transcripts of unknown functions. Further, investigations have suggested that *HsfA2* mitigates oxidative damage caused by heat stress in *Arabidopsis* (Zhang et al. 2009). Evrard et al. (2013) have shown that heat stress-induced *Arabidopsis* mitogen-activated protein kinase 6 (MPK6) specifically targets HsfA2 and thus, participate in the complex regulatory mechanism in heat stress response. Chen et al. (2010) has demonstrated that *Arabidopsis HsfA3* is regulated by *DREB2A (Dehydration-Responsive Element Binding Protein 2A)* and both proteins are the key players in regulating the heat tolerance of *Arabidopsis*.

6.3.2 Salt Stress

The Hsfs have been shown to regulate salt tolerance in plants both positively and negatively (Li et al. 2013; Wu et al. 2018; Yokotani et al. 2008; Hwang et al. 2014). The transgenic *Arabidopsis* expressing *OsHsfA2e* have shown to exhibits tolerance to high-salinity stress (Yokotani et al. 2008). Schmidt et al. (2012) have observed that *OsHsfC1b* participates in ABA-mediated salt stress tolerance in rice and is also entailed during osmotic stress response and plant growth under non-stress conditions (Schmidt et al. 2012). Expression of *Populus euphratica* class A, *PeHsf* has been induced under high saline conditions in leaves and callus cultures. In callus, this up-regulation is significantly inhibited by DPI (an inhibitor of plasma membrane NADPH oxidase) and $LaCl_3$ (an inhibitor of plasma membrane Ca^{2+}-permeable channels). In transgenic tobacco, the improvement in seed germination and root growth has been observed under saline conditions which also exhibited enhanced antioxidant enzyme activities (Shen et al. 2013). The overexpression lines of *AtHsfA6a* were hypersensitive to ABA and exhibited higher tolerance to salt and drought stresses (Hwang et al. 2014). The *ZmHsf04* (*HsfA2*) of maize is strongly induced under heat stress and function in heat/salt stress response (Jiang et al. 2018). This nuclear protein requires AHA2 domain for its transcriptional activity. The *ZmHsf04* over-expressed *Arabidopsis* (Col-0) exhibited thermotolerance, salt tolerance and increase sensitivity to ABA. The transgenic *Arabidopsis* also demonstrated higher expression of heat-specific Hsp genes (*AtHsp25.3-P, AtHsp18.2-CI*, and *AtHsp70B*) and stress-related genes (*AtAPX2* and *AtGolS1*). Pérez-Salamó et al. (2014) have demonstrated the contribution of *Arabidopsis* HsfA4 to enhance the tolerance to salt and oxidative agents. Further, HsfA4 is phosphorylated by MPK3 and MPK6 mediated phosphorylation (Pérez-Salamó et al. 2014). Contrary to this, there are also studies that demonstrate the role of Hsfs in salt sensitivity. For instance, *OsHsfB2b* overexpression in rice significantly decreased drought and salt tolerance, which was enhanced in *OsHsfB2b*-RNAi transgenic rice (Xiang et al. 2013). The ectopically expressed tomato *SlHsfA3* conferred enhanced thermotolerance but salt hypersensitivity in transgenic *Arabidopsis* (Li et al. 2013). Similarly, two heat-inducible *HsfA3* homologs of lily (*LlHsfA3A* and *LlHsfA3B*), when overexpressed in *Arabidopsis* have shown to enhance the thermotolerance and hypersensitivity to salt stress, and implicate proline catabolism in this function (Wu et al. 2018).

6.3.3 Drought Stress

Though, Hsfs are known to offer heat stress tolerance in plants, still, some of the studies do advocate its role under drought conditions. Bechtold et al. (2013) have reported that the overexpression of *Arabidopsis HsfA1b* overexpression improves resistance to drought, and infection. Nonetheless, *HsfA1b* over-expression effect on drought/dehydration tolerance does not entail changes in the expression of *DREB2A* or ABA or dehydration-responsive genes (Bechtold et al. 2013). The chickpea *CarHsfB2* shows no activation due to lack of aromatic, hydrophobic, and acidic amino acid (AHA) motifs. Differential expression of this gene has been observed during various developmental processes (leaf senescence, developing seed, and embryo of germinating seed) and it was also induced by abiotic stress factors (heat, salt, wound and drought), plant growth hormones (IAA, and GA_3) and oxidative (H_2O_2) stress. Among growth regulators, 6-BA inhibited expression and treatments with ABA, MeJA, Et, and SA had no effect on expression. The *CarHsfB2* over-expressed *Arabidopsis* seedlings has exhibited increased tolerance to drought and heat, in which higher expression of stress-responsive genes (*RD22*, *RD26*, and *RD29A*) was also observed under drought and Hsfs (*HsfA2*, *HsfB2a*, and *HsfA7a*) were observed under heat stress (Ma et al. 2016). The *Vitis pseudoreticulata VpHSF1* negatively regulates PEG6000 induced osmotic stress (Peng et al. 2013).

6.3.4 Heavy Metal Stress

The heat shock transcription factors offer heavy metal tolerance particularly cadmium (Cd), a widespread soil pollutant. The wheat *HsfA4a* has been shown to bestow Cd tolerance to a Cd hypersensitive yeast strain. Its closest ortholog in rice *OsHsfA4a* also confers Cd tolerance in yeast (Shim et al. 2009). The *TaHsf4a* over-expression and knocked-down expression in rice plants have revealed the enhanced and reduced tolerance, respectively. Further, DNA binding domain (especially Ala-31 and Leu-42) of Hsf4a has been found to be crucial for Cd tolerance and TaHsf4a mediated Cd tolerance requires metallothionein (Shim et al. 2009). Cai et al. (2017) have observed cadmium tolerance in tomato by *HsfA1*-mediated induction of melatonin biosynthesis. *HsfA1a*-silenced plants have exhibited decreased Cd tolerance, which is enhanced by its overexpression. Further, *HsfA1a*-silenced plants demonstrated a low accumulation of melatonin. Contrary to this, *HsfA1a*-overexpressed lines accumulated melatonin and exhibit induced expression of caffeic acid O-methyltransferase 1 (COMT1) under Cd stress (Cai et al. 2017). Subsequent studies have revealed that HsfA1a binds to COMT1 promoter. In addition, the Hsps induction under Cd stress is compromised in *HsfA1a*-silenced plants which were induced in *HsfA1a*-overexpressing plants under Cd stress. The silencing of *COMT1* in *HsfA1a*-overexpressing plants compromised *HsfA1a*-induced Cd tolerance, melatonin accumulation and also *HsfA1a*-induced expression of *Hsps* (Cai et al. 2017). Crucial role of Hsf/Hsp networks in switchgrass (*Panicum virgatum* L.) Cd tolerance was observed by Song et al. (2018). The comparative transcriptome analysis of Cd-treated switchgrass roots suggested higher expression of majority of Hsfs and Hsps after Cd treatment (Song et al. 2018). Chen et al. (2018) have investigated the contribution of Hsf in heavy metal hyperaccumulator *Sedum alfredii* Hance. Out of 22 *SaHsfs*, 18 were found responsive to Cd stress along with many *SaHsps* in transcriptome under Cd stress. Further, the overexpression of *SaHsfA4a* and *SaHsfA4c* in transgenic yeast demonstrated Cd tolerance and accumulation (Chen et al. 2018).

6.3.5 Cold/Chilling Stress

Zhang et al. (2013) have observed up-regulation of *TaHsf3* in wheat seedlings by high and low temperatures, and to a lesser extent by drought, salt and ABA. Its over-expression in *Arabidopsis* exhibited enhanced tolerance under heat treatment with survival frequency of 75–91% and freezing treatment with 85–95% survival frequency as compared to 25 and 10%, respectively in wild *Arabidopsis* (Zhang et al. 2013). Deng et al. (2018) while working on cold-resistant rubber tree clone ‘93–114’ have identified two cold-inducible *HbHsfA1* and *HbHsfB1* genes. Out of them, HbHsfA1 protein localized in nucleus exhibited strong transcriptional activation in yeast. Further, these genes have demonstrated higher expression in cold-tolerant rubber tree clone ‘93–114’ as compared to cold-sensitive clone ‘Reken501’ and enhanced cold stress tolerance in yeast (Deng et al. 2018).

6.3.6 Anoxia Stress

It has been observed that anoxia induces several Hsps and mild heat acclimatize *Arabidopsis* seedling to anoxia stress. The *AtHsfA2* strongly induced during anoxia and this heat-dependent anoxia acclimation is vanished in an *HsfA2* knockout mutant as well as in *HsfA1a/HsfA1b* double mutant, suggesting their cooperation to confer anoxia tolerance (Banti et al. 2010). Further, overexpression of *HsfA2* demonstrated more tolerance to anoxic and submergence condition (Banti et al. 2010). Such studies demonstrate the possibility of overlapping molecular mechanisms of heat and anoxia tolerance mediated by HsfA2.

6.4 Conclusion

The simple fluctuations in the plants' natural environment may lead plant under stress, which severely impact its productivity/survival. To encounter this, plants devise several mechanism such as phytohormone and other signalling, antioxidant system, osmo-protectants, effective nutrient management, to manage any negative consequences. Apart from this, there are also transcription factors (viz, WRKY, MYB, AP2, etc.) that regulate many of the genes associated with several mechanism of stress management, together called as stress-responsive genes. The Hsfs, a well-claimed master regulator of heat tolerance, is now one of the known accepted TF family found associated with tolerance mechanism to variety of other abiotic stress conditions such as salt, drought, heavy metal, cold, anoxia etc. However, the relevance other than heat is still not much explored and need further attention. Therefore, it is imperative to continue such studies to identify further novel functions to better understand heat shock transcription factors.

Acknowledgements S.M. is thankful to Science & Engineering Research Board (SERB, India) for NPDF (PDF/2016/002133) fellowship. A.K.C. and V.S. acknowledges Central University of Jammu for providing the work facility. V.S. is also thankful to UGC, India for UGC-BSR Research Start Up Grant.

References

Albihlal WS, Obomighie I, Blein T, Persad R, Chernukhin I, Crespi M, Bechtold U, Mullineaux PM (2018) *Arabidopsis* HEAT SHOCK TRANSCRIPTION FACTORA1b regulates multiple developmental genes under benign and stress conditions. J Exp Bot 69(11):2847–2862

Almoguera C, Personat JM, Prieto-Dapena P, Jordano J (2015) Heat shock transcription factors involved in seed desiccation tolerance and longevity retard vegetative senescence in transgenic tobacco. Planta 242(2):461–475

Ashraf MF, Yang S, Wu R, Wang Y, Hussain A, Noman A, Khan MI, Liu Z, Qiu A, Guan D, He S (2018) *Capsicum annuumHsfB2a* positively regulates the response to *Ralstoniasolanacearum* infection or high temperature and high humidity forming transcriptional cascade with *CaWRKY6* and *CaWRKY40*. Plant Cell Physiol 59(12):2608–2623

Baniwal SK, Bharti K, Chan KY, Fauth M, Ganguli A, Kotak S (2004) Heat stress response in plants: a complex game with chaperones and more than twenty heat stress transcription factors. J Biosci 29:471–487

Banti V, Mafessoni F, Loreti E, Alpi A, Perata P (2010) The heat-inducible transcription factor *HsfA2* enhances anoxia tolerance in *Arabidopsis*. Plant Physiol 152(3):1471–1483

Bechtold U, Albihlal WS, Lawson T, Fryer MJ, Sparrow PA, Richard F, Persad R, Bowden L, Hickman R, Martin C, Beynon JL (2013) Arabidopsis HEAT SHOCK TRANSCRIPTION FACTORA1b overexpression enhances water productivity, resistance to drought, and infection. J Exp Bot 64(11):3467–3481

Cai SY, Zhang Y, Xu YP, Qi ZY, Li MQ, Ahammed GJ, Xia XJ, Shi K, Zhou YH, Reiter RJ, Yu JQ (2017) *HsfA1a* upregulates melatonin biosynthesis to confer cadmium tolerance in tomato plants. J Pineal Res 62(2):e12387

Chen H, Hwang JE, Lim CJ, Kim DY, Lee SY, Lim CO (2010) Arabidopsis DREB2C functions as a transcriptional activator of *HsfA3* during the heat stress response. Biochem Biophys Res Commun 401(2):238–244

Chen SS, Jiang J, Han XJ, Zhang YX, Zhuo RY (2018) Identification, expression analysis of the Hsf family, and characterization of class A4 in *Sedum alfredii*Hance under cadmium stress. Int J Mol Sci 19(4):1216

Cheng Q, Zhou Y, Liu Z, Zhang L, Song G, Guo Z, Wang W, Qu X, Zhu Y, Yang D (2015) An alternatively spliced heat shock transcription factor, OsHSFA 2dI, functions in the heat stress-induced unfolded protein response in rice. Plant Biology 17(2):419–429

Damberger FF, Pelton JG, Harrison CJ, Nelson HC, Wemmer DE (1994) Solution structure of the DNA-binding domain of the heat shock transcription factor determined by multidimensional heteronuclear magnetic resonance spectroscopy. Protein Sci 3(10):1806–1821

Deng X, Wang J, Wang J, Tian W (2018) Two *HbHsfA1* and *HbHsfB1* genes from the tropical woody plant rubber tree confer cold stress tolerance in *Saccharomyces cerevisiae*. Braz J Bot 41(3):711–724

Döring P, Treuter E, Kistner C, Lyck R, Chen A, Nover L (2000) The role of AHA motifs in the activator function of tomato heat stress transcription factors HsfA1 and HsfA2. Plant Cell 12(2):265–278

Evrard A, Kumar M, Lecourieux D, Lucks J, von Koskull-Döring P, Hirt H (2013) Regulation of the heat stress response in *Arabidopsis* by MPK6-targeted phosphorylation of the heat stress factor *HsfA2*. Peer J 1:e59

Fragkostefanakis S, Röth S, Schleiff E, Scharf KD (2015) Prospects of engineering thermotolerance in crops through modulation of heat stress transcription factor and heat shock protein networks. Plant Cell Environ 38:1881–1895

Gong B, Yi J, Wu J, Sui J, Khan MA, Wu Z, Zhong X, Seng S, He J, Yi M (2014) LIHSFA1, a novel heat stress transcription factor in lily (*Liliumlongiflorum*), can interact with *LlHSFA2* and enhance the thermotolerance of transgenic *Arabidopsis thaliana*. Plant Cell Rep 33(9):1519–1533

Görlich D, Kutay U (1999) Transport between the cell nucleus and the cytoplasm. Annu Rev Cell Dev Biol 15:607–660

Guo M, Liu JH, Ma X, Luo DX, Gong ZH, Lu MH (2016) The plant heat stress transcription factors (HSFs): structure, regulation, and function in response to abiotic stresses. Front Plant Sci 7:114

Harrison CJ, Bohm AA, Nelson HC (1994) Crystal structure of the DNA binding domain of the heat shock transcription factor. Science 263(5144):224–227

Heerklotz D, Döring P, Bonzelius F, Winkelhaus S, Nover L (2001) The balance of nuclear import and export determines the intracellular distribution and function of tomato heat stress transcription factor *HsfA2*. Mol Cell Biol 21:1759–1768

Hu XJ, Chen D, Lynne Mclntyre C, Fernanda Dreccer M, Zhang ZB, Drenth J, Kalaipandian S, Chang H, Xue GP (2018) Heat shock factor C2a serves as a proactive mechanism for heat protection in developing grains in wheat via an ABA-mediated regulatory pathway. Plant Cell Environ 41(1):79–98

Hwang SM, Kim DW, Woo MS, Jeong HS, Son YS, Akhter S, Choi GJ, Bahk JD (2014) Functional characterization of Arabidopsis HsfA6a as a heat-shock transcription factor under high salinity and dehydration conditions. Plant Cell Environ 37(5):1202–1222

Ikeda M, Ohme-Takagi M (2009) A novel group of transcriptional repressors in *Arabidopsis*. Plant Cell Physiol 50:970–975

Iranbakhsh A, Ardebili NO, Ardebili ZO, Shafaati M, Ghoranneviss M (2018) Non-thermal plasma induced expression of heat shock factor A4A and improved wheat (*Triticumaestivum* L.) growth and resistance against salt stress. Plasma Chem Plasma Process 38(1):29–44

Jiang Y, Zheng Q, Chen L, Liang Y, Wu J (2018) Ectopic overexpression of maize heat shock transcription factor gene ZmHsf04 confers increased thermo and salt-stress tolerance in transgenic Arabidopsis. Acta Physiol Plant 40(1):9

Kotak S, Port M, Ganguli A, Bicker F, Koskull–Döring V (2004) Characterization of C-terminal domains of *Arabidopsis* heat stress transcription factors (Hsfs) and identification of a new signature combination of plant class A Hsfs with AHA and NES motifs essential for activator function and intracellular localization. Plant J 39:98–112

Kumar RR, Goswami S, Singh K, Dubey K, Rai GK, Singh B, Singh S, Grover M, Mishra D, Kumar S, Bakshi S (2018) Characterization of novel heat-responsive transcription factor (*TaHSFA6e*) gene involved in regulation of heat shock proteins (HSPs)—a key member of heat stress-tolerance network of wheat. J Biotechnol 279:1–12

Lang S, Liu X, Xue H, Li X, Wang X (2017) Functional characterization of *BnHSFA4a* as a heat shock transcription factor in controlling the re-establishment of desiccation tolerance in seeds. J Exp Bot 68(9):2361–2375

Li Z, Zhang L, Wang A, Xu X, Li J (2013) Ectopic overexpression of SlHsfA3, a heat stress transcription factor from tomato, confers increased thermotolerance and salt hypersensitivity in germination in transgenic *Arabidopsis*. PLoS ONE 8(1):e54880

Li HC, Zhang HN, Li GL, Liu ZH, Zhang YM, Zhang HM, Guo XL (2015) Expression of maize heat shock transcription factor gene *ZmHsf06* enhances the thermotolerance and drought-stress tolerance of transgenic Arabidopsis. Funct Plant Biol 42(11):1080–1091

Li F, Zhang H, Zhao H, Gao T, Song A, Jiang J, Chen F, Chen S (2018) *Chrysanthemum CmHSFA* 4 gene positively regulates salt stress tolerance in transgenic *chrysanthemum*. Plant Biotechnol J 16(7):1311–1321

Li GL, Zhang HN, Shao H, Wang GY, Zhang YY, Zhang YJ, Zhao LN, Guo XL, Sheteiwy MS (2019) ZmHsf05, a new heat shock transcription factor from *Zea mays* L. improves thermotolerance in *Arabidopsis thaliana* and rescues thermotolerance defects of the athsfa2 mutant. Plant Sci 283:375–384

Liu HC, Liao HT, Charang YY (2011) The role of class A1 heat shock factors (HSFA1s) in response to heat and other stresses in *Arabidopsis*. Plant Cell Environ 34(5):738–751

Liu ZW, Wu ZJ, Li XH, Huang Y, Li H, Wang YX (2016) Identification, classification, and expression profiles of heat shock transcription factors in tea plant (*Camellia sinensis*) under temperature stress. Gene 576:52–59

Lohmann C, Eggers-Schumacher G, Wunderlich M, Schöffl F (2004) Two different heat shock transcription factors regulate immediate early expression of stress genes in *Arabidopsis*. Mol Genet Genomics 271:11–21

Ma H, Wang C, Yang B, Cheng H, Wang Z, Mijiti A, Ren C, Qu G, Zhang H, Ma L (2016) *CarHSFB2*, a class B heat shock transcription factor, is involved in different developmental processes and various stress responses in chickpea (*Cicer arietinum* L.). Plant Mol Biol Rep 34(1):1–14

Mishra SK, Tripp J, Winkelhaus S, Tschiersch B, Theres K, Nover L (2002) In the complex family of heat stress transcription factors, HsfA1 has a unique role as master regulator of thermotolerance in tomato. Genes Dev 16:1555–1567

Mittler R (2006) Abiotic stress, the field environment and stress combination. Trends Plant Sci 11(1):15–19

Nakai A (1999) New aspects in the vertebrate heat stress factor system: *HsfA3* and HsfA4. Cell Stress Chaperones 4:86–93

Nover L, Scharf KD, Gagliardi D, Vergne P, Czarnecka-Verner E, Gurley WB (1996) The Hsf world: classification and properties of plant heat stress transcription factors. Cell Stress Chaperones 1:215–223

Nover L, Bharti K, Döring P, Mishra SK, Ganguli A, Scharf KD (2001) *Arabidopsis* and the heat stress transcription factor world: how many heat stress transcription factors do we need? Cell Stress Chaperones 6(3):177

Ogawa D, Yamaguchi K, Nishiuchi T (2007) High-level overexpression of the *Arabidopsis HsfA2* gene confers not only increased themotolerance but also salt/osmotic stress tolerance and enhanced callus growth. J Exp Bot 58:3373–3383

Peng S, Zhu Z, Zhao K, Shi J, Yang Y, He M, Wang Y (2013) A novel heat shock transcription factor, VpHsf1, from Chinese wild *Vitispseudoreticulata* is involved in biotic and abiotic stresses. Plant Mol Biol Rep 31(1):240–247

Pérez-Salamó I, Papdi C, Rigó G, Zsigmond L, Vilela B, Lumbreras V, Nagy I, Horváth B, Domoki M, Darula Z, Medzihradszky K (2014) The heat shock factor A4A confers salt tolerance and is regulated by oxidative stress and the mitogen-activated protein kinases MPK3 and MPK6. Plant Physiol 165(1):319–334

Personat JM, Tejedor-Cano J, Prieto-Dapena P, Almoguera C, Jordano J (2014) Co-overexpression of two heat shock factors results in enhanced seed longevity and in synergistic effects on seedling tolerance to severe dehydration and oxidative stress. BMC Plant Biol 14(1):56

Scharf KD, Berberich T, Ebersberger I, Nover L (2012) The plant heat stress transcription factor (Hsf) family: structure, function and evolution. Biochim Biophys Acta 1819:104–119

Schmidt R, Schippers JH, Welker A, Mieulet D, Guiderdoni E, Mueller-Roeber B (2012) Transcription factor *OsHsfC1b* regulates salt tolerance and development in *Oryza sativa* ssp. japonica. AoB Plants:pls011

Schramm F, Ganguli A, Kiehlmann E, Englich G, Walch D, von Koskull-Döring P (2006) The heat stress transcription factor HsfA2 serves as a regulatory amplifier of a subset of genes in the heat stress response in Arabidopsis. Plant Mol Biol 60(5):759–772

Schultheiss J, Kunert O, Gase U, Scharf KD, Nover L, Rüterjans H (1996) Solution structure of the DNA-binding domain of the tomato heat-stress transcription factor HSF24. Eur J Biochem 236(3):911–921

Sewelam N, Oshima Y, Mitsuda N, Ohme-Takagi M (2014) A step towards understanding plant responses to multiple environmental stresses: a genome-wide study. Plant Cell Environ 37(9):2024–2035

Shen Z, Ding M, Sun J, Deng S, Zhao R, Wang M, Ma X, Wang F, Zhang H, Qian Z, Hu Y (2013) Overexpression of *PeHSF* mediates leaf ROS homeostasis in transgenic tobacco lines grown under salt stress conditions. Plant Cell Tissue Organ Cult 115(3):299–308

Shi H, Tan DX, Reiter RJ, Ye T, Yang F, Chan Z (2015) Melatonin induces class A1 heat-shock factors (HSFA1s) and their possible involvement of thermotolerance in *Arabidopsis*. J Pineal Res 58(3):335–342

Shim D, Hwang JU, Lee J, Lee S, Choi Y, An G, Martinoia E, Lee Y (2009) Orthologs of the class A4 heat shock transcription factor *HsfA4a* confer cadmium tolerance in wheat and rice. Plant Cell 21(12):4031–4043

Song C, Chung WS, Lim CO (2016) Overexpression of heat shock factor gene *HsfA3* increases galactinol levels and oxidative stress tolerance in *Arabidopsis*. Mol Cells 39(6):477

Song G, Yuan S, Wen X, Xie Z, Lou L, Hu B, Cai Q, Xu B (2018) Transcriptome analysis of Cd-treated switchgrass root revealed novel transcripts and the importance of HSF/HSP network in switchgrass Cd tolerance. Plant Cell Rep 37(11):1485–1497

Vuister GW, Kim SJ, Wu C, Bax A (1994) NMR evidence for similarities between the DNA-binding regions of *Drosophila* melanogaster heat shock factor and the helix-turn-helix and HNF-3/forkhead families of transcription factors. Biochemistry 33(1):10–16

Wang X, Huang W, Yang Z, Liu J, Huang B (2016) Transcriptional regulation of heat shock proteins and ascorbate peroxidase by *CtHsfA2b* from African bermudagrass conferring heat tolerance in *Arabidopsis*. Sci Rep 6:28021

Wang X, Huang W, Liu J, Yang Z, Huang B (2017) Molecular regulation and physiological functions of a novel *FaHsfA2c* cloned from tall fescue conferring plant tolerance to heat stress. Plant Biotechnol J 15(2):237–248

Wei Y, Liu G, Chang Y, He C, Shi H (2018) Heat shock transcription factor 3 regulates plant immune response through modulation of salicylic acid accumulation and signalling in cassava. Mol Plant Pathol 19(10):2209–2220

Wu Z, Liang J, Wang C, Zhao X, Zhong X, Cao X, Li G, He J, Yi M (2018) Overexpression of lily *HsfA3*s in *Arabidopsis* confers increased thermotolerance and salt sensitivity via alterations in proline catabolism. J Exp Bot 69(8):2005–2021

Xiang J, Ran J, Zou J, Zhou X, Liu A, Zhang X, Peng Y, Tang N, Luo G, Chen X (2013) Heat shock factor *OsHsfB2b* negatively regulates drought and salt tolerance in rice. Plant Cell Rep 32(11):1795–1806

Xin H, Zhang H, Zhong X, Lian Q, Dong A, Cao L, Yi M, Cong R (2017) Over-expression of *LlHsfA2b*, a lily heat shock transcription factor lacking trans-activation activity in yeast, can enhance tolerance to heat and oxidative stress in transgenic *Arabidopsis* seedlings. Plant Cell Tissue Organ Cult 130(3):617–629

Xue GP, Sadat S, Drenth J, McIntyre CL (2014) The heat shock factor family from *Triticumaestivum* in response to heat and other major abiotic stresses and their role in regulation of heat shock protein genes. J Exp Bot 65:539–557

Yokotani N, Ichikawa T, Kondou Y, Matsui M, Hirochika H, Iwabuchi M, Oda K (2008) Expression of rice heat stress transcription factor *OsHsfA2e* enhances tolerance to environmental stresses in transgenic *Arabidopsis*. Planta 227(5):957–967

Zhang L, Li Y, Xing D, Gao C (2009) Characterization of mitochondrial dynamics and subcellular localization of ROS reveal that *HsfA2* alleviates oxidative damage caused by heat stress in *Arabidopsis*. J Exp Bot 60(7):2073–2091

Zhang S, Xu ZS, Li P, Yang L, Wei Y, Chen M, Li L, Zhang G, Ma Y (2013) Overexpression of *TaHSF3* in transgenic *Arabidopsis* enhances tolerance to extreme temperatures. Plant Mol Biol Rep 31(3):688–697

Zhu B, Ye C, Lü H, Chen X, Chai G, Chen J, Wang C (2006) Identification and characterization of a novel heat shock transcription factor gene, *GmHsfA1*, in soybeans (*Glycine max*). J Plant Res 119(3):247–256

Zhuang L, Cao W, Wang J, Yu J, Yang Z, Huang B (2018) Characterization and functional analysis of *FaHsfC1b* from *Festucaarundinacea* conferring heat tolerance in *Arabidopsis*. Int J Mol Sci 19(9):2702

Chapter 7
Mode of Communication Between Plants During Environmental Stress

Sameer Dixit, Akanchha Shukla, Santosh Kumar Upadhyay and Praveen Chandra Verma

Abstract Plants are considered to be intelligent enough to communicate with each other. This happens with the help of signals that are emitted by the emitter plant in response to any biotic or abiotic inducer. These signals are perceived by neighboring plant (receiver) and which enables them adapt better to the stress conditions. Plants communicate with the help of biological as well as chemical mediator or sometimes without any mediator. These cross-talks are mainly dedicated to share the information about the invading pathogen, mechanical injuries, and availability of light and nutrient. Generally, emitters are less benefited in all communications still the phenomenon evolves because it benefits the other member of emitter species. Plant communication is beyond the species boundary, both inter and intra species communications occur in nature. A very fine interplay of various gene networks regulates this communication resulting in change of gene expression profiles in both emitter as well as receiver plant. Plant communication might be used in future to protect crops against necrotrophic as well as biotropic pathogens.

Keywords Eavesdropping · Emitter · Receiver · Talking tree · Volatile organic compounds

S. Dixit · P. C. Verma
Council of Scientific and Industrial Research—National Botanical Research Institute (CSIR-NBRI), Rana Pratap Marg, Lucknow, Uttar Pradesh 226001, India

S. Dixit (✉) · A. Shukla
Department of Biology, University of Western Ontario, 1151 Richmond St, London, ON, Canada
e-mail: sameerdxt646@gmail.com

S. K. Upadhyay
Department of Botany, Panjab University, Chandigarh, India

S. P. Singh et al. (eds.), *Molecular Approaches in Plant Biology and Environmental Challenges*, Energy, Environment, and Sustainability,
https://doi.org/10.1007/978-981-15-0690-1_7

7.1 Introduction

A very old proverb states that "*One rotten apple spoils the whole bushel*". This proverb is not related to science but it emphasizes on a much hidden scientific phenomenon that how a fruit influences the ripening of other fruit. It was later found out that ethylene released by a fruit facilitates the ripening of other fruits (Theologis 1992).

Highly evolved animals have well defined and dedicated organ system to communicate with each other. Plants, on the other hand do not have such specialized organs but they are able to communicate with each other. About three decades back two independent reports were documented on plant communication (Baldwin and Schultz 1983; Rhoades 1983). First report demonstrated that how a damaged sugar maple plant releases some signals in the air which changes the biochemistry of nearby undamaged plant which in turn affected the feeding behavior of insects (Baldwin and Schultz 1983). While in second case, authors observed altered leaf quality of uninfested *Salix sitchensis* trees in the response to airborne signal released by the caterpillar infested tree of same species (Rhoades 1983). Although these pioneer reports were criticized due to the lack of reproducibility and proper experimental design yet they were the first footprints in this virgin field which stirred a crucial thought process (Myers and Williams 1984; Bruin et al. 1992). Since then number of researches has been conducted in well-defined and controlled laboratory set up, net house as well as in field. All these experiments introduced the concept of "Talking trees". This term was replaced by "listening tree" which more accurately reflected the process of how environmental cues like light, space and nutrients were received by the plants and helped them in deciding about the growth and abscission of shoots and roots (Ballaré 1999). Similarly, they could perceive (listen) the damage signals occurring in the adjacent plants, and responded by activating their native defense mechanism. Present article summaries the different modes of plant communication and molecules or mediators involved in this phenomenon. We have also tried to understand the molecular mechanisms behind these intriguing observations (Fig. 7.1).

7.2 The Need of Communication in Plant

Generally, emitter plant which releases the signals are received by their neighboring receiver plant. The most common reason of communication between the plants is to spread the information of invading pathogen or insects (Baldwin et al. 2006). However the information of light and sound signals may also be shared between the plants (Ballaré 1999). Sometimes emitted signals are a type of warning by which a plant secures their niche. In maximum cases of communications emitter plant was less benefited because it receives the environmental stress signal and then releases cues for others. For example, the cues emitted by damaged plant not only influence

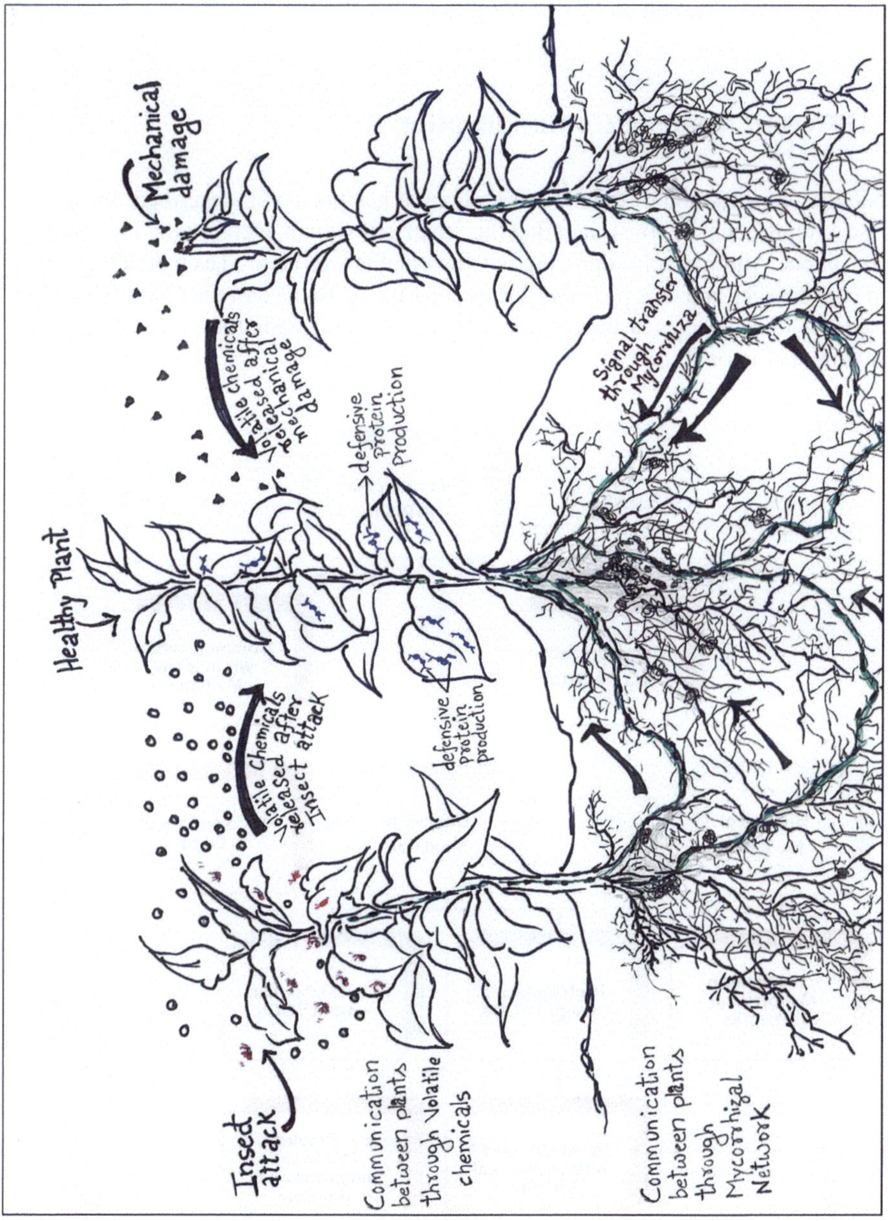

Fig. 7.1 An overview of plant communication

other plants and neighbors but also eavesdroppers to adjust their behaviors (Baldwin et al. 2006). Therefore it cannot be considered as a true communication but it actually represents "eavesdropping". Further, sometimes the information of nutrient or water availability was also passed on by one plant to another.

7.3 Different Mode of Communication

Till date the phenomenon of plant communication has been well described. In several cases, it has been demonstrated that the emanated signals increase the fitness in receiver plants. Alternatively, signaling between plants was favored as it benefits the related conspecifics and thus provides "extended fitness" to the emitter. Communication could be intra-species or inter-species, by using some mediator or occasionally without any mediator (Fig. 7.2).

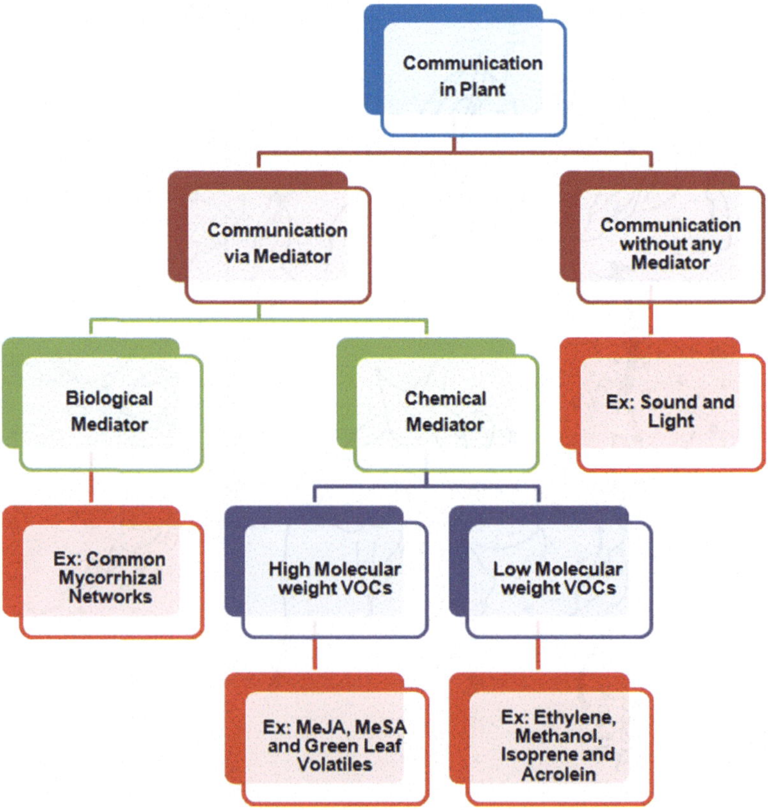

Fig. 7.2 Different modes of plant–plant communication

7.3.1 Communication via Mediator

Mediator is an agent which bridges communication between the emitter and the receiver. Plant to plant communication can take place through biological means as seen in mycorrhizal networks of fungus and plants (Simard et al. 1997) or through chemicals like terpenoids, green-leaf volatiles, methanol, phenylpropanoids, benzenoids, phytohormones, methyl salicylate and jasmonate, hexanal, isoprene, acrolein, methacrolein etc. (Pichersky and Gershenzon 2002). Hence, we can classify mediators as biological and chemical mediators.

7.3.1.1 Biological Mediators

The best studied plant to plant communication via biological mediator is through Mycorrhizal fungus (Simard et al. 1997). The fungal hyphae interconnect with the roots of multiple plants and thus build underground common mycorrhizal networks (CMNs). These CMNs can be as small as to connect two nearby plants or as large as to interconnect the whole forest. Traditionally it was anticipated that these networks are the means of nutrient or water supply across the interconnected plants, but recently it was revealed that these networks are the transfer routes of communication signals for plant defense (Song et al. 2010). It was reported that CMNs mediate plant–plant communication between healthy plants and pathogen-infected tomato plants having leaf blight. Healthy tomato plants connected with diseased plants showed increased disease resistance (Song et al. 2010). Herbivore damage signals could also be transmitted through theses CMNs, it was seen that within 24 h of aphid attack on broad been (*Vicia faba*) signals were transferred in uninfested neighboring plants which activated chemical defenses and made them repell aphids but attractive to their aphids enemies (Babikova et al. 2013). It was also reported that growth of neighbouring plants was suppressed by chemicals produced by marigold which were transported through mycelial networks (Barto et al. 2011). Plant-Mycorrhizal interaction is generally considered as symbiosis between them but now it is emerging as Plant-CMNs-Plant symbiosis.

7.3.1.2 Chemical Mediators

Plant to plant communication through chemical signals has been widely documented (Table 7.1). Chemicals emitted from damaged floral or vegetative parts of plants warn nearby healthy plants to activate defense against the damage. These chemicals are generally volatile organic compounds (VOCs). In stress situations (damage) they are released as complex bouquets. The efflux of VOCs by plant surface is enhanced after insect attack or mechanical wounding. It has been observed in many cases that the mixture of VOCs act as a communicating signal while occasionally only one specific VOC in mixture might work as communicating signal. *Chrysanthemum*

Table 7.1 Chemical signals used in communication by plants

S. no	Compounds	Emitter	Receiver	Type	Effects in neighbouring plants	References
1	Volatile organic compounds	*Salix sitchensis*	*Salix sitchensis*	Intraspecies	Resistance to herbivores	Rhoades (1983)
2	Phenolic compounds	*Populus* x *euroamericana*	*Populus* x *Euroamericana*	Intraspecies	Resistance to herbivores	Baldwin and Schultz (1983)
3	Phenolic compounds	*Acer saccharum*	*Acer saccharum*	Intraspecies	Resistance to herbivores	Baldwin and Schultz (1983)
4	Defensive chemicals	*Alnus rubra*	*Alnus rubra*	Intraspecies	Deterioration of food quality for herbivores	Myers and Williams (1984)
5	Methyl jasmonate	*Artemisia tridentate*	*Lycopersicon esculentum*	Interspecies	Synthesis of proteinase inhibitors	Farmer and Ryan (1990)
6	Jasmonates	*Artemisia tridentate*	*Nicotiana attenuate*	Interspecies	Increased resistance to natural herbivores	Karban et al. (2014)
		Artemisia tridentata	*Artemisia tridentate*	Intraspecies		
7	Airborne chemicals (not specify)	*Gossypium hirsutum*	*Gossypium hirsutum*	Intraspecies	Increased attraction of predatory mites, reduced oviposition by herbivorous mites	Bruin et al. (1992)
8	(Z)-3-hexenyl acetate	*Lycopersicon esculentum*	*Lycopersicon esculentum*	Intraspecies	Improved aphid resistance	Hildebrand et al. (1993)
		Zea mays	*Zea mays*	Intraspecies	Resistance against insect attack	Ton et al. (2007)
		Phaseolus lunatus	*Phaseolus lunatus*	Intraspecies	Enhanced defenses system	Kost and Heil (2006)
9	(Z)-3-hexenylisovalerate	*Phaseolus lunatus*	*Phaseolus lunatus*	Intraspecies	Extrafloral nectar (EFN) secretion	Heil (2008)

(continued)

Table 7.1 (continued)

S. no	Compounds	Emitter	Receiver	Type	Effects in neighbouring plants	References
10	(Z)-3-hexenylbutyrate					
11	mono-, sesqui- and homoterpene	*Alnus glutinosa*	*Alnus glutinosa*	Intraspecies	Reduced herbivory	Dolch and Tscharntke (2000)
12	(Z)-jasmone	*Vicia faba*	*Vicia faba*	Intraspecies	Increased synthesis of VOCs and attraction of predators	Birkett et al. (2000)
		Phaseolus lunatus	*Phaseolus lunatus*	Intraspecies	Defense against herbivores	Heil and Silva Bueno (2007)
13	phenethyl acetate	*Zea mays*	*Zea mays*	Intraspecies	Resistance against insect	Ton et al. (2007)
14	β-ocimene	*Alnus glutinosa*	*Alnus glutinosa*	Intraspecies	Increased activity of oxidative enzymes and proteinase inhibitors	Poveda et al. (2005)
		Antirrhinum Majus	*Arabidopsis thaliana*	Interspecies	Phytotoxic effect	Horiuchi et al. (2007)
		Populus deltoides x *nigra*	*Populus deltoides* x *nigra*	Intraspecies	Elevated defensive responses	Frost et al. (2007)
		Phaseolus lunatu	*Phaseolus lunatus*	Intraspecies	Defense against herbivores	Heil and Silva Bueno (2007)
15	4,8-dimethylnona-1,3,7-triene (DMNT)	*Alnus glutinosa*	*Alnus glutinosa*	Intraspecies	Increased activity of oxidative enzymes and proteinase inhibitors	Dolch and Tscharntke (2000)

(continued)

Table 7.1 (continued)

S. no	Compounds	Emitter	Receiver	Type	Effects in neighbouring plants	References
		Phaseolus lunatus	*Phaseolus lunatus*	Intraspecies	Defenses in neighboring plants by increased EFN secretion	Kost and Heil (2006)
		Zea mays	*Zea mays*	Intraspecies	Elevated defensive responses	Ramadan et al. (2011)
		Populus deltoides x nigra	*Populus deltoides* x nigra	Intraspecies	Elevated defensive responses	Frost et al. (2007)
		Zea mays (maize)	*Zea mays* (maize)	Intraspecies	Resistance against insect attack	Ton et al. (2007)
16	4,8,12-trimethyltrideca-1,3,7,11-tetraene (TMTT)	*Alnus glutinosa*	*Alnus glutinosa*	Intraspecies	Increased activity of oxidative enzymes and proteinase inhibitors	Dolch and Tscharntke (2000)
		Zea mays	*Zea mays*	Intraspecies	Elevated defensive responses	Ramadan et al. (2011)
		Lycopersicon esculentum	*Cuscuta pentagona*	Interspecies	Growth of parasitic plant	Runyon et al. (2006)
		Phaseolus lunatus	*Phaseolus lunatus*	Intraspecies	Defense against herbivores	Heil and Silva Bueno (2007)
17	(Z)-3-Hexenal	*Zea mays*	*Zea mays*	Intraspecies	Priming of JA synthesis and VOCs release	Engelberth et al. (2004)
		Zea mays	*Zea mays*	Intraspecies	Resistance against insect attack	Ton et al. (2007)

(continued)

Table 7.1 (continued)

S. no	Compounds	Emitter	Receiver	Type	Effects in neighbouring plants	References
18	(Z)-3-hexen-1-ol	*Zea mays*	*Zea mays*	Intraspecies	Priming of JA synthesis and VOCs release	Engelberth et al. (2004)
19	(E)-2-hexenal	*Zea mays*	*Zea mays*	Intraspecies	Resistance against insect	Ton et al. (2007)
20	Methyl benzoate	*Antirrhinum Majus*	*Arabidopsis thaliana*	Intraspecies	Phytotoxic effect	Horiuchi et al. (2007)
21	Volatiles (not defined)	*Cirsium arvense*	*Hordeum vulgare*	Interspecies	Becomes significantly less attractive to aphids	Glinwood et al. (2004)
		Cirsium vulgare				
23	Linalool	*Phaseolus lunatus*	*Phaseolus lunatu*	Intraspecies	Defenses in neighboring plants by increased EFN secretion	Kost and Heil (2006)
		Zea mays	*Zea mays*	Intraspecies	Resistance against insect	Ton et al. (2007)
		Phaseolus lunatus	*Phaseolus lunatu*	Intraspecies	Defense against herbivores	Heil and Silva Bueno (2007)
		Zea mays	*Zea mays*	Intraspecies	Elevated defensive responses	Ramadan et al. (2011)
24	α-Pinene	*Lycopersicon esculentum*	*Cuscuta pentagona*	Interspecies	Growth of parasitic plant	Runyon et al. (2006)
25	β-myrcene	*Lycopersicon esculentum*	*Cuscuta pentagona*	Interspecies	Growth of parasitic plant	Runyon et al. (2006)
		Zea mays	*Zea mays*	Intraspecies	Elevated defensive responses	Ramadan et al. (2011)

(continued)

Table 7.1 (continued)

S. no	Compounds	Emitter	Receiver	Type	Effects in neighbouring plants	References
		Antirrhinum Majus	*Arabidopsis thaliana*	Intraspecies	Phytotoxic effect	Horiuchi et al. (2007)
26	2-carene	*Lycopersicon esculentum*	*Cuscuta pentagona*	Interspecies	Growth of parasitic plant	Runyon et al. (2006)
27	p-cymene	*Lycopersicon esculentum*	*Cuscuta pentagona*	Interspecies	Growth of parasitic plant	Runyon et al. (2006)
28	β-phellandrene	*Lycopersicon esculentum*	*Cuscuta pentagona*	Interspecies	Growth of parasitic plant	Runyon et al. (2006)
29	Limonene	*Lycopersicon esculentum*	*Cuscuta pentagona*	Interspecies	Growth of parasitic plant	Runyon et al. (2006)
30	germacrene D	*Populus deltoides* x nigra	*Populus deltoides* x nigra	Intraspecies	Elevated defensive responses	Frost et al. (2007)
31	α-farnesene	*Populus deltoides* x nigra	*Populus deltoides* x nigra	Intraspecies	Elevated defensive responses	Frost et al. (2007)
32	(Z)-3-hexen-1-ylacetate	*Zea mays*	*Zea mays*	Intraspecies	Elevated defensive responses	Ramadan et al. (2011)
33	(E)-3-hexenyl acetate	*Phaseolus lunatus*	*Phaseolus lunatus*	Intraspecies	Extrafloral nectar (EFN) secretion	Heil (2008)
34	β-caryophyllene	*Populus deltoides* x nigra	*Populus deltoides* x nigra	Intraspecies	Elevated defensive responses	Frost et al. (2007)
		Phaseolus lunatus	*Phaseolus lunatu*	Intraspecies	Defense against herbivores	Heil and Silva Bueno (2007)
		Zea mays	*Zea mays*	Intraspecies	Resistance against insect	Ton et al. (2007)

(continued)

Table 7.1 (continued)

S. no	Compounds	Emitter	Receiver	Type	Effects in neighbouring plants	References
		Zea mays	*Zea mays*	Intraspecies	Elevated defensive responses	Ramadan et al. (2011)
35	Indole	*Zea mays*	*Zea mays*	Intraspecies	Resistance against insect	Ton et al. (2007)
		Zea mays	*Zea mays*	Intraspecies	Elevated defensive responses	Ramadan et al. (2011)
36	Geranyl acetate	*Zea mays*	*Zea mays*	Intraspecies	Resistance against insect	Ton et al. (2007)
37	(E)-α-bergamotene	*Zea mays*	*Zea mays*	Intraspecies	Resistance against insect	Ton et al. (2007)
		Zea mays	*Zea mays*	Intraspecies	Elevated defensive responses	Ramadan et al. (2011)
38	(E)-β-farnesene	*Zea mays*	*Zea mays*	Intraspecies	Resistance against insect	Ton et al. (2007)
		Zea mays	*Zea mays*	Intraspecies	Elevated defensive responses	Ramadan et al. (2011)
39	β-sesquiphellandrene	*Zea mays*	*Zea mays*	Intraspecies	Resistance against insect	Ton et al. (2007)
40	(Z)-hexenyl acetate	*Phaseolus lunatus*	*Phaseolus lunatu*	Intraspecies	Defense against herbivores	Heil and Silva Bueno (2007)
41	(E)-2-hexenyl acetate	*Phaseolus lunatus*	*Phaseolus lunatus*	Intraspecies	Extrafloral nectar (EFN) secretion	Heil (2008)
42	5-hexenyl acetate	*Phaseolus lunatus*	*Phaseolus lunatus*	Intraspecies	Extrafloral Nectar (EFN) Secretion	Heil (2008)

(continued)

Table 7.1 (continued)

S. no	Compounds	Emitter	Receiver	Type	Effects in neighbouring plants	References
43	C11 homoterpene	*Phaseolus lunatus*	*Phaseolus lunatu*	Intraspecies	Defense against herbivores	Heil and Silva Bueno (2007)
44	(Z)-3-hexen-1-yl-butyrate	*Phaseolus lunatus*	*Phaseolus lunatu*	Intraspecies	Defense against herbivores	Heil and Silva Bueno (2007)
45	Methyl Salicylate	*Phaseolus lunatus*	*Phaseolus lunatu*	Intraspecies	Defense against herbivores	Heil and Silva Bueno (2007)
46	(E)-bocimene	*Zea mays*	*Zea mays*	Intraspecies	Elevated defensive responses	Ramadan et al. (2011)
47	Herbivore-induced plant volatiles (HIPV)	*Brassica olerace*	*Brassica oleracea*	Intraspecies	Inhibited the growth rate of caterpillars	Peng et al. (2011)
		Vaccinium corymbosum	*Vaccinium corymbosum*	Intraspecies	Induction of direct defenses against gypsy moth	Rodriguez-Saona et al. (2009)
		Hordeum spp.	*Hordeum spp.*	Intraspecies	Decreased attractiveness to aphids and increased attractiveness to parasitoids	Glinwood et al. (2004)
48	Methanol	*Nicotiana benthamiana*	*Nicotiana benthamiana*	Intraspecies	Induction of MIGs and bacterial resistance	Dorokhov et al. (2012)

cinerariaefolium respond to wounding by releasing VOCs mixture into the air and enhance the level of natural insecticide like pyrethrins in neighboring plants by up-regulating their synthesis genes, individual components of VOC mixture does not have same ability (Kikuta et al. 2011). Whereas wound inducible emission of VOCs by *Nicotiana benthamiana* leads to increased bacterial resistance in neighboring plant only due to the presence of Methanol in mixture. Every plant responds differently to each inducer so the specificity of communicating signal might be dependent on (a) the type of inducer and/or (b) the emitter plant (Dorokhov et al. 2012). Out of 48 studies on plant communication, 39 provided evidences that emitted signals are transferred aerially and received by neighboring plants activating a defense response in them (Karban et al. 2014). Plant VOCs can be categorized as high molecular weight (HMW) and low molecular weight (LMW) volatiles. Both types of VOCs work as communication signal.

High Molecular Weight Volatiles

Terpene Alcohols, Methyl Jasmonate (MeJA), Methyl Salicylate (MeSA) and Green Leaf Volatiles (GLVs) are HMW volatiles compounds. They are less volatile, therefore the rate at which they disperse in atmosphere is slow aiding long distance signaling. After release, the VOCs form plumes of higher concentrations (Thistle et al. 2004) and in this process some VOCs are also oxidized. Presumably, whenever a plant is wounded or attacked by herbivore there is emission of MeJA, which has been shown to effectively turn on defense related genes (Creelman and Mullet 1997). Whereas after pathogen infection MeSA is released which serves as a signal for activation of defense genes in neighboring plants (Shulaev et al. 1997). GLV are released by plants immediately after damage but its release can also be induced (Hatanaka 1993; Pare and Tumlinson 1997). GLVs not only stimulate transient JA biosynthesis, but exposure to GLV induces much stronger defense response as seen in studies on corn (Engelberth et al. 2004).

Low Molecular Weight Volatiles

Ethylene, Methanol, Isoprene, Acrolein, Methacrolein, Linalool, Hexenal, Methyl Benzoate and Terpenes etc. are low molecular weight but highly volatile compounds. These volatiles are rapidly diffused in the atmosphere and thus are considered to mediate communication between neighboring plants (Baldwin et al. 2006).

LMW volatiles are generally produced in Shikimic acid-pathway, fatty acid-derived carbon compounds and also from terpenes. Being highly volatile these small VOCs produce defense response in the same plant, as seen in volatiles released from damaged lima bean leaves that induce EFN secretion in undamaged leaves of the same plant. These VOCs include (Z)-hexenylacetate, (Z)-b-ocimene, (E)-b-ocimene, linalool, C11 homoterpene (Heil and Silva Bueno 2007). It has been reported that methyl benzoate emitted from snapdragon inhibited the root growth of *Arabidopsis* growing nearby (Horiuchi et al. 2007). Intraplant signalling through methacrolein was seen in sagebrush, Methacrolein being highly volatile diffuses rapidly (Preston et al. 2004).

Airborne and Underground Chemical Signals

The medium for transfer of the signals is predominantly air, when these mediators are released they first come in contact with local undamaged parts of the plants establishing intraplant communication (Heil and Silva Bueno 2007; Heil 2008). It was seen in corn and cotton that leaves of the same plant amplified volatile compound production as herbivore attacked the plant (Turlings and Tumlinson 1992; Rose et al. 1996). Defense response is also observed when there is egg laying on the leaves by herbivores which generates systemic release of signals (Hilker and Meiners 2002). Herbivore attacked leaves of hybrid poplar (*Populus deltoides*) showed increased defense proteins in neighboring branches which did not had any vascular connections with the damaged leaf. Thus all the above reports suggest that volatile signals are disseminated through air.

Plants have developed a defense system called systemic acquired resistance (SAR), whenever there is a pathogen attack, the response is heightened and this protects the plant against subsequent attack of the same pathogen as well as other pathogens (Vlot et al. 2008). When the signals are dispersed in air they reach neighboring plants which may be of same species or different, other plants eavesdrop on the VOC's signals and prepare themselves for herbivore attack by synthesizing and secreting defense proteins. It was seen that herbivore damage induces the release of volatiles which act as signaling molecules and activate synthesis of plant volatiles dependent on jasmonic acid in neighboring plants. The C6-volatiles like GLVs induced subset of defense-related genes after herbivore attack (Bate and Rothstein 1998). It was seen that C6-Green leaf volatiles triggered VOC emissions in tomato (Farag and Paré 2002), other studies showing similar results were reported (Ruther and Kleier 2005; Arimura et al. 2002; Yan and Wang 2006). In addition, barley (*Hordeum vulgare*) became less aphid infested after the exposure of plants to air from various thistle species (Glinwood et al. 2004).

Plant volatiles not only mediate communication between leaf and stem but also establish communication between roots and shoots without any vascular connection (Erb et al. 2008). (E)-b-caryophyllene, a sesquiterpene, released in maize root was found to attract entomopathogenic nematodes which killed herbivore larvae (Rasmann et al. 2005). This is an indirect defense mechanism. Many studies have been reported for such defense (Neveu et al. 2002; Aratchige et al. 2004). Enhanced production of EFN in cotton plants positively regulated root herbivory and attracted parasitoids for defense (Wackers and Bezemer 2003; Masters et al. 2001). Belowground signals generally affect plant performance whereas aboveground signals are the outcome of plant-insect interactions (Poveda et al. 2005).

7.3.2 *Communication Without Mediator*

Communication between plants is not solely dependent on mediator; they are also able interconnect/communicate without involving any mediator. Evidences that plants use their root system to eavesdrop on their neighboring stressed plant have been documented. Report on *Pisum sativum* subjected to drought like stress conditions concluded that within 15 min of stress induction the plants closed their stomata and their unstressed neighbors also responded similarly. This suggested the presence of some cues that are passed on to the neighbors and warns them for taking protective measures (Falik et al. 2012). Studies on plants using sound as a communication means have also been documented, where young corn plant roots growing in water, made click sounds. It was also seen that roots bended towards the sound source of same frequency. This mode of communication is advantageous over chemical signals as they are more rapidly transported (Gagliano 2013). However, they are transmitted over short distances only.

7.4 Molecular Mechanism Underlying Plant Communications

Receiver plant responds to the communicating signals by modulating a set of genes, this modulation occurs not only in receiver plant but also in emitter plant (Table 7.2). It has been observed that methanol emission after leaf wounding enhances the transcript levels of Methanol Inducible Genes (MIGs) in receiver as well as emitter plant (Dorokhov et al. 2012). Whereas, gene responsible for methanol production is expressed exclusively in emitter plant (Dorokhov et al. 2012). This shows the fine interplay of genes in plant communication. Receiver plant responds differently towards each unique signal given out by emitter plant. For example, MIG-21 is exclusively expressed in response to methanol communication signal (Dorokhov et al. 2012). Similarly, elevated expression of Lipoxygenase (LOX), a key player of JA biosynthesis, was the downstream response in receiver plant after mixture of VOCs were released by mechanical wounding and spider mite infestation in emitter plant (Kikuta et al. 2011; Arimura et al. 2000). However, in some cases common genes were modulated by two different signals. *T. urticae* infested leaf volatiles in lima bean leaves induced basic type of pathogen-related (PR) protein (PR-2 (β-1,3-glucanase) (Arimura et al. 2000). Later it was also characterized as a MIG whose mRNA level were highly upregulated after methanol evaporation (Dorokhov et al. 2012). Wound inducible VOCs not only enhances the PR proteins but they also increases the level of enzyme which plays a role in the bio synthesis of natural insecticides like pyrethrins (Kikuta et al. 2011).

Table 7.2 Genes involved in plant communication

S. no	Gene used in communication	Communicating signal	Inducer	Emitter	Receiver	References
1	PR-2	Jasmonic acid; (E)-b-ocimene; (E)-4,8-dimethyl-1,3,7-nonatriene (DMNT); (E,E)-4,8,12-trimethyl-1,3,7,11-tridecatetraene (TMTT)	Spider mites (*Tetranychus urticae*)	*Phaseolus lunatus*	*Phaseolus lunatus*	Arimura et al. (2000)
2	PR-3					
3	PR-4					
4	Lipoxygenase					
5	Phenylalanine ammonia-lyase					
6	Farnesyl pyrophosphate synthetase					
7	S-adenosylmethionine synthetase					Arimura et al. (2002)
8	1-aminocyclopropane-1-carboxylic acid oxidase					
9	SAM decarboxylase					
10	β-1,3-glucanase	Methanol	*Ralstonia solanacearum*	*Nicotiana benthamiana*	*Nicotiana benthamiana*	Dorokhov et al. (2012)
11	Proteinase inhibitor II					
12	Methanol-inducible gene-21					
13	PME inhibitor					

(continued)

Table 7.2 (continued)

S. no	Gene used in communication	Communicating signal	Inducer	Emitter	Receiver	References
14	Non-cell autonomous pathway protein					
15	1-deoxy-D-xylulose 5-phosphate synthase	(Z)-3-hexena; (E)-2-hexenal; (Z)-3-hexen-1-ol; (Z)-3-hexen-1-yl acetate; (E)-b-farnesene	Wound-induced VOCs	*Chrysanthemum cinerariaefolium*	*Chrysanthemum cinerariaefolium*	Kikuta et al. (2011)
16	Chrysanthemyl diphosphate synthase					
17	13-lipoxygenase					
18	Allene oxide synthase					

7.5 Conclusion and Future Prospects

Plant communication has been observed to evolve in selective evolutionary pressure. It has blazed the trail of plant intelligence between Plant-Neurobiologists. Although our understanding in this area began in 1983 the studies took maximum leap in this century, where gradually, more insightful studies lifted the veils of this communication phenomenon. Plant communication is promoted by both Biotic and abiotic stresses eliciting defense response favoring emitter and receiver plants. However emitter plant is at less benefit than receiver plant. Despite this, plant communication is favored because it benefits the related conspecifics and thus provides "extended fitness" to the emitter. In addition to the receiver plant the communicating signals are also received by eavesdropper due to lack of signal specificity which may evolve in future. Since communication between plants helps growth, survival and health of plants it might serve as promising crop protection strategy against both necrotrophic and biotropic pathogens. We have been continuously working on developing methanol as an insect resistance signal. Apart from methanol, more sensitive or specific emitters can be screened or developed by genetic manipulation. These emitters can be grown intermittently in the crop field where they can act as guards and deliver warning signals to the other plants growing in the field. Genetic manipulation can also be done in unresponsive receiver plant to convert it into active receiver. Communications through sound and electric waves have recently come into picture and need to be explored in future. Thus plant communication holds great promise however, further investigations and validation studies need to be addressed before implementing it in crop protection regime.

Acknowledgements We would like to apologize to all the authors whose work we were unable to cite, because of space constraint, We express gratitude to all the lab members of the PC Verma laboratory for valuable discussions. We also like to acknowledge Dr. Nidhi Thakur and Varsha Srivastava for their constructive comments.

References

Aratchige NS, Lesna I, Sabelis MW (2004) Below-ground plant parts emit herbivore-induced volatiles: olfactory responses of a predatory mite to tulip bulbs infested by rust mites. Exp Appl Acarol 33(1–2):21–30

Arimura G, Ozawa R, Shimoda T, Nishioka T, Boland W, Takabayashi J (2000) Herbivory-induced volatiles elicit defense genes in lima bean leaves. Nature 406(6795):512–515

Arimura G, Ozawa R, Nishioka T, Boland W, Koch T, Kühnemann F, Takabayashi J (2002) Herbivore-induced volatiles induce the emission of ethylene in neighboring lima bean plants. Plant J 29(1):87–98

Babikova Z, Johnson D, Bruce T, Pickett JA, Gilbert L (2013) How rapid is aphid-induced signal transfer between plants via common mycelial networks? Commun Integr Biol 6(6):e25904

Baldwin IT, Schultz JC (1983) Rapid changes in tree leaf chemistry induced by damage: evidence for communication between plants. Science 221(4607):277–279

Baldwin IT, Halitschke R, Paschold A, von Dahl CC, Preston CA (2006) Volatile signaling in plant–plant interactions: "talking trees" in the genomics era. Science 311(5762):812–815

Ballaré CL (1999) Keeping up with the neighbours: phytochrome sensing and other signalling mechanisms. Trends Plant Sci 4(5):201

Barto EK, Hilker M, Müller F, Mohney BK, Weidenhamer JD, Rillig MC (2011) The fungal fast lane: common mycorrhizal networks extend bioactive zones of allelochemicals in soils. PLoS ONE 6(11):e27195

Bate NJ, Rothstein SJ (1998) C6-volatiles derived from the lipoxygenase pathway induce a subset of defense-related genes. Plant J 16(5):561–569

Birkett MA, Campbell CA, Chamberlain K, Guerrieri E, Hick AJ, Martin JL, Matthes M, Napier JA, Pettersson J, Pickett JA, Poppy GM (2000) New roles for cis-jasmone as an insect semiochemical and in plant defense. Proc Natl Acad Sci 97(16):9329–9334

Bruin J, Dicke M, Sabelis MW (1992) Plants are better protected against spider mites after exposure to volatiles from infested conspecifics. Experientia 48:525–529

Creelman RA, Mullet JE (1997) Biosynthesis and action of jasmonates in plants. Annu Rev Plant Physiol Plant Mol Biol 48:355–381

Dolch R, Tscharntke T (2000) Defoliation of alders (*Alnus glutinosa*) affects herbivory by leaf beetles on undamaged neighbours. Oecologia 125(4):504–511

Dorokhov YL, Komarova TV, Petrunia IV, Frolova OY, Pozdyshev DV, Gleba YY (2012) Airborne signals from a wounded leaf facilitate viral spreading and induce antibacterial resistance in neighboring plants. PLoS Pathog 8(4):e1002640

Engelberth J, Alborn HT, Schmelz EA, Tumlinson JH (2004) Airborne signals prime plants against insect herbivore attack. Proc Natl Acad Sci USA 101(6):1781–1785

Erb M, Ton J, Degenhardt J, Turlings TC (2008) Interactions between arthropod-induced aboveground and belowground defenses in plants. Plant Physiol 146(3):867–874

Falik O, Mordoch Y, Ben-Natan D, Vanunu M, Goldstein O, Novoplansky A (2012) Plant responsiveness to root–root communication of stress cues. Ann Bot 110(2):271–280

Farag MA, Paré PW (2002) C6-Green leaf volatiles trigger local and systemic VOC emissions in tomato. Phytochemistry 61(5):545–554

Farmer EE, Ryan CA (1990) Interplant communication: airborne methyl jasmonate induces synthesis of proteinase inhibitors in plant leaves. Proc Natl Acad Sci 87(19):7713–7716

Frost CJ, Appel HM, Carlson JE, De Moraes CM, Mescher MC, Schultz JC (2007) Within-plant signalling via volatiles overcomes vascular constraints on systemic signalling and primes responses against herbivores. Ecol Lett 10(6):490–498

Gagliano M (2013) Green symphonies: a call for studies on acoustic communication in plants. Behav Ecol 24(4):789–796

Glinwood R, Ninkovic V, Pettersson J, Ahmed E (2004) Barley exposed to aerial allelopathyfromthistles (*Cirsium spp.*) becomes less acceptable to aphids. Ecol Entomol 29:188–195

Hatanaka A (1993) The biogeneration of green odour by green leaves. Phytochem 34:1201–1218

Heil M (2008) Indirect defense via tritrophic interactions. New Phytol 178(1):41–61

Heil M, Silva Bueno JC (2007) Within-plant signaling by volatiles leads to induction and priming of an indirect plant defense in nature. Proc Natl Acad Sci USA 104(13):5467–5472

Hildebrand DF, Brown GC, Jackson DM, Hamilton-Kemp TR (1993) Effects of some leaf-emitted volatile compounds on aphid population increase. J Chem Ecol 19(9):1875–1887

Hilker M, Meiners T (2002) The plant's response towards insect egg deposition. In: Hilker M, Meiners T (eds) Chemoecology of insect eggs and egg deposition. Blackwell Verlag, Oxford, pp 205–233

Horiuchi J, Badri DV, Kimball BA, Negre F, Dudareva N, Paschke MW, Vivanco JM (2007) The floral volatile, methyl benzoate, from snapdragon (*Antirrhinum majus*) triggers phytotoxic effects in *Arabidopsis thaliana*. Planta 226(1):1–10

Karban R, Yang LH, Edwards KF (2014) Volatile communication between plants that affects herbivory: a meta-analysis. Ecol Lett 17(1):44–52

Kikuta Y, Ueda H, Nakayama K, Katsuda Y, Ozawa R, Takabayashi J, Hatanaka A, Matsuda K (2011) Specific regulation of pyrethrin biosynthesis in *Chrysanthemum cinerariaefolium* by a blend of volatiles emitted from artificially damaged conspecific plants. Plant Cell Physiol 52(3):588–596

Kost C, Heil M (2006) Herbivore-induced plant volatiles induce an indirect defence in neighbouring plants. J Ecol 94(3):619–628

Masters GJ, Jones TH, Rogers M (2001) Host-plant mediated effects of root herbivory on insect seed predators and their parasitoids. Oecologia 127(2):246–250

Myers JH, Williams KS (1984) Does tent caterpillar attack reduce the food quality for red alder foliage? Oecologia 62:74–79

Neveu N, Grandgirard J, Nenon JP, Cortesero AM (2002) Systemic release of herbivore-induced plant volatiles by turnips infested by concealed root-feeding larvae *Delia radicum* L. J Chem Ecol 28(9):1717–1732

Pare PW, Tumlinson JH (1997) De novo biosynthesis of volatiles induced by insect herbivory in cotton plants. Plant Physiol 114(4):1161–1167

Peng J, Van Loon JJA, Zheng S, Dicke M (2011) Herbivore-induced volatiles of cabbage (*Brassica oleracea*) prime defence responses in neighbouring intact plants. Plant biology 13(2):276–284

Pichersky E, Gershenzon J (2002) The formation and function of plant volatiles: perfumes for pollinator attraction and defense. Curr Opin Plant Biol 5(3):237–243

Poveda K, Steffan-Dewenter I, Scheu S, Tscharntke T (2005) Effects of decomposers and herbivores on plant performance and aboveground plant-insect interactions. Oikos 108:503–510

Preston CA, Laue G, Baldwin IT (2004) Plant–plant signaling: application of trans- or cis-methyl jasmonate equivalent to sagebrush releases does not elicit direct defenses in native tobacco. J Chem Ecol 30(11):2193–2214

Ramadan A, Muroi A, Arimura GI (2011) Herbivore-induced maize volatiles serve as priming cues for resistance against post-attack by the specialist armyworm *Mythimna separata*. J plant Interact 6(2–3):155–158

Rasmann S, Köllner TG, Degenhardt J, Hiltpold I, Toepfer S, Kuhlmann U, Gershenzon J, Turlings TC (2005) Recruitment of entomopathogenic nematodes by insect-damaged maize roots. Nature 434(7034):732–737

Rhoades DF (1983) Responses of alder and willow to attack by tent caterpillars and webworms: evidence for phenomonal sensitivity of willows. In: Hedin PA (ed) Plant resistance to insects, symposium series 208. American Chemical Society, Washington DC, pp 55–68

Rodriguez-Saona CR, Rodriguez-Saona LE, Frost CJ (2009) Herbivore-induced volatiles in the perennial shrub, *Vaccinium corymbosum*, and their role in inter-branch signaling. J Chem Ecol 35(2):163–175

Rose U, Manukian A, Heath RR, Tumlinson JH (1996) volatile semiochemicals released from undamaged cotton leaves (a systemic response of living plants to caterpillar damage). Plant Physiol 111(2):487–495

Runyon JB, Mescher MC, De Moraes CM (2006) Volatile chemical cues guide host location and host selection by parasitic plants. Science 313(5795):1964–1967

Ruther J, Kleier S (2005) Plant-plant signaling: ethylene synergizes volatile emission in *Zea mays* induced by exposure to (Z)-3-hexen-1-ol. J Chem Ecol 31(9):2217–2222

Shulaev V, Silverman P, Raskin I (1997) Airborne signalling by methyl salicylate in plant pathogen resistance. Nature 386:718–721

Simard SW, Perry DA, Jones MD, Myrold DD, Durall DM, Molina R (1997) Net transfer of carbon between ectomycorrhizal tree species in the field. Nature 388:579–582

Song YY, Zeng RS, Xu JF, Li J, Shen X, Yihdego WG (2010) Interplant communication of tomato plants through underground common mycorrhizal networks. PLoS ONE 5(10):e13324

Theologis A (1992) One rotten apple spoils the whole bushel: the role of ethylene in fruit ripening. Cell 70(2):181–184

Thistle HW, Peterson H, Allwine G, Lamb B, Strand T, Holsten EH, Shea PJ (2004) Surrogate pheromone plumes in three forest trunk spaces: composite statistics and case studies. Forest Sci 50(16):610–625

Ton J, D'alessandro M, Jourdie V, Jakab G, Karlen D, Held M, Mauch-Mani B, Turlings TC (2007) Priming by airborne signals boosts direct and indirect resistance in maize. Plant J 49(1):16–26
Turlings TC, Tumlinson JH (1992) Systemic release of chemical signals by herbivore-injured corn. Proc Natl Acad Sci USA 89(17):8399–8402
Vlot AC, Klessig DF, Park SW (2008) Systemic acquired resistance: the elusive signal(s). Curr Opin Plant Biol 11(4):436–442
Wackers FL, Bezemer TM (2003) Root herbivory induces an aboveground indirect defense. Ecol Lett 2003(6):9–12
Yan ZG, Wang CZ (2006) Wound-induced green leaf volatiles cause the release of acetylated derivatives and a terpenoid in maize. Phytochemistry 67(1):34–42

Chapter 8
Molecular Approaches for Combating Multiple Abiotic Stresses in Crops of Arid and Semi-arid Region

Vinod Kumar, Shourabh Joshi, Naveen C. Pant, Punesh Sangwan, Ajar Nath Yadav, Abhishake Saxena and Dharmendra Singh

Abstract Under constantly changing environmental conditions; crop plants are exposed to abiotic stresses, which lead to affect growth and development including the productivity of agricultural crops. Understanding the mechanism of stress at molecular level and improving crop varieties for tolerance to abiotic stress is a challenging task. In Arid and semi-arid regions, the agricultural crops are challenged to multiple abiotic stresses (draught, salinity and heat), simultaneously. The conditions like low annual rainfall, soil salinity, and variable high temperature conditions lead to low agricultural productivity. Exposure to the abiotic stresses induces a complex signaling pathway in different plant species and resulting into variable molecular, biochemical and physiological changes to acclimatize under stress conditions including multiple abiotic stresses simultaneously. In different sections of this chapter, the metabolic changes in plants in response to draught, salinity, and temperature stress is described followed by an insight into plant signaling pathways and possible biochemical, physiological and molecular approaches for alleviating the negative effects of these stresses.

Keywords Plant abiotic stresses · Sustainable agriculture · Crop improvement · Draught

---

V. Kumar (✉)
Biochemistry Laboratory, Forage Section, Department of Genetics and Plant Breeding, CCS Haryana Agricultural University, Hisar 125004, India
e-mail: sangwan.vinod@yahoo.com

S. Joshi
Department of Plant Biotechnology, Agriculture University, Jodhpur 342304, India

N. C. Pant
Department of Plant Biochemistry, SKN Agriculture University, Jobner 342304, India

P. Sangwan
Department of Biochemistry, Eternal University, Baru Sahib 173101, India

A. N. Yadav · A. Saxena · D. Singh
Department of Biotechnology, Eternal University, Baru Sahib 173101, India

S. P. Singh et al. (eds.), *Molecular Approaches in Plant Biology and Environmental Challenges*, Energy, Environment, and Sustainability,
https://doi.org/10.1007/978-981-15-0690-1_8

8.1 Introduction

Plants are constantly exposed to real-time changes in the environmental conditions and the undesirable variations in the growth conditions, which negatively impact economically important crops (Ramegowda and Senthil-Kumar 2015). The abiotic stresses such as salinity, drought and extremely high temperature result into poor plant growth and crop yields These abiotic stresses may occur alone or in combination, causing severe crop yield losses in later stage (Mittler 2006). A major area of world's agricultural land i.e. 19.5% and 45% is affected by salinity and drought, respectively (dos Reis et al. 2012; Flowers and Yeo 1995). A huge part (Nearly 800 million ha) of world's agricultural land is affected by salt (FAO 2009). Under presently changing environmental conditions and climate change, more severe abiotic stresses are predicted elsewhere. Such stresses are supposed to be more pronounced in arid and semi-arid region. Crop plants for arid regions are those that survive and produce in spite of aridity. The major crops of arid region with varying degree of tolerance include cereal grains (Corn, Sorghum, Pearl Millet), grain legumes (Common Bean, Cowpea, Mung Bean), seed spices and various crops of medicinal and aromatic importance (Creswell and Martin 1998). Due course of evolution, however, certain plants have evolved adaptive features to tolerate these stresses conditions (Phukan et al. 2014). Tolerance to abiotic stress is mainly attributed to production certain metabolites, proteins and modulation in gene expression. It has been observed that complex array of signalling pathways are activated that help alleviate the negative effect of abiotic stress. Based on such observations, various molecular approaches have been proposed for developing crop species tolerant to abiotic species. In this chapter, responses of plants for various abiotic stresses, mechanism of signal transduction pathway and potential molecular approaches have been discussed in following sections.

8.2 Effect of Various Abiotic Stresses on Crop Plants

Plant growth and development are affected by a variety of environmental factors like relative humidity, light, temperature, availability of water, mineral nutrients, etc. (Koyro et al. 2012). Plants require an optimum level of environmental factors which in turn have a pronounced effect on the growth and development of the plants. Intensity and/or quantity of each abiotic factor decide the overall effect on the plant at physiological and molecular level. Deviation from optimal conditions is regarded as abiotic stress which adversely affects yield and/or productivity (Rahneshan et al. 2018). Since multiple environmental factors are involved in plant abiotic stresses, the rapid change in environmental conditions is likely to override the adaptive potential of plants which in turn will severely affect the production and yield of various crops.

8.2.1 Drought

Water is the single most important constituent of any living organism including plants. It comprises of more than 90% of the fresh weight of plant tissues. Water availability is one of the major constraints affecting plant yield and productivity (Boyer 1982). It is also an important factor governing the distribution of various plant species across the globe. Arid or semiarid region constitutes around 35% of the world's available land with inadequate precipitation unable to sustain plant life. Moreover, the precipitation is rarely uniform. Yield loss due to drought in different crops probably exceeds losses from all other abiotic stresses, and may lead to around 50% yield loss depending upon the severity and duration of drought. The effects of drought on plants are in the form of morphological to molecular changes, which are evident at all critical stages of plant growth and development. The most pronounced effect of drought stress is on germination and plant establishment (Harris et al. 2002). Plants exhibits a multitude of physiological and biochemical responses towards drought, thus making drought tolerance a complex phenomenon. Drought stress reduces water-use efficiency and disturbs plant water relations, thus affecting leaf size, stems elongation and root proliferation. CO_2 assimilation during photosynthesis in leaves is significantly reduced mainly due to stomatal closure, formation of reactive oxygen species (ROS) and corresponding membrane damage, altered enzyme activity especially of those involved in CO_2 fixation and partitioning. Inflated metabolic flux through alternative means (photorespiration) especially in C3 plants increases the oxidative damage in the plant tissues especially due to generation of ROS. Oxidative damage caused by reactive oxygen species to various cellular constituents is the major deterrents to plant growth and development under drought stress. Plants in arid and semi-arid regions have developed a wide range of mechanisms to withstand drought stress. Some of the major mechanisms are exemplified by increase in diffusive resistance to reduce water loss, smaller, succulent leaves to compensate transpiration loss, increased water uptake and its efficient use. Drought tolerance in crop plants can be managed by adoption of suitable strategies viz., screening for drought tolerance, selective breeding utilizing marker assisted selection and engineering plants with drought resistance traits.

Water deficit at the critical physiological stages affects both quality and quantity of the harvest. Physiological processes including cellular growth and differentiation are severely affected by drought stress. Severe water deficit inhibits cell elongation in higher plants due to interruption in water flow from the xylem to the surrounding cells. Impaired cell division, cell expansion and elongation results in reduced plant height, leaf area and crop growth under severe drought (Kaya et al. 2006; Hussain et al. 2008). During drought stress, it is very important to access various parameter such as drought severity, duration of stress, plant growth condition, plants responses, and interaction of drought with other abiotic factors at that duration. Many physiological processes which affect crop yield also respond to water deficit. Thus, it is difficult to interpret and ascertain the plant responses and the means by which plants regulate the change in physiological processes and how such processes work in a combined way.

8.2.2 Salt Stress

Soil salinity is one of the major constraints today because it limits crop productivity and restricts cultivation on effected lands. Soil with higher content of soluble salts is considered as saline i.e. when the osmotic pressures exerted by ions are equivalent to that exerted by 40 mM NaCl (0.2 MPa). An estimated 6% and 20% of the total land and world's irrigated land, respectively, are affected by soil salinity (Turan et al. 2012). Many arid and semi-arid regions of the world contain Around the world, the soil and water resources of many arid and semi-arid regions are too saline and unfit for cultivation of most of the crops. Salinity affects crop plants through osmotic imbalance, ion specific effects and oxidative damage (Rani et al. 2019). A negative correlation exists between soil salinity and crop productivity. Salinity affects plant growth and development apart from restricting the use of land for cultivation. The physiological drought caused by decreased soil water potential and affected soil porosity is a consequence of soil salinity. Moreover the problem of soil salinity is aggravated by irrigation with saline water and is typically a major problem in tropical regions, where excessive water is lost through transpiration. High salt content have pronounced effect on physiology of the plants, with effects usually being observed at the levels of cellular and whole plant (Murphy et al. 2003). The main effect of salt stress is osmotic imbalance due to high ionic strength in rhizospheric region, which restricts water absorption, thus severely limiting plant growth and development. The salt stress also exhibits secondary effects usually caused by ionic disequilibrium, which results in inactivation of various enzymes, nutrient imbalance and ROS induced oxidative stress. ROS further damages plants by increased lipid peroxidation, DNA damage and inhibition of photosynthesis (Turan and Tripathy 2013). An excessive cellular amount of Na^+ and Cl^- like ions within plants led to enzyme inhibition especially nitrate reductase (Khan et al. 2011), Rubisco, phosphoenolpyruvate carboxylase, and associated metabolic dysfunctions including damage of photosynthetic apparatus (Soussi et al. 1998). Plants usually develop various physiological and biochemical mechanisms in order to survive in soils with high salt concentration i.e., accumulating ions into various tissue compartments, vacuoles and older leaves. In most plants, during water absorption by plant roots from the soil, Na^+ and Cl^- are excluded effectively (Munns 2005). Plants produce osmolytes like glycine betaine, trehalose or proline in response to osmotic stress due to high salinity which protect them from dehydration, protein denaturation and associated metabolic dysfunctions. Considering the adverse effects of salinity on growth and development of plants, identification, conservation and development of salt tolerant plant genetic resources is of great importance to cater the requirement of productivity and sustainable agriculture.

8.2.3 High Temperature Stress

High ambient temperature is one of the most critical factors affecting yield and corresponding crop productivity among various environmental variables affecting plant growth and development. An increase by 0.2 °C per decade in global temperature is predicted, with corresponding increase in temperatures by 1.8–4.0 °C than the current level by 2100 ((IPCC) IPoCC 2007). Heat stress often causes a multifarious and adverse effect on plant growth and development, apart from significant effect on various physiological processes and in turn corresponding yield of the crop plants. One of the main effects of high temperature (HT) stress is ROS induced oxidative damage (Hasanuzzaman et al. 2013). Plants mitigate the effect of high temperature by alteration in their metabolism in response to HT stress, particularly by enhancing production of compatible solutes, maintenance of cellular turgidity by osmotic adjustment, and modification of antioxidant defense system to re-establish cellular redox balance thus maintaining homeostasis for optimal growth and development (Janská et al. 2010). Alterations in gene expression at molecular level, particularly of genes involved in direct protection from HT stress viz., genes for osmoprotectants, detoxifying enzymes, transporters, and regulatory proteins serves as a means for protection against HT stress. The HT stress induces modification in various biochemical and physiological processes usually by altered gene expression to develop heat tolerance for acclimation and/or adaptation (Moreno and Orellana 2011). The stability of different proteins, membranes, RNA species and cytoskeleton is affected by heat stress apart from altered efficiency of various enzymatic and non-enzymatic antioxidants. The plant response to HT stress may vary with magnitude, duration and plant type. At extremely high temperature, cellular damage and cell death becomes evident, leading to a catastrophic effect on cellular integrity (Ahuja et al. 2010). Heat stress affects all aspects of plant growth and development severely reducing plant growth and yield (McClung and Davis 2010).

8.3 Common Effector Molecules and Signal Transduction Mechanism Under Multiple Abiotic Stresses

During its life-cycle, plants often exposed to multiple abiotic stresses while growing in arid and semi-arid regions, and leading to accumulation of changes at biochemical and molecular level and fixes as physiological adaptations. Plants have both unique and shared responses towards multiple abiotic stress, which comes together at field conditions, mostly. The knowledge of both shared and individual responses of plants towards multiple abiotic stresses can be utilized to develop broad spectrum stress tolerant cultivars.

8.3.1 Plants Response to Individual and Interactive Abiotic Stress Condition

The stress induced response production of reactive oxygen species is generally a common stress induced response under various abiotic stress conditions. At the onset of concurrent drought and salinity stresses, enhanced transpiration is considered as a negative factor as it might intensify the stress effects due to increased salt uptake and higher water loss by plants (Mittler 2006). It also creates a physiological water deficit and decreased carbon metabolism in plants due to reduced CO_2 diffusion in chloroplast. During the combined heat and drought stress, some plants accumulate sucrose instead of proline (Rizhsky et al. 2002). Furthermore, compartmentation and transportation of ions in plants is also affected as a consequence of combined effect of high temperature and salt stress (Munns 2005).

The interaction between two stress conditions can either be additive when they occur in combined (e.g. drought and salinity) or antagonistic to each other, if occurs individually (Pandey et al. 2015). The stomatal conductance and net photosynthetic rate are decreased while oxidative damage, reduction of chlorophyll b, accumulation of Na^+ in roots are characteristic features of combined drought and salinity as compared to any one kind of stress (Pandey et al. 2015; Ahmed et al. 2013; Rivero et al. 2014). In contrast to heat and salinity stress, the accumulation of Na^+ is more preferred in shoots during salinity stress alone (Ahmed et al. 2013). This change in accumulation pattern of Na^+ under heat and salt stress is supposed to enhance salinity stress tolerance (Rivero et al. 2014). Similarly, higher accumulation of glycine betaine and trehalose over proline has been observed as a feature of combined heat and salinity stress over salinity stress alone. The enhanced production of proline during salinity stress is attributed to the increased activity of the enzyme1-pyrroline-5-carboxylatesynthase (P5CS) which useds glutamate as substrate. However, under combined stress, activity of P5CS gets hampered and activity of ornithineamino-transferase (OAT) get increased, and proline synthesis is occurred from ornithine through ornithine aminotransferase (Krell et al. 2007; Verslues and Sharma 2010). The accumulation of such metabolites is suggested to enhance the tolerance of plants under concurrent stresses (Rivero et al. 2014).

8.3.2 Physio-Biochemical Response of Plants to Various Abiotic Stress

Drought, salinity and heat stresses are mostly concomitant and might led to turgor loss as a result of induced osmotic stress. During stress, cellular membrane may become disorganized, proteins get denature and plants often produce excess levels of reactive oxygen species resulting in oxidative damage. Ultimately, inhibition of photosynthesis, altered metabolism and damaged cellular structures lead to negatively affect the plant growth. The stress induced response in cells includes modification of cell wall architecture, cell cycle/division, membrane system and osmotic adjustment for maintenance of turgor by production of compatible solutes, and maintaining the ionic balance of cells through removing excessive ROSs. Under effect of multiple abiotic stresses, while some effects/responses are common, others are specific i.e. increased level of amino acids, sugars, and sugar alcohols to different stress conditions. Interestingly, proline accumulates during drought and salt stress while not during heat stress. In other case, organic acids and intermediates of TCA-cycle decreased in glycophytes after salt stress, but increased in response to drought or temperature stress. Plants synthesize various osmoprotectants such as amino acids (proline), amines (γ-aminobutyric acid and glycine betaine) and carbohydrates (fructans, monosaccharides and disaccharides, oligosaccharides of trehalose and raffinose family, and polysaccharides e.g. starch) in order to develop resistance to abiotic stress mainly drought (Bhattacharya 2018).

8.3.3 Molecular Response of Plants to Various Abiotic Stresses

During the abiotic stress condition plant expresses its stress inducible mechanism as functional and regulatory proteins. Functional proteins include membrane encoded proteins, various enzymes involved in osmolyte biosynthesis, detoxifying enzymes and proteins involved in protection of cellular macromolecules as LEA protein, molecular chaperons, anti-freezing protein etc. Regulatory proteins cover various genes encoding transcription factors and protein kinases, central to the signal transduction machinery (Xiong and Zhu 2001). A complex array of molecular components involving several metabolic pathways are involved for regulating the abiotic stress tolerance in plants (Rodríguez et al. 2005). As a part of common regulatory mechanism to respond under various abiotic stresses, the induced Ca^{2+} influx activates a MAP-kinase enzyme cascade and phytohormone signaling (e.g., Abscisic acid, Ethylene, Jasmonic acid and Salicylic acid). Induction of molecular chaperons and antioxidant enzymes are other responsive measures. Targeting the above molecular components and adaptive strategies for enhancing abiotic stress tolerance in crop plants could have significant effects (Zhu 2002). The primary signals under draught and salinity cause osmotic and osmotic as well as ionic stress, respectively. Whereas, secondary signals might cause severe oxidative damage leading to metabolic dysfunctions and damage of various cellular biomolecules. As a characteristic feature, abscisic acid accumulation in turn induces several adaptive responses in plants (Zhu 2002).

8.3.4 *Plant Signal Transduction Mechanism Under the Influence of Abiotic Stresses*

As the plant senses any of the environmental stress through sensor molecules, a signal transduction cascade gets initiated (or suppressed) through modulation of the activity of various transcription factors which, in turn, modify the expression of respective stress responsive genes. A schematic representation of signal transduction and plant response at physiological, biochemical and molecular level under abiotic stress conditions is given in Fig. 8.1. During drought, salinity and cold stress influx of Ca^{2+} ions into cytoplasm takes place through ligand-sensitive Ca^{2+} channels, which controls internal Ca^{2+} release and are one type of sensor for these stresses. Receptor like kinases (RLKs) consists of an extra cellular domain which when receives signals, autophosphorylate cytoplasmic histidine residue and transfer phosphoryl moiety to asparate, which may constitute part of sensor protein or separate protein. These sensors either couples with (MAPK) or directly phosphorylates specific targets leading to initiation of cellular responses for particular stress. Phosphorylation and dephosphorylation of protein are the most common signaling molecules (Xiong and Zhu 2001; Rodríguez et al. 2005). For environmental stress i.e. drought, salinity and cold stress, three types of signal transduction networks have been suggested as follows (Xiong and Zhu 2002);

i. Oxidative stress signaling (based on MAP-kinase cascade)
ii. Ca^{2+} dependent signaling (based on activation of LEA-type genes)
iii. Ca^{2+} dependent SOS signaling.

8.3.4.1 Oxidative Stress Signaling

During various abiotic stresses, plant produces reactive oxygen species (ROS) as hydroxyl radicals, hydrogen peroxide and superoxide, which not only causes cellular damages to the plants but also inhibits photosynthesis extensively. In response to these ROS signals plant develops various protective mechanism to scavenge these toxic compounds and lower down their deleterious impact. One of the protective mechanisms includes activation of antioxidants including various enzymes such as catalase, superoxide dismutase, glutathione reductase, ascorbate peroxidase and production of other molecules including ascorbic acid, glutathione, carotenoids, anthocyanins and tocopherol (Xiong and Zhu 2002). MAPK pathways mediates ROS signaling and responsible for over production of osmolytes that play important role in osmotic adjustment by activating receptors/sensors such as protein tyrosine kinases, G-proteincoupled receptors, and two-component histidine kinases. MAP Kinase pathways are likely to be conserved in all eukaryotes and transduces signals from cell surface to nucleus. The core MAPK cascades get sequentially activated by an upstream kinase. In a sequential phosphorylation event, the activated MAP kinase kinase kinase (MAPKKK) phosphorylates (on serine and threonine residues) a MAP

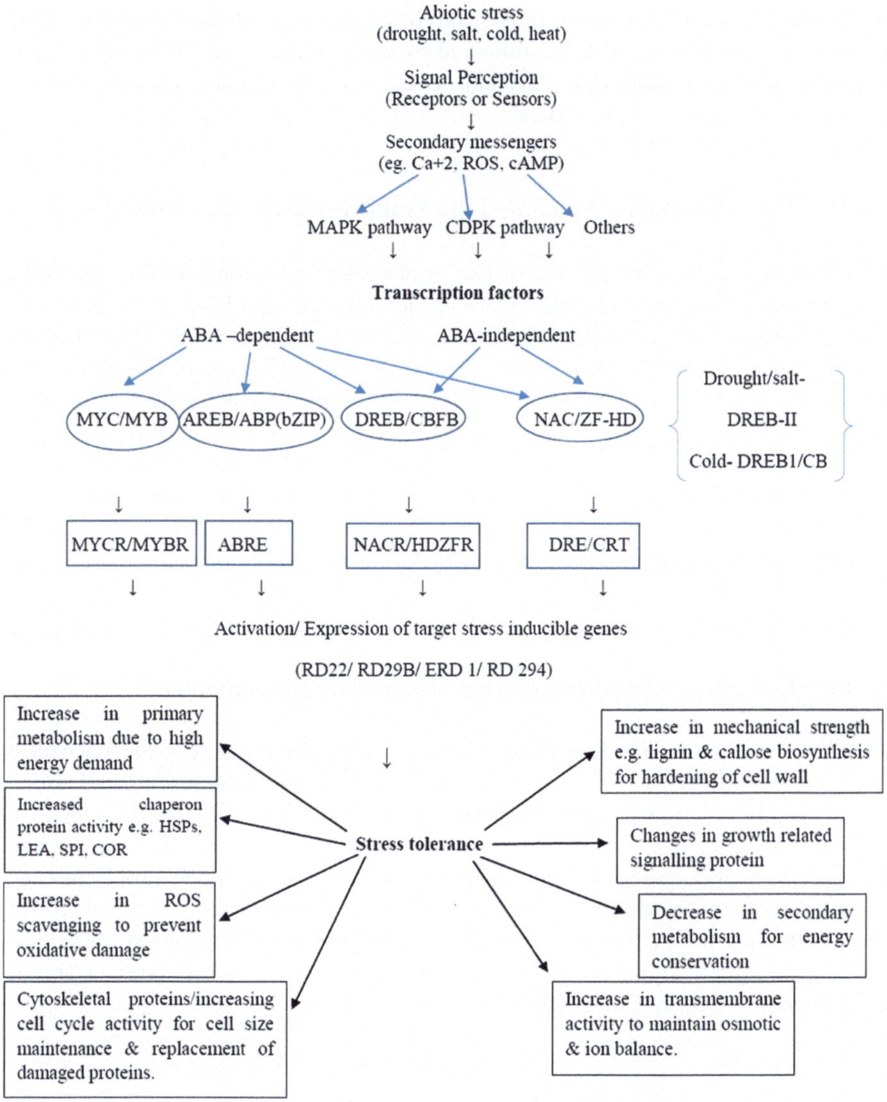

Fig. 8.1 A schematic representation of signal transduction and plant response at physiological, biochemical and molecular level under abiotic stress conditions (modified from Onaga and Wydra 2016; Lata et al. 2011)

kinase kinase (MAPKK) which in turn phosphorylates (on threonine and tyrosine residues) a MAP kinase (MAPK) followed by entry of activated MAPK in nucleus to activate further signaling mechanism or regulating gene expression by directly activating transcription factors (Rodríguez et al. 2005; Xiong and Zhu 2002).

8.3.4.2 Ca^{2+} Dependent Signaling Late Embryogenesis Abundant

In this signaling type, the release of Ca^{2+} (an intracellular ion with low cytosolic concentration under normal stage) from intracellular storage leads to its increased cellular concentration in response to stress signals. Thereafter, the Calcium Dependent Protein Kinases (CDPKs) (serine/threonine protein kinases having C-terminal calmodulin-like Ca^{2+} binding domain) senses Ca^{2+} influx and activates late embryogenesis abundant (LEA)-type genes responsible for activating the pathways to repair the damage caused by abiotic stress (Xiong and Zhu 2002; Serrano et al. 2003). The examples of LEA-type stress responsive genes are dehydration-responsive element (DRE)/C-repeat (CRT) class genes. The proteins encoded by LEA genes prevents protein denaturation and maintain their structure and integrity of membrane (Xiong and Zhu 2002).

8.3.4.3 Ca^{2+} Dependent Salt Overlay Sensitive (SOS) Signaling

The SOS signaling is calcium dependent mechanism. SOS pathway triggers at excess extra or intra cellular Na^+ concentration which induces cytoplasmic Ca^{2+} and affects Na^+, K^+ and H^+ ion transporters. The ionic aspect of salt stress is signaled via the SOS pathway where a calcium responsive SOS_3-SOS_2 protein kinase complex controls the expression and activity of ion transporters such as SOS_1. SOS_3 family of Ca^{2+} binding proteins is an important group of Ca^{2+} sensors in plants (Serrano et al. 2003). The CDPKs and the SOS_3 family of Ca^{2+} sensors play a major role in coupling this universal inorganic signal to specific protein phosphorylation cascades in abiotic stress signaling (Xiong et al. 2002).

8.3.5 Gene Expression Regulation and Transcription Factors

Several genes are involved in abiotic stress responses and regulate gene expression with the help of transcription factors (TFs). Transcription factor are regulatory protein which binds to the specific sites of DNA and alter expression of the target genes. These transcription factors belong to multigene families and often respond differently to various stress or, sometimes share common transcription factors. Overlapped gene-expression profiles can be seen during multiple abiotic stress condition. Therefore, these diverse abiotic stresses often activate similar cell signaling pathways and cellular responses. Various sets of *cis* and trans-acting factors are involved in stress-responsive transcription which are controlled by both ABA-dependent and -independent regulatory systems (Seki et al. 2001; Chen et al. 2002; Fowler and Thomashow 2002; Knight and Knight 2001).

Molecular and genomic analysis suggests different stress responsive transcriptional regulatory mechanism, some of them are ABA dependent and others are not. In ABA regulated genes, ABA-response elements (ABREs) play a key role. During drought stress, the role of two *cis* acting promoter elements, ABRE and dehydration response element (DRE) is pivotal in regulating gene expression. The regulatory role of DRE is also observed in low temperature and salt stress responses (Xiong and Zhu 2002; Mahalingam et al. 2003). The stress-inducible TFs include members of the DREB family, ethylene-responsive element binding factor (ERF) family, basic-domain leucine zipper (bZIP) family, zinc-finger family, basic helix-loophelix (bHLH) family and homeodomain TFs family. These TFs could regulate various stress inducible genes cooperatively or independently. The DREB1A transcription factor specifically interacts with the DRE and induces the expression of stress tolerance genes. ABRE is a major cis-acting element in ABA-responsive gene expression (Su et al. 1998). Similarly, DREB2A and DREB2B transcription factors get activated during osmotic stress (Xiong and Zhu 2002). The specific interaction among transcription factors and target genes of complex nature as multiple TFs might regulate a group of genes or a single gene also (Su et al. 1998).

8.4 Combating Abiotic Stresses Using Physiological and Biochemical Approaches

Various biochemical, physiological and microbiological factors/approaches as potential targets for combating abiotic stresses in plants have been studied. A brief description to these has been presented in following subsections.

8.4.1 Role of Proline and Glycine Betaine

Osmolytes are important plant metabolites for maintaining osmotic pressure in plant tissues. Two major osmolytes i.e. Proline and Glycine betaine (GB) are known for enhancing adaptive responses of plants under varying types of abiotic stresses. Although it's still a matter of debate whether increased level and accumulation of these two is a product of stress or an adaptive strategy by plants, a large amount of literature is available in support of these assumptions. In addition, varying level of tolerance to stresses was shown by exogenous application of these osmolytes on different crop species. Furthermore, transgenic plants with enhanced content of these metabolites were reported to have considerable degree of tolerance to stresses (Ashraf and Foolad 2007; Hayat et al. 2012).

8.4.2 Role of Plant Antioxidants in Improving Plant Tolerance to Stress

As mentioned in Sect. 8.3.4.1 the enzymatic and non-enzymatic antioxidants play critical role in the detoxification of ROSs (Ahmad et al. 2016). Increased activities of the mentioned antioxidant enzymes under stress conditions help in efficient detoxification of the generated ROS and protect DNA from oxidative damage, lipid peroxidation and membrane stability along with other positive effects to varying extent in different plant species. In this context, it is proposed that plant tolerance to abiotic stress may be enhanced by increased antioxidant profile of plant.

8.4.3 Role of Mineral Nutrition and Management of Nutrient Supply

In plants, numerous functions of mineral elements have been reported such as electron carriers, activation of enzymes, component of structure of various biomolecules, provides osmotic balance and maintain charge balance in cells (Waraich et al. 2011). Literature evidences towards significant contribution of mineral nutrients in management of abiotic stress in plants revealed an enormous opportunity to deal with environment induced stress. Along with recommended fertilizers applications, targeted studies using potassium (K; essential and most abundant cation in plants) and zinc (Zn) minerals have been reported with positive effect on plant resistance to stress (Wang et al. 2013). Overexpression of K transporter also positively regulates stress response of rice plants (Chen et al. 2017). Application of K may act via increasing membrane stability, osmolyte management and regulating stomatal closure for enhancing plant tolerance (Wang et al. 2004). Although K plays a key role in multiple plant growth and developmental aspects, a deeper understanding of mechanism of K in whole plant-stress response is required which will in turn help in recommendation of optimal K requirement to tackle stresses in plants (Wang et al. 2013). In some recent and relevant studies, Silicon application has been a matter of great concern because of its beneficial application towards alleviating negative effects of abiotic stresses (Chen et al. 2011; Romero-Aranda et al. 2006; Liang et al. 2007). Silicon-mediated alleviation of stress in plants is mediated via enhancement in gas exchange attributes and photosynthetic pigments; changes in mineral nutrients uptake and plant growth and biomass and enhancement in antioxidant defense system (Liang et al. 2007).

8.4.4 Role of Plant Growth Hormones

It is well known that plants produce various growth regulators at different developmental stages and play crucial role in adapting plants in stress conditions (Peleg and Blumwald 2011). Ethylene and abscisic acid are produced at maturity and stress conditions, respectively, as a response of metabolic changes in plants. There are different reports of combating abiotic stresses in plants by use of Brassinosteroids (BRs) (Zargar et al. 2019) and Salicylic acid (SA) (Khan et al. 2015). As indicated by name, BRs are of immense importance in abiotic stress conditions due to their involvements in multiple plant growth developmental activities such as morphogenesis and regulating expression of hundreds of important genes, cell division etc. (Fariduddin et al. 2014; Choudhary et al. 2012). For improving plant abiotic stress-tolerance, significance of SA has got recognition due to SA-mediated regulation of major plant-metabolic processes. The underlying mechanism of this SA-induced tolerance is however unclear (Khan et al. 2015).

8.5 Combating Abiotic Stresses Using Molecular Approaches

At molecular level, the approaches for combating abiotic stress are divided into pre- and post-genomic era. In pre-genomic era, the forward genetics was the major approach for identification of abiotic stress (strong light, UV, high and low temperatures, freezing, drought, salinity, heavy metals and hypoxia) tolerant genes. Marker assisted selection (MAS) was the broadly used method for confirming transfer of tolerant gene or region in desired plants e.g. incorporation of HKT allele from *Triticum monococcum* to durum wheat for improving salinity tolerance (Flowers and Yeo 1995; James et al. 2012). Molecular markers were also used in transfer of genes *Kin1* in *Arabidopsis*, *OsB28* in rice, *PKABA1* in wheat for salinity, cold and drought stress (Winicov 1998). In addition, other molecular approaches such as induced mutations, somaclonal variations, differential display and differentiation screening have been used but reported to have limited success in understanding mechanism and finding candidates of abiotic stress tolerance.

The post-genomic era is focusing on functional genomics approaches for fighting abiotic stresses. In functional genomics, function of gene and intergenic regions of the genome identified. Also, how such genes or regions on a "genome-wide" scale contribute to different biological processes can be studied. Functional genomics focuses on the dynamic expression of gene products in a specific context, for example, at a specific developmental stage or during a drought condition. Functional genomics mainly utilize sequence-based, hybridization-based and gene inactivation-based or genome-editing based methodologies to study the abiotic stress tolerance in a collective manner on a genome-wide scale, which can be further utilized for elucidation of abiotic stress networks.

8.5.1 Sequencing Based Approaches

Expressed Sequence Tag (EST) sequencing method was the earliest genome wide sequence-based approach and has been successfully used for several stresses using different tissues and developmental stages to identify several specific and stress-responsive transcripts. However these approaches suffers from one major issue that is they under-represent rare transcripts or unexpressed transcripts under certain conditions. EST analysis required cDNA library preparation and offer easier strategy for gene discovery and genome annotation. ESTs are indexed in National Center for Biotechnology Information (NCBI) dbEST database from various crops and plant species of arid and semi-arid regions (Ergen and Budak 2009). Further, the information on gene number and gene families playing significant roles in abiotic stress responses can be established by clustering the sequences of ESTs obtained through respective stress-treated cDNA libraries (Li et al. 2014a, b). Extensive efforts have been made to compare abundance of expressed ESTs in Arabidopsis, rice and halophytes such as *Thellungiellahalophila* and *Spartina alterniflora* (Wang et al. 2004; Baisakh et al. 2008).

Next approach is Serial Analysis of Gene Expression (SAGE) in which oligo(dT)-trapped mRNA is reverse transcribed to cDNA from which small tags are excised to form concatemers and sequenced, providing quantification of gene expression under different stress conditions. SAGE coupled with next-generation sequencing technologies, emerged as a high throughput, sensitive and cost-effective approach (Cheng et al. 2013). Additionally, by combining 5′ RACE (Rapid amplification of cDNA ends) transcription start sites in key differentially expressed genes were also identified (Wei et al. 2004). Studies using SAGE in plants not only revealed new expressed regions in the plant genome but also implied their novel functions including stress response in crops (Cheng et al. 2013).

Another tag sequencing-based approach is Massively Parallel Signature Sequencing (MPSS) in which tagged PCR products obtained from cDNA are amplified to produce hundred thousand of PCR products ligated to microbeads and sequenced (Kudapa et al. 2013). This method enables the parallel analysis of millions of transcripts on a genome-wide scale (Akpınar et al. 2013). The high throughput analysis and longer tags in MPSS can detect novel transcripts particularly in non-model plant or crop species lacking whole genome sequence (Hamilton and Robin Buell 2012). Expression data for several genotypes of economically important crops such as soybean, maize and rice, is publicly available in plant MPSS database (http://mpss.udel.edu/) (Nakano et al. 2006). Further, the advancements in NGS platforms have expanded genome-wide sequence expression analysis through RNA sequencing (RNA-seq) and exome sequencing (Sánchez-León et al. 2012; Mo et al. 2018). The digital gene expression system through RNA- and micro-RNA sequencing has been demonstrated in crops under abiotic stress conditions in different tissues (Pandey et al. 2014).

8.5.2 Hybridization Based Approaches

In post-genomic era differential gene expression in response to abiotic stress has been analysed alternatively through hybridization-based approaches using gene chips and microarray technology. However, the methods represent targeted approach with closed system and required prior sequences of gene and transcripts to design probes (Mehboob-ur-Rahman et al. 2016). During abiotic stresses plants respond and adapt by altering physiological and biochemical processes leading to altered responses of thousands of genes analysed through customised microarray chips (Gul et al. 2016). In microarray, DNA sequences representing genes of a plant are placed on microchips and used as substrates for hybridization for quantifying expression of different genes in a sample, thus provide complete quantitative information about relative expression of genes corresponding to their response under various abiotic stresses (Wu et al. 2015). This approach has major technical limitations including cross-hybridization and background noise that affect microarray analysis of stress responsive genes. Extensive microarray studies for different stresses have been reported in Arabidopsis (Richards et al. 2012), rice (Jung and An 2013), wheat (Quijano et al. 2015), corn (Allardyce et al. 2013), soybean (Le et al. 2012), tomato (Martínez-Andújar et al. 2012) and expression data available in public domain (Source: www.genevestigator.com/gv/plant.jsp).

More advanced hybridisation-based approach is GeneChip Genome Array used to detect expression of several genes at the same time in the whole genome (Wu et al. 2015). In contrast to microarray, several hundred thousand oligonucleotides are synthesized using photolithography on miniature gene chips (Joshi et al. 2012). In pigeon pea using this technique genome-wide expression pattern characterised for drought response and candidate drought responsive genes were identified (Saxena et al. 2011).

8.5.3 Gene Inactivation-Based Approaches

Using the high throughput whole genome sequencing methods large number of differentially expressed genes or transcripts of different pathways and processes can be identified involved in abiotic stress response in various crop species, however their molecular function and role is unknown. Functional characterization of differentially expressed genes achieved through two main approaches based on gene inactivation, namely T-DNA insertion mutation and TILLING (Targeted Induced Local Lesions In Genomes).

TILLING is used to generate novel mutant alleles that enable high-throughput genome-wide analysis of point mutations in target genomes including diploids and allohexaploids for crop improvement (Lee et al. 2014; Chen et al. 2014). The TILLING populations can be traditionally screened for phenotypic or genotypic variations under abiotic stresses (De Lorenzo et al. 2009). Further, more advanced modified

TILLING method called EcoTilling is used to identify SNPs and small indels in polyploid species for differentiating among alleles of paralogous and homologous genes (Akpınar et al. 2013). This method has been used in rice for identifying SNPs involved in salinity stress tolerance, in barley for identifying INDELs and SNPs involved in drought stress tolerance and in chickpea for identifying allelic variants having different components of transcription factor genes (Negrao et al. 2011; Cseri et al. 2011; Bajaj et al. 2016).

T-DNA insertional mutagenesis is more popular method utilized as a tool to identify function of gene in Arabidopsis and other crop plants (Jung and An 2013). Random insertion of T-DNA fragments in either exon or intron results in the target gene inactivation. A large number of sequence-indexed T-DNA insertion lines for Arabidopsis (https://www.arabidopsis.org/portals/mutants/stockcenters.jsp) and rice (RiceGE, http://signal.salk.edu/cgi-bin/RiceGE) are available in public domain libraries. The T-DNA insertion mutants are used for elucidating metabolic/signalling pathways and for functional analysis of genes in plants (Gao and Zhao 2013).

RNA interference (RNAi) technology is a method for suppression of gene expression. In this method, micro RNA (miRNA) and small interfering RNA (siRNA) are used for silencing of gene through RNA Induced Silencing Complex (RISC) mediated target gene degradation. This method has been used for confirming the functional aspects of SOS_2 (a serine/threonine type protein kinase) and SOS_3 (a calcium binding protein) in *Arabidopsis*, rice, wheat and *Brassica* (Kumar et al. 2009; Yang et al. 2009; Kushwaha et al. 2011; Feki et al. 2014). Under salinity stress SOS_3 interacts with SOS_2 after receiving cytoplasmic calcium signals which further activates SOS_1 (Na^+/H^+ antiporter gene) leading to homeostasis in tolerant plant (Sharma et al. 2015).

8.5.4 Genome Editing Based Approaches

Genome editing offer targeted mutation, INDEL and sequence modifications to a predetermined location within the genome contrary to TILLING and T-DNA insertion based random mutation or random gene silencing (Strange and Petolino 2012). Most commonly used genome editing tools are Zinc finger nucleases (ZFNs), transcriptional activator-like effector nucleases (TALENs) and clustered regularly interspaced short palindromic repeat (CRISPR)-Cas9 (CRISPR-associated nuclease9) (Kumar and Jain 2014). ZFNs are failed to provide desired specificity of gene editing in plants (Gaj et al. 2012). Further, TALENs have emerged as an alternative to ZFNs, providing targeted gene editing at very high success rate through double-strand breaks. TALENs have limitation of delivery in recombinant adeno-associated viruses (AAV) due to their large size (Gaj et al. 2013). The cutting-edge technology for targeted genome editing is type II clustered regularly interspaced short palindromic repeat (CRISPR)/Cas9 system from Streptococcus pyogenes (Bortesi and Fischer 2015). The sgRNA was designed against target gene which guides Cas9 nuclease to target DNA leading to double strand breaks to initiate non-homologous end joining and homologous recombination repair pathways, resulting in genome modifications (Zhang et al. 2016). The results can be analysed for loss-of-function, gain-of-function and gene expression in spatio-temporal manner. The application of CRISPR/Cas9 in understanding gene function, gene regulatory networks and engineering abiotic stress tolerance in crop was reviewed in Khatodia et al. (Khatodia et al. 2016).

8.6 Conclusion

In last few decades, extensive work has been carried out to get a deeper insight into plant abiotic stress, its consequences and mechanism. Plants respond to abiotic stress through changes in activities of various enzymes (e.g. Kinases, Phosphatases and Antioxidant enzymes) involved in regulatory pathways, production of certain metabolites (including osmolytes, growth hormones, ROSs, proline, HSPs etc.) and altered gene expression via involvement of signaling proteins including transcription factors. Plants, in arid and semi-arid region experience multiple abiotic stresses (Drought, Salinity and Temperature) simultaneously. Thus, there may be considerable degree of cross talks among the signaling cascades initiated by different stresses, implying that mechanism involved in regulation of plant responses under multiple biotic stresses is complex. For obvious reasons, it is difficult to predict the exact response every time in continuously changing environmental condition in addition to variable experimental conditions. These instances however, provide opportunities to develop strategies to counteract negative effect of abiotic stress by targeting either one or more effective response mechanism. It will be interesting to unveil common regulatory networks operating in regulation of multiple abiotic stresses and thereby to identify molecular targets for development of crop plants tolerant to multiple abiotic stresses. Comprehensive systems based approaches involving, metabolomics, phenomics and genomics should pave the way to understand the intricacies of mechanisms of abiotic stress response at the finer level.

References

Ahmad P, Rasool S, Gul A, Sheikh SA, Akram NA, Ashraf M, Kazi A, Gucel S (2016) Jasmonates: multifunctional roles in stress tolerance. Front Plant Sci 7:813

Ahmed IM, Cao F, Zhang M, Chen X, Zhang G, Wu F (2013) Difference in yield and physiological features in response to drought and salinity combined stress during anthesis in Tibetan wild and cultivated barleys. PLoS ONE 8:e77869

Ahuja I, de Vos RC, Bones AM, Hall RD (2010) Plant molecular stress responses face climate change. Trends Plant Sci 15:664–674

Akpınar BA, Lucas SJ, Budak H (2013) Genomics approaches for crop improvement against abiotic stress. Sci World J

Allardyce JA, Rookes JE, Hussain HI, Cahill DM (2013) Transcriptional profiling of Zea mays roots reveals roles for jasmonic acid and terpenoids in resistance against Phytophthora cinnamomi. Funct Integr Genomics 13:217–228

Ashraf M, Foolad M (2007) Roles of glycine betaine and proline in improving plant abiotic stress resistance. Environ Exp Bot 59:206–216

Baisakh N, Subudhi PK, Varadwaj P (2008) Primary responses to salt stress in a halophyte, smooth cordgrass (*Spartina alterniflora* Loisel.). Funct Integr Genomics 8:287–300

Bajaj D, Srivastava R, Nath M, Tripathi S, Bharadwaj C, Upadhyaya HD, Tyagi AK, Parida SK (2016) EcoTILLING-based association mapping efficiently delineates functionally relevant natural allelic variants of candidate genes governing agronomic traits in chickpea. Front Plant Sci 7:450

Bhattacharya A (2018) Abiotic stress and plant physiology: Vol. 01: metabolic activities. New India Publishing Agency, India

Bortesi L, Fischer R (2015) The CRISPR/Cas9 system for plant genome editing and beyond. Biotechnol Adv 33:41–52

Boyer JS (1982) Plant productivity and environment. Science 218:443–448

Chen W, Provart NJ, Glazebrook J, Katagiri F, Chang H-S, Eulgem T, Mauch F, Luan S, Zou G, Whitham SA (2002) Expression profile matrix of Arabidopsis transcription factor genes suggests their putative functions in response to environmental stresses. Plant Cell 14:559–574

Chen W, Yao X, Cai K, Chen J (2011) Silicon alleviates drought stress of rice plants by improving plant water status, photosynthesis and mineral nutrient absorption. Biol Trace Elem Res 142:67–76

Chen L, Hao L, Parry MA, Phillips AL, Hu YG (2014) Progress in TILLING as a tool for functional genomics and improvement of crops. J Integr Plant Biol 56:425–443

Chen G, Liu C, Gao Z, Zhang Y, Jiang H, Zhu L, Ren D, Yu L, Xu G, Qian Q (2017) OsHAK1, a high-affinity potassium transporter, positively regulates responses to drought stress in rice. Front Plant Sci 8:1885

Cheng CK, Au CH, Wilke SK, Stajich JE, Zolan ME, Pukkila PJ, Kwan HS (2013) 5′-Serial analysis of gene expression studies reveal a transcriptomic switch during fruiting body development in *Coprinopsis cinerea*. BMC Genom 14:195

Choudhary SP, Yu J-Q, Yamaguchi-Shinozaki K, Shinozaki K, Tran L-SP (2012) Benefits of brassinosteroid crosstalk. Trends Plant Sci 17:594–605

Creswell R, Martin FW (1998) Dry land farming: crops & techniques for arid regions. ECHO technical note, Fort Myers, Fl, USA

Cseri A, Cserháti M, Von Korff M, Nagy B, Horváth GV, Palágyi A, Pauk J, Dudits D, Törjék O (2011) Allele mining and haplotype discovery in barley candidate genes for drought tolerance. Euphytica 181:341

De Lorenzo L, Merchan F, Laporte P, Thompson R, Clarke J, Sousa C, Crespi M (2009) A novel plant leucine-rich repeat receptor kinase regulates the response of *Medicago truncatula* roots to salt stress. Plant Cell 21:668–680

dos Reis SP, Lima AM, de Souza CRB (2012) Recent molecular advances on downstream plant responses to abiotic stress. Int J Mol Sci 13:8628–8647

Ergen NZ, Budak H (2009) Sequencing over 13000 expressed sequence tags from six subtractive cDNA libraries of wild and modern wheats following slow drought stress. Plant, Cell Environ 32:220–236

FAO (2009) Declaration of the world summit on food security, 16–18 Nov 2009, Rome

Fariduddin Q, Yusuf M, Ahmad I, Ahmad A (2014) Brassinosteroids and their role in response of plants to abiotic stresses. Biol Plant 58:9–17

Feki K, Quintero FJ, Khoudi H, Leidi EO, Masmoudi K, Pardo JM, Brini F (2014) A constitutively active form of a durum wheat Na+/H+ antiporter SOS1 confers high salt tolerance to transgenic Arabidopsis. Plant Cell Rep 33:277–288

Flowers T, Yeo A (1995) Breeding for salinity resistance in crop plants: where next? Funct Plant Biol 22:875–884

Fowler S, Thomashow MF (2002) Arabidopsis transcriptome profiling indicates that multiple regulatory pathways are activated during cold acclimation in addition to the CBF cold response pathway. Plant Cell 14:1675–1690

Gaj T, Guo J, Kato Y, Sirk SJ, Barbas CF III (2012) Targeted gene knockout by direct delivery of zinc-finger nuclease proteins. Nat Methods 9:805

Gaj T, Gersbach CA, Barbas CF III (2013) ZFN, TALEN, and CRISPR/Cas-based methods for genome engineering. Trends Biotechnol 31:397–405

Gao Y, Zhao Y (2013) Epigenetic suppression of T-DNA insertion mutants in Arabidopsis. Mol Plant 6:539–545

Gul A, Ahad A, Akhtar S, Ahmad Z, Rashid B, Husnain T (2016) Microarray: gateway to unravel the mystery of abiotic stresses in plants. Biotechnol Lett 38:527–543

Hamilton JP, Robin Buell C (2012) Advances in plant genome sequencing. Plant J 70:177–190

Harris D, Tripathi R and Joshi A (2002) On-farm seed priming to improve crop establishment and yield in dry direct-seeded rice. Direct seeding: research strategies and opportunities. International Research Institute, Manila, Philippines, pp 231–240

Hasanuzzaman M, Nahar K, Alam M, Roychowdhury R, Fujita M (2013) Physiological, biochemical, and molecular mechanisms of heat stress tolerance in plants. Int J Mol Sci 14:9643–9684

Hayat S, Hayat Q, Alyemeni MN, Wani AS, Pichtel J, Ahmad A (2012) Role of proline under changing environments: a review. Plant Signaling Behav 7:1456–1466

Hussain M, Malik M, Farooq M, Ashraf M, Cheema M (2008) Improving drought tolerance by exogenous application of glycinebetaine and salicylic acid in sunflower. J Agron Crop Sci 194:193–199

(IPCC) IPoCC (2007) Climate change 2007–The physical science basis. In Contribution of Working Group I to the Fourth Assessment Report of the Intergovernmental Panel on Climate Change

James RA, Blake C, Zwart AB, Hare RA, Rathjen AJ, Munns R (2012) Impact of ancestral wheat sodium exclusion genes Nax1 and Nax2 on grain yield of durum wheat on saline soils. Funct Plant Biol 39:609–618

Janská A, Maršík P, Zelenková S, Ovesná J (2010) Cold stress and acclimation–what is important for metabolic adjustment? Plant Biol 12:395–405

Joshi R, Karan R, Singla-Pareek SL, Pareek A (2012) Microarray technology. In: Gupta AK, Guptha SM, Pareek A (eds) Biotechnology in medicine and agriculture: principles and practices. IK International Publishing House Pvt. Ltd., India, pp 273–296

Jung K-H, An G (2013) Functional characterization of rice genes using a gene-indexed T-DNA insertional mutant population. Rice Protocols. Springer, Heidelberg, pp 57–67

Kaya MD, Okçu G, Atak M, Cıkılı Y, Kolsarıcı Ö (2006) Seed treatments to overcome salt and drought stress during germination in sunflower (*Helianthus annuus* L.). Eur J Agron 24:291–295

Khan T, Mazid M, Mohammad F (2011) A review of ascorbic acid potentialities against oxidative stress induced in plants. J Agrobiology 28:97–111

Khan MIR, Fatma M, Per TS, Anjum NA, Khan NA (2015) Salicylic acid-induced abiotic stress tolerance and underlying mechanisms in plants. Front Plant Sci 6:462

Khatodia S, Bhatotia K, Passricha N, Khurana S, Tuteja N (2016) The CRISPR/Cas genome-editing tool: application in improvement of crops. Front Plant Sci 7:506

Knight H, Knight MR (2001) Abiotic stress signalling pathways: specificity and cross-talk. Trends Plant Sci 6:262–267

Koyro H-W, Ahmad P, Geissler N (2012) Abiotic stress responses in plants: an overview. Environmental adaptations and stress tolerance of plants in the era of climate change. Springer, Heidelberg, pp 1–28

Krell A, Funck D, Plettner I, John U, Dieckmann G (2007) Regulation of proline metabolism under salt stress in the psychrophilic diatom *Fragilariopsis cylindrus* (Bacillariophyceae) 1. J Phycol 43:753–762

Kudapa H, Ramalingam A, Nayakoti S, Chen X, Zhuang W-J, Liang X, Kahl G, Edwards D, Varshney RK (2013) Functional genomics to study stress responses in crop legumes: progress and prospects. Funct Plant Biol 40:1221–1233

Kumar V, Jain M (2014) The CRISPR–Cas system for plant genome editing: advances and opportunities. J Exp Bot 66:47–57

Kumar G, Purty RS, Sharma MP, Singla-Pareek SL, Pareek A (2009) Physiological responses among Brassica species under salinity stress show strong correlation with transcript abundance for SOS pathway-related genes. J Plant Physiol 166:507–520

Kushwaha HR, Kumar G, Verma PK, Singla-Pareek SL, Pareek A (2011) Analysis of a salinity induced BjSOS3 protein from *Brassica* indicate it to be structurally and functionally related to its ortholog from *Arabidopsis*. Plant Physiol Biochem 49:996–1004

Lata C, Yadav A, Prasad M (2011) Role of plant transcription factors in abiotic stress tolerance. Abiotic stress response in plants-physiological, biochemical and genetic perspectives, IntechOpen

Le DT, Nishiyama R, Watanabe Y, Tanaka M, Seki M, Yamaguchi-Shinozaki K, Shinozaki K, Tran L-SP (2012) Differential gene expression in soybean leaf tissues at late developmental stages under drought stress revealed by genome-wide transcriptome analysis. PLoS ONE 7:e49522

Lee LS, Till BJ, Hill H, Huynh OA, Jankowicz-Cieslak J (2014) Mutation and mutation screening. Cereal genomics. Springer, Heidelberg, pp 77–95

Li F, Fan G, Wang K, Sun F, Yuan Y, Song G, Li Q, Ma Z, Lu C, Zou C (2014a) Genome sequence of the cultivated cotton *Gossypium arboreum*. Nat Genet 46:567

Li J, Sun X, Yu G, Jia C, Liu J, Pan H (2014b) Generation and analysis of expressed sequence tags (ESTs) from halophyte Atriplex canescens to explore salt-responsive related genes. Int J Mol Sci 15:11172–11189

Liang Y, Sun W, Zhu Y-G, Christie P (2007) Mechanisms of silicon-mediated alleviation of abiotic stresses in higher plants: a review. Environ Pollut 147:422–428

Mahalingam R, Gomez-Buitrago A, Eckardt N, Shah N, Guevara-Garcia A, Day P, Raina R, Fedoroff NV (2003) Characterizing the stress/defense transcriptome of Arabidopsis. Genome Biol 4:R20

Martínez-Andújar C, Pluskota WE, Bassel GW, Asahina M, Pupel P, Nguyen TT, Takeda-Kamiya N, Toubiana D, Bai B, Górecki RJ (2012) Mechanisms of hormonal regulation of endosperm cap-specific gene expression in tomato seeds. Plant J 71:575–586

McClung CR, Davis SJ (2010) Ambient thermometers in plants: from physiological outputs towards mechanisms of thermal sensing. Curr Biol 20:R1086–R1092

Mehboob-ur-Rahman Rahmat Z, Gul M, Zafar Y (2016) Plant functional genomics: approaches and applications. In: Khan MS, Khan IA, Barh D (eds) Applied molecular biotechnology: the next generation of genetic engineering, 1st edn. CRC Press, Boca Raton, pp 157–186

Mittler R (2006) Abiotic stress, the field environment and stress combination. Trends Plant Sci 11:15–19

Mo Y, Howell T, Vasquez-Gross H, de Haro LA, Dubcovsky J, Pearce S (2018) Mapping causal mutations by exome sequencing in a wheat TILLING population: a tall mutant case study. Mol Genet Genomics 293:463–477

Moreno AA, Orellana A (2011) The physiological role of the unfolded protein response in plants. Biol Res 44:75–80

Munns R (2005) Genes and salt tolerance: bringing them together. New Phytol 167:645–663

Murphy LR, Kinsey ST, Durako MJ (2003) Physiological effects of short-term salinity changes on Ruppia maritima. Aquat Bot 75:293–309

Nakano M, Nobuta K, Vemaraju K, Tej SS, Skogen JW, Meyers BC (2006) Plant MPSS databases: signature-based transcriptional resources for analyses of mRNA and small RNA. Nucleic Acids Res 34:D731–D735

Negrao S, Almadanim C, Pires I, McNally K, Oliveira M (2011) Use of EcoTILLING to identify natural allelic variants of rice candidate genes involved in salinity tolerance. Plant Genet Resour 9:300–304

Onaga G, Wydra K (2016) Advances in plant tolerance to abiotic stresses. Plant Genomics, IntechOpen

Pandey R, Joshi G, Bhardwaj AR, Agarwal M, Katiyar-Agarwal S (2014) A comprehensive genome-wide study on tissue-specific and abiotic stress-specific miRNAs in *Triticum aestivum*. PLoS ONE 9:e95800

Pandey P, Ramegowda V, Senthil-Kumar M (2015) Shared and unique responses of plants to multiple individual stresses and stress combinations: physiological and molecular mechanisms. Front Plant Sci 6:723

Peleg Z, Blumwald E (2011) Hormone balance and abiotic stress tolerance in crop plants. Curr Opin Plant Biol 14:290–295

Phukan UJ, Mishra S, Timbre K, Luqman S, Shukla RK (2014) Mentha arvensis exhibit better adaptive characters in contrast to Mentha piperita when subjugated to sustained waterlogging stress. Protoplasma 251:603–614

Quijano CD, Brunner S, Keller B, Gruissem W, Sautter C (2015) The environment exerts a greater influence than the transgene on the transcriptome of field-grown wheat expressing the Pm3b allele. Transgenic Res 24:87–97

Rahneshan Z, Nasibi F, Moghadam AA (2018) Effects of salinity stress on some growth, physiological, biochemical parameters and nutrients in two pistachio (*Pistacia vera* L.) rootstocks. J Plant Interact 13:73–82

Ramegowda V, Senthil-Kumar M (2015) The interactive effects of simultaneous biotic and abiotic stresses on plants: mechanistic understanding from drought and pathogen combination. J Plant Physiol 176:47–54

Rani S, Sharma MK, Kumar N (2019) Impact of salinity and zinc application on growth, physiological and yield traits in wheat. Curr Sci 00113891:116

Richards CL, Rosas U, Banta J, Bhambhra N, Purugganan MD (2012) Genome-wide patterns of Arabidopsis gene expression in nature. PLoS Genet 8:e1002662

Rivero RM, Mestre TC, Mittler R, Rubio F, Garcia-Sanchez F, Martinez V (2014) The combined effect of salinity and heat reveals a specific physiological, biochemical and molecular response in tomato plants. Plant, Cell Environ 37:1059–1073

Rizhsky L, Liang H, Mittler R (2002) The combined effect of drought stress and heat shock on gene expression in tobacco. Plant Physiol 130:1143–1151

Rodríguez M, Canales E, Borrás-Hidalgo O (2005) Molecular aspects of abiotic stress in plants. Biotecnología Aplicada 22:1–10

Romero-Aranda MR, Jurado O, Cuartero J (2006) Silicon alleviates the deleterious salt effect on tomato plant growth by improving plant water status. J Plant Physiol 163:847–855

Sánchez-León N, Arteaga-Vázquez M, Alvarez-Mejía C, Mendiola-Soto J, Durán-Figueroa N, Rodríguez-Leal D, Rodríguez-Arévalo I, García-Campayo V, García-Aguilar M, Olmedo-Monfil V (2012) Transcriptional analysis of the Arabidopsis ovule by massively parallel signature sequencing. J Exp Bot 63:3829–3842

Saxena RK, Cui X, Thakur V, Walter B, Close TJ, Varshney RK (2011) Single feature polymorphisms (SFPs) for drought tolerance in pigeonpea (Cajanus spp.). Funct Integr Genomics 11:651–657

Seki M, Narusaka M, Abe H, Kasuga M, Yamaguchi-Shinozaki K, Carninci P, Hayashizaki Y, Shinozaki K (2001) Monitoring the expression pattern of 1300 Arabidopsis genes under drought and cold stresses by using a full-length cDNA microarray. Plant Cell 13:61–72

Serrano R, Gaxiola R, Ríos G, Forment J, Vicente O, Ros R (2003) Salt stress proteins identified by a functional approach in yeast. Monatshefte für Chemie/Chemical Monthly 134:1445–1464

Sharma R, Mishra M, Gupta B, Parsania C, Singla-Pareek SL, Pareek A (2015) De novo assembly and characterization of stress transcriptome in a salinity-tolerant variety CS52 of Brassica juncea. PLoS ONE 10:e0126783

Soussi M, Ocana A, Lluch C (1998) Effects of salt stress on growth, photosynthesis and nitrogen fixation in chick-pea (*Cicer arietinum* L.). J Exp Bot 49:1329–1337

Strange TL, Petolino JF (2012) Targeting DNA to a previously integrated transgenic locus using zinc finger nucleases. Transgenic plants. Springer, Heidelberg, pp 391–397

Su J, Shen Q, Ho T-HD, Wu R (1998) Dehydration-stress-regulated transgene expression in stably transformed rice plants. Plant Physiol 117:913–922

Turan S, Tripathy BC (2013) Salt and genotype impact on antioxidative enzymes and lipid peroxidation in two rice cultivars during de-etiolation. Protoplasma 250:209–222

Turan S, Cornish K, Kumar S (2012) Salinity tolerance in plants: breeding and genetic engineering. Aust J Crop Sci 6:1337

Verslues PE, Sharma S (2010) Proline metabolism and its implications for plant-environment interaction. The Arabidopsis Book/American Society of Plant Biologists 8

Wang Z-l, Li P-h, Fredricksen M, Gong Z-z, Kim C, Zhang C, Bohnert HJ, Zhu J-K, Bressan RA, Hasegawa PM (2004) Expressed sequence tags from *Thellungiella halophila*, a new model to study plant salt-tolerance. Plant Sci 166:609–616

Wang M, Zheng Q, Shen Q, Guo S (2013) The critical role of potassium in plant stress response. Int J Mol Sci 14:7370–7390

Waraich EA, Ahmad R, Ashraf M (2011) Role of mineral nutrition in alleviation of drought stress in plants. Aust J Crop Sci 5:764

Wei C-L, Ng P, Chiu KP, Wong CH, Ang CC, Lipovich L, Liu ET, Ruan Y (2004) 5′ Long serial analysis of gene expression (LongSAGE) and 3′ LongSAGE for transcriptome characterization and genome annotation. Proc Natl Acad Sci 101:11701–11706

Winicov I (1998) New molecular approaches to improving salt tolerance in crop plants. Ann Bot 82:703–710

Wu M, Chen A, Wang Z, Zhang Z, Wang C, Li F, Wei P, Wang R, Luo Z, Wei C (2015) Plant microarray for gene expression profiling and their application. J Agric Technol 11:93

Xiong L, Zhu JK (2001) Abiotic stress signal transduction in plants: molecular and genetic perspectives. Physiol Plant 112:152–166

Xiong L, Zhu JK (2002) Molecular and genetic aspects of plant responses to osmotic stress. Plant, Cell Environ 25:131–139

Xiong L, Schumaker KS, Zhu J-K (2002) Cell signaling during cold, drought, and salt stress. Plant Cell 14:S165–S183

Yang Q, Chen Z-Z, Zhou X-F, Yin H-B, Li X, Xin X-F, Hong X-H, Zhu J-K, Gong Z (2009) Overexpression of SOS (Salt Overly Sensitive) genes increases salt tolerance in transgenic Arabidopsis. Mol Plant 2:22–31

Zargar SM, Mahajan R, Bhat JA, Nazir M, Deshmukh R (2019) Role of silicon in plant stress tolerance: opportunities to achieve a sustainable cropping system. 3 Biotech 9:73

Zhang B, Yang X, Yang C, Li M, Guo Y (2016) Exploiting the CRISPR/Cas9 system for targeted genome mutagenesis in petunia. Sci Rep 6:20315

Zhu J-K (2002) Salt and drought stress signal transduction in plants. Annu Rev Plant Biol 53:247–273

Chapter 9
Applications of Landscape Genetics to Study the Effect of Varying Landscapes and Environmental Challenges in Plant Populations

Akshay Nag, Anshu Alok and Kashmir Singh

Abstract Evolutionary processes like adaptation, dispersal and genetic drift play an important role in shaping a geographical habitat of the organisms and species. To study these processes in a geographical context is fundamental for the biogeographical studies. Nevertheless, the introduction of new molecular methods allowed a stronger assessment of the associations between the patterns of genetic diversity between populations and the micro-evolutionary processes governing them. Traditionally the use of molecular methods has been applied to study the evolutionary biology, phylogenetics and population structure of the plant species to formulate conservation programs for declining populations and these approaches are collectively referred to as *Molecular Ecology*. Recently a more robust approach, which has emerged as an altogether separate discipline known as *Landscape Genetics* has been used for studying the effects of changing environment due to human intervention, habitat fragmentation and effect of varying landscapes. This approach takes help from the disciplines of *Molecular Ecology, Population Genetics and Biogeography*, aided by the recent advances in simulation and remote sensing technology. Landscape genetics has recently evolved as a discipline, through which, we can incorporate the effect of local habitats and connectivity of the landscapes to analysis of gene flow. It also allows us to have a better understanding of local adaptation processes by helping to develop novel hypothesis on impending selection forces. It plays ever more significant role in species conservation and management studies. Recent advances in molecular tools and statistic has also aided researchers to device more precise strategies for these type of studies. The improvements in sequencing technologies have profoundly enhanced our ability to study genetic variation in wild species, which has opened up new and unparalleled opportunities for genetic analysis in conservation biology. In this chapter we will have an insight on the basic aspects of this discipline, tools and methods used in it and the advent of molecular biology techniques, which serve as the backbone of the research projects in this discipline.

A. Nag · A. Alok · K. Singh (✉)
Department of Biotechnology, Panjab University, South Campus, Sector 25, Chandigarh, India
e-mail: kashmirbio@pu.ac.in

S. P. Singh et al. (eds.), *Molecular Approaches in Plant Biology and Environmental Challenges*, Energy, Environment, and Sustainability,
https://doi.org/10.1007/978-981-15-0690-1_9

9.1 Introduction

The most basic level of diversity in the biological systems is Genetic variation and it is a prerequisite to ascertain the variability of species, populations, and ecosystems. Genetic diversity is also important for the fitness and endurance of individuals, the feasibility of populations, and for the species to be able to adapt themselves to change in ecological conditions (Allendorf and Luikart 2009; Frankham et al. 2002). Therefore, conservation of genetic diversity is very important, and researchers in various disciplines, like ecology, evolution, and conservation, are concerned to understand the elements that shape genetic variation patterns in nature. The concepts of ascertaining genetic diversity are available for more than 100 years, but during those times, laboratory techniques were not available to quantify genes or DNA. Due to these constraints, considerable amount of the initial research by population geneticists was theoretical and conceptual. This scenario changed after the structure of DNA was discovered and even more so after PCR (polymerase chain reaction) was developed. Thanks to PCR, it was now possible to obtain large amounts of DNA even from very little samples, and this technique transformed many research areas, including medicine, forensics, genetic engineering, and population genetics.

The human activities have left an enormous impact on environment and the biological diversity and human-induced fragmentation and loss of habitats were found to be its major cause (Wilcove et al. 1986) For the populations to survive in the changing landscapes, the competence to move between remaining patches of habitat was recognized as crucial for the conservation of species and populations in fragmented landscapes (Levins 1969; Hanski 1998). The effects of climate change and ever fluctuating environmental conditions also became a pivotal issue of landscape ecology, which developed into separate discipline in the 1980s (Naveh and Lieberman 2013). Simultaneous developments in so many research areas led scientists to combine the methods and concepts of population genetics and landscape ecology to evaluate the effect of environmental heterogeneity on genetic diversity and gene flow (Pamilo 1988; Merriam et al. 1989; Manicacci et al. 1992; Gaines et al. 1997). However, it was not until 2003, when Manel et al. formally introduced the discipline of "landscape genetics" (Manel et al. 2003). Following this introduction, many new approaches were introduced for analyzing landscape genetic data (Murphy et al. 2008; Guillot et al. 2005). Presently, landscape genetics is still a highly dynamic and fast evolving field. Novel approaches and methods are regularly coming up and answers to new problems are being sought as a result of improvements in theoretical as well as technological knowhow. Landscape genetics as an altogether different discipline is having a rapid growth, which is really thrilling and inspiring, but it also offers incredible challenges.

9.1.1 *Definition*

The term 'landscape genetics' was proposed by Manel et al. (2003) and recently a comprehensive definition has been proposed by Balkenhol et al. (2015) as "***a research that combines population genetics, landscape ecology, and spatial analytical techniques to explicitly quantify the effects of landscape composition, configuration, and matrix quality on microevolutionary processes, such as gene flow, drift, and selection, using neutral and adaptive genetic data.***"

9.2 Landscape Ecology: The Basics

Ecology as a discipline can be defined as the interaction between species and their environment (Arthur 1935). The varying habitats and environments in which organisms dwell and interact are spatially structured at various scales (Levin 1992). Landscape ecology studies deal with understanding the consequences of landscape, geographical location and distance on the organisms and the interactions between organisms and the environment (Wiens et al. 1993).

9.2.1 *Landscape*

From an ecological point of view, a landscape can be seen as a matrix of various interacting ecological processes and patterns at a particular scale. From a spatial point of view, a landscape can be thought of as a region that is structurally diverse in at least one of the factors of interest (Turner 2005), whereas from an organism's point of view, it can be viewed as a diverse distribution of conditions and resources, like the conditions which govern an ecological position or niche. However, it should be remembered that the size of the landscape doesn't necessarily define it, which implies that the landscapes are defined by the spatially heterogeneous matrix like rivers, valleys, mountains, dense forests, barren patch etc. it also varies according to the perspective of the phenomenon under observation, which is generally the population genetic structure and the events that are responsible for it, like gene flow, genetic drift or adaptive evolution (Manel et al. 2003; Segelbacher et al. 2010).

9.2.2 *Effect of Landscapes on Genetic Processes*

The habitats and the environments where organisms live vary spatially or three dimensionally through measures, and these have an effect on the behavior and evolution of the organism, which leads to the higher-level processes of adaptive evolution, gene flow and population dynamics (Wiens et al. 1993; Johnson et al. 1992). The matrix, structure and arrangement of the landscape components are crucial in defining the distribution of the populations and gene flow. These components also interact with different genetic characteristics and affect the genetic fitness. Particularly, a disturbance to habitat stability may affect the ecological processes essential for population perseverance (Fahrig 2003). There are so many means through which landscape composition and structure affects population processes, the major factors governing the population processes are as follows: (1) effects of area (2) effects of edge and (3) effects of isolation.

9.2.2.1 Area Effects

Most of the species require a minimum area of their habitat to withstand and meet all the requirements to support their life history (Robbins et al. 1989). Theoretically, each population requires a threshold level of habitat, or minimum area below which the populations are not able to sustain themselves (Fahrig 2003; Flather and Bevers 2002; Hill and Caswell 1999). The volume of this threshold habitat, varies from species to species and depends upon specific characteristics related to the development and life-history of the populations (Vance et al. 2003; With and King 1999). Consequently, the effects of loss of habitat on a particular species or a particular population of the species depends upon the interaction of its ecological demands and capabilities to cope up or adapt to the extent of loss in habitat in the area adjoining the landscape (Fahrig 2003; McGarigal and Cushman 2002; Schmiegelow and Mönkkönen 2002). Loss of habitat and its fragmentation changes the dynamics of availability of resources and this might have an effect on the behavior of individuals and their interaction with the altering habitat, which will subsequently change the competence of the species to secure the minimum requirement of resources needed to reproduce and persist (Wiens et al. 1993; Mangel and Clark 1996).

9.2.2.2 Edge Effects

The boundaries of the habitat, known as the edges may act as a buffer zone for the adaptive behavior of an individual and subsequently the populations. These edges have one of the strongest impacts on the processes affecting population genetics of the organism. Edges also have a role to play in affecting the movement and fitness of the organism. The edges which are formed by natural discontinuities or gaps in geographical or geophysical features are referred to as “inherent” edges, whereas those developed mostly due to human induced disturbances are known as “induced” edges. These type of edges might be comparatively everlasting features of the landscape, produced by gaps in basic factors like land–water interface, or these might be temporary features of a landscape, induced by turbulences e.g., deforestation due to harvesting of timber or forest fire. As discussed earlier, the edges are the buffer zone and are created as result of combination of climatic and human induced or natural disturbances. These events often produce marked changes in the structure, composition and texture of landscapes along the edges. Changes in the climatic factors like light, moisture, temperature and wind affect seedling production, growth and survival of various forest plant species (Wales 1972; Gates and Mosher 1981; de Casenave et al. 1995). Some species get benefited from the altered microclimate along the edges whereas others do not. Some species adapt to the new microclimate, others perish (Chen et al. 1992).

9.2.2.3 Isolation Effects

The gene flow among the populations depends upon the heterogeneity of the landscape and the species under observation. Some abiotic factors may act as barrier for one species, while not so for others. Therefore, the effect of landscape heterogeneity acts as one of the crucial drivers to shape the spatial genetic structure of the populations and it decides the dispersal patterns and subsequent isolation of individuals and native populations. When heterogeneity (i.e. differences in landscape texture like valleys, trenches, canyons, rivers, deserts, mountains etc.) in a landscape increases, movement of organisms gets affected, which results in varying levels of isolation which depends on the richness, dissemination, distribution and dispersal abilities of the species (McCoy and Mushinsky 1999; Rukke 2000; Virgós 2001; Bender et al. 1998; Tischendorf et al. 2003).

9.2.3 Population Dynamics and Dispersal

Connectivity between populations depends upon the following aspects: (1) the structure, organization and resistance of the landscape, (2) population density and extent of its distribution, and (3) the distribution of the organism (Cushman and Landguth 2012). Most of the conventional theory of population genetics is centered on some presumptions where we assume that the populations are separately constrained, and internally panmictic. When no evolutionary force like non-random mating, selection, mutation, gene flow, meiotic drive or random genetic drift acts upon the populations, the allele frequency and genotypic frequencies of a population will stay constant in the subsequent generations (Hartl et al. 1997). These assumptions are not followed by any natural populations though. In reality, one or more of these assumptions are bound to be violated by the populations. Its effect on populations depends on the assumptions which get violated. Non-random mating commonly results in inbreeding due to which homozygosity for all the genes increases, which leads to deviations from Hardy-Weinberg ratios. One of the other major causes is the absence of panmixia and greater possibility of mating with individuals in the proximity, which in turn is theorized as isolation-by-resistance or isolation-by distance (IBD). IBD is a highly common phenomenon observed in studies of landscape genetics. Furthermore, genetic migration or gene flow offers genetic connectivity between two or more populations via long distance dispersal.

9.2.4 Influence of Landscapes on Genetic Variation

Historically, studies in population genetics have been concentrated on studying the genetic structure of populations and modifications in genetic variation produced as a result of the mutation, selection, gene flow and drift. Genetic variation can be defined as the differences in the DNA sequence at the same locus. The loci on which these variations are observed may be under selection (non-neutral loci/adaptive loci) or may not be under selection (neutral loci). The major causes of genetic variation found in a population or a species are–genetic drift, mutation, gene flow, and selection (Allendorf and Luikart 2009; Hedrick 1995). As a result of mutation, new alleles or genetic variants are created because of errors in DNA replication. Selection (mostly natural) has an impact on genetic variation, as it increases the frequencies of alleles in favor of a specific environment and decreases the frequencies of alleles that are less favorable in that environment. Due to Gene flow, alleles are moved between populations and due to these phenomenon populations tend to get genetically more similar. Genetic drift is another evolutionary force acting upon the populations, which causes the change in allele frequencies as alleles get passed on from generation to generation. This happens due to random sampling effects (Wright 1931). Thus, number of breeding individuals have an influence on genetic drift.

Genetic drift has a direct association with the effective population size (N_e). The concept of N_e was put forward by Wright (1931) to describe the effects of genetic drift. N_e can be termed as the number of breeding individuals in a model population, which have genetic contribution in the next generation. In simple words, N_e can be explained as the number of individuals successful in passing on their genes to the next generation. Populations having higher N_e will have high genetic variation across generations, since more new alleles are being produced due to mutation as a result of larger number of recombination events. Large N_e would also mean that fewer alleles are lost at the hands of genetic drift, because of larger sample size of breeders in each generation. N_e also influences the genetic drift and selection in natural populations (Wright 1931). If N_e decreases, the genetic drift causes a decrease in the effects of selection and it may result in the loss of important adaptive genetic variation (Hedrick 1995).

There are a lot of ways through which landscapes have an influence on N_e and on the distribution and amount of genetic variation. First of all, the nature of the landscape will decide whether or not individuals or populations reside in a particular area or not. It influences the habitat of an organism and the number of individuals living in a particular location. Consequently, landscape has an effect on the amount of genetic variation. Moreover, the landscape also has an influence on the amount of movement, dispersal and gene flow which has a direct effect on N_e and genetic variation.

9.2.5 Gene Flow

The phenomenon of movement of genes among populations, via successful mating is known as Gene flow. High amounts of gene flow mean that the populations tend to get homogenized which results in similar allele frequencies. On the other hand, if gene flow is low, it results in isolation of population and genetic differentiation. Consecutively, if N_e is small, genetic drift will come into action, which will cause genetic differentiation through arbitrary modifications in allele frequencies. Nevertheless, theoretically only a single migrant in each generation is all that is needed to overcome genetic drift (Hedrick 1995).

9.3 Overview of Dna Types and Molecular Methods

9.3.1 Types of DNA

The DNA found in the plant cells is of three main types: nuclear DNA (nDNA), chloroplast DNA (cpDNA), and mitochondrial DNA (mtDNA). MtDNA gets inherited uniparentally and generally follows maternal inheritance (Gillham 1974; Birky 1978), however, paternal inheritance has also been documented in some conifer species (Neale et al. 1989). Nuclear DNA is the DNA which is present inside the nucleus and has biparental mode of ingeritance. On the other hand, the cpDNA is the DNA found inside the chloroplasts. The mode of its inheritance differs from species to species. It can be paternal, maternal, or biparental (Gillham 1974; Neale et al. 1989). On an average, the N_e of nDNA loci is approximately four times higher than the N_e of both mtDNA and cpDNA, since these are generally uniparentally inherited and are haploid.

In population genetics, the research question is influenced by the type of DNA, the mode of inheritance and the mutation rate. Generally, in landscape genetic studies nDNA loci are used more often (90%) than cpDNA or mtDNA loci (Storfer et al. 2010) since they deliver comparatively enhanced resolution for identifying genotypic variations in the recent history.

9.3.2 Molecular Methods

The two major kinds of molecular approaches involving nDNA, also known as molecular marker systems are codominant and dominant methods. In codominant systems e.g. SNP (single-nucleotide polymorphism) or SSR (Simple Sequence repeats or microsatellite analysis) analysis, we can visualize both alleles in the form of bands on DNA gel electrophoresis, at a particular locus (AA vs. AB). While, in case of dominant loci methods, such as RAPD (random amplified polymorphic DNA) or

AFLP (amplified fragment length polymorphism), heterozygous loci could not be identified as we can see only one band per locus. The data of the dominant loci are generally scored in a binary (1/0) format, where 0 means absence and 1 means presence. The most commonly used approaches are summarized in Box 1, as it is not possible to describe the in depth account of these methods.

Box 1. Molecular methods: an overview Allozymes—Allozymes are used to detect allelic variations of proteins coded by genes.

SSR—Microsatellite analyses or simple sequence repeats (SSRs), (also known as short tandem repeats (STRs) or "μsats" are tandemly repeating sequences of 2–6 base.

AFLP analysis—AFLP is a method which combines the power of PCR and precision of restriction enzymes for analyzing the genetic variation and knowledge of the DNA sequence is not a prior requirement. The amplified fragments of DNA are separated on the basis of their size on a polyacrylamide gel and visualized using fluorescence or autoradiography (Vos et al. 1995).

SNP analysis—It is a comparatively new method of molecular analysis which analyses single base pair polymorphisms in various loci across the genome. This method is not very frequently used in landscape genetics, but the its application is predicted to increase quickly in the coming years because of its larger coverage of genome and possibility to be automated for high throughput analyses (Garvin et al. 2010; Morin et al. 2004).

9.3.3 *Measures of Genetic Diversity*

9.3.3.1 Metrics of Population Analysis

Genetic diversity assessment is the most important objective of landscape genetic studies. It plays the pivotal role in assessing the spatial structure and extent of relatedness within or between the populations under study. There are multiple ways to quantify genetic diversity. When we consider populations for analysis of genetic diversity, there are four main measures of genetic diversity for codominant loci: (i) allele number, (ii) allelic richness (iii) proportion of polymorphic loci and (iv) heterozyosity.

Genetic diversity can be assessed by using the average proportion of alleles per locus (A); but, value of A and the confidence levels vary significantly with the change in sample size. Using Allelic richness (A_r) can solve this problem. yY using rarefaction or subsampling techniques, differences in sample size could be normalized to assess A_r, (Kalinowski 2004; Leberg 2008; El Mousadik and Petit 1996).

Heterozygosity has two main components: observed heterozygosity and expected. Observed heterozygosity (H_o) can be defined as the ratio of individuals which are heterozygous at a particular locus. Expected heterozygosity (H_e) is measured by observing the assumptions of Hardy-Weinberg equilibrium and is calculated by using observed frequency of alleles (Nei 1978). Values of H_o and H_e are assessed for every locus in a population and then their average is calculated across all the loci.

Polymorphism (P) is calculated as the number of loci which contain more than one allele i.e. polymorphic. P is used less frequently for landscape genetic studies polymorphic loci are more informative for measuring diversity.

A_r is the one of the most sensitive components for identifying recent losses in genetic diversity (Allendorf and Luikart 2009). When the location of the populations under observation is not very clear and the sampling of the accessions is more continuous, an alternative approach is to group the individuals into clusters, which are genetically most similar in view of the population structure and then calculate these metrics for each of these clusters (Shirk and Cushman 2011).

For cpDNA and mtDNA sequence data, there is less variation and the sequences corresponding to these types of DNA are more or less conserved. Therefore, the main diversity metrics to be considered are haplotype number, nucleotide diversity and haplotypic diversity. The term haplotype has been derived from haploid genotype and it is simply the haploid form of a genotype. Haplotypic diversity (h), or gene diversity, is the expected heterozygosity in haploid form. Nucleotide diversity (π) is the average difference in the nucleotides at individual loci along DNA sequence. In case of dominant loci, generally the average number of bands per population are considered to calculate diversity, but equivalents of H_e and π can also be estimated and used under particular assumptions (Bonin et al. 2007).

9.3.3.2 Genetic Structure

Evaluation of genetic structure between populations and individuals is the main objective of a landscape genetic analysis. Panmictic populations or completely randomly mating populations have no significant variations in the allele frequencies and gene flow estimates also tend to get finite. However, the patterns of gene flow are governed by environmental variables, landscape patterns as well as geographic distance between populations. Over the course of generations, these factors generally limit the gene flow and as a result specific patterns of subdivision in populations and genetic structure is produced. The visualization of these patterns can be done using measures of genetic structure or genetic distance, barrier detection methods and Bayesian clustering analysis.

Wright's F statistics developed by Sewall Wright is the oldest and most popular metric to assess genetic structure. F statistics are used to evaluate the genetic variation partitioning within a species (Wright 1931). There are three different derivatives of F-statistics, i.e. (1) among the total population (T) or all populations being considered, (2) the subpopulation (S) or individuals and (3) within a subpopulation (I). F_{ST}, which is most widely used to define the genetic structure of the populations, represents genetic differentiation at subpopulation-level w.r.t. the total population. It ranges from 0 which means population is in panmixia to 1 which means complete genetic isolation. The other two F-statistics, F_{IS} and F_{IT}, which are also standard genetic estimates are also used to measure the genetic structure. For more thorough information on Wright's F-statistics, see Meirmans and Hedrick (2011).

After Wright, numerous F_{ST} equivalents were proposed for the estimation of genetic differentiation among populations. G_{ST}, was developed by Nei (1972), which is similar to F_{ST} and moreover its extension. It considers several alleles at a particular locus in place of only two, as proposed by Wright's model. Another metric i.e. analysis of variance (ANOVA) approach was developed by Weir and Cockerham (1984) to estimate an F_{ST} analog theta (θ).

Slatkin (1995) developed R_{ST} specially for loci like SSRs with high numbers of alleles and heterozygosity. However this metric is not recommended now, as very few loci fall under the assumptions of this model (Balloux et al. 2000) (Meirmans and Hedrick 2011). More recently, additional F_{ST} analogs like G'_{ST} and G''_{ST} and substitute metrics like Jost's D have been introduced to improve the analyses for samples having a small number of populations with high heterozygosity, or those likely to be out of Hardy-Weinberg's equilibrium (Hedrick 1995; Meirmans and Hedrick 2011; Jost 2008). Despite of plethora of metrics developed through the course of time, F_{st} and G_{st} are still the most widely used metrics for assessing the population structure till date.

To assess these metrics, there are various freely accessible software, which can be used to analyze the landscape genetic data and assess different statistics and metrics corresponding to gene flow between populations, like: GENEPOP (Rousset 2008), FSTAT (Goudet 1995) and GENALEX (Peakall and Smouse 2006). A large number software packages based on R platform are also available for population genetics calculations, like: gstudio (Dyer 2014), pegas, and genetics (Warnes 2015).

9.3.4 Bayesian Clustering Methods

Apart from the above discussed traditional approaches to evaluate the genetic structure, there is another approach used, known as Bayesian clustering analysis. In these methods, instead of whole populations, an individual is treated as basic unit of diversity assessment. Genetic lineage from source populations is estimated on the basis of the number of distinct genetic clusters generated by analyzing individuals. On the basis of generated clusters, group affiliation is allocated to individuals and genotypes (Pritchard et al. 2000). This approach of evaluating the genetic structure to define populations on the basis of individuals and multilocus genotypes is incorporated in the software STRUCTURE, and since its introduction, this approach has become one of the most frequently used methodologies in population genetics and landscape genetic studies. Other Bayesian based clustering methods have also been introduced which are based on individuals in the software such as TESS (Chen et al. 2007), BAPS5 (Corander et al. 2008) and GENELAND (Guillot et al. 2005).

In case of moderate to high genetic differentiation between subpopulations, these methodologies could be predominantly effective for describing populations and identifying landscape barriers (Chen et al. 2007; Blair et al. 2012; Latch and Rhodes 2005; Safner et al. 2011).

9.3.5 Barrier Detection Methods

Barriers play defining role in structuring the genetic populations through a landscape and their identification is essential from a conservation and management point of view. Barriers induce genetic isolation between populations and as a result of it genetic diversity is lost during the course of generations through genetic drift. Bayesian clustering and edge detection are the two main approaches used to identify barriers in landscape genetic and population genetic studies (Safner et al. 2011). In the Bayesian clustering approaches, the most likely groups of genetically associated individuals are assessed on the basis of their multilocus genotypes.

On the other hand, the Edge detection methods could be of use identify barriers between groups of individuals or genotypes from different populations. For example, Monmonier's maximum difference algorithm (Monmonier 2010) and Wombling (Womble 1951) basically assign putative barriers to geographic zones where the differences in allele frequencies or genetic distance is found to be large even over short geographic distances.

9.4 Genomic Approaches in Landscape Genetics

The discovery of structure of DNA revolutionized the biological research. Gradual advances in the DNA technologies led to rapid modifications of the available methods (Table 9.1) to quantify processes behind the science of genetics. Population genetics as a discipline got revolutionized with the advances in the molecular biology techniques, especially genomics. The current advances of NGS platforms has more or less transformed the field landscape genetics by providing extraordinary statistical power to assess the estimates of population genetic factors such as gene flow and population structure (Safner et al. 2011; Coop et al. 2010).

NGS technologies has made it possible to utilize a massive number of loci in population genetics. These huge datasets have arrived with the challenges of data handling and analyses. This sub discipline of landscape genetics, in which NGS technologies are used and the subsequent data is analyzed using complex computational tools has been termed as "landscape genomics" (Joost et al. 2007; Eckert et al. 2010; Manel et al. 2010a). At the heart of the landscape genomics approach is to study large sets of genetic loci ranging from hundreds to thousands, distributed

Table 9.1 Features of major approaches to landscape genomics

S. No.	Method	Statistics	Approach	Inference
1	QTL mapping	Interval mapping, ANOVA (marker regression), maximum likelihood	Large numbers of markers are surveyed and then mapped across the genome. Mapping populations are used to identify the genomic regions and establish association with the phenotype	Identification of genomic regions, which are associated with a phenotype by establishing a significant correlation with the trait
2	GWAS	Search for linkage disequilibrium between alleles and phenotype linkage disequilibrium (LD) between phenotype and genotype	Large number of markers, even larger than QTLs. One set of association panel can be used to establish association with many traits	Identification of a locus (SNP)linked with different traits
3	Candidate gene	Genotyping of individuals with different phenotypes, correlation of genotype with phenotype establishing correlation of genotype to phenotype	Prior knowledge of loci from other studies of same or other species and search for the association of differences in alleles to phenotype	allelic differences may lead to their association with the trait, which can subsequently establish association with the environment
4	Exomics, transcriptomics	Comparison of copy number of mRNA in different phenotypes to study the effect of a particular environmental conditions	NGS base high throughput approach of sequencing cDNA libraries	Differences in the phenotype may be associated the differences in the gene expression

in the genomes of individuals from different habitats or having different traits or phenotypes to determine the genomic regions under selection (Bonin et al. 2007; Stinchcombe and Hoekstra 2008).

9.5 Effect of Climate Change and Spatial Data Collection

Landscape genomics has made it possible to incorporate a great number of environmental as well as biotic factors and their effect on the populations (Coop et al. 2010; Manel et al. 2010a). These can range from various climatic features to assess the effect of changing climate on the spatial distribution of adaptive genetic variation (Manel et al. 2010b).

Connectivity between the populations is one of the highly important aspects for long survival of a species, because the habitat is changing at a very rapid speed and many species might not get any time to evolve through the process of natural selection. If we want to devise any management and conservation strategies for various species, the amount of genetic variance and population structure is the only resource that we have to work with. In the rapidly changing environments, the most primary concern of any species, which have to survive is the amount of genetic variation found in different populations. As investigations of landscape fragmentation effects the processes that have operated since a long time, it is highly likely that the pattern of the environmental factors affecting genetic connectivity may change dramatically (García-Ramos and Rodríguez 2002).

To study the effects of changing climates on the structure of the populations, it's necessary to have high quality climate data to test the hypothesis of the spatial distribution and genetic variation patterns in the populations. To achieve the statistical significance from the analyses, it's imperative to have environmental data for the past several years. There are GIS databases, where environmental data are readily available, but the biggest challenge in collecting fresh high resolution GIS data is that, it is highly time consuming and a bottleneck for the new studies. Hence these already available GIS datasets serve as goldmines for landscape geneticists. These data sets include various important constituents like climatic vegetation and geological variables e.g., WorldClim, http://www.worldclim.org and The Global Map Project, http://www.iscgm.org/cgi-bin/fswiki/wiki.cgi. Many global environmental data sets are freely available now due to an increase in the environmental data from weather stations and remote sensors. More information on the GIS resources is available in (Manel et al. 2010a). These databases serve as valuable substitute in the absence of local environmental data.

9.6 Conclusion and Future Perspectives

Current is an era of fast growth in landscape genetics, as we are coming through novel possibilities of data analysis and different idea and methods to carry out the landscape genetic studies. It is not surprising that new researchers in this field get confused or overwhelmed to an extent by noticing the current complexities and challenges in landscape genetics and the variety of research approaches. They might feel sometimes helpless due to the plethora of statistical approaches currently followed in this field. This complexity is bound to increase even more in the coming years. If we just consider the advances in whole genome sequencing and bioinformatics, we can observe that these advances are leading to a transformation which offers incredible prospects and challenges to this field. The methods for analyzing and examining the landscape genetic data is getting sophisticated day by day (Kranstauber et al. 2014; Merow et al. 2014). To take full advantage of these opportunities, we have to efficiently combine the current research in landscape genetics, that has focused on unfolding the population structure and associating it with landscape attributes, with the advances in the fields of genomics, bioinformatics, and other evolving approaches in experimental and field settings (Cushman and Landguth 2012). To accomplish this, landscape geneticists have to get more aware of the advances in the NGS platforms and the computational analyses, which are associated with these technologies so that results can be generalized and their inferences could be explored across scales of biological organization, from nucleotides to ecosystems.

References

Allendorf FW, Luikart G (2009) Conservation and the genetics of populations. Wiley. ISBN 1444309056

Arthur G (1935) Tansley the use and abuse of vegetational concepts and terms. Ecology 16:284–307

Balkenhol N, Cushman S, Storfer A, Waits L (2015) Landscape genetics: concepts, methods, applications. John Wiley & Sons. ISBN 1118525280

Balloux F, Brunner H, Lugon-Moulin N, Hausser J, Goudet J (2000) Microsatellites can be misleading: an empirical and simulation study. Evolution (N. Y) 54:1414–1422

Bender DJ, Contreras TA, Fahrig L (1998) Habitat loss and population decline: a meta-analysis of the patch size effect. Ecology 79:517–533

Birky CW (1978) Transmission genetics of mitochondria and chloroplasts. Annu Rev Genet 12:471–512

Blair C, Weigel DE, Balazik M, Keeley ATH, Walker FM, Landguth E, Cushman S, Murphy M, Waits L, Balkenhol N (2012) A simulation-based evaluation of methods for inferring linear barriers to gene flow. Mol Ecol Resour 12:822–833

Bonin A, Ehrich D, Manel S (2007) Statistical analysis of amplified fragment length polymorphism data: a toolbox for molecular ecologists and evolutionists. Mol Ecol 16:3737–3758

Chen J, Franklin JF, Spies TA (1992) Vegetation responses to edge environments in old-growth douglas-fir forests. Ecol Appl 2:387–396

Chen C, Durand E, Forbes F, François O (2007) Bayesian clustering algorithms ascertaining spatial population structure: a new computer program and a comparison study. Mol Ecol Notes 7:747–756

Coop G, Witonsky D, Di Rienzo A, Pritchard JK (2010) Using environmental correlations to identify loci underlying local adaptation. Genetics 185:1411–1423
Corander J, Marttinen P, Sirén J, Tang J (2008) Enhanced bayesian modelling in BAPS software for learning genetic structures of populations. BMC Bioinform 9:539
Cushman SA, Landguth EL (2012) Multi-taxa population connectivity in the Northern Rocky Mountains. Ecol Modell 231:101–112
de Casenave JL, Pelotto J, Protomastro J (1995) Edge-interior differences in vegetation structure and composition in a Chaco semi-arid forest, Argentina. Ecol Manag 72:61–69
Dyer R (2014) GSTUDIO: analyses and functions related to the spatial analysis of genetic marker data. R package version 1.3
Eckert AJ, Bower AD, González-Martínez SC, Wegrzyn JL, Coop G, Neale DB (2010) Back to nature: ecological genomics of loblolly pine (Pinus taeda, Pinaceae). Mol Ecol 19:3789–3805
El Mousadik A, Petit RJ (1996) High level of genetic differentiation for allelic richness among populations of the argan tree [*Argania spinosa* (L.) Skeels] endemic to Morocco. Theor Appl Genet 92:832–839
Fahrig L (2003) Effects of habitat fragmentation on biodiversity. Annu Rev Ecol Evol Syst 34:487–515
Flather C, Bevers M (2002) Patchy reaction-diffusion and population abundance: the relative importance of habitat amount and arrangement. Am Nat 159:40–56
Frankham R, Briscoe DA, Ballou JD (2002) Introduction to conservation genetics. Cambridge university press, ISBN 0521639859
Gaines M, Diffendorfer J, Tamarin RH (1997) The effects of habitat fragmentation on the genetic structure of small mammal populations. J Hered 88:194–304
García-Ramos G, Rodríguez D (2002) Evolutionary speed of species invasions. Evolution (N.Y) 56:661–669
Garvin MR, Saitoh K, Gharrett AJ (2010) Application of single nucleotide polymorphisms to non-model species: a technical review. Mol Ecol Resour 10:915–934
Gates J, Mosher JA (1981) A functional approach to estimating habitat edge width for birds. Am Midl Nat 189–192
Gillham NW (1974) Genetic analysis of the chloroplast and mitochondrial genomes. Annu Rev Genet 8:347–391
Goudet J (1995) FSTAT (version 1.2): a computer program to calculate F-statistics. J Hered 86:485–486
Guillot G, Mortier F, Estoup A (2005) GENELAND: a computer package for landscape genetics. Mol Ecol Notes 5:712–715
Hanski I (1998) Metapopulation dynamics. Nature 396:41
Hartl DL, Clark AG, Clark AG (1997) Principles of population genetics Sinauer Associates, vol 116. Sunderland, MA
Hedrick PW (1995) Gene flow and genetic restoration: the Florida panther as a case study. Conserv Biol 9:996–1007
Hill MF, Caswell H (1999) Habitat fragmentation and extinction thresholds on fractal landscapes. Ecol Lett 2:121–127
Johnson A, Wiens J, Milne B (1992) TO crist animal movements and population dynamics in heterogeneous landscapes. Landsc Ecol 7:63–75
Joost S, Bonin A, Bruford MW, Després L, Conord C, Erhardt G, Taberlet P (2007) A spatial analysis method (SAM) to detect candidate loci for selection: towards a landscape genomics approach to adaptation. Mol Ecol 16:3955–3969
Jost L (2008) GST and its relatives do not measure differentiation. Mol Ecol 17:4015–4026
Kalinowski ST (2004) Counting alleles with rarefaction private alleles and hierarchical sampling designs. Conserv Genet 5:539–543
Kranstauber B, Safi K, Bartumeus F (2014) Bivariate Gaussian bridges: directional factorization of diffusion in Brownian bridge models. Mov Ecol 2:5

Latch E, Rhodes OE (2005) The effects of gene flow and population isolation on the genetic structure of reintroduced wild turkey populations: agenetic signatures of source populations. Conserv Genet 6:981–997
Leberg PL (2008) Estimating allelic richness: effects of sample size and bottlenecks. Mol Ecol 11:2445–2449
Levin SA (1992) The problem of pattern and scale in ecology: the Robert H. MacArthur Award Lecture Ecology 73:1943–1967
Levins R (1969) Some demographic and genetic consequences of environmental heterogeneity for biological control. Am Entomol 15:237–240
Manel S, Schwartz MK, Luikart G, Taberlet P (2003) Landscape genetics: combining landscape ecology and population genetics. Trends Ecol Evol 18:189–197
Manel S, Poncet BN, Legendre P, Gugerli F, Holderegger R (2010a) Common factors drive adaptive genetic variation at different spatial scales in Arabis alpina. Mol Ecol 19:3824–3835
Manel S, Joost S, Epperson BK (2010b) Perspectives on the use of landscape genetics to detect genetic adaptive variation in the field. Mol Ecol 19:3760–3772
Mangel M, Clark CW (1996) Towards a unifield foraging theory. Ecology 67:1127–1138
Manicacci D, Olivieri I, Perrot V, Atlan A, Gouyon P-H, Prosperi J-M, Couvet D (1992) Landscape ecology: population genetics at the metapopulation level. Landsc Ecol 6:147–159
McCoy ED, Mushinsky HR (1999) Habitat fragmentation and the abundances of vertebrates in the Florida scrub. Ecology 80:2526–2538
McGarigal K, Cushman SA (2002) Comparative evaluation of experimental approaches to the study of habitat fragmentation effects. Ecol Appl 12:335–345
Meirmans PG, Hedrick PW (2011) Assessing population structure: FST and related measures. Mol Ecol Resour 11:5–18
Merow C, Dahlgren J, Metcalf CJE (2014) Advancing population ecology with integral projection models: a practical guide. Methods Ecol Evol 5:99–110
Merriam G, Kozakiewicz M, Tsuchiya E, Hawley K (1989) Barriers as boundaries for metapopulations and demes of Peromyscus leucopus in farm landscapes. Landsc Ecol 2:227–235
Monmonier MS (2010) Maximum-difference barriers: an alternative numerical regionalization method. Geogr Anal 5:245–261
Morin P, Luikart G, Wayne RK (2004) SNPs in ecology, evolution and conservation. Trends Ecol Evol 19:208–216
Murphy MA, Evans JS, Cushman SA, Storfer A (2008) Representing genetic variation as continuous surfaces: an approach for identifying spatial dependency in landscape genetic studies. Ecography (Cop.) 31:685–697
Naveh Z, Lieberman AS (2013) Landscape ecology: theory and application. Springer Science & Business Media, ISBN 1475723318
Neale DB, Marshall KA, Sederoff RR (1989) Chloroplast and mitochondrial DNA are paternally inherited in Sequoia sempervirens. In: Don Endl D (ed) Proceedings of the National Academy of Sciences, vol 86, pp 9347–9349
Nei M (1972) Genetic distance between populations. Am Nat 106:283–292
Nei M (1978) Estimation of average heterozygosity and genetic distance from a small number of individuals. Genetics 89:583–590
Pamilo P (1988) Genetic variation in heterogeneous environments. In: Proceedings of the annales zoologici fennici. JSTOR, pp 99–106
Peakall R, Smouse PE (2006) Genalex 6: genetic analysis in Excel. Population genetic software for teaching and research. Mol Ecol Notes 6:288–295
Pritchard J, Stephens M, Donnelly P (2000) Inference of population structure using multilocus genotype data. Genetics 155:945–959
Robbins C, Dawson D, Dowell BA (1989) Habitat area requirements of breeding forest birds of the middle Atlantic states. Wildl Monogr 3–34
Rousset F (2008) Genepop'007: a complete re-implementation of the Genepop software for Windows and Linux. Mol Ecol Resour 8:103–106

Rukke BA (2000) Effects of habitat fragmentation: increased isolation and reduced habitat size reduces the incidence of dead wood fungi beetles in a fragmented forest landscape. Ecography (Cop.) 23:492–502
Safner T, Miller M, McRae B, Fortin MJ (2011) Comparison of Bayesian clustering and edge detection methods for inferring boundaries in landscape genetics. Int J Mol Sci 12:865–889
Schmiegelow F, Mönkkönen M (2002) Habitat loss and fragmentation in dynamic landscapes: avian perspectives from the boreal forest. Ecol Appl 12:375–389
Segelbacher G, Cushman SA, Epperson BK, Fortin MJ, Francois O, Hardy OJ, Holderegger R, Taberlet P, Waits LP, Manel S (2010) Applications of landscape genetics in conservation biology: concepts and challenges. Conserv Genet 11:375–385
Shirk AJ, Cushman SA (2011) sGD: software for estimating spatially explicit indices of genetic diversity. Mol Ecol Resour 11:922–934
Slatkin M (1995) A measure of population subdivision based on microsatellite allele frequencies. Genetics 139:457–462
Stinchcombe J, Hoekstra HE (2008) Combining population genomics and quantitative genetics: finding the genes underlying ecologically important traits. Heredity (Edinb) 100:158
Storfer A, Murphy MA, Spear SF, Holderegger R, Waits LP (2010) Landscape genetics: where are we now? Mol Ecol 19:3496–3517
Tischendorf L, Bender DJ, Fahrig L (2003) Evaluation of patch isolation metrics in mosaic landscapes for specialist versus generalist dispersers. Landsc Ecol 18:41–50
Turner MG (2005) Landscape ecology: what is the state of the science? Annu Rev Ecol Evol Syst 36:319–344
Vance MD, Fahrig L, Flather CH (2003) Effect of reproductive rate on minimum habitat requirements of forest-breeding birds. Ecology 84:2643–2653
Virgós E (2001) Role of isolation and habitat quality in shaping species abundance: a test with badgers (*Meles meles* L.) in a gradient of forest fragmentation. J Biogeogr 28:381–389
Vos P, Hogers R, Bleeker M, Reijans M (1995) AFLP: a new technique for DNA fingerprinting. Nucleic Acids Res 23:4407–4414
Wales BA (1972) Vegetation analysis of north and south edges in a mature oak-hickory forest. Ecol Monogr 42:451–471
Warnes G (2015) With contributions from Gregor Gorjanc; Friedrich Leisch and Michael Man. genetics: Population Genetics. R package version 1.3. 8.1
Weir BS, Cockerham CC (1984) Estimating F-statistics for the analysis of population structure. Evolution (N.Y) 38:1358–1370
Wiens J, Chr N, Van Horne B, Ims RA (1993) Ecological mechanisms and landscape ecology. Oikos 66:369–380
Wilcove DS, McLellan CH, Dobson AP (1986) Habitat fragmentation in the temperate zone. Conserv Biol 6:237–256
With K, King AW (1999) Extinction thresholds for species in fractal landscapes. Conserv Biol 13:314–326
Womble WH (1951) Differential systematics. Science (80-.) 114:315–322
Wright S (1931) Evolution in Mendelian populations. Genetics 16:97

Chapter 10
Arsenic in Rice Grain: Role of Transporters in Arsenic Accumulation

Manish Tiwari

Abstract Arsenic toxicity emerged as one of the major challenges for the human being localized in a certain region of the globe, including the eastern part of India, Bangladesh, and China. Rice, a staple crop growing in an arsenic contaminated area, accumulates high arsenic in the grains. The consumption of such rice grains associated with several detrimental effects on kidney, liver, cancer, and cardio-vascular diseases in human. Being as food crop of half of the world population, many reports showed that eating of arsenic-containing rice grains is one the major routes of arsenic toxicity to human. Given that contribution of arsenic toxicity by using arsenic-containing rice grains, various studies have been conducted in past years to understand the molecular mechanism of arsenic metabolism in rice, and greatly increases our knowledge about key molecular players. In this chapter, the role of membrane-bound transporters, which mediate arsenic uptake from soil and cellular as well as long-distance transport has been reviewed in the light of recent reports.

Keywords Arsenic · Rice · Transporter · Heavy metals · Detoxification · Acr · Lsi

10.1 Introduction

Arsenic (As) is a well-known heavy metal, and chemically it is metalloid in nature. It is ubiquitously present in the soil, water, and terrestrial environment, however, due to possessing unique geological rock beds, population living in East Asian countries facing severe As toxicity. Arsenic is proved to be a potent carcinogenic element leading to various form of cancer in the human body. Based on a large number of available reports, it is now consensus that As act as a tumor promoter by interfering with several molecular mechanisms including epigenetic changes such as DNA methylation. The chronic exposure of As causes several health ailments, including skin, kidney, liver, cardiovascular, and respiratory diseases in the human population. Contrary to

M. Tiwari (✉)
Council of Scientific and Industrial Research—National Botanical Research Institute (CSIR-NBRI), Rana Pratap Marg, Lucknow, Uttar Pradesh 226001, India
e-mail: maneeshtiwari11@yahoo.co.in

S. P. Singh et al. (eds.), *Molecular Approaches in Plant Biology and Environmental Challenges*, Energy, Environment, and Sustainability,
https://doi.org/10.1007/978-981-15-0690-1_10

this, acute As poisoning can develop several symptoms such as headaches, confusion, diarrhea as well as drowsiness. The lung, skin, kidney, and liver are the main organs affected by As poisoning primarily. Concerning to historical backdrop of As toxicity, it falls around 3000 BC when human learned to make alloy like bronze by adding As into copper (Harper 1987).

There are numerous sources of As toxicity to the human being, for instance, As laden drinking water, food, soil, and air. Apart from this, few anthropogenic activities are also responsible for As toxicity. The use of As-containing pesticides is one the main reason of As addition to the environment which leaves high As into the soil. In the United States of America, chromated copper arsenate was frequently used in wood preservatives for protecting the wood from fungi and insects during mid of 20th century which caused millions acre of land are contaminated with As. Furthermore, it is also evidenced that people engaged in mining and smelting of copper (Cu) ores are much prone to lung cancer by inhalation of airborne As contaminants resulting As concentration up to (0.000003 mg/m3) in the air (Lagerkvist and Zetterlund 1994).

As mentioned earlier, the manifestation of high concentration of As in soil may result by natural or various anthropogenic activities. It is believed that due to possessing specific arsenopyrite rich bed in Earth crust, surface areas of the Indian subcontinent are heavily polluted with As, and increased As the level in the soil surface. The water percolation through such rock beds elevated As content in ground water in the water cycle. It is also important to note that ground water is the main source of drinking water as well as for agricultural practices. The As toxicity in these affected area draw much attention and extensively debated areas since the past three decades. A plethora of reports is now available showing heavy contamination of ground water with As in Ganges-delta in India and Bangladesh. The inhabitant of this region mostly depends on the river, shallow dig wells and aquifers for their drinking water and household purposes. Water from these resources is loaded with a high amount of As compared to the permissible level assigned by WHO (2001). Since, As is considered as a promising carcinogen and its exposure becomes a potential health threat to humans as well as environment.

Interestingly, excess of As in the soil and drinking water leads to the entry of As into the food chain and extend the As toxicity range globally (Verbruggen et al. 2009). According to an estimate, As poisoning affects the health of nearly ten million people in South Asia and Bengal delta region (Smith and Steinmaus 2000; Brammer and Ravenscroft 2009; Tripathi et al. 2007). There are two major forms of soil As that is inorganic species like arsenite (AsIII) and arsenate (AsV) and organic form monomethylarsonic acid (MMA) and dimethylarsinic acid (DMA). It is important to mention here that inorganic from is much toxic and main cause of As toxicity.

Rice is an annual plant and mostly grown in the tropical region of the world. It is a major cultivated crop in As-contaminated areas and accumulates higher As in the grains (Zhao et al. 2010). A number of studies suggest that nutritional intake of As via consuming rice grain has a substantial contribution on As toxicity apart from drinking water. Studies also have shown that As in rice grain is one of the leading routes to human exposure of As (Mondal et al. 2008; Tuli 2010). It has been indicated that population living in an area far away from As contaminated sites also expose with As by dietary sources. In addition to human, other life forms also suffered enormously with this toxic metalloid in various ways, for instance, irrigating paddy field in Bangladesh using of As-containing ground-water resulted in a considerable loss in paddy yield (Panaullah 2009). Moreover, reports demonstrate that As affected nutraceutical quality of rice by changing mineral content in the rice grains cultivated in As-contaminated sites (Dwivedi 2010a). The human exposure of As has been described as the largest mass poisoning of a population in the history, and it is further extrapolated that As in paddy is considered as neglected cancer risk (Stone 2008).

10.2 Arsenic in Soil

Considering the abundance, As ranked as 20th most prevalent metal of the Earth crust. It is present in nearly 3 mg kg^{-1} on average concentration (Cullen and Reimer 1989). It is found that more than 200 commercial mineral ores reported in Earth crust namely arsenopyrites (FeAsS), realgar (As_4S_4) and orpiment (As_2S_3) contain a high percentage of As (Zhao et al. 2010). As naturally enters into the environment by weathering of rocks, volcanic emissions, and hot spring discharges. The percolation of surface water via arsenopyrites rich bed to the ground water seems to be the main in reason of high As in water bodies at least in Indian and Bangladesh scenario. Some anthropogenic activities like ores mining, smelting, use of herbicide, pesticide and wood preservatives are equally responsible for As release in environment and consequentially reaches into the food chain through the hydro-geo-biological cycle (Tripathi et al. 2007). Majority of As present in the soil exists as inorganic species like AsIII and AsV (Fig. 10.1a). The organic species are also present in the soil but in less quantity comparing with inorganic species. The abundance of As species in the soil is determined by redox ability of soil, for example, AsV is a predominant inorganic species in aerobic environment whereas AsIII is predominantly found in anaerobic soil. These inorganic species of As are inter-convertible to each other either through abiotic factors such as redox and pH of soil solution or chemical reactions catalyzed by microbial released enzymes in the rhizosphere. Some minerals (manganese oxide;

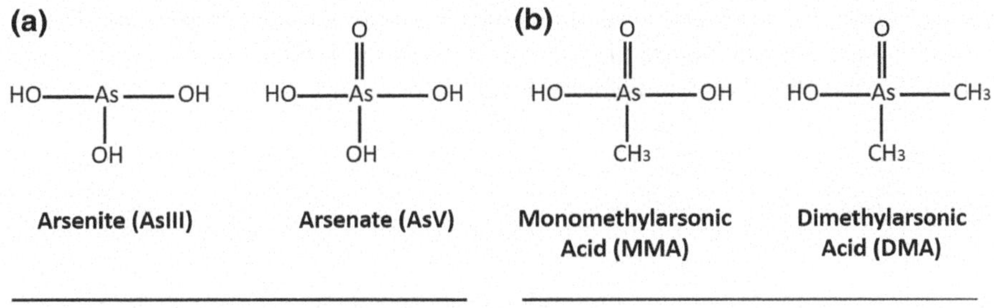

Fig. 10.1 Main As species in soil, (**a**) inorganic forms and (**b**) organic forms

MnO) present in the soil also catalyzes oxidation reaction of AsIII in the aerobic soil while acidic environment resulted by dissolved sulfide act as reductant and favors the reduction of AsV into AsIII in anaerobic soil (Oscarson et al. 1981; Rochette et al. 2000).

In addition to inorganic forms of As, some organic species of As are present in very less amount in the soil, for instance, MMA, DMA, and other trace forms (Fig. 10.1a). The conversion from inorganic to organic form by methylation is mainly achieved by microorganisms and algae (Bentley and Chasteen 2002). The MMA and DMA are the main constituents of the As based herbicides and pesticides. Therefore, the use of these chemicals added more organo-As compounds in the soil, ultimately leading to As accumulation.

10.3 Arsenic Toxicity

10.3.1 Human

Arsenic is a highly toxic environmental pollutant which causes chronic and epidemic effects on human health across the globe (Nordstrom 2002). The As crisis is common in South-east Asian countries where millions of people are affected through water and crop contamination. The As level in ground water exceeds up to safe limit of 10 $\mu g\ L^{-1}$ as prescribed by WHO and over 100 million people are supposed to encounter with As through drinking water in coastal belt of Bangladesh, India, and China (Brammer and Ravenscroft 2009; Mukherjee et al. 2006; Rosen and Liu 2009). Due to having potent carcinogenic potential, inorganic species of As are grouped in the class-1 type of carcinogens (IARC 2004). Also, As ingestion associated links lung, kidney, nervous disorders and skin lesions in humans. Contrary to reports indicating the harmful properties of As, As is commonly used as a therapeutic drug in ancient Chinese medical practices (Liu et al. 2013). Apart from this, a series of recent studies report the substantial virtue of anti-cancerous activity of AsIII for curing acute promyelocytic leukemia patient (Zhang et al. 2010; Nasr et al. 2008; Hu et al. 2009). These studies also described and successfully dissected the molecular mechanism of the anti-cancerous property of AsIII. It is a bigger paradox of nature that the same compound retains carcinogenic as well as anti-cancerous properties. It seems that carcinogenic nature of As may be due to high intake of As via different sources compared to the recommended level.

10.3.2 Plant

The plant and animal share similar cellular and biochemical reactions both at the molecular level. The presence of As found to interfere with many vital physiological and biochemical processes in the plant. In rice, major symptoms associated with As are the reduction of germination, biomass, root and shoot elongation that eventually led to low crop yield (Shri et al. 2009; Rahman 2012). In addition, As damaged the cellular ultra-structure and changed the homeostasis of reactive oxygen species (ROS) (Liu et al. 2013). A comparative proteomic profiling of rice shoots revealed several proteins, including those involved in redox and protein metabolism are differentially expressed under As stress (Liu 2013). By utilizing a large number of rice accessions growing in the Indian subcontinent, studies have suggested that a differential As speciation occurred in different tissues and grains of these rice varieties (Burko et al. 2011; Bhattacharya et al. 2012). It is further noticed that As content in rice grain varies significantly and associated with phosphate requirement and iron (Fe) concentration in soil (Mukherjee et al. 2006; Rahman 2011). In a nice report, Dwivedi et al. (2010a) screened a large number of rice varieties by considering As accumulation as a trait, and revealed the incidence of a strong correlation between As accumulation and stress-responsive amino acids and antioxidative response (Dwivedi 2010b).

In general, As is considered as biologically non-essential to the plants. Nevertheless, one study revealed that AsIII could drive anoxygenic photosynthesis in hot spring photosynthetic bacteria (Kulp et al. 2008). The high concentration of As impedes the growth of rice and *Arabidopsis,* unlike to cadmium (Cd), which causes chlorotic symptoms (Song et al. 2010). Studies at the molecular level indicate that As strongly interfere with salient biochemical processes. Due to the common physicochemical property to phosphate, AsV can compete with phosphate in many biochemical reactions. These reactions include reduction of NAD^+, citric acid cycle, and generation of ATP. The AsV can be incorporated in earlier steps of certain metabolic pathways instead of phosphate. However, it cannot process similar to phosphate in further downstream steps because of precise enzyme-substrate specificity. Nonetheless, under the presence of adequate phosphate, AsV toxicity is less pronounced in plants in contrast to AsIII (Wang 2002). Intriguingly, AsIII toxicity in the plant is differed from AsV and is considered more toxic than AsV. It shows high affinity with cysteine residue and can form the strong covalent bonds with closely spaced thiolates (cysteine) of peptides and imidazolium nitrogen (histidine) which are largely responsible for determining the structure of proteins. The binding of AsIII to the cysteine residues involved in the formation of active sites of enzymes primarily impairs the enzyme activity. Similarly, disulfide bridge formations between the polypeptides are also prevented. Hence, the tertiary structure of proteins can be distorted (Hughes 2002). The direct binding of AsIII to cellular proteins resulted in the dysfunction or inactivation of proteins, and ultimately, metabolic processes are inhibited.

10.4 Arsenic in Rice Grain

Rice is a staple crop of half of the world's population, especially in East Asia. The evolutionary study reveals that rice was originated from wild rice, *Oryza rufipogon,* and both Indica and japonica varieties are evolved from a common ancestor (Huang et al. 2012). Based on genetic divergence of the genome of the rice varieties, it is now clear that domestication of rice occurred in Chinese province followed by spread towards South-east Asia and rest of the world (Huang et al. 2012). As discussed in the earlier section, As ingestion via As containing rice grain increases the risk of As toxicity at another level. Paddy is the main crop cultivated in this region and use of As polluted water for irrigation purposes alleviate manifold As content in the grains. This ecological process is known as biological magnification. According to an estimate, use of As laden groundwater for irrigation of paddy alone can insert more than 1000 tons of As to the soil each year in Bangladesh alone (Ali 2003). This is one of the main reasons for high As the level in paddy fields which may range from 4 to 8 mg kg^{-1} soil and can reach up to 83 mg kg^{-1} in the dry season (Abedin 2002). The As in paddy fields is readily absorbed by rice plants and efficiently translocated to the grains and ultimately resulted around 10-fold higher As content than other cereal grains grown in same conditions (Tuli 2010; Williams et al. 2007; Khan et al. 2010). The high level of As in rice grains constitutes a threat to the local inhabitant, relying mostly on rice for their diet (Zhao et al. 2010). Interestingly, presence of inorganic As species is high in rice of Asian countries as compared to Californian rice which has more percentage of methylated As, and these methylated species are considered least toxic to human beings (Carey et al. 2010). Hence, it appears that edaphic factors are the presumably crucial regulator for As metabolism in rice and discrepancy in As content of rice grown in various geographical regions is decided by these factors.

It is also a fact that naturally grown rice from every part of the world contains As on an average of 80–200 μg kg^{-1} as reported by a global survey (Zavala and Duxbury 2008). The As level in rice grain can elevate multifold when cultivated in either As polluted sites or irrigated with polluted water. Therefore, it is plausible that the presence of As in rice grain is an unavoidable natural phenomenon; however, it can be managed by developing rice varieties that accumulate less amount in grain.

In general, plants vary considerably in their ability to translocate heavy metals in aerial tissues from the soil, and this can be denoted as 'transfer factor' (TF). The plant having a least TF is often referred to as 'extruder'. They have evolved certain mechanisms to avoid uptake of As and restricted the mobilization from root to above ground plant parts. On the other hand, hyperaccumulators like *Pteris vittata* can accumulate As up to 2% of their dry weight, and in this case, TF exceeds more than one. More importantly, TF of rice for As reaches to nearly one that simply depicts efficient root to shoot transport of As in rice. High As content in the above ground tissue of rice may be due to two reasons first; huge amount of silicon (Si) required for proper growth and for achieving this rice has evolved very efficient Si uptake and transport system (Ma et al. 2006), AsIII shares Si transporters and readily distributed in aerial portion of rice; second, AsIII is a predominant inorganic As species exist in anaerobic conditions and submerged paddy fields with water provide the similar environment that enhances the bioavailability and accumulation of AsIII in rice grain (Zhao et al. 2010).

10.5 Arsenic Transport in Plant

Given that impact of As toxicity, numerous attempts have been made to understand uptake and transport of As. However, biochemical and molecular mechanisms underlying the As transport and assimilation in rice and in other plants are still missing. At present, information about As uptake and transport are confined to few members of the *NIP* gene family and phosphate transporters in plants (Ma et al. 2008; Shin et al. 2004). In general, plants exhibit the least selectivity in between the essential and non-essential elements at the time of uptake from the soil solution. It has been recognized that transporters engage in the acquisition of essential nutrients are mainly responsible for the mobilization of non-essential metals. By sharing a similar structure and chemical properties, non-essential metals are often succeeded to mimic the transporters meant for the uptake of essential nutrients. Research has shown that different As species such as AsIII, AsV, MMA and DMA can able to move inside the root via transporters of essential elements. The details of such mechanisms are discussed in the following sub-sections.

10.5.1 Arsenite Transport

As mentioned earlier, AsIII (H_3AsO_3) exists in an undissociated state in soil and readily absorb by root cells under normal pH range (>94% undissociated at pH <8.0) (Zhao et al. 2010). The AsIII moves in the root are mainly through aquaporin that is Lsi1 and Lsi2 in rice (Ma et al. 2008). Aquaporin is a newly characterized transporter family commonly formed channels in the membrane of root cells to facilitate quick absorption of water in the plant. Several essential and non-essential metals are also get absorbed in the dissolved state with the water. The Lsi1 and Lsi2 are well known characterized members for AsIII transport. The Lsi1 causes AsIII influx in the cell; however, the efflux activity of Lsi1 was also reported later in rice. The research has shown that aquaporin can perform a bidirectional movement of molecules across the membrane Zhao et al. (2010). The direction of movement is determined by the concentration gradient of molecules across the membrane. If the outside concentration is low, then extrusion activity is operational, however, if the condition is a reverse influx of substrate take place. It has been demonstrated that when AsV is abundant in soil, and after entry into the epidermal cells, it gets reduced in AsIII normally through enzymatic reduction. Due to this, AsIII concentration rises inside cells, and efflux of AsIII from the cell started. In this condition, plants export AsIII outside environment through Lsi1 according to concentration gradient (Zhao et al. 2010). This observation was also seen in yeast when the expression of the same NIP caused susceptibility in yeast to the AsIII whereas conferred tolerance upon AsV addition in the media due to efflux of AsIII (Bienert et al. 2008).

In general, these aquaporins form tetramers within the membrane, and each monomer constitutes a channel that consists of six transmembrane helices. The selectivity of any NIP for particular solute or compound depends on a narrow region of the channel which provides the passage and allowing the movement of a single molecule at a time (Wallace et al. 2002). The polarity of amino acid lying this region determines the selectivity of solutes, for instance, hydrophobic amino acids in aquaglyceroporins stabilize the hydrophobic backbone of glycerol and also permitting transport of metalloid and other compounds chemically similar with glycerol.

10.5.2 Arsenate Acquisition

The AsV exists as oxyanions ($H_2AsO_4^-$ or $HAsO_4^-$) under normal pH range (pH 4–8) under oxidizing condition. These oxyanions are the chemical analog of a corresponding phosphate anion. Thus, AsV uptake in plants mainly occurs via phosphate transporters. It is interesting to note that several mutants of phosphate transporter have been identified during a genetic screen based on AsV tolerance (Shin et al. 2004). The AsV repressed activation of the genes responsible for phosphate uptake in *Arabidopsis*, and this response reflect a regulatory mechanism of plants to protect from the extreme toxicity of As (Catarecha et al. 2007). Most of the bacterial phosphate transporters act as secondary active transporters which couples with Na^+ for its transport activity (Virkki et al. 2007). Sharing of phosphate transporters by AsV not only confined in plants but extends in other domains of life also, for instance, phosphate transporters PiT-1 and PiT-2 have been shown to facilitate AsV uptake in *Escherichia coli*, *Saccharomyces cerevisiae* and human (Persson et al. 1999).

10.5.3 Organo-Arsenic Transport

The MMA and DMA are major As organic species in soil. These organo-arsenic species are produced by the microbial and algal enzymatic actions. Due to non-polar chemical nature, it is assumed that these forms can simply absorb through diffusion across the membrane (Li et al. 2009). Detail liposome-mediated transfer assay of MMA and DMA are estimated to be $1.4 \times 10^{-13}$and 4.5×10^{-11} cm s^{-1} respectively. This observation suggests that uptake of these species in the root cells through simple diffusion is a very slow process, and other possible route may also exist (Li et al. 2009). While the role of Lsi1 and Lsi2 has been established in the AsIII translocation, comparative analytical estimation of MMA (AsV) and DMA (AsV) in rice *lsi1* mutant exhibited that MMA and DMA content in the mutant was reduced to nearly 80% of MMA and 50% of DMA relative to wild plant. At the same time, the mutation in *Lsi2* had a very little effect on MMA and DMA amount in comparison to WT plants (Li et al. 2009). Consistent to this, transport assay in *Xenopus* oocyte revealed MMA level was increased in oocyte and prooved that Lsi1 can directly facilitate the movement of these organo-As compounds similar to AsIII in the rice.

10.6 Arsenic Metabolism in Plant

The considerable number of research has been done to know the fate of As species once enter into the cell. Upon the entry of As species, a series of multiple biochemical reactions involved for metabolizing As. Finally, these compounds lose toxic nature and get sequestered into either of the cellular reservoirs or distributed in various tissues (Fig. 10.2). The knowledge of As assimilation and partitioning in plants, including rice, is still developing (Ye et al. 2010). Consistent to separate route for the uptake of each As species, they metabolized through the same pathway except the additional step of AsV reduction in case of AsV. There are several reports showed the abundance of AsIII in plant tissues when fed with AsV and suggest that reduction of AsV is the foremost step (Pickering et al. 2000). This kind of reduction capacity is found in root as well as in shoot tissues. Arsenic reductase (Acr) is the key enzyme, initially identified in prokaryotes and represents tyrosine phosphate phosphatase class, catalyzes this reaction. Glutathione and glutaredoxin serve as proton donor during the reaction. Several homologs of *Acr* are now characterized in *Arabidopsis* (Dhankher et al. 2002), *Holcus* (Duan et al. 2007) and rice (Ellis et al. 2006). Now, it is a known fact that toxicity caused by AsV is actually due to AsIII because AsV gets quickly converted into AsIII upon cellular intake.

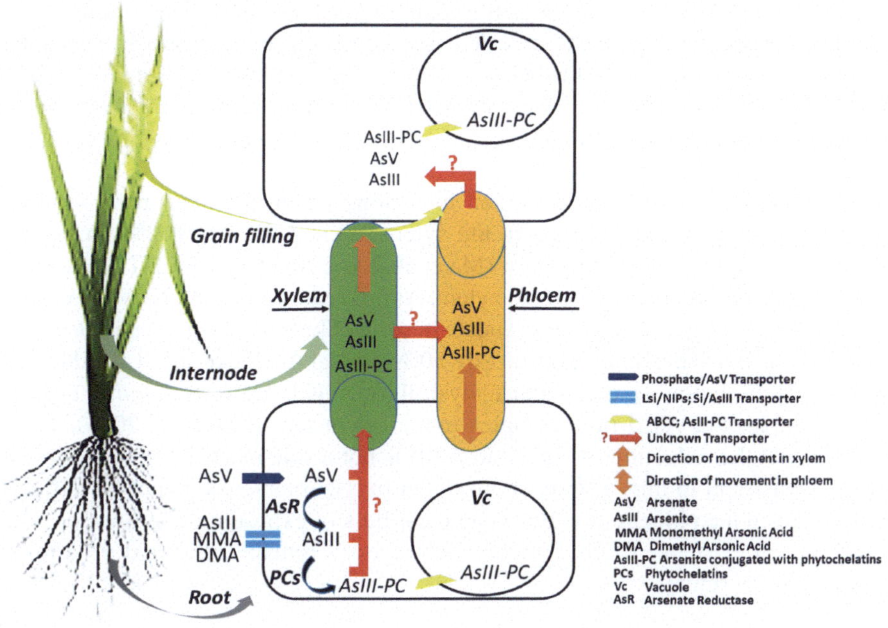

Fig. 10.2 Schematic representation of uptake, assimilation and long distance distribution of Arsenic forms in rice tissues

In general, AsIII is shown to be more toxic compared to AsV. The AsIII shows a high degree of affinity with sulfhydryl (-SH) group of proteins and distorts protein structure and function upon binding to them. After reaching into cytosol directly through either Lsi1 or upon reduction with Acr, AsIII binds to cellular thiols (glutathione-GSH and phytochelatin-PC) and AsIII detoxified (Pickering et al. 2000). At the time of AsIII exposure, expression of many genes are induced which are known to be involved in thiol biosynthesis, therefore, reflecting high demand of thiols during As stress in the rice (Ahsan et al. 2008; Chakrabarty et al. 2009). The PCs or glutathione-conjugated AsIII are then sequestered in the vacuoles.

Moreover, thiol bound AsIII is also translocated toward stellar direction via silicic acid transporters, namely Lsi1 and Lsi2 for xylem loading (Fig. 10.2). Vacuolar compartmentation of AsIII is energy-dependent process and supposed to mediate by a different class of active transporters. Yeast vacuolar transporter, Ycf1p, is a member of ABC transporter family which transport the glutathiones conjugated AsIII (Ghosh et al. 1999). Complexation of glutathione with AsIII is the major phenomena present in yeasts, but none of this kind of transporter is reported in plants till date. Two ABCC-type transporters are characterized by *Arabidopsis* as the major vacuolar AsIII-phytochelatin complex transporters engaged in AsIII detoxification (Song et al. 2010). Earlier it has been revealed that these transporters specifically sequester the AsIII into vacuoles, while later investigation revealed they could also sequester the PC-bind cadmium (Cd) and mercury (Hg) into vacuoles (Park et al. 2012).

MMA and DMA enter by the same route utilized by AsIII. The kinetic study of these organic species of As revealed that DMA moves with much slow rate, nonetheless, very efficiently loaded in xylem compared to MMA. Similar to AsV, MMA(V) is partly reduced to the MMA(III) in the root of rice. Interestingly, it has been found that only MMA(V) can translocate into shoot via xylem mediated transport (Li et al. 2009).

Systematic studies of As speciation in various organ and plant parts indicate that the major As species present in the xylem sap is AsIII (80–100%) in several plants when supplemented with AsV (Ma et al. 2008; Xu et al. 2007). These results illustrate that before loading into the xylem, AsV is reduced into AsIII by enzymatic reduction. Contrary to this, xylem exudate of plants like wheat, barley, and *Brassica* retain AsV in much higher proportion (40–50%) of total As (Pickering et al. 2000; Xu et al. 2007). Furthermore, elemental analysis through ICP-MS of *Helianthus annuus* showed the lack of AsIII-thiol complex in the xylem sap (Raab et al. 2005). Thus, it seems that the reduction of AsV into AsIII is not a sole mechanism found in the plants. Consistent to this, a study showed that free inorganic As, mainly As(III), is transported in the phloem of castor bean regardless of exposure of As species. The possibility of existence of other metabolic processes remains alive and need to be explored in the future.

10.7 Root to Shoot Translocation of as in Plant

Plants have evolved certain measures to overcome the toxicity of As for instance, compartmentation of thiol chelated-As in vacuoles of root cells and dispersion of As species in different plant tissues and organs. After complexation with thiols, AsIII loses toxicity, and long-distance movement is ceased as well. The xylem and phloem mediated transport systems are the major channel of long-distance translocation (Fig. 10.2). The lack of thiol bound AsIII moiety in xylem sap during HPLC coupled ICP-MS analysis verified this hypothesis (Raab et al. 2005). Based on the potential of As accumulation, certain plants are considered as 'hyperaccumulator' for example, *Thlaspi*, *Brassica* and some fern like *Pteris*. Members of fern family pteridaceae are generally potent As hyperaccumulator which can accumulate 1–1000 mg-kg^{-1} of dry body weight. It is important to note that hyperaccumulator plants have very efficient transport systems, including vacuolar or long distance transport. However, the underlying transporters responsible for efficient translocation remains to be elusive.

The detail investigation of phloem sap exhibited the presence of certain ligand molecules such as nicotianamine, glutathione (GSH), and phytochelatins (PCs) in many plants. Nicotianamine is primarily meant for binding with Fe, Cu, Mn, and Zn and facilitate translocation by increasing solubility of metals (Curie et al. 2009). The detection of GSH and PCs in the phloem is the most astonishing fact. These biomolecules have more affinity with Cd, Hg, and some metalloids like As (Mendoza-Cozatl 2008; Shukla 2012). The PCs are, in general, known for maintaining cellular metal homeostasis by binding and facilitating the vacuolar sequestration of heavy metals. However, occurrences of PCs in phloem sap suggest that they may participate in the long-distance movement as well (Mendoza-Cozatl 2008). Studies supported this notion; for instance, *AtPCS1* (At5g44070) is highly expressed in companion cells of phloem. Companion cells and sieve elements are perforated to each other through plasmodesmata, and thus PCs and GSH synthesized in these cells are likely to move in phloem for further transport to sink tissues (Chen et al. 2006). But, transporters through which these metals loaded into phloem is largely unknown except a low cation transporter (LCT1) was characterized in rice for phloem loading of Cd (Uraguchi et al. 2011).

10.8 Role of Transporters in Arsenic Uptake and Movement

There are many taxa spread in the plant kingdom that retain the ability to survive under the extreme metal concentrations referred to as 'metallophytes.' These plants can be subdivided into two groups; excluder and hyperaccumulator. The excluders are those metallophytes which had evolved a mechanism for restricted metal uptake as well as least root to shoot translocation and maintain constant metal content in aerial tissues irrespective of metals concentration in soil. On the other hand, hyperaccumulator has the exceptional potential of storing metals in above ground tissues, for example,

Thlaspi caerulescens, *T. praecox*, *Arabidopsis halleri*, *Pteris*, and *Sedum alfredii* (Baker et al. 1994). The As hyperaccumulators are confined to certain members of fern family Pteridaceae like *Pteris vittata* and *Pteris cretica*. The frond of *Pteris* is found to accumulate As as much as 0.1% of body weight (Raab et al. 2004).

It has been long sought the molecular mechanism underpinning hyperaccumulation behavior of *Pteris*. A lot of progress has been achieved in the past few years by using these plants as a model system. In connection to As hyperaccumulation potential of *Pteris*, an efflux transporter, *PvACR3*, was identified which encoded a protein similar to the bacterial ACR3. The ACR3 of yeast is known for efflux of As outside to the cells while localization study reveals that PvACR3 was found on tonoplast of gametophyte and effluxes As into the vacuoles (Indriolo et al. 2012). Knocking down the expression of *PvACR3* through RNAi resulted in sensitivity towards As in *Pteris*. Interestingly, homologs of this *PvACR3* are missing in the higher plants (Gymnosperm and Angiosperm) during the course of evolution. An important ABC-transporter characterized for the export of thiol conjugated As into vacuoles of *Arabidopsis* (Song et al. 2010). From all these studies, the role of transporters behind metal tolerance and accumulation is well established, and it is amenable that transporters are mainly responsible for determining tolerance and hyperaccumulation trait in plants. Even though the identification of additional As transporter is thought to be crucial for understanding the mechanisms and thereby could be a potential candidate gene for genetic modification of plants in the future. The summary of As transporters are described in the following sub-section:

10.8.1 Lsi

The Lsi protein is the first transporter characterized for AsIII transport in rice. The genetic study identified that *Lsi1* and *Lsi2* (low silicon1 & 2) efficiently efflux Si towards stele region for the xylem loading of Si due to having a unique proximal and distal orientation (Ma et al. 2006, 2007). Lsi1 and Lsi2 belong to NIP (nodulin26 like intrinsic protein) subfamily of plant aquaporin (Maurel et al. 2008). It has been found that size of AsIII (diameter 4.11 Å) resembles very much to Si (silicic acid in diameter 4.38 Å) in terms of diameter and tetrahedron structure (Ma et al. 2008). These observations indicate that the transport system of Si responsible for Si uptake may be utilized by AsIII transport in rice and which was proved (Ma et al. 2008). The mutation of *Lsi1* in rice is alone resulted in 57% decrease in AsIII quantity compared to the wild plant in the short term performed assay establishing the role of *Lsi1* for the entry of AsIII in rice (Ma et al. 2008). Mutation in another OsNIP3.2 significantly reduced As concentration in the roots of rice. The synchrotron-based detail analysis showed that reduction in As accumulation occurred in the stele of the lateral roots in the mutant compared to wild plant (Chen et al. 2017). The observation that members of the NIP family can transport AsIII led the investigation of such capability in other members as well. After that, many NIPs from various source organisms, for instance, *Arabidopsis*, rice and lotus were identified for their inherent

AsIII permeability behavior either in a heterologous system, oocyte of *XenopusI,* yeast or in plant (Bienert et al. 2008; Isayenkov and Maathuis 2008; Kamiya and Fujiwara 2009). Besides AsIII, NIPs can also be permeable for antimonite (Sb), boron (B), germanium (Ge) and some other neutral molecule such as urea, glycerol, and formamide. The recent study demonstrates that overexpression of OsNIP1;1 or OsNIP3;3 in rice decreased root-to-shoot translocation of AsIII (Sun et al. 2018).

10.8.2 ABCC

OsABCC1 is a member of the C-type ATP-binding cassette transporters family in *Oryza sativa*. This protein is recently investigated for role in the detoxification and reduction of As in rice grains. The cellular localization experiment at basal and upper nodes suggested that OsABCC1 was localized to the phloem region of vascular bundles. The knockout mutant of OsABCC1 resulted in a considerable reduction for As tolerance in rice (Song et al. 2014). It was proposed that OsABCC1 inhibits As translocation to the grains by sequestering As in the vacuoles of the phloem companion cells at nodes in rice.

10.8.3 NRAMP

The NRAMP was foremost characterized in mouse (Vidal et al. 1993). NRAMPs form a highly conserved family of integral membrane proteins which are mostly engaged in divalent metal transport. NRAMPs have been identified in a wide range of organisms, including bacteria, fungi, plants, and animals (Cellier et al. 1995). The function of these NRAMPs are elucidated in the transport of divalent nutrient cations, such as Fe, Mn, and Zn. Ironically, Fe transport behavior of these proteins was studied in the plants, but Mn transport ability was demonstrated later (Cailliatte et al. 2010). The role of *Arabidopsis* NRAMPs is also well documented in Cd transport either through heterologous expression in yeast or plants. OsNRAMP1, an iron transporter, participate in cellular Cd translocation and its overexpression consequences high amount of Cd in the leaves of rice. Similarly, the role of OsNRAMP5 has been elucidated in Mn and Cd uptake and distribution in rice grain (Ishimaru et al. 2012). An exciting study revealed that expression of OsNRAMP1 and cellular localization in Arabidopsis displayed As transport and tolerance (Tiwari et al. 2014). The detail investigation demonstrated that OsNRAMP1 facilitated As loading in xylem and proposed that OsNRAMP1 can be a good candidate gene for developing rice varieties with low As in grains through gene manipulation, and ultimately lowered the risk of food chain As contamination.

10.8.4 MATE

Multidrug and Toxic compound Extrusion proteins (MATE) are a group of secondary active transporters which utilize electrochemical gradient across membrane maintained by ATPases for transport activity (Kuroda and Tsuchiya 2009). The MATE transporters engage in intracellular channeling of secondary metabolites, especially flavonoids and alkaloids in vacuoles. A member of *MATE* gene family, FRD3, maintains Fe homeostasis in *Arabidopsis* through efflux of citrate outside the pericycle cells. This activity of FRD3 increases Fe mobilization in aerial tissues (Takanashi 2013; Yokosho et al. 2009). It assumes that MATE proteins are playing a big role in diverse physiological functions in plant kingdom seems yet to be generated. The investigation of OsMATE1 and OsMATE2 showed that it alters development, stress responses, and pathogen susceptibility in Arabidopsis as well as leading to As sensitivity in Arabidopsis (Tiwari et al. 2014). The knocking down of RNAi OsMATE2 using endosperm-specific mitigates As accumulation in rice grains and transgenic lines demonstrated a significant reduction of both OsMATE2 transcripts (~38–87%) and grain As content (36.9–47.8%) compared to wild type plant (Das et al. 2018).

10.9 Conclusions

Our knowledge about As translocation into rice grains, including xylem and phloem transport, is increasing day by day. It is well obvious that phloem transport is more responsible for As loading in grain. Transporters responsible for As as well as other nutrients in xylem loading and phloem unloading process are largely unknown. The identification and such transporters and their manipulation could be a strategy to restrict As loading in rice grain. The detailed analysis of As in rice grain revealed that DMA and AsIII are main As species found in grains (Lombi et al. 2009). It has also been reported that DMA is translocated much efficiently in grains compared to inorganic As species (Carey et al. 2010). The comparative study of xylem and phloem contribution about As filling in grain revealed that phloem mediated transport accounts for 90% of AsIII and 55% of DMA to the grain. It has been shown that girdling of phloem significantly lowers As content in grains. The better understanding of As uptake and translocation in rice enable the researcher to develop varieties with less As in grains via gene editing or molecular breeding in the near future.

Acknowledgements Department of Biotechnology, Govt. of India is acknowledged for funding assistance in the form of Ramalingaswami Fellowship (Grant No. BT/HRD/35/02/2006).

References

Abedin MJ et al (2002) Arsenic accumulation and metabolism in rice (*Oryza sativa* L.). Environ Sci Technol 36(5):962–968

Ahsan N et al (2008) Comparative proteomic study of arsenic-induced differentially expressed proteins in rice roots reveals glutathione plays a central role during As stress. Proteomics 8(17):3561–3576

Ali M et al (2003) Fate of arsenic extracted with groundwater. In M Ahmed, ed, Fate of Arsenic in the Environment. Intl Training Network, Dhaka, Bangladesh, pp 7–20

Baker AJM, Reeves RD, Hajar ASM (1994) Heavy metal accumulation and tolerance in British populations of the metallophyte Thlaspi caerulescens J. & C. Presl (Brassicaceae), in New Phytol. United States. pp 61–68

Bentley R, Chasteen TG (2002) Microbial methylation of metalloids: arsenic, antimony, and bismuth. Microbiol Mol Biol Rev 66(2):250–271

Bhattacharya P et al (2012) In vitro assessment on the impact of soil arsenic in the eight rice varieties of West Bengal. J Hazard Mater, India

Bienert GP et al (2008) A subgroup of plant aquaporins facilitate the bi-directional diffusion of As(OH)3 and Sb(OH)3 across membranes. BMC Biol 6:26

Brammer H, Ravenscroft P (2009) Arsenic in groundwater: a threat to sustainable agriculture in South and South-east Asia. Environ Int 35(3):647–654

Burko Y et al (2011) From organelle to organ: ZRIZI MATE-Type transporter is an organelle transporter that enhances organ initiation. Plant Cell Physiol 52(3):518–527

Cailliatte R et al (2010) High-affinity manganese uptake by the metal transporter NRAMP1 is essential for Arabidopsis growth in low manganese conditions. Plant Cell 22(3):904–917

Carey AM et al (2010) Grain unloading of arsenic species in rice. Plant Physiol 152(1):309–319

Catarecha P et al (2007) A mutant of the Arabidopsis phosphate transporter PHT1;1 displays enhanced arsenic accumulation. Plant Cell 19(3):1123–1133

Cellier M et al (1995) Nramp defines a family of membrane proteins. Proc Natl Acad Sci U S A 92(22):10089–10093

Chakrabarty D et al (2009) Comparative transcriptome analysis of arsenate and arsenite stresses in rice seedlings. Chemosphere 74(5):688–702

Chen A, Komives EA, Schroeder JI (2006) An improved grafting technique for mature Arabidopsis plants demonstrates long-distance shoot-to-root transport of phytochelatins in Arabidopsis. Plant Physiol 141(1):108–120

Chen Y et al (2017) The Nodulin 26-like intrinsic membrane protein OsNIP3;2 is involved in arsenite uptake by lateral roots in rice. J Exp Bot 68(11):3007–3016

Cullen WR, Reimer KJ (1989) Arsenic speciation in the environment. Chem Rev 89:713–764

Curie C et al (2009) Metal movement within the plant: contribution of nicotianamine and yellow stripe 1-like transporters. Ann Bot 103(1):1–11

Das N et al (2018) Expression of rice MATE family transporter OsMATE2 modulates arsenic accumulation in tobacco and rice. Plant Mol Biol 98(1–2):101–120

Dhankher OP et al (2002) Engineering tolerance and hyperaccumulation of arsenic in plants by combining arsenate reductase and gamma-glutamylcysteine synthetase expression. Nat Biotechnol 20(11):1140–1145

Duan GL et al (2007) A CDC25 homologue from rice functions as an arsenate reductase. New Phytol 174(2):311–321

Dwivedi S et al (2010a) Arsenic affects mineral nutrients in grains of various Indian rice (Oryza sativa L.) genotypes grown on arsenic-contaminated soils of West Bengal. Protoplasma 245(1–4):113–124

Dwivedi S et al (2010b) Arsenate exposure affects amino acids, mineral nutrient status and antioxidants in rice (*Oryza sativa* L.) genotypes. Environ Sci Technol 44(24)::9542–9549

Ellis DR et al (2006) A novel arsenate reductase from the arsenic hyperaccumulating fern Pteris vittata. Plant Physiol 141(4):1544–1554

Ghosh M, Shen J, Rosen BP (1999) Pathways of As(III) detoxification in Saccharomyces cerevisiae. Proc Natl Acad Sci U S A 96(9):5001–5006

Harper M (1987) Possible toxic metal exposure of prehistoric bronze workers. Br J Ind Med 44(10):652–656

Hu J et al (2009) Long-term efficacy and safety of all-trans retinoic acid/arsenic trioxide-based therapy in newly diagnosed acute promyelocytic leukemia. Proc Natl Acad Sci U S A 106(9):3342–3347

Huang X et al (2012) A map of rice genome variation reveals the origin of cultivated rice. Nature 490(7421):497–501

Hughes MF (2002) Arsenic toxicity and potential mechanisms of action. Toxicol Lett 133(1):1–16

IARC (2004) IARC monographs on the evaluation of carcinogenic risks to humans some drinking-water disinfectants and contaminants, including arsenic. Vienna IARC 84(0):p. 0

Indriolo E et al (2012) A vacuolar arsenite transporter necessary for arsenic tolerance in the arsenic hyperaccumulating fern Pteris vittata is missing in flowering plants. Plant Cell 22(6):2045–2057

Isayenkov SV, Maathuis FJ (2008) The Arabidopsis thaliana aquaglyceroporin AtNIP7;1 is a pathway for arsenite uptake. FEBS Lett 582(11):1625–1628

Ishimaru Y et al (2012) OsNRAMP5, a major player for constitutive iron and manganese uptake in rice. Plant Signal Behav 7(7):763–766

Kamiya T, Fujiwara T (2009) Arabidopsis NIP1;1 transports antimonite and determines antimonite sensitivity. Plant Cell Physiol 50(11):1977–1981

Khan KA et al (2010) Arsenic bioavailability to rice is elevated in Bangladeshi paddy soils. Environ Sci Technol 44(22):8515–8521

Kulp TR et al (2008) Arsenic(III) fuels anoxygenic photosynthesis in hot spring biofilms from Mono Lake. California Sci 321(5891):967–970

Kuroda T, Tsuchiya T (2009) Multidrug efflux transporters in the MATE family. Biochim Biophys Acta 1794(5):763–768

Lagerkvist BJ, Zetterlund B (1994) Assessment of exposure to arsenic among smelter workers: a five-year follow-up. Am J Ind Med 25(4):477–488

Li RY et al (2009) The rice aquaporin Lsi1 mediates uptake of methylated arsenic species. Plant Physiol 150(4):2071–2080

Liu Y et al (2013) Comparative proteomic analysis of rice shoots exposed to high arsenate. J Integr Plant Biol

Liu XJ et al (2013) Arsenic speciation in Chinese Herbal Medicines and human health implication for inorganic arsenic. Environ Pollut 172:149–154

Lombi E et al (2009) Speciation and distribution of arsenic and localization of nutrients in rice grains. New Phytol 184(1):193–201

Ma JF et al (2006) A silicon transporter in rice. Nature 440(7084):688–691

Ma JF et al (2007) Genotypic difference in silicon uptake and expression of silicon transporter genes in rice. Plant Physiol 145(3):919–924

Ma JF et al (2008) Transporters of arsenite in rice and their role in arsenic accumulation in rice grain. Proc Natl Acad Sci U S A 105(29):9931–9935

Maurel C et al (2008) Plant aquaporins: membrane channels with multiple integrated functions. Annu Rev Plant Biol 59:595–624

Mendoza-Cozatl DG et al (2008) Identification of high levels of phytochelatins, glutathione and cadmium in the phloem sap of Brassica napus. A role for thiol-peptides in the long-distance transport of cadmium and the effect of cadmium on iron translocation. Plant J 54(2):249–259

Mondal P, Majumder CB, Mohanty B (2008) Effects of adsorbent dose, its particle size and initial arsenic concentration on the removal of arsenic, iron and manganese from simulated ground water by Fe3+ impregnated activated carbon. J Hazard Mater 2008:702 (Netherlands)

Mukherjee A et al (2006) Arsenic contamination in groundwater: a global perspective with emphasis on the Asian scenario. J Health Popul Nutr 24(2):142–163

Nasr R et al (2008) Eradication of acute promyelocytic leukemia-initiating cells through PML-RARA degradation. Nat Med 14(12):1333–1342

Nordstrom DK (2002) Public health. Worldwide occurrences of arsenic in ground water. Science 296(5576):2143–2145

Oscarson DW et al (1981) Oxidative power of Mn(IV) and Fe(III) oxides with respect to As(IIII) in terrestrial and aquatic environments. Nature 291:50–51

Panaullah GM et al (2009) Arsenic toxicity to rice (*Oryza sativa* L.) in Bangladesh. Plant Soil 39:317–331

Park J et al (2012) The phytochelatin transporters AtABCC1 and AtABCC2 mediate tolerance to cadmium and mercury. Plant J 69(2):278–288

Persson BL et al (1999) Phosphate permeases of Saccharomyces cerevisiae: structure, function and regulation. Biochim Biophys Acta 1422(3):255–272

Pickering IJ et al (2000) Reduction and coordination of arsenic in Indian mustard. Plant Physiol 122(4):1171–1177

Raab A, Feldmann J, Meharg AA (2004) The nature of arsenic-phytochelatin complexes in Holcus lanatus and Pteris cretica. Plant Physiol. 2004: United States. pp 1113–1122

Raab A et al (2005) Uptake, translocation and transformation of arsenate and arsenite in sunflower (Helianthus annuus): formation of arsenic-phytochelatin complexes during exposure to high arsenic concentrations. New Phytol 168(3):551–558

Rahman MA et al (2011) Effect of external iron and arsenic species on chelant-enhanced iron bioavailability and arsenic uptake in rice (*Oryza sativa* L.). Chemosphere 84(4):439–445

Rahman MM et al (2012) Phytotoxicity of arsenate and salinity on early seedling growth of rice (*Oryza sativa* L.): a threat to sustainable rice cultivation in South and South-East Asia. Bull Environ Contam Toxicol 88(5):695–702

Rochette EA et al (2000) Kinetics of arsenate reduction by dissolved sulfide. Environ Sci Technol 34:4714–4720

Rosen BP, Liu Z (2009) Transport pathways for arsenic and selenium: a minireview. Environ Int 35(3):512–515

Shin H et al (2004) Phosphate transport in Arabidopsis: Pht1;1 and Pht1;4 play a major role in phosphate acquisition from both low- and high-phosphate environments. Plant J 39(4):629–642

Shri M et al (2009) Effect of arsenic on growth, oxidative stress, and antioxidant system in rice seedlings. Ecotoxicol Environ Saf 72(4):1102–1110

Shukla D et al (2012) Expression of phytochelatin synthase from aquatic macrophyte Ceratophyllum *demersum* L. enhances cadmium and arsenic accumulation in tobacco. Plant Cell Rep 31(9):1687–1699

Smith AH, Steinmaus CM (2000) Arsenic in urine and drinking water. In: Environ health perspect. 2000: United States. pp A494–A495

Song WY et al (2010) Arsenic tolerance in Arabidopsis is mediated by two ABCC-type phytochelatin transporters. Proc Natl Acad Sci U S A 107(49):21187–21192

Song WY et al (2014) A rice ABC transporter, OsABCC1, reduces arsenic accumulation in the grain. Proc Natl Acad Sci U S A 111(44):15699–15704

Stone R (2008) Food safety. Arsenic and paddy rice: a neglected cancer risk? Science 321(5886):184–185

Sun SK et al (2018) Decreasing arsenic accumulation in rice by overexpressing OsNIP1;1 and OsNIP3;3 through disrupting arsenite radial transport in roots. New Phytol 219(2):641–653

Takanashi K et al (2013) LjMATE1: a citrate transporter responsible for iron supply to the nodule infection zone of Lotus japonicus. In: Plant Cell Physiol. 2013: Japan, pp 585–94

Tiwari M et al (2014a) Expression in Arabidopsis and cellular localization reveal involvement of rice NRAMP, OsNRAMP1, in arsenic transport and tolerance. Plant, Cell Environ 37(1):140–152

Tiwari M et al (2014b) Expression of OsMATE1 and OsMATE2 alters development, stress responses and pathogen susceptibility in Arabidopsis. Sci Rep 4:3964

Tripathi RD et al (2007) Arsenic hazards: strategies for tolerance and remediation by plants. Trends Biotechnol 25(4):158–165

Tuli R et al (2010) Recent advances in arsenic accumulation and metabolism in rice. Mol Breeding (26):307–323

Uraguchi S et al (2011) Low-affinity cation transporter (OsLCT1) regulates cadmium transport into rice grains. Proc Natl Acad Sci U S A 108(52):20959–20964
Verbruggen N, Hermans C, Schat H (2009) Mechanisms to cope with arsenic or cadmium excess in plants. Curr Opin Plant Biol England. pp 364–372
Vidal SM et al (1993) Natural resistance to infection with intracellular parasites: isolation of a candidate for Bcg. Cell 73(3):469–485
Virkki LV et al (2007) Phosphate transporters: a tale of two solute carrier families. Am J Physiol Renal Physiol 293(3):F643–F654
Wallace IS et al (2002) Functional selectivity for glycerol of the nodulin 26 subfamily of plant membrane intrinsic proteins. FEBS Lett 523(1–3):109–112
Wang J et al (2002) Mechanisms of arsenic hyperaccumulation in Pteris vittata. Uptake kinetics, interactions with phosphate, and arsenic speciation. Plant Physiol 130(3):1552–1561
WHO(2001) Arsenic in drinking water. WHO, Geneva
Williams PN et al (2007) Greatly enhanced arsenic shoot assimilation in rice leads to elevated grain levels compared to wheat and barley. Environ Sci Technol 41(19):6854–6859
Wolfe-Simon F et al (2010) A bacterium that can grow by using arsenic instead of phosphorus. Science 332(6034):1163–1166
Xu XY, McGrath SP, Zhao FJ (2007) Rapid reduction of arsenate in the medium mediated by plant roots. New Phytol 176(3):590–599
Ye WL et al (2010) Arsenic speciation in phloem and xylem exudates of castor bean. Plant Physiol 154(3):1505–1513
Yokosho K et al (2009) OsFRDL1 is a citrate transporter required for efficient translocation of iron in rice. Plant Physiol 149(1):297–305
Zavala YJ, Duxbury JM (2008) Arsenic in rice: I. Estimating normal levels of total arsenic in rice grain. Environ Sci Technol 42(10):3856–3860
Zhang XW et al (2010) Arsenic trioxide controls the fate of the PML-RARalpha oncoprotein by directly binding PML. Science 328(5975):240–243
Zhao FJ, McGrath SP, Meharg AA (2010a) Arsenic as a food chain contaminant: mechanisms of plant uptake and metabolism and mitigation strategies. Annu Rev Plant Biol 61:535–559
Zhao FJ et al (2010b) Accumulation, distribution, and speciation of arsenic in wheat grain. Environ Sci Technol 44(14):5464–5468
Zhao FJ et al (2010c) The role of the rice aquaporin Lsi1 in arsenite efflux from roots. New Phytol 186(2):392–399

Chapter 11
Metabolic Engineering of Stress Protectant Secondary Metabolites to Confer Abiotic Stress Tolerance in Plants

Deepak Ganjewala, Gurminder Kaur and Nidhi Srivastava

Abstract Plants produce vivid secondary metabolites playing essential roles in adaptation to changing environment, management of abiotic stresses including drought, temperature, salinity etc. Secondary metabolites such as isoprenoids, carotenoids and flavonoids are highly specialized in alleviating effects of abiotic stresses. In plants, biosynthesis and accumulation of these secondary metabolites rapidly changes during stress to overcome adverse effects of abiotic stresses. Hence, there is a strong correlation between secondary metabolites and abiotic stress tolerance potential of plants. Metabolic engineering offers promises to enhance the production of secondary metabolites thereby improving plant's tolerance to various environmental stresses. It has been successfully used in many plants such as Arabidopsis (*Arabidopsis thaliana*), tobacco (*Nicotiana tabacum*), sweet potato (*Ipomea batata*), alfaalfa (*Medicago sativa*), glasswort (*Salicornia europaea*) and Brassica (*Brassica napus*) to enhance the levels of secondary metabolites in order to confer abiotic stress tolerance. The present chapter deals with stress protectant secondary metabolites produced by plants as well as elaborately discuss metabolic engineering to improve their producion in some important plants to make them stress tolerant.

Keywords Abiotic stress · Secondary metabolites · Metabolic engineering · Isoprenoid · Carotenoid · Flavonoid · Abscisic acid

---

D. Ganjewala (✉) · N. Srivastava
Amity Institute of Biotechnology, Amity University Uttar Pradesh, Sector 125, Noida, Uttar Pradesh 201303, India
e-mail: deepakganjawala73@yahoo.com

G. Kaur
Institute of Biosciences and Technology, Shri Ramswaroop Memorial University, Lucknow-Deva Road, Barabanki, Uttar Pradesh 225003, India

S. P. Singh et al. (eds.), *Molecular Approaches in Plant Biology and Environmental Challenges*, Energy, Environment, and Sustainability,
https://doi.org/10.1007/978-981-15-0690-1_11

Abbreviations

AACT	Acetoacetyl-CoA thiolase
ABA	Abscisic acid
chy-β	*β*-carotene hydroxylase
CRISPR/Cas9	Clustered Regularly Interspaced Short Palindromic Repeats
DFR	Dihydroflavonol-4-reductase
GPPS	Geranylgeranyl pyrophosphate synthase
HMGR	3-hydroxy-3-methylglutaryl-CoA reductase
IPP	Isopentenyl diphosphate
lcy-β	Lycopene *β*-cyclase
MEP	Methylerythritol-4-phosphate
MVA	Mevalonic acid
Or	Orange gene
PSY	Phytoene synthase
PTDS	PCR-based two-step DNA synthesis
PYR/PYL/RCAR	Pyrabactin resistance1/PYR1-like/regulatory component of ABA receptor
ROS	Reactive oxygen species
UV	Ultra violet

11.1 Introduction

Abiotic stress can be defined as any negative impact on plants, animals and microbes caused by a nonliving factor in environment. Studies of abiotic stresses have always been in the focus of researchers due to its adverse effects on survival and productivity of plants (Vickers et al. 2009). A number of studies carried out so far have provided vast knowledge of various processes, metabolites and pathways involved in different types of stresses in plants. Advanced molecular biology techniques greatly facilitated our understanding of molecular processes underlying stress tolerance mechanisms of plants by providing many new breakthroughs. In this regard, two popularly known molecular biology techniques, genetic engineering and metabolic engineering have been proven to be very useful conferring abiotic stress tolerance to plants through manipulating expressions of the genes of the secondary metabolite biosynthetic pathways. In this chapter, we have discussed in detail about various stress prtotectant secondary metabolites and metabolic engineering approaches used to enhance their production to confer abiotic stress tolerance in plants.

Plant's tolerance to a number of environmental stresses involves a highly complex mechanism comprising of numerous processes, which is not yet fully elucidated and understood. Several abiotic factors e.g. drought, soil salinity, light, temperature, air pollution and mechanical damage may trigger abiotic stress directly affecting plant's performance and productivity (Vickers et al. 2009a). Among all, drought (low water availability), salt (high salinity), and heat (high temperatures) are the most threatening stresses. According to published reports, all such stresses drastically reduce (70%) the productivity of staple food crops (Kaur et al. 2008; Mantri et al. 2012; Bromham et al. 2013). Also, they are responsible for increasing the production of reactive oxygen species (ROS), which adversely affect antioxidant defenses of plants making their survival much difficult under such conditions (Miller et al. 2010). Nevertheless, abiotic stress factors can potentially influence almost all physiological, biochemical, and molecular processes from an early stage of seed germination to the maturity, and eventually cause severe losses in the economic yield of crop plants. As per the report by Zhao et al. (2013) ~20% of the irrigated soils worldwide are affected by salt stress. Salt stress is reported to severely limit the productivity and distribution of major plants worldwide (Tuteja 2007). Under salinity, levels of Na^+ are excessively very high, which poses several harmful effects on plants including osmotic stress, ionic toxicity and related oxidative stress. All stresses together decline the rate of photosynthesis and thus the productivity (Zhu 2002; Oh et al. 2000). Some plants have evolved counter-mechanisms like developmental regulation of stress protectants, detoxification, homeostasis and osmotic adjustment to adapt them to salt and drought stresses (Bohnert et al. 1995; Zhu 2001). However, a large number of commercial plants are highly susceptible to drought, salt and heat stresses, hence need to be converted to more stress tolerant against all such stresses. Our studies in tomato (*Lycopercocon esculentum*) indicated that silicon supplementation help to retain moisture in the soil for long time and this approach may be beneficial for areas with low water availability (Malhotra et al. 2016a, b). Silicon supplementation has been suggested for modifying antioxidant potential of the tomato, which is linked with drought tolerance (Malhotra et al. 2017).

For the development of new plants with improved abiotic stress tolerance we need deeper knowledge and understanding of mechanisms underlying different stresses and how plants do response to these conditions. Currently, we know about the changes that take place in cellular, biochemical and molecular processes during stressed situation. In addition, many genes related to abiotic stresses have been identified as well as studied for their expression patterns under stressed condition (Bhatnagar-Mathur et al. 2008). Prior knowledge of the biochemistry, molecular biology and genetics is highly desirable for metabolic engineering of any biochemical pathways which is intended to improve abiotic stress tolerance of plants (Bhatnagar-Mathur et al. 2008; Varshney et al. 2012). In this context, available cloned genes of biochemical pathways have been proven to be very useful for metabolic engineering. A number of reports indicated that the molecular manipulation of genes, such as those encoding antioxidant enzymes (Natwar et al. 2014; Wu et al. 2014; Zhang et al. 2014), transcription factors (Tang et al. 2014; Xu et al. 2014; Min et al. 2013; Sun et al. 2014), and ion transporters (Muhammad et al. 2014; Yarra et al. 2012) offers attractive targets for developing transgenic plants with abiotic stress tolerant trait. Several reports have underlined importance of secondary metabolites such as isoprenoids, carotenoids, phenolics and flavonoids for improving plant's tolerance against drought, salt, heat, light and other stresses. Thus, by increasing the production of these secondary metabolites through metabolic engineering seems to be a very useful strategy to enhance stress tolerance level of plants. Readily available cloned genes of various secondary metabolic pathways have facilitated metabolic engineering of the secondary biosynthetic pathways for the better management of abiotic stresses in plants. Rapidly increasing demand of food of burgeoning population has always been a challenge before biotechnologists. The food probem can be resolved to a large extent by improving abiotic stress tolerance in plants using molecular biology techniques and metabolic engineering. Review of literature revealed that molecular biology techniques were used for manipulation of genes involved in the synthesis of metabolites like mannitol, glycine betaine and heat shock proteins in order to enhance plant's tolerance against many negative environmental factors (Chen Tony and Murata 2002; Wani et al. 2013). Recently, Pandey and Arora (Pande and Arora 2017) have elaborately discussed molecular strategies to develop transgenic crops resistant against specific abiotic stresses. In the present chapter, we have discussed metabolic engineering of the secondary metabolite biosynthesis pathways for improving tolerance power of plants against various types of stresses. For this we have analyzed reports, reviews and research papers published till date on metabolic engineering of the secondary metabolite pathways conferring abiotic stress tolerance. To our knowledge, this is the first chapter of its kind providing valuable information to researchers working in this area.

11.2 Abiotic Stress Protectant Secondary Metabolites

Secondary metabolic pathways often derived from the primary metabolic pathways upon initial gene duplication are frequently restricted to specific taxonomic groups and play a major role in the plant and environment interaction (Mazid et al. 2011). Secondary metabolites play many essential roles in plant's life such as in adaptation of plants to the changing environment and abiotic stress conditions. Among major classes of secondary metabolites, flavonoids and phenolics are reported to be the most important classes of stress protectant metabolites. Here, we have specifically discussed secondary metabolites with stress protectant properties. Structures of the some stress protectant secondary metabolites are presented in Fig. 11.1.

11.2.1 Isoprenoids

Isoprenoids represent one of the largest classes of secondary metabolites (Fig. 11.1). They are produced via two (Fig. 11.2) separate biochemical pathways: cytosolic acetate-MVA and plastidic methylerythritol-4-phosphate (MEP) pathways (Ganjewala et al. 2009; Kirby and Keasling 2009). In plants, isoprenoids have antioxidant functions during various stresses (Vickers et al. 2009; Bajda et al. 2009; Soto et al. 2011). Many volatiles in the isoprenoid category are reported to play similar stress protectant roles in plants (Vickers et al. 2009). However, the molecular biology behind the stress protectant nature of the isoprenoids is yet to be elucidated. Stress protectant nature of isoprenoids have tremendous biotechnological prospects (Aharoni et al. 2005). Isoprene a C_5 gaseous secondary metabolite (Fig. 11.1) displayed protective effect against ozone fumigation and heat shock as well as ability to scavenge singlet oxygen (Edreva et al. 2008). Emission of isoprene provides a protective effect under high light, drought and ozone conditions in plants. Isoprene alters the ROS response system of plants which has been suggested to protect plants from abiotic stresses (Vickers et al. 2009a, b). Unfortunately, most of the crop plants do not emit

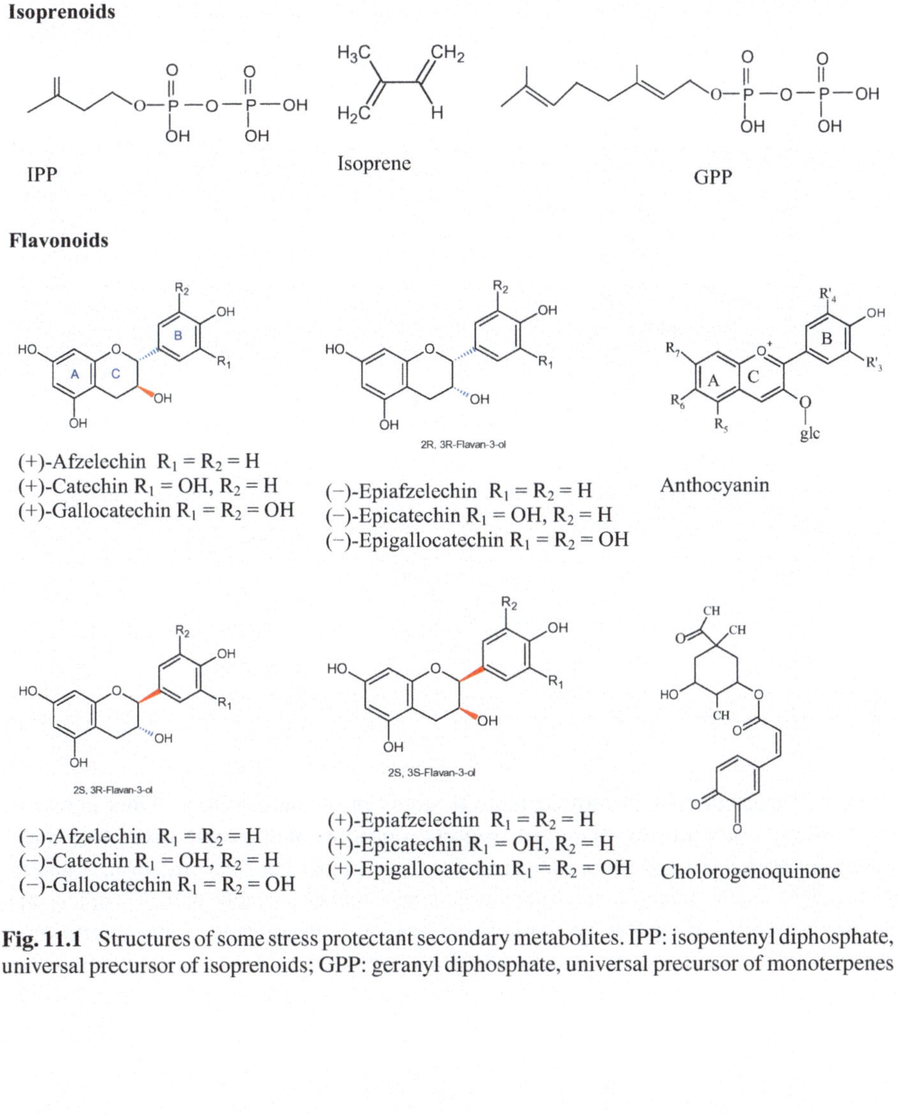

Fig. 11.1 Structures of some stress protectant secondary metabolites. IPP: isopentenyl diphosphate, universal precursor of isoprenoids; GPP: geranyl diphosphate, universal precursor of monoterpenes

Carotenoids

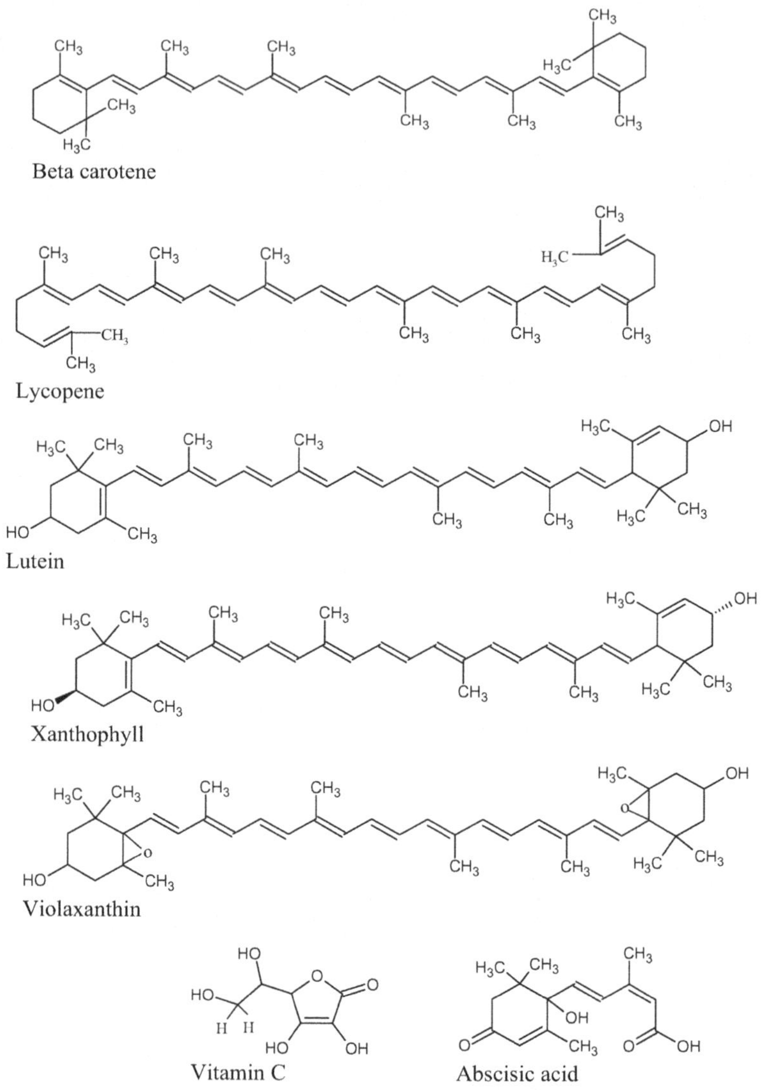

Fig. 11.1 (continued)

isoprene, hence devoid of simple natural stress protectant system. Thus converting non-isoprene emitting plants to isoprene emitting plants their stress tolerance potential can be leveled upto isoprene emitting plants. In this direction, Vickers et al. (2009, 2011) have successfully engineered tobacco plants with a single constitutively expressed *IspS* gene, which were capable of emitting isoprene and protect themselves against ozone stress. Similarly, transgenics *Arabidopsis* plants were developed by the transfer of *IspS* gene which showed resistance to thermal stress

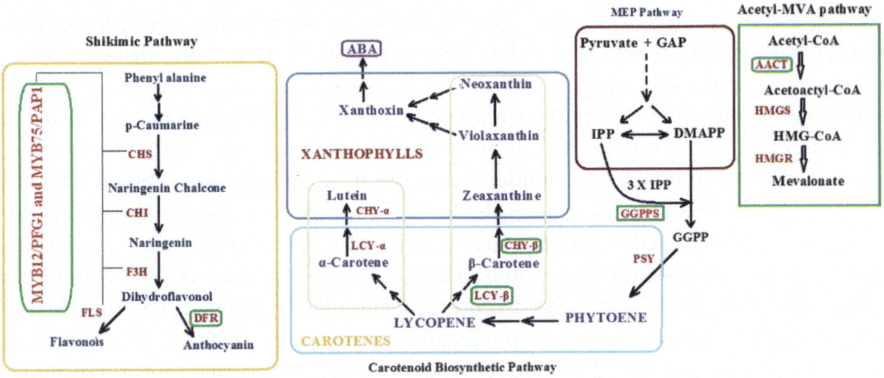

Fig. 11.2 Schematic diagram of secondary metabolic pathways induced during abiotic stress condition in plants. Overexpression of genes depicted in green box, AACT (acetoacetyl-CoA thiolase), GGPS (geranylgeranyl pyrophosphate synthase), LCY-β (lycopene cyclase), CHY-β (carotene hydroxylase) and DFR (dihydroflavonol 4-reductase) results in the enhancement of abiotic stress tolerance. Overexpression of regulatory gene *MYB12* also upregulate the genes involved in the flavonoid and ABA biosynthesis

(Loivamäki et al. 2007; Sasaki et al. 2007). Apart from the *IspS* several other genes have stress protectant effects depending on the type and extent of specific abiotic stresses. For example, in alfalfa an enzyme acetoacetyl-CoA thiolase of acetate-MVA pathway has important function in adaptation to thermal stress (Soto et al. 2011).

11.2.2 Phenolics and Flavonoids

Flavonoids are known as ‘specialized metabolites’ because plants synthesize species-specific flavonoids (Harborne and Williams 2000). Flavonoids comprises of polyphenolic compounds containing a 15-carbon skeleton (C6–C3–C6) made up of a heterocyclic benzopyran ring, an aromatic ring and the phenyl ring in their structures. Benzopyran ring has asymmetric carbon while phenyl ring has attached hydroxyl groups. Flavonoids could be distinguished from each other based on benzopyran and phenyl ring into subgroups, such as flavonols, isoflavones, flavanones, flavan-3-ols, proanthocyanidins, and anthocyanins (Fig. 11.1). Flavonoids possesses strong antioxidant properties, which has been attributed for biotic and abiotic stress tolerance capabilities of plants (Buer et al. 2010; Dixon et al. 2013). A strong correlation found between the amount of flavonoids and stress tolerance capabilities of plants. Published reports have clearly indicated protective effects of flavonoids including anthocyanins against oxidative stress generated due to adverse environmental conditions (Kim et al. 2012). Several transcription factors, including MYB, bHLH and

WD40 have been reported to play important roles in the biosynthesis of flavonoids (Grotewold 2006; Lepiniec et al. 2006; Feller et al. 2011; Qiu et al. 2014).

Previously published reports and review articles have discussed phenolics and flavonoids in context to their stress protectant roles in plants (Ferdinando et al. 2012; Mierziak et al. 2014; Ganjewala 2016; Yogendra and Nirmaljit 2017). In tobacco, a brown pigment produced by condensation of chlorogenoquinone (Fig. 11.1) and proteins is reported to control the spread of stress-induced tissue damage. Also, polyamines accumulated in tobacco have powerful ROS scavenging ability that safeguards tobacco plants from water stress damages. Phenylamides produced in beans helps them fight against heat stress. In cotton, accumulation of anthocyanins in leaves suffering from Na/K imbalance is attributed for scavenging of ROS. Accumulation of anthocyanins is correlated and the ROS scavenging capacity of cotton plants under drought stress condition.

11.2.3 Carotenoids

Carotenoids (Fig. 11.1) are the largest class of pigments widely distributed in plants, algae, fungi and cyanobacteria (Cunningham and Gantt 1998). They are fat soluble terpenoids having multiple conjugated double bonds responsible for their characteristic colors in yellow to red range (Bramley 1997). In plants, carotenoids are biosynthesized from the isopentenyl diphosphate (IPP) produced from the MEP pathway (Fig. 11.2) present in chloroplasts of photosynthetic tissues and chromoplasts of fruits and flowers (Galpaz et al. 2006; Chen et al. 2015).

Carotenoids provide potent nutritional benefits to human and animal health because animals and humans are unable to synthesize vitamin A (Fraser and Bramley 2004). They are major accessory pigments of light harvesting system and potential scavengers of ROS (such as triplet chlorophyll and singlet oxygen derived from excess light energy) produced in the photosynthetic apparatus (Tao et al. 2007; Dall'Osto et al. 2007; Andrade-Souza et al. 2011). Also, carotenoids guards plants against oxidative stress as non-enzymatic antioxidants by scavenging ROS generated due to excess excitation energy from chlorophyll during photosynthesis (Shi et al. 2015). Antioxidant nature of carotenoids helps to maintain the redox state of the cell and facilitate proper functioning of the cell under abiotic stress. Recent studies have revealed that increase in carotenoid contents is associated with high light conditions, UV irradiation, and salt stress (Shi et al. 2015). Carotenoids as being powerful antioxidant and important nutrients their higher concentration in plants will certainly increases the tolerance power of plants to the negative factors in two ways: i) by impairing ROS production and ii) acting synergistically with other antioxidants such as vitamins E and C. In tomato, under the condition of high salt concentration level of vitamin C, lycopene and β-carotene (Fig. 11.1) goes up to 35% which has been correlated with enhanced antioxidant capacity (Krauss et al. 2005). In plants, carotenoid serve as precursor for the biosynthesis of abscisic acid (ABA), which is an important signaling molecule in a variety of developmental processes and adaptive stress responses to environmental stimuli (Nambara and Marion-Poll 2005;

Taylor et al. 2005). Abscisic acid regulates numerous processes in plants including seed dormancy and the plant responses to low water availability.

11.3 Changes in Metabolic Pathways During Abiotic Stress

Plants have multiple biochemical pathways for the synthesis as well as degradation of the myriad of primary and secondary metabolites (Nascimento and Fett-Neto 2010). The major primary metabolic pathways lead to the synthesis of carbohydrates or sugars, amino acids and ammonium compounds like polyamines and glycinebetaine. Due to osmoprotective properties of these primary metabolites they have been found highly effective in overcoming negative effects of various abiotic stresses in plants. Hence, by increasing the biosynthesis of osmoprotectants metabolites in crop plants through metabolic engineering may result in improvement of their performances under stressed conditions (Rathinasabapathi 2000; Rontein et al. 2002). Previously, several review articles have been published focused on applications of genetic and metabolic engineering to enhance the production of betaines and other compatible solutes to improve abiotic stress tolerance in plans (Chen Tony and Murata 2002; Wani et al. 2013; Lawlor 2013). Besides, several metabolic pathways of antioxidant enzymes impairing ROS production have also been engineered to elevate plant's tolerance. Like primary metabolites, many secondary metabolites viz., isoprenoids, carotenoids, flavonoids and ABA have demonstrated tremendous stress protectant potential. In plants, it is found that oxidative stresses caused by heat, salt, drought, flooding, UV light, or temperature extremes triggers biosynthesis of secondary metabolites. At present, biochemical pathways of the biosynthesis and regulation of major classes of secondary metabolites are well established (Fig. 11.2), and they offer an attractive target for metabolic engineering in order to enhance the synthesis of diverse secondary metabolites useful in the management of abiotic stresses in plants (Nascimento and Fett-Neto 2010). Biochemical pathways leading to synthesis of cellulose and suberin have similar prospects in abiotic stresses tolerance as they helps plants in adaptation during stress conditions (Wang et al. 2011; Franke et al. 2012). Earlier, several articles have been published on metabolic pathways, their importance in abiotic stress tolerance, and development of commercially important stress tolerant transgenic crop plants (Jain 2013; Bakhsh and Hussain 2015).

11.4 Metabolic Engineering of Secondary Metabolites

Applications of metabolic engineering to improve abiotic stress tolerance are highly promising (Aharoni et al. 2005). However, we need knowledge of all enzymatic steps of a biochemical pathway, gene encoding all the enzymes as well as processes and factors involved in regulation of the pathway for metabolic engineering. Currently, we have knowledge of genes encoding enzymes of the primary and secondary metabolic pathways producing stress protectant metabolites from a number of plant species. The number of such genes mitigating various types of stress tolerance in plants are rapidly increasing. In addition to cloned genes, several transcriptional factors with roles in abiotic stress tolerance have been identified. Abiotic stress tolerance of many plants has been improved by altering expressions of genes involved in the biosynthesis of mannitol, glycine betaine and heat shock proteins. Transcription factors, like DREB1 (dehydration-responsive element-binding protein), MAPK (Mitogen activated protein kinases), WRKY, etc. have also been modified to confer abiotic stress tolerance in plants. The WRKY name has been coined from the highly conserved 60 amino acid long WRKY domains of the TFs, which contain a conserved heptapeptide motif WRKYGQK at the N-terminus and a novel zinc-finger-like motif at the C-terminus. Transcription factors WRKY are one of the largest families of transcriptional regulators exclusively found in plants. They have various biological functions such as in disease resistance, abiotic stress management, nutrient scarcity, senescence, seed and trichome development, embryogenesis, as well as additional developmental and hormone-controlled processes (Bakshi and Oelmüller 2014). As reported, accumulation of isoprenoids, carotenoids and flavonoids are correlated with different types of stress tolerance of plants thus any increase in the production of such metabolites will improve tolerance potential of plants. At present, most of the genes of acetate-MVA, MEP, carotenoids and phenyl propanoid pathways have been cloned. Despite all these developments, only limited efforts have been devoted so far in the direction of improving stress tolerance of plants through manipulating secondary metabolic pathways. Till now, transgenics of only some plants such as Arabidopsis (*A. thaliana*) (Davison et al. 2002), tobacco (*N. tabacum*) (Gotz et al. 2002), sweet potato (*I. batata*) (Wagner et al. 2002), alfalfa (*M. sativa* L) (Soto et al. 2011; Han et al. 2008), glasswort (*S. europaea*) (Chen et al. 2011) and Brassica (*B. napus*) (Kim et al. 2012) have been developed. Table 11.1 provided information on metabolic engineering of secondary metabolic pathways.

In many cases, in plants several long and complex metabolic pathways may involve in the biosynthesis and regulation of secondary metabolites. Also, many important traits of plants are regulated by multiple genes. Transformation of plants with multigene controlled traits has always been very difficult to perform. However, new developments in genetic engineering techniques allowed transformation of plants with multigene controlled trait. Multigene transformation is a conventional approach for development of transgenic plants with more determined phenotypes representing more complex examples of metabolic engineering (Naqvi et al. 2009). Most advanced molecular biology techniques such as antisense gene, RNAi and CRISPR/Cas9 (Clustered Regularly Interspaced Short Palindromic Repeats) have revolutionized genome editing in plants. A review by Jain (2013) provides knowledge of CRISPR-Cas9 system to understand the biology of abiotic stress tolerance in plants. In plants, CRISPR-Cas9 system has opened lots of option for genome editing in various biological contexts including management of the environmental stresses. Previous studies have provided valuable information on plants performances to regulate abiotic stress in a timely and quantitative manner (Lopez-Arredondo et al.

Table 11.1 Abiotic stress tolerance genes of secondary metabolic pathways

Pathway	Gene	Source plant	Transformed plant	Stress protectant metabolite	Expression	Stress tolerance	References
Flavonoid biosynthetic pathway	*MYB*12	*Arabidopsis thaliana*	*Arabidopsis*	Flavonoids and ABA	Overexpression	Drought and salt	Wang et al. (2016)
Flavonoid biosynthetic pathway	*MYB*12/*PFG*1 or *MYB75*/*PAP*1	*Arabidopsis thaliana*	*Arabidopsis*	Flavonoids and ABA	Overexpression	Oxidative and drought	Nakabayashi et al. (2014)
Flavonoid biosynthetic pathway	*DFR-OX B*	*Arabidopsis thaliana*	*Brassica napus*	Anthocyanin	Overexpression	Drought and salt stress	Kim et al. (2012)
Carotenoid biosynthetic pathway	β-*LCY*1	*Nicotiana tabacum*	*Nicotiana tabacum*	Carotenoids	Overexpression	Drought and salt stress	Shi et al. (2015)
Carotenoid biosynthetic pathway	*Or*	*Ipomoea batatas* cv. Sinhwangmi	*Ipomoea batatas* cv. Yulmi (Ym)	β-carotene, lutein, carotenoids	Overexpression	Salt stress	Kim et al. (2013a, b)
Carotenoid biosynthetic pathway	β-*LCY*1	Salicornia europaea L.	*Arabidopsis thaliana and Nicotiana benthamiana*	Carotenoids	Inhibition	Salt stress	Chen et al. (2011)
Carotenoid biosynthetic pathway	*CHY*-β	*Ipomoea batatas* cv. Sinhwangmi	*Ipomoea batatas* cv. Yulmi (Ym)	β-cryptoxanthin and zeaxanthin, β-carotene	Inhibition (down regulation)	Salt stress	Kim et al. (2012)
IPP biosynthetic pathway	*GGPS*	*Ipomoea batatas*	*Arabidopsis thaliana*	Carotenoid	Overexpression	Osmotic stress	Chen et al. (2015)

2016). In the following section metabolic engineering of isoprenoid, flavonoid and carotenoid biosynthesis has been discussed.

11.4.1 Manipulation of Isoprenoids

Isoprenoids have roles in physiological adaptation of plants to low temperature and high salt conditions. In alfalfa, isoprenoids play important roles during osmotic stress (Soto et al. 2011). Alfalfa is a particularly abiotic stress-sensitive crop; its yield is reduced by 50% under osmotic stress. An enzyme acetoacetyl-CoA thiolase (MsAACT) known as thiolase-II of acetate-MVA pathway regulates isoprenoid biosynthesis during salt stress adaptation in alfalfa has been cloned (Soto et al. 2011). Transgenic Alfalfa over-expressing MsAACT1 showed enhanced production of squalene without altering the 3-hydroxy-3-methylglutaryl-CoA reductase (HMGR) activity in salt-stress condition. Inhibition of acetate-MVA pathway by mevastatin and vitamin C (a potent antioxidant) has suggested that this pathway has crucial roles in abiotic stress-inducible antioxidant defense in plants.

11.4.2 Manipulation of Flavonoids

Several researchers have successfully attempted engineering of the flavonoid biosynthetic pathways (Butelli et al. 2008, 2012; Nakabayashi et al. 2014). Several studies have revealed link between flavonoid accumulation and tolerance to UV-B irradiation in plants (Stracke et al. 2010; Kusano et al. 2011). However, the exact mechanism by which flavonoid help in coping up with UV-B irradiation is yet to be elucidated. Butelli et al. (2008, 2012) have reported that in *Arabidopsis* and other plants the biosynthesis of flavonoid can be manipulated by altering expressions of flavonoid biosynthetic genes. Till date 35 genes encoding enzymes or transcription factors involved in the processing and production of flavonoids and anthocyanins have been identified (Yonekura-Sakakibara et al. 2008). Of the 35, the flavonol regulators MYB12/PFG1 (production of flavonol glycoside S1), MYB11/PFG2, and MYB111/PFG3 regulate expression of four genes, chalcone synthase (CHS), chalcone isomerase (CHI), flavanone 3-hydroxylase (F3H), flavonoid 3′-hydroxylase (F3′H), and flavonol synthase (FLS) (Stracke et al. 2010) involved in the initial stages biosynthesis, whereas the anthocyanin regulators MYB75/PAP1 (production of anthocyanin pigment1), MYB90/PAP2, MYB113, and MYB114 control expression of dihydroflavonol reductase (DFR) and leucoanthocyanidin dioxygenase/anthocyanidin synthase (LDOX/ANS) which are involved in the later stages of the biosynthesis of flavonoids (Gonzalez et al. 2008; Dubos et al. 2010). In *A. thaliana* oxidative and drought tolerance has been improved by over accumulation of flavonoids with radical scavenging activity, which mitigates oxidative and drought stress (Nakabayashi et al. 2014). Analysis of metabolome and transcriptome profiles under the oxidative and drought stress of wild-type, single over expressers of MYB12/PFG1 (production of flavonol glycosides1) or MYB75/PAP1 (production of anthocyanin pigment1), double over expressers of MYB12 and PAP1, transparent testa4 (tt4) as a flavonoid-deficient mutant, and flavonoid-deficient MYB12 or PAP1 overexpressing lines (obtained by crossing tt4 and the individual MYB over expresser) have revealed over accumulation of flavonoid, which was a key for enhanced tolerance to such stresses (Nakabayashi et al. 2014). Over accumulation of flavonoids and anthocyanins has also been confirmed by antioxidant assay (Nakabayashi et al. 2014). Regulators of flavonoid biosynthesis (e.i. MYB12/PFG1 (production of flavonol glycosides1), MYB11/PFG2, and MYB111/PFG3) play essential role in abiotic stress tolerance in plants including *A. thaliana*. The *AtMYB12* has been found to regulate flavonoids accumulation and abiotic stress tolerance in *A. thaliana* (Wang et al. 2016). Up-regulation of the gene *AtMYB12* has been particularly shown to enhance synthesis of flavonoids in several plant species (Wang et al. 2016). The *AtMYB12* gene was chemically synthesized through the PTDS (PCR-based two-step DNA synthesis) and used to develop transgenic *Arabidopsis* plants overexpressing *AtMYB12*. Such transgenic *Arabidopsis* plants showed markedly increased accumulation of flavonoids and tolerance to salt and drought stresses (Wang et al. 2016). These studies have suggested that *MYB12* gene is a promising target to increase accumulation of flavonoids and thus the abiotic stress tolerance in plants. Accumulation of higher amount of anthocyanins in plants is also correlated with plants ability to tolerate salt stress. Kim et al. (2012) have achieved high accumulation of anthocyanins in *B. napus* by overexpression of *A. thaliana* dihydroflavonol-4-reductase (At*DFR*) to significantly enhance salt stress tolerance. Development of stress resistant *Brassica* species is highly desirable in view of its rapidly increasing cultivation area worldwide. Similar to *Brassica*, overexpression of *AtDFR* gene in other commercially important plants may be very useful to modulate and/or confer salinity and drought stress tolerance. However, currently our knowledge about biochemical and molecular mechanisms underlying accumulation of flavonoid under stressed situation is limited hence how it enhances stress tolerance of plants is also not clear.

11.4.3 Manipulation of Carotenoids

Like flavonoids, carotenoids are specialized in mediating abiotic stresses hence its metabolic engineering has found immense scope in management of abiotic stresses in plants. Despite this fact, only limited efforts have been fueled into manipulation of carotenoid biosynthesis so far in few plants, such as Arabidopsis, tobacco and sweet potato. However, most elaborate work on metabolic engineering of the carotenoid biosynthesis has been carried out in sweet potato as it is a most valuable industrial crop and good source of antioxidant carotenoids (Chen et al. 2011; Wang et al. 2016). Kang et al. (2017) have elaborately discussed biosynthesis, accumulation and catabolism of carotenoids focusing on prospects of metabolic engineering of carotenoids in sweet potato. Studies have revealed that the production of carotenoids can be improved through regulated expression of a few genes such as *β*-carotene hydroxylase (*chy-β*) and lycopene *β*-cyclase (*lcy-β*) (Wang et al. 2016; Kang et al. 2017; Kim et al. 2013a, b). The *chy-β* encodes an enzyme *β*-carotene hydroxylase (CHY-*β*), which is a key regulatory enzyme in the *β*–*β*-branch of carotenoid biosynthesis catalyzing hydroxylation of both *β*-carotene to *β*-cryptoxanthin and *β*-cryptoxanthin to zeaxanthin (Wang et al. 2016). The *lcy-β* encodes an enzyme lycopene *β*-cyclase (LCY-*β*), which acts at the branch point of the carotenoid biosynthetic pathway where it catalyzes cyclization of the lycopene. Wang et al. (2016) have cloned a partial cDNA encoding CHY-*β* from the storage roots of orange-fleshed sweet potato (cv. Shinhwangmi). Down-regulation of *chy-β* gene in cultured transgenic sweet potato cells has been found to increase levels of *β*-carotene and total carotenoids. Also, the cloned CHY-*β* was used to make an RNA interference-Ib*chy-β* construct and introduced into cultured cells of white-fleshed sweet potato (cv. Yulmi) to inhibit activity of *chy-β* gene (Wang et al. 2016). Inhibition of *chy-β* gene resulted in an increased synthesis of *β*-carotene and total carotenoids, which in turn enhanced the salt stress tolerance potential of sweet potato (Wang et al. 2016). In sweet potato, the major carotenoid identified is violaxanthin, which mediate plant's response to different stresses. Violaxanthin is down-stream of *β*-carotene and is a good candidate to control the up-stream pathway. Therefore, manipulation of violaxanthin biosynthesis in white-fleshed sweet potato cv. Yulmi (Ym) callus by metabolic engineering approach has successfully enhanced abiotic stress tolerance. Engineering of a bacterial *chy-β* gene in tobacco and *Arabidopsis* has improved their tolerance level against UV, and high light stress, respectively (Davison et al. 2002; Gotz et al. 2002). The reason for improvement in stress tolerance levels of tobacco and *Arabidopsis* was the overexpression of bacterial *chy-β* gene which accelerated the biosynthesis of xanthophyll carotenoids.

Similarly, transgenic plants of *Arabidopsis* and tobacco displaying high tolerance to salt stress have been developed by introducing *lcy-β* gene from *S. europaea*. In *S. europaea*, inhibition of the lycopene cyclization step in the carotenoid biosynthetic pathway affects the salt suitability of the plants (Chen et al. 2011). Transgenic plants of tobacco (*Nicotiana benthami*ana) produced by introducing Se*lcy-β* and At*lcy-β* showed no significant differences in ectopic expression of Se*lcy-β* and At*lcy-β*. A Shi et al. (2015) have cloned and characterized a *lcy-β* gene from *N. tabacum* referred as Nt*lcy-β*. Their study further revealed that expression of Nt*lcy-β* markedly varied with developmental stages, in response to salt, drought, and ABA treatment (Shi et al. 2015). Over expression of Nt*lcy-β* gene in *N. tabacum* significantly increases accumulation of carotenoids and plant's tolerance to salt and drought stress. In *Arabidopsis*, over expression of *lcy-β* gene is reported to increase the level of lutein and the rate of non-photochemical quenching induction under high light (Pogson and Rissler 2000). Improved salt stress tolerance of transgenic *Arabidopsis* can be attributed to high accumulation of carotenoids, which impairs ROS thus protects photosynthesis system under salt stress (Chen et al. 2011). In transgenic calli of sweet potato down-regulation of *lcy-β* gene increase carotenoid synthesis and enhance salt stress tolerance (Kim et al. 2013a). In contrast to transgenic plants over expressing Nt*lcy-β*, transgenic plants where expression of Nt*lcy-β* was completely suppressed using RNAi approach showed low production of carotenoid, chlorophyll, and ABA thus making plants highly susceptible to salt and drought stress (Shi et al. 2015).

Besides *chy-β* and *lcy-β*, other genes such as orange gene (*Or*), geranylgeranyl pyrophosphate synthase (*GGPS*), and phytoene synthase gene (*PSY*) are associated with carotenoid accumulation and salt stress tolerance in plants (Chen et al. 2015; Han et al. 2008; Kim et al. 2013b). First time, *Or* gene was cloned from storage roots of orange-fleshed sweet potato cv. Sinhwangmi and termed as (Ib*Or*) (Kim et al. 2013a). It is an open reading frame in the 942 bp cDNA encoding a 313-amino acid protein containing a cysteine-rich zinc finger domain. In transgenic white-fleshed sweet potato cv. Yulmi, over expression of *Or* gene enhance *β*-carotene, lutein, and total carotenoids contents, which in turn increase antioxidant activity and thereby tolerance to salt stress (Kim et al. 2013a, b). A gene *GGPS* encoding geranylgeranyl pyrophosphate synthase (GGPS) has been cloned from the storage roots of sweet potato and named as IbGGPS (Chen et al. 2015). Enzyme GGPS catalyzes formation of geranylgeranyl pyrophosphate, which is a key step in biosynthetic pathway of carotenoids and many other terpenes. Studies showed that expression of Ib*GPPS* is related with plant's tolerance to different stresses. In *A. thaliana*, expression of Ib*GPPS* protects plants from photo-oxidative stress (Xie et al. 2012), whereas, in sweet potato enhances biosynthesis of carotenoids thereby protecting plants from osmotic stress. Also, over expression of phytoene synthase gene (Se*PSY*) from *S. europaea* in *Arabidopsis* conferred tolerance against salt and oxidative stresses (Han et al. 2008).

Carotenoid derived other metabolites like ABA (Fig. 11.1) is known to regulate a number of processes including plant's response to drought condition. Previous studies have revealed that ABA binds with small soluble protein receptors namely, PYR/PYL/RCAR (pyrabactin resistance1/PYR1-like/regulatory component of ABA receptor) found in the cytoplasm and nucleus of the plant cell. The names of proteins were derived from pyrabactin resistance/pyrabactin-like or regulatory components of ABA receptor. Pyrabactin is a synthetic ABA that helps plant to cope up with drought condition by selectively inhibiting seed germination. In their non-bonded state the ABA receptor proteins exist in dimeric form. There are six members of the PYR/PYL/RCAR family of proteins (PYR1/RCAR11: PYL1/RCAR12, PYL2/RCAR14, PYL3/RCAR13, PYL8/RCAR3, PYL9/RCAR1), which binds a protein phosphatase 2C in the presence of ABA. Based on the structures of ABA receptors they have been kept under the START superfamily of ligand-binding proteins. Studies have revealed that constitutive over expression of ABA receptors in plants may be very useful to improve drought tolerance. However, over expression of ABA may have some negative impact on yield under non-stress conditions, suggesting that precise regulation of the activity of individual or multiple ABA receptors is very important in achieving enhanced drought tolerance without compromising with plant yield. Transgenic *Arabidopsis* with drought tolerance has been developed using an engineered PYR1 ABA-receptor that can be activated by non-herbicidal agrochemical and inducer of ABA responses (Lopez-Arredondo et al. 2016). Some chemical herbicides in plants are reported to activate or repress the receptors that sense the stress or the signaling pathways triggered by hormones, which mediate the corresponding stress responses. In future, applications of such chemicals may be an alternative approach to actively control tolerance to abiotic stress conditions in plants.

11.5 Conclusion

Previously, several reviews and chapters have discussed stress protectants metabolites such as mannitol, glycine betaine, heat shock proteins and transcription factors as well as metabolic engineering of these metabolites to some extent. However, no chapter or review was avaialble on secondary metabolites specialized in mitigating different types of stresses and their metabolic engineering. The present chapter discussed stress protectant potential of vivid secondary metabolites and engineering of their biosynthetic pathways to improve plant's tolerance levels. In plants, levels of secondary metabolites such as isoprenoids, carotenoids and flavonoids strikingly change during stressed condition highlighting their importance in the management of stresses. Metabolic engineering has been proven to be very useful in increasing production of secondary metabolites. Engineering of multiple enzymatic reactions of the same or different metabolic pathways is more fruitful approach than engineering of single gene to generate abiotic stress tolerant plants. Significance of metabolic engineering is rapidly increasing to impart unique characteristics traits in plants to deal with changing environmental conditions. Thorough investigations of genes and metabolic pathways are fundamental to understand their precise roles during abiotic stresses and to improve abiotic stress tolerance via metabolic engineering. In this regard, Arabidopsis is a very useful model system. It is evident from this chapter that most of the efforts on metabolic engineering have been focused on sweet potato. However, there is a need to devote more such efforst in industrially important other crops susceptible to various abiotic stresses in order to improve their abiotic stress tolerance. Several examples presented in this chapter would be very helpful for researchers to develop stress tolerant transgenics plants. In the past several decades scientific advances in areas of molecular biology faciliated and enriched our understanding of secondary metabolite biosynthesis and regulation in plants. Applications of genome/transcriptome sequencing in stress sensitive crops may provide breakthroughs for the discovery of the new/novel genes of secondary metabolic pathways related to abiotic stress tolerance. Also, recent gene editing technique CRISPER-Cas9 can be useful for over expression or suppression of the abiotic stress related genes. Currently, this approach is being used to generate abiotic stress tolerant transgenic sweet potato. Although metabolic engineering is the most straightforward approach to manipulate gene expressions, transgenic plants have several complex issues mainly concerned with population.

Acknowledgements The corresponding author of this chapter is highly grateful to Dr. Ashok Kumar Chauhan, Founder President and Dr. Atul Chauhan, Chancellor, Amity University Uttar Pradesh, Noida, India for providing necessary support and facilities. Support and help from the lab members in literature survey is duly acknowledged.

References

Aharoni A, Jongsma MA, Bouwmeester HJ (2005) Volatile science? Metabolic engineering of terpenoids in plants. Trends Plant Sci 10:594–602

Andrade-Souza V, Costa MGC, Chen CX, Gmitter FG Jr, Costa MA (2011) Physical location of the carotenoid biosynthesis genes Psy and β-Lcy in *Capsicum annuum* (Solanaceae) using heterologous probes from *Citrus sinensis* (Rutaceae). Genet Mol Res 10:404–409

Bajda A, Konopka-Postupolska D, Krzymowska M et al (2009) Role of polyisoprenoids in tobacco resistance against biotic stresses. Physiol Plant 135:351–364

Bakhsh A, Hussain T (2015) Engineering crop plants against abiotic stress: current achievements and Prospects. Emir J Food Agric 27:24–39

Bakshi M, Oelmüller R (2014) WRKY transcription factors: jack of many trades in plants. Plant Signal Behav 9:e27700

Bhatnagar-Mathur P, Vadez V, Sharma KK (2008) Transgenic approaches for abiotic stress tolerance in plants: retrospect and prospects. Plant Cell Rep 27:411–424

Bohnert HJ, Nelson DE, Jensen RG (1995) Adaptations to environmental stresses. Plant Cell 7:1099–1111

Bramley PM (1997) The regulation and genetic manipulation of carotenoid biosynthesis in tomato fruit. Pure Appl Chem 69:2159–2162

Bromham L, Saslis-Lagoudakis CH, Bennett TH, Flowers TJ (2013) Soil alkalinity and salt tolerance: adapting to multiple stresses. Biol Lett 9:1744

Buer CS, Imin N, Djordjevic MA (2010) Flavonoids: new roles for old molecules. J Integr Plant Biol 52:98–111

Butelli E, Titta L, Giorgio M et al (2008) Enrichment of tomato fruit with health-promoting anthocyanins by expression of select transcription factors. Nat Biotechnol 26:1301–1308

Butelli E, Licciardello C, Zhang Y, Liu J, Mackay S, Bailey P, Reforgiato-Recupero G, Martin C (2012) Retrotransposons control fruit-specific, cold-dependent accumulation of anthocyanins in blood oranges. Plant Cell 24:1242–1255

Chen Tony HH, Murata N (2002) Enhancement of tolerance of abiotic stress by metabolic engineering of betaines and other compatible solutes. Curr Opin Plant Biol 5:250–257

Chen X, Han H, Jiang P, Nie L, Bao H, Fan P, Lv S, Feng J, Li Y (2011) Transformation of b-lycopene cyclase genes from *Salicornia europaea* and *Arabidopsis* conferred salt tolerance in Arabidopsis and Tobacco. Plant Cell Physiol 52(5):909–921

Chen W, He S, Liu D, Patil GB, Zhai H, Wang F et al (2015) A sweetpotato geranylgeranyl pyrophosphate synthase gene, *IbGGPS*, increases carotenoid content and enhances osmotic stress tolerance in *Arabidopsis thaliana*. PLoS ONE 10(9):e0137623

Cunningham FX, Gantt E (1998) Genes and enzymes of carotenoid biosynthesis in plants. Annu Rev Plant Physiol Plant Mol Biol 49:557–583

Dall'Osto L, Fiore A, Cazzaniga S, Giuliano G, Bassi R (2007) Different roles of α- and β-branch xanthophylls in photosystem assembly and photoprotection. J Biol Chem 282:35056–35068

Davison PA, Hunter CN, Horton P (2002) Overexpression of beta-carotene hydroxylase enhances stress tolerance in Arabidopsis. Nature 418:203–206

Dixon RA, Liu C, Jun JH (2013) Metabolic engineering of anthocyanins and condensed tannins in plants. Curr Opin Biotechnol 24:329–335

Dubos C, Stracke R, Grotewold E, Weisshaar B, Martin C, Lepiniec L (2010) MYB transcription factors in arabidopsis. Trends Plant Sci 15:573–581

Edreva A, Velikova V, Tsonev T, Dagnon A, Gurel L, Gesheva AE (2008) Stress-protective role of secondary metabolites: diversity of functions and mechanisms. Gen Appl Plant Physiol 34:67–78

Feller A, Machemer K, Braun EL, Grotewold E (2011) Evolutionary and comparative analysis of MYB and bHLH plant transcription factors. Plant J 66:94–116

Ferdinando MD, Brunetti C, Fini A, Tattini M (2012) Flavonoids as antioxidants in plants under abiotic stresses. In: Ahmad P, Prasad M (eds) Abiotic stress responses in plants. Springer, New York

Franke RB, Dombrink I, Schreiber L (2012) Suberin goes genomics: use of a short living plant to investigate a long lasting polymer. Front Plant Sci 3:4–5

Fraser PD, Bramley PM (2004) The biosynthesis and nutritional uses of carotenoids. Prog Lipid Res 43:228–265

Galpaz N, Ronen G, Khalfa Z, Zamir D, Hirschberg J (2006) A chromoplast-specific carotenoid biosynthesis pathway is revealed by cloning of the tomato white-flower locus. Plant Cell 18:1947–1960

Ganjewala D (2016) Secondary metabolite credentials and biological properties of *Lifchi Chinensis*. In: Kumar M, Kumar V, Prasad R, Varma A (eds) The litchi: biotechnology. Springer, Heidelberg, pp 213–242

Ganjewala D, Kumar S, Luthra R (2009) An account of cloned genes of methyl-erythritol-4-phosphate pathway of isoprenoid biosynthesis in Plants. Curr Issues Mol Biol S1:34–44

Gonzalez A, Zhao M, Leavitt JM, Lloyd AM (2008) Regulation of the anthocyanin biosynthetic pathway by the TTG1/bHLH/Myb transcriptional complex in Arabidopsis seedlings. Plant J 53:814–827

Gotz T, Sandmann G, Romer S (2002) Expression of a bacterial carotene hydroxylase gene (crtZ) enhances UV tolerance in tobacco. Plant Mol Biol 50:129–142

Grotewold E (2006) The genetics and biochemistry of floral pigments. Annu Rev Plant Biol 57:761–780

Han H, Li Y, Zhou S (2008) Overexpression of phytoene synthase gene from *Salicornia europaea* alters response to reactive oxygen species under salt stress in transgenic Arabidopsis. Biotech Lett 30:1501–1507

Harborne JB, Williams CA (2000) Advances in flavonoid research since 1992. Phytochem 55:481–504

Jain M (2013) Emerging role of metabolic pathways in abiotic stress tolerance. J Plant Biochem Physiol 1:2–3

Kang L, Park SC, Ji CY, Kim HS, Lee HS, Kwak SS (2017) Metabolic engineering of carotenoids in transgenic sweetpotato. Breed Sci 67:27–34

Kaur G, Kumar S, Nayyar H, Upadhyaya HD (2008) Cold stress injury during the pod-filling phase in Chickpea (*Cicer arietinum* L.): effects on quantitative and qualitative components of seeds. J Agron Crop Sci 194:457–464

Kim SH, Ahn YO, Ahn MJ, Lee HS, Kwak SS (2012) Down-regulation of β-carotene hydroxylase increases β-carotene and total carotenoids enhancing salt stress tolerance in transgenic cultured cells of sweetpotato. Phytochem 74:69–78

Kim SH, Kim YH, Ahn YO, Ahn MJ, Jeong JC, Lee HS et al (2013a) Down regulation of the lycopene ϵ-cyclase gene increases carotenoid synthesis via the β-branch-specific pathway and enhances salt-stress tolerance in sweetpotato transgenic calli. Physiol Plant 147:432–442

Kim SH, Ahn YO, Ahn MJ, Jeong JC, Lee HS, Kwak SS (2013b) Cloning and characterization of an Orange gene that increases carotenoid accumulation and salt stress tolerance in transgenic sweetpotato cultures. Plant Physiol Biochem 70:445–454

Kirby J, Keasling JD (2009) Biosynthesis of plant isoprenoids: perspectives for microbial engineering. Annu Rev Plant Biol 60:335–355

Krauss S, Schnitzler WH, Grassmann J, Woitke M (2005) The influence of different electrical conductivity values in a simplified recirculating soilless system on inner and outer fruit quality characteristics of tomato. J Agric Food Chem 54:441–448

Kusano M, Tohge T, Fukushima A, Kobayashi M, Hayashi N, Otsuki H, Kondou Y, Goto H, Kawashima M, Matsuda F, Niida R, Matsui M, Saito K, Fernie AR (2011) Metabolomics reveals comprehensive reprogramming involving two independent metabolic responses of Arabidopsis to UV-B light. Plant J 67:354–369

Lawlor DW (2013) Genetic engineering to improve plant performance under drought: physiological evaluation of achievements, limitations, and possibilities. J Exp Bot 64:83–108

Lepiniec L, Debeaujon I, Routaboul JM, Baudry A, Pourcel L et al (2006) Genetics and biochemistry of seed flavonoids. Annu Rev Plant Biol 57:405–430

Loivamäki M, Gilmer F, Fischbach RJ, Sorgel C, Bachl A, Walter A, Schnitzler JP (2007) Arabidopsis, a model to study biological functions of isoprene emission? Plant Physiol 144:1066–1078

Lopez-Arredondo D, Gonzalez-Morales SI, Bello-Bello E, Alejo-Jacuinde G, Herrera L (2016) Engineering food crops to grow in harsh environments. F1000Res 4(F1000 Faculty Rev):651

Malhotra C, Kapoor RT, Ganjewala D (2016a) Alleviation of abiotic and biotic stresses in plants by silicon supplementation. Sci Agric 13:59–73

Malhotra C, Kapoor RT, Ganjewala D (2016b) Protective role of sodium silicate against water stress in *Lycopersicon esculentum* mill. Int J Pharm Biosci 7:909–917

Malhotra C, Kapoor RT, Ganjewala D, Singh NB (2017) Sodium silicate mediated response of antioxidative defense system in *Lycopersicon esculentum* mill. under water stress. Int J Phytomed 9:364–378

Mantri N, Patade V, Penna S, Ford R (2012) Abiotic stress responses in plants: present and future. In: Ahmad P, Prasad MNV (eds) Abiotic stress responses in plants. Springer, New York, pp 1–19

Mazid M, Khan TA, Mohammad F (2011) Effect of abiotic stress on synthesis of secondary plant products: a critical review. Agric Rev 32:172–182

Mierziak J, Kostyn K, Kulma A (2014) Flavonoids as important molecules of plant interactions with the environment. Molecules 19:16240–16265

Miller G, Suzuki N, Ciftci-Yilmaz S, Mittler R (2010) Reactive oxygen species homeostasis and signaling during drought and salinity stresses. Plant Cell Environ 33:453–467

Min DH, Zhao Y, Huo DY, Li LC, Chen M, Xu ZS, Ma YZ (2013) Isolation and identification of a wheat gene encoding a zinc finger protein (TaZnFP) responsive to abiotic stresses. Acta Physiol Planta 35:1597–1604

Muhammad R, Khurram S, Rashid A, Moddassir A, Imran H, Shahid M, Gerald AB, Asir AS (2014) Cloning and characterization of Na^+/H^+ antiporter (LfNHX1) gene from a halophyte grass *Leptochloa fusca* for drought and salt tolerance. Mol Biology Report 41:1669–1682

Nakabayashi R, Yonekura-Sakakibara K, Urano K, Suzuki M, Yamada Y, Nishizawa T, Matsuda F, Kojima M, Sakakibara H, Shinozaki K, Michael AJ, Tohge T, Yamazaki M, Saito K (2014) Enhancement of oxidative and drought tolerance in Arabidopsis by over accumulation of antioxidant flavonoids. Plant J 77:367–379

Nambara E, Marion-Poll A (2005) Abscisic acid biosynthesis and catabolism. Annu Rev Plant Biol 56:165–185

Naqvi S, Zhu C, Farre G, Ramessar K, Bassie L, Breitenbach J et al (2009) Transgenic multivitamin corn through biofortification of endosperm with three vitamins representing three distinct metabolic pathways. Proc Natl Acad Sci USA 106:7762–7767

Nascimento NC, Fett-Neto AG (2010) Plant secondary metabolism and challenges in modifying its operation: an overview. In: Fett-Neto A (ed) Plant secondary metabolism engineering, vol 643. Methods in molecular biology (methods and protocols). Humana Press, Totowa

Natwar S, Avinash M, Bhavanath J (2014) Over-expression of the peroxisomal ascorbate peroxidase (SbpAPX) gene cloned from *Halophyte Salicornia* brachiata confers salt and drought stress tolerance in transgenic Tobacco. Mar Biotechnol 16:321–332

Oh SK, Kim IJ, Shin DH, Yang J, Kang H, Han KH (2000) Cloning, characterization, and heterologous expression of a functional geranylgeranyl pyrophosphate synthase from sunflower (*Helianthus annuus* L.). J Plant Physiol 157:535–542

Pande A, Arora S (2017) Molecular strategies for development of abiotic stress tolerance in plants. Cell Cell Life Sci J 2:000112

Pogson BJ, Rissler HM (2000) Genetic manipulation of carotenoid biosynthesis and photoprotection. Philos Trans R Soc Lond Ser B 355:1395–1403

Qiu J, Sun S, Luo S, Zhang J, Xiao X et al (2014) Arabidopsis AtPAP1 transcription factor induces anthocyanin production in transgenic *Taraxacum brevicorniculatum*. Plant Cell Rep 33:669–680

Rathinasabapathi B (2000) Metabolic engineering for stress tolerance: installing osmoprotectant synthesis pathways. Ann Bot 86:709–716

Rontein D, Basset G, Hanson AD (2002) Metabolic engineering of osmoprotectant accumulation in plants. Metab Eng 4:49–56

Sasaki K, Saito T, Lamsa M, Oksman-Caldentey KM, Suzuki M, Ohyama K, Muranaka T, Ohara K, Yazaki K (2007) Plants utilize isoprene emission as a thermotolerance mechanism. Plant Cell Physiol 48:1254–1262

Shi Y, Guo J, Zhang W, Jin L, Liu P, Chen X, Li F, Wei P, Li Z, Li W, Wei W, Zheng Q, Chen Q, Zhang J, Lin F, Qu L, Snyder JH, Wang R (2015) Cloning of the lycopene β-cyclase gene in *Nicotiana tabacum* and its overexpression confers salt and drought tolerance. Int J Mol Sci 16:30438–30457

Soto G, Stritzler M, Lisi C, Alleva K, Pagano ME, Ardila F, Mozzicafreddo M, Cuccioloni M, Angeletti M, Ayub ND (2011) Acetoacetyl-CoA thiolase regulates the mevalonate pathway during abiotic stress adaptation. J Exp Bot 62:5699–5711

Stracke R, Favory JJ, Gruber H, Bartelniewoehner L, Bartels S, Binkert M, Funk M, Weisshaar B, Ulm R (2010a) The Arabidopsis bZIP transcription factor HY5 regulates expression of the PFG1/MYB12 gene in response to light and ultraviolet-B radiation. Plant Cell Environ 33:88–103

Stracke R, Jahns O, Keck M, Tohge T, Niehaus K, Fernie AR, Weisshaar B (2010b) Analysis of PRODUCTION OF FLAVONOL GLYCOSIDES-dependent flavonol glycoside accumulation in *Arabidopsis thaliana* plants reveals MYB11-, MYB12- and MYB111-independent flavonol glycoside accumulation. New Phytol 188:985–1000

Sun PP, Zhu XF, Huang XS, Liu JH (2014) Overexpression of a stress-responsive MYB transcription factor of *Poncirus trifoliata* confers enhanced dehydration tolerance and increases polyamine biosynthesis. Plant Physiol Biochem 78:71–79

Tang LL, Cai H, Zhai H, Luo X, Wang ZY, Cui L, Bai X (2014) Overexpression of *Glycine soja* WRKY20 enhances both drought and salt tolerance in transgenic alfalfa (*Medicago sativa* L.). Plant Cell Tissue Organ Culture 118:77–86

Tao N, Hu Z, Liu Q, Xu J, Cheng Y, Guo L et al (2007) Expression of phytoene synthase gene (Psy) is enhanced during fruit ripening of Cara Cara navel orange (*Citrus sinensis* Osbeck). Plant Cell Rep 26:837–843

Taylor IB, Sonneveld T, Bugg TDH, Thompson AJ (2005) Regulation and manipulation of the biosynthesis of abscisic acid, including the supply of xanthophyll precursors. J Plant Growth Regul 24:253–273

Tuteja N (2007) Abscisic acid and abiotic stress signaling. Plant Signal Behav 2:135–138

Varshney RK, Ribaut JM, Buckler ES, Tuberosa R, Rafalski JA, Langridge P (2012) Can genomics boost productivity of orphan crops? Nat Biotechnol 30:1172–1176

Vickers CE, Gershenzon J, Lerdau MT, Loreto F (2009a) A unified mechanism of action for volatile isoprenoids in plant abiotic stress. Nat Chem Biol 5:283–291

Vickers CE, Possell M, Cojocariu CI, Velikova VB, Laothawornkitkul J, Ryan A, Mullineaux PM, Nicholas Hewitt C (2009b) Isoprene synthesis protects transgenic tobacco plants from oxidative stress. Plant Cell Environ 32:520–531

Vickers CE, Possell M, Laothawornkitku J, Ryan AC, Hewitt CN, Mullineaux PM (2011) Isoprene synthesis in plants: lessons from a transgenic tobacco model. Plant Cell Environ 34:1043–1053

Wagner T, Windhovel U, Romer S (2002) Bansformation of tobacco with a mutated cyanobacterial phytoene desaturase gene confers resistance to bleaching herbicides. Z Naturforsch C 57:671–679

Wang ZY, Xiong L, Li W, Zhu JK, Zhu J (2011) The plant cuticle is required for osmotic stress regulation of abscisic acid biosynthesis and osmotic stress tolerance in Arabidopsis. Plant Cell 23:1971–1984

Wang F, Kong W, Wong G, Fu L, Peng R, Li Z, Yao Q (2016) AtMYB12 regulates flavonoids accumulation and abiotic stress tolerance in transgenic *Arabidopsis thaliana*. Mol Genet Genomics. https://doi.org/10.1007/s00438-016-1203-2

Wani SH, Singh NB, Haribhushan A, Mir JI (2013) Compatible solute engineering in plants for abiotic stress tolerance—role of glycine betaine. Curr Genom 14:157–165

Wu G, Wang G, Ji J, Gao H, Guan W, Wu J, Guan C, Wang Y (2014) Cloning of a cytosolic ascorbate peroxidase gene from *Lycium chinense* Mill. and enhanced salt tolerance by overexpressing in tobacco. Gene 543:85–92

Xie Y, Xu D, Cui W, Shen W (2012) Mutation of Arabidopsis Hy1 causes Uv-C hypersensitivity by impairing carotenoid and flavonoid biosynthesis and the down-regulation of antioxidant defence. J Exp Bot 63:3869–3883

Xu Q, He Q, Li S, Tian Z (2014) Molecular characterization of StNAC2 in potato and its overexpression confers drought and salt tolerance. Acta Physiol Plant 36:1841–1851

Yarra R, He SJ, Abbagani S, Ma B, Bulle M, Zhang WK (2012) Overexpression of a wheat Na^+/H^+ antiporter gene (TaNHX2) enhances tolerance to salt stress in transgenic tomato plants (*Solanum lycopersicum* L.). Plant Cell Tissue Organ Cult 111:49–57

Yogendra KM, Nirmaljit K (2017) Antioxidants: Environmental stress mitigating metabolites. Curr Trends Biomed Eng Biosci 5:555660

Yonekura-Sakakibara K, Tohge T, Matsuda F, Nakabayashi R, Takayama H, Niida R, Watanabe-Takahashi A, Inoue E, Saito K (2008) Comprehensive flavonol profiling and transcriptome coexpression analysis leading to decoding gene-metabolite correlations in Arabidopsis. Plant Cell 20:2160–2176

Zhang DY, Yang HL, Li XS, Li HY, Wang YC (2014) Overexpression of *Tamarix albiflonum* TaMnSOD increases drought tolerance in transgenic cotton. Mol Breed 34:1–11

Zhao Q, Zhang H, Wang T, Chen SX, Dai SJ (2013) Proteomicsbased investigation of salt-responsive mechanisms in plant roots. J Proteomics 82:230–253

Zhu JK (2001) Plant salt tolerance. Trends Plant Sci 6:66–71

Zhu JK (2002) Salt and drought stress signal transduction in plants. Annu Rev Plant Biol 53:247–273

Chapter 12
An Update on Molecular Strategies of Transgenic Rice Tolerance to Abiotic Stresses

P. Hima Kumari, K. Venkatesh, S. Krupanidhi and S. Anil Kumar

Abstract As stalkless in nature, plants always encounter either abiotic or biotic stresses or both all through the life. Abiotic stresses causes severe damage and results in reduced yield. Salt, drought, heat, and cold are the important abiotic stresses, which derails the plant metabolism. With the increasing salinity, drought, heat, and decreasing natural resources; development of plant resilience for these stresses is the prime requisite. The improved varieties can be produced either by breeding or by transgenic approach. With transgenic approach, it would be easy to obtain the desired varieties in less time compared to conventional plant breeding, a time-consuming method. Rice is the third largest produced crop and its cultivation requires more water than other cereal crops. Development of rice transgenics that can withstand abiotic stresses coupled with high yield and early maturation is the need of the hour. The genetically modified rice showed improved stress tolerance to different abiotic stresses with an increase in yield. The present work describes the development of transgenic rice tolerant to various abiotic stresses alongside the early maturation and increase in yield.

Keywords Abiotic stress · Cold stress · Drought stress · Heat stress · Rice · Salt stress · Transgenics

12.1 Introduction

All through the life, plants encounter different abiotic and biotic stresses due to sedentary in nature. Abiotic stresses account for half of the agricultural yield loss (Qin et al. 2011). Salt, drought, heat, and cold are the most important abiotic stresses

P. Hima Kumari · K. Venkatesh
Department of Genetics, Osmania University, Hyderabad 500007, India

S. Krupanidhi · S. Anil Kumar (✉)
Department of Biotechnology, Vignan's Foundation for Science, Technology and Research, Vadlamudi, Guntur, Andhra Pradesh 522213, India
e-mail: anilkumarou@gmail.com

S. P. Singh et al. (eds.), *Molecular Approaches in Plant Biology and Environmental Challenges*, Energy, Environment, and Sustainability,
https://doi.org/10.1007/978-981-15-0690-1_12

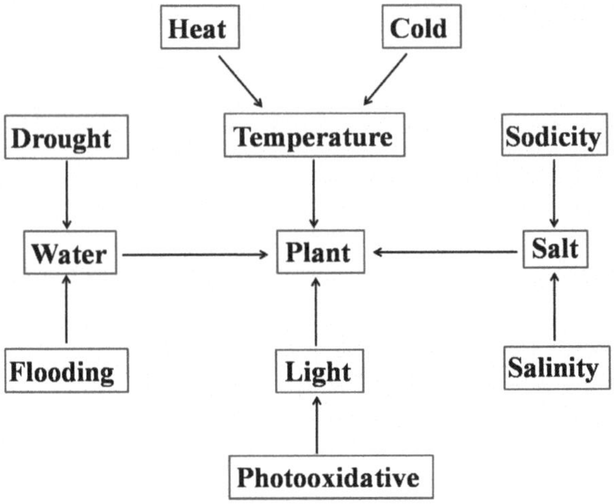

Fig. 12.1 Shematic representation of affect of different abiotic stresses on plant

that cause reduced yields (Fig. 12.1). Across the globe, drought coupled with salinity affects 64 and 6% of cultivated land and leads to severe yield loss (Boyer 1982; Cramer et al. 2011). Combined together, drought and heat stresses cause 50% of the yield loss (Wang et al. 2003). Increased salt stress is toxic in nature and prevents the water uptake by lowering the water potential. Plants maintain electro-neutrality by exclusion of Na^+ at root level or by sequestering into the vacuole through sodium proton antiporters (Mäser et al. 2001; Munns and Tester 2008). During cold conditions, rigidification of cell membrane occurs resulting in the lowering of the activity of all the cellular processes (Los and Murata 2004). The generation of reactive oxygen species (ROS) occurs during abiotic stresses, toxic for cellular metabolism (Miller et al. 2010). These stress factors are interconnected and alter the plant metabolism which causes the plant death (Rodríguez et al. 2005). Plants sense the stresses and have developed mechanisms to overcome the negative effects of stresses (Atkinson and Urwin 2012; Hu and Xiong 2014).

Rice, staple food of Asian countries and ranks third in production. It is an annual cop, but perennial in tropical areas, and can produce ratoon crop for 30 years. Rice provides 130 calories per 100 g and contains high water content (68%) followed by carbohydrates (28%), proteins (3%), and negligible fat content. To prevent malnutrition, the World Health Organization, have developed several biofortification programs of rice like iron, vitamin A, folic acid, thiamine for the development of complete nutritional food. Rice is highly suitable for high rain fall and low-cost labour areas. Rice requires more amount of water all through their growth, a challenge for water shortage and low rain-fed areas. Rice is more susceptible to cold compared to other cereals. Drought stress in rice reduces the germination rate, vegetative and reproduction growth (Manickavelu et al. 2006). Salinity stress inhibits the crop production by the sequestration of Na^+ and Cl^- resulting in the change in electroneutrality of the cell and ion homeostasis which impairs all the metabolic and physiological processes (Munns and Tester 2008).

Similarly, high temperature or heat stress hampers the biochemical, morphological, anatomical and physiological processes by inactivating the enzymes, and damaging the membranes (Wahid et al. 2007). In rice, low temperature or chilling affects the different developmental stages and its affect is more during the seedling, delays flowering, reduces the number of spikelets, booting stages, and poor grain filling (Chinnusamy et al. 2007; Ma et al. 2015).

12.2 Development of Transgenic Rice Tolerant to Various Abiotic Stresses

To meet the increasing world's population, decreasing natural resources and change in the climate, climate resilience crops are required which can resists the abiotic and biotic stresses (Lesk et al. 2016). Transgenic rice showed increased tolerance to salt, drought, heat, and cold stresses with improved yields (Chang et al. 2017; Fang et al. 2019; Moon et al. 2019; Tang et al. 2019). The role of molecular approaches of transgenic rice tolerant to drought, salt, and cold stresses has been reviewed (Sahebi et al. 2018; Reddy et al. 2017; Yasuda 2017). Roy et al. (2011) reviewed the role of genetic engineering for the development of abiotic stress tolerant crops (Roy et al. 2011). In this chapter, we have updated the gaps identified in the field of transgenic rice plants developed for tolerance to different abiotic stresses like salt, drought, heat, and cold stresses that have not been covered earlier.

12.2.1 Development of Transgenic Rice for Salt Tolerance

Transgenics developed by overexpression of transgenes showed enhanced stress tolerance compared to wild-type (WT) plants (Table 12.1). Birada et al. (2018) developed Salt Responsive Protein 3-1 (*SaSRP3-1*) gene and pyramided transgenic plants with *SaSRP3-1* and *SaVHAc1* (Vacuolar H^+-ATPase subunit c1) genes. During the reproductive stage, both lines showed better grain filling, while pyramided lines showed an improved yield. GWAS (Genome Wide Association Study) in rice showed 93 candidate genes associated with salinity tolerance (Yu et al. 2017). *Thellungiella halophila* inorganic pyrophosphatase (*ThPP1*) transgenics exhibited improved alkaline stress tolerance than WT plants (He et al. 2017). To dispose damaged/harmful cells during abiotic and biotic stresses plants have developed mechanisms like programmed cell death (PCD) or apoptosis (Del Pozo and Lam 2007; Khurana et al. 2005). Overexpression of antiPCD genes in plants improved abiotic stress tolerance. *SfIAP* (*Spodoptera frugiperda* inhibitor of apoptosis) rice transgenics showed enhanced salt tolerance, maintained photosynthetic rates, relative water content (RWC), cell viability, and reduced cell death (Hoang et al. 2014). Transcription factors (TFs) play a vital role in imparting salt stress tolerance. Under salt

Table 12.1 Non-exhaustive list of transgenic rice developed for various abiotic stresses

Gene/trascription factor	Tolerance of transgenic plant to	Reference
OsbZIP46 and SAPK6	Drought and temperature stresses	Chang et al. (2017)
OsFTL10	Drought stress	Fang et al. (2019)
OsDREB1G	Cold stress	Moon et al. (2019)
OsMYB6	Salt and drought stresses	Tang et al. (2019)
SaSRP3-1 and SaVHAc1	Salt stress	Biradar et al. (2018)
ZAT6	Salt stress	Tang and Luo (2018)
IDS1	Salt stress	Cheng et al. (2018)
ScMYBAS1	Drought stress	Peixoto-Junior et al. (2018)
OsCLC1	Salt stress	Um et al. (2018)
OsDIRP1	Drought and cold stresses	Cui et al. (2018)
OsbZIP72	Drought stress	Abreu et al. (2018)
OsbZIP23	Drought stress	Srivastava et al. (2017)
OsDRAP1	Drought stress	Huang et al. (2018b)
DREB and PIF	Drought stress	Kudo et al. (2017)
OsDRZ1	Drought stress	Yuan et al. (2018)
EDT1	Drought stress	Wu et al. (2019)
OsCTZFP8	Cold stress	Jin et al. (2018)
DaCBF4	Cold stress	Byun et al. (2018)
LsEm1	Salt and drought stresses	Xiang et al. (2018a)
RPL23A	Drought and salt stresses	Moin et al. (2017)
EhEm1	Salt and drought stresses	Xiang and Man (2018)
OsGS1;1 and OsGS2	Drought and salt stresses	James et al. (2018)
AtDREB1A	Drought and cold stresses	Latha et al. (2019)
Rab7	Drought and heat stresses	El-Esawi and Alayafi (2019)
TlOsm	Drought, salt, and cold stresses	Le et al. (2018)
CcCDR	Drought, salt, and cold stresses	Sunitha et al. (2016)
OsbZIP42	Drought stress	Joo et al. (2019)
OsHSP50.2	Drought stress	Xiang et al. (2018b)
OsLG3	Drought stress	Xiong et al. (2018)
OsMADS25	Salt stress	Xu et al. (2018)
OsRACK1A	Involved in regulation of salt stress	Zhang et al. (2018)
OsJAZ1	Drought stress	Fu et al. (2017)
OsNAC6	Drought stress	Lee et al. (2017)
OsASR5	Drought stress	Li et al. (2017b)
GA2 oxidase	Drought stress	Lo et al. (2017)
OsMAPK3	Chilling stress	Zhang et al. (2017)

stress, transgenic rice lines overexpressing barley calcium-sensor calcineurin B-like (*CBL*) protein showed water protection in vivo, accumulated higher proline, and decreased sodium uptake (Guo et al. 2016). TFs such as basic region/leucine zipper (bZIP), WRKY, AP2/ERF, and MYB improved salt tolerance in many plant species. Overexpression of *ZAT6* (Zinc finger of *Arabidopsis thaliana* 6) under salt stress increased abscisic acid (ABA), GA, antioxidant enzymes (CAT, APOX, GR, and SOD), and decreased lipid peroxidation. Additionally Ca^{2+}-dependent protein kinase genes *OsCPK*9 and *OsCPK*25 were expressed by seven folds under stress condition (Tang and Luo 2018). Cheng et al. (2018) reported an AP2/ERF (apetala2/ethylene response factor) TF indeterminate spikelet 1 (*IDS1*) negatively regulates salt tolerance in rice (Cheng et al. 2018). Reports on the utilisation of CRISPR/Cas9 systems for genome editing in rice is scanty. Cas9-OsRR22-gRNA targeting *OsRR22* gene generated 9 mutant plants with 6 mutations identified from 14 T_0 transgenics. T_2 homozygous mutant lines during seedling stage exhibited enhanced salt tolerance than the WT plants (Zhang et al. 2019).

12.2.2 Development of Transgenic Rice Tolerant to Drought Stress

Transgenic rice developed by overexpression of drought-related genes displayed enhanced tolerance to drought conditions (Table 12.1). Flowering Locus T-like (*FTL*) genes are the genetic factors for flowering. About 13 *OsFTL* genes have been reported in rice including *Hd3a* (*OsFTL2*) and *RFT1* (*OsFTL3*). *OsFTL10* overexpression in rice showed enhanced drought tolerance, early flowering, upregulation of downstream genes such as *OsMADS15* and modulated Ehd1 and OsMADS5 by feedback mechanism. It shares functional similarly with Hd3a in rice (Fang et al. 2019). Rice transgenic plants generated using sugarcane R2R3-MYB genes (*ScMYBAS1-2* and *ScMYBAS1-3*) showed better growth and RWC under drought stress conditions. *ScMYBAS1-3* exhibited higher biomass, while *ScMYBAS1-2* showed decreased biomass compared to WT plants (Peixoto-Junior et al. 2018). Overexpression of C4 photosynthesis enzymes- Maize-specific pyruvate orthophosphate dikinase (*PPDK*), maize C4-specific phosphoenolpyruvate carboxylase (*PCK*) showed enhanced drought tolerance individually and in combination in rice. Transgenic plants exhibited increased leaf photosynthetic rate, and higher grain yields when phosphoenolpyruvate carboxylase (*PEPC*) and carbonic anhydrase (*CA*) genes were overexpressed (Gu et al. 2013). Anion channel proteins such as plant chloride channels (CLCs) play key roles in ion homeostasis and cell turgor. CLCs are localized on plasma and organellar membrane imparts abiotic stress tolerance. In rice, five CLC genes have been reported; *OsCLC1* and *OsCLC2* are tonoplast specific (Nakamura et al. 2006). *OsCLC1* transgenic rice showed increased expression of *DREB1A, OsbHLH148,* better growth, tiller number, grain yield, and improved drought tolerance (Um et al. 2018). The small peptide drought tolerance 11 (*OsDT11*) is a cystein-rich peptide (*CRP*) and displayed drought tolerance in rice (Li et al. 2017a). *OsDSSR1* rice displayed improved drought stress tolerance by reduced ABA sensitivity, increased free proline, soluble sugars, superoxide dismutase and ascorbate peroxidase activities. *OsSodCc2* and *OscAPX* expression along with other stress-related genes was also noticed (Cui et al. 2018). RNA-binding proteins such as DEAD-box RNA helicase (RHs) play a crucial key role in abiotic stress adaptation (Lee and Kang 2016; Nawaz and Kang 2017). *OsSUV3* and Thermotolerant Growth Required1 (*TOGR1*) overexpression also showed drought tolerance in rice (Tuteja et al. 2013; Wang et al. 2016a).

During water stress, differential expression of phospholipase D (*PLDα1*) was noticed under drought stress in rice (Singh et al. 2012). *PLDα1* transgenic recovered from intense water deficit, showed partial recovered productivity, maintained photosynthesis, decreased oxidative stress, and delayed senescence (Abreu et al. 2018). Small ubiquitin-like modifier (SUMO) conjugation responds to heat stress in rice. Over-expression of *OsSCE1*in rice decreased drought tolerance and knockdown OsSCE1line showed marginal increase in drought tolerance (Srivastava et al. 2017; Nurdiani et al. 2018). Abscisic acid (ABA) plays an important role in stress tolerance. Five 9-*cis*-epoxycarotenoid dioxygenase (*NCED*) genes were identified in rice associated with ABA biosynthesis (Zhu et al. 2009). *OsNCED1* serves as the housekeeping gene and suppressed under water stress (Ye et al. 2011). *OsNCED2* and *OsNCED3* are associated with delayed seed germination (Zhu et al. 2009; Song et al. 2012). *OsNCED3*, *OsNCED4*, and *OsNCED5* were expressed under drought stress (Teng et al. 2014; Zhang et al. 2015). *OsNCED3* rice plants displayed drought tolerance with improved levels of ABA (Huang et al. 2018a).

Dehydration responsive element binding (DREB) member genes regulate stress-responsive genes by interacting with DRE and GCC-box *cis* element (Liu et al. 1998; Dubouzet et al. 2003). There are two DREB TFs: DREB1 (*OsDREB1A*, *OsDREB1B*, *OsDREB1C*, and *OsDREB1D*) and DREB2 (*OsDREB2A*, *OsDREB2B*, *OsDREB2C*, *OsDREB2E*, and *OsABI4*) were reported in the rice genome (Matsukura et al. 2010; Srivastav et al. 2010). Transgenic rice plants overexpressing DREB2-like TF *OsDRAP1* (Drought Responsive AP2/EREBP gene) displayed enhanced drought tolerance with better water conductivity (Huang et al. 2018b). *OsDREB1E* and *OsDREB1G* enhanced drought tolerance (Chen et al. 2008). *OsDRAP1* interacted with many genes/proteins such as *OsCBSX3* which could regulate downstream drought stress-related genes. Water deficit and exogenous ABA-induced *OsDREB2A* expression enhanced drought tolerance (Cui et al. 2011). Rice coexpressing phytochrome-interacting factor-like 1 (*OsPIL1*), and *DREB1A* showed tolerance to drought stress with hypocotyl elongation (Kudo et al. 2017). *OsDRZ1* (novel zinc finger protein 1) transgenic lines showed increased seedling drought tolerance, proline content, antioxidant enzyme activities and plant growth (Yuan et al. 2018). *PeaT1*rice showed increased expression of 32 up-regulated and 25 down-regulated drought stress-responsive proteins (Shi et al. 2018). Transcription Factor *WUSCHEL Homeobox OsWOX13*-overexpressing rice lines under a water deficit specific promoter (Rab21) showed drought resistance and early flowering (Minh-Thu et al. 2018). Wu et al. (2019) reported a novel drought-tolerant

EDT1 (Enhanced drought tolerance 1) gene, belonging to bZIP transcription factors. *EDT1*- overexpressing rice showed drought tolerance and up-regulated the stress -responsive genes like *OsbZIP12*, *SNAC1*, *OsLEA3*, *OsbZIP16*, *OsbZIP10*, and *OsABI2* (Wu et al. 2019).

12.2.3 Transgenic Rice Developed for Tolerance to Cold Stress

Rice is sensitive to cold stress during all growth phases (Zhu 2016). Cold stress during early vegetative stages hampers growth of seedling, plant height, and reduces the number of tillers (Tian et al. 2011). It also hampers the reproductive phase by sterile pollen, poor grain filling, and decreased seed setting rates resulting in the reduced yield (Xu et al. 2008; Ye et al. 2009; Zhang et al. 2011a). For chilling, *COLD1-RGA1* (chilling tolerance divergence 1, COLD1; rice G-protein α subunit1, RGA1) function has been evaluated in rice (Ma et al. 2015). OsbHLH002 directly targets *OsTPP1 (Trehalose-6-phosphate phosphatase1)* whereas OsMAPK3 under chilling stress phosphorylates OsbHLH002 which increase *OsTPP1* expression and trehalose content which imparts cold stress tolerance (Jin et al. 2018). Overexpression of the tomato *SlCZFP1* zinc finger protein in rice displayed improved tolerance to cold (Table 12.1) (Zhang et al. 2011b). *OsCTZFP8* transgenic lines showed enhanced cold tolerance during reproductive stage along with higher yield accompanied with increase in fertile pollen and seed setting than WT plants (Jin et al. 2018). The overexpression of *OsDREB1F* (drought responsive element binding) displayed enhanced tolerance to cold stress in rice and *Arabidopsis* (Wang et al. 2008). TF plays a crucial role in cold tolerance. C-repeat binding factor (CBF)/DREB is the key TF for cold tolerance (Thomashow 2010). *OsDREB1G* a cold inducible gene in leaf, root, and node when overexpressed confer cold tolerance in rice (Moon et al. 2019; Mao and Chen 2012). Overexpression of *CBF* homolog *OsDREB1A* in rice improved cold tolerance (Ito et al. 2006). Transgenic rice expressing *CBF* genes from rice (*OsDREB1*), barley (*HvCBF4*), and maize (*ZmCBF3*) displayed cold tolerance (Ito et al. 2006; Oh et al. 2007; Xu et al. 2011). Transgenic plants generated using *DaCBF4* and *DaCBF7* exhibited enhanced cold tolerance (Byun et al. 2015, 2018). Functional characterization of rice NAC transcription factor, *ONAC095* acts as a negative regulator of drought, which may be used for imparting cold tolerance (Huang et al. 2016).

12.2.4 Transgenic Rice Developed for Tolerance to Heat Stress

Heat-shock proteins (HSPs) play key roles as chaperons to impart abiotic stress tolerance. However, the molecular mechanism remains unexplored. Heat and osmotic stresses upregulates the expression of *OsHSP50.2*. *OsHSP50.2* transgenic rice lines displayed decreased water loss, electrolyte leakage, MDA, and increased proline and SOD activity (Xiang et al. 2018a). The LEA family with diverse multidomain and multifunctional proteins confer abiotic stress tolerance (Magwanga et al. 2018; Mertens et al. 2018). Hu et al. (2019) isolated a novel *LEA5C* gene and *OsLea14-A* genes from rice and its overexpression in rice displayed enhanced tolerance to dehydration, salinity, $CuSO_4$, and $HgCl_2$ stresses (Table 12.1) (Hu et al. 2019).

12.3 Pleiotropic Effects of Transgenes for Multiple Stress Tolerance

TFs acts as master regulators of abiotic stress tolerance in plants and induce stress-tolerance responsive gene expression. Several MYB TFs are involved in plant growth, development, and stress response (Butt et al. 2017). Ribosomal Protein Large (*RPL*) subunit gene (*RPL23A*) transgenic lines displayed tolerance to both salt and drought stresses with improved fresh weight, root length, proline, and chlorophyll contents (Moin et al. 2017). Transgenic rice expressing *EhEm1* (*Eutrema halophilum*) showed improved germination, increased survival rates, lowered chlorophyll damage, low MDA and high proline content. *EhEm1* rice lines showed tolerance to both salt and drought stresses. Additionally, the transgenic plants exhibited upregulation of stress-related *OsCDPK* genes (Xiang and Man 2018). *OsSGL* (stress tolerance and grain length) overexpressing rice plants showed enhanced cell division, grain filling, grain length, attributing to higher yield (Wang et al. 2016b). Rab play a vital role in several plants processes such as developmental, stress tolerance, and intracellular trafficking pathways (Agarwal et al. 2009). Rab7 protein exhibited exemplary tolerance to abiotic stresses (Pereira-leal and Seabra 2001). Mazel et al. (2004) reported that AtRab7-related proteins are confined to vacuolar membrane, while in soybean they are localized in tonoplasts and endosomes (Mazel et al. 2004). In rice, OsRab5a played a prominent role in nutrient uptake in roots, endosome transport and protein trafficking (Wang et al. 2002, 2010; Fukuda et al. 2011). *OsRab7B3* improved leaf senescence (Pitakrattananukool et al. 2012) and *OsRab7* improved salt tolerance (Peng et al. 2014). Transgenic rice lines harbouring rice *Rab7* (*OsRab7*) showed increased survival rate, chlorophyll, gas-exchanges, soluble protein, soluble sugar, proline, RWC and antioxidant enzymes activities. On the contrary, H_2O_2, electrolyte leakage, and MDA were decreased compared to WT. Chloroplastic *OsGS2* transgenic rice lines displayed elevated tolerance to salt (Hoshida et al. 2000), while cytosolic *OsGS1;1* overexpression showed enhanced oxidative stress tolerance (Lee et al. 2013). Co-expression of *OsGS1;1* and *OsGS2* in transgenic rice improved osmotic and salt tolerance and conferred limited tolerance to glufosinate. The transgenic lines showed increased fresh weight, proline, chlorophyll, RWC, photosynthetic and agronomic performance, and decreased electrolyte leakage and MDA. Transgenic lines displayed enhanced grain filling rates and subsequently improved yields under drought and salinity stress condition (James et al. 2018).

In rice, about 158 NAC family members have been reported (Jin et al. 2014). Transgenic plants harbouring NAC TFs (*OsNAC6*), *OsNAC10* (Jeong et al. 2013), and *SNAC1* (Hu et al. 2006) displayed improved drought and salt tolerance and also biotic stress tolerance (Nakashima et al. 2007). Finger millet NAC transcription factor *EcNAC67* transgenic rice lines showed drought and salt tolerance. During salt stress, transgenic lines displayed increased root and shoot biomass and recovered well after stress imposition. During drought stress, transgenic lines retained higher RWC, and less reduction in grain yield and spikelet sterility compared to WT plants (Rahman et al. 2016). NAC (*NAM/ATAF1/2/CUC2*) family of TFs are specific to plants with 149 members in rice (Gong et al. 2004; Xiong et al. 2005; Kadier et al. 2017). *OsNAC45* transgenic lines showed enhanced salt and drought tolerance by decreasing ROS, increasing lignin in roots (Yu et al. 2018). Transgenic rice lines overexpressing *ONAC002* (*SANC1/OsNAC9*), *ONAC048* (*SNAC2/OsNAC6*), *ONAC009* (*OsNAC5*), *ONAC122* (*OsNAC10*), *ONAC045*, or *ONAC058* (*OsNAP*) showed enhanced drought and salt tolerance (Jeong et al. 2010; 2013; Hu et al. 2006, 2008; Nakashima et al. 2007; Zheng et al. 2009; Takasaki et al. 2010; Song et al. 2011; Redillas et al. 2012; Chen et al. 2014; Liang et al. 2014). A new stress-responsive rice NAC gene *ONAC022* localized in the nucleus and its expression is induced by drought, high salinity, and ABA. *ONAC022*- transgenic rice lines showed improved drought tolerance showed increased survival ratios, reduced water loss and transpiration, decreased stomata opening and improved proline and soluble sugars and better growth than WT plants under drought stress condition in the greenhouse. Transgenic plants promoted salt tolerance and accumulated less Na^+ in the roots and shoots under salt stress in hydroponics. ABA biosynthetic genes, signalling and regulatory genes, and late stress-responsive genes were reported to be upregulated in transgenic rice plants (Hong et al. 2016). *OsMYB6* transgenic rice plants displayed both drought and salt stress tolerance with high abiotic stress-related gene expressions, proline, lower relative electrolyte leakage, and MDA content under stress (Tang et al. 2019). Transgenic expression of *AtDREB1A* in rice contributed to enhanced drought and salt tolerance in transgenic rice lines (Latha et al. 2019). Overexpression of *OsDIRP1* in rice showed decreased tolerance to drought and salt stresses with improved cold stress tolerance. This implies that the *OsDIRP1* negatively regulates drought and salt stresses and positively regulates cold stress (Cui et al. 2018). ROS-scavenging enzymes (*OsCATA*, *OsCATB*, *OsAPX2*, *OsSOD-Cu/Zn*) and stress tolerance genes (*OsLEA3*, *OsRD29A*, *OsSNAC1*, *OsSNAC2*, *OsDREB2A*, *OsDREB2B*, *OsRAB16A*, *OsRAB16C*) were up-regulated in transgenic rice. *OsRab7* expression enhanced grain yield, drought and heat tolerance by regulating osmolytes, antioxidants and abiotic stress-related genes (El-Esawi and Alayafi 2019). Some genes and TFs exert tolerance to more than one stress when overexpressed in rice. *TlOsm* rice displayed enhanced tolerance to salt, drought, and cold stresses. The transgenic lines also showed increased plant growth, water content, and enhanced survival rate (Le et al. 2018). In plants, methylglyoxal levels rise under abiotic stresses, a cytotoxic metabolite above normal concentrations (Kaur et al. 2014). Glyoxalase transgenic rice plants detoxified methylglyoxal, showed protected chloroplast and mitochondria, enhanced photosynthesis, lowered oxidative damage and conferred tolerance against salt, drought, heat, and fungal stresses (Gupta et al. 2018). Transgenic rice plants expressing *Cajanus cajan* cold and drought regulatory protein-encoding gene (*CcCDR*) showed drought, salt, and cold tolerance by modulation of ABA-dependent and-independent signalling-pathway genes. The CcCDR rice displayed improved plant biomass. They also produced large panicles and increased grain filling (Sunitha et al. 2016). Harpin-encoding genes impart both disease resistance and abiotic stress tolerance. Harpin-encoding gene hrf1-transgenic rice lines showed improved ABA signalling, promoted stomatal closure, increased RWC, ABA content, proline content, soluble sugars, decreased water loss and enhanced ROS-scavenging efficiency, and expression of stress-responsive genes such as *OsLEA3-1*, *OsP5CS*, *Mn-SOD*, and *NM_001074345* (Zhang et al. 2011). Transgenic rice expressing *ZFP245*conferred tolerance to salt, drought, cold, and oxidative stresses (Huang et al. 2009). Similarly, rice zinc finger protein *OsCOIN* displayed enhanced tolerance to chilling, salt, and drought stresses (Liu et al. 2007). The overexpression of *OsDREB1F* (drought responsive element binding), *OsDREB1A* and *B* showed tolerance to salt, drought, and cold stresses in rice and *Arabidopsis* (Dubouzet et al. 2003; Wang et al. 2008; Ito et al. 2006). Overexpression of *CBF* homolog *OsDREB1A* in rice improved drought, high salt, and cold tolerance (Ito et al. 2006).

12.4 Conclusion

In conclusion, we have reviewed the different strategies for the development of transgenic rice tolerant to abiotic stresses. To combat different abiotic stresses, expression of trangenes or transcription factors that confer tolerance to multiple stresses is needed. Rice trangenics expressing the transgenes and transcription factors displayed enhanced stress tolerance with a concomitant increase in the yield. Early maturation of rice transgenics was also noticed.

Acknowledgements PHK acknowledges UGC for DR. D. S. Kothari Postdoctoral Fellowship Scheme [F.4-2/2006 (BSR)/BL/14-15/0392]. SAK acknowledges DST-SERB for the National-Postdoctoral Fellowship (PDF/2015/000929).

References

Abreu FRM, Dedicova B, Vianello RP, Lanna AC, de Oliveira JAV, Vieira AF, Brondani C (2018) Overexpression of a phospholipase (*OsPLDα1*) for drought tolerance in upland rice (*Oryza sativa L*). Protoplasma 255:1751–1761. https://doi.org/10.1007/s00709-018-1265-6

Agarwal P, Reddy MK, Sopory SK, Agarwal PK (2009) Plant rabs: characterization functional diversity and role in stress tolerance. Plant Mol Biol Rep 27:417–430. https://doi.org/10.1007/s11105-009-0100-9

Atkinson NJ, Urwin PE (2012) The interaction of plant biotic and abiotic stresses: from genes to the field. J Exp Bot 63:3523–3543. https://doi.org/10.1093/jxb/ers100

Biradar H, Karan R, Subudhi PK (2018) Transgene pyramiding of salt responsive protein 3-1 (*SaSRP3-1*) and *SaVHAc1* from *Spartina alterniflora* L enhances salt tolerance in rice. Front Plant Sci 9:1304. https://doi.org/10.3389/fpls.2018.01304

Boyer JS (1982) Plant productivity and environment. Science 218:443–448. https://doi.org/10.1126/science.218.4571.443

Butt HI, Yang Z, Gong Q, Chen E, Wang X, Zhao G, Li F (2017) *GaMYB85* an R2R3 MYB gene in transgenic *Arabidopsis* plays an important role in drought tolerance. BMC Plant Biol 17:142. https://doi.org/10.1186/s12870-017-1078-3

Byun MY, Lee J, Cui LH, Kang Y, Oh TK, Park H, Kim WT (2015) Constitutive expression of *DaCBF7* an antarctic vascular plant *Deschampsia antarctica* CBF homolog resulted in improved cold tolerance in transgenic rice plants. Plant Sci 236:61–74. https://doi.org/10.1016/j.plantsci.2015.03.020

Byun MY, Cui LH, Lee J, Park H, Lee A, Kim WT, Lee H (2018) Identification of rice genes associated with enhanced cold tolerance by comparative transcriptome analysis with two transgenic rice plants overexpressing *DaCBF4* or *DaCBF7* isolated from antarctic flowering plant *Deschampsia Antarctica*. Front Plant Sci 9:601. https://doi.org/10.3389/fpls.2018.00601

Chang Y, Nguyen BH, Xie Y, Xiao B, Tang N, Zhu W, Xiong L (2017) Co-overexpression of the constitutively active form of *OsbZIP46* and ABA-Activated Protein Kinase *SAPK6* improves drought and temperature stress resistance in rice. Frontiers in Plant Science 8:1102. https://doi.org/10.3389/fpls.2017.01102

Chen JQ, Meng XP, Zhang Y, Xia M, Wang XP (2008) Over-expression of *OsDREB* genes lead to enhanced drought tolerance in rice. Biotech Lett 30:2191–2198. https://doi.org/10.1007/s10529-008-9811-5

Chen X, Wang Y, Lv B, Li J, Luo L, Lu S, Ming F (2014) The NAC family transcription factor OsNAP confers abiotic stress response through the ABA pathway. Plant Cell Physiol 55:604–619. https://doi.org/10.1093/pcp/pct204

Cheng X, Zhang S, Tao W, Zhang X, Liu J, Sun J, Chen T (2018) Indeterminate spikelet1 recruits histone deacetylase and a transcriptional repression complex to regulate rice salt tolerance. Plant Physiol 178:824–837. https://doi.org/10.1104/pp.18.00324

Chinnusamy V, Zhu J, Zhu JK (2007) Cold stress regulation of gene expression in plants. Trends Plant Sci 12:444–451. https://doi.org/10.1016/j.tplants.2007.07.002

Cramer GR, Urano K, Delrot S, Pezzotti M, Shinozaki K (2011) Effects of abiotic stress on plants: a systems biology perspective. BMC Plant Biol 11:163. https://doi.org/10.1186/1471-2229-11-163

Cui M, Zhang W, Zhang Q, Xu Z, Zhu Z, Duan F, Wu R (2011) Induced over-expression of the transcription factor OsDREB2A improves drought tolerance in rice. Plant Physiol Biochem 49:1384–1391. https://doi.org/10.1016/j.plaphy.2011.09.012

Cui LH, Min HJ, Byun MY, Oh HG, Kim WT (2018) OsDIRP1 a putative RING E3 ligase plays an opposite role in drought and cold stress responses as a negative and positive factor respectively in rice (*Oryza sativa* L.). Front Plant Sci 9:1797. https://doi.org/10.3389/fpls.2018.01797

Del Pozo O, Lam E (2007) Expression of the *Baculo virus* p35 protein in tobacco affects cell death progression and compromises N gene-mediated disease resistance response to tobacco mosaic virus. Mol Plant Microbe Interact 16:485–494. https://doi.org/10.1094/mpmi.2003.16.6.485

Dubouzet JG, Sakuma Y, Ito Y, Kasuga M, Dubouzet EG, Miura S, Yamaguchi-Shinozaki K (2003) *OsDREB* genes in rice *Oryza sativa* L encode transcription activators that function in drought-high-salt- and cold-responsive gene expression. Plant J 33:751–763. https://doi.org/10.1046/j.1365-313x.2003.01661.x

El-Esawi MA, Alayafi AA (2019) Overexpression of rice *Rab7* gene improves drought and heat tolerance and increases grain yield in rice. (*Oryza sativa* L). Genes 10:56. https://doi.org/10.3390/genes10010056

Fang M, Zhou Z, Zhou X, Yang H, Li M, Li H (2019) Overexpression of *OsFTL10* induces early flowering and improves drought tolerance in *Oryza sativa* L. Peer J 7:e6422. https://doi.org/10.7717/peerj.6422

Fu J, Wu H, Ma S, Xiang D, Liu R, Xiong L (2017) *OsJAZ1* attenuates drought resistance by regulating JA and ABA signaling in rice. Front Plant Sci 8:2108. https://doi.org/10.3389/fpls.2017.02108

Fukuda M, Satoh-Cruz M, Wen L, Crofts AJ, Sugino A, Washida H, Okita TW, Ogawa M, Kawagoe Y, Maeshima M, Kumamaru T (2011) The Small GTPase Rab5a is essential for intracellular transport of proglutelin from the golgi apparatus to the protein storage vacuole and endosomal membrane organization in developing rice endosperm. Plant Physiol 57:632–644. https://doi.org/10.1104/pp.111.180505

Gong W, Shen YP, Ma LG, Pan Y, Du YL, Wang DH, Yang JY, Hu LD, Liu XF, Dong CX, Ma L (2004) Genome-wide ORFeome cloning and analysis of *Arabidopsis* transcription factor genes. Plant Physiol 135:773–782. https://doi.org/10.1104/pp.104.042176

Gu JF, Qiu M, Yang JC (2013) Enhanced tolerance to drought in transgenic rice plants overexpressing C_4 photosynthesis enzymes. Crop J 1:105–114. https://doi.org/10.1016/j.cj.2013.10.002

Guo W, Chen T, Hussain N, Zhang G, Jiang L (2016) Characterization of salinity tolerance of transgenic rice lines harboring HsCBL8 of wild barley (*Hordeum spontanum*) line from Qinghai-Tibet plateau. Front Plant Sci 7:1678. https://doi.org/10.3389/fpls.2016.01678

Gupta BK, Sahoo KK, Ghosh A, Tripathi AK, Anwar K, Das P, Singla-Pareek SL (2018) Manipulation of glyoxalase pathway confers tolerance to multiple stresses in rice. Plant Cell Environ 41:11861200. https://doi.org/10.1111/pce.12968

He R, Yu G, Han X, Han J, Li W, Wang B, Cheng X (2017) *ThPP1* gene encodes an inorganic pyrophosphatase in *Thellungiella halophila* enhanced the tolerance of the transgenic rice to alkali stress. Plant Cell Rep 36:1929–1942. https://doi.org/10.1007/s00299-017-2208-y

Hoang TML, Williams B, Khanna H, Dale J, Mundree SG (2014) Physiological basis of salt stress tolerance in rice expressing the anti-apoptotic gene *SfIAP*. Funct Plant Biol 41:1168–1177. https://doi.org/10.1071/fp13308

Hong Y, Zhang H, Huang L, Li D, Song F (2016) Overexpression of a stress-responsive NAC transcription factor gene ONAC022 improves drought and salt tolerance in rice. Front Plant Sci 7:4. https://doi.org/10.3389/fpls.2016.00004

Hoshida H, Tanaka Y, Hibino T, Hayashi Y, Tanaka A, Takabe T, Takabe T (2000) Enhanced tolerance to salt stress in transgenic rice that overexpresses chloroplast glutamine synthetase. Plant Mol Biol 43:103–111. https://doi.org/10.1023/A:1006408712416

Hu H, Xiong L (2014) Genetic engineering and breeding of drought-resistant crops. Ann Rev Plant Biol 65:715–741

Hu H, Dai M, Yao J, Xiao B, Li X, Zhang Q, Xiong L (2006) Overexpressing a *NAM ATAF* and *CUC* (NAC) transcription factor enhances drought resistance and salt tolerance in rice. Proc Natl Acad Sci 103:12987–12992. https://doi.org/10.1073/pnas.0604882103

Hu H, You J, Fang Y, Zhu X, Qi Z, Xiong L (2008) Characterization of transcription factor gene SNAC2 conferring cold and salt tolerance in rice. Plant Mol Biol 67:169–181. https://doi.org/10.1007/s11103-008-9309-5

Hu T, Liu Y, Zhu S (2019) Overexpression of *OsLea14-A* improves the tolerance of rice and increases Hg accumulation under diverse stresses. Environ Sci Pollut Res 26:10537. https://doi.org/10.1007/s11356-019-04464-z

Huang J, Sun SJ, Xu DQ, Yang X, Bao YM, Wang ZF, Zhang H (2009) Increased tolerance of rice to cold drought and oxidative stresses mediated by the overexpression of a gene that encodes the zinc finger protein ZFP245. Biochem Biophys Res Commun 389:556–561. https://doi.org/10.1016/j.bbrc.2009.09.032

Huang L, Hong Y, Zhang H, Li D, Song F (2016) Rice NAC transcription factor *ONAC095* plays opposite roles in drought and cold stress tolerance. BMC Plant Biol 16:203. https://doi.org/10.1186/s12870-016-0897-y

Huang Y, Guo Y, Liu Y, Zhang F, Wang Z, Wang H, Chen L (2018a) 9-cis-Epoxycarotenoid Dioxygenase 3 regulates plant growth and enhances multi-abiotic stress tolerance in rice. Front Plant Sci 9:162. https://doi.org/10.3389/fpls.2018.00162

Huang L, Wang Y, Wang W, Zhao X, Qin Q, Sun F, Li Z (2018b) Characterization of transcription factor gene *OsDRAP1* conferring drought tolerance in rice. Front Plant Sci 9:94. https://doi.org/10.3389/fpls.2018.00094

Ito Y, Katsura K, Maruyama K, Taji T, Kobayashi M, Seki M, Yamaguchi-Shinozaki K (2006) Functional analysis of rice DREB1/CBF-type transcription factors involved in cold-responsive gene expression in transgenic rice. Plant Cell Physiol 47:141–153. https://doi.org/10.1093/pcp/pci230

James D, Borphukan B, Fartyal D, Ram B, Singh J, Manna M, Reddy MK (2018) Concurrent overexpression of *OsGS1;1* and *OsGS2* genes in transgenic rice (*Oryza sativa* L.): impact on tolerance to abiotic stresses. Front Plant Sci 9:786. https://doi.org/10.3389/fpls.2018.00786

Jeong JS, Kim YS, Baek KH, Jung H, Ha SH, Do Choi Y, Kim JK (2010) Root-specific expression of osnac10 improves drought tolerance and grain yield in rice under field drought conditions. Plant Physiol 153:185–197. https://doi.org/10.1104/pp.110.154773

Jeong JS, Kim YS, Redillas MCFR, Jang G, Jung H, Bang SW, Kim JK (2013) OsNAC5 overexpression enlarges root diameter in rice plants leading to enhanced drought tolerance and increased grain yield in the field. Plant Biotechnol J 11:101–114. https://doi.org/10.1111/pbi.12011

Jin J, Zhang H, Kong L, Gao G, Luo J (2014) PlantTFDB 30: A portal for the functional and evolutionary study of plant transcription factors. Nucleic Acids Res 42:D1182–D1187. https://doi.org/10.1093/nar/gkt1016

Jin Y-M, Piao,R Yan Y-F, Chen M, Wang L, He H, Lin X-F (2018) Overexpression of a new zinc finger protein transcription *OsCTZFP8* improves cold tolerance in rice. Int J Genomics 2018:1–13. https://doi.org/10.1155/2018/5480617

Joo J, Lee YH, Song SI (2019) OsbZIP42 is a positive regulator of ABA signaling and confers drought tolerance to rice. Planta 249:1521–1533. https://doi.org/10.1007/s00425-019-03104-7

Kadier Y, Zu Y, Dai Q, Song G, Lin S, Sun Q, Lu M (2017) Genome-wide identification classification and expression analysis of NAC family of genes in sorghum [*Sorghum bicolor* (L.) Moench]. Plant Growth Regul 83:301–312. https://doi.org/10.1007/s10725-017-0295-y

Kaur C, Ghosh A, Pareek A, Sopory SK, Singla-Pareek SL (2014) Glyoxalases and stress tolerance in plants. Biochem Soc Trans 2014:485–490. https://doi.org/10.1042/bst20130242

Khurana SMP, Pandey SK, Sarkar D, Arunachalam C (2005) Apoptosis in plant disease response: a close encounter of the pathogen kind. Curr Sci 10:740–752

Kudo M, Kidokoro S, Yoshida T, Mizoi J, Todaka D, Fernie AR, Yamaguchi-Shinozaki K (2017) Double overexpression of *DREB* and *PIF* transcription factors improves drought stress tolerance and cell elongation in transgenic plants. Plant Biotechnol J 15:458–471. https://doi.org/10.1111/pbi.12644

Latha GM, Raman KV, Lima JM, Pattanayak D, Singh AK, Chinnusamy V, Mohapatra T (2019) Genetic engineering of indica rice with *AtDREB1A* gene for enhanced abiotic stress tolerance. Plant Cell Tissue Organ Cult 136:173–188. https://doi.org/10.1007/s11240-018-1505-7

Le TTT, Williams B, Mundree SG (2018) An osmotin from the resurrection plant *Tripogon loliiformis* (*TlOsm*) confers tolerance to multiple abiotic stresses in transgenic rice. Physiol Plant 162:13–34. https://doi.org/10.1111/ppl.12585

Lee K, Kang H (2016) Emerging roles of RNA-binding proteins in plant growth development and stress responses. Mol Cells 39:179–185. https://doi.org/10.14348/molcells.2016.2359

Lee HJ, Abdula SE, Jang DW, Park SH, Yoon UH, Jung YJ, Cho YG (2013) Overexpression of the glutamine synthetase gene modulates oxidative stress response in rice after exposure to cadmium stress. Plant Cell Rep 32:1521–1529. https://doi.org/10.1007/s00299-013-1464-8

Lee DK, Chung PJ, Jeong JS, Jang G, Bang SW, Jung H, Kim JK (2017) The rice OsNAC6 transcription factor orchestrates multiple molecular mechanisms involving root structural adaptions and nicotianamine biosynthesis for drought tolerance. Plant Biotechnol 15:754–764. https://doi.org/10.1111/pbi.12673

Lesk C, Rowhani P, Ramankutty N (2016) Influence of extreme weather disasters on global crop production. Nature 529:84–87. https://doi.org/10.1038/nature16467

Li X, Han H, Chen M, Yang W, Liu L, Li N, Chu Z (2017a) Overexpression of *OsDT11* which encodes a novel cysteine-rich peptide enhances drought tolerance and increases ABA concentration in rice. Plant Mol Biol 93:21–34. https://doi.org/10.1007/s11103-016-0544-x

Li J, Li Y, Yin Z, Jiang J, Zhang M, Guo X, Li Z (2017b) *OsASR5* enhances drought tolerance through a stomatal closure pathway associated with ABA and H_2O_2 signalling in rice. Plant Biotechnol J 15:183–196. https://doi.org/10.1111/pbi.12601

Liang C, Wang Y, Zhu Y, Tang J, Hu B, Liu L, Chu C (2014) *OsNAP* connects abscisic acid and leaf senescence by fine-tuning abscisic acid biosynthesis and directly targeting senescence-associated genes in rice. Proc Natl Acad Sci 111:10013–10018. https://doi.org/10.1073/pnas.1321568111

Liu Q, Kasuga M, Sakuma Y, Abe H, Miura S, Yamaguchi-Shinozaki K, Shinozaki K (1998) Two transcription factors DREB1 and DREB2 with an EREBP/AP2 DNA binding domain separate two cellular signal transduction pathways in drought- and low-temperature-responsive gene expression respectively in *Arabidopsis*. Plant Cell 10:1391–1406. https://doi.org/10.1105/tpc.10.8.1391

Liu K, Wang L, Xu Y, Chen N, Ma Q, Li F, Chong K (2007) Overexpression of *OsCOIN* a putative cold inducible zinc finger protein increased tolerance to chilling salt and drought and enhanced proline level in rice. Planta 226:1007–1016. https://doi.org/10.1007/s00425-007-0548-5

Lo SF, Ho THD, Liu YL, Jiang MJ, Hsieh KT, Chen KT, Yu LC, Lee MH, Chen CY, Huang TP, Kojima M (2017) Ectopic expression of specific GA 2 oxidase mutants promotes yield and stress tolerance in rice. Plant Biotechnol J 15:850–864. https://doi.org/10.1111/pbi.12681

Los DA, Murata N (2004) Membrane fluidity and its roles in the perception of environmental signals. Biochim Biophys Acta 1666:142–157. https://doi.org/10.1016/j.bbamem.2004.08.002

Ma Y, Dai X, Xu Y, Luo W, Zheng X, Zeng D, Chong K (2015) COLD1 confers chilling tolerance in rice. Cell 160:1209–1221. https://doi.org/10.1016/j.cell.2015.06.046

Magwanga RO, Lu P, Kirungu JN, Cai X, Zhou Z, Wang X, Liu F (2018) Whole genome analysis of cyclin dependent kinase (CDK) gene family in cotton and functional evaluation of the role of CDKF4 gene in drought and salt stress tolerance in plants. Int J Mol Sci 19:2625. https://doi.org/10.3390/ijms19092625

Manickavelu A, Nadarajan N, Ganesh SK, Gnanamalar RP, Babu RC (2006) Drought tolerance in rice: morphological and molecular genetic consideration. Plant Growth Regul 50:121–138. https://doi.org/10.1007/s10725-006-9109-3

Mao D, Chen C (2012) Colinearity and similar expression pattern of rice DREB1s reveal their functional conservation in the cold-responsive pathway. PLoS ONE 7:e47275. https://doi.org/10.1371/journal.pone.0047275

Mäser P, Thomine S, Schroeder JI, Ward JM, Hirschi K, Sze H (2001) Phylogenetic relationships within cation transporter families of *Arabidopsis*. Plant Physiol 126:1646–1667. https://doi.org/10.1104/pp.126.4.1646

Matsukura S, Mizoi J, Yoshida T, Todaka D, Ito Y, Maruyama K, Yamaguchi-Shinozaki K (2010) Comprehensive analysis of rice DREB2-type genes that encode transcription factors involved in the expression of abiotic stress-responsive genes. Mol Genet Genomics 283:185–196. https://doi.org/10.1007/s00438-009-0506-y

Mazel A, Leshem Y, Tiwari BS, Levine A (2004) Induction of salt and osmotic stress tolerance by overexpression of an intracellular vesicle trafficking protein AtRab7 (AtRabG3e). Plant Physiol 134:118–128. https://doi.org/10.1104/pp.103.025379

Mertens J, Aliyu H, Cowan DA (2018) LEA proteins and the evolution of the WHy domain. Appl Environ Microbiol 84:e00539–e00518. https://doi.org/10.1128/aem.00539-18

Miller G, Suzuki N, Ciftci-Yilmaz S, Mittler R (2010) Reactive oxygen species homeostasis and signalling during drought and salinity stresses. Plant, Cell Environ 33:453–467. https://doi.org/10.1111/j.1365-3040.2009.02041.x

Minh-Thu P-T, Kim JS, Chae S, Jun KM, Lee G-S, Kim D-E, Kim Y-K (2018) A *WUSCHEL*homeobox transcription factor *OsWOX13* enhances drought tolerance and triggers early flowering in rice. Mol Cells 41:781. https://doi.org/10.14348/molcells.2018.0203

Moin M, Bakshi A, Madhav MS, Kirti PB (2017) Expression profiling of ribosomal protein gene family in dehydration stress responses and characterization of transgenic rice plants overexpressing *RPL23A* for water-use efficiency and tolerance to drought and salt stresses. Front Chem 5:97. https://doi.org/10.3389/fchem.2017.00097

Moon S-J, Min MK, Kim J-A, Kim DY, Yoon IS, Kwon TR, Kim B-G (2019) Ectopic Expression of *OsDREB1G* a member of the *OsDREB1* subfamily confers cold stress tolerance in rice. Front Plant Sci 10:297. https://doi.org/10.3389/fpls.2019.00297

Munns M, Tester M (2008) Mechanisms of salinity tolerance. Annu Rev Plant Biol 59:651–681. https://doi.org/10.1146/annurev.arplant.59.032607.092911

Nakamura A, Fukuda A, Sakai S, Tanaka Y (2006) Molecular cloning functional expression and subcellular localization of two putative vacuolar voltage-gated chloride channels in rice (*Oryza sativa* L.). Plant Cell Physiol 47:32–42. https://doi.org/10.1093/pcp/pci220

Nakashima K, Tran LSP, Van Nguyen D, Fujita M, Maruyama K, Todaka D, Hayashi Ito Y, Shinozaki K. Yamaguchi-Shinozaki K, N (2007) Functional analysis of a NAC-type transcription factor OsNAC6 involved in abiotic and biotic stress-responsive gene expression in rice. Plant J 51:617–630. https://doi.org/10.1111/j.1365-313x.2007.03168.x

Nawaz G, Kang H (2017) Chloroplast- or mitochondria-targeted DEAD-Box RNA Helicases play essential roles in organellar RNA metabolism and abiotic stress responses. Front Plant Sci 8:871. https://doi.org/10.3389/fpls.2017.00871

Nurdiani D, Widyajayantie D, Nugroho S (2018) *OsSCE1* encoding SUMO E2-conjugating enzyme involves in drought stress response of *Oryza sativa* rice. Science 25:73–81. https://doi.org/10.1016/j.rsci.2017.11.002

Oh SJ, Kwon CW, Choi DW, Song SI, Kim JK (2007) Expression of barley *HvCBF4* enhances tolerance to abiotic stress in transgenic rice. Plant Biotechnol J 5:646–656. https://doi.org/10.1111/j.1467-7652.2007.00272.x

Peixoto-Junior RF, de Andrade LM, dos Santos Brito M, Nobile PM, Martins APB, Carlin SD, Ribeiro RV, de Souza Goldman MH, de Oliveira JFNC, de Oliveira Figueira AV, Creste S (2018) Overexpression of ScMYBAS1 alternative splicing transcripts differentially impacts biomass accumulation and drought tolerance in rice transgenic plants. PLoS ONE 13:e0207534. https://doi.org/10.1371/journal.pone.0207534

Peng X, Ding X, Chang T, Wang Z, Liu R, Zeng X, Cai Y, Zhu Y (2014) Overexpression of a vesicle trafficking gene *OsRab7* enhances salt tolerance in rice. Sci World J 2014:1–7. https://doi.org/10.1155/2014/483526

Pereira-leal JB, Seabra MC (2001) Evolution of the Rab family of small GTP-binding proteins. J Mol Biol 313:889–901. https://doi.org/10.1006/jmbi.2001.5072

Pitakrattananukool S, Kawakatsu T, Anuntalabhochai S, Takaiwa F (2012) Overexpression of *OsRab7B3* a small GTP-binding protein gene enhances leaf senescence in transgenic rice. Biosci Biotechnol Biochem 76:1296–1302. https://doi.org/10.1271/bbb.120050

Qin F, Shinozaki K, Yamaguchi-Shinozaki K (2011) Achievements and challenges in understanding plant abiotic stress responses and tolerance. Plant Cell Physiol 52:1569–1582. https://doi.org/10.1093/pcp/pcr106

Rahman H, Ramanathan V, Nallathambi J, Duraialagaraja S, Raveendran M (2016) Over-expression of a *NAC67* transcription factor from finger millet (*Eleusine coracana* L) confers tolerance against salinity and drought stress in rice. BMC Biotechnol 16:35. https://doi.org/10.1186/s12896-016-0261-1

Reddy INBL, Kim BK, Yoon IS, Kim KH, Kwon TR (2017) Salt tolerance in rice: focus on mechanisms and approaches. Rice Sci 24:123–144. https://doi.org/10.1016/j.rsci.2016.09.004

Redillas MCFR, Jeong JS, Kim YS, Jung H, Bang SW, Choi YD, Kim JK (2012) The overexpression of *OsNAC9* alters the root architecture of rice plants enhancing drought resistance and grain yield under field conditions. Plant Biotechnol J 10:792–805. https://doi.org/10.1111/j.1467-7652.2012.00697.x

Rodríguez M, Canales E, Borras-Hidalgo O (2005) Molecular aspects of abiotic stress in plants. Biotecnología Aplicada 22:1–10

Roy B, Noren SK, Mandal AB, Basu AK (2011) Genetic engineering for abiotic stress tolerance in agricultural crops. Biotechnology 10:1–22. https://doi.org/10.1016/b978-0-12-813066-7.00001-2

Sahebi M, Hanafi MM, Rafii MY, Mahmud TMM, Azizi P, Osman M, Abiri R, Taheri S, Kalhori N, Shabanimofrad M, Miah G (2018) Improvement of drought tolerance in rice (*Oryza sativa* L.): genetics, genomic tools, and the *WRKY* gene family. Biomed Res Int 3158474. https://doi.org/10.1155/2018/3158474

Shi F, Yang X, Zeng H, Guo L, Qiu D (2018) Label-free quantitative proteomic analysis revealed a positive effect of ectopic over-expression of PeaT1 from *Alternaria tenuissima* on rice (*Oryza sativa*) response to drought. 3 Biotech 8:480. https://doi.org/10.1007/s13205-018-1507-5

Singh A, Pandey A, Baranwal V, Kapoor S, Pandey GK (2012) Comprehensive expression analysis of rice phospholipase D gene family during abiotic stresses and development. Plant Signal Behav 7:855–874. https://doi.org/10.4161/psb.20385

Song SY, Chen Y, Chen J, Dai XY, Zhang WH (2011) Physiological mechanisms underlying OsNAC5-dependent tolerance of rice plants to abiotic stress. Planta 234:331–345. https://doi.org/10.1007/s00425-011-1403-2

Song S, Dai X, Zhang WH (2012) A rice F-box gene *OsFbx352* is involved in glucose-delayed seed germination in rice. J Exp Bot 63:5559–5568. https://doi.org/10.1093/jxb/ers206

Srivastav A, Mehta S, Lindlof A, Bhargava S (2010) Over-represented promoter motifs in abiotic stress-induced *DREB* genes of rice and sorghum and their probable role in regulation of gene expression. Plant Signal Behav 5:775–784. https://doi.org/10.4161/psb.5.7.11769

Srivastava AK, Zhang C, Caine RS, Gray J, Sadanandom A (2017) Rice SUMO protease overly tolerant to salt 1 targets the transcription factor *OsbZIP23* to promote drought tolerance in rice. Plant J 92:1031–1043. https://doi.org/10.1111/tpj.13739

Sunitha M, Srinath T, Reddy VD, Rao KV (2016) Expression of cold and drought regulatory protein (*CcCDR*) of pigeonpea imparts enhanced tolerance to major abiotic stresses in transgenic rice plants. Planta 245:1137–1148. https://doi.org/10.1007/s00425-017-2672-1

Takasaki H, Maruyama K, Kidokoro S, Ito Y, Fujita Y, Shinozaki K, Nakashima K (2010) The abiotic stress-responsive NAC-type transcription factor *OsNAC5* regulates stress-inducible genes and stress tolerance in rice. Mol Genet Genomics 284:173–183. https://doi.org/10.1007/s00438-010-0557-0

Tang W, Luo C (2018) Overexpression of zinc finger transcription factor ZAT6 enhances salt tolerance. Open Life Sci 13:431–445. https://doi.org/10.1515/biol-2018-0052

Tang Y, Bao X, Zhi Y, Wu Q, Guo Y, Yin X, Liu K (2019) Overexpression of a MYB Family Gene *OsMYB6* increases drought and salinity stress tolerance in transgenic rice. Front Plant Sci 10:168. https://doi.org/10.3389/fpls.2019.00168

Teng K, Li J, Liu L, Han Y, Du Y, Zhang J, Zhao Q (2014) Exogenous ABA induces drought tolerance in upland rice: the role of chloroplast and ABA biosynthesis-related gene expression on photosystem II during PEG stress. Acta Physiol Plant 36:2219–2227. https://doi.org/10.1007/s11738-014-1599-4

Thomashow MF (2010) Molecular basis of plant cold acclimation: insights gained from studying the CBF cold response pathway. Plant Physiol 154:571–577. https://doi.org/10.1104/pp.110.161794

Tian Y, Zhang H, Pan X, Chen X, Zhang Z, Lu X et al (2011) Overexpression of ethylene response factor TERF2 confers cold tolerance in rice seedlings. Transgenic Res 20:857–866. https://doi.org/10.1007/s11248-010-9463-9

Tuteja N, Sahoo RK, Garg B, Tuteja R (2013) OsSUV3 dual helicase functions in salinity stress tolerance by maintaining photosynthesis and antioxidant machinery in rice (*Oryza sativa* L cv IR64). Plant J 76:115–127. https://doi.org/10.1111/tpj.12277

Um TY, Lee S, Kim JK, Jang G, Do Choi Y (2018) Chloride channel 1 promotes drought tolerance in rice leading to increased grain yield. Plant Biotechnol Rep 12:283–293. https://doi.org/10.1007/s11816-018-0492-9

Wahid A, Gelani S, Ashraf M, Foolad MR (2007) Heat tolerance in plants: an overview. Environ Exp Bot 61:199–223. https://doi.org/10.1016/j.envexpbot.2007.05.011

Wang X, Xia M, Chen Q, Wu Z, Wu P (2002) Identification of a new small GTP-binding protein gene OsRab5a genomic organization and expression pattern analysis during nitrate supply and early nutrient starvation in rice (*Oryza sativa* L) root. Plant Sci 163:273–280. https://doi.org/10.1016/s0168-9452(02)00096-1

Wang W, Vinocur B, Altman A (2003) Plant responses to drought salinity and extreme temperatures: towards genetic engineering for stress tolerance. Planta 218:1–14. https://doi.org/10.1007/s00425-003-1105-5

Wang Q, Guan Y, Wu Y, Chen H, Chen F, Chu C (2008) Overexpression of a rice *OsDREB1F* gene increases salt drought and low temperature tolerance in both *Arabidopsis* and rice. Plant Mol Biol 67:589–602. https://doi.org/10.1007/s11103-008-9340-6

Wang Y, Ren Y, Liu X, Jiang L, Chen L, Han X, Jin M, Liu S, Liu F, Lv J et al (2010) *OsRab5a* regulates endomembrane organization and storage protein trafficking in rice endosperm cells. Plant J 64:812–824. https://doi.org/10.1111/j.1365-313x.2010.04370.x

Wang D, Qin B, Li X, Tang D, Zhang Y, Cheng Z, Xue Y (2016a) Nucleolar DEAD-Box RNA Helicase TOGR1 regulates thermotolerant growth as a pre-rRNA chaperone in rice. PLoS Genet 12:e1005844. https://doi.org/10.1371/journal.pgen.1005844

Wang M, Lu X, Xu G, Yin X, Cui Y, Huang L, Xia X (2016b) *OsSGL* a novel pleiotropic stress-related gene enhances grain length and yield in rice. Sci Rep 6:38157. https://doi.org/10.1038/srep38157

Wu T, Zhang M, Zhang H, Huang K, Chen M, Chen C, Du X (2019) Identification and characterization of EDT1 conferring drought tolerance in rice. J Plant Biol 62:39–74. https://doi.org/10.1007/s12374-018-0203-7

Xiang D, Man L (2018) *EhEm1* a novel Em-like protein from *Eutrema halophilum* confers tolerance to salt and drought stresses in rice. Mol Breeding 38:17. https://doi.org/10.1007/s11032-017-0750-5

Xiang DJ, Man LL, Zhang CL, Peng-Liu Li ZG, Zheng GC (2018a) A new Em-like protein from *Lactuca sativaLsEm1* enhances drought and salt stress tolerance in *Escherichia coli* and rice. Protoplasma 255:1089–1106. https://doi.org/10.1007/s00709-018-1207-3

Xiang J, Chen X, Hu W, Xiang Y, Yan M, Wang J (2018b) Overexpressing heat-shock protein OsHSP502 improves drought tolerance in rice. Plant Cell Rep 37:1585–1595. https://doi.org/10.1007/s00299-018-2331-4

Xiong Y, Liu T, Tian C, Sun S, Li J, Chen M (2005) Transcription factors in rice: a genome-wide comparative analysis between monocots and eudicots. Plant Mol Biol 59:191–203. https://doi.org/10.4137/ebo.s9369

Xiong H, Yu J, Miao J, Li J, Zhang H, Wang X, Li Z (2018) Natural variation in OsLG3 Increases drought tolerance in rice by inducing ROS scavenging. Plant Physiol 178:451–467. https://doi.org/10.1104/pp.17.01492

Xu LM, Zhou L, Zeng YW, Wang FM, Zhang HL, Shen SQ, Li ZC (2008) Identification and mapping of quantitative trait loci for cold tolerance at the booting stage in a *japonica* rice near-isogenic line. Plant Sci 174:340–347. https://doi.org/10.1016/j.plantsci.2007.12.003

Xu M, Li L, Fan Y, Wan J, Wang L (2011) ZmCBF3 overexpression improves tolerance to abiotic stress in transgenic rice (*Oryza sativa*) without yield penalty. Plant Cell Rep 30:1949. https://doi.org/10.1007/s00299-011-1103-1

Xu N, Chu Y, Chen H, Li X, Wu Q, Jin L, Huang J (2018) Rice transcription factor OsMADS25 modulates root growth and confers salinity tolerance via the ABA-mediated regulatory pathway and ROS scavenging. PLoS Genet 14:e1007662. https://doi.org/10.1371/journal.pgen.1007662

Yasuda H (2017) Cross-tolerance to thermal stresses and its application to the development of cold tolerant rice. JAQR 51:99–105. https://doi.org/10.6090/jarq.51.99

Ye C, Fukai S, Godwin I, Reinke R, Snell P, Schiller J, Basnayake J (2009) Cold tolerance in rice varieties at different growth stages. Crop Pasture Sci 60:328–338. https://doi.org/10.1071/cp09006

Ye N, Zhu G, Liu Y, Li Y, Zhang J (2011) ABA controls H_2O_2 accumulation through the induction of OsCATB in rice leaves under water stress. Plant Cell Physiol 52:689–698. https://doi.org/10.1093/pcp/pcr028

Yu J, Zao W, He Q, Kim TS, Park YJ (2017) Genome-wide association study and gene set analysis for understanding candidate genes involved in salt tolerance at the rice seedling stage. Mol Genet Genomics 292:1391–1403. https://doi.org/10.1007/s00438-017-1354-9

Yu S, Huang A, Li J, Gao L, Feng Y, Pemberton E, Chen C (2018) OsNAC45 plays complex roles by mediating POD activity and the expression of development-related genes under various abiotic stresses in rice root. Plant Growth Regul 84:519–531. https://doi.org/10.1007/s10725-017-0358-0

Yuan X, Huang P, Wang R, Li H, Lv X, Duan M, Huang J (2018) A zinc finger transcriptional repressor confers pleiotropic effects on rice growth and drought tolerance by down-regulating stress-responsive genes. Plant Cell Physiol 59:2129–2142. https://doi.org/10.1093/pcp/pcy133

Zhang X, Guo X, Lei C, Cheng Z, Lin Q, Wang J, Wan J (2011a) Overexpression of *SlCZFP1* a novel TFIIIA-type zinc finger protein from tomato confers enhanced cold tolerance in transgenic *Arabidopsis* and rice. Plant Mol Biol Rep 29:185–196. https://doi.org/10.1007/s11105-010-0223-z

Zhang L, Xiao S, Li W, Feng W, Li J, Wu Z, Shao M (2011b) Overexpression of a Harpin-encoding gene hrf1 in rice enhances drought tolerance. J Exp Bot 62:4229–4238. https://doi.org/10.1093/jxb/err131

Zhang DP, Zhou Y, Yin JF, Yan XJ, Lin S, Xu WF, Liang JS (2015) Rice G-protein subunits qPE9-1 and RGB1 play distinct roles in abscisic acid responses and drought adaptation. J Experim Botany 66:6371–6384. https://doi.org/10.1093/jxb/erv350

Zhang Z, Li J, Li F, Liu H, Yang W, Chong K, Xu Y (2017) OsMAPK3 Phosphorylates OsbHLH002/OsICE1 and inhibits its ubiquitination to activate OsTPP1 and enhances rice chilling tolerance. Dev Cell 43:731–743. https://doi.org/10.1016/j.devcel.2017.11.016

Zhang D, Wang Y, Shen J, Yin J, Li D, Gao Y, Liang J (2018) *OsRACK1A* encodes a circadian clock-regulated WD40 protein negatively affect salt tolerance in rice. Rice 11:45. https://doi.org/10.1186/s12284-018-0232-3

Zhang A, Liu Y, Wang F, Li T, Chen Z, Kong D, Luo L (2019) Enhanced rice salinity tolerance via CRISPR/Cas9-targeted mutagenesis of the *OsRR22* gene. Mol Breeding 39:47. https://doi.org/10.1007/s11032-019-0954-y

Zheng X, Chen B, Lu G, Han B (2009) Overexpression of a NAC transcription factor enhances rice drought and salt tolerance. Biochem Biophys Res Commun 379:685–989. https://doi.org/10.1016/j.bbrc.2008.12.163
Zhu JK (2016) Abiotic stress signaling and responses in plants. Cell 167:313–324. https://doi.org/10.1016/j.cell.2016.08.029
Zhu G, Ye N, Zhang J (2009) Glucose-induced delay of seed germination in rice is mediated by the suppression of ABA catabolism rather than an enhancement of ABA biosynthesis. Plant Cell Physiol 50:644–651. https://doi.org/10.1093/pcp/pcp022

Chapter 13
An Update on the Applications of CRISPR/Cas9 Technology in Tomato

Sushmita, Gurminder Kaur and Praveen Chandra Verma

Abstract A genome is an organism's complete set of DNA, including all of its genes. Each genome comprises all the information required to form and sustain the organism. Introducing any change in this genome might help us introduce the desired characters in the organism. Current genome editing technologies enable us to produce very efficient and specific nucleotide modifications. This becomes possible with the availability of some effective molecules and the DNA repair machinery of the cells. The lesions produced can cause localized nucleotide changes through inaccurate, nonhomologous end joining or desired nucleotide changes through homologous recombination with a donor DNA template. Till now there are three kinds of genome editing machinery complexes, the ZFNs (Zinc Finger Nucleases), TALENs (Transcription Activator-Like Effector Nucleases) and the most current one, CRISPR (Clustered regularly interspaced short palindromic repeats)/Cas9 (CRISPR-associated protein 9) RNA-guided Nucleases. In the following chapter, we discuss all the traits of tomato, edited using CRISPR-cas9 genome editing technology.

Keywords CRISPR/Cas9 · Tomato · Mutants

13.1 Introduction

In the present world, nearly one billion of the population suffers from chronic malnutrition. With the estimate of the world population explosion to 9 billion by 2050, the agricultural techniques would require to improve drastically. The future crops must have higher yields, better in nutritional quality and at the same time would require

Sushmita · P. C. Verma (✉)
Council of Scientific and Industrial Research—National Botanical Research Institute (CSIR-NBRI), Rana Pratap Marg, Lucknow, Uttar Pradesh 226001, India
e-mail: praveencverma@nbri.res.in

Academy of Scientific and Innovative Research (AcSIR), Ghaziabad 201002, India

G. Kaur
Faculty of Bioscience, Institute of Bioscience and Technology, Shri Ramswaroop Memorial University, Lucknow-Deva Road, Barabanki, Uttar Pradesh 225003, India

S. P. Singh et al. (eds.), *Molecular Approaches in Plant Biology and Environmental Challenges*, Energy, Environment, and Sustainability,
https://doi.org/10.1007/978-981-15-0690-1_13

fewer inputs. The conventional breeding technologies surely promise us some results but the period of time which it requires is tremendous. So, we certainly require some alternatives to improve our agricultural productivity. This is where the genome editing technologies come to the rescue, as they promise us with some beneficial trait incorporation in the crops within limited period of time. There are a few genome editing technologies developed till now which are based on site specific nucleases (SSNs) (Chen and Gao 2014) Zinc Finger Nucleases (ZFNs) (Kim et al. 1996) and Transcription activator like effector nucleases (TALENS) (Christian et al. 2010) were the two most used techniques in the field of genome editing since last two decades, but with the discovery of CRISPR/Cas9 and its repurposing to be used as a genome editing tool, built the hopes for a better agricultural scenario (Jinek et al. 2012). It has been proven to be the most influential of all other genome editing tools used until now. It has great potential for incorporating novel traits into important crops. Naturally, it is a bacterial immune mechanism against phage infection and plasmid transfer. It has been repurposed as a strong RNA-directed DNA targeting tool for genome editing, epigenetic modulation, genome imaging, and transcriptional perturbation (Jiang and Doudna 2017). This technology permits geneticists to very accurately modify any genomic sequence with the aid of a short guide RNA.

CRISPR–Cas is presently classified into six different types. From these six types, type II is largely widely used for the purpose of genome-engineering (Makarova et al. 2015). Finding of vital machinery and mechanism of the type II CRISPR system were central to its utilization as a genome-engineering technique. This includes the illustration that *Streptococcus thermophilus* can be used exclusively to slice the double-stranded DNA, directed by the endonuclease, Cas9 (Barrangou et al. 2007; Garneau et al. 2010). A short sequence of DNA present next to the RNA-binding site, called the protospacer-adjacent motif (PAM), is the mechanism through which CRISPR/Cas differentiates self and non-self genome (Mojica et al. 2009); a small transactivating CRISPR RNA (tracrRNA), guides the post-transcriptional processing and maturation of the CRISPR RNA (crRNA) with the aid of sequence complementarity (Deltcheva et al. 2011). With all these findings, the CRISPR/Cas9 from *S. thermophilus* system was found to be able to function in *E. coli* and provide resistance against foreign infecting genes (Sapranauskas et al. 2011).

13.1.1 An Array of Cas9 Activities

Cas9 can be used in its original form (wtCas9) to create double-strand DNA breaks. This double-stranded break is then repaired mostly via the nonhomologous end joining (NHEJ) pathway. The NHEJ pathway introduces minor insertions or deletions (indels) that result in CRISPR knockout (CRISPR–ko) (Shalem et al. 2014; Wang et al. 2014; Koike-Yusa et al. 2014). A different approach in the manipulation of cas9 is by inactivating the nuclease domain of Cas9, creating a dead Cas9 (dCas9) (Qi et al. 2013). This dCas9 can then be appended with additional domains, either activating for transcriptional activation (CRISPRa) (Gilbert et al. 2013) or inhibitory

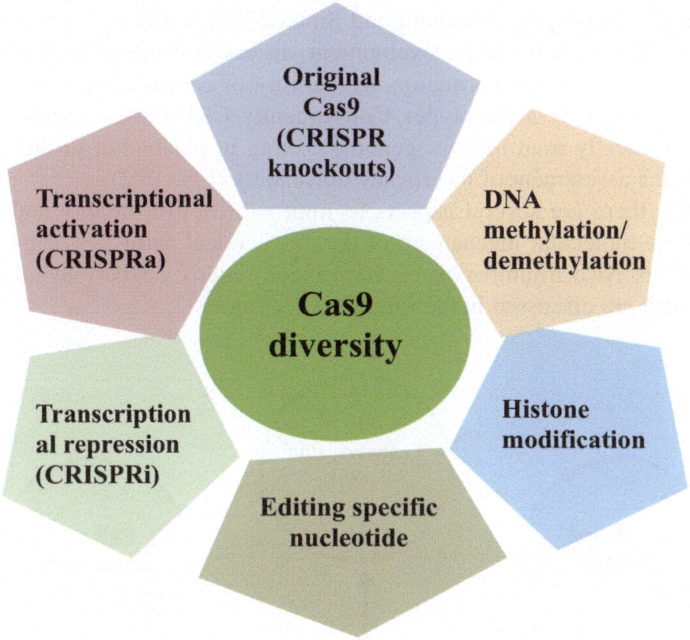

Fig. 13.1 Cas9: a versatile tool for genome engineering

domains to hinder transcription (CRISPRi) (Mali et al. 2013; Gilbert et al. 2014; Konermann et al. 2015; Rajagopal et al. 2016). Also, some domains are found to alter the DNA methylation status (Vojta et al. 2016; Choudhury et al. 2016), modify histones (Hilton et al. 2015; Kearns et al. 2015; Kwon et al. 2017; Polstein et al. 2015; Thakore et al. 2015) or precisely edit particular nucleotides (Komor et al. 2016; Hess et al. 2016; Ma et al. 2016) (Fig. 13.1).

13.2 Uses of CRISPR/Cas9 in Crop Improvement

The earliest reports of CRISPR/Cas9 genome editing in plants emerged in the year 2013. The application was successful in both stable and transient expressions in transgenic. The application of CRISPR/Cas was found to be efficient in the models plants like *Arabidopsis* (Li et al. 2013) and *Nicotiana benthamiana* (Nekrasov et al. 2013) as well as for major crop species, rice (Thuesombat et al. 2014), sorghum (Jiang et al. 2013), and wheat (Wang et al. 2014). More reports came for the development of stable transgenic lines of Arabidopsis and rice in which the mutations were found to be heritable in the next generations too (Feng et al. 2014; Zhang et al. 2014). CRISPR related reports on tomato also became common as tomato (*Solanum lycopersicum*) is the principal vegetable crop all over the world and an essential component of a healthy diet (Kimura and Sinha 2008; Willcox et al. 2003). It has become a model system for plant development, disease resistance, and fruit ripening because of the ease of transformation, availability of complete genome sequence, and the presence of several genotypes. Consequently, CRISPR/Cas9 has become the most enthusiastically used tool for genome editing in plants, but further trials are required for the assessment of its efficacy universally (Fig. 13.2).

In tomato, there are several aspects to improve the fruit quality and to better understand the molecular mechanism of the biochemical pathways being operated inside the plants. Also, a number of genes have been edited for experimental purposes. Some of which are discussed in the following section.

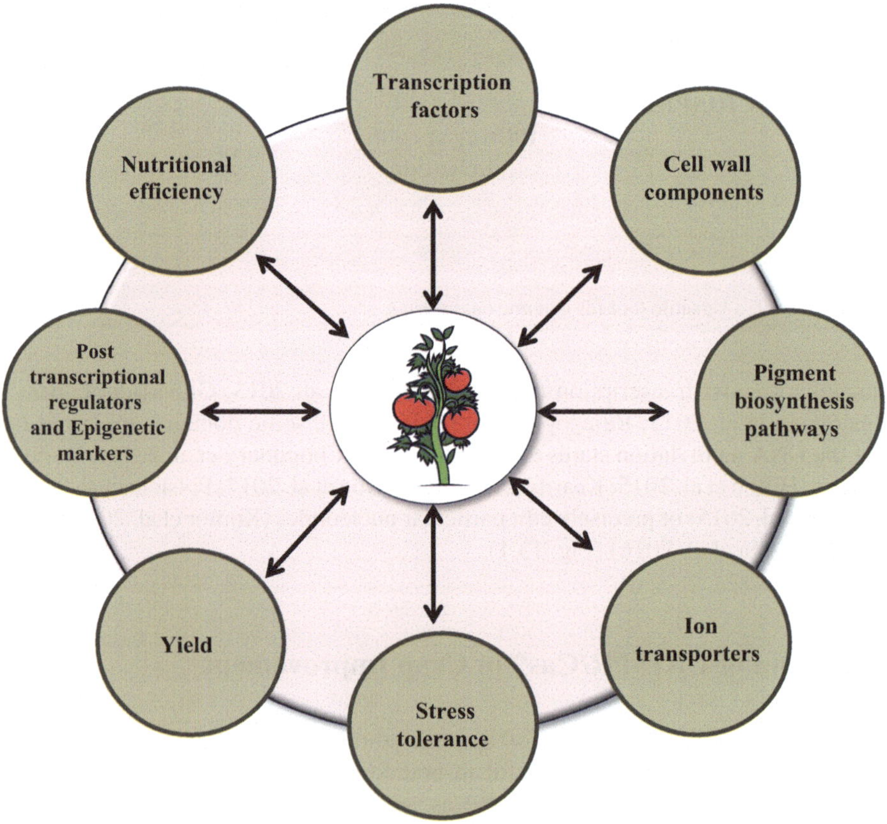

Fig. 13.2 Manipulated traits in tomato using CRISPR/Cas9

13.2.1 Experimental

Plant growth. The CRISPR/Cas9 technology was used to target mutations in the DELLA domain of the PROCERA gene in tomato. A number of loss-of-function mutations and a dominant dwarf mutation with the deletion of an amino acid in the DELLA domain were obtained. At the seedling stage, the heterozygotes were found to have an intermediary phenotype, however in the adult stage heterozygotes were found to be as dwarfed as homozygotes (Tomlinson et al. 2019) (Table 13.1).

Mutation stability. To test the efficacy and stability of CRISPR/Cas9 in tomato, phytochrome interacting factor (*SlPIF4*) and phytoene desaturase (*SlPDS*) were mutated. The mutations were found to be stable in the first and the later generations (Pan et al. 2016) (Table 13.1).

Leaf morphology. Tomato homolog of Arabidopsis ARGONAUTE7 (*SlAGO7*) is essential for the biogenesis of transacting short interfering RNAs. These RNAs are designated to regulate organ polarity by posttranscriptionally silencing the Auxin Response Factor (*ARF*) (Husbands et al. 2015). Loss of function mutation of ARGONAUTE7 was created with CRISPR/Cas9, which produced reduced levels of the transacting short interfering RNAs and thus lessened ARF mRNA degradation. This resulted in formation of initially formed leaflets without petioles and later-formed leaves without laminae (Brooks et al. 2014) (Table 13.1).

13.2.2 Applications

Scientists have also been curious to explore some of the agriculturally important traits in tomato to improve its productivity, stress tolerance and also to introduce new traits. A number of related parameters which have been manipulated are herein reviewed.

Manipulating stress tolerance. During the process of growth and development, a plant experiences a number of stresses in its environment. A number of stress parameters that have been explored using CRISPR/Cas9. To interpret the role of *SlCBF1* (C-repeat/dehydration-responsive element binding factors), *slcbf1* mutants were generated using the CRISPR/Cas9. *slcbf1* mutants possessed higher electrolyte leakage and activities of antioxidant enzymes in comparison to the wild type plants. The *slcbf1* mutants had reduced tolerance to chilling conditions, thus depicting the role of *SlCBF1* in tomato chilling tolerance (Li et al. 2018). Another gene, namely ***NPR1*** (nonexpressor of pathogenesis-related gene 1) has been explored widely for its role in defense response against pathogens in plants; nevertheless its role in abiotic stress is not much explored. A study with the *slnpr1* mutants generated through CRISPR/Cas9 showed strong induction of ***SlNPR1*** by drought stress. The *slnpr1* mutants were more droughts sensitive. The stomatal aperture was enlarged, electrolytic leakage, reactive oxygen species levels were increased and the antioxidant enzyme activities were decreased in comparison to the wild type plants. Also, the

Table 13.1 Applications of CRISPR/Cas9 in tomato

Sl no.	Group	Trait manipulated	Gene targeted	References
[A]	**Experimentals**			
		Plant growth	*DELLA*	(Tomlinson et al. 2019)
		Mutation stability	*SlPDS, SlPIF*	(Pan et al. 2016)
		Leaf morphology	*AGO7*	(Brooks et al. 2014)
[B]	**Applications**			
	Stress	Chilling tolerance	*CBF1*	(Li et al. 2018)
		Drought tolerance	*NPR1*	(Li et al. 2019)
	Transcription factors	Fruit ripening	*RIN*	(Ito et al. 2015; Ito et al. 2017)
			NOR like 1	(Gao et al. 2018)
			NOR, CNR	(Gao et al. 2019)
	Cell wall components	Fruit ripening	*PL, TBG4, PG2a*	(Wang et al. 2019)
	Pigment biosynthesis pathways	Carotenoid biosynthesis	*PSY1, CrtR-b2*	(D'Ambrosio et al. 2018)
		Carotenoid biosynthesis	*PSY1, CRTISO*	(Dahan-Meir et al. 2018)
		Anthocyanin biosynthesis	*DFR*	(Danilo et al. 2018)
	Nutritional efficiency	γ-Aminobutyric acid (GABA) biosynthesis	*GAD2, GAD3*	(Nonaka et al. 2017)
		γ-Aminobutyric acid (GABA) shunt	*GABA-TP1, GABA-TP3, CAT9, SSADH*	(Li et al. 2018)
	Post-transcriptional regulators and epigenetic makers	Fruit ripening	ORRM4	(Yang et al. 2017)
		Fruit ripening	*DML2*	(Lang et al. 2017)
		Fruit ripening	lncRNA1459	(Li et al. 2018)
		Fruit ripening	lncRNA2155	(Yu et al. 2018)
	Disease resistance	Powdery mildew	*Mlo1*	(Nekrasov et al. 2017)
		Tomato yellow leaf curl virus	*REP, CP*	(Fonseca et al. 2009)
		Bacterial speck disease	*JAZ2*	(Ortigosa et al. 2019)
	Yield	Inflorescence	*TMF, BOP*	(Xu et al. 2016)
		Photoperiodism	*SP5G*	(Soyk et al. 2017)
		Inflorescence	*J2, EJ2*	(Soyk et al. 2017)

(continued)

Table 13.1 (continued)

Sl no.	Group	Trait manipulated	Gene targeted	References
	Crop management	Herbicide resistance	*ALS*	(Veillet et al. 2019)
	Parthenocarpy	Fruit characteristic	*IAA9*	(Ueta et al. 2017)
	Transporter	Phosphate transporter	*PHO1;1*	(Zhao et al. 2018)
	Domestication	Fruit size, number and lycopene content	*SP, O, FW2.2, CycB FAS and MULT*	(Zsögön et al. 2018)

expression of drought related genes *SlGST*, *SlDHN*, and *SlDREB* were decreased showing reduced drought tolerance by the *slnpr1* mutants (Li et al. 2019) (Table 13.1).

Manipulating the transcription factors. Targeting genes encoding transcription factors involved in the ripening pathways has been till now the most widely used technique to alter fruit ripening. But they are also known to regulate other ripening-related processes like change in different pigments and flavors in the fruit. Thus, mutating the transcription factor genes also brings influence a number of changes which occur at the time of ripening like-change in pigment synthesis and change in flavour. Targeting the *RIN* (ripening inhibitor) gene, encoding a MADS-box transcription factor regulating fruit ripening, produced incomplete ripening, confirming the major regulatory role of *RIN* in ripening activation, as RIN was already proved to be a ripening activator (Ito et al. 2015). The long believed role of RIN was contradicted with the new study presenting stimulation of ripening of fruits in the absence of *RIN*. The CRISPR mutants of *RIN* knockouts could not repress the initiation of ripening. Additionally, inactivation of the *rin* mutant allele moderately restored the initiation of ripening. Consequently, *RIN* is not essential for the induction of ripening, and *rin* is a gain-of-function mutation and not a null mutation (Ito et al. 2017). NOR-like1 inhibition led to delayed ripening with reduced ethylene production, reduced lycopene accumulation, and chlorophyll loss. The ripening-related genes like genes of cell wall metabolism (*SlPL*, *SlPG2a*, *SlEXP1*, and *SlCEL2*), ethylene biosynthesis (*SlACS2*, *SlACS4*) color formation (*SlSGR1*, *Sl*Ggpps2) were found to be targets of NOR-like1. It was thus found that a NAC domain transcription factor NOR-like1 is also played the role of the positive regulator in tomato fruit ripening (Gao et al. 2018). More ripening related transcription factors, NAC NON-RIPENING (NOR) and a SBP-box COLORLESS NON-RIPENING (CNR), were mutated using CRISPR/Cas9 genome editing technology. The mutated plants showed only delayed or partial non-ripening phenotypes, unlike the original mutants which showed strong non-ripening phenotypes (Gao et al. 2019) (Table 13.1).

Manipulating the cell wall. Commanding the rate of softening of the fruits for augmenting shelf life is a key target for researchers for genetically engineered tomato fruits. With the objective of achieving the same without affecting other aspects of ripening, ripening-induced genes encoding the pectin-decomposing enzymes like pectate lyase (*PL*), β-galactanase (*TBG4*), and polygalacturonase 2a (*PG2a*) were mutagenized with the help of CRISPR/Cas9 genome editing technology. The mutant fruits generated were compared for their physiochemical properties showed that only mutations in pectate lyase (*PL)* gene produced firmer fruits while mutations in β-galactanase (*TBG4*), and polygalacturonase 2a (*PG2a*) influenced the color and weight of fruits (Wang et al. 2019) (Table 13.1).

Manipulating pigment biosynthesis pathways. To test the efficacy of CRISPR/Cas9 in tomato, two vital genes of the carotenoid biosynthesis pathway, *PSY1* and *CrtR-b2* were knocked out. The results proved that the CRISPRCas9 system could be a potent method for the creation of valuable mutations in tomato, which could be implemented further in breeding programs (D'Ambrosio et al. 2018). In another gene targeting experiment, efficient gene targeting was shown to be accomplished without the use of any selection markers or reporters. The following carotenoid biosynthesis pathway genes were selected-phytoene synthase 1 (*PSY1*) and carotenoid isomerase (*CRTISO*) (Dahan-Meir et al. 2018). Combining the CRISPR-Cas system with the rolling circle replicon of bean yellow dwarf virus and optimizing the method for targeted mutagenesis as well as gene replacement in tomato, presented an approach to insert a gene or a segment of the gene. The WVA106 genotype was reported with deletion of 1013 bp in the *DFR* (dihydroflavonol 4-reductase) gene, and accurate insertion of transgenes by HDR-DNA repair was acquired at the *DFR* locus. This resulted in the renewal of a functional *DFR* gene and the re-establishment of biosynthesis of anthocyanin, providing a strategy for targeted transgene insertion in tomato (Danilo et al. 2018) (Table 13.1).

Manipulating the nutritional efficiency. γ-Aminobutyric acid (GABA), is a non-proteinogenic amino acid and the tomato fruits are found to contain high levels of GABA. GABA has hypotensive effects, i.e., it has enhanced blood pressure lowering effects. The key enzyme in the biosynthesis of GABA is Glutamate decarboxylase (GAD). The C terminal of this enzyme has a domain which is autoinhibitory and regulates enzymatic function. The deletion of this domain increases the activity of GAD. Out of the five *GAD* genes present in tomato genome (*SlGAD1–5*), two of them, i.e., *SlGAD2* and *SlGAD3* are expressed at the time of fruit development. Deleting the autoinhibitory domain of *SlGAD2* and *SlGAD3* using CRISPR/Cas9 technology increased GABA accumulation as shown by the experiments. The results provide a strategy for the enhancement nutritional efficiency in other crops as well as medicinally important plants too using CRISPR/Cas9-based genetic engineering (Nonaka et al. 2017). In another study, γ-aminobutyric acid (GABA) shunt was manipulated targeting five key genes using a multiplex pYLCRISPR/Cas9 system, thus resulting in enhanced GABA accumulation in leaves and fruits. This study established a system for multisite knockout mutations (Li et al. 2018) (Table 13.1).

Manipulating post-transcriptional regulators and epigenetic makers. Mutating RNA editing factor *SlORRM4* (of the ORRM family) with CRISPR/Cas9 genome editing in tomato plants also showed significantly delayed fruit ripening, ethylene production as well as respiratory rates in comparison to the wild type plants. *SlORRM4* was found to be localized in the mitochondria, and its loss of function interrupted RNA editing of a good number of important mitochondrial genes, thereby affecting the association of core subunits in the respiratory chain complex. The findings surely enhanced our understanding of the complex correlation of fruit ripening and RNA editing factors (Yang et al. 2017). In another interesting study, they had generated CRISPR based mutants of *SlDML2* tomato fruits which failed to ripen, depicting its critical role in ripening. It was found that the role of DNA demethylation is important for activation of many genes related to fruit ripening including genes of flavor; fruit pigment synthesis, cell wall hydrolysis, ethylene synthesis and signaling as well as repression of ripening inhibited genes (Lang et al. 2017). To decipher the function of lncRNA1459 in the ripening of tomato fruits, lncRNA1459 knock-out mutants were produced using the CRISPR/Cas9 system. The knockouts showed a significant delay in ripening, decreased lycopene accumulation, and ethylene production in comparison with wild type fruits. Overall the results proved that lncRNA1459 is a positive regulator in the process of fruit ripening (Li et al. 2018). One more long noncoding RNA lncRNA2155 was knocked out using CRISPR/Cas9 for deciphering its role in ripening. The mutants were observed to ripen lately (Yu et al. 2018) (Table 13.1).

Manipulating disease resistance. A study targeting *MILDEW RESISTANT LOCUS O 1* (*Mlo1*) in tomato through CRISPR/Cas9 produced a new variety of tomato, Tomelo. Tomelo was resistant against powdery mildew causing fungal pathogen *Oidium neolycopersici* (Nekrasov et al. 2017). Another study targeting the genome of tomato yellow leaf curl virus (TYLCV) with CRISPR/Cas9 at the sequences encoding replicase (*Rep*) and coat protein (*CP*) caused effective viral interference in the transgenic tomato and tobacco plants. This study unlocked the doors to engineer virus resistance in other crops too (Fonseca et al. 2009). The existence of antagonism between the salicylate and jasmonate defense pathways hurdles to achieve broad spectrum pathogen resistance in crops. *Pseudomonas syringae* Pto DC3000 which cause the bacterial speck disease in tomato manipulates the hormonal crosstalk by producing coronatine (COR), which mimics the bioactive jasmonic acid-isoleucine (JA-Ile) (Tashkandi et al. 2018). This activation of JA pathway thus inhibits SA defense pathway and reopens stomata to smoothen bacterial invasion and growth in the apoplast. *SlJAZ2* is a major co-receptor of COR in the stomatal guard cells. *SlJAZ2* was mutated with CRISPR/Cas9 to produce dominant JAZ2 repressors, which lacked the C-terminal Jas domain (*SlJAZ2*Δ*jas)* which barred stomatal reopening by COR. This made the plant resistant to *Pseudomonas syringae*, Pto DC3000 resistance without altering its resistance against the necrotrophic fungal pathogen *Botrytis cinerea*, which is a causal agent of the tomato gray mold (Ortigosa et al. 2019) (Table 13.1).

Manipulating the yield. The plant inflorescence determines its flower production and also the plant productivity. Eliminating the *SlBOP* through CRISPR/Cas9 approach produced pleiotropic effects and most prominently simplification of inflorescences into single flowers, similar to the *tmf* mutants. The defects in the flowering severed in higher order mutants of *slbop* and *tmf* mutants. This study showed that the meristem maturation is mediated by the transcription factor TERMINATING FLOWER (TMF), and its interaction with other BOPs (BLADE-ON-PETIOLE) (Xu et al. 2016). The expression of *SELF-PRUNING 5G* (*SP5G*) is increased in long days in wild species of tomato, but not in the ones which are cultivated. CRISPR/Cas9-engineered mutations were induced in *SP5G*, which produced rapid flowering and more compact growth of field tomato plants. This resulted in rapid fruit production and improved yield (Soyk et al. 2017). In a different study, two of the MADS Box transcription factors JOINTLESS 2 and ENHANCED JOINTLESS 2 were engineered using CRISPR/Cas9 technology for bypassing the negative epistasis and finding the detailings of the genetics of tomato inflorescence (Soyk et al. 2017) (Table 13.1).

Manipulating herbicide resistance. CRISPR/Cas9 was used to target *acetolactate synthase* (*SlALS*) gene in tomato as well potato by Target-AID (target-activation-induced cytidine deaminase), a Cytidine base editor (CBE) which directs a C-to-T base conversion. Efficient cytidine base edition was done which conferred chlorsulfuron-resistant plants. In summary, this study established an efficient and economical approach for the production of mutated and transgene free plants (Veillet et al. 2019) (Table 13.1).

Manipulating parthenocarpic characteristics. CRISPR based somatic mutations were produced in tomato (*Solanum lycopersicum*) *IAA9* (*SlIAA9*) which a key gene controlling parthenocarpy. The *SlIAA9* knock out plants showed simple leaves instead of compound leaves which are found in wild type plants, and fruit development was initiated before fertilization, causing parthenocarpy. The mutation was found to be heritable in the further generations, and thus the developed system could be used to develop parthenocarpic fruits in tomato as well as many other cultivars (Ueta et al. 2017) (Table 13.1).

Manipulating a transporter. CRISPR/Cas9 based *SlPHO1;1* knockout mutant showed phenotypes of Pi starvation, like an increase in fresh weight of root and decrease in fresh weight of shoot. The mutants contained more amount of soluble Pi in roots and less amount of soluble Pi in the shoot, they also accumulated more anthocyanin. These results proved the role of *SlPHO1;1* in Pi transport in the tomato (Zhao et al. 2018) (Table 13.1).

Domestication of Wild tomato. In a breakthrough study, a wild tomato *Solanum pimpinellifolium* was edited at six loci that are central for productivity and yield in the contemporary tomato crop varieties. The gene loci targeted with multiplex CRISPR/Cas9 approach were general plant growth habit (*SELFPRUNING*) (Pnueli et al. 1998), fruit size (*FASCIATED* and *FRUIT WEIGHT 2.2*) (Frary et al. 2000; Liu et al. 2002), shape (*OVATE*) (Xu et al. 2015) and number (*MULTIFLORA*) (Lippman et al. 2008), and nutritional quality (*LYCOPENE BETA CYCLASE*) (Ronen et al. 2000). No mutations were recovered in either *FASCIATED* (*FAS*) or *MULTIFLORA* (*MULT*). The engineered S. *pimpinellifolium* fruits were three fold increased fruit size, ten folds increased fruit number and 500% more lycopene accumulation in the fruits. These results surely paved the way for exploring the genetic diversity present in the wild plants (Zsögön et al. 2018) (Table 13.1).

13.3 Conclusion

This massive era of genome editing promises us a better understanding of the genetic world. Of all the genome editing technologies explored so far, CRISPR/Cas9 technology shows the maximum capabilities. The ease of use and cost effectiveness, strengthen its potential. With the use of targetable endonuclease as a tool for genetic manipulation and analysis, practical applications are being explored in the areas of increased yield and disease resistance in major crop plants. In model plants, increasing analyses of standard gene characterization to demystify the genes of lesser explored species, provide hopes for a more comprehensible scientific world. From manipulating fruit texture by targeting the cell wall degrading enzymes to even modifying the histones, CRISPR/Cas9 has been used to manipulate diverse genes in tomato too. These studies definitely give us valuable insights into the function of the mutated genes, transcription factors, non-coding RNAs, RNA editing factors, etc., as mutating those, changes particular phenotypes or characters in the plant itself with comparison to the wild type plants.

Moreover, CRISPR/Cas9 technology also promises us to fulfill increasing food demand to the ever increasing world population by increasing the yield as well as increasing the resistance against the harmful pests. With the escalating CRISPR studies, we can surely anticipate a revolution in the genetic world. As scientists, it's our responsibility to balance between science and society.

References

Barrangou R, Fremaux C, Deveau H, Richards M, Boyaval P, Moineau S, Romero DA, Horvath P (2007) CRISPR provides acquired resistance against viruses in prokaryotes. Science 315(5819):1709–1712

Brooks C, Nekrasov V, Lippman ZB, Van Eck J (2014) Efficient gene editing in tomato in the first generation using the clustered regularly interspaced short palindromic repeats/CRISPR-associated9 system. Plant Physiol 166(3):1292–1297. https://doi.org/10.1104/pp.114.247577

Chen K, Gao C (2014) Targeted genome modification technologies and their applications in crop improvements. Plant Cell Rep 33(4):575–583. https://doi.org/10.1007/s00299-013-1539-6

Choudhury SR, Cui Y, Lubecka K, Stefanska B, Irudayaraj J (2016) CRISPR-dCas9 mediated TET1 targeting for selective DNA demethylation at BRCA1 promoter. Oncotarget 7(29):46545. https://doi.org/10.18632/oncotarget.10234

Christian M, Cermak T, Doyle EL, Schmidt C, Zhang F, Hummel A, Bogdanove AJ, Voytas DF (2010) Targeting DNA double-strand breaks with TAL effector nucleases. Genetics 186(2):757–761. https://doi.org/10.1534/genetics.110.120717

D'Ambrosio C, Stigliani AL, Giorio G (2018) CRISPR/Cas9 editing of carotenoid genes in tomato. Transgenic Res 27(4):367–378. https://doi.org/10.1007/s11248-018-0079-9

Dahan-Meir T, Filler-Hayut S, Melamed-Bessudo C, Bocobza S, Czosnek H, Aharoni A, Levy AA (2018) Efficient in planta gene targeting in tomato using geminiviral replicons and the CRISPR/Cas9 system. Plant J 95(1):5–16. https://doi.org/10.1111/tpj.13932

Danilo B, Perrot L, Botton E, Nogué F, Mazier M (2018) The DFR locus: a smart landing pad for targeted transgene insertion in tomato. PLoS ONE 13(12):e0208395. https://doi.org/10.1371/journal.pone.0208395

Deltcheva E, Chylinski K, Sharma CM, Gonzales K, Chao Y, Pirzada ZA, Eckert MR, Vogel J, Charpentier E (2011) CRISPR RNA maturation by trans-encoded small RNA and host factor RNase III. Nature 471(7340):602. https://doi.org/10.1038/nature09886

Feng Z, Mao Y, Xu N, Zhang B, Wei P, Yang DL, Wang Z, Zhang Z, Zheng R, Yang L, Zeng L (2014) Multigeneration analysis reveals the inheritance, specificity, and patterns of CRISPR/Cas-induced gene modifications in Arabidopsis. Proc Natl Acad Sci 111(12):4632–4637. https://doi.org/10.1073/pnas.1400822111

Fonseca S, Chini A, Hamberg M, Adie B, Porzel A, Kramell R, Miersch O, Wasternack C, Solano R (2009) (+)-7-iso-Jasmonoyl-L-isoleucine is the endogenous bioactive jasmonate. Nat Chem Biol 5(5):344. https://doi.org/10.1038/nchembio.161

Frary A, Nesbitt TC, Frary A, Grandillo S, Van Der Knaap E, Cong B, Liu J, Meller J, Elber R, Alpert KB, Tanksley SD (2000) fw2.2: a quantitative trait locus key to the evolution of tomato fruit size. Science 289(5476):85–88. https://doi.org/10.1126/science.289.5476.85

Gao Y, Wei W, Zhao X, Tan X, Fan Z, Zhang Y, Jing Y, Meng L, Zhu B, Zhu H, Chen J (2018) A NAC transcription factor, NOR-like1, is a new positive regulator of tomato fruit ripening. Horticult Res 5. https://doi.org/10.1038/s41438-018-0111-5

Gao Y, Zhu N, Zhu X, Wu M, Cai-Zhong J, Grierson D, Luo Y, Shen W, Zhong S, Fu DQ, Qu G (2019) Diversity and redundancy of the ripening regulatory networks revealed by the fruitENCODE and the new CRISPR/Cas9 CNR and NOR mutants. Horticult Res 6(1):39. https://doi.org/10.1038/s41438-019-0122-x

Garneau JE, Dupuis MÈ, Villion M, Romero DA, Barrangou R, Boyaval P, Fremaux C, Horvath P, Magadán AH, Moineau S (2010) The CRISPR/Cas bacterial immune system cleaves bacteriophage and plasmid DNA. Nature 468(7320):67. https://doi.org/10.1038/nature09523

Gilbert LA, Larson MH, Morsut L, Liu Z, Brar GA, Torres SE, Stern-Ginossar N, Brandman O, Whitehead EH, Doudna JA, Lim WA (2013) CRISPR-mediated modular RNA-guided regulation of transcription in eukaryotes. Cell 154(2):442–451. https://doi.org/10.1016/j.cell.2013.06.044

Gilbert LA, Horlbeck MA, Adamson B, Villalta JE, Chen Y, Whitehead EH, Guimaraes C, Panning B, Ploegh HL, Bassik MC, Qi LS (2014) Genome-scale CRISPR-mediated control of gene repression and activation. Cell 159(3):647–661. https://doi.org/10.1016/j.cell.2014.09.029

Hess GT, Frésard L, Han K, Lee CH, Li A, Cimprich KA, Montgomery SB, Bassik MC (2016) Directed evolution using dCas9-targeted somatic hypermutation in mammalian cells. Nat Methods 13(12):1036. https://doi.org/10.1038/nmeth.4038

Hilton IB, D'ippolito AM, Vockley CM, Thakore PI, Crawford GE, Reddy TE, Gersbach CA (2015) Epigenome editing by a CRISPR-Cas9-based acetyltransferase activates genes from promoters and enhancers. Nat Biotechnol 33(5):510. https://doi.org/10.1038/nbt.3199

Husbands AY, Benkovics AH, Nogueira FT, Lodha M, Timmermans MC (2015) The ASYMMETRIC LEAVES complex employs multiple modes of regulation to affect adaxial-abaxial patterning and leaf complexity. Plant Cell 27(12):3321–3335. https://doi.org/10.1105/tpc.15.00454

Ito Y, Nishizawa-Yokoi A, Endo M, Mikami M, Toki S (2015) CRISPR/Cas9-mediated mutagenesis of the RIN locus that regulates tomato fruit ripening. Biochem Biophys Res Commun 467(1):76–82. https://doi.org/10.1016/j.bbrc.2015.09.117

Ito Y, Nishizawa-Yokoi A, Endo M, Mikami M, Shima Y, Nakamura N, Kotake-Nara E, Kawasaki S, Toki S (2017) Re-evaluation of the rin mutation and the role of RIN in the induction of tomato ripening. Nat Plants 3(11):866. https://doi.org/10.1038/s41477-017-0041-5

Jiang F, Doudna JA (2017) CRISPR–Cas9 structures and mechanisms. Annu Rev Biophys 46:505–529. https://doi.org/10.1146/annurev-biophys-062215-010822

Jiang SY, Ma Z, Vanitha J, Ramachandran S (2013) Genetic variation and expression diversity between grain and sweet sorghum lines. BMC Genomics 14(1):18. https://doi.org/10.1186/1471-2164-14-18

Jinek M, Chylinski K, Fonfara I, Hauer M, Doudna JA, Charpentier E (2012) A programmable dual-RNA-guided DNA endonuclease in adaptive bacterial immunity. Science 337(6096):816–821. https://doi.org/10.1126/science.1225829

Kearns NA, Pham H, Tabak B, Genga RM, Silverstein NJ, Garber M, Maehr R (2015) Functional annotation of native enhancers with a Cas9–histone demethylase fusion. Nat Methods 12(5):401. https://doi.org/10.1038/nmeth.3325

Kim YG, Cha J, Chandrasegaran S (1996) Hybrid restriction enzymes: zinc finger fusions to Fok I cleavage domain. Proc Natl Acad Sci 93(3):1156–1160. https://doi.org/10.1073/pnas.93.3.1156

Kimura S, Sinha N (2008) Tomato (*Solanum lycopersicum*): a model fruit-bearing crop. Cold Spring Harbor Protoc 2008(11):pdb-emo105. https://doi.org/10.1101/pdb.emo105

Koike-Yusa H, Li Y, Tan EP, Velasco-Herrera MDC, Yusa K (2014) Genome-wide recessive genetic screening in mammalian cells with a lentiviral CRISPR-guide RNA library. Nat Biotechnol 32(3):267

Komor AC, Kim YB, Packer MS, Zuris JA, Liu DR (2016) Programmable editing of a target base in genomic DNA without double-stranded DNA cleavage. Nature 533(7603):420

Konermann S, Brigham MD, Trevino AE, Joung J, Abudayyeh OO, Barcena C, Hsu PD, Habib N, Gootenberg JS, Nishimasu H, Nureki O (2015) Genome-scale transcriptional activation by an engineered CRISPR-Cas9 complex. Nature 517(7536):583. https://doi.org/10.1038/nature14136

Kwon DY, Zhao YT, Lamonica JM, Zhou Z (2017) Locus-specific histone deacetylation using a synthetic CRISPR-Cas9-based HDAC. Nat Commun 8:15315. https://doi.org/10.1038/ncomms15315

Lang Z, Wang Y, Tang K, Tang D, Datsenka T, Cheng J, Zhang Y, Handa AK, Zhu JK (2017) Critical roles of DNA demethylation in the activation of ripening-induced genes and inhibition of ripening-repressed genes in tomato fruit. Proc Natl Acad Sci 114(22):E4511–E4519. https://doi.org/10.1073/pnas.1705233114

Li JF, Norville JE, Aach J, McCormack M, Zhang D, Bush J, Church GM, Sheen J (2013) Multiplex and homologous recombination-mediated genome editing in Arabidopsis and *Nicotiana benthamiana* using guide RNA and Cas9. Nat Biotechnol 31(8):688. https://doi.org/10.1038/nbt.2654

Li R, Zhang L, Wang L, Chen L, Zhao R, Sheng J, Shen L (2018a) Reduction of tomato-plant chilling tolerance by CRISPR–Cas9-mediated SlCBF1 mutagenesis. J Agric Food Chem 66(34):9042–9051. https://doi.org/10.1021/acs.jafc.8b02177

Li R, Li R, Li X, Fu D, Zhu B, Tian H, Luo Y, Zhu H (2018b) Multiplexed CRISPR/Cas9-mediated metabolic engineering of γ-aminobutyric acid levels in *Solanum lycopersicum*. Plant Biotechnol J 16(2):415–427. https://doi.org/10.1111/pbi.12781

Li R, Fu D, Zhu B, Luo Y, Zhu H (2018c) CRISPR/Cas9-mediated mutagenesis of lncRNA1459 alters tomato fruit ripening. Plant J 94(3):513–524. https://doi.org/10.1111/tpj.13872

Li R, Liu C, Zhao R, Wang L, Chen L, Yu W, Zhang S, Sheng J, Shen L (2019) CRISPR/Cas9-mediated SlNPR1 mutagenesis reduces tomato plant drought tolerance. BMC Plant Biol 19(1):38. https://doi.org/10.1186/s12870-018-1627-4

Lippman ZB, Cohen O, Alvarez JP, Abu-Abied M, Pekker I, Paran I, Eshed Y, Zamir D (2008) The making of a compound inflorescence in tomato and related nightshades. PLoS Biol 6(11):e288. https://doi.org/10.1371/journal.pbio.0060288

Liu J, Van Eck J, Cong B, Tanksley SD (2002) A new class of regulatory genes underlying the cause of pear-shaped tomato fruit. Proc Natl Acad Sci 99(20):13302–13306. https://doi.org/10.1073/pnas.162485999

Ma Y, Zhang J, Yin W, Zhang Z, Song Y, Chang X (2016) Targeted AID-mediated mutagenesis (TAM) enables efficient genomic diversification in mammalian cells. Nat Methods 13(12):1029. https://doi.org/10.1038/nmeth.4027

Makarova KS, Wolf YI, Alkhnbashi OS, Costa F, Shah SA, Saunders SJ, Barrangou R, Brouns SJ, Charpentier E, Haft DH, Horvath P (2015) An updated evolutionary classification of CRISPR–Cas systems. Nat Rev Microbiol 13(11):722. https://doi.org/10.1038/nrmicro3569

Mali P, Aach J, Stranges PB, Esvelt KM, Moosburner M, Kosuri S, Yang L, Church GM (2013) CAS9 transcriptional activators for target specificity screening and paired nickases for cooperative genome engineering. Nat Biotechnol 31(9):833. https://doi.org/10.1038/nbt.2675

Mojica FJ, Díez-Villaseñor C, García-Martínez J, Almendros C (2009) Short motif sequences determine the targets of the prokaryotic CRISPR defence system. Microbiology 155(3):733–740. https://doi.org/10.1099/mic.0.023960-0

Nekrasov V, Staskawicz B, Weigel D, Jones JD, Kamoun S (2013) Targeted mutagenesis in the model plant *Nicotiana benthamiana* using Cas9 RNA-guided endonuclease. Nat Biotechnol 31(8):691. https://doi.org/10.1038/nbt.2655

Nekrasov V, Wang C, Win J, Lanz C, Weigel D, Kamoun S (2017) Rapid generation of a transgene-free powdery mildew resistant tomato by genome deletion. Sci Rep 7(1):482. https://doi.org/10.1038/s41598-017-00578-x

Nonaka S, Arai C, Takayama M, Matsukura C, Ezura H (2017) Efficient increase of γ-aminobutyric acid (GABA) content in tomato fruits by targeted mutagenesis. Sci Rep 7(1):7057. https://doi.org/10.1038/s41598-017-06400-y

Ortigosa A, Gimenez-Ibanez S, Leonhardt N, Solano R (2019) Design of a bacterial speck resistant tomato by CRISPR/Cas9-mediated editing of Sl JAZ 2. Plant Biotechnol J 17(3):665–673. https://doi.org/10.1111/pbi.13006

Pan C, Ye L, Qin L, Liu X, He Y, Wang J, Chen L, Lu G (2016) CRISPR/Cas9-mediated efficient and heritable targeted mutagenesis in tomato plants in the first and later generations. Sci Rep 6:24765. https://doi.org/10.1038/srep24765

Pnueli L, Carmel-Goren L, Hareven D, Gutfinger T, Alvarez J, Ganal M, Zamir D, Lifschitz E (1998) The SELF-PRUNING gene of tomato regulates vegetative to reproductive switching of sympodial meristems and is the ortholog of CEN and TFL1. Development 125(11):1979–1989 PMID: 9570763

Polstein LR, Perez-Pinera P, Kocak DD, Vockley CM, Bledsoe P, Song L, Safi A, Crawford GE, Reddy TE, Gersbach CA (2015) Genome-wide specificity of DNA binding, gene regulation, and chromatin remodeling by TALE- and CRISPR/Cas9-based transcriptional activators. Genome Res 25(8):1158–1169. https://doi.org/10.1101/gr.179044.114

Qi LS, Larson MH, Gilbert LA, Doudna JA, Weissman JS, Arkin AP, Lim WA (2013) Repurposing CRISPR as an RNA-guided platform for sequence-specific control of gene expression. Cell 152(5):1173–1183. https://doi.org/10.1016/j.cell.2013.02.022

Rajagopal N, Srinivasan S, Kooshesh K, Guo Y, Edwards MD, Banerjee B, Syed T, Emons BJ, Gifford DK, Sherwood RI (2016) High-throughput mapping of regulatory DNA. Nat Biotechnol 34(2):167. https://doi.org/10.1038/nbt.3468

Ronen G, Carmel-Goren L, Zamir D, Hirschberg J (2000) An alternative pathway to β-carotene formation in plant chromoplasts discovered by map-based cloning of Beta and old-gold color mutations in tomato. Proc Natl Acad Sci 97(20):11102–11107. https://doi.org/10.1073/pnas.190177497

Sapranauskas R, Gasiunas G, Fremaux C, Barrangou R, Horvath P, Siksnys V (2011) The *Streptococcus thermophilus* CRISPR/Cas system provides immunity in *Escherichia coli*. Nucleic Acids Res 39(21):9275–9282. https://doi.org/10.1093/nar/gkr606

Shalem O, Sanjana NE, Hartenian E, Shi X, Scott DA, Mikkelsen TS, Heckl D, Ebert BL, Root DE, Doench JG, Zhang F (2014) Genome-scale CRISPR-Cas9 knockout screening in human cells. Science 343(6166):84–87. https://doi.org/10.1126/science.1247005

Soyk S, Müller NA, Park SJ, Schmalenbach I, Jiang K, Hayama R, Zhang L, Van Eck J, Jiménez-Gómez JM, Lippman ZB (2017a) Variation in the flowering gene SELF PRUNING 5G promotes day-neutrality and early yield in tomato. Nat Genet 49(1):162. https://doi.org/10.1038/ng.3733

Soyk S, Lemmon ZH, Oved M, Fisher J, Liberatore KL, Park SJ, Goren A, Jiang K, Ramos A, van der Knaap E, Van Eck J (2017). Bypassing negative epistasis on yield in tomato imposed by a domestication gene. Cell 169(6):1142–1155. https://www.ncbi.nlm.nih.gov/pubmed/28528644

Tashkandi M, Ali Z, Aljedaani F, Shami A, Mahfouz MM (2018) Engineering resistance against tomato yellow leaf curl virus via the CRISPR/Cas9 system in tomato. Plant Signal Behav 13(10):e1525996. https://doi.org/10.1080/15592324.2018.1525996

Thakore PI, D'ippolito AM, Song L, Safi A, Shivakumar NK, Kabadi AM, Reddy TE, Crawford GE, Gersbach CA (2015) Highly specific epigenome editing by CRISPR-Cas9 repressors for silencing of distal regulatory elements. Nat Methods 12(12):1143. https://doi.org/10.1038/nmeth.3630

Thuesombat P, Hannongbua S, Akasit S, Chadchawan S (2014) Effect of silver nanoparticles on rice (*Oryza sativa* L. cv. KDML 105) seed germination and seedling growth. Ecotoxicol Environ Saf 104:302–309. https://doi.org/10.24297/jaa.v4i1.4295

Tomlinson L, Yang Y, Emenecker R, Smoker M, Taylor J, Perkins S, Smith J, MacLean D, Olszewski NE, Jones JD (2019) Using CRISPR/Cas9 genome editing in tomato to create a gibberellin-responsive dominant dwarf DELLA allele. Plant Biotechnol J 17(1):132–140. https://doi.org/10.1111/pbi.12952

Ueta R, Abe C, Watanabe T, Sugano SS, Ishihara R, Ezura H, Osakabe K (2017) Rapid breeding of parthenocarpic tomato plants using CRISPR/Cas9. Sci Rep 7(1):507. https://doi.org/10.1038/s41598-017-00501-4

Veillet F, Perrot L, Chauvin L, Kermarrec MP, Guyon-Debast A, Chauvin JE, Nogué F, Mazier M (2019) Transgene-free genome editing in tomato and potato plants using Agrobacterium-mediated delivery of a CRISPR/Cas9 cytidine base editor. Int J Mol Sci 20(2):402. https://doi.org/10.3390/ijms20020402

Vojta A, Dobrinić P, Tadić V, Bočkor L, Korać P, Julg B, Klasić M, Zoldoš V (2016) Repurposing the CRISPR-Cas9 system for targeted DNA methylation. Nucleic Acids Res 44(12):5615–5628. https://doi.org/10.1093/nar/gkw159

Wang T, Wei JJ, Sabatini DM, Lander ES (2014a) Genetic screens in human cells using the CRISPR-Cas9 system. Science 343(6166):80–84. https://doi.org/10.1126/science.1246981

Wang S, Wong D, Forrest K, Allen A, Chao S, Huang BE, Maccaferri M, Salvi S, Milner SG, Cattivelli L, Mastrangelo AM (2014b) Characterization of polyploid wheat genomic diversity using a high-density 90 000 single nucleotide polymorphism array. Plant Biotechnol J 12(6):787–796. https://doi.org/10.1111/pbi.12183

Wang D, Samsulrizal NH, Yan C, Allcock NS, Craigon J, Blanco-Ulate B, Ortega-Salazar I, Marcus SE, Bagheri HM, Fons LP, Fraser PD (2019) Characterization of CRISPR mutants targeting genes modulating pectin degradation in ripening tomato. Plant Physiol 179(2):544–557. https://doi.org/10.1104/pp.18.01187

Willcox JK, Catignani GL, Lazarus S (2003) Tomatoes and cardiovascular health. Crit Rev Food Sci Nutr. https://doi.org/10.1080/10408690390826437

Xu C, Liberatore KL, MacAlister CA, Huang Z, Chu YH, Jiang K, Brooks C, Ogawa-Ohnishi M, Xiong G, Pauly M, Van Eck J (2015) A cascade of arabinosyltransferases controls shoot meristem size in tomato. Nat Genet 47(7):784. https://doi.org/10.1038/ng.3309

Xu C, Park SJ, Van Eck J, Lippman ZB (2016) Control of inflorescence architecture in tomato by BTB/POZ transcriptional regulators. Genes Dev 30(18):2048–2061. https://doi.org/10.1101/gad.288415.116

Yang Y, Zhu G, Li R, Yan S, Fu D, Zhu B, Tian H, Luo Y, Zhu H (2017) The RNA editing factor SlORRM4 is required for normal fruit ripening in tomato. Plant Physiol 175(4):1690–1702. https://doi.org/10.1104/pp.17.01265

Yu T, Tzeng DT, Li R, Chen J, Zhong S, Fu D, Zhu B, Luo Y, Zhu H (2018) Genome-wide identification of long non-coding RNA targets of the tomato MADS box transcription factor RIN and function analysis. Ann Bot 123(3):469–482. https://doi.org/10.1093/aob/mcy178

Zhang H, Zhang J, Wei P, Zhang B, Gou F, Feng Z, Mao Y, Yang L, Zhang H, Xu N, Zhu JK (2014) The CRISPR/Cas9 system produces specific and homozygous targeted gene editing in rice in one generation. Plant Biotechnol J 12(6):797–807. https://doi.org/10.1111/pbi.12200

Zhao P, You Q, Lei M (2018) A CRISPR/Cas9 deletion into the phosphate transporter SlPHO1;1 reveals its role in phosphate nutrition of tomato seedlings. Physiol Plant. https://doi.org/10.1111/ppl.12897

Zsögön A, Čermák T, Naves ER, Notini MM, Edel KH, Weinl S, Freschi L, Voytas DF, Kudla J, Peres LEP (2018) De novo domestication of wild tomato using genome editing. Nat Biotechnol. https://doi.org/10.1038/nbt.4272

Chapter 14
Transgenic Approaches for Enhancement of Salinity Stress Tolerance in Plants

Shweta Jha

Abstract Worldwide food security and sustainable development are alarming issues with respect to climate change. Abiotic stresses negatively influence plant growth and development thus reducing the crop yield and productivity. Amidst the abiotic stresses, salinity stress is the major problem in agriculture lands. There is a rapid increase in salt-affected areas due to insufficient rainfall, imperfect irrigation system and water contamination, resulting in entry of salts in the soil. While traditional approaches used to handle the situations have limitations, current agricultural practices must seek tailored solutions to meet the demands of growing population. To generate climate-smart crops, genetic engineering is an important tool that allows to introduce distinct genetic changes without abolishing native traits, is faster, more effective and applicable to a wide range of species. It has been proved that expression of foreign gene(s) promotes a higher level of salt-tolerance in heterologous plant systems. Till date, several genes have been transferred in plants to increase salinity-stress tolerance, which are involved in synthesis of stress-mitigating compounds, antioxidant enzymes, regulatory proteins and signaling pathways proteins, ion transporters, etc. However, our knowledge about regulatory mechanisms of the salinity tolerance is still enigmatic. In the present chapter, current progress in transgenic approaches and the potential of transgenic plants for enhancement of salinity stress tolerance are reviewed and summarized.

Keywords Abiotic stress · Antioxidant enzymes · Compatible solutes · Genetic engineering · Ion homeostasis · Salinity stress tolerance · Stress signaling · Transgenic plants

---

S. Jha (✉)
Plant Functional Genomics Lab, Biotechnology Unit, Department of Botany (UGC-CAS), J.N.V. University, Jodhpur, Rajasthan 342001, India
e-mail: jha.shweta80@gmail.com

S. P. Singh et al. (eds.), *Molecular Approaches in Plant Biology and Environmental Challenges*, Energy, Environment, and Sustainability,
https://doi.org/10.1007/978-981-15-0690-1_14

14.1 Introduction

Various environmental stresses (both abiotic and biotic) adversely influence growth and development of plants, resulting in drastic reduction of plant productivity. Salinity stress is one of the major abiotic stress causing deleterious impact on plant survival and yield potential. There is a rapid increase in the salt-affected areas due to scanty rainfall, poor/contaminated irrigation system, deforestation and weathering of rocks (Munns and Tester 2008). In arid and semi-arid regions, higher rate of evaporation and less precipitation are responsible for loss of water, leading to deposition and accumulation of heavy amounts of salt in the soil over a certain period, thereby reducing soil porosity and transport of water. Increased salinization of arable land has global detrimental effects. Soil salinity is expected to have deleterious effect on soil fertility, including plant physiology and metabolism, that takes a major toll on growth and productivity of crop plants, ultimately reducing food production.

According to the Food and Agriculture Organization of the United States (FAO), 6–20% of total irrigated land of the world is adversely affected by salinity, causing significant yield reduction in major agricultural crops like rice, wheat and maize (Food and Agriculture Organization of the United Nations 2011). Moreover, the world population is expected to cross 9.5 billion by 2050 and the global food demand is predicted to increase by 70–85% (Food and Agriculture Organization of the United Nations 2017). To deal with the global hunger of exponentially growing population, a "Green to Gene Revolution" is required to achieve food and nutrition security worldwide. For this, concerted research effort and entanglement of new approaches are needed for more sustainable development of agriculture, via integration of major environmental stress tolerance traits in important crop plants.

Current scientific advances present substantial potential to develop climate-smart crops that can grow in adverse environmental conditions with sustainable yield and productivity. Genetic engineering is an effective tools for generating climate-resilient crops that can be used for a wider range of plant species (Dhankher and Foyer 2018). It provides a more promising approach to characterize specific gene function by plant transformation. The present chapter reviews the transgenic approaches used so far to develop salinity stress tolerance in plants, along with salient examples of key genes overexpressed in homologous or heterologous environment of plant cells.

14.2 Mechanisms of Salt Stress Response in Plants

Plant abiotic stress responses are highly intricate and include many genes. Due to its multigenic nature, it is quite more difficult to completely understand the salinity stress response in plants. However, in the past few decades, enormous efforts have been made to gain a better understanding of molecular mechanisms of plant salt stress response (Munns and Tester 2008; Deinlein et al. 2014).

Plants constantly face harsh climatic conditions. They cannot "fly-away" from such unfavorable conditions due to their sessile nature. In order to avoid or tolerate stress condition, they have developed several adaptive strategies including upregulation of stress-responsive genes involved in diverse physiological and biochemical processes related to cellular metabolism. In the first step, stress signals are perceived by membrane receptors, leading to activation of a whole cascade of signaling pathway that ultimately modulates the expression of stress-responsive genes in a synchronized manner. The physiological response of salinity stress is similar to water deficit condition or drought stress. Most plants show a drastic reduction in photosynthetic rate, causing wilting, reduced plant growth or even death under salt stress (Munns and Tester 2008). High sodium ion concentration in soil decreases soil osmotic potential, thereby limiting the water availability, resulting in osmotic stress to plants and loss of turgidity. In addition, excess of cations, mainly Na^+, causes ion toxicity to most plants, and adversely affects enzyme activity and membrane stability, and enhances production of Reactive oxygen species (ROS). Glycophytes and halophytes differ greatly in their tolerance towards salinity. Glycophytes are salt-sensitive plants that can grow at very low salt concentrations, whereas halophytes are extremely salt-tolerant plants and are well capable to grow at salt concentrations equivalent to seawater, as they are able to effectively exclude Na^+ ions by roots while taking up water from the soil even at higher salinities.

14.2.1 Biphasic Response

Plant response to salinity is biphasic, and can be divided into an early osmotic phase and a late ionic phase. Osmotic phase is rapid and hinders the growth of young leaves, whereas ionic phase is slower and cause senescence of mature leaves due to sodium ion accumulation (Munns and Tester 2008). Osmotic phase starts as soon as the salt concentration in the soil crosses the threshold level. Due to the osmotic effect of the salt around plant roots, a significant reduction occurs in shoot growth rate, leaf development, size of leaves, lateral bud development, cell expansion and number of tillers/no. of branches. The outcome of osmotic stress is depicted as quick retardation in expansion of younger leaves and stomatal closure/lowering of stomatal conductance of older leaves. This response of plants towards osmotic stress is quite similar to water deficit response. The mechanism that causes shoot and leaf growth inhibition under osmotic stress is still unknown. It might be regulated by long distance signaling molecules, like Abscisic acid (ABA). Gibberellins (GAs) and DELLA proteins are also good candidates for regulation of this response mechanism.

The second ionic phase of salinity response initiates after accumulation of sodium ions to a higher toxic concentration in mature leaves, resulting in enhanced senescence and death of older leaves, and further drastic reduction in the rate of shoot growth (Munns and Tester 2008). For most plants, both osmotic and ionic phases are isolated in time, except under highly saline conditions when a huge amount of salts may accumulate in leaves. At low or moderate salt stress conditions, ionic stress exerts much lower and delayed effect on growth and development of plants, as compared to osmotic stress. The mechanism by which salts put toxic effects is not precisely known. Salts may accumulate in the apoplast causing the cells to lose water, or in the cytosol causing protein denaturation and inhibition of enzyme activity involved in cellular metabolism, or they may exert a toxic effect on photosynthetic machinery in the chloroplast.

14.2.2 Strategies to Combat Salt Stress in Plants

Plants respond and tolerate to salinity stress by three distinct mechanisms, as described in the following sections.

14.2.2.1 Osmotic Stress Tolerance

Plants sense immediately the salt stress, either extracellularly at the plasma membrane, or intracellularly (The perception mechanism is still unknown). The best-characterized salt signaling pathway is SOS (Salt overlay sensitive) pathway (Quintero et al. 2002). In this pathway, perception of excessive sodium ions somehow induces a rapid increment in the cytosolic free Ca^{2+} concentrations. This elevated levels of Ca^{2+} is further sensed by a calcium-binding calcineurin B-like protein (CBL4, or SOS3), that subsequently interact with a Ser/Thr type CBL-interacting protein kinase (CIPK24, or SOS2) (Halfter et al. 2000). Activated CBL4/CIPK24 (SOS3/SOS2) complex is then activates the plasma membrane bound putative Na^+/H^+ antiporter (SOS1) and other transporters either through phosphorylation, or at posttranslational level, resulting in ion homeostasis and subsequent salinity stress tolerance (Fig. 14.1).

The decline in the photosynthesis rate due to osmotic stress causes greater synthesis of reactive oxygen species (ROS). These ROS species are highly toxic to the cell and also known to be involved in stress signaling. This increase subsequently lead to enhanced synthesis and activity of ROS scavengers (both antioxidant compounds like ascorbate, glutathione and anti-oxidative enzymes like catalase, SOD, APX, etc.), to balance the redox homeostasis in the cell. The biochemical pathways for such important stress-response mechanisms are very much complex, and include several genes encoding many isoforms of enzymes involved in ROS detoxification (Munns and Tester 2008).

14.2.2.2 Ion Exclusion

The main site of sodium ion toxicity is plant leaves. In order to avoid the accumulation of toxic Na^+ ions in the leaves, roots excludes the excess ions by altering the ion transport, or reduced long distance transport of Na^+ ions to shoots. Na^+/H^+ antiporters

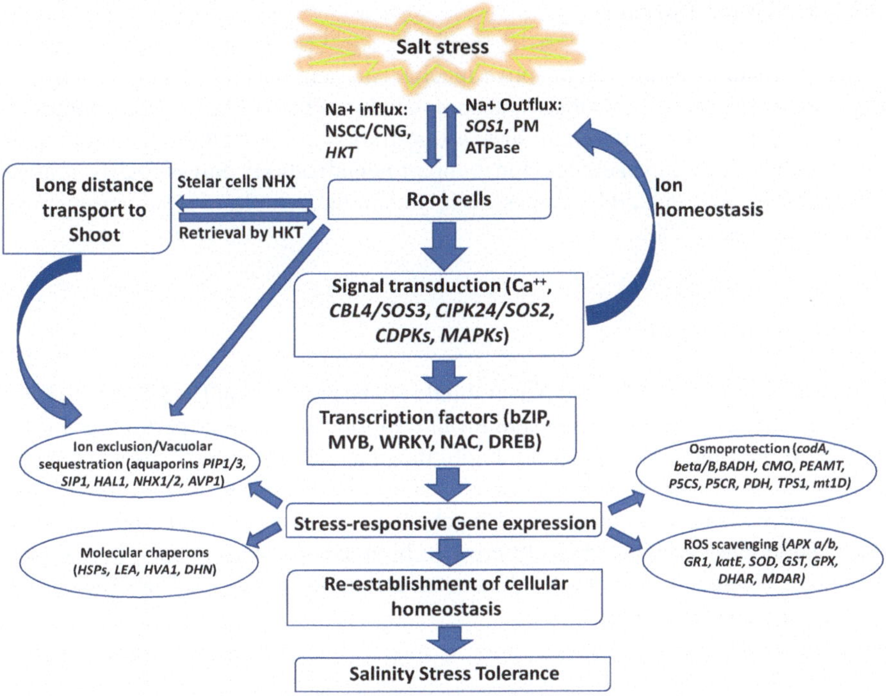

Fig. 14.1 Molecular mechanism of salinity stress response in plants. Candidate genes exploited for genetic engineering of plants for enhanced salt tolerance are shown in brackets

are the key players for this process that act mainly to efflux Na^+ ions out of the cell. Na^+ influx in roots is mediated by nonselective cation channels, like cyclic nucleotide–gated channels, or some members of the high-affinity K^+ transporter (HKT) family (Munns and Tester 2008). Most of the Na^+ entering in the root is efflux out by ATP-dependent plasma membrane Na^+/H^+ antiporters, like SOS1. Some other Na^+-translocating ATPases also play role in this process. Remaining Na^+ can be sequestered into the root cell vacuoles, or transported to the shoot via xylem, by stelar cells plasma membrane Na^+/H^+ antiporter activity (Fig. 14.1).

High-affinity K^+ transporters (HKT), such as *Arabidopsis* HKT1;1 and rice and wheat HKT1;5 have been shown to play important role in retrieval of Na^+ ions from the xylem before reaching to the shoot. The HKT gene family is grouped into two subfamilies. The members of subfamily-1 function as low affinity Na^+ uniporters, and contain a specific serine residue in the "reactive pore loop" region of the protein. While members of subfamily-2 mainly act as Na^+/K^+ symporters, in which serine is substituted by a glycine (Munns and Tester 2008).

14.2.2.3 Tissue Tolerance

Tissue tolerance is clearly an adaptive mechanism, and plants (mostly halophytes) adapt two strategies for tolerance of accumulated ions to avoid toxic concentrations at intracellular level in different tissues: sequestration or intracellular partitioning of excess sodium ions into vacuoles (Ion compartmentalization), and synthesis of large concentrations of compatible solutes/osmolytes in the cytoplasm (Osmoprotection) (Munns and Tester 2008).

Ion Compartmentalization

The salt concentration in the cytoplasm should not be greater than 10–30 mM. To keep this concentration optimal, excess sodium and chloride ions must be partitioned inside the cell, usually through sequestration into the plant cell vacuoles. This mechanism takes place by the help of tonoplast Na^+/H^+ antiporters (e.g. NHX family). Ion compartmentalization is an essential mechanism for salt stress tolerance in most of the halophytes, which can easily grow in high saline environments. In contrast, glycophytes cannot tolerate high salt concentrations since they are less efficient for this mechanism (Staal et al. 1991).

Synthesis of Compatible Solutes/Osmolytes

Elevated vacuolar ion concentrations would require accumulation and synthesis of higher concentrations of compatible solutes (i.e. organic solutes that are compatible and do not inhibit the metabolic activities of the cell) in the cytosol and other subcellular organelles (like chloroplast), to maintain their osmotic potential and optimum volume (a process called as osmotic adjustment). These compounds include amino acids and their derivatives (proline), soluble sugars (Trehalose, fructans), sugar alcohols/polyols (mannitol, sorbitol, ononitol, pinitol), quaternary ammonium compounds (glycine betaine), etc. At low concentrations, these organic solutes act as osmoprotectants and stabilize the cellular membranes, tertiary structure of proteins and their function, and other subcellular components. These compatible solutes were also found to accumulate at high concentrations under drought and cold stress conditions, which function to retain water in the cell in order to balance the osmotic pressure. These compounds may also act as scavengers of harmful ROS species (Shen et al. 1997). The synthesis of these compounds requires sufficient energy and may impose growth penalty, but allow the plant to survive under high salt stress condition.

14.3 Genetic Engineering Approaches to Improve Salinity Stress Tolerance in Plants

In the twentieth century, the conventional breeding programs have been widely used to enhance various agronomic traits (Flowers 2004). However, a very little progress has been made so far for developing abiotic stress tolerant crop plants through traditional breeding approaches. The main limiting factor of plant breeding is that it relies on the use of germplasm of the same or closely related species that have less genetic differences in their gene pool. Many times it is hard to transfer desirable gene to domesticated crop varieties from its wild relatives by crossing due to reproductive barriers, and undesirable traits of donor species also transfer along with the gene of interest. In addition, Breeding strategies require considerable amount of time and labor. In recent times, marker-assisted selection (MAS) and quantitative trait loci (QTL) analysis have been used to enhance tolerance to abiotic stress and to improve growth and productivity of crop plants. However, cost and efficiency of MAS technique for analysis of phenotypic traits, and accurate experimental designing for QTL analysis are the major concerns.

On the other end, Genetic engineering allows us to select and manipulate any desirable gene of interest and easily transfer it to related or non-related plant species. Transgenic approaches are being used extensively worldwide for improving abiotic stress-tolerance in plants (Wani et al. 2017; Paul and Roychoudhury 2018; Singla-Pareek et al. 2001). Recent developments in transgenic research for enhancing salinity stress tolerance in plants are discussed in following sections.

It has been shown that the overexpression of some candidate genes results in a higher level of salinity stress tolerance. These genes can be categorized in three groups: genes involved in direct protection by producing functional proteins or stress-adaptive compounds, genes for regulatory and signaling pathways, and genes involved in ion homeostasis by Na^+ exclusion or transport (Fig. 14.1).

14.3.1 Genes Involved in Direct Protection (Functional Proteins/Stress-Adaptive Compounds)

14.3.1.1 Osmolytes

As discussed in earlier sections, accumulation of large concentration of compatible solutes/osmolytes in cytosol or chloroplast is an important strategies to combat the detrimental impact of salt stress in plants. Several studies have confirmed essential function of these compounds in salinity stress tolerance, by manipulating the genes for their synthesis or metabolism. The main enzyme for proline biosynthesis is Δ1-pyrroline-5-carboxylate synthetase (P5CS), which catalyzes the conversion of glutamate to Δ1-pyrroline-5-carboxylate that is further reduced to proline. Several studies have described the overexpression of *P5CS* gene rendered tolerance for salt stress in transgenic *Arabidopsis*, rice, wheat, potato, etc. (Table 14.1) (Kavi Kishor et al. 1995; Karthikeyan et al. 2011; Hong et al. 2000; Hmida-Sayari et al. 2005). Transgenic tobacco plants overexpressing *P5CS* gene from moth bean accumulated 10-18 fold more proline as compared to untransformed control plants, and showed better biomass production and reproductive development under osmotic stress conditions (Kavi Kishor et al. 1995). In a separate study, transgenic rice plants overexpressing the same *P5CS* gene demonstrated 2.5-fold more proline accumulation than control plants, and exhibited enhanced growth under salt stress condition (Zhu et al. 1998). Recently Surekha et al. (2014) expressed *P5CSF129A* from *Vigna aconitifolia* in transgenic Pigeon pea, which resulted in enhanced accumulation of proline in the cells and subsequent tolerance to salinity. In another study, antisense suppression of Proline dehydrogenase (*ProDH*) reduced proline degradation in *Arabidopsis* and conferred tolerance to freezing and salinity stress (Nanjo et al. 1999).

Glycine betaine (GB) is synthesized in many halophytes and prokaryotes as an adaptive strategy for salt stress tolerance. In bacteria, it is synthesized as one step process using Choline oxidase (CodA) enzyme, which directly converts choline to GB. Whereas in plants it is a two-step process, involving two enzymes—Choline monooxygenease (CMO) and Betaine aldehyde dehydrogenase (BADH). To date, several studies have been performed for overexpression of *codA* gene from *Arthrobacter globiformis* in transgenic rice, *Arabidopsis*, Brassica, alfalfa and Eucalyptus plants, which do not synthesize GB naturally (Sakamoto et al. 1998; Kathuria et al. 2009; Hayashi et al. 1997; Prasad et al. 2000; Kikuchi et al. 2009). Recently Wei et al. (2017) overexpressed the same *codA* gene in tomato that led to enhanced biosynthesis of GB and alleviated salt-induced K^+ efflux and enhanced salt tolerance. Similarly *betA* and *betB* genes from *E. coli* encoding choline dehydrogenase have been transferred in tobacco resulting in improved tolerance to salinity and low temperature due to GB accumulation (Holmstrom et al. 2000). These studies are summarized in Table 14.1. In some studies, *codA* gene was either targeted to cytosol or chloroplast of the rice plant cells. The transgenic lines harboring chloroplast-targeted COD were more tolerant to high salt and cold stress, as compared to cytosol-targeted COD plants (Sakamoto et al. 1998). The results suggested that introduction of GB biosynthetic pathway in chloroplast may lead to more efficient protection of the photosystem II complex of photosynthetic machinery against salt, thus conferring better protection against abiotic stress, compared to introduction of the pathway in the cytosol of rice plant cells. Glycine betaine did not leak into the cytosol after its synthesis and accumulated in the chloroplasts of ChlCOD rice plants, where it had a protective effect on photosystem II complex against inactivation under salt stress, mainly through stabilization of structure and function of proteins (Sakamoto et al. 1998).

There are several reports describing genetic engineering of GB biosynthesis using *CMO* or *BADH* gene from halophytes, either alone or in combination with each other, into non-GB accumulating plant species like wheat, carrot, *Arabidopsis*, tobacco, tomato, sweet potato, maize and trifoliate orange (Table 14.1). Heterologous expression of these genes resulted in enhanced GB accumulation in transgenic plants,

Table 14.1 Comprehensive list of transgenes encoding functional proteins or stress-adaptive compounds for direct protection from salinity stress tolerance

Gene transferred	Product	Source organism	Target plant	Transgene effect	Reference(s)
Compatible solutes					
Glycine betaine synthesis					
codA	Choline oxidase	*Arthrobacter globiformis*	*Eucalyptus camaldulensis*	Enhanced GB accumulation, reduced oxidative stress	(Kikuchi et al. 2009)
codA	Choline oxidase	*Arthrobacter globiformis*	*Arabidopsis thaliana*	Enhanced tolerance to salt and cold stress	(Hayashi et al. 1997)
codA	Choline oxidase	*Arthrobacter globiformis*	*Oryza sativa*	Enhanced GB accumulation and tolerance to salt and cold	(Sakamoto et al. 1998; Kathuria et al. 2009; Mohanty et al. 2002)
codA	Choline oxidase	*Arthrobacter globiformis*	*Brassica juncea*	Salt tolerance	(Prasad et al. 2000)
codA	Choline oxidase	*Arthrobacter globiformis*	*Solanum lycopersicum*	Enhanced salt tolerance	(Wei et al. 2017)
cod A	Choline oxidase	*Arthrobacter globiformis*	*Medicago sativa*	Enhanced salt and drought tolerance	(Li et al. 2014)
betA	Choline dehydrogenase	*E. coli*	*Brassica oleracea var. capitata*	Salt tolerance	(Bhattacharya et al. 2004)
betA and betB	Choline dehydrogenase and Betaine-aldehyde dehydrogenase	*E. coli*	*Nicotiana tabacum*	Improved tolerance to salinity and low temperature	(Holmstrom et al. 2000)
AhBADH	Betaine aldehyde dehydrogenase	*Atriplex hortensis*	*Triticum aestivum*	Salt tolerance	(Guo et al. 2000)

(continued)

Table 14.1 (continued)

Gene transferred	Product	Source organism	Target plant	Transgene effect	Reference(s)
ALDH10A8 and ALDH10A9	Betaine aldehyde dehydrogenase	*Arabidopsis thaliana*	*Arabidopsis thaliana*	Salt tolerance	(Missihoun et al. 2015)
BADH	Betaine aldehyde dehydrogenase	*Daucus carota*	*Daucus carota*	Salt tolerance	(Kumar et al. 2004)
BADH	Betaine aldehyde dehydrogenase	*Spinacia oleracea*	*Ipomoea batatas*	salt, oxidative stress, and low temperature tolerance	(Fan et al. 2012)
BADH	Betaine aldehyde dehydrogenase	*Atriplex hortensis*	*Solanum lycopersicum*	Salt tolerance	(Jia et al. 2002)
SeNHX1 and a BADH	vacuolar Na^+/H^+ antiporter gene and betaine aldehyde dehydrogenase	*Salicornia europaea and Atriplex hortensis*	*Nicotiana tabacum*	Salt tolerance	(Zhou et al. 2008)
BADH	Betaine aldehyde dehydrogenase	*Spinacia oleracea*	*Solanum tuberosum*	Enhanced drought and salinity tolerance	(Zhang et al. 2011)
BADH	Betaine aldehyde dehydrogenase	*Atriplex hortensis*	*Oryza sativa*	Salt tolerance	(Guo et al. 1997)
ScBADH	Betaine aldehyde dehydrogenase	*Suaeda corniculata*	*Arabidopsis thaliana*	Enhanced drought and salinity tolerance	(Wang et al. 2016a)
BADH	Betaine aldehyde dehydrogenase	*Atriplex hortensis*	*Poncirus trifoliata*	Salt tolerance	(Fu et al. 2011)
BADH	Betaine aldehyde dehydrogenase	*Atriplex micrantha*	*Zea mays*	Enhanced salinity tolerance	(Di et al. 2015)

(continued)

Table 14.1 (continued)

Gene transferred	Product	Source organism	Target plant	Transgene effect	Reference(s)
AhCMO	Choline monooxygenase	*Atriplex hortensis*	*Nictiana tabacum*	Salt- and drought-stress tolerance	(Shen et al. 2001)
CMO	Choline monooxygenase	*Spinacia oleracea*	*Oryza sativa*	Enhanced tolerance to salt stress and temperature stress	(Shirasawa et al. 2006)
CMO	Choline monooxygenase	*Salicornia europaea*	*Nictiana tabacum*	Enhanced tolerance to salt stress	(Wu et al. 2010)
SoCMO and SoBADH	Choline monooxygenase and betaine aldehyde dehydrogenase	*Spinacia oleracea*	*Lolium perenne*	Enhanced salt stress tolerance and induced dwarfism	(Bao et al. 2011)
BvCMO	Choline monooxygenase	*Beta vulgaris*	*Nicotiana tabacum (plastid)*	salt and drought tolerance	(Zhang et al. 2008)
CMO	Choline monooxygenase	*Suaeda liaotungensis*	*Nicotiana tabacum*	Enhanced tolerance to salt stress	(Li et al. 2003)
ZmPEAMT1	Phosphoethanolamine N-methyltransferase (choline biosynthesis)	*Zea mays*	*Arabidopsis thaliana*	Enhanced the salt tolerance	(Wu et al. 2007)
Proline biosynthesis					
P5CSF129A	Δ1-pyrroline-5-carboxylate synthetase	*Vigna aconitifolia*	*Cajanus cajan*	Enhanced proline accumulation and salt tolerance	(Surekha et al. 2014)

(continued)

Table 14.1 (continued)

Gene transferred	Product	Source organism	Target plant	Transgene effect	Reference(s)
P5CS	Δ1-pyrroline-5-carboxylate synthetase	*Vigna aconitifolia*	*Nicotiana tabacum*	Osmotic stress tolerance	(Kavi Kishor et al. 1995; Hong et al. 2000)
P5CS	Δ1-pyrroline-5-carboxylate synthetase	*Vigna aconitifolia*	*Oryza sativa*	Salt tolerance	(Karthikeyan et al. 2011; Zhu et al. 1998; Su and Wu 2004)
P5CS	Δ1-pyrroline-5-carboxylate synthetase	*Vigna aconitifolia*	*Triticum aestivum*	Salt tolerance	(Sawahel and Hassan 2002)
P5CS	Δ1-pyrroline-5-carboxylate synthetase	*Arabidopsis thaliana*	*Solanum tuberosum*	Salt tolerance	(Hmida-Sayari et al. 2005)
PRODH (antisense suppression)	Proline dehydrogenase	*Arabidopsis thaliana*	*Arabidopsis thaliana*	Improved tolerance to freezing and salinity stress	(Nanjo et al. 1999)
TaP5CR	Δ1-pyrroline-5-carboxylate reductase	*Triticum aestivum*	*Arabidopsis thaliana*	Salt tolerance	(Ma et al. 2008)
Other osmoprotectants					
ots A and ots B	Trehalose-6-phosphate synthase and trehalose-6-phosphate phosphatase	*E. coli*	*Oryza sativa*	Enhanced trehalose accumulation and Salt, drought and cold tolerance	(Garg et al. 2002)
ScTPS1	Trehalose-6-phosphate synthase	*Saccharomyces cerevisiae*	*Arabidopsis thaliana*	Salt tolerance	(Miranda et al. 2007)

(continued)

Table 14.1 (continued)

Gene transferred	Product	Source organism	Target plant	Transgene effect	Reference(s)
TPS1-TPS2	Trehalose-6-phosphate synthase	*Saccharomyces cerevisiae*	*Medicago sativa*	Salt tolerance	(Suarez et al. 2009)
TPS1	Trehalose-6-phosphate synthase	*Saccharomyces cerevisiae*	*Solanum lycopersicum*	Salt tolerance	(Cortina and Culianez-Macia 2005)
OsTPS1	Trehalose-6-phosphate synthase	*Oryza sativa*	*Oryza sativa*	High salinity and drought tolerance	(Li et al. 2011)
OsTPP1	Trehalose-6-phosphate phosphatase	*Oryza sativa*	*Oryza sativa*	High salinity and drought tolerance	(Ge et al. 2008)
mt1D	Mannitol-1-phosphate dehydrogenase	*E. coli*	*Arabidopsis thaliana*	High salt tolerance	(Thomas et al. 1995)
mt1D	Mannitol-1-phosphate dehydrogenase	*E. coli*	*Nicotiana tabacum*	Salt and dehydration tolerance	(Karakas et al. 1997)
mtlD	Mannitol-1-phosphate dehydrogenase	*E. coli*	*Triticum aestivum*	Enhanced mannitol accumulation leading to salt and drought tolerance	(Abebe et al. 2003)
mt1D and gutD	Mannitol-1-phosphate dehydrogenase and glucitol-6-phosphate dehydrogenase	*E. coli*	*Pinus taeda*	Salt tolerance	(Tang et al. 2005)
mt1D	Mannitol-1-phosphate dehydrogenase	*E. coli*	*Populus nigra x Populus tomentosa*	Salt tolerance	(Lin et al. 2016)
MIPS	L-myo-inositol-1-phosphate synthase	*Potresia coarctata*	*Nicotiana tabacum*	Tolerance up to 300 mM NaCl	(Majee et al. 2004)

(continued)

Table 14.1 (continued)

Gene transferred	Product	Source organism	Target plant	Transgene effect	Reference(s)
IMT1	Myo-inositol O-methyl transferase	*Mesembryantkemum crystallinum*	*Nicotiana tabacum*	Increased D-ononitol synthesis and salt and drought tolerance	(Sheveleva et al. 1997)
Lea proteins					
HVA1	Late embryogenesis abundant protein	*Hordeum vulgare*	*Morus indica*	drought, salinity and cold stress	(Checker et al. 2012)
HVA1	Late embryogenesis abundant protein	*Hordeum vulgare*	*Morus indica*	Tolerance to salinity and water stress	(Lal et al. 2008)
HVA1	Late embryogenesis abundant protein	*Hordeum vulgare*	*Oryza sativa*	tolerance to water deficit and salt stress	(Xu et al. 1996)
Me-LEAn4	Late embryogenesis abundant protein	*Brassica napus*	*Brassica campestris*	Enhanced salt and drought tolerance	(Park et al. 2005)
Me-LEAn4	Late embryogenesis abundant protein	*Brassica napus*	*Lactuca sativa*	Increased tolerance to salt- and water-deficit stress	(Park et al. 2005)
DHN-5	Dehydrin	*Triticum aestivum*	*Arabidopsis thaliana*	Enhanced tolerance to salt and osmotic stress	(Brini et al. 2007)
OsLEA4	Late embryogenesis abundant protein	*Oryza sativa*	*Oryza sativa*	Drought, high salt and heavy metal stress tolerance	(Hu et al. 2016)
Molecular chaperons					
Hsp21	Heat shock protein	*Arabidopsis thaliana*	*Arabidopsis thaliana*	Oxidative stress tolerance	(Härndahl et al. 1999)

(continued)

Table 14.1 (continued)

Gene transferred	Product	Source organism	Target plant	Transgene effect	Reference(s)
AtHSP17.6A	Heat shock protein	*Arabidopsis thaliana*	*Arabidopsis thaliana*	Osmotolerance	(Sun et al. 2001)
DnaK1	Molecular chaperons	*Aphanothece halophytice*	*Nicotiana tabacum*	Resistance to salt stress	(Sugino et al. 1999)
Antioxidant enzymes					
katE	Catalase	*E. coli*	*Oryza sativa*	Enhanced production of catalase, increased salt tolerance	(Moriwaki et al. 2008; Prodhan et al. 2008)
katE	Catalase	*E. coli*	*Nicotiana tabacum*	Resistance to salt stress	(Al-Taweel et al. 2007)
katE + bet	Catalase, choline dehydrogenase and betaine-aldehyde dehydrogenase	*E. coli*	*synechococcus sp. PCC 7942*	Enhanced production of catalase thereby increased salt tolerance	(Kaku et al. 2000)
BcGR1	Cytosolic glutathione reductase	*Chinese cabbage (Brassica campestris)*	*Nicotiana tabacum*	Reduced oxidative stress and enhanced drought and salt tolerance	(Lee and Jo 2004)
APX	Ascorbate peroxidase	*A. thaliana*	*Nicotiana tabacum (chloroplast)*	Tolerance to salt stress and water deficit	(Badawi et al. 2004)
APX	Ascorbate peroxidase	*Pisum sativum*	*Solanum lycopersicum*	Salt tolerance	(Wang et al. 2005)
APXa and APXb	Ascorbate peroxidase	*Oryza sativa*	*A. thaliana*	Salt tolerance	(Lu et al. 2007)
APX	Thylakoid-bound ascorbate peroxidase	*Solanum lycopersicum*	*Nicotiana tabacum*	Salt and osmotic tolerance	(Sun et al. 2010)

(continued)

Table 14.1 (continued)

Gene transferred	Product	Source organism	Target plant	Transgene effect	Reference(s)
GmALDH7	Antiquitin-like aldehyde dehydrogenase family7 member	*Glycine max*	*A. thaliana*; *Nicotiana tabacum*	Improved tolerance to drought, salinity and oxidative stress	(Rodrigues et al. 2006)
Cu/Zn-SOD	Cu/Zn superoxide dismutase	*Pisum sativum*	*N. tabacum*	Oxidative stress tolerance	(SenGupta et al. 1993a, b)
Cu/Zn-SOD, APX and DHAR	Cu/Zn superoxide dismutase, ascorbate peroxidase, dehydroascorbate reductase	*Pisum sativum/Homo sapiens*	*N. tabacum*	Salt and oxidative tolerance	(Lee et al. 2007)
Cu/Zn-SOD	Cu/Zn superoxide dismutase	*Avicennia marina*	*Oryza sativa*	Salt, drought, and oxidative tolerance	(Prashanth et al. 2008)
Cu/Zn-SOD and APX	Cu/Zn superoxide dismutase, ascorbate peroxidase	*Pisum sativum*	*N. tabacum*	Salt and osmotic stress tolerance	(Lee et al. 2010)
Fe-SOD	Fe superoxide dismutase	*A. thaliana*	*N. tabacum*	Enhancement of oxidative stress tolerance	(Van Camp et al. 1996)
Mn-SOD	Mn superoxide dismutase	*A. thaliana*	*A. thaliana*	Salt tolerance	(Wang et al. 2004)
PaSOD and RaAPX	Cu/Zn-superoxide dismutase, ascorbate peroxidase	*Potentilla atrosanguinea, Rheum australe*	*Solanum tuberosum*	Enhanced lignification and starch biosynthesis with improved salt stress tolerance	(Shafi et al. 2017)

(continued)

Table 14.1 (continued)

Gene transferred	Product	Source organism	Target plant	Transgene effect	Reference(s)
GST and GPX	Glutathione S-transferase, glutathione peroxidase	*N. tabacum*	*N. tabacum*	Enhanced salt tolerance	(Roxas et al. 1997)
OsGSTL2	Lambda class of glutathione S-transferase	*Oryza sativa*	*A. thaliana*	Tolerance for heavy metals and other abiotic stresses like cold, osmotic stress and salt	(Kumar et al. 2013)
GST	Glutathione S-transferase	*Suaeda salsa*	*A. thaliana*	Salt tolerance	(Qi et al. 2010)
DHAR	Dehydroascorbate reductase	*Oryza sativa*	*A. thaliana*	Salt tolerance	(Ushimaru et al. 2006)
MDAR	Monodehydroascorbate reductase	*A. thaliana*	*N. tabacum*	Salt, ozone, and PEG tolerance	(Eltayeb et al. 2007)
AmMDAR	Chloroplastic monodehydroascorbate reductase	*Avicennia marina*	*N. tabacum*	Salt tolerance	(Kavitha et al. 2010)
Glyoxalase pathway					
GLYI	Glyoxalase-I	*Brassica juncea*	*O. sativa L. (japonica and indica rice)*	Enhanced salt tolerance	(Verma et al. 2005)
GLYI	Glyoxalase-I	*Brassica juncea*	*Vigna mungo*	Enhanced salt tolerance	(Bhomkar et al. 2008)
GLYI + GLYII	Glyoxalase-I and II	*Brassica juncea, O. sativa*	*N. tabacum*	Enhanced salt tolerance	(Yadav et al. 2005)

(continued)

Table 14.1 (continued)

Gene transferred	Product	Source organism	Target plant	Transgene effect	Reference(s)
Brassica GLYI + *Pennisetum GLYII*	Glyoxalase-I and II	*Brassica juncea*, *Pennisetum glaucum*	*Solanum lycopersicum*	Enhanced salt tolerance	(Alvarez Viveros et al. 2013)
GLYII	Glyoxalase-II	*O. sativa*	*N. tabacum*	Enhanced salt tolerance	(Singla-Pareek et al. 2003)
Polyamines					
ADC	Arginine decarboxylase	*Avena sativa*	*O. sativa*	Salt tolerance	(Roy and Wu 2001)
ODC	Ornithine decarboxylase	*Mus musculus*	*N. tabacum*	Salt tolerance	(Roy and Wu 2001)
SAMDC	S-adenosyl methionine decarboxylase	*Tritordeum*	*O. sativa*	Salt tolerance	(Roy and Wu 2002)
SAMDC	S-adenosylmethionine decarboxylase	*Homo sapiens*	*N. tabacum*	Salt and osmotic tolerance	(Waie and Rajam 2003)
SPDS	Spermidine synthase	*Cucurbita ficifolia*	*A. thaliana*	Multiple abiotic stress tolerance	(Kasukabe et al. 2004)
SPDS	Spermidine synthase	*Malus domestica*	*Pyrus*	Multiple abiotic stress tolerance	(Wen et al. 2008)
Other stress-responsive genes					
SmCP	Cysteine protease	*Salix matsudana*	*A. thaliana*	Salt tolerance	(Zheng et al. 2018)

(continued)

Table 14.1 (continued)

Gene transferred	Product	Source organism	Target plant	Transgene effect	Reference(s)
OsIF	Intermediate filament	*O. sativa*	*O. sativa*	Stabilizes photosynthetic machinery and yield under salinity and heat stress	(Soda et al. 2018)
OsCKX2 (knockdown)	Cytokinin oxidase	*O. sativa*	*O. sativa*	Reduces yield penalty under salinity stress condition	(Joshi et al. 2018)
AtPLAT1	Polycystin, lipoxygenase, alpha-toxin and triacylglycerol lipase) domain protein1	*A. thaliana*	*Nicotiana tabacum cv. Samsun NN and Nicotiana benthamiana*	Cold, drought and salt stress tolerance	(Hyun et al. 2015)
PagGIa, *PagGIb and PagGIc* (downregulation)	GIGANTEA-like genes flowering time regulator	*Populus alba* × *Populus glandulosa*	*A. thaliana*	Vigorous growth, higher biomass and enhanced salt stress tolerance	(Ke et al. 2017)
GmGER 9	Germin	*Glycine max*	*N. tabacum*	Enhanced salt stress tolerance	(Lu et al. 2010)
VvSDIR1	Salt- and drought-induced RING finger 1	*Vitis vinifera*	*Nicotiana tabacum*	Enhanced oxidative stress tolerance, methyl viologen, NaCl, and polyethylene glycol tolerance	(Tak and Mhatre 2013)

(continued)

Table 14.1 (continued)

Gene transferred	Product	Source organism	Target plant	Transgene effect	Reference(s)
MdY3IP1	Nucleus-encoded thylakoid protein, Ycf3-interacting protein	*Malus domestica*	*A. thaliana*	Early-flowering and enhanced salt tolerance	(Yu et al. 2018)
PsSEO-F1	ATP-independent phloem motor protein (forisome)	*Pisum sativum*	*Nicotiana tabacum*	Enhanced salt tolerance	(Srivastava et al. 2016)
PtDRS1	Drought sensitive 1	*Populus trichocarpa*	*Populus deltoides* × *Populus euramericana*	Enhanced salt and drought tolerance	(Mohammadi et al. 2018)
AtFC1	Ferrochelatase-1, terminal enzyme of heme biosynthesis	*A. thaliana*	*A. thaliana*	Resistance to high salinity	(Zhao et al. 2017)
UGT76E11	Uridine diphosphate glycosyltransferase	*A. thaliana*	*A. thaliana*	Enhanced tolerance to salinity and drought	(Li et al. 2018)
UGT76E12	Uridine diphosphate glycosyltransferase	*A. thaliana*	*A. thaliana*	Enhanced tolerance to salinity and drought	(Chen et al. 2019a)
Osmotin gene	Osmotin	*Nicotiana tabacum*	*Nicotiana tabacum*	Osmotic stress tolerance	(Barthakur et al. 2001)

thereby increased abiotic stress tolerance (Shen et al. 2001; Wu et al. 2010; Zhang et al. 2008; Guo et al. 2000; Kumar et al. 2004; Fan et al. 2012; Wang et al. 2016a; Di et al. 2015). Bao et al. (2011) reported that simultaneous expression of Spinach *CMO* and *BADH* genes contributed to enhanced salt stress tolerance and induced dwarfism in transgenic *Lolium perenne*. Improved salt tolerance in tobacco plants was reported by co-transformation of a betaine synthesis gene *BADH* from *Atriplex hortensis* and a vacuolar Na^+/H^+ antiporter gene *SeNHX1* from halophyte *Salicornia europaea* (Zhou et al. 2008). In addition, Wu et al. (2007) reported the role of maize phosphoethanolamine N-methyltransferase (*PEAMT*) for choline biosynthesis (which is a precursor for synthesis of GB), by overexpression in *Arabidopsis*.

Trehalose is a disaccharide and act as osmoprotectants, thus provide tolerance to multiple abiotic stresses including salinity. The *TPS1* gene encoding Trehalose-6-phosphate synthase, which is the main enzyme for Trehalose biosynthesis was transferred to rice, tomato, *Arabidopsis* and alfalfa, conferring tolerance to high salinity and drought (Miranda et al. 2007; Suarez et al. 2009; Cortina and Culianez-Macia 2005; Li et al. 2011). Garg et al. (2002) reported that heterologous expression of bacterial *ots A* and *ots B* genes in rice resulted in increased trehalose accumulation and high tolerance to multiple abiotic stresses in transgenic plants.

The accumulation of polyols like mannitol, myo-inositol, ononitol or pinitol was also found to be directly correlated with abiotic stress tolerance, and the overexpression of genes of their biosynthesis (e.g. bacterial *mt1D* for Mannitol-1-phosphate dehydrogenase) rendered enhanced salt stress tolerance to transgenic plants of tobacco, *Arabidopsis*, hybrid Poplar, wheat and Pine (Tang et al. 2005; Lin et al. 2016; Abebe et al. 2003; Thomas et al. 1995). Transgenic *Arabidopsis* plants exhibited an increase in seed germination on salt-supplemented medium, and transgenic tobacco plants showed enhanced root biomass in the presence of salt. Overexpression of the gene encoding mannose-6-phosphate reductase from celery resulted in enhanced mannitol accumulation in transgenic *Arabidopsis* plants, and increase in growth rate and photosynthesis under saline condition, but not drought stress (Sickler et al. 2007). Majee et al. (2004) reported overexpression of a novel *MIPS* gene encoding L-myo-inositol-1-phosphate synthase from a halophytic wild rice *Porteresia coarctata* in transgenic tobacco, which were able to tolerate up to 300 mM of NaCl concentration as compared to control plants. Similarly, transgenic tobacco plants expressing *imt1* gene encoding Myo-inositol O-methyl transferase resulted in enhanced accumulation of D-ononitol and conferred salt and drought tolerance (Sheveleva et al. 1997).

14.3.1.2 ROS Scavengers/Antioxidant Enzymes

As discussed earlier, the onset of abiotic stress leads to production of ROS species, causing oxidative stress. These have to be removed immediately by cellular antioxidant system (enzymes or antioxidant compounds) to protect the cell from toxic effect of free radicals. To date, several genes encoding anti-oxidative enzymes have been transferred to transgenic plants, as presented in Table 14.1. Overexpression of the bacterial catalase gene (*katE*) was found to enhance salinity stress tolerance in transgenic indica rice cultivar, whereas the same gene protected d1 protein translation during exposure of transgenic tobacco plants to strong light under salt-stressed condition (Moriwaki et al. 2008; Al-Taweel et al. 2007). Superoxide dismutase (SOD) is the first enzyme of ROS scavenging pathway, which converts the harmful superoxide radicals to less toxic peroxide species that is further reduced to oxygen and water by the action of catalase and peroxidases. Great efforts have been executed to develop salt stress tolerance in plants by transfer of various isoforms of genes encoding SOD. SenGupta et al. (1993a, b) reported that transgenic tobacco plants overexpressing *Cu/Zn-SOD* showed enhanced tolerance to oxidative stress by retaining maximum photosynthesis rate under stress condition. Other isoforms like Fe and Mn-SOD were also enhanced salt stress tolerance in transgenic *Arabidopsis* and tobacco plants (Van Camp et al. 1996; Wang et al. 2004). In another study, co-expression of SOD with other anti-oxidative enzymes like APX resulted in more effective scavenging of ROS and tolerance to multiple abiotic stresses in transgenic tobacco and potato (Lee et al. 2007, 2010; Shafi et al. 2017). Seeds of transgenic tobacco plants exhibited enhanced seed longevity and germination rates under stress condition (Lee et al. 2010). Whereas transgenic potato showed enhancement in lignification and starch biosynthesis with improved tolerance to salt stress (Shafi et al. 2017). Similarly other ROS scavenging enzymes like GST, GPX, GR, APX, DHAR, MDAR were overexpressed either singly, or in combination with each other, and provided remarkable tolerance to salt and osmotic stress in transgenic *Arabidopsis*, tobacco, tomato, pea and rice, as shown in Table 14.1 (Kavitha et al. 2010; Ushimaru et al. 2006; Qi et al. 2010; Roxas et al. 1997; Sun et al. 2010; Lee and Jo 2004). In a recent study, expression of a rice Lambda class of glutathione S-transferase (*OsGSTL2*) in *Arabidopsis* was shown to enhance tolerance to heavy metal as well as other abiotic stresses like cold, osmotic stress and salt (Kumar et al. 2013).

14.3.1.3 Polyamines

The level of polyamines (putrescin, spermine, spermidine, etc.) increases under abiotic stress condition, and they bind and stabilize cellular macromolecules like nucleic acids, proteins, cellular membranes, etc. due to their polycationic nature. They can also act as ROS scavengers and help to maintain ion homeostasis in the cell (Galston and Kaur-Sawhney 1990). Several reports have described polyamines as potential candidates for engineering salinity stress tolerance in plants (Table 14.1). The genes encoding major enzymes for polyamine biosynthesis Arginine decarboxylase (*ADC*), ornithine decarboxylase (*ODC*), S-Adenosyl methionine decarboxylase (*SAMDC*) and spermidine synthase (*SPDS*) from different organisms (mouse, human, Tritordeum and plants) have been transferred to *Arabidopsis*, rice, tobacco and pear, conferring tolerance to multiple abiotic stresses (Roy and Wu 2001; Roy and Wu 2002; Waie and Rajam 2003; Wen et al. 2008; Kasukabe et al. 2004). The expression of oat *ADC* under ABA-inducible promoter has been reported to enhance biomass in transgenic rice plants under salt stress condition, due to increase in putrescine levels (Roy and Wu 2001). Transgenic expression of human *SAMDC* increased the polyamine (spermidine and putrescine) biosynthesis in tobacco plants, conferring them tolerance to salt and osmotic stress (Waie and Rajam 2003).

14.3.1.4 Glyoxalase Pathway Enzymes

Methylglyoxal (MG) is a cytotoxic compound produced due to imposition of salinity stress, and can be detoxified by glyoxalase pathway enzymes—GLYI and GLYII in a two-step process. Prevention from MG accumulation by genetic manipulation of the two-step glyoxalase pathway is a promising strategy to increase stress tolerance in plants. To date, these enzymes have been over expressed in transgenic rice, tobacco and black gram, either alone or in combination with each other, and conferred enhanced salt tolerance to transgenic plants up to 200 mM NaCl (Singla-Pareek et al. 2003, 2006, 2008; Yadav et al. 2005; Bhomkar et al. 2008). Alvarez Viveros et al. (2013) co-expressed the Brassica *GLYI* and pearl millet *GLYII* in tomato that conferred salt tolerance to transgenic plants by decreasing oxidative stress. In a recent study, transgenic rice plants expressing both enzymes of glyoxalase pathway showed reduction in yield penalty and enhanced tolerance for multiple abiotic stresses, due to more efficient removal of MG, less ROS accumulation, and better protection of photosynthesis (Gupta et al. 2017).

14.3.1.5 LEA Proteins

Late Embryogenesis Abundant proteins (LEA) normally express in reproductive tissues during late seed development, but their expression also induces in vegetative tissues in adverse environmental conditions, by an unknown mechanism they protect the plants from adverse effect of these stresses. Several attempts have been performed to increase the salt tolerance in transgenic plants by overexpression of these LEA proteins (Table 14.1). For instance, constitutive expression of barley group-3 LEA protein—HVA1 led to accumulation of the protein in roots and leaves of transgenic rice plants and increased tolerance to drought and salinity (Xu et al. 1996). The same gene was overexpressed in mulberry and conferred tolerance to multiple abiotic stresses (Checker et al. 2012; Lal et al. 2008). A wheat LEA-related gene *DHN-5* encoding dehydrin protein conferred the osmotic and salt stress tolerance to *Arabidopsis* plants (Brini et al. 2007). In a recent study by Hu et al. (2016), *OsLEA4* overexpression resulted in enhanced tolerance to drought, high salt and heavy metal stress in transgenic rice plants.

14.3.1.6 Heat Shock Proteins

Heat shock proteins (HSPs) belong to a large gene family, some members are constitutively expressed and required for normal cellular functions like protein folding (act as molecular chaperons), while others are well known to be induced under abiotic stress conditions, and provide thermotolerance to plants. There are some experimental evidences for enhancement in osmotic and oxidative stress tolerance when these HSPs are overexpressed in transgenic *Arabidopsis* (Sun et al. 2001; Härndahl et al. 1999). More research efforts are needed to characterize this family for salinity stress tolerance.

14.3.1.7 Other Stress-Responsive Functional Proteins

A vast majority of novel stress-responsive genes have been identified and functionally characterized in different plant systems, whose expression were increased under different abiotic stress conditions. For instance, a cysteine protease gene from Salix conferred increased salt-stress tolerance, when overexpressed in transgenic *Arabidopsis* (Zheng et al. 2018). In a recent study, a cross-talk between abiotic stress and hormone metabolism was established, and knockdown of rice inflorescence meristem-specific cytokinin oxidase (*CKX*) resulted in increased cytokinin levels and enhanced rice grain yields under salt stress condition (Joshi et al. 2018). Soda et al. (2018) reported the role of rice intermediate filament (*OsIF*) for stabilization of photosynthetic machinery and yield under salinity and heat stress conditions. Some recent studies reported role of glycosyltransferase for improving abiotic stress tolerance. Overexpression of uridine diphosphate glycosyltransferase (*UGT76E12*) gene resulted in enhanced flavonoid accumulation in *Arabidopsis*, that subsequently led to better seed germination and growth under salt, mannitol and ABA treatments (Chen et al. 2019a; Li et al. 2018). Ferrochelatase-1 is a terminal enzyme of heme biosynthesis, which is induced under salt stress. Overexpression of *AtFC1* resulted in improved salt-stress tolerance by limiting sodium accumulation in transgenic *Arabidopsis* (Zhao et al. 2017). In another study, transgenic poplar hybrid plants (*Populus deltoides* × *Populus euramericana*) overexpressing drought sensitive 1 (*PtDRS1*) were shown to develop enhanced salt and drought tolerance (Mohammadi et al. 2018). Recently Yu et al. (2018) reported early-flowering and improved salt-stress tolerance in *Arabidopsis* by overexpression of apple nucleus-encoded thylakoid protein, Ycf3-interacting protein (*MdY3IP1*), while ectopic expression of ATP-independent pea phloem motor protein (forisome) (*PsSEO-F1*) in tobacco also resulted in increased salinity stress tolerance (Srivastava et al. 2016). There are several other genes such as GIGANTEA-like genes flowering time regulator (*GI a/b/c*), Polycystin, Lipoxygenase, Alpha-toxin and Triacylglycerol lipase) domain protein1 (*PLAT1*), salt- and drought-induced RING finger 1 (*SDIR1*), germin proteins and osmotin, which enhanced the tolerance to abiotic stresses when overexpressed in transgenic plants of rice, tobacco and *Arabidopsis*, as shown in Table 14.1 (Ke et al. 2017; Hyun et al. 2015; Lu et al. 2010; Tak and Mhatre 2013; Barthakur et al. 2001). The mechanism of action of all of these protein is still unknown.

14.3.2 Genes for Regulatory Proteins and Signaling Intermediates

The first response of plants to external environmental stimuli like excessive salt is signal perception, and relay of this information via signal transduction pathway that eventually leads to the final physiological response. In prokaryotes, two-component histidine kinase system act as osmosensor. Plants also contain similar type of osmosensory system. Urao et al. (1999) cloned a transmembrane histidine kinase like osmosensor (*AtHK1*) from *Arabidopsis*, which has sequence and function similarity to Yeast osmosensor—SLN1.

The downstream components of salt stress-signaling pathway includes genes for SOS pathway (*CBL4*, *CIPK24* and PM Na^+/H^+ antiporter as *SOS3*, *SOS2* and *SOS1*), Ca^{2+}-dependent protein kinases (*CDPK*s), Mitogen-activated protein kinases (*MAPK*s) and various Transcription factors (like *MYB*, *WRKY*, *NAC*, etc.), which ultimately result in end response like expression of stress-responsive genes. All these signaling pathways are very much complex, either specific to salt stress or constitutes an interconnected network, providing opportunity for cross-talk among different types of stresses. These signaling intermediate and regulatory proteins may act as potential candidates for engineering salt stress tolerance in plants. The transgenic approach for developing salt stress tolerance in plants via modulation of gene expression for salt-overlay sensitive (SOS) pathway (*SOS2* and *SOS3*) was successfully demonstrated in *Arabidopsis*, tobacco and maize, as shown in Table 14.2 (Pardo et al. 1998; Wang et al. 2007; Cheong et al. 2010; Tripathi et al. 2009; Yang et al. 2009). Transgenic approaches that utilized *SOS1* gene (plasma membrane Na^+/H^+ antiporter) to enhance salt stress have been described in Sect. 14.3.3 and Table 14.3.

In response to rapid increase in cytosolic Ca^{2+}, a cascade of protein phosphorylation and dephosphorylation initiates. The Ca^{2+}-dependent protein kinases (CDPKs) and MAP kinases are important components of this cascade, which might play important role as response regulator. Overexpression of a single gene encoding CDPK is reported to confer multiple abiotic stress tolerance (salt, drought and cold) in transgenic rice plants (Saijo et al. 2000). To date, various genes encoding MAP kinases (MAPKKK, MAPKK and MAPK) were cloned and overexpressed in transgenic *Arabidopsis*, rice and tobacco (Table 14.2), which have shown to provide enhanced tolerance to salinity and drought (Kovtun et al. 2000; Kumar and Sinha 2013; Xiong and Yang 2003; Zhang et al. 2011; Kong et al. 2011).

14.3.2.1 Transcription Factors

Transcription factors (TFs) are one of the most important regulators, that interact with the specific conserved cis-elements in the promoter region of stress-responsive genes, and in this way directly regulate the expression of a broad spectrum of downstream target genes. Due to multigenic nature of salt stress response, overexpression of TFs is a promising strategy for enhancing salt-stress tolerance in plants, since multiple

Table 14.2 Comprehensive list of transgenes encoding regulatory proteins or signaling intermediates involved in salinity stress tolerance

Gene transferred	Product	Source organism	Target plant	Transgene effect	Reference(s)
Transcription factors					
bZIP TFs					
ABP9	Basic leucine zipper transcription factor	*Zea mays*	*Gossypium hirsutum*	Enhanced tolerance to salt and drought	(Wang et al. 2017)
HaHB1 and AtHB13	Homeodomain-leucine zipper transcription factors	*Helianthus annuus, A. thaliana*	*A. thaliana*	Drought and salinity stress tolerance	(Cabello and Chan 2012)
OsbZIP71	Basic leucine zipper transcription factor	*Oryza sativa*	*Oryza sativa*	Drought and salinity	(Liu et al. 2014)
LrbZIP	Basic leucine zipper transcription factor	*Nelumbo nucifera (root)*	*N. tabacum*	Salinity stress tolerance	(Cheng et al. 2013)
TabZIP60	Basic leucine zipper transcription factor	*Triticum aestivum*	*A. thaliana*	Drought, salt, and freezing tolerance	(Zhang et al. 2015b)
GmbZIP1	Basic leucine zipper transcription factor	*Glycine max*	*N. tabacum, A. thaliana*	Drought, salinity, and cold	(Gao et al. 2011)
ZmbZIP72	Basic leucine zipper transcription factor	*Zea mays*	*A. thaliana*	Drought and salinity	(Ying et al. 2012)
YAP1	Basic leucine zipper transcription factor	*Saccharomyces cerevisiae*	*A. thaliana*	Enhanced salt and oxidative stress tolerance	(Zhao et al. 2009)

(continued)

Table 14.2 (continued)

Gene transferred	Product	Source organism	Target plant	Transgene effect	Reference(s)
MYB TFs					
MdoMYB121	MYB transcription factor	*Malus domestica*	*Solanum lycopersicum, Malus domestica*	High salinity, drought, and cold stresses	(Cao et al. 2013)
LcMYB1	MYB transcription factor	*Leymus chinensis*	*A. thaliana*	Salinity tolerance	(Cheng et al. 2013)
TaMYB3R1	MYB transcription factor	*Triticum aestivum*	*A. thaliana*	Drought and salinity tolerance	(Cai et al. 2015)
OsMYB91	MYB transcription factor	*Oryza sativa*	*Oryza sativa*	Salinity tolerance	(Zhu et al. 2015)
WRKY TFs					
GhWRKY34	Group III of cotton WRKY family TF	*Gossypium hirsutum*	*A. thaliana*	Enhances tolerance to salt	(Zhou et al. 2015)
MtWRKY76	WRKY transcription factor	*Medicago truncatula*	*Medicago truncatula*	Drought and salinity tolerance	(Liu et al. 2016)
ZmWRKY58	WRKY transcription factor	*Zea mays*	*Oryza sativa*	Drought and salinity tolerance	(Cai et al. 2014)
GhWRKY39	WRKY transcription factor	*Gossypium hirsutum*	*N. tabacum*	Salinity tolerance, disease resistance	(Shi et al. 2014)
AtbHLH17 and AtWRKY28	Basic helix-loop-helix and WRKY transcription factor	*A. thaliana*	*A. thaliana*	Salinity tolerance	(Babitha et al. 2013)

(continued)

Table 14.2 (continued)

Gene transferred	Product	Source organism	Target plant	Transgene effect	Reference(s)
TaWRKY93	WRKY transcription factor	*Triticum aestivum*	*A. thaliana*	Salinity, drought, and low temperature tolerance	(Qin et al. 2015)
NAC TFs					
PgNAC21	NAC transcription factor	*Pennisetum glaucum*	*A. thaliana*	Enhances tolerance to salt	(Shinde et al. 2019)
OsNAC6	NAC transcription factor	*Oryza sativa*	*Oryza sativa*	Improved tolerance to dehydration, high-salt stresses and blast disease	(Nakashima et al. 2007)
SNAC2	Stress-responsive NAC transcription factor	*Oryza sativa L. ssp japonica* (upland rice IRA109)	*Oryza sativa*	Enhanced drought and salt tolerance	(Hu et al. 2008)
BnNAC5	NAC transcription factor	*Brassica napus*	*A. thaliana*	Enhanced salt tolerance	(Zhong et al. 2012)
ONAC063	NAC transcription factor	*Oryza sativa*	*A. thaliana*	Salinity and osmotic tolerance	(Yokotani et al. 2009)
GmNAC20	NAC transcription factor	*Glycine max*	*A. thaliana*	Salinity and freezing tolerance	(Hao et al. 2011)
TaNAC29	NAC transcription factor	*Triticum aestivum*	*A. thaliana*	Drought and salinity tolerance	(Huang et al. 2015)
SNAC1	Stress-responsive NAC transcription factor	*Oryza sativa*	*Triticum aestivum*	Drought and salinity tolerance	(Saad et al. 2013)
OsNAP	NAC transcription factor	*Oryza sativa*	*Oryza sativa*	Cold, salinity, and drought tolerance	(Chen et al. 2014)

(continued)

Table 14.2 (continued)

Gene transferred	Product	Source organism	Target plant	Transgene effect	Reference(s)
DREB TFs					
SsDREB	Dehydration-responsive element-binding protein	*Suaeda salsa*	*Nicotiana tabacum*	Enhanced drought and salt tolerance	(Zhang et al. 2015a)
SlDREB1	Dehydration-responsive element-binding protein	*Solanum tuberosum*	*Solanum tuberosum*	Enhanced salt tolerance	(Bouaziz et al. 2015)
LcDREB2	Dehydration-responsive element-binding protein	*Leymus chinensis*	*A. thaliana*	Salinity tolerance	(Peng et al. 2013)
VrDREB2A	Dehydration-responsive element-binding protein	*Vigna radiata*	*A. thaliana*	Drought and salinity tolerance	(Chen et al. 2016)
LcERF054	Ethylene-responsive element-binding protein	*Lotus corniculatus*	*A. thaliana*	Salinity tolerance	(Sun et al. 2014)
TaERF3	Ethylene-responsive element-binding protein	*Triticum aestivum*	*Triticum aestivum*	Drought and salinity tolerance	(Rong et al. 2014)

(continued)

Table 14.2 (continued)

Gene transferred	Product	Source organism	Target plant	Transgene effect	Reference(s)
Signaling molecules					
OsMKK6	Mitogen-activated protein kinase kinase 6	*Oryza sativa*	*Oryza sativa (indica cultivar var. Pusa Basmati-1)*	Enhanced salt tolerance	(Kumar and Sinha 2013)
AtMKK2	Mitogen-activated protein kinase kinase 2	*A. thaliana*	*A. thaliana*	Enhanced drought and salt tolerance	(Teige et al. 2004)
MAPK5	Mitogen-activated protein kinase 5	*Oryza sativa*	*Oryza sativa*	Enhanced tolerance to drought, salt and cold stresses	(Xiong and Yang 2003)
MAPK44	Mitogen-activated protein kinase 44	*Oryza sativa*	*Oryza sativa*	Salt tolerance	(Jeong et al. 2006)
AtMEK1	MAPK kinase 1	*A. thaliana*	*A. thaliana*	Salt and dehydration tolerance	(Xing et al. 2007)
ZmMKK4	MAPK kinase kinase 4	*Zea mays*	*A. thaliana*	Salt and cold tolerance	(Kong et al. 2011)
GhMPK2	Mitogen-activated protein kinase 2	*Gossypium hirsutum*	*N. tabacum*	Salt and drought tolerance	(Zhang et al. 2011)
OsMSR2	Calmodulin-like protein	*Oryza sativa*	*A. thaliana*	Salt and drought tolerance	(Xu et al. 2011)
ZmCBL4/SOS3	Calcineurin B like protein-4 (salt overly sensitive protein 3)	*Zea mays*	*A. thaliana*	Enhanced salinity tolerance	(Wang et al. 2007)

(continued)

Table 14.2 (continued)

Gene transferred	Product	Source organism	Target plant	Transgene effect	Reference(s)
CBL5	Calcineurin B like protein-5 (calcium sensor)	*A. thaliana*	*A. thaliana*	Salt/osmotic or drought stress tolerance	(Cheong et al. 2010)
CIPK6	CBL-interacting protein kinase 6	*Cicer arietinum*	*N. tabacum*	Salt tolerance	(Tripathi et al. 2009)
SOS1, *SOS2*, *SOS3*	Salt overly sensitive proteins 1/2/3	*A. thaliana*	*A. thaliana*	Salt tolerance	(Yang et al. 2009)
AtCPK6	CBL-interacting protein kinase 6	*A. thaliana*	*A. thaliana*	Salt and drought tolerance	(Xu et al. 2010)
NDPK2	Nucleoside diphosphate kinase 2	*A. thaliana*	*A. thaliana*	Salt, cold and oxidative stress tolerance	(Moon et al. 2003)
SAPK4	SNF1-type serine-threonine protein kinase	*Oryza sativa*	*Oryza sativa*	Salt stress tolerance	(Diédhiou et al. 2008)
ZmSAPK8	SnRK2 protein kinase	*Zea mays*	*A. thaliana*	Salt tolerance	(Ying et al. 2011)
TaSnRK2.8	Sucrose nonfermenting 1-related protein kinase2	*Triticum aestivum*	*A. thaliana*	Tolerance to drought, salt and low temperature	(Zhang et al. 2010)
OsSIK1	Receptor-like kinase	*Oryza sativa*	*Oryza sativa*	Cold, drought and salt tolerance	(Ouyang et al. 2010)

(continued)

Table 14.2 (continued)

Gene transferred	Product	Source organism	Target plant	Transgene effect	Reference(s)
SAP1/11, OsRLCK253	A20/AN1 zinc-finger containing stress associated proteins 1/11, receptor-like cytoplasmic kinase	*Oryza sativa*	*A. thaliana*	Salt, water deficit tolerance	(Giri et al. 2011)
AtSAP13	A20/AN1 zinc-finger containing stress-associated protein 13	*A. thaliana*	*A. thaliana*	Multiple abiotic stresses tolerance	(Dixit et al. 2017)
AtPP2CG1	Protein phosphatase 2C group 1	*A. thaliana*	*A. thaliana*	Enhanced salt tolerance	(Liu et al. 2012)
OsPP1a	Protein phosphatase 1a	*Oryza sativa*	*Oryza sativa*	Salt stress tolerance	(Liao et al. 2016)

Table 14.3 Comprehensive list of transgenes encoding proteins involved in ion homeostasis during salinity stress

Gene transferred	Product	Source organism	Target plant	Transgene effect	Reference(s)
Ion homeostasis					
Aquaporins					
SiPIP3;1 and SiSIP1;1	Plasma membrane intrinsic proteins and small basic intrinsic proteins	*Setaria italica*	*Saccharomyces cerevisiae*	Dehydration and salt stress tolerance	(Singh et al. 2019)
MaPIP1;1	Plasma membrane intrinsic protein	*Musa acuminata*	*A. thaliana*	Enhanced water uptake, salt and drought tolerance	(Xu et al. 2014)
Ion transporters					
SOS1	Plasma membrane Na^+/H^+ antiporter	*A. thaliana*	*A. thaliana*	Enhanced salinity tolerance	(Shi et al. 2003)
CcSOS1	Plasma membrane Na^+/H^+ antiporter (salt overly sensitive 1)	*Chrysanthemum crassum*	*Saccharomyces cerevisiae*	Enhanced salinity tolerance	(Song et al. 2012)
HtSOS1	Plasma membrane Na^+/H^+ antiporter (salt overly sensitive 1)	*Helianthus tuberosus*	*Saccharomyces cerevisiae and Oryza sativa L., ssp. Japonica.cv. Nipponbare)*	Enhanced salinity tolerance	(Li et al. 2014)
SOS1-RNAi	Plasma membrane Na^+/H^+ antiporter (salt overly sensitive 1)	*Thellungiella halophila*	*Thellungiella halophila*	Loss of halophytism	(Oh et al. 2009)
SOS1-RNAi	Plasma membrane Na^+/H^+ antiporter (salt overly sensitive 1)	*Solanum lycopersicum*	*Solanum lycopersicum*	Loss of salt tolerance	(Olías et al. 2009)

(continued)

Table 14.3 (continued)

Gene transferred	Product	Source organism	Target plant	Transgene effect	Reference(s)
SOD2	Plasma membrane Na^+/H^+ antiporter	*Schizosaccharomyces pombe*	*Oryza sativa*	Increased salt and drought tolerance	(Zhao et al. 2006)
HAL1	HALotolerance protein	*Saccharomyces cerevisiae*	*Cucumis melo*	Enhanced salt tolerance	(Bordas et al. 1997)
HAL1	HALotolerance protein	*Saccharomyces cerevisiae*	*Solanum lycopersicum*	Enhanced salt tolerance	(Rus et al. 2001; Gisbert et al. 2000)
AtNHX1	Vacuolar Na^+/H^+ antiporter	*A. thaliana*	*A. thaliana*	Enhanced salt tolerance	(Apse et al. 1999)
AtNHX1	Vacuolar Na^+/H^+ antiporter	*A. thaliana*	*Gossypium arboreum*	Enhanced drought and salt tolerance	(He et al. 2005)
AtNHX1	Vacuolar Na^+/H^+ antiporter	*A. thaliana*	*Solanum lycopersicum*	Enhanced salt tolerance	(Zhang and Blumwald 2001; Leidi et al. 2010)
AtNHX1	Vacuolar Na^+/H^+ antiporter	*A. thaliana*	*Brassica napus*	Enhanced salt tolerance	(Zhang et al. 2001)
AtNHX1	Vacuolar Na^+/H^+ antiporter	*A. thaliana*	*Triticum aestivum*	Enhanced salt tolerance	(Xue et al. 2004)
AtNHX1	Vacuolar Na^+/H^+ antiporter	*A. thaliana*	*Festuca arundinacea*	Enhanced salt tolerance	(Zhao et al. 2007)
AtNHX1	Vacuolar Na^+/H^+ antiporter	*A. thaliana*	*Fagopyrum esculentum*	Enhanced salt tolerance	(Chen et al. 2008)
AtNHX3	Vacuolar Na^+/H^+ antiporter	*A. thaliana*	*Beta vulgaris*	Enhanced salt tolerance	(Liu et al. 2008)
AtNHX1	Vacuolar Na^+/H^+ antiporter	*A. thaliana*	*Actinidia deliciosa*	Enhanced salt tolerance	(Tian et al. 2010)

(continued)

Table 14.3 (continued)

Gene transferred	Product	Source organism	Target plant	Transgene effect	Reference(s)
AtNHX1	Vacuolar Na^+/H^+ antiporter	*A. thaliana*	*Zea mays*	Enhanced salt tolerance	(Xiao-Yan et al. 2004)
AtNHX1	Vacuolar Na^+/H^+ antiporter	*A. thaliana*	*Arachis hypogaea*	Salt tolerance	(Asif et al. 2011)
AmNHX2	Vacuolar Na^+/H^+ antiporter	*Ammopiptanthus mongolicus*	*A. thaliana*	Salt and drought tolerance	(Wei et al. 2011)
AgNHX1	Vacuolar Na^+/H^+ antiporter	*Atriplex gmelini*	*Oryza sativa*	Enhanced salt tolerance	(Ohta et al. 2002)
GhNHX1	Vacuolar Na^+/H^+ antiporter	*Gossipium hirsutum*	*N. tabacum*	Salt tolerance	(Wu et al. 2004)
NHX1	Vacuolar Na^+/H^+ antiporter	*Hordeum brevisubulatum*	*N. tabacum*	Salt tolerance	(Lu et al. 2005)
NHX2	Vacuolar Na^+/H^+ antiporter	*Hordeum vulgare*	*Solanum tuberosum*	Salt tolerance	(Bayat et al. 2010)
NHX1	Vacuolar Na^+/H^+ antiporter	*Pennisetum glaucum*	*Brassica juncea*	Salt tolerance	(Rajagopal et al. 2007)
NHX1	Vacuolar Na^+/H^+ antiporter	*Pennisetum glaucum*	*Oryza sativa*	Salt tolerance	(Verma et al. 2007)
AeNHX1	Vacuolar Na^+/H^+ antiporter	*Agropyron elongatum*	*A. thaliana/Festuca arundinacea*	Salt tolerance	(Qiao et al. 2007)
NHX1	Vacuolar Na^+/H^+ antiporter	*Aleuropus littoralis*	*N. tabacum*	Salt tolerance	(Zhang et al. 2008)
NHX1	Vacuolar Na^+/H^+ antiporter	*Salicornia brachiata*	*N. tabacum*	Salt tolerance	(Jha et al. 2010)
SsNHX1	Vacuolar Na^+/H^+ antiporter	*Salsola soda*	*Medicago sativa*	Salt tolerance	(Li et al. 2010)
NHX1	Vacuolar Na^+/H^+ antiporter	*Thellungiella halophila*	*A. thaliana*	Salt tolerance	(Wu et al. 2009)
HcNHX1	Vacuolar Na^+/H^+ antiporter	*Halostachys caspica*	*A. thaliana*	Salt tolerance	(Guan et al. 2011)

(continued)

Table 14.3 (continued)

Gene transferred	Product	Source organism	Target plant	Transgene effect	Reference(s)
NHX	Vacuolar Na^+/H^+ antiporter	*Oryza sativa*	*Oryza sativa*	Enhanced salt tolerance	(Amin et al. 2016)
PgNHX1 and AtAVP1	Vacuolar Na^+/H^+ antiporter and H^+-pyrophosphatase	*Pennisetum glaucum, A. thaliana*	*Solanum lycopersicum*	Enhanced salinity tolerance	(Bhaskaran and Savithramma 2011)
AtNHX1 and AtSOS1	Vacuolar and plasma membrane Na^+/H^+ antiporter	*A. thaliana*	*A. thaliana*	Enhanced tolerance to combined of heat and salt stresses	(Pehlivan et al. 2016)
TaNHX1 and TVP1	Vacuolar Na^+/H^+ antiporter and H^+ pyrophosphatase	*Triticum aestivum*	*A. thaliana*	Enhanced drought and salt tolerance	(Brini et al. 2007)
SsNHX1 and AtAVP1	Vacuolar Na^+/H^+ antiporter and vacuolar H^+ pyrophosphatase	*Arabidopsis, Suaeda salsa*	*Oryza sativa*	Salt tolerance better than single gene alone	(Zhao et al. 2006)
AVP1	Vacuolar H^+ pyrophosphatase	*A. thaliana*	*Saccharum officinarum* (cv. CP-77-400)	Drought and salinity stress tolerance	(Kumar et al. 2014)
AVP1	Vacuolar H^+ pyrophosphatase	*A. thaliana*	*A. thaliana*	Drought- and salt tolerance	(Gaxiola et al. 2001)
AVP1	Vacuolar H^+ pyrophosphatase	*A. thaliana*	*Medicago sativa*	Salt tolerance	(Bao et al. 2009)
AVP1	Vacuolar H^+ pyrophosphatase	*A. thaliana*	*Gossypium hirsutum* (cv. Coker 312)	Salt and drought tolerance	(Pasapula et al. 2011)
AVP1	Vacuolar H^+ pyrophosphatase	*A. thaliana*	*Agrostis stolonifera*	Enhanced salt tolerance	(Li et al. 2010)

(continued)

Table 14.3 (continued)

Gene transferred	Product	Source organism	Target plant	Transgene effect	Reference(s)
H^+-*PPase*	Membrane H^+-pyrophosphatase	*Rhodospirillum rubrum*	*N. tabacum*	Salt tolerance	(D'yakova et al. 2006)
SsVP1	Vacuolar H^+ pyrophosphatase	*Saueda salsa*	*A. thaliana*	Salt and drought tolerance	(Guo et al. 2006)
ScVP	Vacuolar H^+ pyrophosphatase	*Suaeda corniculate*	*A. thaliana*	Salt, saline-alkali and drought tolerance	(Liu et al. 2011)
TsVP	Vacuolar H^+ pyrophosphatase	*Thellungiella halophila*	*N. tabacum*	Enhanced salinity tolerance	(Gao et al. 2006)
TsVP	Vacuolar H^+ pyrophosphatase	*Thellungiella halophila*	*Zea mays*	Enhanced salinity tolerance	(Li et al. 2008)
TsVP	Vacuolar H^+ pyrophosphatase	*Thellungiella halophila*	*Gossypium hirsutum* (cv. Luyuan 890)	Enhanced salinity tolerance	(Lv et al. 2008)
GmHKT1	High-affinity potassium transporter	*Glycine max*	*N. tabacum*	Enhanced tolerance to salt stress	(Chen et al. 2011)
SbVPPase	Vacuolar H^+ pyrophosphatase	*Sorghum bicolor*	*Bacopa monnieri*	Enhanced tolerance to salt stress	(Ahire et al. 2018)
PpENA1	Sodium pumping ATPase	*Physcomitrella patens*	*Oryza sativa*	Salt tolerance	(Jacobs et al. 2011)

stress-responsive genes can be regulated simultaneously by modulation of expression of a single TF. Meanwhile, TFs themselves are regulated at transcriptional and post-translational level, thereby constitutes a complex regulatory network for salinity stress signal transduction (Wang et al. 2016b). In the last few decades, considerable research efforts have been made to enhance salinity stress tolerance in model and crop plants through genetic engineering of different TFs, such as bZIP, MYB, WRKY, NAC, AP2/EREBP, etc., which are shown to be involved in plant abiotic stress responses. But still a limited number of TFs have been well characterized up to now, and most of the TFs family members have yet to be identified. Some recent studies are summarized in the following section and in Table 14.2.

AP2/EREBP Transcription Factor

AP2/EREBP constitute a large family of TFs in plants, and contains a characteristic highly conserved DNA-binding domain (ERF or AP2/ethylene-responsive element-binding factor binding domain) that specifically interacts with DRE/CRT (dehydration-responsive element/C-repeat element) cis-acting elements in the promoter of downstream target genes. The members of this family have been shown to play important role not only in various growth and developmental processes in plants, but also in abiotic/biotic stress responses (Wang et al. 2016b). AP2/EREBP family is further classified into four major sub-families: DREB (dehydration-responsive element-binding protein), ERF, AP2 (Apetala 2) and RAV (related to ABI3/VP1). DREB subfamily is further sub-divided into two groups: DREB1/CBF (C-repeat binding factor) that is commonly involved in cold-stress response, and DREB2 that is induced by heat, salt and drought stress. The DREB and ERF subfamilies have been thoroughly investigated for their role in plant stress responses. Overexpression of *DREB2A* from mung bean resulted in improved tolerance to drought and high-salt in transgenic *Arabidopsis* (Chen et al. 2016), while DREB TFs from other species like Suaeda, potato and sheep-grass conferred enhanced salinity and oxidative-stress tolerance in transgenic tobacco, potato and *Arabidopsis*, respectively, without hampering normal growth and development of the transgenic plants (Zhang et al. 2015a; Bouaziz et al. 2015; Peng et al. 2013). Recently, the ERF TFs from wheat and Lotus have been shown to promote salt and drought stress tolerance when overexpressed in transgenic wheat and *Arabidopsis* (Rong et al. 2014; Sun et al. 2014). Some AP2/ERF TFs are involved in various hormonal signaling pathway (like ethylene, jasmonic acid, salicylic acid), and may function both in biotic and abiotic stresses (Wang et al. 2016b).

MYB Transcription Factor

MYB is also a large family of TFs widely distributed in plants, which contains highly conserved MYB DNA-binding domain repeats at N-terminus and a variable activation domain at C-terminus. This family is divided into four sub-classes (on the basis of no. of MYB repeats): 1R-MYB, R2R3-MYB, R1R2R3-MYB and 4R-MYB. In plants, R2R3-MYB having 2 MYB-repeats is the most common group of TFs. The members of MYB TF family participate in various cellular metabolic reactions and cell-cycle, hormonal metabolism and stress-signal transduction pathway (Wang et al. 2016b). To date, a variety of MYB TFs have been shown to be induced by different abiotic stresses, and functionally characterized for their involvement in salt stress response (Table 14.2). A recent study indicated increase in tolerance to high salinity in transgenic rice overexpressing R2R3-type MYB gene—*OsMYB91* (Zhu et al. 2015). Overexpression of apple *MYB121* resulted in tolerance to multiple abiotic stresses, including salinity, in transgenic tomato (Cao et al. 2013). Whereas wheat *MYB3R1* exhibited a pleiotropic effects on development and osmotic stress response in transgenic *Arabidopsis* (Cai et al. 2015). Very recently Chen et al. (2019b) reported the role of *MdMYB46* to enhance salt and osmotic stress tolerance in transgenic *Arabidopsis* and apple. This TF was shown to directly bind to the promoter of lignin biosynthesis-related gene and other stress-responsive genes, thus not only promoting lignin deposition and secondary cell wall biosynthesis, but also directly activating stress responsive signals in transgenic plants (Chen et al. 2019b).

WRKY Transcription Factor

Like other TF families, WRKY TFs also comprise a large superfamily of plants, characterized by the presence of conserved WRKY DNA-binding domain and C2H2 or C2HC zinc-finger motif at N- and C-terminus, respectively. It is further divided into three sub-classes: Group I that contains 2 WRKY domains and 1 C2H2 zinc-finger motif, Group II members contain 1 WRKY domain and 1 C2H2 zinc-finger motif, and Group III that contains 1 WRKY domain and 1 C2HC zinc-finger motif. The members of WRKY TF family are found to be involved in various vegetative and reproductive plant growth and development processes, as well as responses to multiple biotic and abiotic stresses (Wang et al. 2016b). To date, several members of this family have been functionally characterized, for example, overexpression of a group III member of cotton WRKY family—*WRKY34* in transgenic *Arabidopsis* resulted in enhanced tolerance for salt stress (Zhou et al. 2015). In a separate study by Liu et al. (2016), overexpression of Medicago *WRKY76* exhibited significant salt and drought stress tolerance in transgenic plants. Co-expression of *Arabidopsis WRKY28* and *bHLH17* conferred enhanced resistance to salt stress in transgenic *Arabidopsis* plants (Babitha et al. 2013). Similarly, other WRKY members from wheat, cotton and maize also conferred enhanced tolerance to multiple abiotic/biotic stresses in transgenic *Arabidopsis*, tobacco and rice (Qin et al. 2015; Shi et al. 2014; Cai et al. 2014).

NAC Transcription Factor

The members of NAC (NAM-No apical meristem, ATAF-*Arabidopsis* transcription activation factor, CUC-cup-shaped cotyledon) superfamily are widely distributed in plants, and contain a highly conserved DNA-binding NAC-domain at their N-terminus, and a variable transcriptional activator/repressor domain at C-terminus. They can form homo or heterodimers with other proteins, by the help of N-terminus NAC domain (Wang et al. 2016b). Like other TFs, the members of NAC TF family are known to be involved in various plant growth and development processes like cell division, apical meristem formation, flower development, etc. These are also found to be differentially expressed in a tissue or developmental-stage specific manner, in response to multiple abiotic or biotic stress, and can be potential candidates for genetic engineering for stress tolerance. In a recent study, Pearl millet stress-responsive NAC transcription factor *PgNAC21* conferred significant increase in salt stress tolerance, when overexpressed in transgenic *Arabidopsis* (Shinde et al. 2019). Overexpression of *OsNAC6* resulted in improved tolerance to dehydration and high-salinity stress as well as blast disease in transgenic rice plants (Nakashima et al. 2007). Other NAC family members were also shown to exhibit enhanced salt and drought stress tolerance in transgenic plants of wheat, rice and *Arabidopsis*, via ABA-dependent or ABA-independent pathway (Saad et al. 2013; Huang et al. 2015; Chen et al. 2014).

bZIP Transcription Factor

Basic leucine zipper (bZIP) TFs comprise a large superfamily, which is widely distributed in plants. This family is characterized by the presence of a basic highly conserved N-terminus DNA-binding bZIP domain, and a C-terminus leucine rich motif that is responsible for dimerization process. The members of this family have been shown to play important role in various growth and developmental processes, as well as responses to multiple abiotic stress (Wang et al. 2016b). These are very less functionally characterized plant TFs, and shown to regulate the expression of downstream target genes, possibly in ABA-dependent manner. Cabello and Chan (2012) have shown that overexpression of homeodomain-leucine zipper transcription factors *HaHB1* and *AtHB13* conferred tolerance to drought and salinity stresses in transgenic *Arabidopsis*, via the induction of membrane-stabilizing proteins. Recently, overexpression of a maize bZIP transcription factor (*ABP9*) resulted in enhanced tolerance to salt and drought in transgenic cotton (Wang et al. 2017). Various other members of bZIP TF family were also shown to enhance multiple abiotic stress tolerances in transgenic rice and tobacco (Table 14.2).

In addition to the TF families described above, some novel zinc finger proteins (ZFPs) were also reported to be involved in abiotic stress responses (Xu et al. 2008; Sun et al. 2010), whose overexpression resulted in increased tolerance to salt and drought stress in transgenic rice plants. In a recent study, rice transcription factor *OsMADS25* have been shown to modulate root growth and confers salinity tolerance via the ABA-mediated regulatory pathway and ROS scavenging (Xu et al. 2018).

14.3.2.2 Other Signaling Molecules

As we know, protein phosphorylation and dephosphorylation (executed by protein kinases and phosphatases, respectively) constitute important steps of the stress signal transduction pathway inside the cell. Besides MAP kinases, some other protein kinases like Nucleoside diphosphate kinase 2 (NDPK2), Sucrose nonfermenting 1-related protein kinase2 (SnRK2.8), SNF1-type serine-threonine protein kinase (SAPK4), SnRK2 protein kinase (SAPK8), Receptor-like kinase (RLK, SIK1) and receptor-like cytoplasmic kinase (RLCK253) have been shown to play important role in providing tolerance to multiple abiotic stresses in transgenic plants of rice and *Arabidopsis* (Moon et al. 2003; Diédhiou et al. 2008; Ying et al. 2011; Zhang et al. 2010; Ouyang et al. 2010; Giri et al. 2011). These kinases may directly interact with the stress-induced MAPKs, or other regulatory molecules like A20/AN1 zinc-finger containing stress-associated Proteins (SAP) to maintain cellular redox homeostasis. The *Arabidopsis* and rice genome contain 14 and 18 members of SAP gene family, respectively. These novel SAP proteins showed differential expression in response to multiple environmental stresses (such as cold, salt, drought, wounding, heavy metals, etc.), and exhibited improved tolerance to salt, drought stress and toxic metals like arsenic (As), cadmium (Cd), and zinc (Zn), as well as pathogen (*Pseudomonas syringii*) attack in transgenic rice, *Arabidopsis* and tobacco (Giri et al. 2011; Tyagi et al. 2014; Dansana et al. 2014; Dixit et al. 2017). The SAP proteins from *Arabidopsis*, wheat and rice have been suggested to possess E3 Ubiquitin ligase activity and may act as redox-sensor (Kang et al. 2011; Zhang et al. 2017; Ströher et al. 2009; Jha et al. unpublished data). In addition to phosphorylation, dephosphorylation of proteins is also important for relay of stress signals in plants. In this context, protein phosphatase-1a (*OsPP1a*) and protein phosphatase 2C Group 1 (*AtPP2CG1*) were shown to positively regulate salinity stress tolerance in transgenic rice and *Arabidopsis*, in ABA-dependent manner (Liao et al. 2016; Liu et al. 2012).

14.3.3 Genes for Ion Homeostasis

Under the salt stress condition, plants try to maintain ion homeostasis inside the cell by exclusion of excessive toxic ions or compartmentalize them into vacuoles. Sodium/proton transporters (Na^+/H^+ antiporter, or NHX) are the key proteins for these mechanisms. Efficient vacuolar sequestration of Na^+ ions is the major strategy adapted by the halophytes, which helps them to survive under highly saline conditions. The *Arabidopsis* transgenic plants constitutively overexpressing *AtNHX1* showed an increased tolerance to NaCl up to 200 mM. They were able to grow better than untransformed control plants under saline condition, due to high transcript, protein levels and activity of Na^+/H^+ antiporter in transgenic plants (Apse et al. 1999). In another important study by Zhang and Blumwald (Zhang and Blumwald 2001), the same gene was overexpressed in transgenic tomato plants that exhibited growth, flowering and fruiting at salt concentrations up to 200 mM. Most importantly, they accumulated excessive NaCl in foliage but not in fruits, maintaining the fruit quality of transgenic tomato. Leidi et al. (2010) reported potassium compartmentation in vacuoles of transgenic tomato plants, mediated by *AtNHX1* exchanger. To date, *NHX1* gene from different plant species other than *Arabidopsis* have been overexpressed in different homologous or heterologous plant systems, like cotton, wheat, maize, rice, brassica, etc. (Wu et al. 2004; Jha et al. 2010; Amin et al. 2016; Li et al. 2010; Rajagopal et al. 2007), the detailed account of which is presented in Table 14.3.

In another study, the yeast ion transporter *HAL1* was overexpressed in tomato and cucurbita (Bordas et al. 1997; Rus et al. 2001; Gisbert et al. 2000). Transgenic tomato plants exhibited improved salt tolerance and increase in fruit yield as compared to untransformed plants, due to high K^+/Na^+ ratio under salt stress condition (Rus et al.

2001). Vacuolar H^+-translocating pyrophosphatase (AVP1) are also known to contribute to proton electrochemical potential gradient that energizes Na^+ ions sequestration into the vacuoles. Overexpression of *AtAVP1* has been reported in several transgenic plant systems like sugarcane, alfalfa and cotton, providing enhanced salt and drought tolerance (Kumar et al. 2014; Pasapula et al. 2011; Bao et al. 2009). Similar H(+)-pyrophosphatase gene(s) from halophytes *Thellungiella halophila* and *Suaeda* spp. were transferred to *Arabidopsis*, maize, cotton, tobacco showing improvement in growth and photosynthetic performance and enhanced salt stress tolerance over non-transgenic control plants (Liu et al. 2011; Gao et al. 2006; Li et al. 2008; Lv et al. 2008). Co-expression of *NHX1* gene from pearl millet or *Suaeda* spp. along with the gene encoding *AtAVP1* resulted in more efficient sequestration of sodium ions, reduced cytosolic Na^+ concentrations and improved tissue tolerance in transgenic tomato and rice (Bhaskaran and Savithramma 2011; Zhao et al. 2006). In a recent study, the vacuolar H^+-pyrophosphatase gene (*VPPase*) from Sorghum conferred salt tolerance in transgenic Brahmi (Ahire et al. 2018). The sodium pumping ATPase (*ENA1*) from a moss showed higher biomass in transgenic rice under salinity stress condition (Jacobs et al. 2011). Moreover, overexpression of a novel soybean gene encoding high-affinity potassium transporter (*HKT1*) modulated Na^+ and K^+ transport and enhanced salt tolerance in transgenic tobacco plants (Chen et al. 2011). Recently overexpression of *HKT1;1* from hybrid poplar (*Populus deltoides* × *Populus euramericana*) is reported to improve salt tolerance in *Populus davidiana* × *Populus bolleana*, along with better relative growth rate, and improved efficiency of antioxidant systems in transgenic plants (Xu et al. 2018). The gene for *SOS1* encoding a putative plasma membrane Na^+/H^+ antiporter was isolated from *Arabidopsis*, Chrysanthemum and Jerusalem artichoke and functionally characterized for imparting salinity stress tolerance (Shi et al. 2003; Song et al. 2012; Li et al. 2014), which was further supported by RNA interference studies for *SOS1*, resulting in loss of salt tolerance in tomato and *Thellungiella* (Olías et al. 2009; Oh et al. 2009).

Aquaporins or water channels are specialized proteins that regulate the water flux across the plasma membrane. Plants contain various cell-type or tissue-specific isoforms of aquaporins: plasma membrane intrinsic proteins (PIPs), tonoplast intrinsic proteins (TIPs), NOD26-like intrinsic proteins (NIPs), and small basic intrinsic proteins (SIPs). Some experimental evidences showed their role in biotic and abiotic stress responses. In a recent study, overexpression of *PIP3;1* and *SIP1;1* from foxtail millet suggested their involvement in abiotic stress response (Singh et al. 2019), whereas overexpression of a banana aquaporin gene *PIP1;1* enhanced water uptake in *Arabidopsis* leading to salt and drought tolerance (Xu et al. 2014).

14.4 Conclusions and Future Perspective

To satisfy the requirements of exponentially increasing population of the world, there is an urgent requirement to transform the current farming practices to achieve sustainable food security at global level. Future research programs should focus on the challenges imposed by different environmental stresses, along with developing novel strategies to combat the impact of those abiotic and biotic stresses on agriculture.

In the last few years, there has been immense advancements in the field of transgenic technology to raise stress-tolerant plants. Despite the fact that salinity is a quite difficult trait to manipulate due to its multigenic nature, genetic engineering of a single gene is proving to be a useful tool in achieving significant salt stress tolerance in plants. This chapter summarizes the recent progress in genetic engineering of salt stress-responsive genes, and their possible use to develop salt-stress tolerance in transgenic plants. The use of new tools and techniques in genetic engineering, such as gene pyramiding and RNAi, as well as the recent breakthroughs in molecular biology like Genome editing by CRISPR/Cas9 technology have facilitated to understand the accurate function of stress-responsive genes, thus precise engineering of stress tolerance in plants (Nongpiura et al. 2016).

Although most of the studies have been performed in model plants such as *Arabidopsis* and tobacco under controlled greenhouse conditions, but they provide a clear vision about the key role of important genes in response to salt stress. However, results from these studies in model species cannot be extrapolated to crop plants. Rice and tomato are advocated as model crop plants due to availability of genome information and well-established transformation protocols. But still there is a critical requirement to assess the function of promising genes in other non-model economically important crop plants and validation under field conditions, as in nature various environmental stresses act together in a combined way, thus provoking the deleterious impact to a large extent, in comparison to a single stress factor usually studied under controlled laboratory conditions.

A comprehensive understanding of molecular mechanism of stress response is imperative for identification of promising genes, which can be used for genetic engineering for salt stress tolerance in plants. Substantial progress has been made in this field by means of functional genomics tools (Nongpiura et al. 2016). With the advent of fast and cost-effective next generation sequencing (NGS) technologies for whole genome and transcriptome sequencing, as well as advancement of other "omics" approaches like proteomics and metabolomics, it is now possible to decipher the molecular mechanism of salinity stress response in different plants, and to identify the promising genes responsible for imparting stress tolerance in salt-tolerant genotypes. These genes from wild tolerant genotype could further be explored for their tolerance potential by overexpression in sensitive crop plants. For example, two rice genotypes with contrasting tolerance to salinity stress (salt-sensitive IR64 and salt tolerant Pokkali) were used to unravel the differentially expressed proteins involved in salt stress tolerance (Lakra et al. 2017). A similar study was conducted in our lab using comparative proteomics and transcriptomics approaches for two pearl millet genotypes with contrasting salt-stress tolerance. The differences in proteome or transcriptome profiles of these two genotypes were investigated, and some novel candidate genes have been identified that may be responsible for imparting stress tolerance (Jha et al., unpublished data). Similarly, we have also identified some promising genes from a xero-halophytic shrub—Atriplex using the similar approaches (Jha et al., unpublished data). The genes or proteins identified in these studies would be further characterized for developing salt-stress tolerance in important crop plants.

Although manipulation of single gene has great potential for raising stress tolerant transgenic plants, concerns have been raised for using single gene for engineering salinity stress tolerance due to its multigenic quantitative nature. The use of TFs might be a potential solution to overcome this problem, as they act as master regulator to regulate a vast array of stress-responsive genes simultaneously. But still the regulatory mechanism of individual TF and its upstream and downstream interacting partners are not recognized completely. Future research efforts are needed to decipher these regulatory mechanisms and protein interactions.

The constitutive overexpression of some genes may cause adverse impact on vegetative or reproductive development in transgenic plants such as dwarfism, delayed flowering and reduction in yield in unstressed condition. The use of stress-inducible promoter may overcome this problem, which can protect the cell from toxic effect of certain transgene products, and to save the energy currency of the cell used for synthesis of these compounds when they are not required in unstressed conditions.

In field conditions, the plants are exposed to multiple abiotic and biotic stresses simultaneously. Therefore, there is a need to identify commonly regulated genes required for universal stress response and represents common points of cross-talk among different signaling pathways. Genetic manipulation of those genes would be a more potent strategy for developing multiple stress tolerance in plants as compared to manipulation of gene(s) for individual signaling pathway.

Until now several key genes have been identified and functionally characterized for salinity stress tolerance in plants. Nevertheless, there is a need to design tissue and cell-type specific expression of transgene, and evaluation of other important parameters (like yield) of transgenic plants. With the increase in availability of plant genome sequences, significant progress has been made to unravel the salt stress signaling pathways, but there is still much more to be determined. For instance, there is a need to determine the exact role of small RNA (micro RNA) and Long non-coding RNA (LncRNA), which are now emerging as novel regulators for plant abiotic stress responses.

Taken together, future research should focus on absolute understanding of regulatory mechanisms of plant abiotic stress responses and their cross-talks, which would help to generate precisely engineered crop plants for multiple abiotic stress-tolerance with better productivity and yield.

Acknowledgements The author thankfully acknowledges Department of Science and Technology, Government of India for DST-SERB Young Scientist grant (SB/YS/LS-39/2014); University Grants Commission, Government of India for UGC-BSR Start-up research grant (F.30-50/2014/BSR); UGC-CAS program and DST-FIST program in Department of Botany, J.N.V. University, Jodhpur, India.

Conflict of Interest The author declares no financial or commercial conflict of interest.

References

Abebe T, Guenzi AC, Martin B, Cushman JC (2003) Tolerance of mannitol-accumulating transgenic wheat to water stress and salinity. Plant Physiol 131:1748–1755

Ahire ML, Anil Kumar S, Punita DL, Mundada PS, Kavi Kishor PB, Nikam TD (2018) The vacuolar proton pyrophosphatase gene (*SbVPPase*) from the *Sorghum bicolor* confers salt tolerance in transgenic Brahmi [*Bacopa monnieri* (L.) Pennell]. Physiol Mol Biol Plants 24(5):809–819. https://doi.org/10.1007/s12298-018-0586-4

Al-Taweel K, Iwaki T, Yabuta Y, Shigeoka S, Murata N, Wadano A (2007) A bacterial transgene for catalase protects translation of d1 protein during exposure of salt-stressed tobacco leaves to strong light. Plant Physiol 145:258–265

Alvarez Viveros MF, Inostroza-Blancheteau C, Timmermann T, González M, Arce-Johnson P (2013) Overexpression of *GlyI* and *GlyII* genes in transgenic tomato (*Solanum lycopersicum* Mill.) plants confers salt tolerance by decreasing oxidative stress. Mol Biol Rep 40:3281–3290

Amin U, Biswas S, Elias SM, Razzaque S, Haque T, Malo R, Seraj ZI (2016) Enhanced salt tolerance conferred by the complete 2.3 kb cDNA of the rice vacuolar Na^+/H^+ antiporter gene compared to 1.9 kb coding region with 5′ UTR in transgenic lines of rice. Front Plant Sci 7:14

Apse MP, Aharon GS, Snedden WA, Blumwald E (1999) Salt tolerance conferred by overexpression of a vacuolar Na^+/H^+ antiporter in *Arabidopsis*. Science 285:1256–1258

Asif MA, Zafar Y, Iqbal J, Iqbal MM, Rashid U, Ali GM et al (2011) Enhanced expression of *AtNHX1*, in transgenic groundnut (*Arachis hypogaea* L.) improves salt and drought tolerance. Mol Biotechnol 49(3):250–256. https://doi.org/10.1007/s12033-011-9399-1

Babitha KC, Ramu SV, Pruthvi V, Mahesh P, Nataraja KN, Udayakumar M (2013) Co-expression of *AtbHLH17* and *AtWRKY28* confers resistance to abiotic stress in *Arabidopsis*. Transgenic Res 22:327–341. https://doi.org/10.1007/s11248-012-9645-8

Badawi GH, Kawano N, Yamauchi Y, Shimada E, Sasaki R, Kubo A, Tanaka K (2004) Overexpression of ascorbate peroxidase in tobacco chloroplasts enhances the tolerance to salt stress and water deficit. Physiol Plant 121:231–238

Bao AK, Wang SM, Wu GQ, Xi JJ, Zhang JL, Wang CM (2009) Overexpression of the *Arabidopsis* H^+-*PPase* enhanced resistance to salt and drought stress in transgenic alfalfa (*Medicago sativa* L.). Plant Sci 176:232–240

Bao Y, Zhao R, Li F, Tang W, Han L (2011) Simultaneous expression of *Spinacia oleracea* chloroplast choline monooxygenase (*CMO*) and betaine aldehyde dehydrogenase (*BADH*) genes contribute to dwarfism in transgenic *Lolium perenne*. Plant Mol Biol Rep 29:379–388. https://doi.org/10.1007/s11105-010-0243-8

Barthakur S, Babu V, Bansal KC (2001) Overexpression of osmotin induces proline accumulation and confers tolerance to osmotic stress in transgenic tobacco. J Plant Biochem Biotechnol 10:31–37

Bayat F, Shiran B, Belyaev DV, Yur'eva NO, Sobol'kova GI, Alizadeh H et al (2010) Potato plants bearing a vacuolar Na^+/H^+ antiporter *HvNHX2* from barley are characterized by improved salt tolerance. Russ J Plant Physiol 57:696–706

Bhaskaran S, Savithramma DL (2011) Co-expression of *Pennisetum glaucum* vacuolar Na^+/H^+ antiporter and *Arabidopsis* H^+-pyrophosphatase enhances salt tolerance in transgenic tomato. J Exp Bot 62(15):5561–5570. https://doi.org/10.1093/jxb/err237

Bhattacharya RC, Maheswari M, Dineshkumar V, Kirti PB, Bhat SR, Chopra VL (2004) Transformation of *Brassica oleracea* var. capitata with bacterial *betA* gene enhances tolerance to salt stress. Sci Horticult 100:215–227

Bhomkar P, Upadhyay CP, Saxena M, Muthusamy A, Prakash NS, Poggin K, Hohn T, Sarin NB (2008) Salt stress alleviation in transgenic *Vigna mungo* L. Hepper (blackgram) by overexpression of the *glyoxalase I* gene using a novel Cestrum yellow leaf curling virus (*CmYLCV*) promoter. Mol Breed 22:169–181

Bordas M, Montesinos C, Debauza M, Salvador A, Roig LA, Serrano R, Moreno V (1997) Transfer of the yeast salt tolerance gene HAL1 to *Cucumis melo* L. cultivars and in vitro evaluation of salt tolerance. Transgenic Res 6:41–50

Bouaziz D, Jbir R, Charfeddine S (2015) The *StDREB1* transcription factor is involved in oxidative stress response and enhances tolerance to salt stress. Plant Cell Tissue Organ Cult 121:237. https://doi.org/10.1007/s11240-014-0698-7

Brini F, Hanin M, Lumbreras V, Amara I, Khoudi H, Hassairi A, Pages M, Masmoudi K (2007a) Overexpression of wheat dehydrin *DHN-5* enhances tolerance to salt and osmotic stress in *Arabidopsis thaliana*. Plant Cell Rep 26:2017–2026

Brini F, Hanin M, Mezghani I, Berkowitz GA, Masmoudi K (2007b) Overexpression of wheat Na^+/H^+ antiporter *TNHX1* and H^+-pyrophosphatase *TVP1* improve salt- and drought-stress tolerance in *Arabidopsis thaliana* plants. J Exp Bot 58:301–308

Cabello JV, Chan RL (2012) The homologous homeodomain-leucine zipper transcription factors *HaHB1* and *AtHB13* confer tolerance to drought and salinity stresses via the induction of proteins that stabilize membranes. Plant Biotechnol J 10:815–825. https://doi.org/10.1111/j.1467-7652.2012.00701.x

Cai R, Zhao Y, Wang Y, Lin Y, Peng X, Li Q et al (2014) Overexpression of a maize *WRKY58* gene enhances drought and salt tolerance in transgenic rice. Plant Cell Tissue Organ Cult 119:565–577. https://doi.org/10.1007/s11240-014-0556-7

Cai H, Tian S, Dong H, Guo C (2015) Pleiotropic effects of *TaMYB3R1* on plant development and response to osmotic stress in transgenic *Arabidopsis*. Gene 558:227–234. https://doi.org/10.1016/j.gene.2014.12.066

Cao Z-H, Zhang S-Z, Wang R-K, Zhang R-F, Hao Y-J (2013) Genome wide analysis of the apple MYB transcription factor family allows the identification of *MdoMYB121* gene conferring abiotic stress tolerance in plants. PLoS ONE 8(7):e69955. https://doi.org/10.1371/journal.pone.0069955

Checker VG, Chhibbar AK, Khurana P (2012) Stress-inducible expression of barley *Hva1* gene in transgenic mulberry displays enhanced tolerance against drought, salinity and cold stress. Transgenic Res 21:939–957. https://doi.org/10.1007/s11248-011-9577-8

Chen LH, Zhang B, Xu ZQ (2008) Salt tolerance conferred by overexpression of *Arabidopsis* vacuolar Na^+/H^+ antiporter gene *AtNHX1* in common buckwheat (*Fagopyrum esculentum*). Transgenic Res 17:121–132

Chen H, He H, Yu D (2011) Overexpression of a novel soybean gene modulating Na^+ and K^+ transport enhances salt tolerance in transgenic tobacco plants. Physiol Plant 141(1):11–18. https://doi.org/10.1111/j.1399-3054.2010.01412.x

Chen X, Wang Y, Lv B, Li J, Luo L, Lu S et al (2014) The NAC family transcription factor *OsNAP* confers abiotic stress response through the ABA pathway. Plant Cell Physiol 55:604–619. https://doi.org/10.1093/pcp/pct204

Chen H, Liu L, Wang L, Wang S, Cheng X (2016) *VrDREB2A*, a DREB-binding transcription factor from *Vigna radiata*, increased drought and high-salt tolerance in transgenic *Arabidopsis thaliana*. J Plant Res 129:263–273

Chen L, Wang W-S, Liu Q, Dong R-R, Li Y-J, Chen T-T, Hou B-K (2019a) Overexpression of the UGT76E12 gene modulates seed germination, growth, and response to NaCl, mannitol, and abscisic acid. Biol Plant 63:328–334. https://doi.org/10.32615/bp.2019.038

Chen K, Song M, Guo Y, Liu L, Xue H, Dai H, Zhang Z (2019b) *MdMYB46* could enhance salt and osmotic stress tolerance in apple by directly activating stress responsive signals. Plant Biotechnol J. https://doi.org/10.1111/pbi.13151 (Epub ahead of print)

Cheng L, Li S, Hussain J, Xu X, Yin J, Zhang Y et al (2013a) Isolation and functional characterization of a salt responsive transcriptional factor, *LrbZIP* from lotus root (*Nelumbo nucifera* Gaertn). Mol Biol Rep 40:4033–4045. https://doi.org/10.1007/s11033-012-2481-3

Cheng L, Li X, Huang X, Ma T, Liang Y, Ma X et al (2013b) Overexpression of sheepgrass R1-MYB transcription factor *LcMYB1* confers salt tolerance in transgenic *Arabidopsis*. Plant Physiol Biochem 70:252–260. https://doi.org/10.1016/j.plaphy.2013.05.025

Cheong YH, Sung SJ, Kim BG, Pandey GK, Cho JS, Kim KN et al (2010) Constitutive overexpression of the calcium sensor *CBL5* confers osmotic or drought stress tolerance in *Arabidopsis*. Mol Cells 29:159–165

Cortina C, Culianez-Macia FA (2005) Tomato abiotic stress enhanced tolerance by trehalose biosynthesis. Plant Sci 169:75–82

D'yakova EV, Rakitin AL, Kamionskaya AM, Baikov AA, Lahti R, Ravin NV et al (2006) A study of the effect of expression of the gene encoding the membrane H^+ pyrophosphatase of *Rhodospirillum rubrum* on salt resistance of transgenic tobacco plants. Doklady Biol Sci 409:346–348

Dansana PK, Kothari KS, Vij S, Tyagi AK (2014) *OsiSAP1* overexpression improves water-deficit stress tolerance in transgenic rice by affecting expression of endogenous stress-related genes. Plant Cell Rep 33:1425–1440. https://doi.org/10.1007/s00299-014-1626-3

Deinlein U, Stephan AB, Horie T, Luo W, Xu G, Schroeder JI (2014) Plant salt-tolerance mechanisms. Plant Sci 19(6):371–379. https://doi.org/10.1016/j.tplants.2014.02.001

Dhankher OP, Foyer CH (2018) Climate resilient crops for improving global food security and safety. Plant Cell Environ 41:877–884. https://doi.org/10.1111/pce.13207

Di H, Tian Y, Zu H, Meng X, Zeng X, Wang Z (2015) Enhanced salinity tolerance in transgenic maize plants expressing a *BADH* gene from *Atriplex micrantha*. Euphytica 206:775–783. https://doi.org/10.1007/s10681-015-1515-z

Diédhiou CJ, Popova OV, Dietz KJ, Golldack D (2008) The SNF1-type serine-threonine protein kinase *SAPK4* regulates stress-responsive gene expression in rice. BMC Plant Biol 8:49

Dixit A, Tomar P, Vaine E, Abdullah H, Hazen S, Dhankher OP (2017) A stress-associated protein, *AtSAP13*, from *Arabidopsis thaliana* provides tolerance to multiple abiotic stresses. Plant Cell Environ 41:1171–1185

Eltayeb AE, Kawano N, Badawi GH, Kaminaka H, Sanekata T, Shibahara T et al (2007) Overexpression of monodehydroascorbate reductase in transgenic tobacco confers enhanced tolerance to ozone, salt and polyethylene glycol stresses. Planta 225:1255–1264

Fan W, Zhang M, Zhang H, Zhang P (2012) Improved tolerance to various abiotic stresses in transgenic sweet potato (*Ipomoea batatas*) expressing spinach betaine aldehyde dehydrogenase. PLoS ONE 7(5):e37344. https://doi.org/10.1371/journal.pone.0037344

Flowers TJ (2004) Improving crop salt tolerance. J Exp Bot 55(396):307–319. https://doi.org/10.1093/jxb/erh003

Food and Agriculture Organization of the United Nations (2011) The state of the world's land and water resources for food and agriculture; managing systems at risk. http://www.fao.org/3/a-i1688e.pdf. Accessed 14 May 2019

Food and Agriculture Organization of the United Nations (2017) The future of food and agriculture—trends and challenges. Food and Agriculture Organization of the United Nations, Rome

Fu X-Z, Khan EU, Hu S-S, Fan Q-J, Liu J-H (2011) Overexpression of the betaine aldehyde dehydrogenase gene from *Atriplex hortensis* enhances salt tolerance in the transgenic trifoliate orange (*Poncirus trifoliata* L. Raf.). Environ Exp Bot 74:106–113. https://doi.org/10.1016/j.envexpbot.2011.05.006

Galston AW, Kaur-Sawhney R (1990) Polyamines in plant physiology. Plant Physiol 94:406–410

Gao F, Gao Q, Duan X, Yue G, Yang A, Zhang J (2006) Cloning of an H^+-*PPase* gene from *Thellungiella halophila* and its heterologous expression to improve tobacco salt tolerance. J Exp Bot 57(12):3259–3270. https://doi.org/10.1093/jxb/erl090

Gao SQ, Chen M, Xu ZS, Zhao CP, Li L, Xu HJ et al (2011) The soybean *GmbZIP1* transcription factor enhances multiple abiotic stress tolerances in transgenic plants. Plant Mol Biol 75:537–553. https://doi.org/10.1007/s11103-011-9738-4

Garg AK, Kim JK, Owens TG, Ranwala AP, Choi YC, Kochian LV, Wu RJ (2002) Trehalose accumulation in rice plants confers high tolerance levels to different abiotic stresses. Proc Natl Acad Sci USA 99:15898–15903. https://doi.org/10.1073/pnas.252637799

Gaxiola RA, Li J, Undurraga S, Dang LM, Allen GJ, Alper SL, Fink GR (2001) Drought- and salt-tolerant plants result from overexpression of the *AVP1* H^+-pump. Proc Natl Acad Sci USA 98:11444–11449

Ge LF, Chao DY, Shi M, Zhu MZ, Gao JP, Lin HX (2008) Overexpression of the trehalose-6-phosphate phosphatase gene *OsTPP1* confers stress tolerance in rice and results in the activation of stress responsive genes. Planta 228:191–201

Giri J, Vij S, Dansana PK, Tyagi AK (2011) Rice A20/AN1 zinc-finger containing stress-associated proteins (SAP1/11) and a receptor-like cytoplasmic kinase (OsRLCK253) interact via A20 zinc-finger and confer abiotic stress tolerance in transgenic *Arabidopsis* plants. New Phytol 191:721–732. https://doi.org/10.1111/j.1469-8137.2011.03740.x

Gisbert C, Rus AM, Bolarin MC, Lopez-Coronado M, Arrillaga I, Montesinos C et al (2000) The yeast *HAL1* gene improves salt tolerance of transgenic tomato. Plant Physiol 123:393–402

Guan B, Hu Y, Zeng Y, Wang Y, Zhang F (2011) Molecular characterization and functional analysis of a vacuolar Na^+/H^+ antiporter gene (*HcNHX1*) from *Halostachys caspica*. Mol Biol Rep 38:1889–1899

Guo Y, Zhang L, Xiao G, Cao S, Gu D, Tian W, Chen S (1997) Expression of betaine aldehyde dehydrogenase gene and salinity tolerance in rice transgenic plants. Sci China C Life Sci 40(5):496–501. https://doi.org/10.1007/BF03183588

Guo BH, Zhang YM, Li HJ, Du LQ, Li YX, Zhang JS et al (2000) Transformation of wheat with a gene encoding for the betaine aldehyde dehydrogenase (BADH). Acta Bot Sin 42:279–283

Guo S, Yin H, Zhang X, Zhao F, Li P, Chen S et al (2006) Molecular cloning and characterization of a vacuolar H^+-pyrophosphatase gene, *SsVP*, from the halophyte *Suaeda salsa* and its overexpression increases salt and drought tolerance of *Arabidopsis*. Plant Mol Biol 60:41–50

Gupta B, Sahoo K, Ghosh A, Tripathi A, Anwar K, Das P et al (2017) Manipulation of glyoxalase pathway confers tolerance to multiple stresses in rice. Plant Cell Environ 41:1186–1200

Halfter U, Ishitani M, Zhu JK (2000) The *Arabidopsis* SOS2 protein kinase physically interacts with and is activated by the calcium-binding protein SOS3. Proc Natl Acad Sci USA 97:3735–3740

Hao YJ, Wei W, Song QX, Chen HW, Zhang YQ, Wang F et al (2011) Soybean NAC transcription factors promote abiotic stress tolerance and lateral root formation in transgenic plants. Plant J 68:302–313. https://doi.org/10.1111/j.1365-313X.2011.04687.x

Härndahl U, Hall RB, Osteryoung KW, Vierling E, Bornman JF, Sundby C (1999) The chloroplast small heat shock protein undergoes oxidation-dependent conformational changes and may protect plants from oxidative stress. Cell Stress Chaperones 4:129–138

Hayashi H, Alia L, Mustardy P, Deshnium MI, Murata N (1997) Transformation of *Arabidopsis thaliana* with the *codA* gene for choline oxidase; accumulation of glycine betaine and enhanced tolerance to salt and cold stress. Plant J 12:133–142

He C, Yan J, Shen G, Fu L, Holaday AS, Auld D, Blumwald E, Zhang H (2005) Expression of an *Arabidopsis* vacuolar sodium/proton antiporter gene in cotton improves photosynthetic performance under salt conditions and increases fiber yield in the field. Plant Cell Physiol 46:1848–1854

Hmida-Sayari A, Gargouri-Bouzid R, Bidani A, Jaoua L, Savoure A, Jaoua S (2005) Overexpression of Δ^1-pyrroline-5-carboxylate synthetase increases proline production and confers salt tolerance in transgenic potato plants. Plant Sci 169:746–752

Holmstrom KO, Somersalo S, Mandal A, Palva TE, Welin B (2000) Improved tolerance to salinity and low temperature in transgenic tobacco producing glycine betaine. J Exp Bot 51:177–185

Hong Z, Lakkineni K, Zhang K, Verma DPS (2000) Removal of feedback inhibition of Δ1-pyrroline-5-carboxylate synthetase results in increased proline accumulation and protection of plants from osmotic stress. Plant Physiol 122:1129–1136

Hu H, You J, Fang Y, Zhu X, Qi Z, Xiong L (2008) Characterization of transcription factor gene *SNAC2* conferring cold and salt tolerance in rice. Plant Mol Biol 67:169–181. https://doi.org/10.1007/s11103-008-9309-5

Hu T, Zhu S, Tan L, Qi W, He S, Wang G (2016) Overexpression of *OsLEA4* enhances drought, high salt and heavy metal stress tolerance in transgenic rice (*Oryza sativa* L.). Environ Exp Bot 123:68–77

Huang Q, Wang Y, Li B, Chang J, Chen M, Li K et al (2015) *TaNAC29*, a NAC transcription factor from wheat, enhances salt and drought tolerance in transgenic *Arabidopsis*. BMC Plant Biol 15:268. https://doi.org/10.1186/s12870-015-0644-9

Hyun T-K, Albacete A, Graaff E, Eom S-H, Großkinsky D-K, Böhm H, Janschek U, Rim Y, Ali WW, Kim S-Y, Roitsch T (2015) The *Arabidopsis* PLAT domain protein1 promotes abiotic stress tolerance and growth in tobacco. Transgenic Res 24:651–663. https://doi.org/10.1007/s11248-015-9868-6

Jacobs A, Ford K, Kretschmer J, Tester M (2011) Rice plants expressing the moss sodium pumping ATPase *PpENA1* maintain greater biomass production under salt stress. Plant Biotechnol J 9:838–847

Jeong MJ, Lee SK, Kim BG, Kwon TR, Cho WS, Park YT et al (2006) A rice (*Oryza sativa* L.) MAP kinase gene, *OsMAPK44*, is involved in response to abiotic stresses. Plant Cell Tissue Organ Cult 85:151–160

Jha A, Joshi M, Yadav NS, Agarwal PK, Jha B (2010) Cloning and characterization of the *Salicornia brachiate* Na^+/H^+ antiporter gene *SbNHX1* and its expression by abiotic stress. Mol Biol Rep 38:1965–1973

Jia G-X, Zhu Z-Q, Chang F-Q, Li Y-X (2002) Transformation of tomato with the *BADH* gene from Atriplex improves salt tolerance. Plant Cell Rep 21:141–146. https://doi.org/10.1007/s00299-002-0489-1

Joshi R, Sahoo KK, Tripathi AK, Kumar R, Gupta BK, Pareek A, Singla-Pareek SL (2018) Knockdown of an inflorescence meristem-specific cytokinin oxidase – *OsCKX2* in rice reduces yield penalty under salinity stress condition. Plant Cell Environ 41:936–946. https://doi.org/10.1111/pce.12947

Kaku N, Hibino T, Meng Y, Tanaka Y, Araki E, Takabe T (2000) Effects of overexpression of *Escherichia coli katE* and *bet* genes on the tolerance for salt stress in a fresh water Cyanobacterium *synechococcus* sp. PCC 7942. Plant Sci 159:281–288

Kang M, Fokar M, Abdelmageed H, Allen RD (2011) *Arabidopsis* SAP5 functions as a positive regulator of stress responses and exhibits E3 ubiquitin ligase activity. Plant Mol Biol 75(4–5):451–466. https://doi.org/10.1007/s11103-011-9748-2

Karakas B, Ozias Akins P, Stushnoff C, Suefferheld M, Rieger M (1997) Salinity and drought tolerance of mannitol accumulating transgenic tobacco. Plant Cell Environ 20:609–616

Karthikeyan A, Pandian SK, Ramesh M (2011) Transgenic indica rice cv. ADT 43 expressing a Δ1-pyrroline-5-carboxylate synthetase (*P5CS*) gene from *Vigna aconitifolia* demonstrates salt tolerance. Plant Cell Tissue Organ Cult 107:383–395

Kasukabe Y, He L, Nada K, Misawa S, Ihara I, Tachibana S (2004) Overexpression of spermidine synthase enhances tolerance to multiple environmental stresses and upregulates the expression of various stress-regulated genes in transgenic *Arabidopsis thaliana*. Plant Cell Physiol 45:712–722

Kathuria H, Giri J, Nataraja KN, Murata N, Udayakumar M, Tyagi AK (2009) Glycine betaine-induced water-stress tolerance in *codA*-expressing transgenic indica rice is associated with up-regulation of several stress responsive genes. Plant Biotechnol J 7(6):512–526. https://doi.org/10.1111/j.1467-7652.2009.00420.x

Kavi Kishor PB, Hong Z, Miao GH, Hu CAA, Verma DPS (1995) Overexpression of Δ^1-pyrroline-5-carboxylate synthetase increase proline production and confers osmotolerance in transgenic plants. Plant Physiol 108:1387–1394

Kavitha K, George S, Venkataraman G, Parida A (2010) A salt-inducible chloroplastic monodehydroascorbate reductase from halophyte *Avicennia marina* confers salt stress tolerance on transgenic plants. Biochimie 92:1321–1329

Ke Q, Kim H-S, Wang Z, Ji C-Y, Jeong J-C, Lee H-S, Choi Y-I, Xu B, Deng X, Yun D-J, Kwak S-S (2017) Down-regulation of *GIGANTEA-like* genes increases plant growth and salt stress tolerance in poplar. Plant Biotechnol J 15:331–343. https://doi.org/10.1111/pbi.12628

Kikuchi A, Yu X, Shimazaki T, Kawaoka A, Ebinuma H, Watanabe KN (2009) Allelopathy assessments for the environmental biosafety of the salt-tolerant transgenic *Eucalyptus camaldulensis*, genotypes codA 12-5B, codA 12-5C, and codA 20C. J Wood Sci 55:149–153. https://doi.org/10.1007/s10086-008-1007-z

Kong X, Pan J, Zhang M, Xing X, Zhou Y, Liu Y et al (2011) *ZmMKK4*, a novel group C mitogen-activated protein kinase kinase in maize (*Zea mays*), confers salt and cold tolerance in transgenic *Arabidopsis*. Plant Cell Environ 34:1291–1303

Kovtun Y, Chiu WL, Tena G, Sheen J (2000) Functional analysis of oxidative-stress activated mitogen-activated protein kinase cascade in plants. Proc Natl Acad Sci USA 97:2940–2945

Kumar K, Sinha AK (2013) Overexpression of constitutively active mitogen activated protein kinase kinase 6 enhances tolerance to salt stress in rice. Rice 6:25. https://doi.org/10.1186/1939-8433-6-25

Kumar S, Dhingra A, Daniell H (2004) Plastid-expressed betaine aldehyde dehydrogenase gene in carrot cultured cells, roots, and leaves confers enhanced salt tolerance. Plant Physiol 136:2843–2854. www.plantphysiol.org/cgi/doi/10.1104/pp.104.045187

Kumar S, Asif MH, Chakrabarty D, Tripathi RD, Dubey RS, Trivedi PK (2013) Expression of a rice Lambda class of glutathione S-transferase, *OsGSTL2*, in *Arabidopsis* provides tolerance to heavy metal and other abiotic stresses. J Hazard Mater 248–249:228–237. https://doi.org/10.1016/j.jhazmat.2013.01.004

Kumar T, Uzma Khan MR, Abbas Z, Ali GM (2014) Genetic improvement of sugarcane for drought and salinity stress tolerance using *Arabidopsis* vacuolar pyrophosphatase (*AVP1*) gene. Mol Biotechnol 56:199–209. https://doi.org/10.1007/s12033-013-9695-z

Lakra N, Kaur C, Anwar K, Pareek S, Pareek A (2017) Proteomics of contrasting rice genotypes: identification of potential targets for raising crops for saline environment. Plant Cell Environ 41:947–969

Lal S, Gulyani V, Khurana P (2008) Overexpression of *HVA1* gene from barley generates tolerance to salinity and water stress in transgenic mulberry (*Morus indica*). Transgen Res 17:651–663

Lee H, Jo J (2004) Increased tolerance to methyl viologen by transgenic tobacco plants that overexpress the cytosolic glutathione reductase gene from *Brassica campestris*. J Plant Biol 47:111–116

Lee YP, Kim SH, Bang JW, Lee HS, Kwak SS, Kwon SY (2007) Enhanced tolerance to oxidative stress in transgenic tobacco plants expressing three antioxidant enzymes in chloroplasts. Plant Cell Rep 26:591–598

Lee YP, Baek KH, Lee HS, Kwak SS, Bang JW, Kwon SY (2010) Tobacco seeds simultaneously overexpressing Cu/Zn superoxide dismutase and ascorbate peroxidase display enhanced seed longevity and germination rates under stress conditions. J Exp Bot 61:2499–2506

Leidi EO, Barragán V, Rubio L, El-Hamdaoui A, Ruiz MT, Cubero B, Fernández JA, Bressan RA, Hasegawa PM, Quintero FJ, Pardo JM (2010) The *AtNHX1* exchanger mediates potassium compartmentation in vacuoles of transgenic tomato. Plant J 61(3):495–506

Li Q-L, Liu D-W, Gao X-R, Su Q (2003) Cloning of cDNA encoding choline monooxygenase from *Suaeda liaotungensis* and salt tolerance of transgenic tobacco. Acta Bot Sin 45(2):242–247

Li B, Wei AY, Song CX, Li N, Zhang JR (2008) Heterologous expression of the *TsVP* gene improves the drought resistance of maize. Plant Biotechnol J 6:146–159

Li W, Wang D, Jin T, Chang Q, Yin D, Xu S et al (2010a) The vacuolar Na^+/H^+ antiporter gene *SsNHX1* from the halophyte *Salsola soda* confers salt tolerance in transgenic alfalfa (*Medicago sativa* L.). Plant Mol Biol Rep 29:278–290

Li Z, Baldwin CM, Hu Q, Liu H, Luo H (2010b) Heterologous expression of *Arabidopsis* H^+-pyrophosphatase enhances salt tolerance in transgenic creeping bentgrass (*Agrostis stolonifera* L.). Plant Cell Environ 33:272–289

Li HW, Zang BS, Deng XW, Wang XP (2011) Overexpression of the trehalose-6-phosphate synthase gene *OsTPS1* enhances abiotic stress tolerance in rice. Planta 234:1007–1018

Li Q, Tang Z, Hu Y, Yu L, Liu Z, Xu G (2014a) Functional analyses of a putative plasma membrane Na^+/H^+ antiporter gene isolated from salt tolerant *Helianthus tuberosus*. Mol Biol Rep 41:5097–5108. https://doi.org/10.1007/s11033-014-3375-3

Li H, Wang Z, Ke Q, Ji CY, Jeong JC, Lee H-S, Lim YP, Xu B, Deng X-P, Kwak S-S (2014b) Overexpression of *codA* gene confers enhanced tolerance to abiotic stresses in alfalfa. Plant Physiol Biochem 85:31–40. https://doi.org/10.1016/j.plaphy.2014.10.010

Li Q, Yu HM, Meng XF, Lin JS, Li YJ, Hou BK (2018) Ectopic expression of glycosyltransferase *UGT76E11* increases flavonoid accumulation and enhances abiotic stress tolerance in *Arabidopsis*. Plant Biol 20:10–19

Liao YD, Lin KH, Chen CC, Chiang CM (2016) *Oryza sativa* protein phosphatase 1a (*OsPP1a*) involved in salt stress tolerance in transgenic rice. Mol Breed 36:22

Lin KH, Lin MY, Wang TC, Wu YS (2016) Efficient Agrobacterium-mediated transformation of *Populus nigra* × *Populus tomentosa* expressing the *MtlD* gene improves salt tolerance. Taiwan J Forest Sci 31:19–35

Liu H, Wang Q, Yu M, Zhang Y, Wu Y, Zhang H (2008) Transgenic salt-tolerant sugar beet (*Beta vulgaris* L.) constitutively overexpressing an *Arabidopsis thaliana* vacuolar Na^+/H^+ antiporter gene, *AtNHX3*, accumulates more soluble sugar but less salt in storage root. Plant Cell Environ 31:1325–1334

Liu L, Wang Y, Wang N, Dong Y-Y, Fan X-D, Liu X-M, Yang J, Li H-Y (2011) Cloning of a vacuolar H^+-pyrophosphatase gene from the halophyte *Suaeda corniculata* whose heterologous overexpression improves salt, saline-alkali and drought tolerance in *Arabidopsis*. J Integr Plant Biol 53(9):731–742. https://doi.org/10.1111/j.1744-7909.2011.01066.x

Liu X, Zhua Y, Zhai H, Cai H, Ji W, Luo X, Li J, Bai X (2012) *AtPP2CG1*, a protein phosphatase 2C, positively regulates salt tolerance of *Arabidopsis* in abscisic acid-dependent manner. Biochem Biophys Res Commun 422(4):710–715. https://doi.org/10.1016/j.bbrc.2012.05.064

Liu C, Mao B, Ou S, Wang W, Liu L, Wu Y et al (2014) *OsbZIP71*, a bZIP transcription factor, confers salinity and drought tolerance in rice. Plant Mol Biol 84:19–36. https://doi.org/10.1007/s11103-013-0115-3

Liu L, Zhang Z, Dong J, Wang T (2016) Overexpression of *MtWRKY76* increases both salt and drought tolerance in *Medicago truncatula*. Environ Exp Bot 123:50–58. https://doi.org/10.1016/j.envexpbot.2015.10.007

Lu SY, Jing YX, Shen SH, Zhao HY, Ma LQ, Zhou XJ et al (2005) Antiporter gene from *Hordeum brevisubulatum* (Trin.) link and its overexpression in transgenic tobaccos. J Int Plant Biol 47:343–349

Lu ZQ, Liu DL, Liu SK (2007) Two rice cytosolic ascorbate peroxidases differentially improve salt tolerance in transgenic *Arabidopsis*. Plant Cell Rep 26:1909–1917

Lu M, Han Y-P, Gao J-G, X W-J, Li W-B (2010) Identification and analysis of the germin-like gene family in soybean. BMC Genomics 11:620. https://doi.org/10.1186/1471-2164-11-620

Lv SL, Zhang KW, Gao Q, Lian LJ, Song YJ, Zhang JR (2008) Overexpression of an H^+-*PPase* gene from *Thellungiella halophila* in cotton enhances salt tolerance and improves growth and photosynthetic performance. Plant Cell Physiol 49:1150–1164

Ma L, Zhou E, Gao L, Mao X, Zhou R, Jia J (2008) Isolation, expression analysis and chromosomal location of *P5CR* gene in common wheat (*Triticum aestivum* L.). S Afr J Bot 74:705–712

Majee M, Maitra S, Dastidar KG, Pattnaik S, Chatterjee A, Hait NC et al (2004) A novel salt-tolerant L-myo-inositol-1-phosphate synthase from *Porteresia coarctata* (Roxb.) Tateoka, a halophytic wild rice, molecular cloning, bacterial overexpression, characterization, and functional introgression into tobacco-conferring salt tolerance phenotype. J Biol Chem 279:28539–28552

Miranda JA, Avonce N, Suárez R, Thevelein JM, van Dijck P, Iturriaga G (2007) A bifunctional TPS-TPP enzyme from yeast confers tolerance to multiple and extreme abiotic-stress conditions in transgenic *Arabidopsis*. Planta 226:1411–1421

Missihoun TD, Willée E, Guegan J-P, Berardocco S, Shafiq MR, Bouchereau A, Bartels D (2015) Overexpression of *ALDH10A8* and *ALDH10A9* genes provides insight into their role in glycine betaine synthesis and affects primary metabolism in *Arabidopsis thaliana*. Plant Cell Physiol 56(9):1798–1807. https://doi.org/10.1093/pcp/pcv105

Mohammadi K, Movahedi A, Maleki SS, Sun W, Zhang J, Almasi Zadeh Yaghuti A, Nourmohammadi S, Zhuge Q (2018) Functional analysis of overexpressed *PtDRS1* involved in abiotic stresses enhances growth in transgenic poplar. Plant Physiol Biochem 126:22–31. https://doi.org/10.1016/j.plaphy.2018.01.023

Mohanty A, Kathuria H, Ferjani A, Sakamoto A, Mohanty P, Murata N, Tyagi AK (2002) Transgenics of an elite indica rice variety Pusa Basmati 1 harbouring the *codA* gene are highly tolerant to salt stress. Theor Appl Genet 106:51–57

Moon H, Lee B, Choi G, Shin D, Prasad DT, Lee O et al (2003) NDP kinase 2 interacts with two oxidative stress activated MAPKs to regulate cellular redox state and enhances multiple stress tolerance in transgenic plants. Proc Natl Acad Sci USA 100:358–363

Moriwaki T, Yamamoto Y, Aida T, Funahashi T, Shishido T, Asada M et al (2008) Overexpression of the *Escherichia coli* catalase gene, *katE*, enhances tolerance to salinity stress in the transgenic indica rice cultivar, BR5. Plant Biotechnol Rep 2:41–46

Munns R, Tester M (2008) Mechanisms of salinity tolerance. Annu Rev Plant Biol 59:651–681. https://doi.org/10.1146/annurev.arplant.59.032607.092911

Nakashima K, Tran LS, Van Nguyen D, Fujita M, Maruyama K, Todaka D, Ito Y, Hayashi N, Shinozaki K, Yamaguchi-Shinozaki K (2007) Functional analysis of a NAC-type transcription factor *OsNAC6* involved in abiotic and biotic stress-responsive gene expression in rice. Plant J 51:617–630. https://doi.org/10.1111/j.1365-313X.2007.03168.x

Nanjo T, Kobayashia M, Yoshibab Y, Kakubaric Y, Yamaguchi-Shinozaki K, Shinozaki K (1999) Antisense suppression of proline degradation improves tolerance to freezing and salinity in *Arabidopsis thaliana*. FEBS Lett 461:205–210

Nongpiura RC, Singla-Pareek S, Pareek A (2016) Genomics approaches for improving salinity stress tolerance in crop plants. Curr Genomics 17:343–357

Oh DH, Leidi E, Zhang Q, Hwang SM, Li Y, Quintero FJ et al (2009) Loss of halophytism by interference with SOS1 expression. Plant Physiol 151:210–222

Ohta M, Hayashi Y, Nakashima A, Hamada A, Tanaka A, Nakamura T et al (2002) Introduction of a Na^+/H^+ antiporter gene from *Atriplex gmelini* confers salt tolerance to rice. FEBS Lett 532:279–282

Olías R, Eljakaoui Z, Li J, Morales PAD, Marín-manzano MC, Pardo JM et al (2009) The plasma membrane Na^+/H^+ antiporter SOS1 is essential for salt tolerance in tomato and affects the partitioning of Na^+ between plant organs. Plant Cell Environ 32:904–916

Ouyang SQ, Liu YF, Liu P, Lei G, He SJ, Ma B, Zhang WK, Zhang JS, Chen SY (2010) Receptor-like kinase OsSIK1 improves drought and salt stress tolerance in rice (*Oryza sativa*) plants. Plant J 62:316–329. https://doi.org/10.1111/j.1365-313X.2010.04146.x

Pardo JM, Reddy MP, Yang S, Maggio A, Huh GH, Matsumoto T, Coca MA et al (1998) Stress signaling through Ca^{2+}/calmodulin-dependent protein phosphatase calcineurin mediates salt adaptation in plants. Proc Natl Acad Sci USA 16:9681–9686

Park BJ, Liu Z, Kanno A, Kameya T (2005a) Genetic improvement of Chinese cabbage for salt and drought tolerance by constitutive expression of a *B. napus* LEA gene. Plant Sci 169:553–558

Park BJ, Liu Z, Kanno A, Kameya T (2005b) Increased tolerance to salt- and water-deficit stress in transgenic lettuce (*Lactuca sativa* L.) by constitutive expression of LEA. Plant Growth Regul 45:165–171

Pasapula V, Shen G, Kuppu S, Paez-Valencia J, Mendoza M, Hou P et al (2011) Expression of an *Arabidopsis* vacuolar H^+-pyrophosphatase gene (*AVP1*) in cotton improves drought and salt tolerance and increases fibre yield in the field conditions. Plant Biotechnol J 9(1):88–99. https://doi.org/10.1111/j.1467-7652.2010.00535.x

Paul S, Roychoudhury A (2018) Transgenic plants for improved salinity and drought tolerance. In: Gosal SS, Wani SH (eds) Biotechnologies of crop improvement, vol 2. Transgenic approaches. Springer, Switzerland, pp 141–181. https://doi.org/10.1007/978-3-319-90650-8_7

Pehlivan N, Sun L, Jarrett P, Yang X, Mishra N, Chen L, Kadioglu A, Shen G, Zhang H (2016) Co-overexpressing a plasma membrane and a vacuolar membrane sodium/proton antiporter significantly improves salt tolerance in transgenic *Arabidopsis* plants. Plant Cell Physiol 57(5):1069–1084. https://doi.org/10.1093/pcp/pcw055

Peng X, Zhang L, Zhang L, Liu Z, Cheng L, Yang Y et al (2013) The transcriptional factor *LcDREB2* cooperates with *LcSAMDC2* to contribute to salt tolerance in *Leymus chinensis*. Plant Cell Tissue Organ Cult 113:245–256. https://doi.org/10.1007/s11240-012-0264-0

Prasad KVSK, Sharmila P, Kumar PA, Saradhi PP (2000) Transformation of *Brassica juncea* (L.) Czern with bacterial codA gene enhances its tolerance to salt stress. Mol Breed 6:489–499

Prashanth SR, Sadhasivam V, Parida A (2008) Over expression of cytosolic copper/zinc superoxide dismutase from a mangrove plant *Avicennia marina* in indica rice var Pusa Basmati-1 confers abiotic stress tolerance. Transgenic Res 17:281–291

Prodhan SH, Hossain A, Nagamiya K, Komamine A, Morishima H (2008) Improved salt tolerance and morphological variation in indica rice (*Oryza sativa* L.) transformed with a catalase gene from *E. coli*. Plant Tissue Cult Biotechnol 18:57–63

Qi YC, Liu WQ, Qiu LY, Zhang SM, Ma L, Zhang H (2010) Overexpression of glutathione S-transferase gene increases salt tolerance of *Arabidopsis*. Russ J Plant Physiol 57:233–240

Qiao WH, Zhao XY, Li W (2007) Overexpression of *AeNHX1*, a root-specific vacuolar Na^+/H^+ antiporter from *Agropyron elongatum*, confers salt tolerance to *Arabidopsis* and Festuca plants. Plant Cell Rep 26:1663–1672

Qin Y, Tian Y, Liu X (2015) A wheat salinity-induced WRKY transcription factor *TaWRKY93* confers multiple abiotic stress tolerance in *Arabidopsis thaliana*. Biochem Biophys Res Commun 464:428–433. https://doi.org/10.1016/j.bbrc.2015.06.128

Quintero FJ, Ohta M, Shi HZ, Zhu J-K, Pardo JM (2002) Reconstitution in yeast of the *Arabidopsis* SOS signaling pathway for Na^+ homeostasis. Proc Natl Acad Sci USA 99:9061–9066

Rajagopal D, Agarwal P, Tyagi W, Singla-Pareek S, Reddy MK, Sopory SK (2007) *Pennisetum glaucum* Na^+/H^+ antiporter confers high level of salinity tolerance in transgenic *Brassica juncea*. Mol Breed 19:137–151

Rodrigues SM, Andrade MO, Gomes APS, Damatta FM, Baracat-Pereira MC, Fontes EPB (2006) *Arabidopsis* and tobacco plants ectopically expressing the soybean antiquitin-like *ALDH7* gene display enhanced tolerance to drought, salinity, and oxidative stress. J Exp Bot 57:1909–1918

Rong W, Qi L, Wang A, Ye X, Du L, Liang H et al (2014) The ERF transcription factor *TaERF3* promotes tolerance to salt and drought stresses in wheat. Plant Biotechnol J 12:468–479. https://doi.org/10.1111/pbi.12153

Roxas VP, Smith RK Jr, Allen ER, Allen RD (1997) Overexpression of glutathione S-transferase/glutathione peroxidase enhances the growth of transgenic tobacco seedlings during stress. Nat Biotechnol 15:988–991

Roy M, Wu R (2001) Arginine decarboxylase transgene expression and analysis of environmental stress tolerance in transgenic rice. Plant Sci 160:869–875

Roy M, Wu R (2002) Overexpression of S-adenosylmethionine decarboxylase gene in rice increases polyamine level and enhances sodium chloride-stress tolerance. Plant Sci 163:987–992

Rus AM, Estaño MT, Gisbert C, Garcia-Sogo B, Serrano R, Caro M, Moreno V, Bolarín MC (2001) Expressing the yeast *HAL1* gene in tomato increases fruit yield and enhances K^+/Na^+ selectivity under salt stress. Plant Cell Environ 24:875–880

Saad AS, Li X, Li HP, Huang T, Gao CS, Guo MW et al (2013) A rice stress-responsive NAC gene enhances tolerance of transgenic wheat to drought and salt stresses. Plant Sci 203–204:33–40. https://doi.org/10.1016/j.plantsci.2012.12.016

Saijo Y, Hata S, Kyozuka J, Shimamoto K, Izui K (2000) Overexpression of a single Ca^{2+}-dependent protein kinase confers both cold and salt/drought tolerance on rice plants. Plant J 23:319–327

Sakamoto A, Alia H, Murata N (1998) Metabolic engineering of rice leading to biosynthesis of glycinebetaine and tolerance to salt and cold. Plant Mol Biol 38:1011–1019

Sawahel WA, Hassan AH (2002) Generation of transgenic wheat plants producing high levels of the osmoprotectants proline. Biotechnol Lett 24:721–725

SenGupta A, Heinen JL, Holaday AS, Burke JJ, Allen RD (1993a) Increased resistance to oxidative stress in transgenic plants that overexpress chloroplastic Cu/Zn superoxide dismutase. Proc Natl Acad Sci USA 90:1629–1633

SenGupta A, Webb RP, Holaday AS, Allen RD (1993b) Overexpression of superoxide dismutase protects plants from oxidative stress. Plant Physiol 103:1067–1073

Shafi A, Pal AK, Sharma V, Kalia S, Kumar S, Ahuja PS, Singh AK (2017) Transgenic potato plants overexpressing SOD and APX exhibit enhanced lignification and starch biosynthesis with improved salt stress tolerance. Plant Mol Biol Rep 35:504–518

Shen V, Jensen RG, Bonhert HJ (1997) Increased resistance to oxidative stress in transgenic plants by targeting mannitol biosynthesis in chloroplasts. Plant Physiol 113:1177–1183

Shen YG, Du BX, Zhang JS, Chen SY (2001) Cloning and characterization of *CMO* gene from *Atriplex hortensis*. Sheng wu gong cheng xue bao (Chin J Biotechnol) 17(1):1–6

Sheveleva E, Chmara W, Bohnert HJ, Jensen RG (1997) Increased salt and drought tolerance by D-ononitol production in transgenic *Nicotiana tabacum* L. Plant Physiol 115:1211–1219

Shi H, Lee BH, Wu SJ, Zhu JK (2003) Overexpression of a plasma membrane Na^{+}/H^{+} antiporter gene improves salt tolerance in *Arabidopsis thaliana*. Nat Biotechnol 21:81–85

Shi W, Liu D, Hao L, Wu CA, Guo X, Li H (2014) *GhWRKY39*, a member of the WRKY transcription factor family in cotton, has a positive role in disease resistance and salt stress tolerance. Plant Cell Tissue Organ Cult 118:17–32. https://doi.org/10.1007/s11240-014-0458-8

Shinde H, Dudhate A, Tsugama D, Gupta SK, Liu S, Takano T (2019) Pearl millet stress-responsive NAC transcription factor *PgNAC21* enhances salinity stress tolerance in *Arabidopsis*. Plant Physiol Biochem 135:546–553. https://doi.org/10.1016/j.plaphy.2018.11.004

Shirasawa K, Takabe T, Takabe T, Kishitani S (2006) Accumulation of glycine betaine in rice plants that overexpress *choline monooxygenase* from spinach and evaluation of their tolerance to abiotic stress. Ann Bot 98(3):565–571. https://doi.org/10.1093/aob/mcl126

Sickler CM, Edwards GE, Kiirats O, Gao Z, Loescher W (2007) Response of mannitol producing *Arabidopsis thaliana* to abiotic stress. Funct Plant Biol 34:382–391

Singh RK, Shweta S, Muthamilarasan M, Rani R, Prasad M (2019) Study on aquaporins of *Setaria italica* suggests the involvement of *SiPIP3;1* and *SiSIP1;1* in abiotic stress response. Funct Integr Genomics 1–10. https://doi.org/10.1007/s10142-018-00653-0

Singla-Pareek S, Reddy MK, Sopory SK (2001) Transgenic approach towards developing abiotic stress tolerance in plants. Proc Ind Natl Sci Acad (PINSA) B67(5):265–284

Singla-Pareek S, Reddy MK, Sopory SK (2003) Genetic engineering of the glyoxalase pathway in tobacco leads to enhanced salinity tolerance. Proc Natl Acad Sci USA 100:14672–14677

Singla-Pareek S, Yadav SK, Pareek A, Reddy MK, Sopory SK (2006) Transgenic tobacco overexpressing glyoxalase pathway enzymes grow and set viable seeds in zinc-spiked soils. Plant Physiol 140:613–623

Singla-Pareek S, Yadav SK, Pareek A, Reddy MK, Sopory SK (2008) Enhancing salt tolerance in a crop plant by overexpression of *glyoxalase II*. Transgenic Res 17:171–180

Soda N, Gupta BK, Anwar K, Sharan A, Govindjee Singla-Pareek S, Pareek A (2018) Rice intermediate filament, *OsIF*, stabilizes photosynthetic machinery and yield under salinity and heat stress. Sci Rep 8(4072):1–13. https://doi.org/10.1038/s41598-018-22131-0

Song A, Lu J, Jiang J, Chen S, Guan Z, Fang W, Chen F (2012) Isolation and characterisation of *Chrysanthemum crassum SOS1*, encoding a putative plasma membrane Na^{+}/H^{+} antiporter. Plant Biol (Stuttg) 14(5):706–713. https://doi.org/10.1111/j.1438-8677.2011.00560.x

Srivastava VK, Raikwar S, Tuteja R, Tuteja N (2016) Ectopic expression of phloem motor protein pea forisome *PsSEO-F1* enhances salinity stress tolerance in tobacco. Plant Cell Rep 35(5):1021–1041. https://doi.org/10.1007/s00299-016-1935-9

Staal M, Maathuis FJM, Elzenga JTM, Overbeek JHM, Prins HBA (1991) Na^{+}/H^{+} antiport activity in tonoplast vesicles from roots of the salt-tolerant *Plantago maritima* and the salt-sensitive *Plantago media*. Physiol Plant 82:179–184

Ströher E, Wang XJ, Roloff N, Klein P, Husemann A, Dietz KJ (2009) Redox-dependent regulation of the stress-induced zinc-finger protein SAP12 in *Arabidopsis thaliana*. Mol Plant 2(2):357–367. https://doi.org/10.1093/mp/ssn084

Su J, Wu R (2004) Stress-inducible synthesis of proline in transgenic rice confers faster growth under stress conditions than that with constitutive synthesis. Plant Sci 166:941–948

Suarez R, Calderón C, Iturriaga G (2009) Improved tolerance to multiple abiotic stresses in transgenic alfalfa accumulating trehalose. Crop Sci 49:1791–1799

Sugino M, Hibino T, Tanaka Y, Nii N, Takabe T (1999) Overexpression of *DnaK* from a halotolerant cyanobacterium *Aphanothece halophytice* acquires resistance to salt stress in transgenic tobacco plants. Plant Sci 146:81–88

Sun W, Bernard C, van de Cotte B, Van Montagu M, Verbruggen N (2001) *At-HSP17.6A*, encoding a small heat-shock protein in *Arabidopsis*, can enhance osmotolerance upon overexpression. Plant J 27:407–415

Sun WH, Duan M, Yang DFSS, Meng QW (2010a) Over-expression of *StAPX* in tobacco improves seed germination and increases early seedling tolerance to salinity and osmotic stresses. Plant Cell Rep 29:917–926. https://doi.org/10.1007/s00299-010-0878-9

Sun S-J, Guo S-Q, Yang X, Bao Y-M, Tang H-J, Su H, Huang J, Zhang H-S (2010b) Functional analysis of a novel Cys2/His2-type zinc finger protein involved in salt tolerance in rice. J Exp Bot 61:2807–2818. https://doi.org/10.1093/jxb/erq120

Sun ZM, Zhou ML, Xiao XG, Tang YX, Wu YM (2014) Genome-wide analysis of AP2/ERF family genes from *Lotus corniculatus* shows *LcERF054* enhances salt tolerance. Funct Integr Genomics 14:453–466. https://doi.org/10.1007/s10142-014-0372-5

Surekha C, Kumari KN, Aruna LV, Suneetha G, Arundhati A, Kavi Kishor PB (2014) expression of the *Vigna aconitifolia P5csf129a* gene in transgenic pigeonpea enhances proline accumulation and salt tolerance. Plant Cell Tissue Organ Cult 116:27–36. https://doi.org/10.1007/s11240-013-0378-z

Tak H, Mhatre M (2013) Molecular characterization of *VvSDIR1 f*rom *Vitis vinifera* and its functional analysis by heterologous expression in *Nicotiana tabacum*. Protoplasma 250:565–576. https://doi.org/10.1007/s00709-012-0442-2

Tang W, Peng X, Newton RJ (2005) Enhanced tolerance to salt stress in transgenic loblolly pine simultaneously expressing two genes encoding mannitol-1-phosphate dehydrogenase and glucitol-6-phosphate dehydrogenase. Plant Physiol Biochem 43:139–146

Teige M, Scheikl E, Eulgem T, Doczi R, Ichimura K, Shinozaki K, Dangl JL, Hirt H (2004) The MKK2 pathway mediates cold and salt stress signaling in *Arabidopsis*. Mol Cell 15:141–152. https://doi.org/10.1016/j.molcel.2004.06.023

Thomas JC, Sepahi M, Arendall B, Bohnert HJ (1995) Enhancement of seed germination in high salinity by engineering mannitol expression in *Arabidopsis thaliana*. Plant Cell Environ 18:801–806

Tian N, Wang J, Xu ZQ (2010) Overexpression of Na^+/H^+ antiporter gene *AtNHX1* from *Arabidopsis thaliana* improves the salt tolerance of kiwifruit (*Actinidia deliciosa*). S Afr J Bot 77:160–169

Tripathi V, Parasuraman B, Laxmi A, Chattopadhyay D (2009) CIPK6, a CBL-interacting protein kinase is required for development and salt tolerance in plants. Plant J 58:778–790

Tyagi H, Jha S, Sharma M, Giri J, Tyagi AK (2014) Rice SAPs are responsive to multiple biotic stresses and overexpression of *OsSAP1*, an A20/AN1 zinc-finger protein, enhances the basal resistance against pathogen infection in tobacco. Plant Sci 225:68–76. https://doi.org/10.1016/j.plantsci.2014.05.016

Urao T, Yakubov B, Satoh R, Yamaguchi-Shinozaki K, Seki M, Hirayama T, Shinozaki K (1999) A transmembrane hybrid-type histidine kinase in *Arabidopsis* functions as an osmosensor. Plant Cell 11:1743–1754

Ushimaru T, Nakagawa T, Fujioka Y, Daicho K, Naito M, Yamauchi Y et al (2006) Transgenic *Arabidopsis* plants expressing the rice dehydroascorbate reductase gene are resistant to salt stress. J Plant Physiol 163:1179–1184

Van Camp W, Capiau K, Van Montagu M, Inze D, Slooten L (1996) Enhancement of oxidative stress tolerance in transgenic tobacco plants overproducing Fe-superoxide dismutase in chloroplasts. Plant Physiol 112:1703–1714

Verma M, Verma D, Jain RK, Sopory SK, Wu R (2005) Overexpression of *glyoxalase-I* gene confers salinity tolerance in transgenic japonica and indica rice plants. Rice Genet Newslett 22:58–62

Verma D, Singla-Pareek S, Rajagopal D, Reddy MK, Sopory SK (2007) Functional validation of a novel isoform of Na^+/H^+ antiporter from *Pennisetum glaucum* for enhancing salinity tolerance in rice. J Biosci 32:621–626

Waie B, Rajam MV (2003) Effect of increased polyamine biosynthesis on stress responses in transgenic tobacco by introduction of human S-adenosylmethionine gene. Plant Sci 164:727–734
Wang Y, Ying Y, Chen J, Wang X (2004) Transgenic *Arabidopsis* overexpressing Mn-SOD enhanced salt-tolerance. Plant Sci 167:671–677
Wang Y, Wisniewski ME, Meilan R, Webb R, Fuchigami L, Boyer C (2005) Overexpression of cytosolic ascorbate peroxidase in tomato (*Lycopersicon esculentum*) confers tolerance to chilling and salt stress. J Am Soc Horticult Sci 130:167–173
Wang M, Gu D, Liu T, Wang Z, Guo X, Hou W et al (2007) Overexpression of a putative maize calcineurin B-like protein in *Arabidopsis* confers salt tolerance. Plant Mol Biol 65:733–746
Wang FW, Wang ML, Guo C, Wang N, Li XW, Chen H, Dong YY, Chen XF, Wang ZM, Li HY (2016a) Cloning and characterization of a novel betaine aldehyde dehydrogenase gene from *Suaeda corniculata*. Genet Mol Res 15(2). https://doi.org/10.4238/gmr.15027848
Wang H, Wang H, Shao H, Tang X (2016b) Recent advances in utilizing transcription factors to improve plant abiotic stress tolerance by transgenic technology. Front Plant Sci 7:67. https://doi.org/10.3389/fpls.2016.00067
Wang C, Lu G, Hao Y, Guo H, Guo Y, Zhao J, Cheng H (2017) *ABP9*, a maize bZIP transcription factor, enhances tolerance to salt and drought in transgenic cotton. Planta 246(3):453–469. https://doi.org/10.1007/s00425-017-2704-x
Wani SH, Dutta T, Neelapu NRR, Surekha C (2017) Transgenic approaches to enhance salt and drought tolerance in plants. Plant Gene 11:219–231. https://doi.org/10.1016/j.plgene.2017.05.006
Wei Q, Guo YJ, Cao HM, Kuai BK (2011) Cloning and characterization of an *AtNHX2*-like Na^+/H^+ antiporter gene from *Ammopiptanthus mongolicus* (Leguminosae) and its ectopic expression enhanced drought and salt tolerance in *Arabidopsis thaliana*. Plant Cell Tissue Organ Cult 105:309–316
Wei D, Zhang W, Wang C, Meng Q, Li G, Chen THH, Yang X (2017) Genetic engineering of the biosynthesis of glycine betaine leads to alleviate salt-induced potassium efflux and enhances salt tolerance in tomato plants. Plant Sci 257:74–83. https://doi.org/10.1016/j.plantsci.2017.01.012
Wen XP, Pang XM, Matsuda N, Kita M, Inoue H, Hao YJ et al (2008) Over-expression of the apple spermidine synthase gene in pear confers multiple abiotic stress tolerance by altering polyamine titers. Transgenic Res 17:251–263
Wu CA, Yang GD, Meng QW, Zheng CC (2004) The cotton *GhNHX1* gene encoding a novel putative tonoplast Na^+/H^+ antiporter plays an important role in salt stress. Plant Cell Physiol 45:600–607
Wu S, Yu Z, Wang F, Li W, Ye C, Li J, Tang J, Ding J, Zhao J, Wang B (2007) Cloning, characterization, and transformation of the phosphoethanolamine N-methyltransferase gene (*ZmPEAMT1*) in maize (*Zea mays* L.). Mol Biotechnol 36:102–112. https://doi.org/10.1007/s12033-007-0009-1
Wu C, Gao X, Kong X, Zhao Y, Zhang H (2009) Molecular cloning and functional analysis of a Na^+/H^+ antiporter gene *ThNHX1* from a halophytic plant *Thellungiella halophila*. Plant Mol Biol Rep 27:1–12
Wu S, Su Q, An LJ (2010) Isolation of choline monooxygenase (*CMO*) gene from *Salicornia europaea* and enhanced salt tolerance of transgenic tobacco with CMO genes. Indian J Biochem Biophys 47(5):298–305
Xiao-Yan Y, Ai-Fang Y, Ke-Wei Z, Ju-Ren Z (2004) Production and analysis of transgenic maize with improved salt tolerance by the introduction of *AtNHX1* gene. Acta Bot Sin 46:854–861
Xing Y, Jia W, Zhang J (2007) *AtMEK1* mediates stress-induced gene expression of *CAT1* catalase by triggering H_2O_2 production in *Arabidopsis*. J Exp Bot 58:2969–2981
Xiong L, Yang Y (2003) Disease resistance and abiotic stress tolerance in rice are inversely modulated by an abscisic acid-inducible mitogen-activated protein kinase. Plant Cell 5:745–759
Xu D, Duan X, Wang B, Hong B, Ho THD, Wu R (1996) Expression of a late embryogenesis abundant protein gene, *HVA1*, from barley confers tolerance to water deficit and salt stress in transgenic rice. Plant Physiol 110:249–257
Xu DQ, Huang J, Guo SQ, Yang X, Bao YM, Tang HJ, Zhang HS (2008) Overexpression of a TFIIIA-type zinc finger protein gene *ZFP252* enhances drought and salt tolerance in rice (*Oryza sativa* L.). FEBS Lett 582:1037–1043

Xu J, Tian YS, Peng RH, Xiong AS, Zhu B, Jin XF et al (2010) *AtCPK6*, a functionally redundant and positive regulator involved in salt/drought stress tolerance in *Arabidopsis*. Planta 231:1251–1260
Xu GY, Rocha PSCF, Wang ML, Xu ML, Cui YC, Li LY et al (2011) A novel rice calmodulin-like gene, *OsMSR2*, enhances drought and salt tolerance and increases ABA sensitivity in *Arabidopsis*. Planta 234:47–59. https://doi.org/10.1007/s00425-011-1386-z
Xu Y, Hu W, Liu J, Zhang J, Jia C, Miao H, Xu B, Jin Z (2014) A banana aquaporin gene, *MaPIP1;1*, is involved in tolerance to drought and salt stresses. BMC Plant Biol 14:59
Xu N, Chu Y, Chen H, Li X, Wu Q, Jin L et al (2018a) Rice transcription factor *OsMADS25* modulates root growth and confers salinity tolerance via the ABA-mediated regulatory pathway and ROS scavenging. PLoS Genet 14(10):e1007662. https://doi.org/10.1371/journal.pgen.1007662
Xu M, Chen C, Cai H, Wu L (2018b) Overexpression of *PeHKT1;1* improves salt tolerance in populus. Genes 9(475):1–15. https://doi.org/10.3390/genes9100475
Xue ZY, Zhi DY, Xue GP, Zhang H, Zhao YH, Xia GM (2004) Enhanced salt tolerance of transgenic wheat (*Triticum aestivum* L.) expressing a vacuolar Na^+/H^+ antiporter gene with improved grain yields in saline soils in the field and a reduced level of leaf Na^+. Plant Sci 167:849–859
Yadav SK, Singla-Pareek S, Ray M, Reddy MK, Sopory SK (2005) Transgenic tobacco plants overexpressing glyoxalase enzymes resist an increase in methylglyoxal and maintain higher reduced glutathione levels under salinity stress. FEBS Lett 579:6265–6271
Yang Q, Chen ZZ, Zhou XF, Yin HB, Li X, Xin XF et al (2009) Overexpression of *SOS* (*salt overly sensitive*) genes increases salt tolerance in transgenic *Arabidopsis*. Mol Plant 2:22–31
Ying S, Zhang DF, Li HY, Liu YH, Shi YS, Song YC et al (2011) Isolation and characterization of a maize *SnRK2* protein kinase gene confers enhanced salt tolerance in transgenic *Arabidopsis*. Plant Cell Rep 30(9):1683–1699. https://doi.org/10.1007/s00299-011-1077-z
Ying S, Zhang DF, Fu J, Shi YS, Song YC, Wang TY et al (2012) Cloning and characterization of a maize bZIP transcription factor, *ZmbZIP72*, confers drought and salt tolerance in transgenic *Arabidopsis*. Planta 235:253–266. https://doi.org/10.1007/s00425-011-1496-7
Yokotani N, Ichikawa T, Kondou Y, Matsui M, Hirochika H, Iwabuchi M et al (2009) Tolerance to various environmental stresses conferred by the salt-responsive rice gene *ONAC063* in transgenic *Arabidopsis*. Planta 229:1065–1075. https://doi.org/10.1007/s00425-009-0895-5
Yu J-Q, Wang J-H, Sun C-H, Zhang Q-Y, Hu D-G, Hao Y-J (2018) Ectopic expression of the apple nucleus encoded thylakoid protein *MdY3IP1* triggers early-flowering and enhanced salt tolerance in *Arabidopsis thaliana*. BMC Plant Biol 18:18. https://doi.org/10.1186/s12870-018-1232-6
Zhang HX, Blumwald E (2001) Transgenic salt-tolerant tomato plants accumulate salt in foliage but not in fruit. Nat Biotechnol 19:765–768
Zhang HX, Hodson JN, Williams JP, Blumwald E (2001) Engineering salt-tolerant Brassica plants: characterization of yield and seed oil quality in transgenic plants with increased vacuolar sodium accumulation. Proc Natl Acad Sci USA 98:12832–12836
Zhang J, Tan W, Yang X-H, Zhang H-X (2008a) Plastid-expressed choline monooxygenase gene improves salt and drought tolerance through accumulation of glycine betaine in tobacco. Plant Cell Rep 27:1113–1124. https://doi.org/10.1007/s00299-008-0549-2
Zhang GH, Su Q, An LJ, Wu S (2008b) Characterization and expression of a vacuolar Na^+/H^+ antiporter gene from the monocot halophyte *Aeluropus littoralis*. Plant Physiol Biochem 46:117–126
Zhang H, Mao X, Wang C, Jing R (2010) Overexpression of a common wheat gene *TaSnRK2.8* enhances tolerance to drought, salt and low temperature in *Arabidopsis*. PLoS ONE 5(12):e16041. https://doi.org/10.1371/journal.pone.0016041
Zhang L, Xi D, Li S, Gao Z, Zhao S, Shi J et al (2011a) A cotton group C MAP kinase gene, *GhMPK2*, positively regulates salt and drought tolerance in tobacco. Plant Mol Biol 77(1–2):17–31. https://doi.org/10.1007/s11103-011-9788-7
Zhang N, Si H-J, Wen G, Du H-H, Liu B-L, Wang D (2011b) Enhanced drought and salinity tolerance in transgenic potato plants with a BADH gene from spinach. Plant Biotechnol Rep 5:71–77. https://doi.org/10.1007/s11816-010-0160-1

Zhang X, Liu X, Wu L, Yu G, Wang X, Ma H (2015a) The *SsDREB* transcription factor from the succulent halophyte *Suaeda salsa* enhances abiotic stress tolerance in transgenic tobacco. Int J Genomics 13. Article ID 875497. https://doi.org/10.1155/2015/875497

Zhang L, Zhang L, Xia C, Zhao G, Liu J, Jia J et al (2015b) A novel wheat bZIP transcription factor, *TabZIP60*, confers multiple abiotic stress tolerances in transgenic *Arabidopsis*. Physiol Plant 153:538–554. https://doi.org/10.1111/ppl.12261

Zhang N, Yin Y, Liu X, Tong S, Xing J, Zhang Y, Pudake RN, Izquierdo EM, Peng H, Xin M, Hu Z et al (2017) The E3 ligase *TaSAP5* alters drought stress responses by promoting the degradation of DRIP proteins. Plant Physiol 175:1878–1892. https://doi.org/10.1104/pp.17.01319

Zhao FY, Zhang XJ, Li PH, Zhao YX, Zhang H (2006a) Co-expression of the *Suaeda salsa SsNHX1* and *Arabidopsis AVP1* confer greater salt tolerance to transgenic rice than the single *SsNHX1*. Mol Breed 17:341–353

Zhao FY, Guo SL, Zhang H, Zhao YX (2006b) Expression of yeast *SOD2* in transgenic rice results in increased salt tolerance. Plant Sci 170:216–224

Zhao JS, Zhi DY, Xue ZY (2007) Enhanced salt tolerance of transgenic progeny of tall fescue (*Festuca arundinacea*) expressing a vacuolar Na^+/H^+ antiporter gene from *Arabidopsis*. J Plant Physiol 164:1377–1383

Zhao J, Guo S, Chen S, Zhang H, Zhao Y (2009) Expression of yeast *YAP1* in transgenic *Arabidopsis* results in increased salt tolerance. J Plant Biol 52:56–64

Zhao W-T, Feng S-J, Li H, Faust F, Kleine T, Li L-N, Yang Z-M (2017) Salt stress-induced *FERROCHELATASE 1* improves resistance to salt stress by limiting sodium accumulation in *Arabidopsis thaliana*. Sci Rep 7(1):14737. https://doi.org/10.1038/s41598-017-13593-9

Zheng L, Chen S, Xie L, Lu Z, Liu M, Han X, Qiao G, Jiang J, Zhuo R, Qiu W (2018) Overexpression of cysteine protease gene from *Salix matsudana* enhances salt tolerance in transgenic *Arabidopsis*. Environ Exp Bot 147:53–62

Zhong H, Guo QQ, Chen L, Ren F, Wang QQ, Zheng Y et al (2012) Two *Brassica napus* genes encoding NAC transcription factors are involved in response to high-salinity stress. Plant Cell Rep 31:1991–2003. https://doi.org/10.1007/s00299-012-1311-3

Zhou S, Chen X, Zhang X, Li Y (2008) Improved salt tolerance in tobacco plants by co-transformation of a betaine synthesis gene *BADH* and a vacuolar Na^+/H^+ antiporter gene *SeNHX1*. Biotechnol Lett 30:369–376. https://doi.org/10.1007/s10529-007-9548-6

Zhou L, Wang NN, Gong SY, Lu R, Li Y, Li XB (2015) Overexpression of a cotton (*Gossypium hirsutum*) WRKY gene, *GhWRKY34*, in *Arabidopsis* enhances salt-tolerance of the transgenic plants. Plant Physiol Biochem 96:311–320. https://doi.org/10.1016/j.plaphy.2015.08.016

Zhu B, Su J, Chang MC, Verma DPS, Fan YL, Wu R (1998) Overexpression of a pyrroline-5-carboxylate synthetase gene and analysis of tolerance to water and salt stress in transgenic rice. Plant Sci 139:41–48

Zhu N, Cheng S, Liu X, Du H, Dai M, Zhou DX et al (2015) The R2R3-type MYB gene *OsMYB91* has a function in coordinating plant growth and salt stress tolerance in rice. Plant Sci 236:146–156. https://doi.org/10.1016/j.plantsci.2015.03.023

Chapter 15
Genome Engineering in Rice: Applications, Advancements and Future Perspectives

Shalini Tiwari and Charu Lata

Abstract Rice (*Oryza sativa* L.) is one of the essential cereal crops for the majority of the world's population. However, we need to ensure a continuous supply and enhanced the productivity of this crop in the purview of global climate change and increasing world population. Several crop improvement strategies, including genetic engineering and molecular breeding, have been routinely used to develop varieties superior in stress tolerance and yield. However, each one of them has limitations. Genome engineering or genome editing using targeted nucleases is recently being deployed as a key strategy to improve rice and other crops which promises a significant improvement in yield without the requirement of additional agricultural land in the future. Targeted genome editing using artificial nucleases has largely revolutionized the field of crops' genome modification. Several studies recently used Zinc-finger nucleases (ZFNs), transcription activator-like effector nucleases (TALENs), and clustered regularly interspaced short palindromic repeat (CRISPR)/CRISPR-associated protein (Cas) to successfully modulate genes in a precise and predictable manner in plants for gene function studies and crop improvement programmes. These techniques open up new prospects to develop improved plant lines by adding important traits or by removing undesirable traits. The ability of these technologies to perform targeted and efficient modifications in genome sequence will undoubtedly lead to novel developments in plants, including crop plants. Moreover, due to the non-insertion of foreign DNA, this technique is socially acceptable and may help to alleviate regulatory issues associated with genetically modified plants. In this review, we describe the recent advancement in the CRISPR/Cas9 system and also highlight the strengths and weaknesses of this technology in comparison to the other two well-established genome editing platforms (ZFNs and TALENs). We have also discussed the small size new protein named CasX, its DNA cleavage characteristics, and its

---

S. Tiwari · C. Lata
Council of Scientific and Industrial Research—National Botanical Research Institute (CSIR-NBRI), Rana Pratap Marg, Lucknow, Uttar Pradesh 226001, India

C. Lata (✉)
CSIR—National Institute of Science Communication and Information Resources, 14 Satsang Vihar Marg, New Delhi 110067, India
e-mail: charulata@niscair.res.in

S. P. Singh et al. (eds.), *Molecular Approaches in Plant Biology and Environmental Challenges*, Energy, Environment, and Sustainability,
https://doi.org/10.1007/978-981-15-0690-1_15

advantages over other CRISPR-Cas genome-editing enzymes. These technologies are mostly used for substitution of targeted gene fragments and insertion of exogenous DNA sequences into specific genomic location in crop plants that offer great potential for genetic improvement and breeding of rice.

Keywords CRISPR · Crop improvement · Genome editing · Nucleases · Rice

15.1 Introduction

Various abiotic stresses, including drought, salinity, heat, cold, flooding, and heavy metal stress, are major threats to crop production worldwide. From the past few decades, researchers have focused on understanding the molecular mechanism of plant adaption and tolerance response to these adverse environment conditions. However, it has been difficult to study complex interconnected signaling pathways that lead to multiple responses to abiotic stresses using traditional approaches due to involvement and complex networking of a large number of genes and gene products in various plant defensive and developmental processes. Various disciplines of functional genomic approaches for crop improvement under stress condition are namely, gene discovery, transcript profiling, altering gene expression by transformation technologies, genetical genomic-approaches, and genome editing through which complex cellular networks of stress perception, signaling, and defense response system can be examined from transcription, through cell complement proteins, to the metabolite profiles of stressed plant tissues (Tiwari et al. 2017a) (Fig. 15.1).

Genome editing tools make specific changes to the DNA of a cell or organism via the introduction of targeted mutation, insertion/deletion (indel), and sequence-specific alteration using customized nucleases. These customized nucleases create site-specific double-strand breaks (DSBs) at the targeted loci of the genome. Typically, these DSBs are repaired mainly by the two intracellular repair pathways, i.e., homology-directed repair/homology recombination (HDR/HR) and nonhomologous end joining (NHEJ) (Jain 2015). NHEJ uses various enzymes to directly join the DNA ends that lead to the introduction of indels. HDR uses a homologous sequence as a template for insertion of desired sequences and to introduce specific point mutations via recombination. ZFNs, TALENs, and CRISPR/Cas are the most commonly used genome editing tools (Jain 2015; Malzahn et al. 2017). Due to the simplicity, the CRISPR-cas system greatly facilitates the study of genome function and engineering under abiotic stress in various plants. This system has opened surplus options for genome editing in various biological perspectives.

In several major crops, recent studies have shown high efficacy of indels in addition to introducing exogenous DNA at a targeted locus, as e.g. rice (*Oryza sativa*), barley (*Hordeum vulgare*), maize (*Zea mays*), bread wheat (*Triticum*), soybean (*Glycine max*), sweet orange (*Citrus sinensis*), and tomato (*Solanum lycopersicum*) (Araki and Ishii 2015). Among cereal crops, rice is a major source of food for a large population around the globe. Numerous research based on targeted nucleases in rice

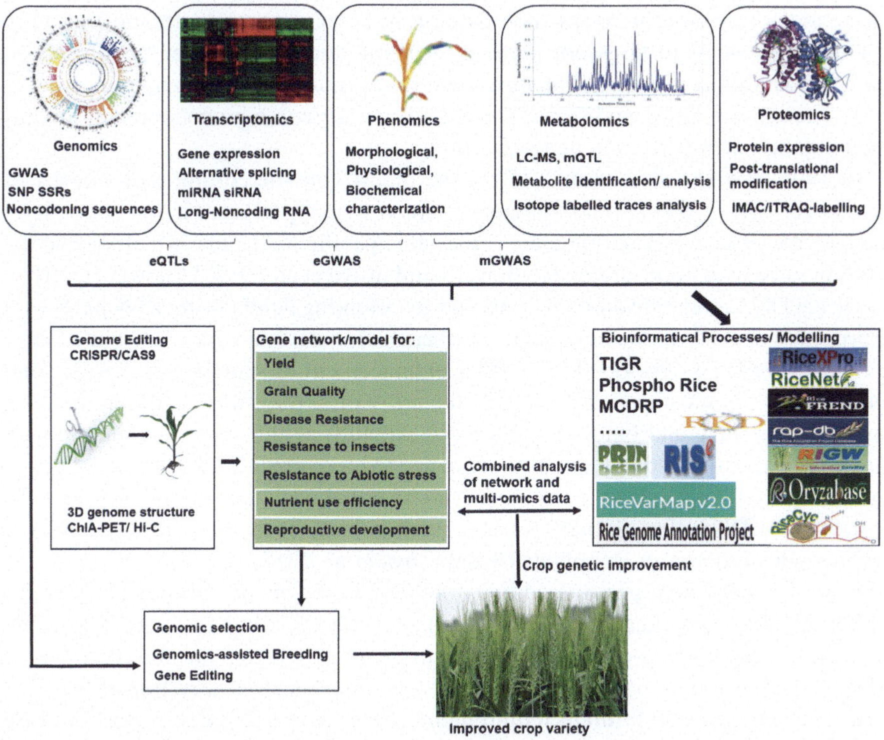

Fig. 15.1 A schematic representation of integrative functional genomic approaches for abiotic stress tolerance in rice

involves general genome editing studies (Malzahn et al. 2017; Lu and Zhu 2017; Li et al. 2016a; Petolino 2015). Several reports have confirmed that NHEJ-mediated indels can lead to both biotic and abiotic stress resistance (Kamburova et al. 2017). Therefore, this review briefly delineates the mechanisms of different genome editing technologies and their use to improve rice crops. The present study also highlights various applications and their advantages in rice using engineered nuclease-based technologies.

15.2 Gene Editing Tools: ZFN, TALEN & CRISPR-Cas System

To understand the complex network of genome structure and function, modern biotechnology has delivered several genome editing tools that allow precise and specific alteration of DNA sequences within the plant genomes in which the ZFNs, TALENs, and CRISPR-Cas are the most common tools used for genome editing.

These tools have been used for modification in plant genomes with application in a wide range of plant kingdom families. Several studies have been conducted on monocot and dicot plants to enhance yield and productivity and endurance capability against environmental stresses. Rice (*Oryza sativa*), a monocot plant, too has been mutated using different genome editing tools.

In 1996, a study showed that ZFNs, the oldest genome editing tool, consist of zinc finger (Zn-finger) protein domains coupled with type II restriction endonuclease enzyme *Fok*I catalytic domains, act as site-specific nucleases, which cleave the DNA in vitro in definite targeted regions (Kamburova et al. 2017; Urnov et al. 2010). To cleave DNA, catalytic domain dimerizes by orienting neighboring ZFN pairs with appropriate spacing at the target site. Furthermore, heterodimerization-related *Fok*I variants have been developed for further enhancement of sequence specificity and reduction of off-site cleavage. In ZFNs, the array of 4–6 Zn-finger protein domains fused to recognize and bind ~3 bp of DNA whereby within each module, domain junctions are optimized to recognize sequence (Moore et al. 2001). The most effective assembled ZFN is from a two-finger archive that each recognizes specific 6-bp sequences of DNA (Petolino 2015). This method became the basis for gene editing in the model and non-model plants (Kamburova et al. 2017).

After ZFNs, a new genome-editing tool TALENs was discovered in 2009 for targeting DNA via engineering transcription activator-like (TAL) effectors (Malzahn et al. 2017). In nature, these effectors are introduced into host plant cells by the type III secretion system of bacterium *Xanthomonas*. This bacterial secretion alters plant host gene expression to fulfill its requirements. In the nucleus, TAL effectors bind the promoter region of a target gene within the range of 60 bp of start codons and lead to activation of the transcription machinery (Malzahn et al. 2017). The central repeat sequences of the DNA binding region of each TAL effector is typically composed of 34 amino acids (Boch and Bonas 2010). Further, research on tobacco discovered another repeat variable di-residue (RVD) on each repeat that governs the binding specificity of nucleotide (Boch et al. 2009). This finding results in the discovery of site-specific nuclease, i.e., TAL effector nuclease (TALEN), based on the fusion of the *Fok*1 nuclease domain with the DNA binding TALE repeats (Mahfouz et al. 2011; Miller et al. 2011; Li et al. 2010). TALENs has been preferred over ZFNs because of their lesser toxicity as well as easier to engineer.

Just after two years of the invention of TALENs, a new genome-editing tool, CRISPR was introduced. CRISPR acts as an adaptive immune system in bacteria and archaebacteria against multiple invasive viruses, and phages (Malzahn et al. 2017; Wiedenheft et al. 2012). The bacteria can protect themselves by cleaving viral DNA using a series of CRISPR associated (Cas) proteins and insert fragments of viral DNA into their genomes. Then bacteria use Cas proteins associated with RNA to create targeted DSBs in invading viral DNA, transcribed from the viral DNA library.

The classification of CRISPR depends on the signature Cas proteins, i.e., Cas protein sequences, gene composition, and organization of the genomic loci (Tong et al. 2019; Van Der Oost et al. 2014). Therefore, the CRISPR system has been divided into two general classes. Class 1 (type I, type III, and type IV systems) involves the functioning of multiple Cas enzymes, while Class 2 (type II, type V, and type VI systems) requires only a single Cas protein to perform its function. Each subtype of Class 2 CRISPR systems utilize a distinctive Cas protein as an effector, e.g., type II use Cas9; type V uses Cas12a (Cpf1), Cas12b (C2c1), Cas12c (C2c3), Cas12d (CasY), and Cas12e (CasX); and type VI uses Cas13a (C2c2), Cas13b, and Cas13c (Koonin et al. 2017).

Cas9 is the most efficient and commonly used CRISPR protein that is composed of HNH and a Ruv C-like endonuclease domains, and each domain cut one strand of DNA. In 2012, Cas9 of *Streptococcus pyogenes* firstly demonstrated to be paired with a synthetic single guide RNA (gRNA) in vitro to create a specific DSB in DNA (Jinek et al. 2012). Shortly after, being a simple design and ease of vector construction, CRISPR-Cas9 was proven as a potent RNA guided single-strand nuclease for genome editing in human as well as in plants (Malzahn et al. 2017).

Among genome editing technologies, CRISPR-Cas9 is more efficient and broadly used in plant improvement applications than TALENs and ZFNs. The most important advantages of CRISPR/Cas9 in comparison to other genome editing technologies is its simplicity and efficiency. The detailed comparison of CRISPR-Cas9, TALENs, and ZFNs are mentioned in Table 15.1.

Table 15.1 Comparison of ZFN, TALEN and CRISPR-Cas9 systems

Characteristics	ZFNs	TALENs	CRISPR/Cas9
Components	Zn finger domains Nonspecific *Fok*I nuclease domain	TALE DNA-binding domains Nonspecific *Fok*I nuclease domain	crRNA,Cas9
Structural Proteins	Dimeric Protein	Dimeric Protein	Monomeric Protein
Catalytic domain	Restriction endonuclease *Fok*I	Restriction endonuclease *Fok*I	RUVC and HNH
Length of target sequence (bp)	24–36	24–59	20–22
Protein engineering steps	Required	Required	Should not be complex to test gRNA
Cloning	Necessary	Necessary	Not necessary
gRNA production	Not applicable	Not applicable	Easy to produce
Mode of action	DSB in target DNA	DSB in target DNA	DSB or SSN in target DNA
Target recognition Efficiency	High	High	High
Mutation rate	High	Middle	Low
Creation of large scale libraries	Impossible	Technically difficult	Possible
Multiplexing	Difficult	Difficult	Possible

15.3 Applications of Genome Editing in Rice Crop Improvement

Genome editing technique exhibits point mutations generation, new genes insertion or large regions deletion of nucleotide sequences and modification or replacement of elements and fragments of genes (Mishra et al. 2018). Even in plants, especially in rice, various types of genome alterations have been used by utilizing ZFN, TALEN, and CRISPR/Cas systems (Table 15.2). These technologies have broad applications for developing new varieties of crops, to increase their production, yield, nutritional value, and resistant to abiotic and biotic stresses (Mishra et al. 2018). Therefore, genome modification technologies have been used in plant breeding for various purposes i.e. (1) Make small modifications to gene function, (2) Insert point mutations similar to natural SNPs, (3) Insert foreign genes, (4) Gene knockout, and pyramiding (5) Activation or repression of gene expression, and (6) Epigenetic editing (Kamburova et al. 2017).

For example, the use of ZFN in *O. sativa* could efficiently cleave and stimulate mutation at SSIVa locus to affect plant height, grain filling, and starch content (Jung et al. 2018). In plants, ZFN has also been used for the targeted modification of endogenous malate dehydrogenase (MDH) gene, and increased yields were found in plants with modified MDH (Shukla et al. 2013). Cantos et al. (2014), identified "safe harbor" loci in the *indica* rice genome by using zinc-finger nucleases to induce and repair DNA damage.

In rice, several studies have reported about genes that could be helpful to develop resistance against pathogen attack by using TALEN technology. Li et al. (2012), used TALEN based disruption of rice bacterial blight susceptibility gene *Os11N3* in rice to prevent the virulence strategy of *Xanthomonas oryzae*. Similarly, Blanvillain-Baufumé et al. (2017), reported editing the activities of *SWEET14* by using TALE-nuclease editing for rice resistance to *X. oryzae*. A study by Li et al. (2019), demonstrated that *X. oryzae* (Xoo strain PXO99) T3 translocator *Hpa1* interacts with rice aquaporin *OsPIP1;3* at the plant plasma membrane to regulate TALE translocation from the bacterial cells into the cytosol of plant cells. The TALEN system has also investigated the suppression of the transcription factor *PHR2* by negatively regulating the phosphate starvation response by rice *SPX6* gene (Zhong et al. 2018). During storage, grain deterioration causes serious economic losses as it compromises the quality and seed longevity of rice. TALEN-based lipoxygenase LOX3 mutagenesis was reported to enhance the storage tolerance of rice seeds (Ma et al. 2015a). Yang et al. (2018) reported TALEN-induced genome editing of *OsSWEET11 and OsSWEET15* in rice caryopses that play central roles in seed filling.

Another most common and widely used genome editing tool CRISPR/Cas9 has also been used extensively to modify various genes to increase tolerance against environmental stresses. Several genes including *Gn1a*, *GS3*, *DEP1*, and *IPA1* have been edited separately in rice using the CRISPR/Cas9 that leads to improved phenotypes such as dense erect panicles, enhanced grain number, and grain size (Mishra et al.

Table 15.2 Examples of application of genome editing technologies in rice

Plant species	Target gene	Reference
ZFN mediated genome editing in rice		
Rice	*SSIVa*	Jung et al. (2018)
Rice	Safe harbour loci (Identification)	Cantos et al. (2014)
TALEN mediated genome editing in rice		
Rice	*PIP1;3*	Li et al. (2019)
Rice	*SWEET11, SWEET15*	Yang et al. (2018)
Rice	*SPX6*	Zhong et al. (2018)
Rice	*SWEET14*	Blanvillain-Baufumé et al. (2017)
Rice	*WAXY*	Nishizawa-Yokoi et al. (2016)
Rice	*MST7, MST8, PMS3, CSA, DERF1*	Zhang et al. (2016)
Rice	*ALS*	Li et al. (2016b)
Rice	*LOX3*	Ma et al. (2015a)
Rice	*EPSPS*	Wang et al. (2015a)
Rice	*BADH2*	Shan et al. (2015)
Rice	*DEP1, CKX2, SD1*	Shan et al. (2013a)
Rice	*11N3 (SWEET14)*	Li et al. (2012)
CRISPR-Cas9 mediated genome editing in rice		
Rice	*SPO11-1, REC8, OSD1, MATL*	Xie et al. (2019)
Rice	*NGv1*	Endo et al. (2019)
Rice	*BBM1*	Khanday et al. (2019)
Rice	*GUS, PDS, Chalk5*	Pathak et al. (2019)
Rice	*TOS17*	Saika et al. (2018)
Rice	*eIF4G*	Macovei et al. (2018)
Rice	*PMR*	Liu et al. (2017)
Rice	*MEGs, PEGs*	Yuan et al. (2017)
Rice	*Hd2, Hd4, Hd5*	Li et al. (2017)
Rice	*SBEI, SBEIIB*	Sun et al. (2017)
Rice	*ACT, GST*	Wang et al. (2017)
Rice	*RBOHH*	Yamauchi et al. (2017)
Rice	*EPFL9*	Yin et al. (2017)
Rice	*SWEET11*	Ma et al. (2017)
Rice	*RAV2*	Duan et al. (2016)
Rice	*DMC1A, DMC1B*	Mikami et al. (2016)
Rice	*NAL1, LPA1, LG1, GL1-1*	Hu et al. (2016)
Rice	*DEP1, ROC5*	Zheng et al. (2016)
Rice	*Gn1a, DEP1, GS3, IPA1*	Li et al. (2016c)

(continued)

Table 15.2 (continued)

Plant species	Target gene	Reference
Rice	*ERF922*	Wang et al. (2016)
Rice	*OST2*	Osakabe et al. (2016)
Rice	*CSA*	Li et al. (2016d)
Rice	*RUPO*	Liu et al. (2016a)
Rice	*TMS5*	Zhou et al. (2016)
Rice	*ALS*	Sun et al. (2016)
Rice	*ALS*	Endo et al. (2016)
Rice	*CDKA2, CDKB1, CDKB2*	Endo et al. (2015)
Rice	*MPK1, MPK2, MPK5, MPK6, PDS*	Xie et al. (2015)
Rice	*GSTU, MRP15, ANP, WAXY, 7 FTL genes, and 21 other genes*	Ma et al. (2015b)
Rice	*AOX1a, AOX1b, AOX1c, BEL*	Xu et al. (2015)
Rice	*DsRed (transgene), YSA, PDS, DL*	Mikami et al. (2015)
Rice	*P450, DWD1*	Woo et al. (2015)
Rice	*PDS, BADH2, MPK2, 02g23823*	Wang et al. (2015b)
Rice	*ROC5, SPP, YSA*	Zhang et al. (2014)
Rice	*BEL*	Xu et al. (2014)
Rice	*SWEET11, SWEET13, SWEET1a, SWEET1b, CPS4, CYP99A2, CYP76M5, CYP76M6, KO1, KOL5*	Zhou et al. (2014)
Rice	*MYB1*	Mao et al. (2013)
Rice	*PDS, BADH2, MPK2, 02g23823*	Shan et al. (2013b)
Rice	*SWEET11, SWEET14*	Jiang et al. (2013)
Rice	*CAO1, LAZY1*	Miao et al. (2013)

2018; Li et al. 2016c). A study also reported efficient editing of group I and group II *PYL* genes using this technology, resulting in improved growth and productivity of rice (Miao et al. 2018). Genome editing technologies were also used to produce bacterial leaf blight-resistant plants caused by *Xanthomonas oryzae* pv. *Oryzae* (Li et al. 2012). This technology enhanced plant immune system by successfully modifying the ethylene-responsive factor *OsERF922* gene, resulting in increased resistance to *Magnaporthe oryzae* and salt stress (Wang et al. 2016; Liu et al. 2012). In another effort, Osakabe et al. (2016), obtained new alleles that conferred resistance to salt stress using the CRISPR-induced mutagenesis of *OST2* gene in rice. CRISPR/Cas9-targeted mutagenesis generated novel rice alleles eIF4G that confers resistance to rice tungro spherical virus (Macovei et al. 2018). Similarly, the CRISPR/Cas9 technique was effectively used to acquire a herbicide-resistant crop (Svitashev et al. 2015). Further, editing of the *ALS* gene in rice created chlorsulfuron (herbicide)-resistant mutant rice plant (Sun et al. 2016).

Using the CRISPR/Cas9 technology, researchers are also able to identify the functions of various genes at the molecular level. Alteration using CRISPR/Cas9 of *BBM1* gene (a member of the AP2 family of transcription factors) which is expressed in sperm cells resulted in heritable synthetic asexual-propagation trait thus paving the way for asexual reproduction in rice crop (Khanday et al. 2019). Similarly, Xie et al. (2019), introduced apomixis, a type of asexual reproduction, into rice by mutating *OsSPO11-1*, *OsREC8*, *OsOSD1*, and *OsMATL* through a CRISPR/Cas9 system. Interestingly, a successful example of targeted deletion of rice retrotransposon *Tos17* via CRISPR/Cas9 suggested a novel alternative approach to plant breeding (Saika et al. 2018).

Another area of interest is to develop rice varieties to fulfill increasing public health nutrition problems. High level of Amylase Content (AC) and Resistant Starch (RS) reduce the risk of various diseases, including hypertension, diabetes, and colon cancers that lead to improved human health (Chen et al. 2012). Therefore, the CRISPR/Cas9 was used targeting rice branching enzyme (SBEIIb) to produce amylose content rich rice. SBEIIb mutated rice plants showed a considerable increase in AC and RS content suggesting the significant role of SBEII in determining the fine structure and nutritional properties of starch in rice varieties. (Sun et al. 2017).

15.4 Innovations in Genome Editing

Innovative breeding technology is urgently required to enhanced agricultural production that boosts global access to nutritious foods. For the betterment of rice crop, among various genome editing technologies, CRISPR/Cas system has been used extensively, in which, CRISPR/Cas9 and Cas12a are the only two kinds that are the foundation of this revolutionary technology. Recently, a new protein has been discovered by metagenomic analysis of microbial DNA from groundwater samples, referred to as CasX (also known as Cas12e5). This newly discovered protein, when expressed with cognate crRNAs target the plasmid DNA to, prevents bacterial transformation (Burstein et al. 2017). Biochemical and in vivo data determined the activity of CasX for modification in *Escherichia coli* and human genome (Liu et al. 2019). However, genome editing by application of CasX enzyme in plant system is yet to be performed. Sequence analysis of CasX showed no similarity to other CRISPR-cas enzymes, except for the presence of a RuvC nuclease domain present in both Cas9 and Cas12a enzyme families, transposases, and recombinases (Burstein et al. 2017; Liu et al. 2019). Due to the smaller size of CasX (<1000 amino acids), its DNA cleaving properties and its origin from non-pathogenic microorganisms suggest its significant advantages over other CRISPR/Cas genome-editing enzymes.

Recent advancement in genome editing techniques enables efficient, targeted genome modification of most crop plants, hence encouraging crop improvement. Wang et al. (2019), in his study generated two variants of cas9 (cas9 3.6 and Cas9 3.7) and observed that xCas9 3.7 functioning is better in comparison to xCas9 3.6 variant for editing of rice genome, therefore, indicating the ability of xCas9 variants as versatile tools that will expand the scope of rice genome editing.

15.5 The Need for Genome Editing and Its Social Acceptance

Genome editing systems are efficient tools that are capable of manipulating the genomes of animals as well as plants. Sequence-specific nucleases, including ZFNs, TALENs, and CRISPR/Cas, have been used for effective targeted mutagenesis in a wide range of plant kingdom species (Baltes and Voytas 2015; Voytas 2013). However, delivery of these nucleases using *Agrobacterium* or protoplast transformation may lead to unwanted genetic alterations due to random integration of DNA encoding nuclease into the host genome. This random DNA incorporation into the host genome may lead to unexpected inactivation or alteration of the expression of the host gene that hinders trait improvement and gene function studies. Therefore, methods for modification of plant genomes that do not involve DNA delivery system would have a higher value than other techniques. In plants, this could solve complex problems, including the concern related to transgenic plants (Van Der Oost et al. 2014). This genome editing approach enabled targeted mutation of endogenous sequences of plant cell though avoiding the insertion of foreign DNA into the genome.

15.6 Conclusion and Future Perspective

In conclusion, recently, gene editing tools, especially the CRISPR/Cas9 system, have become more important in defining plant research. It has emerged as the most efficient tool for crop improvement due to its capability to induce mutations at desired targets in the genome with higher precision, effectiveness, and simplicity. Exploring the different functions of any specific gene in the individual cells or the whole organism, various modifications have been introduced in the rice crop plant. Genome editing has also become a powerful tool against environmental stresses, including molecular plant-microbe interactions and disease resistance breeding. Several examples witness applications of genome editing in generating plants that are resistant to various pathogens and tolerance to abiotic stresses. However, the application of genome editing technologies in plant-PGPR (plant growth promoting rhizobacteria) interaction is still unknown. Numerous genes, with the potential to aid the plant-PGPR interaction, have been reported by several researchers (Tiwari et al. 2016, 2017b). Application of the genome editing technology in designing more competent PGPR could also benefit the plant system. Also, the interaction between plants and microbes play an important role in heavy metal uptake and tolerance for the successful survival and growth of plants in contaminated soils (Tiwari and Lata 2018). Use of microbe-derived from genome modification could also take phytoremediation to the next level, enabling the successful reclamation of contaminated soil. Furthermore, genome editing-based epigenetic regulation is also more promising in crop improvement by manipulating DNA methylation, and histone modification as such modifications can be inherited in plants offsprings without altering the genomic sequence. Although this approach target regions of the mammalian cells (Liu et al. 2016b), but such epigenome editing tools are yet to be discovered in the plant system. This advancement in plant system will expand the dimension of the genome editing technologies for improvement of rice as well as other crops. On the other hand, recently discovered and successful application of CasX enzyme in human genome modification also promote the efficient application of this admirable technology in rice and other crop improvements. Thus, this study on genome editing technology highlighted the importance of producing modified crop plants and brought about a revolutionary change in rice enhancement that is essential for food security for meeting the demands and ensuring the need for rice for future generations.

Acknowledgements CL acknowledges "Early Career Research Award (ECRA)" by Science & Engineering Research Board (SERB), Government of India [Grant No. ECR/2017/001593]. Authors are also thankful to the Director, CSIR—National Botanical Research Institute, Lucknow, India, for providing facilities and support for the study.

References

Araki M, Ishii T (2015) Towards social acceptance of plant breeding by genome editing. Trends Plant Sci 20(3):145–149

Baltes NJ, Voytas DF (2015) Enabling plant synthetic biology through genome engineering. Trends Biotechnol 33(2):120–131

Blanvillain-Baufumé S, Reschke M, Solé M, Auguy F, Doucoure H, Szurek B, Meynard D, Portefaix M, Cunnac S, Guiderdoni E, Boch J (2017) Targeted promoter editing for rice resistance to *Xanthomonas oryzae pv. oryzae* reveals differential activities for SWEET 14-inducing TAL effectors. Plant Biotechnol J 15(3):306–317

Boch J, Bonas U (2010) Xanthomonas AvrBs3 family-type III effectors: discovery and function. Annu Rev Phytopathol 48:419–436

Boch J, Scholze H, Schornack S, Landgraf A, Hahn S, Kay S, Lahaye T, Nickstadt A, Bonas U (2009) Breaking the code of DNA binding specificity of TAL-type III effectors. Science 326(5959):1509–1512

Burstein D, Harrington LB, Strutt SC, Probst AJ, Anantharaman K, Thomas BC, Doudna JA, Banfield JF (2017) New CRISPR—Cas systems from uncultivated microbes. Nature 542(7640):237

Cantos C, Francisco P, Trijatmiko KR, Slamet-Loedin I, Chadha-Mohanty PK (2014) Identification of "safe harbor" loci in indica rice genome by harnessing the property of zinc-finger nucleases to induce DNA damage and repair. Front Plant Sci 5:302

Chen L, Magliano DJ, Zimmet PZ (2012) The worldwide epidemiology of type 2 diabetes mellitus-present and future perspectives. Nat Rev Endocrinol 8:228–236. https://doi.org/10.1038/nrendo.2011.183

Duan YB, Li J, Qin RY, Xu RF, Li H, Yang YC, Ma H, Li L, Wei PC, Yang JB (2016) Identification of a regulatory element responsible for salt induction of rice OsRAV2 through ex situ and in situ promoter analysis. Plant Mol Biol 90(1–2):49–62

Endo M, Mikami M, Toki S (2015) Multigene knockout utilizing off-target mutations of the CRISPR/Cas9 system in rice. Plant Cell Physiol 56(1):41–47

Endo M, Mikami M, Toki S (2016) Biallelic gene targeting in rice. Plant Physiol 170(2):667–677

Endo M, Mikami M, Endo A, Kaya H, Itoh T, Nishimasu H, Nureki O, Toki S (2019) Genome editing in plants by engineered CRISPR-Cas9 recognizing NG PAM. Nat Plants 5(1):14

Hu X, Wang C, Fu Y, Liu Q, Jiao X, Wang K (2016) Expanding the range of CRISPR/Cas9 genome editing in rice. Mol Plant 9(6):943–945

Jain M (2015) Function genomics of abiotic stress tolerance in plants: a CRISPR approach. Front Plant Sci 6:375

Jiang W, Zhou H, Bi H, Fromm M, Yang B, Weeks DP (2013) Demonstration of CRISPR/Cas9/sgRNA-mediated targeted gene modification in *Arabidopsis*, tobacco, sorghum and rice. Nucleic Acids Res 41(20):e188–e188

Jinek M, Chylinski K, Fonfara I, Hauer M, Doudna JA, Charpentier E (2012) A programmable dual-RNA–guided DNA endonuclease in adaptive bacterial immunity. Science 337(6096):816–821

Jung YJ, Nogoy FM, Lee SK, Cho YG, Kang KK (2018) Application of ZFN for site directed mutagenesis of rice SSIVa gene. Biotechnol Bioprocess Eng 23(1):108–115

Kamburova VS, Nikitina EV, Shermatov SE, Buriev ZT, Kumpatla SP, Emani C, Abdurakhmonov IY (2017) Genome editing in plants: an overview of tools and applications. Int J Agro

Khanday I, Skinner D, Yang B, Mercier R, Sundaresan V (2019) A male-expressed rice embryogenic trigger redirected for asexual propagation through seeds. Nature 565(7737):91

Koonin EV, Makarova KS, Zhang F (2017) Diversity, classification and evolution of CRISPR-Cas systems. Curr Opin Microbiol 37:67–78

Li T, Huang S, Jiang WZ, Wright D, Spalding MH, Weeks DP, Yang B (2010) TAL nucleases (TALNs): hybrid proteins composed of TAL effectors and FokI DNA-cleavage domain. Nucleic Acids Res 39(1):359–372

Li T, Liu B, Spalding MH, Weeks DP, Yang B (2012) High-efficiency TALEN-based gene editing produces disease-resistant rice. Nat Biotechnol 30(5):390

Li J, Meng X, Zong Y, Chen K, Zhang H, Liu J, Li J, Gao C (2016a) Gene replacements and insertions in rice by intron targeting using CRISPR–Cas9. Nature plants 2(10):16139

Li T, Liu B, Chen CY, Yang B (2016b) TALEN-mediated homologous recombination produces site-directed DNA base change and herbicide-resistant rice. J Genet Genomics 43(5):297–305

Li M, Li X, Zhou Z, Wu P, Fang M, Pan X et al (2016c) Reassessment of the four yield-related genes *Gn1a*, *DEP1*, *GS3*, and *IPA1* in rice using a CRISPR/Cas9 system. Front Plant Sci 7:377. https://doi.org/10.3389/fpls.2016.00377

Li Q, Zhang D, Chen M, Liang W, Wei J, Qi Y, Yuan Z (2016d) Development of japonica photo-sensitive genic male sterile rice lines by editing carbon starved anther using CRISPR/Cas9. J Genet Genomics (Yi chuan xue bao) 43(6):415

Li X, Zhou W, Ren Y, Tian X, Lv T, Wang Z, Fang J, Chu C, Yang J, Bu Q (2017) High-efficiency breeding of early-maturing rice cultivars via CRISPR/Cas9-mediated genome editing. J Genet Genomics (Yi chuan xue bao) 44(3):175

Li P, Zhang L, Mo X, Ji H, Bian H, Hu Y, Majid T, Long J, Pang H, Tao Y, Ma J (2019) Aquaporin PIP1; 3 of rice and harpin Hpa1 of bacterial blight pathogen cooperate in a type III effector translocation. J Exp Bot 70(12):3057–3073

Liu D, Chen X, Liu J, Ye J, Guo Z (2012) The rice ERF transcription factor *OsERF922* negatively regulates resistance to *Magnaporthe oryzae* and salt tolerance. J Exp Bot 63(10):3899–3911

Liu L, Zheng C, Kuang B, Wei L, Yan L, Wang T (2016a) Receptor-like kinase RUPO interacts with potassium transporters to regulate pollen tube growth and integrity in rice. PLoS Genet 12(7):e1006085

Liu XS, Wu H, Ji X, Stelzer Y, Wu X, Czauderna S, Shu J, Dadon D, Young RA, Jaenisch R (2016b) Editing DNA methylation in the mammalian genome. Cell 167(1):233–247

Liu Y, Xu Y, Ling S, Liu S, Yao J (2017) Anther-preferential expressing gene PMR is essential for the mitosis of pollen development in rice. Plant Cell Rep 36(6):919–931

Liu JJ, Orlova N, Oakes BL, Ma E, Spinner HB, Baney KL, Chuck J, Tan D, Knott GJ, Harrington LB, Al-Shayeb B (2019) CasX enzymes comprise a distinct family of RNA-guided genome editors. Nature 566(7743):218

Lu Y, Zhu JK (2017) Precise editing of a target base in the rice genome using a modified CRISPR/Cas9 system. Mol Plant 10(3):523–525

Ma L, Zhu F, Li Z, Zhang J, Li X, Dong J, Wang T (2015a) TALEN-based mutagenesis of lipoxygenase LOX3 enhances the storage tolerance of rice (*Oryza sativa*) seeds. PLoS ONE 10(12):e0143877

Ma X, Zhang Q, Zhu Q, Liu W, Chen Y, Qiu R, Wang B, Yang Z, Li H, Lin Y, Xie Y (2015b) A robust CRISPR/Cas9 system for convenient, high-efficiency multiplex genome editing in monocot and dicot plants. Mol Plant 8(8):1274–1284

Ma L, Zhang D, Miao Q, Yang J, Xuan Y, Hu Y (2017) Essential role of sugar transporter OsSWEET11 during the early stage of rice grain filling. Plant Cell Physiol 58(5):863–873

Macovei A, Sevilla NR, Cantos C, Jonson GB, Slamet-Loedin I, Čermák T, Voytas DF, Choi IR, Chadha-Mohanty P (2018) Novel alleles of rice eIF4G generated by CRISPR/Cas9-targeted mutagenesis confer resistance to rice tungro spherical virus. Plant Biotechnol J 16(11):1918–1927

Mahfouz MM, Li L, Shamimuzzaman M, Wibowo A, Fang X, Zhu JK (2011) De novo-engineered transcription activator-like effector (TALE) hybrid nuclease with novel DNA binding specificity creates double-strand breaks. Proc Natl Acad Sci 108(6):2623–2628

Malzahn A, Lowder L, Qi Y (2017) Plant genome editing with TALEN and CRISPR. Cell Biosci 7(1):21

Mao Y, Zhang H, Xu N, Zhang B, Gou F, Zhu JK (2013) Application of the CRISPR-Cas system for efficient genome engineering in plants. Mol Plant 6(6):2008–2011

Miao J, Guo D, Zhang J, Huang Q, Qin G, Zhang X, Wan J, Gu H, Qu LJ (2013) Targeted mutagenesis in rice using CRISPR-Cas system. Cell Res 23(10):1233

Miao C, Xiao L, Hua K, Zoua C, Zhao Y, Bressanb RA et al (2018) Mutations in a subfamily of abscisic acid receptor genes promote rice growth and productivity. Proc Natl Acad Sci USA 115:6058–6063. https://doi.org/10.1073/pnas.1804774115

Mikami M, Toki S, Endo M (2015) Parameters affecting frequency of CRISPR/Cas9 mediated targeted mutagenesis in rice. Plant Cell Rep 34(10):1807–1815

Mikami M, Toki S, Endo M (2016) Precision targeted mutagenesis via Cas9 paired nickases in rice. Plant Cell Physiol 57(5):1058–1068

Miller JC, Tan S, Qiao G, Barlow KA, Wang J, Xia DF, Meng X, Paschon DE, Leung E, Hinkley SJ, Dulay GP (2011) A TALE nuclease architecture for efficient genome editing. Nat Biotechnol 29(2):143

Mishra R, Joshi RK, Zhao K (2018) Genome editing in rice: recent advances, challenges, and future implications. Front Plant Sci 9

Moore M, Klug A, Choo Y (2001) Improved DNA binding specificity from polyzinc finger peptides by using strings of two-finger units. Proc Natl Acad Sci 98(4):1437–1441

Nishizawa-Yokoi A, Cermak T, Hoshino T, Sugimoto K, Saika H, Mori A, Osakabe K, Hamada M, Katayose Y, Starker C, Voytas DF (2016) A defect in DNA Ligase4 enhances the frequency of TALEN-mediated targeted mutagenesis in rice. Plant Physiol 170(2):653–666

Osakabe Y, Watanabe T, Sugano SS, Ueta R, Ishihara R, Shinozaki K, Osakabe K (2016) Optimization of CRISPR/Cas9 genome editing to modify abiotic stress responses in plants. Sci Rep 6:26685

Pathak B, Zhao S, Manoharan M, Srivastava V (2019) Dual-targeting by CRISPR/Cas9 leads to efficient point mutagenesis but only rare targeted deletions in the rice genome. 3 Biotech 9(4):158

Petolino JF (2015) Genome editing in plants via designed zinc finger nucleases. In Vitro Cell Dev Biol-Plant 51(1):1–8

Saika H, Mori A, Endo M, Toki S (2018) Targeted deletion of rice retrotransposon Tos17 via CRISPR/Cas9. Plant Cell Rep 1–4

Shan Q, Wang Y, Chen K, Liang Z, Li J, Zhang Y, Zhang K, Liu J, Voytas DF, Zheng X, Zhang Y (2013a) Rapid and efficient gene modification in rice and *Brachypodium* using TALENs. Mol Plant 6(4):1365–1368

Shan Q, Wang Y, Li J, Zhang Y, Chen K, Liang Z, Zhang K, Liu J, Xi JJ, Qiu JL, Gao C (2013b) Targeted genome modification of crop plants using a CRISPR-Cas system. Nat Biotechnol 31(8):686

Shan Q, Zhang Y, Chen K, Zhang K, Gao C (2015) Creation of fragrant rice by targeted knockout of the *OsBADH2* gene using TALEN technology. Plant Biotechnol J 13(6):791–800

Shukla V, Gupta M, Urnov F, Guschin D, Jan M, Bundock P (2013) Targeted modification of malate dehydrogenase. WO Patent Publication Number: WO 2013166315 A1

Sun Y, Zhang X, Wu C, He Y, Ma Y, Hou H, Guo X, Du W, Zhao Y, Xia L (2016) Engineering herbicide-resistant rice plants through CRISPR/Cas9-mediated homologous recombination of acetolactate synthase. Mol Plant 9(4):628–631

Sun Y, Jiao G, Liu Z, Zhang X, Li J, Guo X et al (2017) Generation of high-amylose rice through CRISPR/Cas9-mediated targeted mutagenesis of starch branching enzymes. Front Plant Sci 8:298. https://doi.org/10.3389/fpls.2017.00298

Svitashev S, Young JK, Schwartz C, Gao H, Falco SC, Cigan AM (2015) Targeted mutagenesis, precise gene editing, and site-specific gene insertion in maize using Cas9 and guide RNA. Plant Physiol 169(2):931–945

Tiwari S, Lata C (2018) Heavy metal stress, signaling, and tolerance due to plant-associated microbes: an overview. Front Plant Sci 9:452

Tiwari S, Lata C, Chauhan PS, Nautiyal CS (2016) *Pseudomonas putida* attunes morphophysiological, biochemical and molecular responses in *Cicer arietinum* L. during drought stress and recovery. Plant Physiol Biochem 99:108–117

Tiwari S, Lata C, Singh Chauhan P, Prasad V, Prasad M (2017a) A functional genomic perspective on drought signalling and its crosstalk with phytohormone-mediated signalling pathways in plants. Curr Genomics 18(6):469–482

Tiwari S, Prasad V, Chauhan PS, Lata C (2017b) *Bacillus amyloliquefaciens* confers tolerance to various abiotic stresses and modulates plant response to phytohormones through osmoprotection and gene expression regulation in rice. Front Plant Sci 8:1510

Tong Y, Weber T, Lee SY (2019) CRISPR/Cas-based genome engineering in natural product discovery. Nat Prod Rep

Urnov FD, Rebar EJ, Holmes MC, Zhang HS, Gregory PD (2010) Genome editing with engineered zinc finger nucleases. Nat Rev Genet 11(9):636

Van Der Oost J, Westra ER, Jackson RN, Wiedenheft B (2014) Unravelling the structural and mechanistic basis of CRISPR-Cas systems. Nat Rev Microbiol 12(7):479

Voytas DF (2013) Plant genome engineering with sequence-specific nucleases. Annu Rev Plant Biol 64:327–350

Wang M, Liu Y, Zhang C, Liu J, Liu X, Wang L, Wang W, Chen H, Wei C, Ye X, Li X (2015a) Gene editing by co-transformation of TALEN and chimeric RNA/DNA oligonucleotides on the rice *OsEPSPS* gene and the inheritance of mutations. PLoS ONE 10(4):e0122755

Wang C, Shen L, Fu Y, Yan C, Wang K (2015b) A simple CRISPR/Cas9 system for multiplex genome editing in rice. J Genet Genomics 42(12):703–706

Wang F, Wang C, Liu P, Lei C, Hao W, Gao Y, Liu YG, Zhao K (2016) Enhanced rice blast resistance by CRISPR/Cas9-targeted mutagenesis of the ERF transcription factor gene *OsERF922*. PLoS ONE 11(4):e0154027

Wang M, Lu Y, Botella JR, Mao Y, Hua K, Zhu JK (2017) Gene targeting by homology-directed repair in rice using a gemini virus-based CRISPR/Cas9 system. Mol Plant 10(7):1007–1010

Wang J, Meng X, Hu X, Sun T, Li J, Wang K, Yu H (2019) xCas9 expands the scope of genome editing with reduced efficiency in rice. Plant Biotechnol J 17(4):709

Wiedenheft B, Sternberg SH, Doudna JA (2012) RNA-guided genetic silencing systems in bacteria and archaea. Nature 482(7385):331

Woo JW, Kim J, Kwon SI, Corvalán C, Cho SW, Kim H, Kim SG, Kim ST, Choe S, Kim JS (2015) DNA-free genome editing in plants with preassembled CRISPR-Cas9 ribonucleoproteins. Nat Biotechnol 33(11):1162

Xie K, Minkenberg B, Yang Y (2015) Boosting CRISPR/Cas9 multiplex editing capability with the endogenous tRNA-processing system. Proc Natl Acad Sci 112(11):3570–3575

Xie E, Li Y, Tang D, Lv Y, Shen Y, Cheng Z (2019) A strategy for generating rice apomixis by gene editing. J Integr Plant Biol 61(8):911–916

Xu R, Li H, Qin R, Wang L, Li L, Wei P, Yang J (2014) Gene targeting using the *Agrobacterium tumefaciens*-mediated CRISPR-Cas system in rice. Rice 7(1):5

Xu RF, Li H, Qin RY, Li J, Qiu CH, Yang YC, Ma H, Li L, Wei PC, Yang JB (2015) Generation of inheritable and "transgene clean" targeted genome-modified rice in later generations using the CRISPR/Cas9 system. Sci Rep 5:11491

Yamauchi T, Yoshioka M, Fukazawa A, Mori H, Nishizawa NK, Tsutsumi N, Yoshioka H, Nakazono M (2017) An NADPH oxidase RBOH functions in rice roots during lysigenous aerenchyma formation under oxygen-deficient conditions. Plant Cell 29(4):775–790

Yang J, Luo D, Yang B, Frommer WB, Eom JS (2018) SWEET 11 and 15 as key players in seed filling in rice. New Phytol 218(2):604–615

Yin X, Biswal AK, Dionora J, Perdigon KM, Balahadia CP, Mazumdar S, Chater C, Lin HC, Coe RA, Kretzschmar T et al (2017) CRISPR-Cas9 and CRISPR-Cpf1 mediated targeting of a stomatal developmental gene *EPFL9* in rice. Plant Cell Rep. https://doi.org/10.1007/s00299-017-2118-z

Yuan J, Chen S, Jiao W, Wang L, Wang L, Ye W, Lu J, Hong D, You S, Cheng Z, Yang DL (2017) Both maternally and paternally imprinted genes regulate seed development in rice. New Phytol 216(2):373–387

Zhang H, Zhang J, Wei P, Zhang B, Gou F, Feng Z, Mao Y, Yang L, Zhang H, Xu N, Zhu JK (2014) The CRISPR/Cas9 system produces specific and homozygous targeted gene editing in rice in one generation. Plant Biotechnol J 12(6):797–807

Zhang H, Gou F, Zhang J, Liu W, Li Q, Mao Y, Botella JR, Zhu JK (2016) TALEN-mediated targeted mutagenesis produces a large variety of heritable mutations in rice. Plant Biotechnol J 14(1):186–194

Zheng X, Yang S, Zhang D, Zhong Z, Tang X, Deng K, Zhou J, Qi Y, Zhang Y (2016) Effective screen of CRISPR/Cas9-induced mutants in rice by single-strand conformation polymorphism. Plant Cell Rep 35(7):1545–1554

Zhong Y, Wang Y, Guo J, Zhu X, Shi J, He Q, Liu Y, Wu Y, Zhang L, Lv Q, Mao C (2018) Rice SPX6 negatively regulates the phosphate starvation response through suppression of the transcription factor PHR2. New Phytol 219(1):135–148

Zhou H, Liu B, Weeks DP, Spalding MH, Yang B (2014) Large chromosomal deletions and heritable small genetic changes induced by CRISPR/Cas9 in rice. Nucleic Acids Res 42(17):10903–10914

Zhou H, He M, Li J, Chen L, Huang Z, Zheng S, Zhu L, Ni E, Jiang D, Zhao B, Zhuang C (2016) Development of commercial thermo-sensitive genic male sterile rice accelerates hybrid rice breeding using the CRISPR/Cas9-mediated TMS5 editing system. Sci Rep 6:37395

Chapter 16
Secondary Metabolite Pathways in Medicinal Plants: Approaches in Reconstruction and Analysis

Sumya Pathak, Aditya Vikram Agarwal, Parul Agarwal and Prabodh Kumar Trivedi

Abstract Medicinal plants are the abundant source of varied secondary metabolites with important therapeutic properties. These plants and their extracts have been the basis of several traditional medicines and their usage and demand has grown ever since. This ever increasing fetish has attracted plant biologists, who in the past two decades have made enormous efforts towards exploring and engineering the biosynthetic pathways of these sparsely available molecules. Until recently, endeavors to unravel biosynthetic pathways were limited mainly due to limited plant genomics resources. However, recent advancements in generating high-throughput "omics" datasets, computational tools, functional genomics approaches and analytical methods, along with their seamless integration have leads to the explanation of biosynthetic pathways enormous plant bio-active metabolites. Researchers have gone a step ahead in creating alternative sustainable source of these biomolecules through synthetic biology approaches, thereby developing microbial systems producing plant origin bioactive metabolites. Here, we have reviewed the contributions of major biotechnological approaches and their integration towards elucidating, analyzing and reconstructing biosynthetic pathways of bioactive metabolites in plants. We have briefly discussed different approaches that utilize omics datasets to extract biologically relevant knowledge with intentions to build in depth understanding of metabolic models of secondary metabolite biosynthesis.

Keywords Secondary metabolite · -Omics · Trascriptomics · Metabolomics · Abiotic stress · Metabolic pathways

S. Pathak (✉) · A. V. Agarwal
DST—Centre for Policy Research, Babasaheb Bhimrao Ambedkar University, Vidya Vihar, Raebareli Road, Lucknow 226025, India
e-mail: saumyapathak18@gmail.com

P. Agarwal · P. K. Trivedi
Council of Scientific and Industrial Research—National Botanical Research Institute (CSIR-NBRI), Rana Pratap Marg, Lucknow, Uttar Pradesh 226001, India

S. P. Singh et al. (eds.), *Molecular Approaches in Plant Biology and Environmental Challenges*, Energy, Environment, and Sustainability,
https://doi.org/10.1007/978-981-15-0690-1_16

16.1 Introduction

Plants have been the part of traditional medicine system worldwide for the cure of various ailments since the dawn of civilization. The bioactive principal of the most plant based drugs has been characterized as small organic molecules known as secondary metabolites. Plants contain enormous secondary metabolism pathways that are accountable for the synthesis of complex metabolites which play a vital role in moulding the interdependencies and diversity of plant ecosystems. It has long been known that plants allocate substantial metabolic resources towards enhancement of survival strategies in the form of different natural products. Generally, plants synthesize these molecules to magnetize pollinators or seed-dispersing animals, protecting themselves from harmful rays and for their defence against herbivores, microorganisms and other plants (Tatsis and O'Connor 2016). Moreover, they act as signalling molecules within the plant, with other plants and with microbes, herbivores, and predators as well as with other animals to help in pollination or seed dispersion (Robson and Baek 2009). Chemically, these compounds can be divided into different classes including terpenoids, alkaloids, glucosinolates, benzenoids, cynogenic glucosides, phenylpropenoids and flavonoids (Croteau et al. 2000).

Plant derived natural products (also known as herbal products) have laid down the foundation of traditional medicines across the globe and still continue to provide new remedies to mankind. Since the beginning of human civilization, medicinal plants and their juices have been explored for treatment of different diseases by humans for themselves as well as for their livestock. To name a few, products of *Papaver somniferum* (Poppy latex), *Cedrus* species (Cedar), *Glycyrrhiza glabra* (Licorice), *Cupressus sempervirens* (Cypress), *Commiphora* species (Myrrh) and *Withania somnifera* (Winter cherry), have been widely used against diverse ailments. Modern synthetic drugs, which have given birth to the giant pharmaceutical industry, are also based on the chemical structures of plant derived bioactive molecules.

Biosynthesis as well as accumulation of these secondary plant products varied as species- and chemotype-specific manner under tight spatial and temporal regulation of gene expression (Dhar et al. 2015). Accumulation of these bioactive products in plants fluctuates due to cellular, climatic and developmental conditions and many a times the plant species in question may be distributed to specific geographical location. Due to the complex structures of biochemicals, it is generally cumbersome to synthesize complex natural compounds via synthetic chemistry and the entire phenomenon is often economically not feasible. This limits their proper industrial utilization and drug development. Currently, there is a requirement for large amount of medicinal plant parts of appropriate quality, where active ingredients are present in desirable concentrations spurring an ever-increasing global market. These issues limit exploitation of plants for large scale production of economically important secondary metabolites compounds (Shahid et al. 2013).

Recent advancements in next-generation sequencing (NGS) technology, with the new algorithms for bioinformatics analysis of these developed sequence data, have greatly speeded up the process of plant metabolic gene discovery. Use of NGS technologies to establish transcriptome of contrasting plant varieties in terms of secondary plant product biosynthesis has led to the discovery of a number of previously unidentified genes in biosynthetic pathways (Desgagné-Penix et al. 2012; Pathak et al. 2013). These developments have nurtured advancements in the engineering and reconstruction of metabolic pathways. It is of great importance to reconstruct and elucidate the plant secondary metabolic pathways so that complex metabolites can be engineered from simple building blocks.

16.2 Secondary Metabolites: Concept and Classification

Plants are the nature's best chemists as they produce a fascinating group of specialized metabolites based on an array of skeletal structures and functional group combinations. Given their varied structural diversity and the equally astonishing numbers of species in which specialized metabolites are produced, the ever unfolding potential of plant metabolism has long been recognized. Secondary metabolites (SMs) form a unique group in plant metabolism which do not directly involve in growth and development but have essential role in the plant fortune (Oksman-Caldentey and Inzé 2004). More than hundred thousand such metabolites have been discovered and keep on adding to the list everyday but only a small fraction has been characterized (Kliebenstein 2004). It has been increasingly evident that most of the SMs are derived from differential modification of a common backbone structure with major modifications including oxidation, esterification, methylation and glycosylation (Gachon et al. 2005). Although, there has been no evidence of structural or biochemical distinction between primary and secondary metabolites, their functional boundaries have been slackly defined. It is broadly believed that primary products participate in nutrition and essential metabolic processes of the plant where as secondary products influence ecological interactions involving plant pathogen interaction between the plant and its environment (Aharoni and Galili 2011).

Based on their biosynthetic origin, plant SMs can be structurally divided into three major groups: alkaloids, phenlypropanoids, and terpenoids. Other groups including polyketides, glucosinolates and cynogenic glucosides have been found in many species of plants (Hartmann 2007). The biosynthetic pathways of these groups require flux from primary metabolites, catalyzed by enzymes which are still largely unknown and are often quite perplexing (Fig. 16.1)

16.2.1 Phenylpropanoids

Phenylpropanoids, commonly known as plant phenolics, are ubiquitous in plant kingdom and are characterized by a C6C3- moiety derived from the amino acid L-phenylalanine and less frequently L-tyrosine (Rippert et al. 2009). The enormous

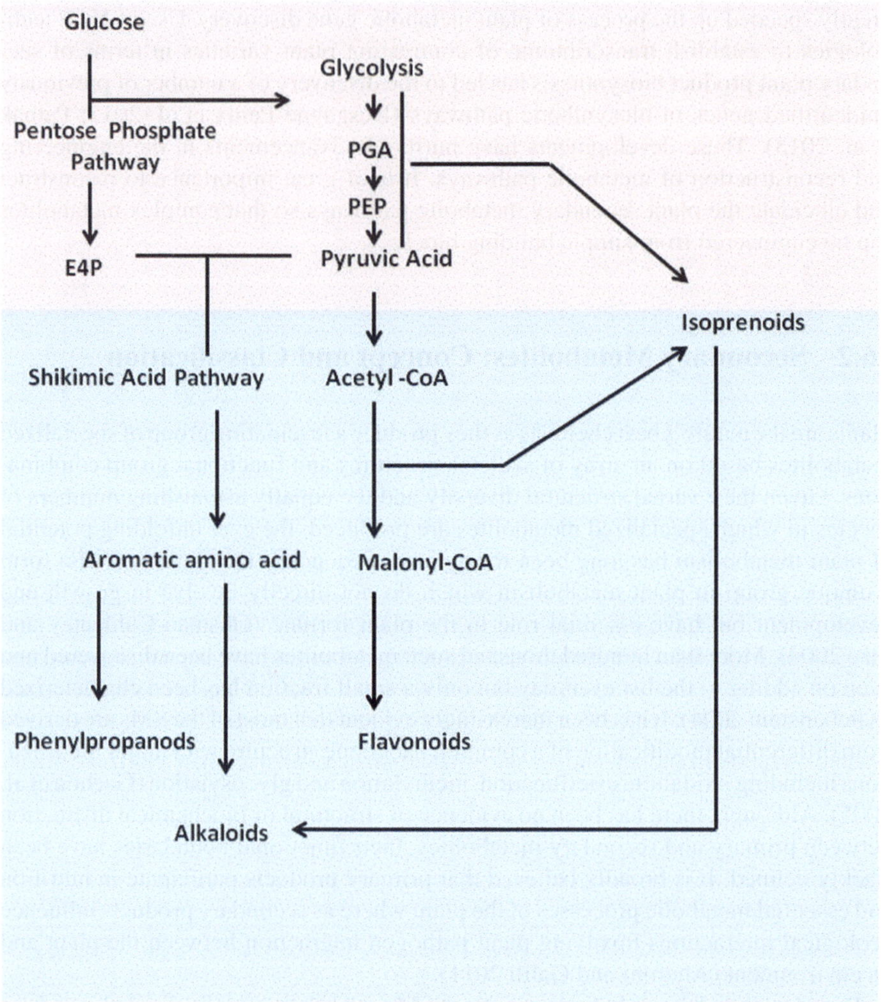

Fig. 16.1 Schematic representation to show interconnection of primary metabolism with pathways leading to the biosynthesis of major classes of seconadray metabolites PGA, phosphoglycerate; PEP, phosphoenol pyruvate; E4P, erythrose-4-phosphate

diversity in this group is the result of various modifications and amplifications in a set of core structures synthesised from shikimate pathway (Saito 2013). The structural genes involved in phenylpropanoid biosynthesis, with special emphasis on lignin and flavonoid formation, regulatory transcription factors, hormonal control of pathway by abiotic treatment and evolution of pathway genes from primary metabolism (Vogt 2010) are reportedly regulated at multiple levels. Phenylammonia lyase (PAL) and tyrosine ammonia lyase (TAL) are involved in directing the carbon flow from upstream shikimate pathway to the various branches distributing flux towards production of specialized molecules by catalysing the non-oxidative deamination of phenylalanine to trans-cinnamate. Complex polymers including lignin, lignin, flavonoids, anthocynins, stilbenes, cutins, suberins etc. are all derived from smaller phenylpropanoid units. A variety of compounds are also involved in plant defence, flower colour, flavours and odours (Boudet 2007).

16.2.2 Alkaloids

Alkaloids represent a group of structurally diverse nitrogen-containing secondary metabolites derived from amino acids such as arginine, ornithine, phenylalanine, lysine, tryptophan or tyrosine. They have a long history in traditional medicine and are an important part of many medicinal plants prompting studies for understanding their biosynthesis over the last century. Alkaloids are known to be produced in about 20% of the flowering plants with more than 12,000 unique molecules known (Kutchan 1995). Although relatively few pathways have been completely elucidated, the results of contemporary research over the past three decades have generated substantially new insights through radio-tracer techniques. Recent investigation of enzymes and genes demonstrate that most of the enzymes are step specific and their expression is controlled within organs, specific cells or organelles inside the cells (De Luca and St Pierre 2000). Based on the structural similarities these have been broadly classified into six groups namely, benzylisoquinoline alkaloids (BIAs), monoterpenoid indole alkaloids (MIAs), tropane, purine, pyrrolizidine and quinolizidine alkaloids (Ziegler and Facchini 2008) (Fig. 16.2).

There are more than 3000 structures identified as MIAs primarily segregated in three tropical plant families Loganiaceae, Apocynaceae and Rubiaceae. The indole alkaloids are derived from tryptophan which synthesised via shikimate pathway with biologically active constituents used as therapeutic agents in medicine, for example, vinblastine and vincristine (Sudžuković et al. 2015). These dimeric alkaloids have been extensively used in the treatment of Hodgkin's disease and leukaemia. Benzylisoquinolines with more than 2500 structures (Fig. 16.2), are found in the families of the superorder Nelumbonaceae, Magnoliids, and Ranunculales. The first committed step in biosynthesis of isoquinoline is the formation of (*S*)-norcoclaurine which is an important precursur of a variety of pathways that lead to a series of diverse structures within this alkaloid group (Ziegler and Facchini 2008). Synthesis of various complex BIAs including morphine, berberine, noscapine and sanguinarine involves a series of isomerisation, methylation and oxidation reactions through a central key branch point intermediate (S)-reticuline (Pathak et al. 2013; Liscombe et al. 2005).

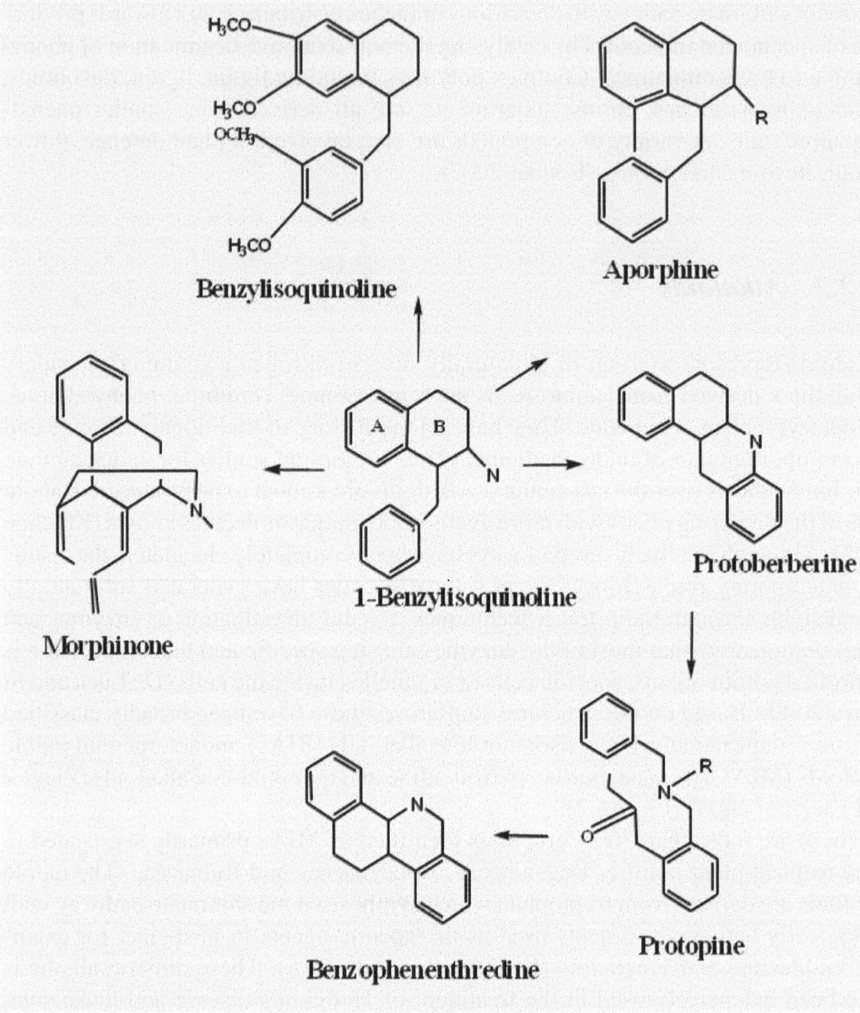

Fig. 16.2 Structural diversity of benzylisoquinolines in plants

16.2.3 Terpenoids

The largest class of plant secondary metabolites are terpenes with over 36,000 individual members reported (Hill 2002) and every year about 1000 novel structures being added in this class of secondary metabolite. Despite such diversity, the class terpenoids has a common mode of biosynthesis i.e. the fusion of isoprene units forming isopentenoid skeleton. The biosynthesis of terpenes can be divided into four stages, the first one being synthesis of isopentenyl diphosphate (IPP) and its allylic isomer dimethy allyl diphosphate (DMAPP), the universal precursors of all terpenes.

These molecules are synthesised via two independent pathways, the cytosolic mevalonate (MVA) pathway and plastid localized 2-C-methyl-D-erythritol-4-phosphate (MEP/DOXP) pathway involving seven enzymatic steps. The second stage commences with condensation of the basic C5 unit, to generate geranyl diphosphate (GPP, C10), farnesyl diphosphate (FPP, C15) and geranylgeranyl diphosphate (GGPP, C20) with three larger prenyl diphosphates. Further, the C10–C20 diphosphates undergo rearrangements and modifications to produce the parent carbon skeletons of each terpene class, marking the third stage. The fourth and conclusive stage includes oxidations, reductions, isomerizations, conjugations and other transformations, thereby generating thousands of distinct terpene metabolites from the same parent skeletons (Kirby and Keasling 2009).

The huge structural diversity of terpenes collimates their functional variability ranging from plant hormones like gibberellins (GA), abscisic acid (ABA), and electron carriers such as plastoquinone and ubiqunone to constituents of membrane structure such as carotenoids (Tholl 2006). Terpenes are also know for defending against herbivores/pathogens (Wittstock and Gershenzon 2002; Wink 2006), acting as attractants for seed dispersing animals and inhibitors for growth of neighbouring plants thereby serving important ecological roles for the plant. Further, many complex terpenes are being used as additive in food, beverages, perfumes, soaps, toothpaste, tobacco and other products as flavour and fragrance agents (Berger 2007).

16.3 Biosynthesis of Secondary Metabolite Is Under Tight Regulation

The biosynthesis and accumulation of secondary metabolites is tissue- and developmental stage-specific and is responsive to various environmental cues, both biotic and abiotic. This strict spacial, temporal and inducible synthesis of these specialized molecules is a result of tight molecular regulation at different levels. An orchestration of a spectrum of biochemical and cellular factors influence biosynthetic flux, transport and storage of SMs at their final site of deposition (Winkel 2004; Yazaki 2005). Developmental factors influence the differentiation of specialized cellular structures at the site of metabolite synthesis and accumulation (Turner et al. 2000). Size and Plant morphology can affect metabolic output. Finally, environmental conditions exert a powerful effect by regulating plant development and metabolism (Broeckling et al. 2005). This multi tier regulation creates a dynamic aspect of biosynthesis and accumulation enabling plants to communicate and respond in order to overcome forthcoming threats.

16.3.1 Abiotic and Biotic Effects

Within a variety of factors in regulation biogenesis of secondary metabolites in plants, two distinct plant stimuli can be determined, being biotic and abiotic factors (Fig. 16.3). Abiotic effects include all physical effects governing habitat, such as light and UV, water availability, and temperature with soil compositions. pre-harvest of *Camellia* sp. (Tea) shoots is regulated by light, resulting in accumulation of phenolic derivatives (flavanols and catechin types) upon overexposure (Ashihara et al. 2008) in arid and semi-arid regions, Drought and salt are major stresses limiting plant growth and productivity (Mahajan and Tuteja 2005). Alterations in intracellular calcium levels and expression levels of certain genes have been implicated in response to osmotic stress and extreme temperature variation related stress (Tuteja 2007). Qualitative and quantitative secondary metabolite profiles have been found to be singular to each chemotype of *Withania*. This feature is time dependent and related to conditions such as season, time of day, tissue age and temperature. Studies based on age, growth phase, temperature as well as lunar cycle dependent variations has put forth a few factors effecting the secondary metabolic profile in *Withania* (Kumar et al. 2012; Tavhare et al. 2016).

Plant hormones have been shown to contribute significantly in regulation of developmental processes and signalling networks. Salicylic acid (SA) and jasmonic acid (JA) enhance tolerance of plants towards different abiotic stresses including osmotic

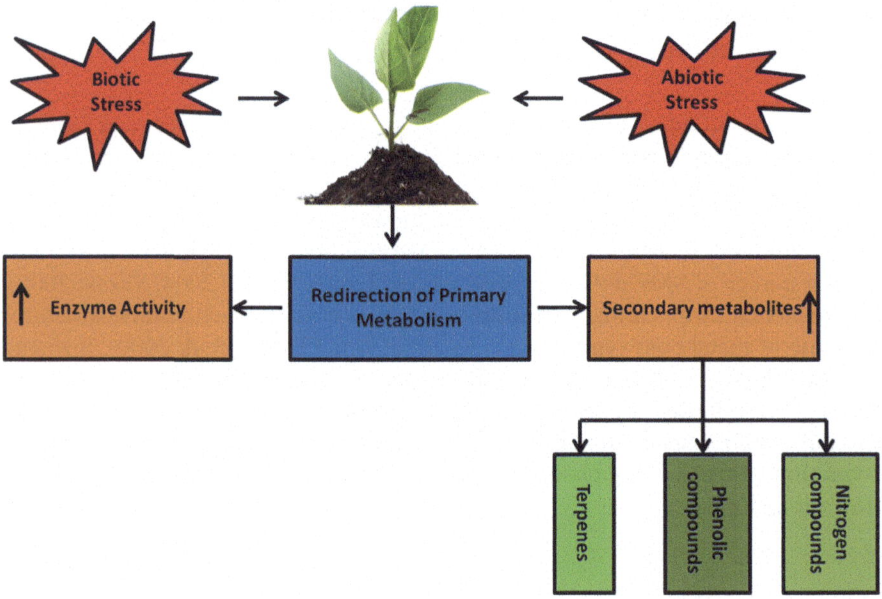

Fig. 16.3 Effect of biotic and abiotic stress on secondary metabolism in plants

stress, drought, salt, heat and UV stress (Horváth et al. 2007; Nazar et al. 2011; Khan et al. 2013). SA and JA have also been show to have antagonistic interactions which affect the expression of pathogen-related (PR) protein genes. MYC2, a bHLH type transcription factor, has been identified as the regulatory centre of JA-signalling pathway, thereby affecting secondary metabolite accumulation in various plant species including *Artemisia annua*, *Nicotiana tabacum* and *Catharanthus roseus* (Sharma et al. 2019). Figure 16.3 depicts the different effect of abiotic and biotic factors on secondary metabolite production. Redirection of primary metabolite into secondary metabolite leads to production of terpenes and nitrogen derivatives. These processes are combinatorial effect of biotic and abiotic effects directly on metabolome of medicinal plant.

Biotic effects have been found out to show more sophisticated interactions with plant biochemistry and physiology than abiotic effects (Briskin 2000). In a simplified sense, it can be understood that biotic effects involve either the plant interactions with microorganisms or plant physiological aspects, as phenology and ontogeny. A highly oxidized tetranortriterpenoid, Azadirachtin-A, is reported to be produced in higher amounts in hairy roots of *Azadiracta indica* during the interaction with, a pathogenic soil bacterium, *A. Rizogenes* (Satdive et al. 2007). An exogenous factor which triggers an intrinsic defence mechanism is herbivory which can include several known secondary metabolites as deterrents. It has been demonstrated for the storage of secoguaianolide type sesquiterpene lactones by caterpillar which is wound induced in *Vernonieae* species such as *Viguiera robusta* (Ambrósio et al. 2008).

16.3.2 Other Regulatory Aspects

Plant molecular responses to environmental stresses involve communication between many gene regulatory networks which are organized in levels of hierarchy (Takahashi et al. 2004) through central, key regulators which are now being elucidated through systems biology and omics approaches. Primary signal in most environmental stresses involve reactive oxygen/nitrogen species (ROS/RNS), hormones like ethylene and abscisic acid (ABA) and transcription factors (TFs). ROS and RNS, through their oxidative effect, form a coordinated network that regulates many plant responses to the environment, often in association with hormones. ABA has been reported as a central regulator of many plant responses, particularly involved in osmotic stress and its signals can be very fast without involving transcriptional activity. Ethylene is involved in many stress responses including ozone, flooding, drought, heat, chilling, wounding and UV-B light exposure (Cramer et al. 2011).

TFs correspond to 5–7% of protein encoding plant genes, as a consequence of dramatic expansion of few families, and have acquired an indispensable position in plant secondary metabolite because of the ability to control multiple steps in a particular metabolic pathway. It has been demonstrated that AtMYB12, one of the flavonol-specific TF, is responsible for several changes in the cellular transcriptome, including modulation in expression of genes involved in the phenylpropanoid pathway, ultimately leading to additional carbon flux for enhanced accumulation of flavonols and several other metabolites (Misra et al. 2010). Post-translational regulation of TFs through the ubiquitin/26S proteasome (UPS) system demonstrates another level of regulation in plants metabolism. Investigations reveal increased anthocynin accumulation in model plants upon loss of 26S proteasome function, suggesting a possible role of UPS in regulating the flavonoid pathway. Micro-RNA and small interfering RNAs (miRNA and siRNA) have also been found as important components of a gene regulatory network (Sharma et al. 2016).

The secondary metabolism of plants changes considerably due to the influence of several abiotic and biotic stress signals. Complete understanding the factors inducing variations in plant secondary metabolism is critical towards extending the knowledge of plant environment interactions and finding alternative strategies to increase productivity of desirably high concentrations of bioactive compounds.

16.4 Gene Mining and Pathway Elucidation

Our trammelled understanding about exploration and regulation of secondary metabolism sets back the development of metabolic engineering strategies to yield medicinally important pharmaceuticals. The primary cause of this obscurity is patchy genomic information from the source plants. Technological advances in the past decade have put forth new insight providing methodological advances in the study of secondary biosynthetic pathways. Sudden surge in development of functional genomics tools including genome/transcriptome sequencing, proteome and metabolome analysis, integrated understanding through systems biology and efficient transformation techniques has led to the discovery of novel genes, identification of gene function and detection of novel pathways of metabolic network. With advancement in these high-throughput techniques, bioinformatics has emerged as an indispensible tool offering a suite of essential approaches for analysis and interpreting huge volumes of raw information generated. Cost effective advanced techniques have made large scale sequencing projects attractive and are frequently being improved to investigate and satiate the lacunas of secondary metabolism. Our understanding for reconstruction of pathway of medicinal plants strengthen by application of different omics profiling methods to acquire a systems-level understanding of plants, and their application to sustainable obtaining of sources for pharmacologically active metabolites (Dhar et al. 2015).

Traditional approaches to functionally characterize genes include creation of various gain- or loss-of-function mutants in plants. Expression of target genes achieve gain of gene function which is driven by strong promoters whereas, ethane methyl sulfonate (EMS) mutagenesis based TILLING approach or T-DNA/transposon insertion induce loss of gene function. Despite a decade full of success stories, these approaches are not being fostered as they target genes in a random manner without any specificity. More recent technologies are aimed at specific gene targeting to induce loss-of-function which can be achieved by targeting either the RNA (RNAi, VIGS) or DNA (CRISPR/Cas9) of a specific gene for alteration or silencing (Teotia et al. 2016).

16.4.1 Transcriptomics

The discovery of secondary biosynthetic pathway genes has been accelerated over the past few years due to revolutionary genomes (DNA) and transcriptome (RNA) sequencing technologies known as next generation sequencing (NGS) technology. NGS is a powerful tool being extensively used in various applications owing to its speed, cost effectiveness and high throughput nature (Mardis 2008; Wang et al. 2009). Sequencing of plant transcriptomes leads to identification of new gene functions, alterations of gene expression with genotype, tissue or physiological changes, as well as tremendous data discovery of Single Nucleotide Polymorphism (SNPs) in a number of model and non model species (Johnson et al. 2012; Schliesky et al. 2012).

Global transcriptome analysis of *C. borivilianum* has been performed to identify genes involved in saponin and flavonoid biosynthesis (Kalra et al. 2013) whereas, comparative analysis of the leaf and root transcriptomes of *Salvia* has revealed steps of tanshinone biosynthetic pathway, such as copalyl diphosphate synthase (SmCPS), kaurene synthase like (SmKSL) and CYP76AH1 (Yang et al. 2013). Ginsenosides, the key active constituents from American ginseng, are of triterpenoid origin, used in various pharmacological remedies. All the genes encoding enzymes involved in the biosynthesis of ginsenoside backbone are identified through transcriptome sequencing of root tissue of American ginseng (Sun et al. 2010). Recently, using NGS, expressed sequence tags (EST) databases of various plants have been developed and used for discovery and illustration of genes involved in secondary metabolite biosynthesis from several other medicinally important plants such as *Artemisia* (Soetaert et al. 2013), *Catharanthus* (Verma et al. 2014), *Azadirachta indica* (Krishnan et al. 2012), *Panax ginseng* (Li et al. 2013), *Papaver somniferum* (Pathak et al. 2013), *Withania somnifera* (Gupta et al. 2013)

16.4.2 Proteomics

Proteomics technologies contribute to the progress made in the knowledge and characterization of plant secondary metabolism which relies on the discovery and characterization of transcription factors and enzymes involved in biosynthesis and regulation (Jacobs et al. 2005; Yang et al. 2012). Numerous transporter and carrier proteins for movement of metabolites also present important targets of proteomic exploration.

Major findings in proteomics about SM have been carried out using elicited cell cultures. Among which polyphenolics biosynthesis, flavonolignan in *Silybummarianum* (Corchete and Bru 2013), isoflavones in *Medicago truncatula* (Lei et al. 2010), and chalcone derivatives in *Boesenbergia rotunda* (Tan et al. 2012) have been analyzed at proteome level under the induction of elicitors. Numerous other protein moieties, in addition to the enzymes of biosynthetic pathway, are potentially involved in modification and movement of end products, e.g. secretory peroxidises (Martinez-Esteso et al. 2009), glutathione-S transferase (Martinez-Esteso et al. 2011), Rab11C and ABC transporter (Corchete and Bru 2013) and laccase (Lei et al. 2010).

The ABC transporters, specialized membrane proteins, have been reported for their indispensible role in metabolite trafficking across cell and organelle membranes (SM transport). The availability of metabolite analysis approaches and species- specific nucleotide databases have proved critical to successfully explore the proteome affecting SM in plants.

16.4.3 Metabolomics

Metabolome can be defined as the global and dynamic metabolic orchestration, of a natural or engineered bio system, to a biological stimuli or genetic manipulation. Study of metabolome i.e. metabolomics, has emerged as a powerful, analytical, and high-through-put tool for research on secondary metabolism by rapid and simultaneous detection, identification, measurement and characterization of many low molecular weight compounds (Breitling et al. 2013). A range of analytical technologies may be used in metabolomics, with chromatography-mass spectrometry (GC-MS and LC-MS) and NMR being the most widely applied techniques (Tugizimana et al. 2013).

Recently Sharma et al. (2016) have assessed the modulation in contents of molecules of various steps of the phenylpropanoid pathway using HPLC coupled with triple quadrupole systems in miR858OX and MIM858 transgenic *Arabidopsis* plants elucidating their role in pathway regulation. Similarly, a fascinating piece of work (Hagel et al. 2015), gave a workflow which is used to study 20 BIA accumulating plants for elucidation of BIA pathway in *Papaver somniferum* L. Advances in plant metabolomics have opened possibilities to understand and predict the behaviour of complex pathways through use of results obtained from modelled and simulated data, in combination with other technologies of functional genomics.

16.4.4 Bioinformatics Approaches

The science of informatics in biology i.e. *Bioinformatics* is a multidisciplinary field that makes use of computers to store, analyse and extract biological meaning from the generated information through integration of statistical algorithms. Development of these computational tools depends on knowledge which is generated from a wide range of disciplines such as mathematics, statistics, computer science, information technology and molecular biology (Sharma and Sarkar 2013). Alike any other scientific venture, the natural product research has experienced a paradigm shift due to availability of recent omics-based complementary approaches. This shift has made computational tools indispensable for keeping pace with the upcoming data and tools. As of today, numerous databases, software and tools are being used to identify, analyse, characterize, maintain and retrieve the huge amount of molecular data on natural product biosynthesis. Two principle strategies (rule-based and rule-independent approaches) for mining pathway for secondary metabolites using computational tools have been recently suggested. These approaches can identify gene clusters encoding known or novel biosynthetic routes for natural products. The number of tools and databases dedicated to natural product biosynthesis has seen a sudden spur in last few years and has led to generation of a one-stop web portal named secondary metabolites bioinformatics portal (SMBP) which links all related tool and data sources (Weber and Kim 2016).

16.4.5 Systems Biology

Exploring across different layers of cell biology is a multi-faceted assignment, with many levels and hierarchies of cellular organization, compartmentalization, and temporal regulation. Systems biology is an emerging approach to understand many critical processes that contribute to cellular organization and dynamics (Mast et al. 2014) by engaging high-throughput experimentation with quantitative analysis and modelling. Systems biology is commonly associated with large-scale "-omics" technologies such as functional genetics, genomics and proteomics which assist in situation identification where a phenotype is expressed by an emergence or unanticipated property of the system (Aitchison and Galitski 2003; Short 2009).

Modern systems biology is a rapidly evolving discipline applying to different levels of biological organization, from molecular sub networks (Albert et al. 2005), to cellular interaction networks (Kim et al. 2009), cells, entire organs (Noble 2002), organisms, and even communities of organisms (Vieites et al. 2009). Areas that have proven particularly rewarding for systems biology include studies of biochemical networks and applications to microorganisms (Güell et al. 2009; Kühner et al. 2009). As already discussed, the biosynthesis of secondary metabolites is highly regulated, and is affected under various circumstances. System biology approach is a holistic method to integrate omics data to these circumstances experienced by plants including environmental ecology, signal transductions and artificial perturbations in the metabolic pathways. This methods, therefore provides an opportunity to fully explore the generation of active ingredients and critical factors related to its accumulation in the plants.

To understand plant systems biology, four distinct network types have been established i.e. gene-to-metabolite interaction, protein-protein interaction, transcriptional-regulatory and gene-regulatory networks. These networks have been characterized for stress responses, plant defence, hormone-induced responses (Goossens et al. 2003; Carrari et al. 2006; Zulak et al. 2007) and have led to the discovery of novel candidate genes in plant species with limited available genome information (Rischer et al. 2006). Systems biology has helped address several practical problems plaguing agriculture practices like quantitative traits, plant stress and defence (Nikiforova et al. 2005; Wang et al. 2006).

16.5 Approaches for Functional Genomics

16.5.1 RNAi

Small RNA based gene silencing which is also known as RNA interference (RNAi) has emerged as an important tool to study gene function because it targets specific genes of known sequences to illustrate their functions in a non-random manner, post-transcriptionally. Interference is part of the normal cellular function as well as an immune response against foreign nucleic acid signatures from viral infections (Roberts et al. 2015) and is induced by exo- or endogenous (i.e., micro RNAs) double stranded RNA (dsRNA) molecules. Inside the cells, dsRNAs are recognized by Dicer family of enzymes and cleaved into short ds fragments of 21–25 bp long siRNAs (Hannon 2002). Intense research has been done in past decade to understand the mechanism of action and exploitation of this phenomenon (Abdurakhmonov et al. 2016) leading to developments, extending from a hairpin structure with inverted repeats to artificial miRNAs (amiRNAs) (Schwab et al. 2006), miRNA target mimicry (TM) (Franco-Zorrilla et al. 2007), miRNA sponge (SP) (Ebert et al. 2007), short tandem target mimic (STTM) (Tang and Tang 2013) and Artificial/synthetic trans-acting siRNAs (atasiRNAs/syn-tasiRNAs) techniques (Zhang 2014).

In context with the gene-specific features and ability to tap the native biosynthetic pathway, there have been a number of investigations aiming application of RNAi to understand plant secondary metabolism. These strategy helps in manipulation of the flavonoid branch of the phenylpropanoid pathway using RNAi has led to numerous novel flower colors, reduced seed coat pigment, altered nodulation, and perturbed biosynthesis. Injection of RNAi construct, targeting chalcone synthase, was demonstrated to reduce anthocyanin biosynthesis and consequently shunting flavonoid metabolism towards production of the hydroxyl cinnamoyl glucose esters, caffeoyl glucose and feruloyl glucose. Suppression of a nicotine N-demethylase gene in tobacco in order to reduce formation of the undesirable compound nornicotine from nicotine is one of the many examples to probe alkaloid metabolism through RNAi (Wagner and Kroumova 2008).

16.5.2 VIGS

Virus-induced gene silencing (VIGS) is a powerful virus-based RNA silencing technique which knocks out expression of a gene without the need to genetically transform the plant. The phenomenon is based on (PTGS) post-transcriptional gene silencing which takes place via a sequence-oriented RNA degradation mechanism that is activated upon sequence similarity between the infecting virus and either a transgene or an endogenous nuclear gene (Kumagai et al. 1995; Ruiz et al. 1998).

VIGS was originally demonstrated in *Nicotiana benthamiana* by silencing Phytoene desaturase (PDS), a carotenoid biosynthetic pathway gene, using viral vector hybridized using tobacco mosaic virus and tomato mosaic virus genome. Delivery of PDS antisense RNA culminated in down regulation of carotenoid accumulation, further leading to its systemic spread throughout the entire plant. Limited levels of photoprotective carotenoid resulted in destruction of chlorophyll, due to photo-oxidation, resulting in a white leafed phenotype. PDS gene homologs have since been used as evidence to monitor the success of VIGS experiment in a large variety of plant species (Becker and Lange 2010).

VIGS has emerged as a preferred reverse-genetic tool to explore gene functions and has been successfully reported in more than 30 plant species including economically important ones like *Solanaceae* and *Rosaceae*. Recently, VIGS has been used for functional characterization of genes catalysis key steps of withanogenesis unravelling their individual contribution (Singh et al. 2015).

VIGS is a quick tool to predict the phenotype of a stable knock-down for specific target gene and therefore is preferably used as a test of choice for species where stable genetic transformation is required due to highly time-consuming option (Chen et al. 2015). VIGS can also be used to elucidate embryo-lethal genes as it acts later in the plant development and is also generally less incorporated with micro- or megaspores (Wege et al. 2007). Exploring full potential of this technique lies in modelling of extensive gene functions for species largely subjective to VIGS.

16.5.3 CRISPR-Cas9 System

Clustered regularly interspaced short palindromic repeats (CRISPR), and associated proteins (Cas) comprise an RNA guided nuclease based system which targets nucleases to specific DNA sequences facilitating genome editing. A structured and precise introduction of DNA double-strand breaks (DSBs) using an engineered nuclease allows reverse genetics, genome engineering and targeted transgene integration experiments. CRISPR/Cas9 system has emerged as a simple, affordable and adoptable tool for genome editing, resulting in extensive research based on the technique which has become known as the ‘CRISPR craze’.

In plants the CRISPR/Cas9 system has been executed using transient expression systems, therefore allows rapid execution and standardization of the method.

Although studies related to perturbation of secondary metabolism are yet to pour in, characterization of gene clusters for antibiotic biosynthesis in actinomycetal genomes has been done using this technique (Tong et al. 2015). CRISPR/Cas9 system has turned out to be a high-throughput functional genomics application, bringing genome editing within the capability of any genome editing laboratory.

16.6 Tools for Enhancement of Secondary Metabolites

It is estimated that a large section of world population depends on traditional remedies to meet their primary health care needs engendering global demand of herbal medicines. About half of all US-FDA approved drugs rendered in the present market are natural products or their analogues. One of the key objectives of plant biotechnology is to develop eco-friendly ways of large scale production of pharmacologically active compounds (Gandhi et al. 2015). Biotechnological advances have immerged as an indispensable tool for multiplication and genetic enhancement of medicinal plants to harness the ever increasing famish of medicinally important secondary metabolites (Table 16.1).

16.6.1 Bioprocessing

Plant secondary metabolites often produced in limited quantities and are accumulate in specific plant parts with different developmental stages, sometimes with exposure to specific stress or as a immune response from plants, or in a specific agro-geo-climatic zone (Chemler and Koffas 2008). Moreover, due to their complex structures, chemical synthesis of these metabolite are rigorous and economically unviable. Industrial scale plant tissue culture technique represents a commercially viable source for production of such phytochemicals to meet the market demands.

16.6.1.1 Suspension Cultures

Suspension cultures are fast growing undifferentiated plant cells or callus in a liquid medium agitated on a rotary shaker. Specific biosynthetic pathways leading to production of cinnamic acid derivatives, anthraquinones, berberines, shikonins, anthocynanins etc. have been reported to work very effectively in suspension cultures (Chiang and Abdullah 2007) however such arrangements show inclination towards production of certain compounds only. Other compounds such as morphinan alkaloids, tropane, quinoline alkaloids, dimeric monoterpene indole alkaloids etc. accumulate just in traces in the suspension cultures irrespective of several optimization efforts including medium engineering and use of elicitors. Numerous pilot- to large-scale

Table 16.1 Various metabolic engineering approaches applied for enhancement of secondary metabolites

S. No.	Source plant	Approach employed	Enhanced compound	Reference
1	*B. vulgaris*	Transient, *Agrobacterium* mediated overexpression of BvDODA1 and CYP76AD1	Betanin and isobetanin	Polturak et al. (2016)
2	*N. tabacum*	Overexpression of AtMYB12 and GmIFS1	Flavonol	Pandey et al. (2014)
3	*N. benthamiana*	Silencing of 5-Epiaristolochene synthase (EAS) and squalene synthase through RNAi	Valencene	Cankar et al. (2015)
4	*S. tuberosum*	Overexpression of WD40-repeat gene (StAN11)	Anthocyanins	Li et al. (2014)
5	*A. thaliana*	Overexpression of *AtDXS*, AtDXR	Aethiopinone	Vaccaro et al. (2014)
6	*S. sclarea*	Overexpression of *SsLPPS*, SsSS	Sclareol	Pan et al. (2014)
7	*P. cablin*	Overexpression of *PcPTS*	Patchoulol	Zhan et al. (2014)
8	*P. bracteatum*	Overexpression of codeinone reductase	Morphine	Sharafi et al. (2013)
9	*Fragaria ananassa*	Overexpression of *FaPHOT2* (phototropin)	Anthocyanins	Kadomura et al. (2013)
10	*N. tabacum*	Overexpression of *AtMYB12*	Flavonol	Pandey et al. (2012)
11	*S. miltiorrhiza*	Overexpression of Allene oxide cyclise	Tanshinone	Gu et al. (2012)
12	*C. acuminate*	Overexpression of allene oxide cyclise	Camptothecin	Pi et al. (2012)
13	*R. cordifolia*	Overexpression of *AtCPK1*	Anthraquinones	Shkryl et al. (2011)
14	*F. koreana*	Silencing of Pinoresinol/lariciresinol reductase	Pinoresinol	Kim et al. (2009)
15	*Z. mays*	Overexpression of feaf color (Lc)	Flavonoids	Li et al. (2007)
16	*S. lycopersicum*	Silencing of de-etiolated1 (*DET1*)	Apocarotenoids and flavonoids	Davuluri et al. (2005)

efforts have not been able to yield results that can resulted to commercial exploitation of tissue cultures for production of these compounds (Gandhi et al. 2015).

16.6.1.2 Organ Culture

Organ culture is a promising alternative for production of such phytochemicals that do not show desired accumulation by suspension culture methods. Several precious metabolites including Morphinan alkaloids of *Papaver somniferum* L. (Papaveraceae), sesquiterpene lactone (artemisinin) of *Artemisia annua* L. (Asteraceae), dimeric indole alkaloid (anhydrovinblastine—a direct precursor of vinblastine and vincristine) of *Catharanthus roseus* (L.) G. Don (Apocynaceae), for instance, is produced in better quantities in shoot culture (Tisserat and Berhow 2009). In comparison to suspension cultures, root cultures have been reported to produce remarkable amounts of several alkaloids i.e. Tropane alkaloids, such as hyoscyamine and scopolamine (Saito and Mizukami 2002).

16.6.1.3 Hairy Root Culture

Agrobacterium rhizogenes infection causes genetic transformation of plant cells resulting in differentiation of cells into hairy roots which can be excised and cultured indefinitely in liquid medium. Transgenic hairy root cultures have made a remarkable entry for the role of plant tissue culture owing to its genetic and biosynthetic stability, faster growth rate and easy maintenance. Expression levels of genes inserted with T-DNA have been correlated to the amount of secondary metabolites produced as well as for production of heterologous proteins (Sharma et al. 2013).

16.6.1.4 Process Optimization

Media optimization plays a crucial role in production of secondary metabolites through plant tissue culture process. The nature of carbon source, nitrogen source, carbon-nitrogen ratio, phosphate levels etc. affect the biomass as well as the production of secondary metabolites. Physical agents including UV-B light and electric current have been demonstrated to enhance the secondary metabolite production too (Ramani and Chelliah 2007; Kaimoyo et al. 2008).

Design of bioreactors has also been taken care of for enhanced production of secondary metabolites by plant cell cultures. Low shear mixing, optimal aeration, sterility and adequate light are some of the general aspects under consideration however choice and scale-up to bioreactors continue to remain a challenge.

16.6.2 Heterologous Production

Plant secondary metabolites via heterologous biosynthesis have been largely investigated in bacteria, yeast and alternative plant species. Engineered microbial species such as *Escherichia coli* and *Saccharomyces cerevisiae* are preferably used due to their quick doubling times compared to the plant species (minutes vs. days), ease of genetic modification, inexpensive carbon sources, as well as various established scale-up technologies (Chang and Keasling 2006; Chang et al. 2007; Roberts 2007). Microbial production routes are expected to overcome the inherent production differences associated with plant suspension cultures. Genes involved in the initial paclitaxel biosynthetic pathway, which contains 19 putative pathway steps (Croteau et al. 2006) have been introduced into *S. cerevisiae* and *E. coli*. Through codon optimization, chimera enzyme fusion and balancing precursor supply high levels of taxadiene-5α-ol were achieved (Ajikumar et al. 2010).

16.6.3 Particle Bombardment Mediated Transformation

The biolistic transformation, developed by Klein et al. (1988) involves a high velocity bombardment of DNA-coated gold/tungsten micro-projectiles into intact cells or tissues. This DNA delivery system employs a simple, efficient, genotype-independent and essentially identical methodology which is notwithstanding to the nature of the target cells and genetic material used. Compared to other techniques this method stands as a highly versatile and adaptable technique, extending its application to a wide range of cells and tissues. During the past two decades, micro-projectile bombardment has become an accepted and regularized method for production of transgenic plants bypassing various difficulties encountered in tissue culture related regeneration methods and *Agrobacterium* host-specificity in several important medicinal species (Franklin et al. 2007). Agrolistic approach can be used to circumvent the convoluted pattern of transgene integration, often revealed by molecular analysis of plants obtained by biolistic transformations (Hansen and Wright 1999).

16.7 Conclusion

Secondary metabolism in plants provide a phenomenal scope for identification, characterization and subsequent utilization of naturally synthesized molecules towards various human needs including medicines, agrochemicals, fragrance and flavor ingredients, food additives and pesticides. As plants still remain a sole sustainable natural resource of many medicinally important secondary metabolites, there is a pressing need to enhance synthesis of secondary metabolites in wild plant species as well as to develop innovative and easy technologies to produce high value bioactive compounds.

Meeting global requirements of these specialized biomolecules require complete understanding of biosynthetic pathways in terms of its intermediate steps; enzymes involved; associated regulatory proteins and optimized physical conditions. However, biogenesis of several important plant secondary metabolites at the level of intermediate steps and their regulation remains sparingly understood. These issues have long been attributed to the lack of functional genomics platforms comprising of developing genomic resources, development of mutants and efficient transformation systems for medicinal plants. However, in the last few years, phenomenal improvements in different omics-data-generation approaches, tools for data analysis through statistical and computational methods, quick functional characterization approaches and analytical techniques have spurred the scenario of phytochemistry.

Precise genome editing tools like CRISPR/Cas have emerged as the most sought after ways to characterize the function of putative biosynthetic genes by opening new horizons in generation of appropriate transgenic lines. The major bottleneck which exists today between the efforts of plant biologists and their endeavors of exploring the riches of secondary metabolites is the inefficient stable-transformation and regeneration methods for major plant species (Dhar et al. 2015). Although systems biology approach provides an entirely new perspective to dodge this issue, current apprehension about this subject at the laboratory scale may not precisely by fitting at the industrial scale.

With enormous data generated across the globe, through various high-throughput approaches, there is urgent need to build strong multi-disciplinary consortia among plant biologist so that major bottlenecks can be dealt with under common developmental objectives.

Acknowledgements SP and AVA acknowledge DST for NPDF as well as STI PDF grant. PA Acknowledge DBT for JRF as well as SRF grant.

References

Abdurakhmonov IY et al (2016) RNA interference for functional genomics and improvement of cotton (*Gossypium* sp.). Front Sci 7

Aharoni A, Galili G (2011) Metabolic engineering of the plant primary–secondary metabolism interface. Curr Opin Biotechnol 22(2):239–244

Aitchison JD, Galitski T (2003) Inventories to insights. J cell Biol 161(3):465–469

Ajikumar PK et al (2010) Isoprenoid pathway optimization for Taxol precursor overproduction in *Escherichia coli*. Science 330(6000):70–74

Albert M-A et al (2005) Experimental and in silico analyses of glycolytic flux control in bloodstream form *Trypanosoma brucei*. J Biol Chem 280(31):28306–28315

Ambrósio SR et al (2008) Constituents of glandular trichomes of *Tithonia diversifolia*: relationships to herbivory and antifeedant activity. Phytochemistry 69(10):2052–2060

Ashihara H, Sano H, Crozier A (2008) Caffeine and related purine alkaloids: biosynthesis, catabolism, function and genetic engineering. Phytochemistry 69(4):841–856

Becker A, Lange M (2010) VIGS—genomics goes functional. Trends Plant Sci 15(1):1–4
Berger RG (2007) Flavours and fragrances: chemistry, bioprocessing and sustainability. Springer Science & Business Media, Berlin
Boudet A-M (2007) Evolution and current status of research in phenolic compounds. Phytochemistry 68(22):2722–2735
Breitling R et al (2013) Metabolomics for secondary metabolite research. Metabolites 3(4):1076–1083
Briskin DP (2000) Medicinal plants and phytomedicines. Linking plant biochemistry and physiology to human health. Plant Physiol 124(2):507–514
Broeckling CD et al (2005) Metabolic profiling of *Medicago truncatula* cell cultures reveals the effects of biotic and abiotic elicitors on metabolism. J Exp Bot 56(410):323–336
Cankar K, Jongedijk E, Klompmaker M, Majdic T, Mumm R, Bouwmeester H et al (2015) (+)-Valencene production in *Nicotiana* benthamiana is increased by down-regulation of competing pathways. Biotechnol J 10(1):180–189
Carrari F et al (2006) Integrated analysis of metabolite and transcript levels reveals the metabolic shifts that underlie tomato fruit development and highlight regulatory aspects of metabolic network behavior. Plant Physiol 142(4):1380–1396
Chang MC et al (2007) Engineering *Escherichia coli* for production of functionalized terpenoids using plant P450s. Nat Chem Biol 3(5):274–277
Chang MCY, Keasling JD (2006) Production of isoprenoid pharmaceuticals by engineered microbes. Nat Chem Biol 2(12):674–681
Chemler JA, Koffas MA (2008) Metabolic engineering for plant natural product biosynthesis in microbes. Curr Opin Biotechnol 19(6):597–605
Chen W et al (2015) MR VIGS: MicroRNA-based virus-induced gene silencing in plants. Methods Protocols, Plant Gene Silencing, pp 147–157
Chiang L, Abdullah MA (2007) Enhanced anthraquinones production from adsorbent-treated *Morinda elliptica* cell suspension cultures in production medium strategy. Process Biochem 42(5):757–763
Corchete P, Bru R (2013) Proteome alterations monitored by DIGE analysis in *Silybum marianum* cell cultures elicited with methyl jasmonate and methyl B cyclodextrin. J Proteomics 85:99–108
Cramer GR et al (2011) Effects of abiotic stress on plants: a systems biology perspective. BMC Plant Biol 11(1):163
Croteau R, Kutchan TM, Lewis NG (2000) Natural products (secondary metabolites). Biochem Mol Biol Plants 24:1250–1319
Croteau R et al (2006) Taxol biosynthesis and molecular genetics. Phytochem Rev 5(1):75–97
Davuluri GR, Van Tuinen, A, Fraser PD, Manfredonia A, Newman R, Burgess D et al (2005) Fruit-specific RNAi-mediated suppression of DET1 enhances carotenoid and flavonoid content in tomatoes. Nat Biotechnol 23(7):890
De Luca V, St Pierre B (2000) The cell and developmental biology of alkaloid biosynthesis. Trends Plant Sci 5(4):168–173
Desgagné-Penix I et al (2012) Integration of deep transcript and targeted metabolite profiles for eight cultivars of opium poppy. Plant Mol Biol 79(3):295–313
Dhar N et al (2015) A decade of molecular understanding of withanolide biosynthesis and in vitro studies in *Withania somnifera* (L.) Dunal: prospects and perspectives for pathway engineering. Front Plant Sci 6
Ebert MS, Neilson JR, Sharp PA (2007) MicroRNA sponges: competitive inhibitors of small RNAs in mammalian cells. Nat Methods 4(9):721–726
Franco-Zorrilla JM et al (2007) Target mimicry provides a new mechanism for regulation of microRNA activity. Nat Genet 39(8):1033–1037
Franklin G, Oliveira M, Dias ACP (2007) Production of transgenic *Hypericum perforatum* plants via particle bombardment-mediated transformation of novel organogenic cell suspension cultures. Plant Sci 172(6):1193–1203

Gachon CM, Langlois-Meurinne M, Saindrenan P (2005) Plant secondary metabolism glycosyltransferases: the emerging functional analysis. Trends Plant Sci 10(11):542–549

Gandhi SG, Mahajan V, Bedi YS (2015) Changing trends in biotechnology of secondary metabolism in medicinal and aromatic plants. Planta 241(2):303–317

Goossens A et al (2003) A functional genomics approach toward the understanding of secondary metabolism in plant cells. Proc Natl Acad Sci 100(14):8595–8600

Güell M et al (2009) Transcriptome complexity in a genome-reduced bacterium. Science 326(5957):1268–1271

Gupta P et al (2013) De novo assembly, functional annotation and comparative analysis of *Withania somnifera* leaf and root transcriptomes to identify putative genes involved in the withanolides biosynthesis. PLoS ONE 8(5):e62714

Gu XC, Chen JF, Xiao Y, Di P, Xuan HJ, Zhou X et al (2012) Overexpression of allene oxide cyclase promoted tanshinone/phenolic acid production in *Salvia miltiorrhiza*. Plant Cell Rep 31(12) 2247–2259

Hagel JM et al (2015) Transcriptome analysis of 20 taxonomically related benzylisoquinoline alkaloid-producing plants. BMC Plant Biol 15(1):227

Hannon GJ (2002) RNA interference. Nature 418(6894):244–251

Hansen G, Wright MS (1999) Recent advances in the transformation of plants. Trends Plant Sci 4(6):226–231

Hartmann T (2007) From waste products to ecochemicals: fifty years research of plant secondary metabolism. Phytochemistry 68(22):2831–2846

Hill R (2002) Dictionary of natural products on CD-ROM, Ed. version 10. Chapman & Hall/CRC, Boca Raton

Horváth E, Szalai G, Janda T (2007) Induction of abiotic stress tolerance by salicylic acid signaling. J Plant Growth Regul 26(3):290–300

Jacobs DI et al (2005) Proteome analysis of the medicinal plant *Catharanthus roseus*. Planta 221(5):690–704

Johnson MT et al (2012) Evaluating methods for isolating total RNA and predicting the success of sequencing phylogenetically diverse plant transcriptomes. PLoS ONE 7(11):e50226

Kadomura-Ishikawa Y, Miyawaki K, Noji S, Takahashi A (2013) Phototropin 2 is involved in blue light-induced anthocyanin accumulation in *Fragaria* x ananassa fruits. J Plant Res 126(6):847–857

Kaimoyo E et al (2008) Sub-lethal levels of electric current elicit the biosynthesis of plant secondary metabolites. Biotechnol Prog 24(2):377–384

Kalra S et al (2013) De novo transcriptome sequencing reveals important molecular networks and metabolic pathways of the plant, *Chlorophytum borivilianum*. PLoS ONE 8(12):e83336

Khan MIR et al (2013) Salicylic acid alleviates adverse effects of heat stress on photosynthesis through changes in proline production and ethylene formation. Plant Signaling Behav 8(11):e26374

Kim PS, Levy D, Lee PP (2009) Modeling and simulation of the immune system as a self-regulating network. Methods Enzymol 467:79–109

Kim HJ, Ono E, Morimoto K, Yamagaki T, Okazawa A, Kobayashi, A, Satake H (2009) Metabolic engineering of lignan biosynthesis in *Forsythia* cell culture. Plant Cell Physiol 50(12):2200–2209

Kirby J, Keasling JD (2009) Biosynthesis of plant isoprenoids: perspectives for microbial engineering. Annu Rev Plant Biol 60:335–355

Klein TM et al (1988) Stable genetic transformation of intact Nicotiana cells by the particle bombardment process. Proc Natl Acad Sci 85(22):8502–8505

Kliebenstein D (2004) Secondary metabolites and plant/environment interactions: a view through *Arabidopsis thaliana* tinged glasses. Plant Cell Environ 27(6):675–684

Krishnan NM et al (2012) A draft of the genome and four transcriptomes of a medicinal and pesticidal angiosperm *Azadirachta indica*. BMC Genom 13(1):1

Kühner S et al (2009) Proteome organization in a genome-reduced bacterium. Science 326(5957):1235–1240

Kumagai M et al (1995) Cytoplasmic inhibition of carotenoid biosynthesis with virus-derived RNA. Proc Natl Acad Sci 92(5):1679–1683

Kumar A et al (2012) Seasonal low temperature plays an important role in increasing metabolic content of secondary metabolites in *Withania somnifera* (L.) Dunal and affects the time of harvesting. Acta physiologiae plantarum 34(5):2027–2031

Kutchan TM (1995) Alkaloid biosynthesis—the basis for metabolic engineering of medicinal plants. Plant Cell 7(7):1059

Lei Z et al (2010) Comparative proteomics of yeast-elicited *Medicago truncatula* cell suspensions reveals induction of isoflavonoid biosynthesis and cell wall modifications. J Proteome Res 9(12):6220–6231

Li C et al (2013) Transcriptome analysis reveals ginsenosides biosynthetic genes, microRNAs and simple sequence repeats in Panax ginseng CA Meyer. BMC Genom 14(1):1

Li H, Flachowsky H, Fischer TC, Hanke MV, Forkmann G, Treutter D et al (2007) Maize Lc transcription factor enhances biosynthesis of anthocyanins, distinct proanthocyanidins and phenylpropanoids in apple (*Malus domestica Borkh.*) Planta 226(5):1243–1254

Li W, Wang B, Wang M, Chen M, Yin JM, Kaleri GM et al (2014) Cloning and characterization of a potato StAN11 gene involved in *Anthocyanin* biosynthesis regulation. J Integr Plant Biol 56(4):364–372

Liscombe DK et al (2005) Evidence for the monophyletic evolution of benzylisoquinoline alkaloid biosynthesis in angiosperms. Phytochemistry 66(11):1374–1393

Mahajan S, Tuteja N (2005) Cold, salinity and drought stresses: an overview. Arch Biochem Biophys 444(2):139–158

Mardis ER (2008) Next-generation DNA sequencing methods. Annu Rev Genomics Hum Genet 9:387–402

Martinez-Esteso M et al (2009) Changes of defense proteins in the extracellular proteome of grapevine (Vitis vinifera cv. Gamay) cell cultures in response to elicitors. J Proteomics 73(2):331–341

Martinez-Esteso M et al (2011) DIGE analysis of proteome changes accompanying large resveratrol production by grapevine (*Vitis vinifera* cv. Gamay) cell cultures in response to methyl-β-cyclodextrin and methyl jasmonate elicitors. J Proteomics 74(8):1421–1436

Mast FD, Ratushny AV, Aitchison JD (2014) Systems cell biology. The Journal of cell biology 206(6):695–706

Misra P et al (2010) Modulation of transcriptome and metabolome of tobacco by Arabidopsis transcription factor, AtMYB12, leads to insect resistance. Plant Physiol 152(4):2258–2268

Nazar R et al (2011) Salicylic acid alleviates decreases in photosynthesis under salt stress by enhancing nitrogen and sulfur assimilation and antioxidant metabolism differentially in two mungbean cultivars. J Plant Physiol 168(8):807–815

Nikiforova VJ et al (2005) Integrative gene-metabolite network with implemented causality deciphers informational fluxes of sulphur stress response. J Exp Bot 56(417):1887–1896

Noble D (2002) Modeling the heart–from genes to cells to the whole organ. Science 295(5560):1678–1682

Oksman-Caldentey K-M, Inzé D (2004) Plant cell factories in the post-genomic era: new ways to produce designer secondary metabolites. Trends Plant Sci 9(9):433–440

Pan XW, Han L, Zhang YH, Chen DF, Simonsen HT (2015) Sclareol production in the moss *Physcomitrella patens* and observations on growth and terpenoid biosynthesis. Plant Biotechnol Rep 9(3):149–159

Pandey A, Misra P, Chandrashekar K, Trivedi PK (2012) Development of AtMYB12-expressing transgenic tobacco callus culture for production of rutin with biopesticidal potential. Plant Cell Rep 31(10):1867–1876

Pandey A, Misra P, Khan MP, Swarnkar G, Tewari MC, Bhambhani S et al (2014) Co-expression of Arabidopsis transcription factor, At MYB 12, and soybean isoflavone synthase, Gm IFS 1, genes in tobacco leads to enhanced biosynthesis of isoflavones and flavonols resulting in osteoprotective activity. Plant Biotechnol J 12(1):69–80

Pathak S et al (2013) Comparative transcriptome analysis using high papaverine mutant of *Papaver somniferum* reveals pathway and uncharacterized steps of Papaverine biosynthesis. PLoS ONE 8(5):e65622

Pennisi E (2013) The CRISPR craze. Science 341(6148):833–836

Pi Y, Jiang K, Lin J, Cao Y, Wang Q, Yang J et al (2012) Effects of over-expression of allene oxide cyclase on camptothecin production by cell cultures of *Camptotheca acuminata*. Afr J Biotechnol 11(24):6535–6541

Polturak G, Breitel D, Grossman N, Sarrion-Perdigones A, Weithorn E, Pliner M et al (2016) Elucidation of the first committed step in betalain biosynthesis enables the heterologous engineering of betalain pigments in plants. New Phytol 210(1):269–283

Ramani S, Chelliah J (2007) UV-B-induced signaling events leading to enhanced-production of catharanthine in *Catharanthus roseus* cell suspension cultures. BMC Plant Biol 7(1):61

Rippert P et al (2009) Tyrosine and phenylalanine are synthesized within the plastids in *Arabidopsis*. Plant Physiol 149(3):1251–1260

Rischer H et al (2006) Gene-to-metabolite networks for terpenoid indole alkaloid biosynthesis in *Catharanthus roseus* cells. Proc Natl Acad Sci 103(14):5614–5619

Roberts AF et al (2015) Biosafety research for non-target organism risk assessment of RNAi-based GE plants. Front Plant Sci 6

Roberts SC (2007) Production and engineering of terpenoids in plant cell culture. Nat Chem Biol 3(7):387–395

Robson B, Baek O (2009) The engines of Hippocrates: From the dawn of medicine to medical and pharmaceutical informatics, vol 5. Wiley New York

Ruiz MT, Voinnet O, Baulcombe DC (1998) Initiation and maintenance of virus-induced gene silencing. Plant Cell 10(6):937–946

Saito K (2013) Phytochemical genomics—a new trend. Curr Opin Plant Biol 16(3):373–380

Saito K, Mizukami H (2002) Plant cell cultures as producers of secondary compounds. Plant Biotechnol Transgenic Plants 77–109

Satdive RK, Fulzele DP, Eapen S (2007) Enhanced production of azadirachtin by hairy root cultures of *Azadirachta indica* A. Juss by elicitation and media optimization. J Biotechnol 128(2):281–289

Schliesky S et al (2012) RNA-seq assembly—are we there yet? Front Plant Sci 3:220

Schwab R et al (2006) Highly specific gene silencing by artificial microRNAs in Arabidopsis. Plant Cell 18(5):1121–1133

Shahid M et al (2013) Recent trends in biotechnology and therapeutic applications of medicinal plants. Springer, Berlin

Sharma A et al (2019) Jasmonate responsive transcription factor WsMYC2 regulates the biosynthesis of triterpenoid withanolides and phytosterol via key pathway genes in *Withania somnifera* (L.) Dunal. Plant Mol Biol 100(4–5):543–560

Sharma D et al (2016) Microrna858 is a potential regulator of phenylpropanoid pathway and plant development. Plant Physiol 171(2):944–959

Sharma P, Padh H, Shrivastava N (2013) Hairy root cultures: a suitable biological system for studying secondary metabolic pathways in plants. Eng Life Sci 13(1):62–75

Sharafi A, Sohi HH, Mousavi A, Azadi P, Khalifani BH, Razavi K (2013) Metabolic engineering of morphinan alkaloids by over-expression of codeinone reductase in transgenic hairy roots of *Papaver bracteatum*, the Iranian poppy. Biotechnol Lett 35(3):445–453

Sharma V, Sarkar IN (2013) Bioinformatics opportunities for identification and study of medicinal plants. Brief Bioinform 14(2):238–250

Shkryl YN, Veremeichik GN, Bulgakov VP, Zhuravlev YN (2011) Induction of anthraquinone biosynthesis in *Rubia cordifolia* cells by heterologous expression of a calcium-dependent protein kinase gene. Biotechnol Bioeng 108(7):1734–1738

Short B (2009) Cell biologists expand their networks. J Cell Biol 186(3):305–311

Singh AK et al (2015) Virus-induced gene silencing of Withania somnifera squalene synthase negatively regulates sterol and defence-related genes resulting in reduced withanolides and biotic stress tolerance. Plant Biotechnol J 13(9):1287–1299

Soetaert SS et al (2013) Differential transcriptome analysis of glandular and filamentous trichomes in *Artemisia annua*. BMC Plant Biol 13(1):1

Sudžuković N, Schinnerl J, Brecker L (2015) Phytochemical meanings of tetrahydro-β-carboline moiety in strictosidine derivatives. Bioorganic Med Chem

Sun C et al (2010) De novo sequencing and analysis of the American ginseng root transcriptome using a GS FLX Titanium platform to discover putative genes involved in ginsenoside biosynthesis. BMC Genom 11(1):262

Takahashi S et al (2004) Monitoring the expression profiles of genes induced by hyperosmotic, high salinity, and oxidative stress and abscisic acid treatment in Arabidopsis cell culture using a full-length cDNA microarray. Plant Mol Biol 56(1):29–55

Tan EC et al (2012) Proteomic analysis of cell suspension cultures of *Boesenbergia rotunda* induced by phenylalanine: identification of proteins involved in flavonoid and phenylpropanoid biosynthesis pathways. Plant Cell Tissue Organ Culture (PCTOC) 111(2):219–229

Tang G, Tang X (2013) Short tandem target mimic: a long journey to the engineered molecular landmine for selective destruction/blockage of microRNAs in plants and animals. Journal of Genetics and Genomics 40(6):291–296

Tatsis EC, O'Connor SE (2016) New developments in engineering plant metabolic pathways. Curr Opin Biotechnol 42:126–132

Tavhare SD, Nishteswar K, Shukla VJ (2016) Effect of seasonal variations on the phytoconstituents of Aśvagandhā wr to lunar cycles. Ancient Sci Life 35(3):150

Teotia S et al (2016) Essential RNA-based technologies and their applications in plant functional genomics. Trends Biotechnol

Tholl D (2006) Terpene synthases and the regulation, diversity and biological roles of terpene metabolism. Curr Opin Plant Biol 9(3):297–304

Tisserat B, Berhow M (2009) Production of pharmaceuticals from Papaver cultivars in vitro. Eng Life Sci 9(3):190–196

Tong Y et al (2015) CRISPR-Cas9 based engineering of actinomycetal genomes. ACS synthetic biology 4(9):1020–1029

Tugizimana F, Piater L, Dubery I (2013) Plant metabolomics: a new frontier in phytochemical analysis. S Afr J Sci 109(5–6):01–11

Turner GW, Gershenzon J, Croteau RB (2000) Development of peltate glandular trichomes of peppermint. Plant Physiol 124(2):665–680

Tuteja N (2007) Chapter Twenty-four-mechanisms of high salinity tolerance in plants. Methods Enzymol 428:419–438

Vaccaro M, Malafronte N, Alfieri M, De Tommasi N, Leone A (2014) Enhanced biosynthesis of bioactive abietane diterpenes by overexpressing AtDXS or AtDXR genes in *Salvia sclarea* hairy roots. Plant Cell, Tissue Organ Cult (PCTOC) 119(1):65–77

Verma M et al (2014) Transcriptome analysis of *Catharanthus roseus* for gene discovery and expression profiling. PLoS ONE 9(7):e103583

Vieites JM et al (2009) Metagenomics approaches in systems microbiology. FEMS Microbiol Rev 33(1):236–255

Vogt T (2010) Phenylpropanoid biosynthesis. Mol Plant 3(1):2–20

Wagner GJ, Kroumova AB (2008) The use of RNAi to elucidate and manipulate secondary metabolite synthesis in plants. In: Current perspectives in microRNAs (miRNA). Springer, Berlin, pp 431–459

Wang D, Amornsiripanitch N, Dong X (2006) A genomic approach to identify regulatory nodes in the transcriptional network of systemic acquired resistance in plants. PLoS Pathog 2(11):e123

Wang Z, Gerstein M, Snyder M (2009) RNA-Seq: a revolutionary tool for transcriptomics. Nat Rev Genet 10(1):57–63

Weber T, Kim HU (2016) The secondary metabolite bioinformatics portal: computational tools to facilitate synthetic biology of secondary metabolite production. Synthetic and Systems Biotechnology 1(2):69–79

Wege S et al (2007) Highly efficient virus-induced gene silencing (VIGS) in California poppy (*Eschscholzia californica*): an evaluation of VIGS as a strategy to obtain functional data from non-model plants. Ann Bot 100(3):641–649

Wink M (2006) Importance of plant secondary metabolites for protection against insects and microbial infections. Nat Occur Bioactive Compounds 3:251–268

Winkel BS (2004) Metabolic channeling in plants. Annu Rev Plant Biol 55:85–107

Wittstock U, Gershenzon J (2002) Constitutive plant toxins and their role in defense against herbivores and pathogens. Curr Opin Plant Biol 5(4):300–307

Yang CQ et al (2012) Transcriptional regulation of plant secondary metabolism. J Integr Plant Biol 54(10):703–712

Yang L et al (2013) Transcriptome analysis of medicinal plant *Salvia miltiorrhiza* and identification of genes related to tanshinone biosynthesis. PLoS ONE 8(11):e80464

Yazaki K (2005) Transporters of secondary metabolites. Curr Opin Plant Biol 8(3):301–307

Zhang ZJ (2014) Artificial trans-acting small interfering RNA: a tool for plant biology study and crop improvements. Planta 239(6):1139–1146

Zhan X, Zhang YH, Chen DF, Simonsen HT (2014) Metabolic engineering of the moss *Physcomitrella patens* to produce the sesquiterpenoids patchoulol and α/β-santalene. Front Plant Sci 5

Ziegler J, Facchini PJ (2008) Alkaloid biosynthesis: metabolism and trafficking. Annu Rev Plant Biol 59:735–769

Zulak KG et al (2007) Gene transcript and metabolite profiling of elicitor-induced opium poppy cell cultures reveals the coordinate regulation of primary and secondary metabolism. Planta 225(5):1085–1106

Chapter 17
Molecular Biology of Glandular Trichomes and Their Functions in Environmental Stresses

Maridul Kundan, Umar Gani, Abhishek Kumar Nautiyal and Prashant Misra

Abstract Glandular trichomes, the epidermal projections over aerial plant parts primarily function in defense against stresses including protection from insects and microbes. These structures are characterized by dedicated genetic machinery for overproduction of structurally diverse secondary metabolites. Certain secondary metabolites of trichome origin display interesting pharmacological activities, and therefore are of immense economical interest as drug, aroma and allelochemicals. For obvious reasons, glandular trichomes have been focus of 'omics' studies, particularly for elucidating molecular basis of such a large scale production of secondary metabolites. In the last decade, next generation sequencing has fueled the development of transcriptome landscapes of glandular trichomes of several medicinal and aromatic plants. Taken together, these studies have started to unravel gene and metabolic networks operating in glandular trichomes, and therefore are potentially useful for identification of novel molecular targets for strategic metabolic engineering of economically important secondary metabolites as well as for development of stress tolerant plant varieties. The present book chapter will update our current knowledge about aspects of glandular trichome biology including its applied value in plant biology.

Keywords Glandular trichomes · Biotic and abiotic stresses · Plant secondary metabolites · Transcriptomics

---

Authors Maridul Kundan, Umar Gani, and Abhishek Kumar Nautiyal are equally contributed.

M. Kundan · U. Gani · A. K. Nautiyal · P. Misra (✉)
Plant Biotechnology Division, CSIR—Indian Institute of Integrative Medicine, Canal Road, Jammu 180001, India
e-mail: pmisra@iiim.ac.in

Academy of Scientific and Innovative Research (AcSIR), New Delhi, India

S. P. Singh et al. (eds.), *Molecular Approaches in Plant Biology and Environmental Challenges*, Energy, Environment, and Sustainability,
https://doi.org/10.1007/978-981-15-0690-1_17

17.1 Introduction

A vast number of Angiosperms possess hair-like epidermal structures which are termed as trichomes. Sometimes, similar structures have also been reported in gymnosperms, bryophytes (Uphof 1962), lichens and algae (Engene et al. 2012). Trichomes are primarily present on surface of leaves and stem, but they are also found on petals, petioles, peduncles and seeds, depending upon the species. Trichomes display tremendous diversity in terms of cellular organization, structure and chemistry. These structures can be broadly divided into two categories—glandular and non glandular, depending upon their morphology and secretion ability (Fahn 2000; Kolb and Muller 2004). Irrespective of the types, all the trichomes originate from epidermal cells. Some of the epidermal cells elongate and modify into unicellular trichomes or may undergo division and specialization to develop into multicellular trichomes. Glandular trichomes, characterized by presence of gland cell(s) or secretory cell(s) have been reported in approximately 30% of all vascular plant species (Fahn 2000) and in a single plant species several types of trichomes (both glandular and non-glandular) have been reported to be present together. Owing to their remarkable ability to biosynthesize, store and secrete a range of secondary metabolites, the glandular trichomes are often referred to as biofactories of specialized metabolites. They secrete a mixture of chemicals that often offer a vast array of uses in the pharmaceutical, pesticides and flavour & fragrance industries, besides playing important role in plant biology. These structures are regarded as suitable systems for studying molecular basis of cellular differentiation and biosynthesis of specialized plant metabolites. For obvious reasons, there have been emerging interests to study the gene expression and metabolism in these structures with main focus on identification of genes involved in biosynthesis of trichome specific natural products. Thus, due to their wide occurrence, usage as development model and many other important functions, glandular trichomes have been of academic and applied significance.

17.2 Morphology and Classification of Glandular Trichomes

Glandular trichomes have a multicellular structure, consisting of a stalk, which is terminated by a glandular head (Turner et al. 2000). They are developed from a single protodermal cell, which following vertical enlargement and multiple divisions develop into a trichome structure. On the basis of structure and cellular organization, glandular trichomes can be subdivided into two major classes, namely peltate and capitate trichomes (Fig. 17.1) (Werker 2000). The peltate trichomes are generally characterized by presence of a short stalk, composed of one or two cells, and large head, comprising of four to eight secretory cells, having a large sub-cuticular space (Turner et al. 2000; Werker 2000). The secretory cells are remarkably active in biosynthesizing metabolites that along with other molecules are transported out

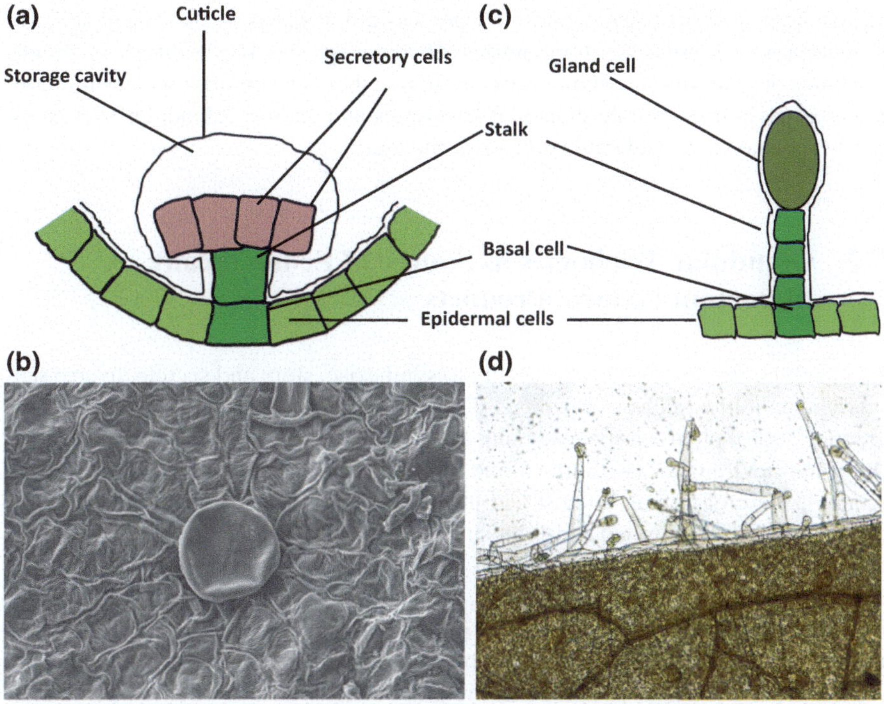

Fig. 17.1 Types of glandular trichomes. **a** and **c** Schematic diagrams showing peltate and capiate trichomes, respectively. **b** Scanning electron microscopic view of leaf surface of a member of family lamiaceae showing peltate trichome (Author's own lab, unpublished picture). **d** Light microscopic view of leaf surface of *N. tabacum* showing glandular trichomes

and stored in a prominent sub-cuticular space. Capitate trichomes are characterized by presence of a single basal cell, which is embedded within the epidermal layer, one or two stalk cells and one or two spherical secretory heads (Werker 2000; Bisio et al. 1999). The *Solanum* sp., for example are characterized by presence of eight types of trichomes. Out of them, four (i.e., type I, IV, VI and VII) are glandular capitate trichomes and the rest (i.e., type II, III, V and VIII) are non-glandular (Glas et al. 2012). The type I and IV glandular trichomes are capitate, whereas type VI and VII appear to display a peltate structure. These types differ in number of stalk and secretory cells. In *S. lycopersicum*, Type I and Type VI are the two abundant types of glandular trichomes (McDowell et al. 2011). Type I trichomes consist of a multicellular stalk with a single, small gland cell at the tip where as Type VI trichomes have a unicellular stalk with a four-cell glandular head. The glandular trichomes of *Nicotiana tabacum, N. sylvestris* and *N. rustica* exhibit very similar features, characterized by presence of 4 to 6 stalk cells and 1 to 6 head cells. In *N. tabacum*, usually, two types of capitate glandular trichomes are found (Shepherd et al. 2005); the larger ones with a long stalk and a chlorophyllous head due to presence of chloroplast, and the smaller ones with a short stalk and non- chlorophyllous head. In contrast to the long trichomes, short trichomes do not possess chloroplasts in their head cells. In the members of Lamiaceae, for example, *Ocimum* sp. and *Mentha piperita* (mint), both types of glandular trichomes are present together. The *Mentha* species has non-glandular trichomes, peltate glandular trichomes and capitate glandular trichomes, present on abaxial and adaxial surfaces of the leaf.

17.3 Glandular Trichomes as Source of Economically Important Natural Products

The glandular trichomes have the ability to synthesise, store and secrete diverse secondary metabolites such as terpenoids (Gershenzon and Dudareva 2007), flavonoids (Treutter 2006), phenylpropenes (Gang et al. 2002), methyl ketones (Fridman et al. 2005) and acyl sugars (Kroumova and Wagner 2003). Many of these secondary metabolites are of human interest and utilized as pharmaceuticals and neutraceuticals (Table 17.1) (Mahmoud and Croteau 2002; Schilmiller et al. 2008). For instance, Lamiaceae, an important aromatic plant family with species such as Basil (*Ocimum basilicum*), lavender (*Lavandula spica*), mint (*Mentha* × *piperita*), oregano (*Origanum vulgare*) and thyme (*Thymus vulgaris*), is renowned for its essential oil, which is produced exclusively in glandular trichomes (Schilmiller et al. 2008). Artemisinin, a sesquiterpene lactone, produced in the glandular trichomes of *Artemisia annua* (a member of Asteraceae family), is used as an effective drug for the treatment of malaria (Duke et al. 1994; Weathers et al. 2011). The glandular trichomes of *Cannabis*

Table 17.1 A list of plants with economically important natural products produced primarily by glandular trichomes

Plant species	Metabolite	Important compound	References
Mentha piperita	Monoterpene	Menthol	Gang et al. (2002)
Salvia sclarea	Diterpene	Sclareol	Moulines et al. (2004), Frija et al. (2011)
Cannabis sativa	Cannabinoids	Tetrahydrocannabinol (THC); Cannabidiol (CBD)	Sirikantaramas et al. (2005), Pellati et al. (2008), Taura et al. (2007)
Humulus lupulus	Terpenes	Humulone	Wang et al. (2008)
Gossypium hirsutum	Sesquiterpene	Gossypol	Mellon et al. (2014)
Artemisia annua	Sesquiterpene	Artemisinin	(Weathers et al. 2011)
Cistus creticus	Diterpene	Labdanum	Attaguile et al. (1995), Demetzos et al. (1997, 2001)
Thymus vulgaris	Monoterpene	Thymol and Carvacrol	Dauqan and Abdullah (2017)
Origanum vulgare	Monoterpene	Carvacrol and Thymol	Sivropoulou et al. (1996)

sativa are source of unique terpeno-phenolic compounds, known as cannabinoids. The Tetrahydrocannabinol (THC)-a psychoactive cannabinoid displays anti-nausea and anti-cancer activities (Sirikantaramas et al. 2005; Pellati et al. 2008) whereas Cannabidiol (CBD)—a non psychoactive cannabinoid has been found to be effective in prevention of neurodegenerative and cardiovascular diseases (Pellati et al. 2008; Taura et al. 2007). Gossypol and other related disesquiterpene produced by the trichomes of *Gossypium hirsutum* (cotton), possessing anti-fungal activities are potential natural pesticides (Mellon et al. 2014; Dayan and Duke 2003). The labdane-type diterpenes, produced in trichomes of *Cistus creticus* (Pink Rock-Rose) trichomes exhibit gastric antiulcer (Attaguile et al. 1995), antifungal, antibacterial and anti-inflammatory activities (Demetzos et al. 1997; Demetzos et al. 2001). *M. piperita* trichomes produce monoterpenes including menthone and menthol (Lange et al. 2000) where as *M. spicata* (spearmint) produces carvone, that gives attribute like odour and taste to the plant and have been used as flavouring agent in food and pharmaceutical preparations (Chauhan et al. 2009). Several species from Solanaceae family like *Solanum lycopersicum*, *S. habrochaites* and *S. pennellii* contain diverse metabolites in their glandular trichomes such as monoterpenes, sesquiterpenes, methylketones, diterpenes and acyl sugars (Antonious 2001; Besser et al. 2009); The glandular trichomes of *Ocimum basilicum* (Basil) secrete phenylpropanoids (Gang et al. 2002); gland exudates of *Medicago sativa* (alfa alfa) contain lipophilic amides (Ranger et al. 2005) and accumulate flavonoids that contributes to the plant's antioxidant properties (Aziz et al. 2005). The glandular trichomes of *Salvia sclarea* (clary sage) trichomes accumulates sclareol—a labdane diterpene, that is used as precursor for Ambrox or Ambroxane, which finds application in flavour and fragrance industry (Moulines et al. 2004; Frija et al. 2011).

17.4 Role of Glandular Trichomes in Plant Biology

Trichomes cover the outermost layer of plant organs such as leaf and stem and thereby are directly exposed to the surroundings to encounter prevailing, changing and often challenging growth conditions. In this regard, these structures can act as a component of physical defense system of plants against insects, pathogens and some abiotic stresses (Fig. 17.2). In addition, owing to the appreciable biosynthetic capabilities for producing secondary metabolites, the glandular trichomes are involved in ecological interactions, including chemical defense against invading pathogens and insects (Runyon et al. 2010; Tian et al. 2012).

17.4.1 Role in Abiotic Stress Tolerance

Abiotic stress conditions trigger an array of morphological, physiological, molecular and biochemical changes that drastically affect plant growth and development and

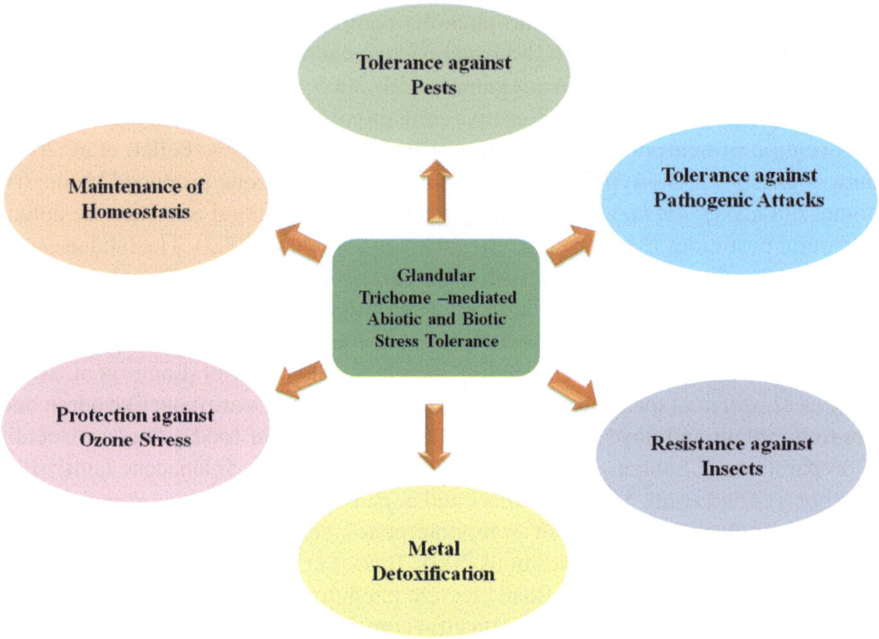

Fig. 17.2 Schematic representation summarizing role of trichomes in stress

remains the major constraint to crop yield. In the recent past, several studies have demonstrated that trichomes under adverse abiotic environmental conditions play a vital role in the plant survival.

17.4.1.1 Role in Heavy Metal Detoxification and Homeostasis

The extensive and burgeoning accumulation of heavy metals (HMs) in the biosphere due to anthropogenic disturbances has become a predicament condition for all forms of life including both plants and animals. HMs are non-biodegradable, inorganic chemical constituents which inflict detrimental effects on plants and animals including humans (Cirlakova 2009). HMs at elevated levels can hamper the functions of several important cellular biomolecules such as DNA, nuclear proteins, enzymes and pigments which can lead to excessive generation of reactive oxygen species (ROS) (Zengin and Munzuroglu 2005; Ali et al. 2013). The increased generation of ROS such as superoxide free radicals ($O_2\cdot-$), hydroxyl free radicals ($OH\cdot-$) or non-free radical species such as singlet oxygen ($^1O_2$) and hydrogen peroxide (H_2O_2) due to disturbance of cellular homeostasis cause an imbalance between ROS generation and scavenging, results in oxidative stress (Sytar et al. 2013). This stress condition implicates serious deteriorative anomalies within plant cells such as DNA damage, protein degradation, redox imbalance, leakage of ions, and disruption of cell components and membranes which ultimately can lead to cell death via programmed cell death (PCD) pathways (Sharma et al. 2012).

Plants have evolved a repertoire of defense strategies to cope with heavy metal stress (Kupper and Kroneck 2005). Trichomes have been implicated in ion homeostasis and heavy metal detoxification process either by serving as storage sites for HMs or as secretion sites of a wide array of secondary metabolites that counteract perilous effects of heavy metal contamination (Hauser 2014). The HMs have been found to be accumulated in the trichomes of both hyper (Sarret et al. 2006; Broadhurst et al. 2004) and non-hyper accumulating (Lavid et al. 2001; Dominguez-Solis et al. 2004) plants. The role of trichomes in heavy metal sequestration and detoxification has been studied in details in *N. tabacum*. On exposure of tobacco plant to the toxic level of cadmium (Cd), trichomes were reported to be the primary sites on the leaf surface, which are engaged in exudation of Cd crystals. Also cysteine synthase overexpressing tobacco lines showed Cd tolerance with 20% less endogenous Cd concentration as compared to wild-type plants and its trichome density has been reported to be 25% higher than the wild-type control plants (Harada and Choi 2008). The sequencing of cDNA libraries corresponding to the trichomes of tobacco, with or without Cd treatment revealed that trichomes are the primary sites of expression of genes encoding for stress related proteins such as antipathogenic T-phylloplanin-like proteins, glutathione peroxidase and many other classes of pathogenesis-related (PR) proteins. Furthermore, the glutathione levels were found to be elevated in the tip cells of trichomes as compared to other cells, reflecting the existence of a well developed sulfur-dependent protective system for heavy metal detoxification. Higher expression of genes encoding metallothionins, functioning in metal tolerance has been documented in trichomes of some plants species such as *Vicia faba* (Foley and Singh 1994). These observations clearly indicate that trichomes have well developed molecular machinery for accumulation, sequestration and exudation of heavy metals.

17.4.1.2 Role in Ozone Stress

During the past century, average tropospheric global ozone (O_3) concentration has drastically elevated and is expectedly increasing further (Hartmann et al. 2014; Oltmans et al. 2013). Several reports suggest that elevated level of atmospheric O_3 inhibits plant growth and development and results in decrease in productivity (Ainsworth et al. 2012; Fares et al. 2013). Certain plants show more tolerance to ozone stress than others but the underlying mechanism is still not well understood (Ainsworth 2017; Feng et al. 2017). In plants O_3 enters mainly via the stomata and reacts with organic molecules in the apoplast and resulting in excessive generation of ROS and ultimately leading to cell damage and cell death (Ainsworth 2017; Cho et al. 2011; Kanagendran et al. 2017). Studies were conducted to investigate the role of glandular and non-glandular trichomes in response to ozone stress. In order to cover a broad range of trichome characteristics, such as trichome density, trichome type etc. twenty-three herbaceous plant species were selected for this study and it was reported that peltate and capitate glandular trichomes showed significant level of tolerance against ozone stress whereas no such resistance was observed in non-glandular trichomes. Also species with lower glandular trichomes on their leaf surface were found to be more vulnerable to ozone stress as compared to those with higher density of glandular trichomes. These results shed light on the possible role of glandular trichome in the reduction of ozone toxicity and may function as chemical barricades that play a key role in neutralizing the toxic O_3 before entering into the apoplast (Li et al. 2017).

Apart from this, there are several reports from many species which proposed that glandular trichomes might play role in tolerance to drought stress. For example, a study in tomato found that overexpression of SlMX1, a MIXTA like MYB transcription factor, led to enhanced trichome density including both glandular and non-glandular trichomes accompanied with increased drought tolerance but yet there are no concrete evidences of the role of glandular trichomes specifically involved in imparting the drought tolerance (Ewas et al. 2016).

17.4.2 Role in Biotic Stress Tolerance

Various biotic factors are the major threats to the productivity of large number of important plant species. Several studies demonstrated that trichomes act as chemical defense barrier against insects, pests, herbivores, fungal infections, and even plants of parasitic behaviour (Tian et al. 2012; Peiffer et al. 2009). Apart from non-glandular trichomes, glandular trichomes synthesize and/or accumulate highly interesting secondary metabolites such as terpenoids, phenylpropenes, methyl ketones (Fridman et al. 2005; Ben-Israel et al. 2009), proteinase inhibitors (Tian et al. 2012) and acyl sugars (Schilmiller et al. 2012; Stout et al. 2012; Xu et al. 2013) and contribute substantially to chemical arsenal of plant defense strategies and thus play a fundamental role in both structural and chemical defense strategy against several herbivory and pathogen attacks and are interesting targets for breeding (Glas et al. 2012; Gruber et al. 2006).

17.4.2.1 Insect Resistance

Recent studies in many tomato wild relatives have reported the correlation of the presence, longevity, density, and size of the type I and the shorter multicellular type IV glandular trichomes with resistance against the whitefly (Firdaus et al. 2013). Previously, studies found that suppression of a glandular trichome specific P450 hydroxylase gene in tobacco led to resistance against aphids. The analysis of P450 suppressed transgenic tobacco plants displayed elevated concentration of cembratriene-ol (CBT-ol) which displayed potent aphidicidal activity (Wang et al. 2001). *NtLTP1*, a glandular-specific lipid transfer protein from tobacco has been implicated in secretion of lipid compounds from trichome heads. The transgenic tobacco lines overexpressing *NtLTP1* displayed enhanced tolerance to aphids (Choi et al. 2012). The investigation of the effect of the *hairless* (*hl*) mutation on trichome density, chemical composition and herbivory resistance in tomato suggested that leaf surface extracts have low levels of sesquiterpene and polyphenolic compound and *hl* mutation causes structural distortion of trichomes in leaf tissue and leads to decreased tolerance against insect herbivory (Kang et al. 2010).

17.4.2.2 Resistance Against Pathogens

Glandular trichomes are often capable of secreting exudates displaying antifungal activities. In a wild potato species (*S. berthaultii*), a trichome exudate was reported to confer resistance to *Phytophthora infestans* (Lai et al. 2000). The disease incidence has shown negative correlation with the density and polyphenol-oxidase activity of short type A trichome bearing a four-lobed membrane-bound gland at their tips. In chickpea (*Cicer arietinum*) the concentration of a highly acidic trichome exudate is critical in response to infection by *Ascochyta rabiei*. Low concentrations of exudates promote germination of *Ascochyta rabiei* conidia whereas high concentrations inhibit its germination (Armstrong-Cho and Gossen 2005). The damaged trichomes have been observed as entry sites for the infection and colonialization of several different fungal pathogens such as *Phoma clematidina* on clematis (Van De Graaf et al. 2002), powdery mildew (*Erysiphe necator*) on grapevine buds (Rumbolz and Gubler 2005), *Botrytis cinerea* on harvested tomato (Charles et al. 2008), and *Beauveria bassiana* on poppy (Landa et al. 2013). A trichome specific glycoprotein known as T-phylloplanin in tobacco was reported to be a potent inhibitor of oomycete *Peronospora tabacina* germination (Kroumova et al. 2007). Also tobacco plants with low expression of phylloplanin are more susceptible to pathogen attacks. Altogether the above mentioned examples clearly provide ample evidences of the active role of trichomes in imparting stress tolerance to several biotic stresses and mediating ecological interactions.

17.5 Omics Approaches for Studying Gene Expression and Function in the Context of Glandular Trichome Biology

Apart from being sources of economically important natural products, trichomes function as physical and chemical defense structures. These structures therefore are interesting systems to understand the molecular basis of secondary metabolism and plant defense. The "Omics" approaches such as transcriptomics, proteomics and metabolomics can provide detailed information about the metabolic and gene regulatory networks operating in trichomes to favour secondary metabolism and defense responses. Initial studies based on Sanger sequencing of cDNA libraries of trichomes provided limited but useful information about the transcriptome landscape of trichomes. Later on, application of next generation sequencing (NGS) revolutionized the area of trichome biology by providing comprehensive information about genes expressing in trichomes (Table 17.2). A dedicated database, TrichOme hosting transcriptomics (ESTs/unigene sequences) and metabolomics (mass-spectrometry-based trichome metabolite profiles) resources of trichomes of a number of plant species is available (Dai et al. 2010) (http://www.planttrichome.org/). The "Omics" approaches have revealed that genes involved in secondary metabolism, defense response, and lipid biosynthesis are enriched in transcriptomes of glandular trichomes. Several medicinal and aromatic plants have been studied for identification and characterization of genes expressing in glandular trichomes (Table 17.3). These studies have been vital for elucidation of molecular basis of biosynthesis of several important natural products as well as for functional attributes of trichomes (Huchelmann et al. 2017; Tissier 2018). In the following heads, the features of gene expression in glandular trichomes have been summarized.

17.5.1 Genes Encoding Enzymes of Secondary Metabolism

Sanger sequencing of cDNA libraries corresponding to the glandular trichomes of *N. tabacum* and *N. sylvestris*, for example, provided platform for identification of genes involved in the trichome specific secondary metabolism leading to cembrenoid and labdanoid diterpenoid biosynthesis (Wang et al. 2001, 2002; Wang and Wagner 2003; Ennajdaoui et al. 2010; Sallaud et al. 2012). A terpene synthase named as cembratriene-ol synthase (CBTS) and CYP71D16, a CYP450 enzyme have been shown to be involved in cembrenoid biosynthesis. The labdanoid biosynthesis, on the other hand is driven by enzymes namely copalyl diphosphate synthase 2 (CPS2) and abienol synthase (ABS). The expression of the genes encoding CBTS, CYP71D16, CPS2 and ABS was reported to be trichome specific and therefore their promoter regions can be used for driving trichome specific gene expression. In addition, the characterization of genes expressing primarily in glandular trichomes led to the discovery of metabolic pathways involved in natural product biosynthesis in several medicinal and aromatic plants. *C. sativa*, for example, biosynthesize bioactive cannabinoids, primarily in the glandular trichomes of female flower. The genes involved in the cannabinoid biosynthesis have been identified using transcriptome resource of glandular trichomes of *C. sativa,* which in turn helped in elucidation of the pathway at molecular level (Sirikantaramas et al. 2005; Taura et al. 2007, 2009; Page and Boubakir 2011).

The terpenoid biosynthesis necessitates substrate supply in the form of isoprene units. Two pathways, namely mevalonate (MVA) and 2-C-methyl-D-erythritol-4-phosphate (MEP) pathway, localized in cytoplasm and plastids, respectively are responsible for generation of isoprene pool. As the glandular trichomes are often enriched in terpenoid class of secondary metabolites, for example, those having monoterpene, diterpene and sesquiterpene backbones, the key genes involved in the MVA and MEP pathways such as DXS and HMGR, respectively are upregulated

Table 17.2 List of summarizing transcriptome sequencing studies on glandular trichomes of different plant species

Species	Trichome type	Transcriptome sequencing approach	ESTs/Unigenes	References
Solanum lycopersicum	Mixed	Sanger	7254 [ESTs]	Besser et al. (2009)
S. lycopersicum	Type I	Sanger	831 [ESTs]	McDowell et al. (2011)
S. lycopersicum	Type VI	NGS	225,000 [ESTs]	McDowell et al. (2011)
S. lycopersicum	Type VII	Sanger	791 [ESTs]	McDowell et al. (2011)
S. lycopersicum	Mixed stems	NGS	278000 [ESTs]	McDowell et al. (2011)
S. lycopersicum	Mixed	NGS	195,377 [ESTs]	Bleeker et al. (2011)
Solanum habrochaites	Mixed	Sanger	2656 [ESTs]	van Der Hoeven et al. (2000), Fei et al. (2004)
S. habrochaites	Type I	Sanger	978 [ESTs]	McDowell et al. (2011)
S. habrochaites	Type IV	Sanger	1425 [ESTs]	McDowell et al. (2011)
S. habrochaites	Type VI	NGS	224000 [ESTs]	McDowell et al. (2011)
S. habrochaites	Mixed leaves	NGS	108,000 [ESTs]	McDowell et al. (2011)
S. habrochaites	Mixed	NGS	182386 [ESTs]	Bleeker et al. (2011)
Solanum pimpinellifolium	Type VI	NGS	227,000 [ESTs]	McDowell et al. (2011)
Solanum pennelllii	Type IV	Sanger	1277 [ESTs]	McDowell et al. (2011)
S. pennelllii	Type VI	Sanger	1137 [ESTs]	McDowell et al. (2011)
S. pennelllii	Mixed leaves	NGS	275000 [ESTs]	McDowell et al. (2011)
Solanum arcanum	Mixed stems	NGS	415,000 [ESTs]	McDowell et al. (2011)
Ocimum basilicum	Peltate	Sanger	4804 [ESTs]	Iijima et al. (2004a)
O. basilicum	Peltate	Sanger	5422 [ESTs]	Iijima et al. (2004b)

(continued)

Table 17.2 (continued)

Species	Trichome type	Transcriptome sequencing approach	ESTs/Unigenes	References
O. basilicum	Peltate	Sanger	7314 [ESTs]	Kapteyn et al. (2007)
O. basilicum	Peltate	Sanger	1344 [ESTs]	Gang et al. (2001)
Medicago sativa	Stem glandular trichomes	Sanger	9659 [ESTs]	Aziz et al. (2005)
Medicago truncatula	Glandular trichomes	Sanger	10,377 [ESTs]	Dai et al. (2010)
Artemisia. annua	Glandular trichomes	NGS	406,044 [ESTs]	Wang et al. (2009)
Humulus lupulus	Glandular trichomes	Sanger	12,665 [ESTs]	Wang et al. (2008)
H. lupulus	Glandular trichomes	Sanger	10,581 [ESTs]	Nagel et al. (2008)
Cannabis sativa	Glandular trichomes from female flower	Sanger	1075 [unigenes]	Marks et al. (2009)
Cistus creticus	Glandular trichomes	Sanger	2022 [ESTs]	Falara et al. (2008)
Mentha × piperita	Peltate	Sanger	1316 [ESTs]	Lange et al. (2000)
Mentha spicata	Peltate glandular trichomes	NGS	25,000 [unigenes]	Jin et al. (2014)
Salvia fruticosa	Glandular trichomes	Sanger	1459 [ESTs]	Chatzopoulou et al. (2010)
Nicotiana tabacum	Glandular trichomes	Sanger	5139 [ESTs]	Cui et al. (2011)
N. tabacum	Glandular trichomes with or without Cd	Sanger	2000 [ESTs]	Harada et al. (2010)
Nicotiana benthamiana	Glandular trichomes	Sanger	6686 [ESTs]	Slocombe et al. (2008)

in glandular trichome as compared to trichome free leaf sample (Glas et al. 2012; Huchelmann et al. 2017; Balcke et al. 2017; Wang et al. 2009). Likewise, genes involved in the biosynthesis of flavonoid and phenylpropanoid class of secondary metabolites have been reported to display higher transcript levels in glandular trichomes.

Table 17.3 List of characterized genes involved in aspects of glandular trichome biology

Plant	Biological functions and metabolic pathways	Genes	References
Nicotiana tabacum	Diterpene transport	*Nt*PDR1	Crouzet et al. (2013)
N. tabacum	Lipid secretion	*Nt*LTP1	Choi et al. (2012)
N. tabacum	Disease defense	T-Phylloplanin	Shepherd et al. (2005), Choi et al. (2012)
N. tabacum	Terpenoid pathway	α-Cembratrienol/β-cembratrienol synthase (CYC-1)	Wang and Wagner (2003)
N. tabacum	Aphid resistance/CBT—diol synthase	CYP71D16	Wang et al. (2001, 2002)
N. sylvestris	Insects resistance/CBT –diol synthase	*Ns*CBTS	Ennajdaoui et al. (2010)
N. tabacum	Labdane diterpene biosynthesis	*Nt*CPS2 *Nt*ABS	Sallaud et al. (2012)
Artemisia annua	artemisinin biosynthetic pathway	*Aa*ORA	Lu et al. (2013)
A. annua	Terpene and lipid biosynthesis	CYP71AV1	Teoh et al. (2006), Polichuk et al. (2010)
A. annua	Terpene and lipid biosynthesis	ALDH1	Teoh et al. (2009)
A. annua	Terpene and lipid biosynthesis	DBR2	Zhang et al. (2008a)
A. annua	Terpene and lipid biosynthesis	Alcohol dehydrogenase 2 (ALDH2)	Polichuk et al. (2010)
A. annua	Terpenoid pathway	Amorpha-4,11-diene synthase (KCS12)	Chang et al. (2000)
A. annua	Terpenoid pathway	Amorpha-4,11-diene synthase (pAC12)	Mercke et al. (2000)
A. annua	Terpenoid pathway	Dihydroartemisinic aldehyde reductase (Red1)	Ryden et al. (2010)
A. annua	Terpene and lipid biosynthesis	2-Alkenal reductase (DBR1)	Zhang et al. (2008b)
A. annua	Terpenoid pathway	β-Caryophyllene synthase (QHS1)	Cai et al. (2002)
A. annua	Terpenoid pathway	β-Farnesene synthase (β-FS)	Picaud et al. (2005)

(continued)

Table 17.3 (continued)

Plant	Biological functions and metabolic pathways	Genes	References
A. annua	Terpenoid pathway	Germacrene A synthase (*Aa*GAS)	Bertea et al. (2006)
A. annua	Terpenoid pathway	(−)-β-Pinene synthase (QH6)	Zhang et al. (2008b)
A. annua	Sesquiterpene β-Caryophyllene Transport	*Aa*PDR3	Cai et al. (2002)
Cannabis sativa	Terpenoid pathway	Olivetol synthase (OLS)	Picaud et al. (2005)
C. sativa	Terpenoid pathway	Aromatic prenyltransferase (PT)	Bertea et al. (2006)
C. sativa	Cannabinoid pathway	Δ1-Tetrahydrocannabinolic acid synthase (THCAS)	Sirikantaramas et al. (2005)
C. sativa	Cannabinoid pathway	Cannabidiolic acid synthase (CBDAS)	Taura et al. (2007)
Mentha citrata	Terpenoid	(−)-Linalool synthase	Crowell et al. (2002)
Mentha. spicata	Terpenoid, flavonoid	MsTPS1 and MsTPS2	Jin et al. (2014)
M. spicata	Terpenoid	Limonene 6-hydroxylase (SM12, CYP71D18)	Lupien et al. (1999)
M. spicata	Terpenoid	Carveol dehydrogenase (ISPD)	Ringer et al. (2005)
Mentha × piperita	Terpenoid	Limonene 3-hydroxylase, PM17, CYP71D13; PM2, CYP71D13	Lupien et al. (1999)
M. piperita	Terpenoid	(+)-Menthofuran synthase (MFS)	Bertea et al. (2001)
M. piperita	Terpenoid	Menthone:(+)-neomenthol reductase (MNR)	Davis et al. (2005)

(continued)

Table 17.3 (continued)

Plant	Biological functions and metabolic pathways	Genes	References
M. piperita	Terpenoid, flavonoid	4-coumarate-CoA ligase, chalcone synthase, chalcone isomerase, flavonoid-3′, 5′-hydroxylase, flavonol-4-reductase, flavonol sulfotransferase, and flavonoid O-methyltransferases	Lange et al. (2000)
Humulus lupulus	Terpenoid pathway	Valerophenone synthase (VPS)	Okada and Ito (2001)
H. lupulus	Terpenoid pathway	Myrcene synthase (HlMTS2)	Wang et al. (2008)
H. lupulus	Terpenoid pathway	β-Caryophyllene/α-Humulene synthase (HlSTS1)	Wang et al. (2008)
H. lupulus	Terpenoid pathway	Germacrene A synthase (HlSTS2)	Wang et al. (2008)
Ocimum basilicum	Terpenoid pathway	Terpinolene synthase (TES)	Iijima et al. (2004b)
O. basilicum	Terpenoid pathway	Geraniol synthase (GES)	Iijima et al. (2004a)
O. basilicum	Terpenoid pathway	Linalool synthase (LIS)	Iijima et al. (2004b)
O. basilicum	Terpenoid pathway	α/β-Selinene synthase (SES)	Iijima et al. (2004a)
O. basilicum	Terpenoid pathway	γ-Cadinene synthase (CDS)	Iijima et al. (2004a)
O. basilicum	Terpenoid pathway	Germacrene D synthase (GDS)	Iijima et al. (2004a)
O. basilicum	Terpenoid pathway	Geraniol/nerol oxidase (GEDH1)	Iijima et al. (2006)
Helianthus anuus	Terpenoid pathway	Germacrene A acid 8β-hydroxylase (CYP71BL1)	Ikezawa et al. (2011)
Cistus creticus	Terpenoid pathway	Geranylgeranyl diphosphate synthase (CcGGDPS1, CcGGDPS2)	Pateraki and Kanellis (2008)
C. creticus	Terpenoid pathway	Copal-8-ol diphosphate synthase (CcCLS)	Falara et al. (2011)

(continued)

Table 17.3 (continued)

Plant	Biological functions and metabolic pathways	Genes	References
Salvia fruticosa	Terpenoid	1,8-Cineole synthase (Sf-CinS1)	Kampranis et al. (2007)
S. pornifera	Terpenoid	Sabinene synthase (Sp-SabS1)	Kampranis et al. (2007)
Solanum habrochaites	Terpenoid	β-Elemene synthase (ShTPS15)	Bleeker et al. (2011)
S. habrochaites	Terpenoid	Germacrene B synthase (SSTLH1)	van Der Hoeven et al. (2000)
S. habrochaites	Terpenoid	Germacrene D synthase (SSTLH2)	van Der Hoeven et al. (2000)
S. habrochaites	Terpenoid	α-Pinene synthase (ShPIS)	Gonzales-Vigil et al. (2012)
S. lycopersicum	Terpenoid	Neryl diphosphate synthase (NDPS1)	Schilmiller et al. (2009)
S. lycopersicum	Terpenoid	1,8-Cineole synthase	Falara et al. (2011)
S. americanum	Defense protein against insect attacks	SaPIN2b	Schluter et al. (2010), Luo et al. (2012)

17.5.2 Gene Involved in Primary Metabolism

The impressively active metabolism requires sufficient primary Carbon flux and energy to support excessive production of secondary metabolites in glandular trichomes. In this regard, transcriptomic and proteomic studies on glandular trichomes suggested modulation of genes involved in primary metabolism as compared to trichome-free leaf. The long glandular trichomes of tobacco contains chlorophyllous head cell and can photosynthesize. Using proteomics approach, a novel Rubisco small subunit (NtRbcS-T), preferentially expressing in head cells of long glandular trichomes of tobacco was identified. NtRbcS-T was implicated in carbon fixation in gland cell having a cellular environment overproducing specialized metabolites along with CO_2 evolution (Laterre et al. 2017). A systems approach involving proteomics, metabolomics and transcriptomics in type VI glandular trichomes and leaves from a cultivated tomato variety (*Solanum lycopersicum* LA4024) provided important insights into carbon flux regulation, source of reducing power and energy to support intensified metabolism in these trichomes (Balcke et al. 2017). It was demonstrated that although type VI glandular trichomes are photosynthetically active, the major carbon flux to support trichome specific metabolism comes from leaf tissue. However, the reducing power and energy generated during photosynthesis can be utilized in driving secondary metabolism (Balcke et al. 2017).

The "Omics" studies on *N. tabacum* and tomato revealed that the genes involved in the metabolism of branched chain amino acids e.g. valine, leucine, isoleucine are upregulated in the glandular trichomes as compared to the trichome-free leaf (Balcke et al. 2017; Jin et al. 2014). These observations are consistent with the role of branched chain amino acids as precursors of acyl-sugars, which are enriched in the glandular trichomes of *N. tabacum* and tomato. The glandular trichomes, in addition, were reported to display higher expression of genes involved in lipid metabolism, especially those concerning with the biosynthesis of polyunsaturated fatty acids and wax (Balcke et al. 2017; Jin et al. 2014; Sallets et al. 2014).

17.5.3 Transporter Genes

Gene expression analysis revealed that the genes encoding transporter proteins belonging to ABC family transporters and Lipid transporter protein (LTP) are preferentially expressed in glandular trichomes. By now, however, only limited number of transporter genes expressing in glandular trichomes have been functionally characterized. An ABCG subfamily transporter gene, named as *NtPDR1*, displaying higher expression in the gland cells of long glandular trichomes of *N.tabacum* has been implicated in transportation of terpenoid compounds such as diterpenoids and sesquiterpenoids (Pierman et al. 2017). In *A. annua*, another ABCG sub-family transporter, namely *AaPDR3*, whose expression is primarily restricted to the glandular trichomes, has been shown to be involved in transportation of β-caryophyllene (Fu et al. 2017). In *N. tabacum*, *NtLTP1*, a gene encoding lipid transporter protein is involved in secretion of lipids from glandular trichomes (Choi et al. 2012).

17.5.4 Genes Involved in Abiotic and Biotic Stresses

The glandular trichomes have been reported to express genes involved in abiotic and biotic stress responses. For example, genes belonging to these functional classes putatively encode PR protein, metallothionein, T-phylloplanin RD22-like BURP domain-containing proteins, and thaumatin-like protein, ascorbate peroxidise, glutathione peroxidase, Fe- superoxide dismutase etc. (Sallets et al. 2014; Marks et al. 2009; Cui et al. 2011). The T-phyllopanin gene, displaying glandular trichome specific expression in *N. tabacum* has been demonstrated to confer defense against pathogens (Shepherd et al. 2005).

17.6 Transcription Factor Genes Involved in the Regulation of Secondary Metabolism in Glandular Trichomes

Transcription factors play central role in regulation of gene expression associated with metabolic pathways. Limited information pertaining to the transcription factors regulating glandular trichome localized metabolism is currently available. Their identification and characterization is, however, crucial for unveiling molecular mechanism involved in the trichome specific expression of structural genes of secondary metabolism. A number of transcription factors belonging to diverse families have been identified with respect to transcriptional regulation of structural genes of artemisinin biosynthesis in *A. annua*. The transcription factors namely AaWRKY1, AaERF1/2, AaORA, AaMYC2, AabZIP1, AaNAC1 and AaSPL2 positively regulate the expression of genes involved in artemisinin biosynthesis (Yu et al. 2012; Lu et al. 2013; Zhang et al. 2015; Jiang et al. 2016; Shen et al. 2016; Lv et al. 2016; 2019). These transcription factors often regulate multiple structural genes simultaneously and therefore play crucial role in fine-tuning of the secondary metabolism under spatial and temporal cues including hormone signaling. In *Mentha spicata*, transcription factor genes namely *MsYABBY5* and *MsMYB* displaying preferential expression in peltate glandular trichomes have been implicated in the regulation of monoterpene production (Wang et al. 2016; Reddy et al. 2017). Another glandular trichome specific transcription factor gene, *Expression* of *terpenoid* 1 (EOT1), belonging to SHI transcription factor family has been demonstrated to regulate terpenoid biosynthesis in *S. lycopersicum* (Spyropoulou et al. 2014).

17.7 Regulatory Genes Involved in Development of Glandular Trichomes

The studies on Arabidopsis have provided detailed information about the molecular players involved in the initiation, development and patterning of non-glandular trichomes (Serna and Martin 2006; Pattanaik et al. 2014). It was established that a complex of three regulatory proteins, viz. MYB, bHLH and WD40 proteins (MBW) plays a central role in the regulation of trichome development in Arabidopsis. In general, the understanding of molecular mechanism of regulation of glandular trichome development is comparatively very limited and it is not clear whether a MBW like regulatory complex is involved in the regulation of glandular trichome development. The studies, conducted so far, have however revealed that transcription factors, mostly belonging to Homeodomain Zipper (HD-Zip), C2H2 Zinc Finger, bHLH and MYB familes are involved in the regulation of glandular trichome development. With the current information, it is also apparent that these regulators might be functionally conserved. Overexpression of *AmMIXTA*, an R2-R3 MYB family transcription factor gene resulted in enhanced density of glandular trichome in *N. tabacum*. Similarly, MIXTA like transcription factors, SlMIXTA1 and AaMIXTA positively regulate glandular trichome development in tomato and *A. annua*, respectively (Ewas et al. 2016; Shi et al. 2018). A HD-Zip transcription factor gene, named as Wooly (*Wo*) has been implicated in the regulation of glandular trichome development in tomato (Yang et al. 2011). The overexpression of mutant allele of tomato *Wo* gene resulted in modified glandular trichome development in *N. tabacum*, suggesting that orthologs of *Wo* gene could be functionally conserved regulators in Solanaceae (Yang et al. 2015). Recently, a C2H2 Zinc finger transcription factor gene, namely Hair (H) has been identified as positive regulator of glandular trichome development in *S. lycopersicum*. It was also demonstrated that H interacts with *Wo* and thereby they might regulate trichome development in a combinatorial manner (Chang et al. 2018). Members of HD-Zip transcription factor families have also been shown to regulate glandular trichome development in *A. annua* (Yan et al. 2017) and *Cucumis sativus* (Liu et al. 2016). Recently, a bHLH transcription factor, SlMYC1 has been identified as regulator of the development of type VI glandular trichomes as well as terpenoid biosynthesis in tomato (Xu et al. 2018).

17.8 Conclusions and Future Outlook

Glandular trichomes belong to defense repertoire of plants against invading pest and pathogens. These structures have also been implicated in conferring tolerance against abiotic stresses such as drought, UV and heavy metal challenge. The impressively active metabolic and genetic machinery enable them to overproduce, accumulate and secrete secondary metabolites that include some economically important natural products. For obvious reasons, glandular trichomes have potential to be developed as production system for economically important secondary metabolites through strategic metabolic engineering involving tools of genome engineering. This approach appears to be attractive to fuel the attempts towards development of alternate production system for those secondary metabolites, which are biosynthesized by rare and endangered plant species. The next generation transcriptome sequencing and other related approaches have been instrumental in identification of genes involved in the trichome specific natural product biosynthesis. However, a detailed understanding of metabolic and gene regulatory networks leading to the biosynthesis and accumulation of metabolites in glandular trichomes awaits further investigations. There is also a need of isolation and characterization of novel and highly active trichome specific promoters. Another important area of research is to understand molecular basis of glandular trichome development, which, as of now, remains poorly understood. The knowledge about molecular regulators of glandular trichome development will be useful in enhancing trichome density and thereby the yield of trichome localized metabolites. Further, glandular trichomes can serve as a rich pool of useful genes which could be potential targets for systematic transgenesis towards development of plants tolerant to environmental stresses. However, by now, only a limited number of genes, particularly those involved in secondary metabolism in trichomes have been characterized. It is therefore desirable to carry out investigations pertaining to the elucidation of functions of other trichome specific genes. Altogether the comprehensive and detailed knowledge of fundamental aspects of trichome biology will provide new leads to plant biologists to exploit the untapped biotechnological potential of trichomes to engineer plants that would exhibit increased resistance to pests and tolerance to many abiotic stresses and also would produce specialized natural compounds of valuable industrial/pharmaceutical potential.

Acknowledgements PM acknowledges funding support by CSIR, New Delhi and Science and Engineering Research Board (SERB), Department of Science and Technology (DST), Government of India for the project under "Scheme for Young Scientists" (SB/YS/LS-305/2013). MK, UG and AKN acknowledge CSIR, DST-INSPIRE and UGC, India respectively for Junior and senior research fellowships. This represents institutional communication number CSIR-IIIM/IPR/0068.

References

Ainsworth EA (2017) Understanding and improving global crop response to ozone pollution. Plant J 90:886–897. https://doi.org/10.1111/tpj.13298

Ainsworth EA, Yendrek CR, Sitch S, Collins WJ, Emberson LD (2012) The effects of tropospheric ozone on net primary productivity and implications for climate change. Ann Rev Plant Biol 63:637–661. https://doi.org/10.1146/annurev-arplant-042110-103829

Ali H, Khan E, Sajad MA (2013) Phytoremediation of heavy metals-concepts and applications. Chemosphere 91:869–881. https://doi.org/10.1016/j.chemosphere.2013.01.075

Antonious GF (2001) Production and quantification of methyl ketones in wild tomato accessions. J Environ Sci Health B 36:835–848. https://doi.org/10.1081/pfc-100107416

Armstrong-Cho C, Gossen BD (2005) Impact of glandular hair exudates on infection of chickpea by Ascochytarabiei. Can J Bot 83:22–27. https://doi.org/10.1139/b04-147

Attaguile G, Caruso A, Pennisi G, Savoca F (1995) Gastroprotective effect of aqueous extract of *Cistus incanus* L. in rats. Pharmacol Res 31:29–32. https://doi.org/10.1016/1043-6618(95)80043-3

Aziz N, Paiva NL, May GD, Dixon RA (2005) Transcriptome analysis of alfalfa glandular trichomes. Planta 221:28–38. https://doi.org/10.1007/s00425-004-1424-1

Balcke GU, Bennewitz S, Bergau N et al (2017) Multi-omics of tomato glandular trichomes reveals distinct features of central carbon metabolism supporting high productivity of specialized metabolites. Plant Cell 29:960–983. https://doi.org/10.1105/tpc.17.00060

Ben-Israel I, Yu G, Austin MB, Bhuiyan N, Auldridge M, Nguyen T et al (2009) Multiple biochemical and morphological factors underlie the production of methyl ketones in tomato trichomes. Plant Physiol 151:1952–1964. https://doi.org/10.1104/pp.109.146415

Bertea CM, Schalk M, Karp F, Maffei M, Croteau R (2001) Demonstration that menthofuran synthase of mint (Mentha) is a cytochrome P450 monooxygenase: cloning, functional expression, and characterization of the responsible gene. Arch Biochem Biophys 390:279–286. https://doi.org/10.1006/abbi.2001.2378

Bertea CM, Voster A, Verstappen FWA, Maffei M, Beekwilder J, Bouwmeester HJ (2006) Isoprenoid biosynthesis in *Artemisia annua*: cloning and heterologous expression of a germacrene A synthase from a glandular trichome cDNA library. Arch Biochem Biophys 446:3012. https://doi.org/10.1016/j.abb.2006.02.026

Besser K, Harper A, Welsby N, Schauvinhold I, Slocombe S, Li Y, Dixon RA, Broun P (2009) Divergent regulation of terpenoid metabolism in the trichomes of wild and cultivated tomato species. Plant Physiol 149:499–514. https://doi.org/10.1104/pp.108.126276

Bisio A, Corallo A, Gastaldo P, Romussi G et al (1999) Glandular hairs and secreted material in *Salvia blepharophylla Brandegee* ex Epling grown in Italy. Ann Bot 83:441–452. https://doi.org/10.1006/anbo.1998.0838

Bleeker PM, Spyropoulou EA, Diergaarde PJ, Volpin H, De Both MT, Zerbe P, Bohlmann J, Falara V, Matsuba Y, Pichersky E, Haring MA, Schuurink RC et al (2011) RNA-seq discovery, functional characterization, and comparison of sesquiterpene synthases from *Solanum lycopersicum* and *Solanum habrochaites* trichomes. Plant Mol Biol 77:323–336. https://doi.org/10.1007/s11103-011-9813-x

Broadhurst CL, Chaney J, Angle JS, Erbe EF, Maugel TK (2004) Nickel localization and response to increasing Ni soil levels in leaves of the nickel hyperaccumulator *Alyssum murale*. Plant Soil 265:225–242. https://doi.org/10.1007/s11104-005-0974-8

Cai Y, Jia JW, Crock J, Lin ZX, Croteau R (2002) A cDNA clone for betacaryophyllene synthase from *Artemisia annua*. Phytochemistry 61:523–529. https://doi.org/10.1016/s0031-9422(02)00265-0

Chang YJ, Song SH, Park SH, Kim SU (2000) Amorpha-4,11-diene synthase of *Artemisia annua*: cDNA isolation and bacterial expression of a terpene synthase involved in artemisinin biosynthesis. Arch Biochem Biophys 383:178–184. https://doi.org/10.1006/abbi.2000.2061

Chang J, Yu T, Yang Q, Li C, Xiong C, Gao S, Xie Q et al (2018) Hair, encoding a single C2H2 zinc-finger protein, regulates multicellular trichome formation in tomato. Plant J. 96:90–102. https://doi.org/10.1111/tpj.14018

Charles MT, Makhlouf J, Arul J (2008) Physiological basis of UV-C induced resistance to Botrytis cinerea in tomato fruit. II. Modification of fruit surface and changes in fungal colonization. Post Harvest Biol Technol 47:21–26. https://doi.org/10.1016/j.postharvbio.2007.05.015

Chatzopoulou FM, Makris AM, Argiriou A et al (2010) EST analysis and annotation of transcripts derived from a trichome-specific cDNA library from *Salvia fruticosa*. Plant Cell Rep 29:523–534. https://doi.org/10.1007/s00299-010-0841-9

Chauhan RS, Kaul MK, Shahi AK, Arun K et al (2009) Chemical composition of essential oils in *Mentha spicata* L. accession [IIIM(J)26] from North-West Himalayan region, India. Ind Crops Prod 29:654–656. https://doi.org/10.1016/j.indcrop.2008.12.003

Cho K, Tiwari S, Agrawal SB, Torres NL, Agrawal M et al (2011) Tropospheric ozone and plants: Absorption, responses, and consequence. Rev Environ Contam Toxicol 212:61–111. https://doi.org/10.1007/978-1-4419-8453-1_3

Choi YE, Lim S, Kim HJ, Han JY, Lee MH, Yang Y, Kim JA, Kim YS (2012) Tobacco NtLTP1, a glandular-specific lipid transfer protein, is required for lipid secretion from glandular trichomes. Plant J 70:480–491. https://doi.org/10.1111/j.1365-313x.2011.04886.x

Cirlakova A (2009) Heavy metals in the vascular plants of Tatra mountains. Oecologia Montana 18:23–26

Crouzet J, Roland J, Peeters E, Trombik T, Ducos E, Nader J, Boutry M (2013) NtPDR1, a plasma membrane ABC transporter from *Nicotiana tabacum*, is involved in diterpene transport. Plant Mol Biol 82:181–192. https://doi.org/10.1007/s11103-013-0053-0

Crowell A, Williams DC, Davis EM, Wildung MR, Croteau R (2002) Molecular cloning and characterization of a new linalool synthase. Arch Biochem Biophys 405:112–121. https://doi.org/10.1016/s0003-9861(02)00348-x

Cui H, Zhang ST, Yang HJ, Ji H, Wang XJ (2011) Gene expression profile analysis of tobacco leaf trichomes. BMC Plant Biol 11:76. https://doi.org/10.1186/1471-2229-11-76

Dai XB, Wang GD, Yang DS, Tang YH, Broun P, Marks MD, Sumner LW, Dixon RA, Zhao PX (2010) TrichOME: a comparative omics database for plant trichomes. Plant Physiol 152:44–54. https://doi.org/10.1104/pp.109.145813

Dauqan EMA, Abdullah A (2017) Medicinal and functional values of thyme (*Thymus vulgaris* L.) Herb J Appl Biol Biotechnol 5:017–022. https://doi.org/10.7324/jabb.2017.50203

Davis EM, Ringer KL, McConkey ME, Croteau R (2005) Monoterpene metabolism. Cloning, expression, and characterization of menthone reductases from peppermint. Plant Physiol 137:873–881. https://doi.org/10.2307/4629731

Dayan FE, Duke SO (2003) Trichomes and root hairs: Natural pesticide factories. Pestic Outlook 14:175–178. https://doi.org/10.1039/b308491b

Demetzos C, Katerinopoulos H, Kouvarakis A, Stratigakis N, Loukis A, Ekonomakis C et al (1997) Composition and antimicrobial activity of the essential oil of *Cistus creticus* subsp. eriocephalus. Planta Med 63:477–479. https://doi.org/10.1055/s-2006-957742

Demetzos C, Dimas K, Hatziantoniou S, Anastasaki T, Angelopoulou D (2001) Cytotoxic and anti-inflammatory activity of labdane and cis-clerodane type diterpenes. Planta Med 67:614–618. https://doi.org/10.1055/s-2001-17362

Dominguez-Solis JR, Lopez-Martin MC, Ager FJ, Ynsa MD, Romero LC, Gotor C (2004) Increased cysteine availability is essential for cadmium tolerance and accumulation in *Arabidopsis thaliana*. Plant Biotech J 2:469–476. https://doi.org/10.1111/j.1467-7652.2004.00092000399

Duke MV, Paul RN, Elsohly HN, Sturtz G, Duke SO (1994) Localization of artemisinin and artemisitene in foliar tissues of glanded and glandless biotypes of *Artemisia annua* L. Int J Plant Sci 155:365–372. https://doi.org/10.2307/2475187

Engene N, Rottacker EC, Kastovsky J, Byrum T, Choi H, Ellisman MH et al (2012) *Moorea producens* gen. nov., sp. nov. and *Moorea bouillonii* comb. nov., tropical marine cyanobacteria rich in bioactive secondary metabolites. Int J Syst Evolution Microbiol 62:1171–1178. https://doi.org/10.1099/ijs.0.033761-0

Ennajdaoui H, Vachon G, Giacalone C et al (2010) Trichome specific expression of the tobacco (*Nicotiana sylvestris*) cembratrien-ol synthase genes is controlled by both activating and repressing cis-regions. Plant Mol Biol 73:673–685. https://doi.org/10.1007/s11103-010-9648-x

Ewas M, Gao YQ, Wang SC et al (2016) Manipulation of SlMXl for enhanced carotenoids accumulation and drought resistance in tomato Sci. Bull. 61:1413–1418. https://doi.org/10.1007/s11434-016-1108-9

Fahn A (2000) Structure and function of secretory cells. Adv Bot Res 31:37–75. https://doi.org/10.1016/s0065-2296(00)31006-0

Falara V, Fotopoulos V, Margaritis T, Anastasaki T, Pateraki I et al (2008) Transcriptome analysis approaches for the isolation of trichome- specific genes from the medicinal plant *Cistus creticus* subsp. *creticus*. Plant Mol. Biol 68:633–651. https://doi.org/10.1007/s11103-008-9399-0

Falara V, Akhtar TA, Nguyen TTH, Spyropoulou EA, Bleeker PM, Schauvinhold I, Matsuba Y, Bonini ME, Schilmiller AL, Last RL, Schuurink RC, Pichersky E (2011) The tomato terpene synthase gene family. Plant Physiol 157:770–789. https://doi.org/10.1104/pp.111.179648

Fares A, Vargas R, Detto M, Goldstein AH, Karlik J, Paoletti E, Vitale M (2013) Tropospheric ozone reduces carbon assimilation in trees: estimates from analysis of continuous flux measurements. Global Change Biol 19:2427–2443. https://doi.org/10.1111/gcb.12222

Fei Z, Tang X, Alba RM, White JA, Ronning CM, Martin GB, Tanksley SD, Giovannoni JJ (2004) Comprehensive EST analysis of tomato and comparative genomics of fruit ripening. Plant J. 40:47–59. https://doi.org/10.1111/j.1365-313x.2004.02188.x

Feng Z, Büker P, Pleijel H, Emberson L, Karlsson PE, Uddling JA (2017) Unifying explanation for variation in ozone sensitivity among woody plants. Glob Change Biol. https://doi.org/10.1111/gcb.13824

Firdaus S, VanHeusden AW, Hidayati N, Supena EDJ, Mumm R, DeVos RCH et al (2013) Identification and QTL mapping of whitefly resistance components in *Solanum galapagense*. Theor Appl Genet 126:1487–1501. https://doi.org/10.1007/s00122-013-2067-z

Foley RC, Singh KB (1994) Isolation of a *Vicia faba* metallothionein like gene: expression in foliar trichomes. Plant Mol Biol 26:435–444. https://doi.org/10.1007/bf00039552

Fridman E, Wang JH, Iijima Y, Froehlich JE, Gang DR et al (2005) Metabolic, genomic, and biochemical analyses of glandular trichomes from the wild tomato species *Lycopersicon hirsutum* identify a key enzyme in the biosynthesis of methylketones. Plant Cell 17:1252–1267. https://doi.org/10.1105/tpc.104.029736

Frija LSMT, Frade RFM, Afonso CAM (2011) Isolation, chemical, and biotransformation routes of Labdane-type diterpenes. Chem Rev 111:4418–4452. https://doi.org/10.1021/cr100258k

Fu X, Shi P, He Q et al (2017) AaPDR3, a PDR transporter 3, is involved in sesquiterpene β-caryophyllene transport in *Artemisia annua*. Front Plant Sci. 8:723. https://doi.org/10.3389/fpls.2017.00723

Gang DR, Wang J, Dudareva N, Nam KH, Simon JE, Lewinsohn E, Pichersky E (2001) An investigation of the storage and biosynthesis of phenylpropenes in sweet basil. Plant Physiol 125:539–555. https://doi.org/10.1104/pp.125.2.539

Gang DR, Lavid N, Zubieta C, Chen F, Beuerle T et al (2002) Characterization of phenylpropene O-methyltransferases from sweet basil: facile change of substrate specificity and convergent evolution within a plant O-methyltransferase family. Plant Cell 14:505–519. https://doi.org/10.1105/tpc.010327

Gershenzon J, Dudareva N (2007) The function of terpene natural products in the natural world. Nat Chem Biol 3:408–414. https://doi.org/10.1038/nchembio.2007.5

Glas JJ, Schimmel BC, Alba JM, Escobar-Bravo R, Schuurink RC, Kant MR (2012) Plant glandular trichomes as targets for breeding or engineering of resistance to herbivores. Int J Mol Sci 13:17077–17103. https://doi.org/10.3390/ijms131217077

Gonzales-Vigil E, Hufnagel DE, Kim J, Last RL Barry CS (2012) Evolution of TPS20-related terpene synthases influences chemical diversity in the glandular trichomes of the wild tomato relative *Solanum habrochaites*. Plant J. https://doi.org/10.1111/j.1365-313x.2012.05040.x

Gruber MY, Wang S, Ethier S, Holowachuk J, Bonham-Smith PC, Soroka J et al (2006) "HAIRY CANOLA"—Arabidopsis GL3 induces a dense covering of trichomes on *Brassica napus* seedlings. Plant Mol Biol 60:679–698. https://doi.org/10.1007/s11103-005-5472-0

Harada E, Choi YE (2008) Investigation of metal exudates from tobacco glandular trichomes under heavy metal stresses using a variable pressure scanning electron microscopy system. Plant Biotechnol. 25:407–411. https://doi.org/10.5511/plantbiotechnology.25.407

Harada E, Kim JA, Meyer AJ, Hell R, Clemens S, Choi YE (2010) Expression profiling of tobacco leaf trichomes identifies genes for biotic and abiotic stresses. Plant Cell Physiol 51:1627–1637. https://doi.org/10.1093/pcp/pcq118

Hartmann DL, Klein Tank AMG, Rusticucci M, Alexander LV, Brönnimann S, Charabi YAR, Zhai P (2014) Observations: atmosphere and surface. In: Stocker TF, Qin D, Hauser MT (eds) Molecular basis of natural variation and environmental control of trichome patterning. Front Plant Sci 5:1–7. https://doi.org/10.3389/fpls.2014.00320

Hauser MT (2014) Molecular basis of natural variation and environmental control of trichome patterning. Front Plant Sci 5:1–7. https://doi.org/10.3389/fpls.2014.00320

Huchelmann A, Boutry M, Charles H (2017) Plant glandular trichomes: natural cell factories of high biotechnological interest. Plant Physiol 175. https://doi.org/10.1104/pp.17.00727

Iijima Y, Davidovich-Rikanati R, Fridman E, Gang DR, Bar E, Lewinsohn E, Pichersky E (2004a) The biochemical and molecular basis for the divergent patterns in the biosynthesis of terpenes and phenylpropenes in the peltate glands of three cultivars of basil. Plant Physiol 136:3724–3736. https://doi.org/10.1104/pp.104.051318

Iijima Y, Gang DR, Fridman E, Lewinsohn E, Pichersky E (2004b) Characterization of geraniol synthase from the peltate glands of sweet basil. Plant Physiol 134:370–379. https://doi.org/10.1104/pp.103.032946

Iijima Y, Wang G, Fridman E, Pichersky E (2006) Analysis of the enzymatic formation of citral in the glands of sweet basil. Arch Biochem Biophys 448:141–149. https://doi.org/10.1016/j.abb.2005.07.026

Ikezawa N, Gopfert J, Nguyen DT, Kim SU, OMaille PE, Spring O, Ro DK (2011) Lettuce costunolide synthase (CYP71BL2) and its homolog (CYP71BL1) from sunflower catalyze distinct regio- and stereoselective hydroxylations in sesquiterpene lactone metabolism. J Biol Chem 286:21601–21611. https://doi.org/10.1074/jbc.m110.216804

Jiang W, Fu X, Pan Q, Tang Y, Shen Q, Lv Z et al (2016) Overexpression of AaWRKY1 Leads to an enhanced content of artemisinin in *Artemisia annua*. Biomed Res Int 2016:9. https://doi.org/10.1155/2016/7314971

Jin J, Panicker D, Wang Q, Kim MJ, Liu J et al (2014) Next generation sequencing unravels the biosynthetic ability of Spearmint (*Mentha spicata*) peltate glandular trichomes through comparative transcriptomics. BMC Plant Biol 14:292. https://doi.org/10.1186/s12870-014-0292-5

Kampranis SC, Ionnidis D, Purvis A, Mahrez W, Ninga E., Katerelos NA, Anssour S, Dunwell JM, Degenhardt J, Makris AM, Goodenough PW, Johnson CB (2007) Rational conversion of substrate and product specificity in a Salvia monoterpene synthase: structural insights into the evolution of terpene synthase function. Plant Cell 19:1994–2005. https://doi.org/10.1105/tpc.106.047779

Kanagendran A, Pazouki L, Li S, Liu B, Kannaste A, Niinemets U (2017) Ozone-triggered surface uptake and stress volatile emissions upon acute ozone exposure scale with foliage surface ozone uptake in ozone resistant *Nicotiana tabacum* 'Wisconsin'. J Exp Bot 69:681–697. https://doi.org/10.1093/jxb/erx431

Kang JH, Shi F, Jones AD, Marks MD, Howe GA (2010) Distortion of trichome morphology by the hairless mutation of tomato affects leaf surface chemistry. J Exp Bot 61:1053–1064. https://doi.org/10.1093/jxb/erp370

Kapteyn J, Qualley AV, Xie ZZ, Fridman E, Dudareva N, Gang DR (2007) Evolution of cinnamate/p-coumarate carboxyl methyltransferases and their role in the biosynthesis of methylcinnamate. Plant Cell 19:3212–3229. https://doi.org/10.1105/tpc.107.054155

Kolb D, Muller M (2004) Light, conventional and environmental scanning electron microscopy of the trichomes of *Cucurbita pepo* subsp. *pepo* var. *styriaca* and histochemistry of glandular secretory products. Ann Bot 94:515–526. https://doi.org/10.1093/aob/mch180

Kroumova AB, Wagner GG (2003) Different elongation pathways in the biosynthesis of acyl groups of trichome exudate sugar esters from various solanaceous plants. Planta 216:1013–1021. https://doi.org/10.1007/s00425-002-0954-7

Kroumova AB, Shepherd RW, Wagner GJ (2007) Impacts of T-phylloplanin gene knockdown and of Helianthus and Datura phylloplanins on *Peronospora tabacina* spore germination and disease potential. Plant Physiol 144:1843–1851. https://doi.org/10.1104/pp.107.097584

Kupper H, Kroneck P (2005) Heavy metal uptake by plants and cyanobacteria. Metal Ions Biol Syst 44:97–144

Lai, A., Cianciolo, V., Chiavarini, S., Sonnino, A.: Effects of glandular trichomes on the development of *Phytophthora infestans* infection in potato (*S. tuberosum*). Euphytica 114, 165–174 (2000). https://doi.org/10.1023/a:1003924318577

Landa BB, Lopez-Díaz C, Jimenez-Fernandez D, Montes-Borrego M et al (2013) In planta detection and monitorization of endophytic colonization by a *Beauveria bassiana* strain using a new-developed nested and quantitative PCR-based assay and confocal laser scanning microscopy. J Invertebr Pathol 114:128–138. https://doi.org/10.1016/j.jip.2013.06.007

Lange BM, Wildung MR, Stauber EJ, Sanchez C, Pouchnik D, Croteau R (2000) Probing essential oil biosynthesis and secretion by functional evaluation of expressed sequence tags from mint glandular trichomes. Proc Natl Acad Sci USA 97:2934–2939. https://doi.org/10.1073/pnas.97.6.2934

Laterre R, Pottier M, Remacle C, Boutry M (2017) Photosynthetic trichomes contain a specific Rubisco with a modified Ph-dependent activity. Plant Physiol 173:2110–2120. https://doi.org/10.1104/pp.17.00062

Lavid N, Barkay Z, Tel-Or E (2001) Planta 212:313. https://doi.org/10.1007/s004250

Li S, Harley PC, Niinemets U (2017) Ozone induced foliar damage and release of stress volatiles is highly dependent on stomatal openness and priming by low level ozone exposure in *Phaseolus vulgaris*. Plant, Cell Environ 40:1984–2003. https://doi.org/10.1111/pce.13003

Liu X, Bartholomew E, Cai Y, Ren H (2016) Trichome-related mutants provide a new perspective on multicellular trichome initiation and development in cucumber (*Cucumis sativus* L.). Front Plant Sci 7:1187. https://doi.org/10.3389/fpls.2016.01187

Lu X, Zhang L, Zhang F, Jiang W, Shen Q, Zhang L et al (2013) *AaORA*, a trichome-specific AP2/ERF transcription factor of *Artemisia annua*, is a positive regulator in the artemisinin biosynthetic pathway and in disease resistance to *Botrytis cinerea*. New Phytol 198:1191–1202. https://doi.org/10.1111/nph.12207

Luo M, Ding LW, Ge ZJ, Wang ZY, Hu BL, Yang XB, Sun QY, Xu ZF (2012) The characterization of SaPIN2b, a plant trichome-localized proteinase inhibitor from *Solanum americanum*. Int J Mol Sci 13:15162–15176. https://doi.org/10.3390/ijms131115162

Lupien S, Karp F, Wildung M, Croteau R (1999) Regiospecific cytochrome P450 limonene hydroxylases from mint (Mentha) species: cDNA isolation, characterization, and functional expression of -4Slimonene- 3-hydroxylase and-4S-limonene-6-hydroxylase. Arch Biochem Biophys 368:181–192. https://doi.org/10.1006/abbi.1999.1298

Lv Z, Wang S, Zhang F, Chen L, Hao X, Pan Q et al (2016) Overexpression of a novel NAC domain-containing transcription factor gene (AaNAC1) Enhances the content of artemisinin and increases tolerance to drought and *Botrytis cinerea* in *Artemisia annua*. Plant Cell Physiol 57:1961–1971. https://doi.org/10.1093/pcp/pcw118

Lv Z, Wang Y, Liu Y, Peng B, Zhang L, Tang K, Chen W (2019) The SPB-box transcription factor AaSPL2 positively regulates artemisinin biosynthesis in *Artemisia annua* L. Front Plant Sci. 10:409. https://doi.org/10.3389/fpls.2019.00409

Mahmoud SS, Croteau RB (2002) Strategies for transgenic manipulation of monoterpene biosynthesis in plants. Trends Plant Sci 7:366–373

Marks MD, Tian L, Wenger JP, Omburo SN, Soto-Fuentes W, He J, Gang DR, Weiblen GD, Dixon RA (2009) Identification of candidate genes affecting delta(9)-tetrahydrocannabinol biosynthesis in *Cannabis sativa*. J Exp Bot 60:3715–3726. https://doi.org/10.1093/jxb/erp210

McDowell ET, Kapteyn J, Schmidt A et al (2011) Comparative functional genomic analysis of Solanum glandular trichome types. Plant Physiol 155:524–539. https://doi.org/10.1104/pp.110.167114

Mellon JE, Dowd MK, Beltz SB, Moore GG (2014) Growth inhibitory effects of gossypol and related compounds on fungal cotton root pathogens. Lett Appl Microbiol 59:161–168. https://doi.org/10.1111/lam.12262

Mercke P, Bengtsson M, Bouwmeester HJ, Posthumus MA, Brodelius PE (2000) Molecular cloning, expression, and characterization of amorpha- 4,11-diene synthase, a key enzyme of artemisinin biosynthesis in *Artemisia annua* L. Arch Biochem Biophys 381:173–180. https://doi.org/10.1006/abbi.2000.1962

Moulines J, Bats JP, Lamidey AM, Da Silva N (2004) About a practical synthesis of Ambrox (R) from sclareol: a new preparation of a ketone key intermediate and a close look at its Baeyer-Villiger oxidation. Helv Chim Acta 87:2695–2705. https://doi.org/10.1002/hlca.200490239

Nagel J, Culley LK, Lu YP, Liu EW, Matthews PD, Stevens JF, Page JE (2008) EST analysis of hop glandular trichomes identifies an O-methyltransferase that catalyzes the biosynthesis of xanthohumol. Plant Cell 20:186–200. https://doi.org/10.1105/tpc.107.055178

Okada Y, Ito K (2001) Cloning and analysis of valerophenone synthase gene expressed specifically in lupulin gland of hop (*Humulus lupulus* L.). Biosci Biotechnol Biochem 65:150–155. https://doi.org/10.1271/bbb.65.150

Oltmans SJ, Lefohn AS, Shadwick D, Harris JM, Scheel HE, Galbally I, Kawasato T (2013) Recent tropospheric ozone changes—a pattern dominated by slow or no growth. Atmos Environ 67:331–351. https://doi.org/10.1016/j.atmosenv.2012.10.057

Page J, Boubakir Z (2011) Aromatic prenyltransferase from Cannabis. PCT patent application WO/2011/017798

Pateraki I, Kanellis AK (2008) Isolation and functional analysis of two *Cistus creticus* cDNAs encoding geranylgeranyl diphosphate synthase. Phytochemistry 69:1641–1652. https://doi.org/10.1016/j.phytochem.2008.02.005

Pattanaik S, Patra B, Singh SK, Yuan L (2014) An overview of the gene regulatory network controlling trichome development in the model plant, *Arabidopsis*. Front Plant Sci 5:259. https://doi.org/10.3389/fpls.2014.00259

Peiffer M, Tooker JF, Luthe DS, Felton GW (2009) Plants on early alert: glandular trichomes as sensors for insect herbivores. New Phytol 184:644–656. https://doi.org/10.1111/j.1469-8137.2009.03002.x

Pellati F, Borgonetti V, Brighenti V, Biagi M, Benvenuti S, Corsi L (2008) *Cannabis sativa* L. and nonpsychoactive cannabinoids: their chemistry and role against oxidative stress, inflammation and cancer. Biomed Res Int. https://doi.org/10.1155/2018/1691428

Picaud S, Olofsson L, Brodelius M, Brodelius PE (2005) Expression, purification, and characterization of recombinant amorpha-4,11-diene synthase from *Artemisia annua* L. Arch. Biochem Biophys 436:215–226. https://doi.org/10.1016/j.abb.2005.02.012

Pierman B, Toussaint F, Bertin A, Levy D, Smargiasso N et al (2017) Activity of the purified plant ABA transporter NtPDR1 is stimulated by diterpenes and sesquiterpenes involved in constitutive and induced defense. J Biol Chem 292:19491–19502. https://doi.org/10.1074/jbc.m117.811935

Polichuk DB, Zhang Y, Reed DW, Schmidt JF, Covello PS (2010) A glandular trichome-specific monoterpene alcohol dehydrogenase from *Artemisia annua*. Phytochemistry 71:1264–1269. https://doi.org/10.1016/j.phytochem.2010.04.026

Ranger CM, Winter RE, Rottinghaus GE, Backus EA, Johnson DW (2005) Mass spectral characterization of fatty acid amides from alfalfa trichomes and their deterrence against the potato leafhopper. Phytochemistry 66:529–541. https://doi.org/10.1016/j.phytochem.2005.01.012

Reddy VA, Wang Q, Dhar N, Kumar N, Venkatesh PN et al (2017) Spearmint R2R3-MYB transcription factor MsMYB negatively regulates monoterpene production and suppresses the expression of geranyl diphosphate synthase large subunit (MsGPPS.LSU). Plant Biotechnol J 15:1105–1119. https://doi.org/10.1111/pbi.12701

Ringer KL, Davis EM, Croteau R (2005) Monoterpene metabolism. Cloning, expression, and characterization of (−)-isopiperitenol/(−)-carveol dehydrogenase of peppermint and spearmint. Plant Physiol 137:863–872. https://doi.org/10.2307/4629730

Rumbolz J, Gubler WD (2005) Susceptibility of grapevine buds to infection by powdery mildew *Erysiphene cator*. Plant Pathol 54:535–548. https://doi.org/10.1111/j.1365-3059.2005.01212.x

Runyon JB, Mescher MC, DeMoraes CM (2010) Plant defenses against parasitic plants show similarities to those induced by herbivores and pathogens. Plant Signal Behav 5:929–931. https://doi.org/10.4161/psb.5.8.11772

Ryden AM, Ruyter-Spira C, Quax WJ, Osada H, Muranaka T, Kayser O, Bouwmeester H (2010) The molecular cloning of dihydroartemisinc aldehyde reductase and its implication in artemisinin biosynthesis in *Artemisia annua*. Planta 76:1778–1783. https://doi.org/10.1055/s-0030-1249930

Sallaud C, Giacalone C, Topfer R, Goepfert S, Bakaher N, Rosti S, Tissier A (2012) Characterization of two genes for the biosynthesis of the labdane diterpene Z-abienol in tobacco (*Nicotiana tabacum*) glandular trichomes. Plant J. 72:1–17. https://doi.org/10.1111/j.1365-313x.2012.05068.x

Sallets A, Beyaert M, Boutry M, Champagne A (2014) Comparative proteomics of short and tall glandular trichomes of *Nicotiana tabacum* reveals differential metabolic activities. J Proteome Res 13:3386–3396. https://doi.org/10.1021/pr5002548

Sarret G, Harada E, Choi YE, Isaure MP, Geoffroy N, Fakra S et al (2006) Trichomes of tobacco excrete zinc as zinc-substituted calcium carbonate and other zinc-containing compounds. Plant Physiol 41:1021–1034. https://doi.org/10.1104/pp.106.082743

Schilmiller AL, Last RL, Pichersky E (2008) Harnessing plant trichome biochemistry for the production of useful compounds. Plant J. 54:702–711. https://doi.org/10.1111/j.1365-313x.2008.03432.x

Schilmiller A, Schauvinhold I, Larson M, Xu R, Charbonneau AL, Schmidt A, Wilkerson C, Last RL, Pichersky E (2009) Monoterpenes in the glandular trichomes of tomato are synthesized from a neryl pyrophosphate precursor rather than geranyl pyrophosphate. Proc Natl Acad Sci USA 106:10865–10870. https://doi.org/10.1073/pnas.0904113106

Schilmiller AL, Charbonneau AL, Last RL (2012) Identification of a BAHD acetyltransferase that produces protective acylsugars in tomato trichomes. Proc Natl Acad Sci USA 109:16377–16382. https://doi.org/10.1073/pnas.1207906109

Schluter U, Benchabane M, Munger A, Kiggundu A, Vorster J, Goulet MC, Cloutier C, Michaud D (2010) Recombinant protease inhibitors for herbivore pest control: a multitrophic perspective. J Exp Bot 61:4169–4183. https://doi.org/10.1093/jxb/erq166

Serna L, Martin C (2006) Trichomes: different regulatory networks lead to convergent structures. Trends Plant Sci 11:274–280. https://doi.org/10.1016/j.tplants.2006.04.008

Sharma P, Bhushan Jha A, Dubey RS, Pessarakli M (2012) Reactive oxygen species, oxidative damage, and antioxidative defense mechanism in plants under stressful conditions. J Bot 26, Article ID 217037. https://doi.org/10.1155/2012/217037

Shen Q, Lu X, Yan T, Fu X, Lv Z, Zhang F et al (2016) The jasmonate-responsive AaMYC2 transcription factor positively regulates artemisinin biosynthesis in *Artemisia annua*. New Phytol 210:1269–1281. https://doi.org/10.1111/nph.13874

Shepherd RW, Bass WT, Houtz RL, Wagner GJ (2005) Phylloplanins of tobacco are defensive proteins deployed on aerial surfaces by short glandular trichomes. Plant Cell 17:1851–1861. https://doi.org/10.2307/4130874

Shi P, Fu X, Shen Q, Liu M, Pan Q, Tang Y et al (2018) The roles of *AaMIXTA1* in regulating the initiation of glandular trichomes and cuticle biosynthesis in *Artemisia annua*. New Phytol 217:261–276. https://doi.org/10.1111/nph.14789

Sirikantaramas S, Taura F, Tanaka Y, Ishikawa Y, Morimoto S, Shoyama Y (2005) Tetrahydrocannabinolic acid synthase, the enzyme contolling marijuana psychoactivity, is secreted into the storage cavity of the glandular trichomes. Plant Cell Physiol 46:1578–1582. https://doi.org/10.1093/pcp/pci166

Sivropoulou A, Papanikolaou E, Nikolaou C, Kokkini S, Lanaras T, Arsenakis M (1996) Antimicrobial and cytotoxic activities of origanum essential oils. J Agric Food Chem 44:1202–1205

Slocombe SP, Schauvinhold I, McQuinn RP et al (2008) Transcriptomic and reverse genetic analyses of branched-chain fatty acid and acyl sugar production in *Solanum pennellii* and *Nicotiana benthamiana*. Plant Physiol 148:1830–1846. https://doi.org/10.1104/pp.108.129510

Spyropoulou EA, Haring MA, Schuurink RC (2014) Expression of Terpenoids 1, a glandular trichome-specific transcription factor from tomato that activates the terpene synthase 5 promoter. Plant Mol Biol 84:345–357. https://doi.org/10.1007/s11103-013-0142-0

Stout JM, Boubakir Z, Ambrose SJ, Purves RW, Page JE (2012) The hexanoyl-CoA precursor for cannabinoid biosynthesis is formed by an acyl-activating enzyme in *Cannabis sativa* trichomes. Plant J. 71:353–365. https://doi.org/10.1111/j.1365-313x.2012.04949.x

Sytar O, Kumar A, Latowski D, Kuczynska P, Strzałka K, Prasad MNV (2013) Heavymetal-induced oxidative damage, defense reactions, and detoxification mechanisms in plants. Acta Physiol Plant 35:985–999. https://doi.org/10.1007/s11738-012-1169-6

Taura F, Sirikantaramas S, Shoyama Y, Yoshikai K et al (2007) Cannabidiolic-acid synthase, the chemotype-determining enzyme in the fiber-type *Cannabis sativa*. FEBS Lett 581:2929–2934. https://doi.org/10.1016/j.febslet.2007.05.043

Taura F, Tanaka S, Taguchi C, Fukamizu T, Tanaka H, Shoyama Y, Morimoto S (2009) Characterization of olivetol synthase, a polyketide synthase putatively involved in cannabinoid biosynthetic pathway. FEBS Lett 583:2061–2066. https://doi.org/10.1016/j.febslet.2009.05.024

Teoh KH, Polichuk DR, Reed DW, Nowak G, Covello PS (2006) *Artemisia annua* L. (Asteraceae) trichome-specific cDNAs reveal CYP71AV1, a cytochrome P450 with a key role in the biosynthesis of the antimalarial sesquiterpene lactone artemisinin. FEBS Lett 580:1411–1416. https://doi.org/10.1016/j.febslet.2006.01.065

Teoh KH, Polichuk DR, Reed DW, Covello PS (2009) Molecular cloning of an aldehyde dehydrogenase implicated in artemisinin biosynthesis in *Artemisia annua*. Botany. 87:635–642. https://doi.org/10.1139/b09-032

Tian D, Tooker J, Peiffer M, Chung SH, Felton GW (2012) Role of trichomes in defense against herbivores: comparison of herbivore response to Woolly and hairless trichome mutants in tomato (*Solanum lycopersicum*). Planta 236:1053–1066. https://doi.org/10.1007/s00425-012-1651-9

Tissier A (2018) Plant secretory structures: more than just reaction bags. Curr Opin Biotechnol 49:73–79. https://doi.org/10.1016/j.copbio.2017.08.003

Treutter D (2006) Significance of flavonoids in plant resistance: a review. Environ Chem Lett 4:147–157. https://doi.org/10.1055/s-2005-873009

Turner GW, Gershenzon J, Croteau RB (2000) Development of peltate glandular trichomes of peppermint. Plant Physiol 124:665–680. https://doi.org/10.1104/pp.124.2.665

Uphof JCT (1962) Plant hairs. In: Encyclopedia of plant anatomy IV, vol 5, pp 1–206

Van De Graaf P, Joseph ME, Chartier-Hollis JM, O'Neill TM (2002) Prepenetration stages in infection of clematis by Phomaclematidina. Plant Pathol 51:331–337. https://doi.org/10.1046/j.1365-3059.2002.00727.x

van Der Hoeven RS, Monforte AJ, Breeden D, Tanksley SD, Steffens JC (2000) Genetic control and evolution of sesquiterpene biosynthesis in *Lycopersicon esculentum* and *L. hirsutum*. Plant Cell 12:2283–2294. https://doi.org/10.1105/tpc.12.11.2283

Wang E, Wagner GJ (2003) Elucidation of the functions of genes central to diterpene metabolism in tobacco trichomes using post transcriptional gene silencing. Planta 216:686–691. https://doi.org/10.1007/s00425-002-0904-4

Wang E, Wang R, DeParasis J, Loughrin JH, Gan S, Wagner GJ (2001) Suppression of a P450 hydroxylase gene in plant trichomeglands enhances natural-product-based aphid resistance. Nat Biotechnol 19:371–374. https://doi.org/10.1038/86770

Wang E, Gan S, Wagner GJ (2002) Isolation and characterization of the CYP71D16 trichome-specific promoter from *Nicotiana tabacum* L. J Exp Bot 53:1891–1897. https://doi.org/10.1093/jxb/erf054

Wang G, Tian L, Aziz N, Broun P, Dai X, He J, King A, Zhao PX, Dixon RA (2008) Terpene biosynthesis in glandular trichomes of hop. Plant Physiol 148:1254–1266. https://doi.org/10.1104/pp.108.125187

Wang W, Wang YJ, Zhang Q, Qi Y, Guo DJ (2009) Global characterization of *Artemisia annua* glandular trichome transcriptome using 454 pyrosequencing. BMC Genom 10:465. https://doi.org/10.1186/1471-2164-10-465

Wang Q, Reddy VA, Panicker D, Mao HZ, Kumar N, Rajan C et al (2016) Metabolic engineering of terpene biosynthesis in plants using a trichome-specific transcription factor MsYABBY5 from spearmint (*Mentha spicata*). Plant Biotechnol J 14:1619–1632. https://doi.org/10.1111/pbi.12525

Weathers PJ, Aesenault PR, Covello PS, McMickle A, Teoh KH, Reed DW (2011) Artemisinin production in *Artemisia annua*: studies in planta and results of a novel delivery method for treating malaria and other neglected diseases. Phytochem Rev 10:173–183. https://doi.org/10.1007/s11101-010-9166-0

Werker E (2000) Trichome diversity and development. Adv Bot Res 31:261–283. https://doi.org/10.1016/s0065-2296(00)31005-9

Xu H, Zhang F, Liu B, Huhman DV, Sumner LW, Dixon RA et al (2013) Characterization of the formation of branched short-chain fattyacid: CoAs for bitter acid biosynthesis in hop glandular trichomes. Mol Plant 6:1301–1317. https://doi.org/10.1093/mp/sst004

Xu J, van Herwijnen ZO, Drager DB, Sui C, Haring MA, Schuurink RC (2018) SlMYC1 regulates type VI glandular trichome formation and terpene biosynthesis in tomato glandular cells. Plant Cell 30:2988–3005. https://doi.org/10.1105/tpc.18.00571

Yan T, Chen M, Shen Q, Li L, Fu X, Pan Q, Tang Y et al (2017) HOMEODOMAIN PROTEIN 1 is required for jasmonate-mediated glandular trichome initiation in *Artemisia annua*. New Phytol 213:1145–1155. https://doi.org/10.1111/nph.14205

Yang C, Li H, Zhang J, Luo Z, Gong P, Zhang C, Li J, Wang T, Zhang Y, Lu YE (2011) A regulatory gene induces trichome formation and embryo lethality in tomato. Proc Natl Acad Sci USA 108:11836–11841. https://doi.org/10.1073/pnas.1100532108

Yang C, Gao Y, Gao S, Yu G, Cheng X, Jiang C, Li H, Ye Z (2015) Transcriptome profile analysis of cell proliferation molecular processes during multicellular trichome formation induced by tomato Wov gene in tobacco. BMC Genom 16:868. https://doi.org/10.1186/s12864-015-2099-7

Yu Z-X, Li J-X, Yang C-Q, Hu W-L, Wang L-J, Chen X-Y (2012) The jasmonate-responsive AP2/ERF transcription factors AaERF1 and AaERF2 positively regulate artemisinin biosynthesis in *Artemisia annua* L. Mol Plant 5:353–365. https://doi.org/10.1093/mp/ssr087

Zengin FK, Munzuroglu O (2005) Effects of some heavy metals on content of chlorophyll, proline and some antioxidant chemicals in bean (*Phaseolus vulgaris* L.) seedlings. Acta Biologica Cracoviensia Series Botanica 47:157–164

Zhang Y, Teoh KH, Reed DW, Maes L, Goossens A, Olson DJH, Ross ARS, Covello PS (2008a) The molecular cloning of artemisinic aldehyde Δ11(13) reductase and its role in glandular trichome-dependent biosynthesis of artemisinin in *Artemisia annua*. J Biol Chem 283:21501–21508. https://doi.org/10.1074/jbc.m803090200

Zhang Y, Teoh KH, Reed DW, Covello PS (2008b) Molecular cloning and characterization of Dbr1, a 2-alkenal reductase from *Artemisia annua*. Botany 87:643–649. https://doi.org/10.1139/b09-033

Zhang F, Fu X, Lv Z, Lu X, Shen Q, Zhang L et al (2015) A basic leucine zipper transcription factor, AabZIP1, connects abscisic acid signaling with artemisinin biosynthesis in *Artemisia annua*. Mol Plant 8:163–175. https://doi.org/10.1016/j.molp.2014.12.004

Chapter 18
Gene Regulatory Networks: Current Updates and Applications in Plant Biology

Parul Gupta and Sunil Kumar Singh

Abstract Gene networks offer a strong perspective for better understanding of gene functions associated with complex biological traits. A gene regulatory network (GRN) is comprised of the regulatory elements which interact together to regulate the transcriptional and translational processes within the cell. GRNs involve the network of genes and proteins and molecular interactions that regulate those genes and proteins. GRNs govern the way a plant responds to various environmental cues. Thus, GRNs are the key to understand the interaction between the plant's genotype and environment. Elucidation of plant GRNs is important for plant resistance and adaptability under external environmental stressors. Serious attempts are being made to characterize these GRNs and modulate them in order to get the desired trait in crop plants. In this chapter, we will discuss the structure, recent advances, and factors influencing the GRNs under various environmental stresses and major challenges for future researches.

Keywords Gene regulatory network · Transcription factor · Cis-regulatory elements · Co-expression network · miRNA · GRN

Abbreviations

ABA1	ABA Deficient 1
ABI	Abscisic Acid-Insensitive
AGL15	Agamous-like 15
AHG3	ABA-hypersensitive Germination 3
AIL3	Aintegumenta-like 3
ANL2	Anthocyaninless 2

P. Gupta (✉)
Department of Botany and Plant Pathology, Oregon State University, Corvallis, USA
e-mail: gupta.parulnbri@gmail.com

S. K. Singh
Department of Biology, University of North Dakota, Grand Forks, USA

S. P. Singh et al. (eds.), *Molecular Approaches in Plant Biology and Environmental Challenges*, Energy, Environment, and Sustainability,
https://doi.org/10.1007/978-981-15-0690-1_18

AP1	Apetala 1
ARAB/ABF	ABA-responsive Element Binding Protein/ABRE Binding Factor
ASG2	Altered Seed Germination 2
bZIP	basic Leucine Zipper
CAL	Cauliflower
CCA1	Circadian Clock Associated 1
CUC	Cup-Shaped Cotyledon
CUC3	Cup-Shaped Cotyledon 3
DOG1	Delay of Germination 1
DREB1/CBF	Dehydration-responsive Element Binding Protein 1/C-Repeat Binding Factor
ESR1	Enhancer of Shoot Regeneration 1
FT	Flowering Locus T
FUS3	Fusca 3
GID1	Gibberellin Receptor 1
HDG11	Homeodomain Glabrous 11
HSFB1	Heat Stress Factor B1
HSP17	Heat Shock Protein 17
JAB	Janus Kinase Binding Protein
KEG1	Keep on Going 1
LEC	Leafy Cotyledon
LFY	Leafy
LHY	Late Elongated Hypocotyl
NAC	NAM, ATAF, and CUC
PDF2	Protodermal Factor 2
PLT3	Plethora 3
RGA	Repressor of Ga1-3
RLT2	Ringlet 2
SCL14	Scarecrow-like 14
SCL18	Scarecrow-like18
SCR	Scarecrow
SEP3	Sepallata 3
TALE	Three Amino Acid Loop Extension
TFL1	Terminal Flower 1
TOC1	Timing of Cab Expression 1
TTG1	Transparent Testa Glabra 1
VAL	VP1/ABI3-LIKE
WER	Werewolf
WOX	WUS/WUS-Related Homeobox

18.1 Introduction

The ability of a cell to maintain cellular processes and response to environmental stimuli depends on changes in gene expression. Alterations in genes expression pattern are a highly controlled process which leads to metabolic and phenotypic changes. The proteins which regulate the transcriptional gene expression are called transcription factors (TFs). Most of the TFs contain a DNA binding domain and other regulatory domain/s. In a simple free-living bacterium such as *Escherichia coli*, >300 TFs and in a model plant species *Arabidopsis thaliana* (*A. thaliana*) >2000 TF genes are reported (Madan Babu and Teichmann 2003; Mitsuda and Ohme-Takagi 2009). Accumulation of genome-wide gene expression datasets is very fast due to advancement in high-throughput next-generation technologies which enable construction and analysis of GRNs to understand the molecular mechanism of complex traits in plants.

A gene regulatory network (GRN) is composed of transcriptional regulators such as TFs and their target genes (TGs). TFs are trans-acting proteins which bind to specific sites in the promoter region of the target genes. TFs coordinate the transcription of many genes as a set of specific factors are expressed in restricted patterns which activate/repress the downstream target genes. The activation and/or repression of target genes is responsible for various cellular processes such as growth, development, and response to environmental stimuli. Molecular-genetic studies of model plant species *A. thaliana* gave the clues about how the developmental processes and environmental adaptations are regulated by GRNs (Chu et al. 2009; Vialette-Guiraud et al. 2016). Promoter region of a gene can have binding sites for multiple transcription factors. There are three different models describing how TFs bind to the promoter region of target genes: (i) Statistical physics model; (ii) Markov-chain model; (iii) Computational model (Chu et al. 2009). As a part of a GRN, a cis-regulatory sequence in the promoter of a gene is required for recruitment of a TF. For activation and/or repression of the gene expression, recruitment of transcriptional activators and/or repressors are also required along with basal transcription factors. Activator and repressor proteins bind to separate cis-regulatory sequences called enhancer and silencer, respectively (Kolovos et al. 2012). This multi-protein-DNA complex dictates the expression pattern of the downstream network of genes. Some regulatory modules control the function of the genes encoding TF proteins. Connections between these particular genes and their associated regulatory modules represent the core networks responsible for various developmental processes (Davidson and Levin 2005). These regulatory modules indirectly influence gene expression representing interactions between microRNA (miRNA) and their target messenger RNA (mRNA). To influence the gene expression, several TFs undergo protein-protein interactions because they bind DNA in the form of homo- or heterodimers, and/or interact with cofactors, basal transcription machinery, and chromatin remodeling factors (Smith and Matthews 2016). Complex interconnection among multiple TFs, target genes, coactivators, and post-transcriptional regulators act together to regulate various cellular processes.

18.1.1 Structure of GRN

A GRN is a graphical representation that shows relationships among genes, which can be a logical relationship (e.g. Boolean network), or a mathematical equation (e.g. algebraic or ordinary differential equations) (Li et al. 2015). In a GRN, nodes are considered as genes and edges represent interactions between genes (Li et al. 2015). TFs act as internal nodes which regulate the activity of terminal nodes (Rebeiz et al. 2015). GRNs are directional and can be visualized under different perspectives: (i) the number of TFs bind to cis-regulatory region of a specific gene; or (ii) the number of cis-regulatory regions recognized by a TF (Ouma et al. 2018). These TFs, target genes, and cis-regulatory elements are considered as basic units and their arrangements in a wiring pattern are termed as network motifs (Ouma et al. 2018; Alon 2007; Chalancon and Babu 2013). GRN modules can be of two types (Fig. 18.1): gene modules (sets of genes regulated by similar TFs); and TF modules (sets of TFs share similar target genes). Similarly, network motifs can also be of two types (Fig. 18.2): feed-forward loops (FFLs); and feed-back loops (FBLs) (Milo et al. 2002; MacNeil and Walhout 2011). In a network motif of three genes, where gene A regulates gene B and gene C is regulated by both gene A and B, is termed as a feed-forward loop (FFL) (Mangan and Alon 2003). The interaction of plant transcription regulator LFY with *AP1* and *CAL* represents FFL transcriptional network motif where LFY controls the meristem identity via the direct activation of two other meristem identity genes, *AP1* and *CAL* (Saddic et al. 2006).

Feed-back loops are two types: (i) positive FBL; and (ii) negative FBL. Positive autoregulation (positive FBL) occurs when a TF activates transcription of its own gene (Fig. 18.2c) and in negative autoregulation (negative FBL) TF suppress its own gene expression (Fig. 18.2d). In a positive FBL, delay in response is an inherited property whereas the negative FBL speeds up the response time in transcription regulatory networks (MacNeil and Walhout 2011; Maeda and Sano 2006; Rosenfeld et al. 2002). In *A. thaliana*, positive autoregulation of MYB23 is required for cell fate

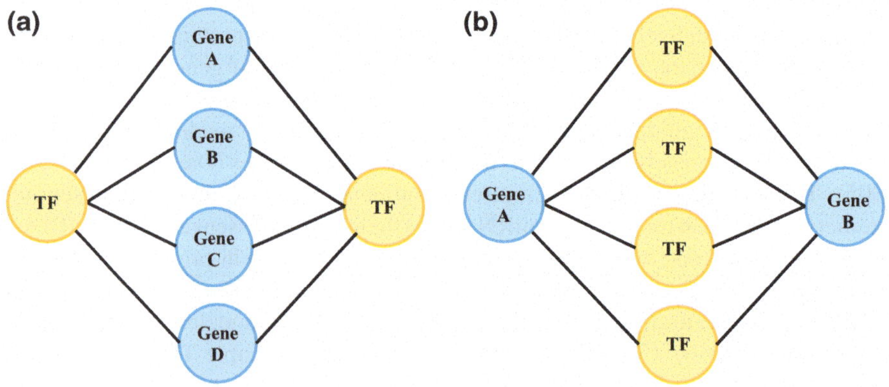

Fig. 18.1 Models for GRN modules. **a** gene module; **b** TF module; solid lines: direct interaction

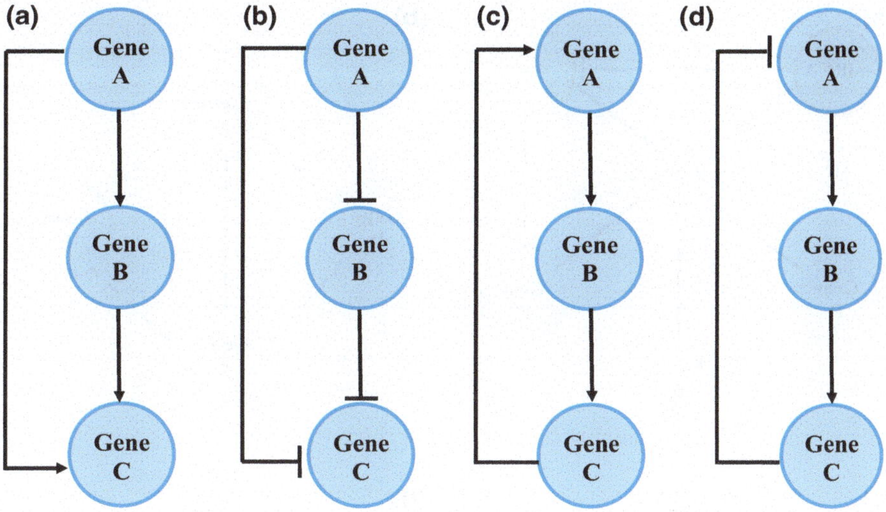

Fig. 18.2 The architecture of network motifs. **a** Coherent feed-forward loop; **b** inherent feed-forward loop; **c** positive feed-back loop; **d** negative feed-back loop; arrows: positive regulation; and T-shape arrows: negative regulation

decisions in root epidermis where expression of *MYB23* is positively regulated by *WER* gene and MYB23 itself (Kang et al. 2009). An example of negative autoregulation can be seen in plant circadian clock where LHY and CCA1 negatively regulate *TOC1* expression but TOC1 positively regulates expression of *LHY* and *CCA1* genes (Alabadí et al. 2001). To induce defense response against pathogen attack, synergistic action of positive and negative FBLs is required to regulates expression of disease resistance genes (Eckardt 2007).

Integrated Network

Integrated GRNs are specified by composite network motifs which include different types of molecular interaction such as protein-protein, DNA-protein, and miRNA-mRNA interactions (Fig. 18.3a). In the integrated GRNs, network motifs consist of combinations of interactions: DNA-protein (between transcription factors and their target genes), protein-protein, genetic, homologous, miRNA-mRNA, and transcription regulatory interactions (Defoort et al. 2018; Yeger-Lotem et al. 2004). Three types of GRNs are inferred based on the type of interaction: (i) association networks; (ii) regulatory networks; and (iii) co-expression networks (Li et al. 2015) (Fig. 18.3b–d). Gene association networks (GANs) are directional and connections can be made only after removing the effects of other genes (Li et al. 2015).

Based on the pan- and core-genome concept, Wirojsirasak et al. (2019) developed pan- and core-gene association networks. In *A. thaliana*, pan-GAN displays an entire gene-gene association system involved in gene regulation, and core-GAN derived from the associated gene-pairs represents the regulation of essential cellular

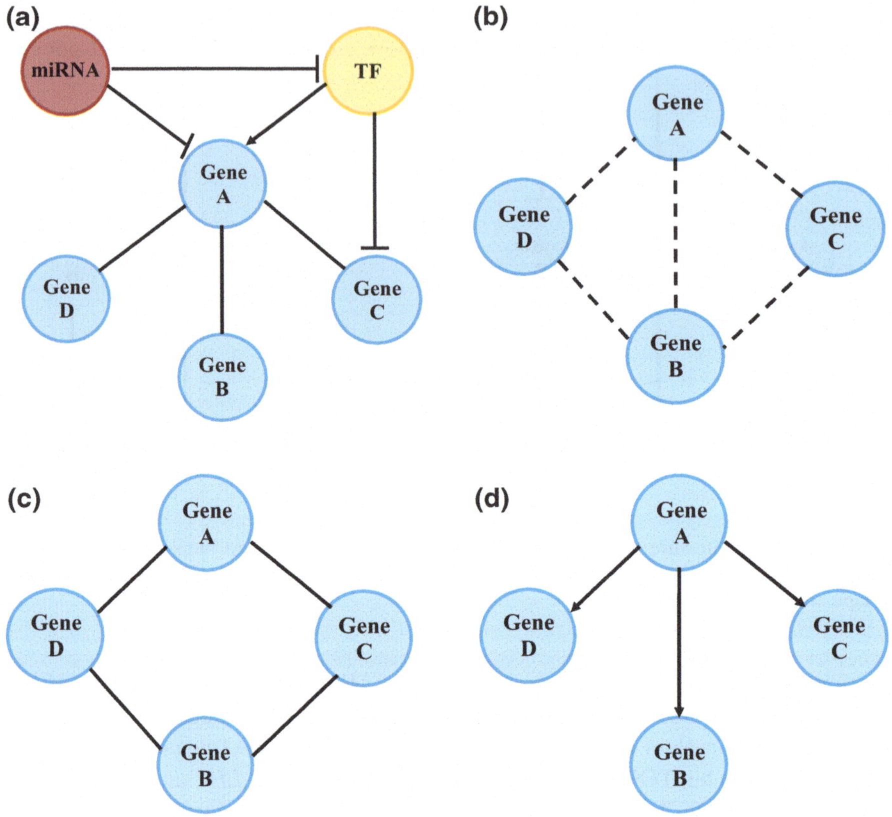

Fig. 18.3 The architecture of GRNs. **a** Models for integrated network; **b** co-expression; **c** association; and **d** regulatory gene networks; dashed lines: direct or indirect interactions; solid lines: direct interaction; arrows: positive regulation; and T-shape arrows: negative regulation

processes (Wirojsirasak et al. 2019). Regulatory networks represent causal relationships between genes and edges with directionality whereas in co-expression network, genes with similar expression profiles have direct or indirect interaction are connected and can be coregulated by common TFs (Li et al. 2015). Some co-expression networks are conserved across plant species indicating several cellular processes are performed by a similar set of orthologous genes in different species (Ficklin and Feltus 2011; Movahedi et al. 2011; Mutwil et al. 2011; Itkin et al. 2013).

Approaches for Predicting Gene Regulatory Networks

GRN prediction is important to show the overall impact of various intrinsic and extrinsic signals to the plant through the relationship between various transcription or regulatory factors and underlying genes. Availability of next-generation sequencing data like whole genomes, transcriptomes, transcription factor bindings, post-translational modifications, chromatin structure, and interaction data has made it easy to look for GRNs in many plant species. Most common approach for predicting the GRNs may include: (i) collecting the data required to identify regulatory factors and functional genes working under a particular intrinsic or extrinsic signal and data analysis; (ii) identifying a connection between regulatory factors and functional genes for the specified signal; and (iii) validating the predicted GRN (Haque et al. 2019).

For gene expression data, RNA-seq is the preferred choice because it provides better coverage, and high resolution along with the information about splice variants and paralogues (van Dam et al. 2017). For GRNs study, a minimum of 20 samples (per treatment or group) with 10 million reads per sample is generally used (Ballouz et al. 2015; Langfelder and Horvath 2008). To identify the co-expression network, RNA-seq data is first quality filtered. The expression is measured for each gene by mapping the reads on genome or assembled transcriptome and quantifying the transcripts after normalization. RNA-seq pipeline from ENCODE project (https://www.encodeproject.org/pages/pipelines/#RNA-seq) can be followed for analyzing the RNA-seq data. Further, co-expression modules are identified by clustering the genes with similar expression pattern in samples under comparison. WGCNA (Langfelder and Horvath 2008) is one of the most widely used tools for clustering which uses hierarchical clustering of co-expressing genes or network. Next, hub gene (usually multiple genes) or highly connected genes are identified for each co-expression module (van Dam et al. 2017). Functional enrichment of the co-expression modules can be determined by using tools like DAVID (Huang et al. 2009), PANTHER (Mi et al. 2013), or g:Profiler (Reimand et al. 2016). These annotated co-expression networks are further used to infer GRNs using tools listed in Table 18.1.

Data for GRNs Prediction

Selection of data types is the most important factor for inferring GRNs. Gene expression data is the basic requisite for predicting GRNs. For dynamic GRNs in different cell types, a combination of spatial expression data with time-series or temporal data is required (de Luis Balaguer et al. 2017). Along with expression data, enhancer-gene linkage data (Hi-C data), histone modification landscape (ChIP-seq data), transcription factor binding site data, transcription factor-gene linkage data (ChIP-seq), quantitative trait loci (QTL) maps, gene splicing data, and information on transcriptionally active regions are used for predicting GRNs. Genotyping and phenotyping data are also used for making networks. These datasets are analyzed individually and in combination to extract information for network prediction.

Models for GRN Prediction

The models used for GRN predictions include data-driven models, probabilistic models, dynamic models, and multi-network models. These models have been reviewed in detail by Huynh-Thu and Sanguinetti (2018). The data-driven models are simple and efficient models containing fully connected weighted networks with an estimation of gene dependencies directly from the data. Data-driven methods include correlation networks (Bin and Steve 2005), information theoretic score based networks, and regression-based methods. The correlation networks are based on the correlated

Table 18.1 Tools available for GRN prediction

Name	Type	Modeling method	Web link
ARACNe (Margolin et al. 2006)	C++	Information theoretic scores	http://califano.c2b2.columbia.edu/aracne/
Banjo (Smith et al. 2005)	Java package	Bayesian network	https://users.cs.duke.edu/~amink/software/banjo/
CatNet (Balov and Salzman 2012)	R package	Bayesian network	https://cran.r-project.org/web/packages/catnet/index.html
CLR (Faith et al. 2007)	MATLAB	Information theoretic scores	http://m3d.mssm.edu/network_inference.html
G1DBN (Lebre 2013)	R package	Dynamic Bayesian	https://cran.r-project.org/web/packages/G1DBN/index.html
GeneNet (Schäfer and Opgen-Rhein 2006)	R package	Gaussian graphical model	https://cran.r-project.org/web/packages/GeneNet/index.html
GENIE3 (Huynh-Thu et al. 2010)	MATLAB/Python/R package	Regression-based	http://www.montefiore.ulg.ac.be/~huynh-thu/software.html
GRACE (Banf and Rhee 2017)	R package	Regression-based	https://github.com/mbanf/GRACE
GRENITS (Morrissey 2011)	R package	Dynamic Bayesian	https://bioconductor.org/packages/release/bioc/html/GRENITS.html
Inferelator (Madar et al. 2009)	R package	Differential equations	http://bonneaulab.bio.nyu.edu/networks.html
Minet (Meyer et al. 2008)	R package	Information theoretic scores	https://www.bioconductor.org/packages/release/bioc/html/minet.html
Netbenchmark (Bellot et al. 2015)	R package	Data-driven	https://www.bioconductor.org/packages/release/bioc/html/netbenchmark.html
TF2Network (Kulkarni et al. 2018)	Web interface	–	http://bioinformatics.psb.ugent.be/webtools/TF2Network/

(continued)

Table 18.1 (continued)

Name	Type	Modeling method	Web link
TGMI (Gunasekara et al. 2018)	R package	Mutual interaction measure	http://sys.bio.mtu.edu/sample_output/TGMI/
TSN1 (Bansal et al. 2006)	MATLAB	Differential equations	http://dibernardo.tigem.it/softwares/time-series-network-identification-tsni
WGCNA (Langfelder and Horvath 2008)	R package	Weighted correlation	https://labs.genetics.ucla.edu/horvath/CoexpressionNetwork/Rpackages/WGCNA/index.html

expression of connected genes but they may predict a false connection between two genes due to the effect of a third gene regulating both of them (Huynh-Thu and Sanguinetti 2018). Information theoretic score based networks consider alternative scores based on information theory and are one of the most used GRN prediction methods (Huynh-Thu and Sanguinetti 2018). Probabilistic models include Gaussian graphical models (Schafer and Strimmer 2005), and Bayesian networks (de Luis Balaguer et al. 2017). Dynamic Bayesian networks (de Luis Balaguer and Sozzani 2017) and differential equation methods (Yoon 2010) are an example of dynamic models.

Resources for Data Analysis and GRN Inference

With the development of NGS techniques, adequate resources are available for analyzing various data types (like, Microarray, RNA-seq, ChIP-seq, ATAC-seq, methylome and Hi-C), identifying co-expressing genes, and integrating this information together to infer the GRNs (van Dam et al. 2017). Similarly, several tools are available (Huynh-Thu and Sanguinetti 2018) for inferring GRNs based on different approaches (models). These tools employ different modeling methods discussed in the previous section. Identification of high confidence interactions within a network is very challenging. As it also depends on the method used for inferring the network, so various methods can be used and compared for their predictions by validating the results. Different models can be evaluated by using simulated gene expression data or synthetic gene circuit (Huynh-Thu and Sanguinetti 2018). The accuracy of a model depends on its ability to identify maximum true positives (high sensitivity or recall) and minimum false positives (high precision). Algorithms can be compared for their effectiveness by using precision-recall curves as described by Huynh-Thu and Sanguinetti (2018). Most frequently used tools are listed in Table 18.1.

Co-expression databases like GeneMANIA (Vlasblom et al. 2015), GENEVESTIGATOR (Zimmermann 2004), and GIANT (Greene et al. 2015) are available which can be used to get information on co-expressing genes in different species. Additionally, tools like Cytoscope (Shannon 2003) and BioLayout (Theocharidis et al. 2009) are also available for visualization of networks.

Applications of GRNs

In addition to predict linkages among genes and regulatory factors, GRNs also provide information about various biological processes and molecular functions of genes.

Hub Gene Analysis

A gene with multiple connections with other genes in a regulatory network is defined as hub gene. Hub genes are the regulators of various biological processes including growth and developmental processes. Genes with high connectivity (hub nodes) along with positive feedback loops can be used to predict potential master switches in a regulatory network (Seo et al. 2009). Analysis of hub genes in rice transcriptional network facilitated the identification of 40 putative targets of 26 *MYB* genes which indicates that co-regulatory network analysis contributes in screening of candidate regulatory factors and their targets under various developmental stages and biotic/abiotic stress conditions (Smita et al. 2015).

Gene Network Regulating Developmental Processes

GRNs involved in developmental processes define the evolution and body structure of an organism and are integrated structure of sub-circuits of each developmental phase and each sub-circuit consists of specific regulatory connections among cis-regulatory transcription factor and their target sites. Growth and development of a plant depend on complex gene network which displays very organized arrangements of genes, proteins, and regulatory factors. A gene co-expression network was used for the identification of hub gene modules involved in reproductive developmental processes in strawberry (Hollender et al. 2014; Shahan et al. 2018; Smith 2018). Genes with most connections are considered as hub genes and most of them are meristem regulators including *WOX*, *AIL3*, *RGA*, *SCR*, *SCL18*, *LFY*, three *NAC* TF genes, *CUC3*, *CUC2*, and five *TALE* genes (Hollender et al. 2014). Co-expression network provides new insights into flower development and fruit setting process in strawberry (Shahan et al. 2018). For growth and development of the bamboo plant, a total of 1896 functional modules related to photosynthesis, hormone biosynthesis, signal transduction, and secondary cell wall biosynthesis were identified (Ma et al. 2018). Gene co-expression network for rice anther development provided the platform for the prediction of gene regulatory relationships during pollen development (Lin et al. 2017). In tea plant, hub genes from 35 co-expression modules are involved in potential regulatory mechanisms of catechins, theanine, and caffeine metabolism (Tai et al. 2018).

For identification of floral organs, ABC combinatorial model of gene expression from *A. thaliana* and *Antirrhinum majus* was used but this model lacked a dynamical explanation of interactions among ABC and non-ABC genes to attain and maintain each primordial floral cell type (Coen and Meyerowitz 1991; Ferrario et al. 2004).

Espinosa-Soto and colleagues (Espinosa-Soto et al. 2004) developed an integrated model of ABC and non-ABC genes of *A. thaliana* to explore the conservation of basic structures of flowers among angiosperms. To identify a regulatory network governing floral transition, a combinatorial approach of experimental data and computational modeling was used. This combinatorial approach showed that *FT* and *TFL1* act through *FD* and a *FD* paralog to regulate the transition and gene *LFY* activates *FD* gene creating a positive feed-back loop (Jaeger et al. 2013). This hub genes network helps in the prediction of how flowering takes place in different genotypes and dynamics of the floral transition. Circadian clock network is the most-studied GRN which constitute a hierarchical structure of transcriptional feed-back loops with regulated protein sets. In a core feed-back loop of the circadian clock, CCA1 and LHY negatively regulate the expression of *TOC1* which in-turn activates the expression of *CCA1* and *LHY* (Lebre 2013; Coen and Meyerowitz 1991). In addition to the core feed-back loop of the circadian clock, two phase-specific feed-back loops: morning loop and evening loop has also been identified (Pruneda-Paz and Kay 2010; Locke et al. 2005; Zeilinger et al. 2006; Locke et al. 2006). The regulatory network involved in floral meristem development and organ differentiation displays miRNA-mediated feed-forward loops consist of floral homeotic TFs and their targets (Chen et al. 2018).

Gene Network Regulating Seed Maturation and Germination

Seed development is a two-phase process: (i) embryo and endosperm morphogenesis; and (ii) seed maturation (Gutierrez et al. 2007). In *A. thaliana*, ABI3, FUS3, LEC2, HAP3 (LEC1) TFs are the master regulators of the seed maturation process and activate a precise set of target genes (Gutierrez et al. 2007; Raz et al. 2001). Many regulatory connections are also observed in these four TFs: HAP3 and LEC2 positively regulate the expression of *ABI3* and *FUS3* genes, and interestingly, ABI3 and FUS3 positively regulate themselves and each other forming positive feed-back loops to maintain their expression during seed development process (Holdsworth et al. 2008; To et al. 2006). Interaction of FUS3 with TTG1 in both maternal and zygotic stage is required for correct seed development (Tsuchiya et al. 2004). Expression of many seed maturation specific genes such as seed storage protein-coding genes are regulated by ABI3, FUS3, and bZIP transcriptional regulators; however, combinatorial interactions are also observed among these TFs (Kroj et al. 2003; Verdier and Thompson 2008; Lara et al. 2003; Xi and Zheng 2011). Co-expression modules detected in two different cultivars of chickpea during seed development displayed connections among WOX9, PDF2, RLT2, ANL2, JAB, HDG11, AGL15, WOX2/3, and SEP3 transcription factors and their target genes related to embryogenesis, cell wall organization, regulation of cell size and post-translational protein modification (Garg et al. 2017).

Seed germination is a complex, multistep and irreversible process which involves the participation of different gene sets and regulators. A condition-dependent network model of global transcriptional interactions for seed germination and dormancy processes is generated to predict the germination-associated function of uncharacterized nodes (Bassel et al. 2011). The key regulators of germination including the positive

(GID1A, GID1C, AHG3) regulators, negative regulators (ABI3, ABI5, ABA1), and major dormancy (DOG1) regulator interact with each other (Holdsworth et al. 2008; Bassel et al. 2011; Bentsink et al. 2010). These transcriptional interactions between antagonistically acting genes may affect the signals required to complete the germination process or to maintain dormancy (Bassel et al. 2011). Abscisic acid (ABA) plays a major role in seed maturation, dormancy, and germination. ABA responses involve changes in gene expression regulated by various transcription factors such as ABI3, ABI4, ABI5, and the bZIP but ABI5 is the key regulator of ABA signaling during and after germination (Finkelstein et al. 2002). Germination regulating interactions are observed not only between ABI3 and ABI5 but also between KEG1 and ABI5 (Nakamura et al. 2001; Stone et al. 2006). SCL14 and ASG2 are the regulators affect the seed germination by interacting with ABI3 and ABI5 (Bassel et al. 2011). VAL family transcription factors are required for repression of the LEC1 transcription factor network and may indirectly regulate gibberellic acid (GA) biosynthesis gene *GA3ox1* during germination (Holdsworth et al. 2008; Suzuki et al. 2007).

Gene Module Detection

Detection of gene modules helps in the identification of arrangement of genes in a network linked with specific biological processes (Segal et al. 2003). The connections between sets of hub genes play an important role in the formation of the module structure of a specific biological process. For detection of functional modules, two approaches have been used: guide-gene approach; and, non-targeted approach. Guide-gene approach is based on a known set of genes used for querying the co-expression network. This approach can be used for identification of gene modules associated with a specific biological function; however, the identified module might be incomplete and/or part of a big complex module (Aoki et al. 2007). The non-targeted approach is computationally intensive, knowledge independent, and relies on the search of modules into the network using graph clustering algorithms (Aoki et al. 2007; Mentzen and Wurtele 2008; Mao et al. 2009). Evolution of functional gene modules can also be identified by integrating the genomic and phylogenetic information (Ruprecht et al. 2017).

Gene Modules Associated with Environmental Stress Response

In response to environmental stress factors, plants modulate their growth, development, and physiology by synergistic actions of local and systemic responses. Using network-based gene clustering approach, 15 functional modules consisting 1,392 interconnected genes in the drought-responsive network in rice were identified. These genes were involved in the biological processes of stimulus-response, photosynthesis, nucleosome assembly, embryonic development, translation, and protein amino acid phosphorylation (Zhang et al. 2012). Using a graph-clustering algorithm, Rice Environment Coexpression Network (RECoN) was built to reveal the modular organization of abiotic-stress responsive genes which consists of 34,792 genes connected by ~18.5 million edges (Krishnan et al. 2017). In *A. thaliana* natural variation in response to drought is different than in response to cold. Non-random nature of the gene-interaction topologies for cold and drought response displayed genetic variation in drought response is more peripheral whereas cold response genes are significantly more central (Des Marais David et al. 2017). This indicates that different parts of transcriptional regulatory networks of drought and cold response are affected during natural selection. An integrated regulatory network, EGRINs, associated with high temperatures, and water deficit was developed using transcriptome, and chromatin-accessibility-based methods. This integrated network consists of regulatory interactions between 113 TFs and their 4,052 target genes responsive to high temperature and water deficit (Wilkins et al. 2016). Involvement of various phytohormone signaling pathways and crosstalk among them plays a very important role in environmental stress response and tolerance in plants (Chinnusamy et al. 2004). The major TFs involved in abiotic stress responsive modules are *NAC* genes, AREB/ABFs and DREB1/CBF which regulate ABA-dependent and ABA-independent gene expression, respectively (Nakashima et al. 2009; Fujita et al. 2013; Nakashima and Yamaguchi-Shinozaki 2013; Mizoi et al. 2013). Wounding triggered by either biotic or abiotic type of stress leads to activation of wound repair and regeneration mechanisms in plants. Regulatory network associated with cellular reprogramming involves 252 TFs where PLT3, ESR1, and HSFB1 are considered as central nodes that have many overlapping targets (Ikeuchi et al. 2018). In this network, AP2/ERFs and LOB/AS2 TFs play a partially overlapping role to regulate downstream genes related to callus formation and organ regeneration. In addition to TFs, small RNAs such as miRNAs and small interfering RNAs (siRNAs), being the regulators of gene expression, act as important players in responses to environmental stresses (Sunkar et al. 2007; Sharma et al. 2019).

Gene Function Analysis

Prediction of the function of a gene/protein using large-scale networks of molecular interactions can be done using two approaches: direct annotation and module-assisted schemes (Sharan et al. 2007). In direct annotation scheme function of a protein can be predicted based on its connections. This approach is utilized by Lee and colleagues (Lee et al. 2010, 2011) to construct RiceNet and AraNet, networks for prediction of gene functions. The module-assisted approach is based on the identification of modules containing genes perform a similar function. This approach was used for functional annotation of rice genes from co-expressed modules using fifteen datasets of gene expression (Childs et al. 2011). Network-based gene function prediction techniques can be useful for understanding the function of scarcely annotated genes/proteins.

Comparative and Integrative Analysis

Comparative network analysis includes similarities and difference between networks which provide insight about evolutionary aspects. Comparative networks can be inter- or intra-species. An interspecies comparative network between dicot model plant *A. thaliana* and monocot model plant rice displays 56–77% homologous genes between these two species (Movahedi et al. 2011; Ma et al. 2005; Vandepoele and de Peer 2005). Similarly, comparative gene co-expression networks among *Populus trichocarpa*, *A. thaliana*, and *Glycine max* revealed the divergence in the regulation of antioxidant and heat shock modules and expansion of HSP17 protein family in *Glycine max* (Weston et al. 2011). A comparison of maize gene co-expression network with the *A. thaliana* network showed conserved gene association patterns (modules involved in cell organization, metabolism, nutrients uptake, and utilization) along with the divergences between the modules involved in developmental pathways (Ma et al. 2017). Co-expression subnetwork derived from a guide-gene approach in rice revealed biological roles of non-coding genome elements such as long non-coding RNA and circular RNA (Yu et al. 2017). Comparison of gene co-expression networks in the two rice genotypes provided insight about gene modules associated with biotic stress response (Wang et al. 2018).

Incorporation of available biological knowledge about the functional association of genes is highly valuable for the reconstruction of gene networks. Wang and colleagues (Wang et al. 2013) developed a pipeline for the integration of prior information from Pathway Commons (PC) and Kyoto Encyclopedia of Genes (KEGG) for gene network reconstruction. Gene networks designed form gene expression data can be integrated with other data types as well for better interpretation of gene functions. Integration of comparative and functional genomics, proteomics, metabolomics and miRNA data with gene expression network is helpful for the identification of genes associated with agriculturally important traits (Sharma et al. 2019; Lee et al. 2010; Hamada et al. 2011; Meng et al. 2011). A statistical framework of quantitative genetic mapping and metabolite–transcript correlation networks can also facilitate the identification of additional genes associated with complex traits (Chan et al. 2011; Wu et al. 2016).

Major Challenges and Future Perspectives

In this chapter, we discussed strategies used for the construction of regulatory networks and how GRNs can provide a better understanding of gene functions and complexity of gene regulation. Several approaches are being used for better interpretation of the gene networks. Integration of various types of data sets, derived from different cell types, tissues, organs or species, not designed for network studies provides a very inappropriate picture of the gene network. To develop reliable networks and inferences, a coherent strategy with the biological question is needed. Proteomics data can be used to infer cellular compartments (Chou and Shen 2007) and molecular functions (Tian and Skolnick 2003) but not the biological processes whereas genomic and transcriptomic data can be used not only to infer molecular functions and biological processes (Lee and Sonnhammer 2003) but also the cellular components (Ryngajllo et al. 2011). Combined approaches utilizing omics datasets, including transcriptomic, proteomic, metabolomic, and epigenomic data will be important to study the relative contribution of different modes of regulation in a multi-module network. Integration of multiple inference methods for different types of datasets is the better way for the inference of gene regulatory networks (Marbach et al. 2012). Factors that play key roles in stem cell regulation and quiescent center function are identified by using an integrated approach of molecular, computational and mathematical biology (de Luis Balaguer et al. 2017). This integrative approach for network construction and analysis will be helpful in the discovery of novel mechanisms and new pathways.

High-throughput experimental validation of regulatory network is also important for elucidation of genes function. Massive parallel DNA sequencing combined with chromatin immunoprecipitation (ChIP-seq) can be a powerful approach for validation of GRNs (Huang et al. 2018). Gene silencing, and gene editing approaches such as virus-induced gene silencing (VIGS), and clustered regularly interspaced short palindromic repeats (CRISPR)/CAS9, respectively can effectively be used for validation of GRNs (Beyene et al. 2017; Rodríguez-Leal et al. 2017). Other approaches used for validation of interaction in network are yeast one- and two-hybrid assays (Brady et al. 2011) combined with electrophoretic mobility-shift assays (Bustos et al. 2010), and tilling array (Zhu et al. 2012). Brooks et al. (2019) developed a combined approach integrating experimental and computational methods to characterize the gene network underlying nitrogen responses in *A. thaliana*. The shortage of freely available online tools which can be used for network-based gene function prediction from variety of datasets is a key issue. Most of the computational methods rely on existing annotations which does not provide exact function of poorly characterized genes. It is important to develop knowledge platforms with the predictive power of network analysis based on integrated omics datasets and genome-wide association studies. GRNs can be used for generation of novel hypothesis, their experimental validation and discovery of gene function. A complex GRN consists of multiple types of interaction data, and displays the topological, evolutionary, and functional properties across the worm and plant species (Defoort et al. 2018). There are some issues need to be addressed in future: (i) "how can the understanding of GRN help in domestication of plants?" (Sun and Dinneny 2018); (ii) "how can epigenomics datasets be integrated with regulatory networks to predict molecular basis of specific traits?" (Li et al. 2015); (iii) "how the adaptive traits can be predicted using GRNs?" (Sun and Dinneny 2018).

References

Alabadí D, Oyama T, Yanovsky MJ, Harmon FG, Más P, Kay SA (2001) Reciprocal regulation between TOC1 and LHY/CCA1 within the arabidopsis circadian clock. Science 293:880–883. https://doi.org/10.1126/science.1061320

Alon U (2007) Network motifs: theory and experimental approaches. Nat Rev Genet 8:450–461. https://doi.org/10.1038/nrg2102

Aoki K, Ogata Y, Shibata D (2007) Approaches for extracting practical information from gene co-expression networks in plant biology. Plant Cell Physiol 48:381–390. https://doi.org/10.1093/pcp/pcm013

Ballouz S, Verleyen W, Gillis J (2015) Guidance for RNA-seq co-expression network construction and analysis: safety in numbers. Bioinformatics 31:2123–2130. https://doi.org/10.1093/bioinformatics/btv118

Balov N, Salzman P (2012) Catnet: categorical Bayesian network inference. R package version, vol 1

Banf M, Rhee SY (2017) Enhancing gene regulatory network inference through data integration with markov random fields. Sci Rep 7:41174. https://doi.org/10.1038/srep41174

Bansal M, Gatta GD, di Bernardo D (2006) Inference of gene regulatory networks and compound mode of action from time course gene expression profiles. Bioinformatics 22:815–822. https://doi.org/10.1093/bioinformatics/btl003

Bassel GW, Lan H, Glaab E, Gibbs DJ, Gerjets T, Krasnogor N, Bonner AJ, Holdsworth MJ, Provart NJ (2011) Genome-wide network model capturing seed germination reveals coordinated regulation of plant cellular phase transitions. PNAS 108:9709–9714. https://doi.org/10.1073/pnas.1100958108

Bellot P, Olsen C, Salembier P, Oliveras-Vergés A, Meyer PE (2015) NetBenchmark: a bioconductor package for reproducible benchmarks of gene regulatory network inference. BMC Bioinform 16:312. https://doi.org/10.1186/s12859-015-0728-4

Bentsink L, Hanson J, Hanhart CJ, Vries HB, Coltrane C, Keizer P, El-Lithy M, Alonso-Blanco C, de Andrés MT, Reymond M, van Eeuwijk F, Smeekens S, Koornneef M (2010) Natural variation for seed dormancy in Arabidopsis is regulated by additive genetic and molecular pathways. PNAS 107:4264–4269. https://doi.org/10.1073/pnas.1000410107

Beyene G, Chauhan RD, Taylor NJ (2017) A rapid virus-induced gene silencing (VIGS) method for assessing resistance and susceptibility to cassava mosaic disease. Virol J 14. https://doi.org/10.1186/s12985-017-0716-6

Bin Z, Steve H (2005) A general framework for weighted gene co-expression network analysis. sagmb 4. https://doi.org/10.2202/1544-6115.1128

Brady SM, Zhang L, Megraw M, Martinez NJ, Jiang E, Yi CS, Liu W, Zeng A, Taylor-Teeples M, Kim D, Ahnert S, Ohler U, Ware D, Walhout AJM, Benfey PN (2011) A stele-enriched gene regulatory network in the Arabidopsis root. Mol Syst Biol 7:459. https://doi.org/10.1038/msb.2010.114

Brooks MD, Cirrone J, Pasquino AV, Alvarez JM, Swift J, Mittal S, Juang C-L, Varala K, Gutiérrez RA, Krouk G, Shasha D, Coruzzi GM (2019) Network Walking charts transcriptional dynamics of nitrogen signaling by integrating validated and predicted genome-wide interactions. Nat Commun 10:1569. https://doi.org/10.1038/s41467-019-09522-1

Bustos R, Castrillo G, Linhares F, Puga MI, Rubio V, Pérez-Pérez J, Solano R, Leyva A, Paz-Ares J (2010) A central regulatory system largely controls transcriptional activation and repression responses to phosphate starvation in Arabidopsis. PLoS Genet 6:e1001102. https://doi.org/10.1371/journal.pgen.1001102

Chalancon G, Babu MM (2013) Structure and evolution of transcriptional regulatory networks. In: Madan Babu M (eds) Bacterial gene regulation and transcriptional networks. Caister Academic Press, UK, p 26

Chan EKF, Rowe HC, Corwin JA, Joseph B, Kliebenstein DJ (2011) Combining genome-wide association mapping and transcriptional networks to identify novel genes controlling glucosinolates in *Arabidopsis thaliana*. PLoS Biol 9:e1001125. https://doi.org/10.1371/journal.pbio.1001125

Chen D, Yan W, Fu L-Y, Kaufmann K (2018) Architecture of gene regulatory networks controlling flower development in *Arabidopsis thaliana*. Nat Commun 9:4534. https://doi.org/10.1038/s41467-018-06772-3

Childs KL, Davidson RM, Buell CR (2011) Gene coexpression network analysis as a source of functional annotation for rice genes. PLoS ONE 6:e22196. https://doi.org/10.1371/journal.pone.0022196

Chinnusamy V, Schumaker K, Zhu J-K (2004) Molecular genetic perspectives on cross-talk and specificity in abiotic stress signalling in plants. J Exp Bot 55:225–236. https://doi.org/10.1093/jxb/erh005

Chou K-C, Shen H-B (2007) Recent progress in protein subcellular location prediction. Anal Biochem 370:1–16. https://doi.org/10.1016/j.ab.2007.07.006

Chu D, Zabet NR, Mitavskiy B (2009) Models of transcription factor binding: sensitivity of activation functions to model assumptions. J Theor Biol 257:419–429. https://doi.org/10.1016/j.jtbi.2008.11.026

Coen ES, Meyerowitz EM (1991) The war of the whorls: genetic interactions controlling flower development. Nature 353:31. https://doi.org/10.1038/353031a0

Davidson E, Levin M (2005) Gene regulatory networks. Proc Natl Acad Sci USA 102:4935. https://doi.org/10.1073/pnas.0502024102

de Luis Balaguer MA, Sozzani R (2017) Inferring gene regulatory networks in the Arabidopsis root using a dynamic Bayesian network approach. In: Kaufmann K, Mueller-Roeber B (eds) Plant gene regulatory networks. Springer, New York, NY, pp 331–348

de Luis Balaguer MA, Fisher AP, Clark NM, Fernandez-Espinosa MG, Möller BK, Weijers D, Lohmann JU, Williams C, Lorenzo O, Sozzani R (2017) Predicting gene regulatory networks by combining spatial and temporal gene expression data in *Arabidopsis* root stem cells. Proc Natl Acad Sci USA 114:E7632–E7640. https://doi.org/10.1073/pnas.1707566114

Defoort J, Van de Peer Y, Vermeirssen V (2018) Function, dynamics and evolution of network motif modules in integrated gene regulatory networks of worm and plant. Nucleic Acids Res 46:6480–6503. https://doi.org/10.1093/nar/gky468

Des Marais David L, Guerrero Rafael F, Lasky Jesse R, Scarpino Samuel V (2017) Topological features of a gene co-expression network predict patterns of natural diversity in environmental response. Proc Roy Soc B: Biol Sci 284:20170914. https://doi.org/10.1098/rspb.2017.0914

Eckardt NA (2007) Positive and negative feedback coordinate regulation of disease resistance gene expression. Plant Cell 19:2700–2702. https://doi.org/10.1105/tpc.107.056226

Espinosa-Soto C, Padilla-Longoria P, Alvarez-Buylla ER (2004) A gene regulatory network model for cell-fate determination during *Arabidopsis thaliana* flower development that is robust and recovers experimental gene expression profiles. Plant Cell 16:2923–2939. https://doi.org/10.1105/tpc.104.021725

Faith JJ, Hayete B, Thaden JT, Mogno I, Wierzbowski J, Cottarel G, Kasif S, Collins JJ, Gardner TS (2007) Large-scale mapping and validation of *Escherichia coli* transcriptional regulation from a compendium of expression profiles. PLoS Biol 5:e8. https://doi.org/10.1371/journal.pbio.0050008

Ferrario S, Immink RG, Angenent GC (2004) Conservation and diversity in flower land. Curr Opin Plant Biol 7:84–91. https://doi.org/10.1016/j.pbi.2003.11.003

Ficklin SP, Feltus FA (2011) Gene coexpression network alignment and conservation of gene modules between two grass species: maize and rice[C][W][OA]. Plant Physiol 156:1244–1256. https://doi.org/10.1104/pp.111.173047

Finkelstein RR, Gampala SSL, Rock CD (2002) Abscisic acid signaling in seeds and seedlings. Plant Cell 14:S15–S45. https://doi.org/10.1105/tpc.010441

Fujita Y, Yoshida T, Yamaguchi-Shinozaki K (2013) Pivotal role of the AREB/ABF-SnRK2 pathway in ABRE-mediated transcription in response to osmotic stress in plants. Physiol Plant 147:15–27. https://doi.org/10.1111/j.1399-3054.2012.01635.x

Garg R, Singh VK, Rajkumar MS, Kumar V, Jain M (2017) Global transcriptome and coexpression network analyses reveal cultivar-specific molecular signatures associated with seed development and seed size/weight determination in chickpea. Plant J 91:1088–1107. https://doi.org/10.1111/tpj.13621

Greene CS, Krishnan A, Wong AK, Ricciotti E, Zelaya RA, Himmelstein DS, Zhang R, Hartmann BM, Zaslavsky E, Sealfon SC, Chasman DI, FitzGerald GA, Dolinski K, Grosser T, Troyanskaya OG (2015) Understanding multicellular function and disease with human tissue-specific networks. Nat Genet 47:569–576. https://doi.org/10.1038/ng.3259

Gunasekara C, Zhang K, Deng W, Brown L, Wei H (2018) TGMI: an efficient algorithm for identifying pathway regulators through evaluation of triple-gene mutual interaction. Nucleic Acids Res 46:e67–e67. https://doi.org/10.1093/nar/gky210

Gutierrez L, Van Wuytswinkel O, Castelain M, Bellini C (2007) Combined networks regulating seed maturation. Trends Plant Sci 12:294–300. https://doi.org/10.1016/j.tplants.2007.06.003

Hamada K, Hongo K, Suwabe K, Shimizu A, Nagayama T, Abe R, Kikuchi S, Yamamoto N, Fujii T, Yokoyama K, Tsuchida H, Sano K, Mochizuki T, Oki N, Horiuchi Y, Fujita M, Watanabe M, Matsuoka M, Kurata N, Yano K (2011) OryzaExpress: an integrated database of gene expression networks and omics annotations in rice. Plant Cell Physiol 52:220–229. https://doi.org/10.1093/pcp/pcq195

Haque S, Ahmad JS, Clark NM, Williams CM, Sozzani R (2019) Computational prediction of gene regulatory networks in plant growth and development. Curr Opin Plant Biol 47:96–105. https://doi.org/10.1016/j.pbi.2018.10.005

Holdsworth MJ, Bentsink L, Soppe WJJ (2008) Molecular networks regulating Arabidopsis seed maturation, after-ripening, dormancy and germination. New Phytol 179:33–54. https://doi.org/10.1111/j.1469-8137.2008.02437.x

Hollender CA, Kang C, Darwish O, Geretz A, Matthews BF, Slovin J, Alkharouf N, Liu Z (2014) Floral transcriptomes in woodland strawberry uncover developing receptacle and anther gene networks. Plant Physiol 165:1062–1075. https://doi.org/10.1104/pp.114.237529

Huang DW, Sherman BT, Lempicki RA (2009) Systematic and integrative analysis of large gene lists using DAVID bioinformatics resources. Nat Protoc 4:44

Huang J, Zheng J, Yuan H, McGinnis K (2018) Distinct tissue-specific transcriptional regulation revealed by gene regulatory networks in maize. BMC Plant Biol 18. https://doi.org/10.1186/s12870-018-1329-y

Huynh-Thu VA, Sanguinetti G (2018) Gene regulatory network inference: an introductory survey. arXiv:1801.04087 [q-bio]

Huynh-Thu VA, Irrthum A, Wehenkel L, Geurts P (2010) Inferring regulatory networks from expression data using tree-based methods. PLoS ONE 5:e12776. https://doi.org/10.1371/journal.pone.0012776

Ikeuchi M, Shibata M, Rymen B, Iwase A, Bågman A-M, Watt L, Coleman D, Favero DS, Takahashi T, Ahnert SE, Brady SM, Sugimoto K (2018) A gene regulatory network for cellular reprogramming in plant regeneration. Plant Cell Physiol 59:770–782. https://doi.org/10.1093/pcp/pcy013

Itkin M, Heinig U, Tzfadia O, Bhide AJ, Shinde B, Cardenas PD, Bocobza SE, Unger T, Malitsky S, Finkers R, Tikunov Y, Bovy A, Chikate Y, Singh P, Rogachev I, Beekwilder J, Giri AP, Aharoni A (2013) Biosynthesis of antinutritional alkaloids in solanaceous crops is mediated by clustered genes. Science 341:175–179. https://doi.org/10.1126/science.1240230

Jaeger KE, Pullen N, Lamzin S, Morris RJ, Wigge PA (2013) Interlocking feedback loops govern the dynamic behavior of the floral transition in Arabidopsis[W][OA]. Plant Cell 25:820–833. https://doi.org/10.1105/tpc.113.109355

Kang YH, Kirik V, Hulskamp M, Nam KH, Hagely K, Lee MM, Schiefelbein J (2009) The MYB23 gene provides a positive feedback loop for cell fate specification in the arabidopsis root epidermis. Plant Cell 21:1080–1094. https://doi.org/10.1105/tpc.108.063180

Kolovos P, Knoch TA, Grosveld FG, Cook PR, Papantonis A (2012) Enhancers and silencers: an integrated and simple model for their function. Epigenetics Chromatin 5:1. https://doi.org/10.1186/1756-8935-5-1

Krishnan A, Gupta C, Ambavaram MMR, Pereira A (2017) RECoN: rice environment coexpression network for systems level analysis of abiotic-stress response. Front Plant Sci 8. https://doi.org/10.3389/fpls.2017.01640

Kroj T, Savino G, Valon C, Giraudat J, Parcy F (2003) Regulation of storage protein gene expression in Arabidopsis. Development 130:6065–6073. https://doi.org/10.1242/dev.00814

Kulkarni SR, Vaneechoutte D, Van de Velde J, Vandepoele K (2018) TF2Network: predicting transcription factor regulators and gene regulatory networks in Arabidopsis using publicly available binding site information. Nucleic Acids Res 46:e31. https://doi.org/10.1093/nar/gkx1279

Langfelder P, Horvath S (2008) WGCNA: an R package for weighted correlation network analysis. BMC Bioinform 9:559. https://doi.org/10.1186/1471-2105-9-559

Lara P, Oñate-Sánchez L, Abraham Z, Ferrándiz C, Díaz I, Carbonero P, Vicente-Carbajosa J (2003) Synergistic activation of seed storage protein gene expression in Arabidopsis by ABI3 and two bZIPs related to OPAQUE2. J Biol Chem 278:21003–21011. https://doi.org/10.1074/jbc.M210538200

Lebre S (2013) G1DBN: a package performing dynamic Bayesian network inference. Version

Lee JM, Sonnhammer ELL (2003) Genomic gene clustering analysis of pathways in eukaryotes. Genome Res 13:875–882. https://doi.org/10.1101/gr.737703

Lee I, Ambaru B, Thakkar P, Marcotte EM, Rhee SY (2010) Rational association of genes with traits using a genome-scale gene network for *Arabidopsis thaliana*. Nat Biotechnol 28:149–156. https://doi.org/10.1038/nbt.1603

Lee I, Seo Y-S, Coltrane D, Hwang S, Oh T, Marcotte EM, Ronald PC (2011) Genetic dissection of the biotic stress response using a genome-scale gene network for rice. PNAS 108:18548–18553. https://doi.org/10.1073/pnas.1110384108

Li Y, Pearl SA, Jackson SA (2015) Gene networks in plant biology: approaches in reconstruction and analysis. Trends Plant Sci 20:664–675. https://doi.org/10.1016/j.tplants.2015.06.013

Lin H, Yu J, Pearce SP, Zhang D, Wilson ZA (2017) RiceAntherNet: a gene co-expression network for identifying anther and pollen development genes. Plant J 92:1076–1091. https://doi.org/10.1111/tpj.13744

Locke JCW, Southern MM, Kozma-Bognár L, Hibberd V, Brown PE, Turner MS, Millar AJ (2005) Extension of a genetic network model by iterative experimentation and mathematical analysis. Mol Syst Biol 1:2005.0013. https://doi.org/10.1038/msb4100018

Locke JCW, Kozma-Bognár L, Gould PD, Fehér B, Kevei É, Nagy F, Turner MS, Hall A, Millar AJ (2006) Experimental validation of a predicted feedback loop in the multi-oscillator clock of *Arabidopsis thaliana*. Mol Syst Biol 2:59. https://doi.org/10.1038/msb4100102

Ma L, Chen C, Liu X, Jiao Y, Su N, Li L, Wang X, Cao M, Sun N, Zhang X, Bao J, Li J, Pedersen S, Bolund L, Zhao H, Yuan L, Wong GK-S, Wang J, Deng XW, Wang J (2005) A microarray analysis of the rice transcriptome and its comparison to Arabidopsis. Genome Res 15:1274–1283. https://doi.org/10.1101/gr.3657405

Ma S, Ding Z, Li P (2017) Maize network analysis revealed gene modules involved in development, nutrients utilization, metabolism, and stress response. BMC Plant Biol 17:131. https://doi.org/10.1186/s12870-017-1077-4

Ma X, Zhao H, Xu W, You Q, Yan H, Gao Z, Su Z (2018) Co-expression gene network analysis and functional module identification in bamboo growth and development. Front Genet 9. https://doi.org/10.3389/fgene.2018.00574

MacNeil LT, Walhout AJM (2011) Gene regulatory networks and the role of robustness and stochasticity in the control of gene expression. Genome Res 21:645–657. https://doi.org/10.1101/gr.097378.109

Madan Babu M, Teichmann SA (2003) Evolution of transcription factors and the gene regulatory network in *Escherichia coli*. Nucleic Acids Res 31:1234–1244

Madar A, Greenfield A, Ostrer H, Vanden-Eijnden E, Bonneau R (2009) The inferelator 2.0: a scalable framework for reconstruction of dynamic regulatory network models. In: 2009 annual international conference of the IEEE Engineering in Medicine and Biology Society. IEEE, Minneapolis, MN, pp 5448–5451

Maeda YT, Sano M (2006) Regulatory dynamics of synthetic gene networks with positive feedback. J Mol Biol 359:1107–1124. https://doi.org/10.1016/j.jmb.2006.03.064

Mangan S, Alon U (2003) Structure and function of the feed-forward loop network motif. Proc Natl Acad Sci USA 100:11980–11985. https://doi.org/10.1073/pnas.2133841100

Mao L, Van Hemert JL, Dash S, Dickerson JA (2009) Arabidopsis gene co-expression network and its functional modules. BMC Bioinform 10:346. https://doi.org/10.1186/1471-2105-10-346

Marbach D, Costello JC, Küffner R, Vega N, Prill RJ, Camacho DM, Allison KR, Kellis M, Collins JJ, Stolovitzky G (2012) Wisdom of crowds for robust gene network inference. Nat Methods 9:796–804. https://doi.org/10.1038/nmeth.2016

Margolin AA, Nemenman I, Basso K, Wiggins C, Stolovitzky G, Favera RD, Califano A (2006) ARACNE: an algorithm for the reconstruction of gene regulatory networks in a mammalian cellular context. BMC Bioinform 7:S7. https://doi.org/10.1186/1471-2105-7-S1-S7

Meng Y, Shao C, Chen M (2011) Toward microRNA-mediated gene regulatory networks in plants. Brief Bioinform 12:645–659. https://doi.org/10.1093/bib/bbq091

Mentzen WI, Wurtele ES (2008) Regulon organization of Arabidopsis. BMC Plant Biol 8:99. https://doi.org/10.1186/1471-2229-8-99

Meyer PE, Lafitte F, Bontempi G (2008) minet: a R/Bioconductor package for inferring large transcriptional networks using mutual information. BMC Bioinform 9:461. https://doi.org/10.1186/1471-2105-9-461

Mi H, Muruganujan A, Casagrande JT, Thomas PD (2013) Large-scale gene function analysis with the PANTHER classification system. Nat Protoc 8:1551–1566. https://doi.org/10.1038/nprot.2013.092

Milo R, Shen-Orr S, Itzkovitz S, Kashtan N, Chklovskii D, Alon U (2002) Network motifs: simple building blocks of complex networks. Science 298:824–827. https://doi.org/10.1126/science.298.5594.824

Mitsuda N, Ohme-Takagi M (2009) Functional analysis of transcription factors in Arabidopsis. Plant Cell Physiol 50:1232–1248. https://doi.org/10.1093/pcp/pcp075

Mizoi J, Ohori T, Moriwaki T, Kidokoro S, Todaka D, Maruyama K, Kusakabe K, Osakabe Y, Shinozaki K, Yamaguchi-Shinozaki K (2013) GmDREB2A;2, a canonical DEHYDRATION-RESPONSIVE ELEMENT-BINDING PROTEIN2-Type transcription factor in soybean, is posttranslationally regulated and mediates dehydration-responsive element-dependent gene expression. Plant Physiol 161:346–361. https://doi.org/10.1104/pp.112.204875

Morrissey ER (2011) GRENITS: gene regulatory network inference using time series, p 5

Movahedi S, Van de Peer Y, Vandepoele K (2011) Comparative network analysis reveals that tissue specificity and gene function are important factors influencing the mode of expression evolution in Arabidopsis and rice1[W]. Plant Physiol 156:1316–1330. https://doi.org/10.1104/pp.111.177865

Mutwil M, Klie S, Tohge T, Giorgi FM, Wilkins O, Campbell MM, Fernie AR, Usadel B, Nikoloski Z, Persson S (2011) PlaNet: combined sequence and expression comparisons across plant networks derived from seven species[W][OA]. Plant Cell 23:895–910. https://doi.org/10.1105/tpc.111.083667

Nakamura S, Lynch TJ, Finkelstein RR (2001) Physical interactions between ABA response loci of Arabidopsis. Plant J 26:627–635. https://doi.org/10.1046/j.1365-313x.2001.01069.x

Nakashima K, Yamaguchi-Shinozaki K (2013) ABA signaling in stress-response and seed development. Plant Cell Rep 32:959–970. https://doi.org/10.1007/s00299-013-1418-1

Nakashima K, Ito Y, Yamaguchi-Shinozaki K (2009) Transcriptional regulatory networks in response to abiotic stresses in Arabidopsis and grasses. Plant Physiol 149:88–95. https://doi.org/10.1104/pp.108.129791

Ouma WZ, Pogacar K, Grotewold E (2018) Topological and statistical analyses of gene regulatory networks reveal unifying yet quantitatively different emergent properties. PLoS Comput Biol 14:e1006098. https://doi.org/10.1371/journal.pcbi.1006098

Pruneda-Paz JL, Kay SA (2010) An expanding universe of circadian networks in higher plants. Trends Plant Sci 15:259–265. https://doi.org/10.1016/j.tplants.2010.03.003

Raz V, Bergervoet JH, Koornneef M (2001) Sequential steps for developmental arrest in Arabidopsis seeds. Development 128:243–252

Rebeiz M, Patel NH, Hinman VF (2015) Unraveling the tangled skein: the evolution of transcriptional regulatory networks in development. Annu Rev Genomics Hum Genet 16:103–131. https://doi.org/10.1146/annurev-genom-091212-153423

Reimand J, Arak T, Adler P, Kolberg L, Reisberg S, Peterson H, Vilo J (2016) g:Profiler—a web server for functional interpretation of gene lists (2016 update). Nucleic Acids Res 44:W83–W89. https://doi.org/10.1093/nar/gkw199

Rodríguez-Leal D, Lemmon ZH, Man J, Bartlett ME, Lippman ZB (2017) Engineering quantitative trait variation for crop improvement by genome editing. Cell 171:470–480.e8. https://doi.org/10.1016/j.cell.2017.08.030

Rosenfeld N, Elowitz MB, Alon U (2002) Negative autoregulation speeds the response times of transcription networks. J Mol Biol 323:785–793. https://doi.org/10.1016/S0022-2836(02)00994-4

Ruprecht C, Proost S, Hernandez-Coronado M, Ortiz-Ramirez C, Lang D, Rensing SA, Becker JD, Vandepoele K, Mutwil M (2017) Phylogenomic analysis of gene co-expression networks reveals the evolution of functional modules. Plant J 90:447–465. https://doi.org/10.1111/tpj.13502

Ryngajllo M, Childs L, Lohse M, Giorgi FM, Lude A, Selbig J, Usadel B (2011) SLocX: predicting subcellular localization of arabidopsis proteins leveraging gene expression data. Front Plant Sci 2. https://doi.org/10.3389/fpls.2011.00043

Saddic LA, Huvermann B, Bezhani S, Su Y, Winter CM, Kwon CS, Collum RP, Wagner D (2006) The LEAFY target LMI1 is a meristem identity regulator and acts together with LEAFY to regulate expression of CAULIFLOWER. Development 133:1673–1682. https://doi.org/10.1242/dev.02331

Schäfer J, Opgen-Rhein R (2006) Reverse engineering genetic networks using the GeneNet package, vol 6, p 4

Schafer J, Strimmer K (2005) An empirical Bayes approach to inferring large-scale gene association networks. Bioinformatics 21:754–764. https://doi.org/10.1093/bioinformatics/bti062

Segal E, Shapira M, Regev A, Pe'er D, Botstein D, Koller D, Friedman N (2003) Module networks: identifying regulatory modules and their condition-specific regulators from gene expression data. Nat Genet 34:166. https://doi.org/10.1038/ng1165

Seo CH, Kim J-R, Kim M-S, Cho K-H (2009) Hub genes with positive feedbacks function as master switches in developmental gene regulatory networks. Bioinformatics 25:1898–1904. https://doi.org/10.1093/bioinformatics/btp316

Shahan R, Zawora C, Wight H, Sittmann J, Wang W, Mount SM, Liu Z (2018) Consensus coexpression network analysis identifies key regulators of flower and fruit development in wild strawberry. Plant Physiol 178:202–216. https://doi.org/10.1104/pp.18.00086

Shannon P (2003) Cytoscape: a software environment for integrated models of biomolecular interaction networks. Genome Res 13:2498–2504. https://doi.org/10.1101/gr.1239303

Sharan R, Ulitsky I, Shamir R (2007) Network-based prediction of protein function. Mol Syst Biol 3:88. https://doi.org/10.1038/msb4100129

Sharma R, Upadhyay S, Bhat B, Singh G, Bhattacharya S, Singh A (2019) Abiotic stress induced miRNA-TF-gene regulatory network: a structural perspective. Genomics. https://doi.org/10.1016/j.ygeno.2019.03.004

Smita S, Katiyar A, Chinnusamy V, Pandey DM, Bansal KC (2015) Transcriptional regulatory network analysis of MYB transcription factor family genes in rice. Front Plant Sci 6. https://doi.org/10.3389/fpls.2015.01157

Smith LM (2018) Identification of woodland strawberry gene coexpression networks. Plant Physiol 178:7–8. https://doi.org/10.1104/pp.18.00880

Smith NC, Matthews JM (2016) Mechanisms of DNA-binding specificity and functional gene regulation by transcription factors. Curr Opin Struct Biol 38:68–74. https://doi.org/10.1016/j.sbi.2016.05.006

Smith VA, Yu J, Smulders T, Hartemink AJ, Jarvis ED (2005) Computational inference of neural information flow networks. PLoS Comput Biol e161. https://doi.org/10.1371/journal.pcbi.0020161.eor (preprint)

Stone SL, Williams LA, Farmer LM, Vierstra RD, Callis J (2006) KEEP ON GOING, a RING E3 ligase essential for Arabidopsis growth and development, is involved in abscisic acid signaling. Plant Cell 18:3415–3428. https://doi.org/10.1105/tpc.106.046532

Sun Y, Dinneny JR (2018) Q&A: how do gene regulatory networks control environmental responses in plants? BMC Biol 16:38. https://doi.org/10.1186/s12915-018-0506-7

Sunkar R, Chinnusamy V, Zhu J, Zhu J-K (2007) Small RNAs as big players in plant abiotic stress responses and nutrient deprivation. Trends Plant Sci 12:301–309. https://doi.org/10.1016/j.tplants.2007.05.001

Suzuki M, Wang HH-Y, McCarty DR (2007) Repression of the LEAFY COTYLEDON 1/B3 regulatory network in plant embryo development by VP1/ABSCISIC ACID INSENSITIVE 3-LIKE B3 genes. Plant Physiol 143:902–911. https://doi.org/10.1104/pp.106.092320

Tai Y, Liu C, Yu S, Yang H, Sun J, Guo C, Huang B, Liu Z, Yuan Y, Xia E, Wei C, Wan X (2018) Gene co-expression network analysis reveals coordinated regulation of three characteristic secondary biosynthetic pathways in tea plant (*Camellia sinensis*). BMC Genomics 19:616. https://doi.org/10.1186/s12864-018-4999-9

Theocharidis A, van Dongen S, Enright AJ, Freeman TC (2009) Network visualization and analysis of gene expression data using BioLayout Express3D. Nat Protoc 4:1535–1550. https://doi.org/10.1038/nprot.2009.177

Tian W, Skolnick J (2003) How well is enzyme function conserved as a function of pairwise sequence identity? J Mol Biol 333:863–882. https://doi.org/10.1016/j.jmb.2003.08.057

To A, Valon C, Savino G, Guilleminot J, Devic M, Giraudat J, Parcy F (2006) A network of local and redundant gene regulation governs Arabidopsis seed maturation. Plant Cell 18:1642–1651. https://doi.org/10.1105/tpc.105.039925

Tsuchiya Y, Nambara E, Naito S, McCourt P (2004) The FUS3 transcription factor functions through the epidermal regulator TTG1 during embryogenesis in Arabidopsis. Plant J 37:73–81. https://doi.org/10.1046/j.1365-313X.2003.01939.x

van Dam S, Võsa U, van der Graaf A, Franke L, de Magalhães JP (2017) Gene co-expression analysis for functional classification and gene-disease predictions. Brief Bioinform. https://doi.org/10.1093/bib/bbw139

Vandepoele K, de Peer YV (2005) Exploring the plant transcriptome through phylogenetic profiling. Plant Physiol 137:31–42. https://doi.org/10.1104/pp.104.054700

Verdier J, Thompson RD (2008) Transcriptional regulation of storage protein synthesis during dicotyledon seed filling. Plant Cell Physiol 49:1263–1271. https://doi.org/10.1093/pcp/pcn116

Vialette-Guiraud ACM, Andres-Robin A, Chambrier P, Tavares R, Scutt CP (2016) The analysis of gene regulatory networks in plant evo-devo. J Exp Bot 67:2549–2563. https://doi.org/10.1093/jxb/erw119

Vlasblom J, Zuberi K, Rodriguez H, Arnold R, Gagarinova A, Deineko V, Kumar A, Leung E, Rizzolo K, Samanfar B, Chang L, Phanse S, Golshani A, Greenblatt JF, Houry WA, Emili A, Morris Q, Bader G, Babu M (2015) Novel function discovery with GeneMANIA: a new integrated resource for gene function prediction in *Escherichia coli*. Bioinformatics 31:306–310. https://doi.org/10.1093/bioinformatics/btu671

Wang Z, Xu W, San Lucas FA, Liu Y (2013) Incorporating prior knowledge into gene network study. Bioinformatics 29:2633–2640. https://doi.org/10.1093/bioinformatics/btt443

Wang A, Shu X, Niu X, Zhao W, Ai P, Li P, Zheng A (2018) Comparison of gene co-networks analysis provide a systems view of rice (*Oryza sativa* L.) response to *Tilletia horrida* infection. PLOS ONE 13:e0202309. https://doi.org/10.1371/journal.pone.0202309

Weston DJ, Karve AA, Gunter LE, Jawdy SS, Yang X, Allen SM, Wullschleger SD (2011) Comparative physiology and transcriptional networks underlying the heat shock response in *Populus trichocarpa*, *Arabidopsis thaliana* and *Glycine max*. Plant Cell Environ 34:1488–1506. https://doi.org/10.1111/j.1365-3040.2011.02347.x

Wilkins O, Hafemeister C, Plessis A, Holloway-Phillips M-M, Pham GM, Nicotra AB, Gregorio GB, Jagadish SVK, Septiningsih EM, Bonneau R, Purugganan M (2016) EGRINs (environmental gene regulatory influence networks) in rice that function in the response to water deficit, high temperature, and agricultural environments. Plant Cell 28:2365–2384. https://doi.org/10.1105/tpc.16.00158

Wirojsirasak W, Kalapanulak S, Saithong T (2019) Pan- and core-gene association networks: integrative approaches to understanding biological regulation. PLoS ONE 14:e0210481. https://doi.org/10.1371/journal.pone.0210481

Wu S, Alseekh S, Cuadros-Inostroza Á, Fusari CM, Mutwil M, Kooke R, Keurentjes JB, Fernie AR, Willmitzer L, Brotman Y (2016) Combined use of genome-wide association data and correlation networks unravels key regulators of primary metabolism in *Arabidopsis thaliana*. PLoS Genet 12:e1006363. https://doi.org/10.1371/journal.pgen.1006363

Xi D-M, Zheng C-C (2011) Transcriptional regulation of seed storage protein genes in Arabidopsis and cereals. Seed Sci Res 21:247–254. https://doi.org/10.1017/S0960258511000237

Yeger-Lotem E, Sattath S, Kashtan N, Itzkovitz S, Milo R, Pinter RY, Alon U, Margalit H (2004) Network motifs in integrated cellular networks of transcription–regulation and protein–protein interaction. Proc Natl Acad Sci USA 101:5934–5939. https://doi.org/10.1073/pnas.0306752101

Yoon MU (2010) Differential equation models and numerical methods for reverse engineering genetic regulatory networks. PhD diss., University of Tennessee, p 164

Yu H, Jiao B, Liang C (2017) High-quality rice RNA-seq-based co-expression network for predicting gene function and regulation. bioRxiv 138040. https://doi.org/10.1101/138040

Zeilinger MN, Farré EM, Taylor SR, Kay SA, Doyle FJ (2006) A novel computational model of the circadian clock in Arabidopsis that incorporates PRR7 and PRR9. Mol Syst Biol 2:58. https://doi.org/10.1038/msb4100101

Zhang L, Yu S, Zuo K, Luo L, Tang K (2012) Identification of gene modules associated with drought response in rice by network-based analysis. PLoS ONE 7:e33748. https://doi.org/10.1371/journal.pone.0033748

Zhu J-Y, Sun Y, Wang Z-Y (2012) Genome-wide identification of transcription factor-binding sites in plants using chromatin immunoprecipitation followed by microarray (ChIP-chip) or sequencing (ChIP-seq). In: Wang Z-Y, Yang Z (eds) Plant signalling networks: methods and protocols. Humana Press, Totowa, NJ, pp 173–188

Zimmermann P (2004) GENEVESTIGATOR. Arabidopsis microarray database and analysis toolbox. Plant Physiol 136:2621–2632. https://doi.org/10.1104/pp.104.046367

Chapter 19
Genomics and Transcriptomics Advance in Plant Sciences

Boas Pucker and Hanna Marie Schilbert

Abstract Recent progress in sequencing technologies facilitates plant science experiments through the availability of genome and transcriptome sequences. Genome assemblies provide details about genes, transposable elements, and the general genome structure. The availability of a reference genome sequence for a species enables and supports numerous wet lab analyses and comprehensive bioinformatic investigations e.g. genome-wide investigations of gene families. After generating a genome sequence, gene prediction and the generation of functional annotations are the major challenges. Although these methods were improved substantially over the last years, incorporation of external hints like RNA-Seq reads is beneficial. Once a high-quality sequence and annotation is available for a species, diversity between accessions can be assessed by re-sequencing. This helps in revealing single nucleotide variants, insertions and deletions, and larger structural variants like inversions and transpositions. Identification of these variants requires sophisticated bioinformatic tools and many of them were developed during past years. Sequence variants can be harnessed for the genetic mapping of traits. Several mapping-by-sequencing approaches were developed to find underlying genes for relevant traits in crops. These genomic approaches are complemented by various transcriptomic methods dominated by a very popular RNA-Seq technology. Transcript abundance is measured via sequencing of the corresponding cDNA molecules. RNA-Seq reads can be subjected to transcriptome assembly or gene expression analysis, e.g. for the identification of transcripts abundance between different tissues, conditions, or genotypes.

Keywords Bioinformatics · Computational biology · Sequencing · Genome assembly · Gene prediction · Read mapping · Variant calling · Single nucleotide variant (SNV) · Insertion/deletion (InDel) · Genotyping-by-sequencing (GBS) · Mapping-by-sequencing (MBS) · RNA-Seq · Transcriptome assembly

---

B. Pucker (✉) · H. M. Schilbert
Genetics and Genomics of Plants, Center for Biotechnology (CeBiTec), Bielefeld University, Bielefeld, Germany
e-mail: bpucker@cebitec.uni-bielefeld.de

S. P. Singh et al. (eds.), *Molecular Approaches in Plant Biology and Environmental Challenges*, Energy, Environment, and Sustainability,
https://doi.org/10.1007/978-981-15-0690-1_19

19.1 Introduction

The genome of an organism determines its phenotype by setting the range of variability for numerous traits. Environmental factors shape the phenotype within this predetermined range. Knowledge about the genome and genes of a species facilitates various biological research projects. Research on *Arabidopsis thaliana* (*A. thaliana*) Columbia–0 was boosted by the availability of the first plant genome sequence (Somssich 2018). The transcriptome of an organism reveals which parts of the genome are 'active' at a certain point in time, under specific conditions, and in a defined cell type. Since the nucleic acid types DNA and RNA have very similar biochemical properties, the investigation of genome and transcriptome can be performed by similar methods. Both omics layers, genomics and transcriptomics, are easily accessible by analytic methods, because general biochemical properties of these nucleic acids are independent from the actual sequence. The intention of this chapter is (1) to describe genomics and transcriptomics workflows which are commonly used in plant research, and (2) to list frequently deployed bioinformatic tools for the analysis steps (Fig. 19.1).

19.2 Sequencing Technologies

Existing sequencing technologies can be grouped into different generations based on their key properties. However, there is disagreement in the literature about this classification system and the assignment of technologies to different generations (Metzker 2010; Shendure et al. 2004; Shendure and Ji 2008; Schadt et al. 2010; Glenn 2011; Quail et al. 2012; Goodwin et al. 2016; Peterson and Arick 2018). Here, we distinguish between three generations: (I) Sanger chain termination sequencing and Maxam Gilbert sequencing as first generation sequencing technologies, (II) Roche/454 pyrosequencing, IonTorrent, Solexa/Illumina, and Beijing Genomics Institute (BGI) sequencing as second generation sequencing technologies, and (III) Single molecule real time sequencing (Pacific Biosciences, PacBio) and nanopore sequencing (Oxford Nanopore Technologies, ONT) as third generation sequencing technologies. Technical details of these sequencing technologies were reviewed elsewhere (Metzker 2009, 2010; Shendure and Ji 2008; Quail et al. 2012; Goodwin et al. 2016; Margulies et al. 2005; Mardis 2008a).

Since the invention of chain termination sequencing (Sanger and Coulson 1975; Sanger et al. 1977), substantial technological advances paved the way for cost reductions. Therefore, broad application of high throughput sequencing (Metzker 2010) and more recently long read sequencing technologies (Li et al. 2017) became possible. Sanger sequencing generates a single read per sample, while other technologies produce large amounts of reads per sample and are hence crucial for many genome sequencing projects. Length of reads produced from Roche 454 pyrosequencing and

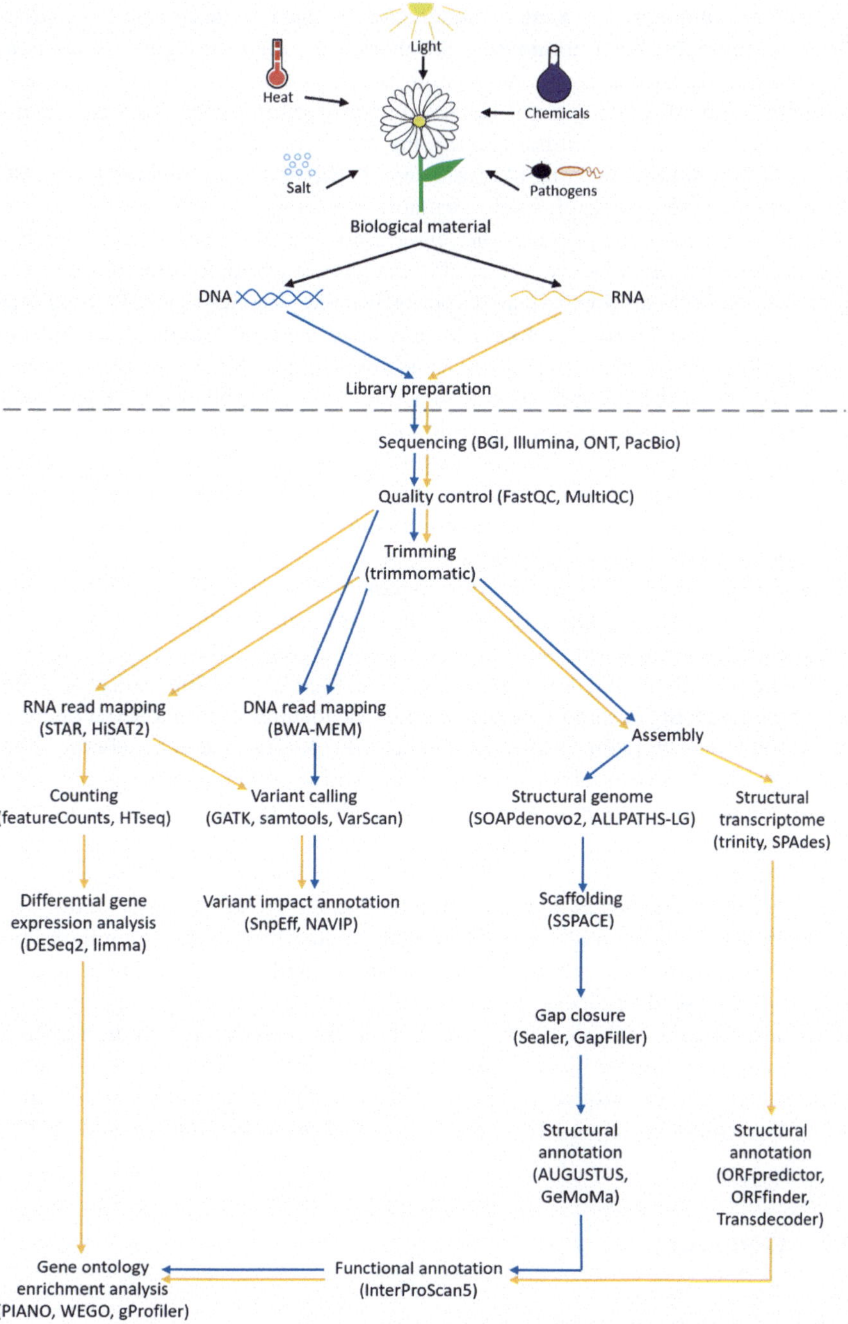

Fig. 19.1 Selected genomics and transcriptomics workflows in plant sciences. These workflows are deployed in many studies in plant research and the listed tools can be applied to perform the displayed steps. Several alternative and additional tools are listed within this chapter

IonTorrent is comparable to Sanger sequencing, but have reduced accuracy. Nevertheless, Illumina has been dominating the market for high throughput sequencing with substantially shorter reads due to high accuracy and low costs of sequencing technology. The BGI became a serious competitor during past years and is now offering the generation of similar sequencing data-sets based on its own technologies. While Illumina sequencing platforms are distributed all around the globe, BGI sequencing technology is exclusively available in China.

Paired-end sequencing provides the opportunity to analyze two ends of the same molecule. Overlapping reads; e.g. 2×300 nt, can be merged, thus leading to a total length of up to <600 nt. Sophisticated approaches like TrueSeq synthetic long reads (McCoy et al. 2014) were developed to maximize the read length of second generation technologies up to several thousand nucleotides. Mate pair reads provide information about the distance of both reads in addition to the mere sequences of both reads. In mate pair sequencing technique, long DNA fragments are modified at their ends, circularized, and fragmented. Fragments with marks are enriched and finally sequenced as paired-end libraries. The size of the initial fragments determines the distance of the two generated reads and can thus be considered as valuable linkage information during genome assembly processes.

However, length of reads generated from mate-pair sequencing is inferior to those generated by Oxford Nanopore Technologies (ONT) and Pacific Biosciences. From ONT, the longest sequenced DNA molecule has been reported to be over 2 Mbp till date (Payne et al. 2018) and the longest single read is close to 1 Mbp (Jain et al. 2018). Dropping sequencing costs and the rise of long read technologies enabled sequencing projects for numerous plant species (Bolger et al. 2014a; Jiao and Schneeberger 2017; Chen et al. 2018). Nevertheless, short reads are still valuable in projects; e.g. RNA-Seq or re-sequencing projects, where a high number of tags is more important than the read length.

In addition to generating extremely long reads at low costs, ONT also provides the first portable sequencers, namely MinION and Flongel, that can be deployed in field applications (Tyler et al. 2018; Pomerantz et al. 2018). Sequencing in the field opens up opportunities, to monitor pathogens in the field accurately (Hu et al. 2019) and to assess the biodiversity (Pomerantz et al. 2018). Real time base calling and the start of downstream analysis before completion of a sequencing run are beneficial when decisions are time critical (Stoiber and Brown 2017). Moreover, it also allows researchers to stop the sequencing process once sufficient data is generated and to commit the remaining sequencing capacity to other projects (Nguyen et al. 2017).

19.3 Genomics

19.3.1 Genome Assembly

Quality control and preprocessing Quality checks via FastQC (Andrews 2010) or MultiQC (Ewels et al. 2016) are usually the first steps to assess the quality of sequencing data. Next, reads need to be preprocessed prior to a de novo assembly, while this is not necessary for other applications like read mapping. Low quality sequences and remaining adapter fragments are removed during the trimming process, e.g. by trimmomatic (Bolger et al. 2014b). Removal of adapter sequences is especially important for de novo genome assemblies, because these sequences can occur in independent reads and cause the miss-join of random sequences into contigs.

Assembly concept A read can only represent a fraction of a complete genome sequence. Hence, intense manual work or the application of sophisticated bioinformatic tools is necessary to reconstruct complete genome sequences based on sequence reads (Mardis 2008b; Chaisson et al. 2009; Myers 2016). Initial sequencing projects involved the cloning of genomic fragments into vectors like bacterial artificial chromosomes (BACs) prior to sequencing. Genome sequences were generated by sequencing several BACs consecutively and combining the BAC sequences almost manually.

Second generation genome assemblies Especially, the rise of high throughput sequencing methods caused shift from manually curated BAC-based high continuity genome sequences towards whole genome shotgun draft assemblies. Dedicated assemblers were developed to harness the full potential of the available data types, for example combinations of paired-end and mate-pair data. SOAPdenovo2 (Luo et al. 2012), ALLPATHS-LG (Gnerre et al. 2011), Platanus (Kajitani et al. 2014), and the proprietary CLC assembler (QIAGEN 2016) are examples for tools which were successfully deployed for the assembly of plant genomes, but there are also many alternatives (Table 19.1). Modification of parameters, especially *k*-mer sizes, should be optimized empirically (Bradnam et al. 2013; Chikhi and Medvedev 2014; Shariat et al. 2014; Salzberg et al. 2012). In addition, the best combination of data from multiple sequencing libraries and sequencing technologies needs to be identified. After the generation of contigs in the assembly process, the information of mate pair and paired-end data-sets can be used to connect contigs to scaffolds without knowing the sequence enclosed between contigs of a scaffold. While some assemblers provide this functionality, dedicated tools like SSPACE (Boetzer et al. 2011) are available. Next, gaps between contigs within a scaffold can be partially closed, e.g. via GapFiller (Boetzer and Pirovano 2012) or Sealer (Paulino et al. 2015). The reduced sequencing costs allowed the assembly of plant genome sequences by single groups (Pucker et al. 2016), but most genome sequences were highly fragmented. More recently, the proprietary NRGene assembler (DeNovoMAGIC^{TM}) and the competing open source alternative TRITEX (Monat et al. 2019) are promising substantially improved assemblies.

Third generation genome assemblies The assembly situation changed again when long reads became available, thus enabling the generation of high continuity genome assemblies for numerous plant species with moderate effort (Michael et al. 2018; Pucker et al. 2019; Copetti et al. 2017; Lightfoot et al. 2017). The technological boost on the sequencing side caused an explosion in the development of novel assemblers and read correction tools which can handle noisy long reads efficiently (Table 19.2).

Table 19.1 Assembler for second generation sequencing data

Name	Availability	Link	References
CLC	Licence required	https://www.qiagenbioinformatics.com/products/clc-main-workbench	QIAGEN (2016)
SOAPdenovo2	Binary available	https://github.com/aquaskyline/SOAPdenovo2	Luo et al. (2012)
Velvet	Installation required	https://github.com/dzerbino/velvet	Zerbino and Birney (2008)
ALLPATHS-LG	Installation required	http://software.broadinstitute.org/allpaths-lg/blog/?page_id=12	Gnerre et al. (2011)
Ray	Installation required	http://denovoassembler.sourceforge.net	Boisvert et al. (2010)
Newbler	Installation required	http://sequencing.roche.com	Margulies et al. (2005)
MaSuRCA	Installation required	https://github.com/alekseyzimin/masurca	Zimin et al. (2013)
SGA	Installation required	https://github.com/jts/sga	Simpson and Durbin (2012)
Platanus	Installation required	http://platanus.bio.titech.ac.jp	Kajitani et al. (2014)

This table is an incomplete list of tools that can be applied for the de novo plant genome assembly based on second generation sequencing data

Table 19.2 Third generation assembler

Name	Availability	Link	References
FALCON	SMRT Link	https://https://pacb.www.pacb.com/training/smrt-link-overview	Chin et al. (2016)
Canu	Installation required	https://https://github.com/marbl/canu	Koren et al. (2017)
Flye	Installation required	https://github.com/fenderglass/Flye	Kolmogorov et al. (2019)
Miniasm	Installation required	https://github.com/lh3/miniasm	Li (2016)
wtdbg2	Installation required	https://github.com/ruanjue/wtdbg2	Ruan and Li (2019)

This table is an incomplete list of tools that can be applied for the de novo plant genome assembly based on third generation sequencing data

FALCON (Chin et al. 2016), Canu (Koren et al. 2017), Flye (Kolmogorov et al. 2019), Miniasm (Li 2016), and wtdbg2 (Ruan and Li 2019) are examples for frequently applied assemblers. Depending on the sequencing coverage and repeat content, the computational costs of assemblies can be high. Several hundred CPU hours, some hundred GB of RAM, and several TB of disc space are often required to assemble plant genomes. Assembled contigs can be joined into scaffolds based on additional information like genetic linkage (Pucker et al. 2019; Gan et al. 2016), optical mapping information, e.g. from Bionano Genomics and OptGen (Jiao et al. 2017; Lin et al. 2012; Tang et al. 2015), and Hi-C (Jiao et al. 2017; Burton et al. 2013; Phillippy 2017). Genetic linkage can rely on molecular markers measured in the lab (Pucker et al. 2019) or on sequencing of multiple individual plants of a segregating population by a high throughput method (Gan et al. 2016). Optical mapping is a size estimation of large DNA fragments which are generated by enzymatic restriction digest and cut site specific coloring with fluorescent dyes. Hi-C measures the 3D distances of genomic loci and assumes that neighboring sequences are also likely to be co-located in 2D.

Due to the high error rate in long reads, raw assemblies require several polishing steps. Firstly, long reads are aligned for correction, e.g. via BLASR (Chaisson and Tesler 2012) and minimap2 (Li 2018). Arrow (Chin et al. 2016) can be applied to polish assemblies based on PacBio reads, while nanopolish (Loman et al. 2015) is the best choice for ONT reads. Secondly, highly accurate short reads are mapped to the assembly to further correct the sequence in single copy regions. Paired-end or mate pair reads provide higher specificity during the mapping compared to single end reads. BWA-MEM (Li 2013) is a suitable read mapping tool and Pilon (Walker et al. 2014) can be used for the detection and correction of assembly errors. Iterative rounds of correction are possible. There is still an ongoing debate about the optimal number of polishing rounds that should be performed (Koren et al. 2017; Vaser et al. 2017). Since the most frequent error types are insertions/deletions, open reading frames are often affected by apparent frameshifts and premature stop codons. Therefore, the contribution of polishing approaches can be benchmarked based on an increase/decrease of frameshifts and premature stop codons in protein encoding genes. The optimal number of correction rounds can be determined by minimizing the number of these variants.

Assembly validation After combining reads into contigs, the correctness of these connections needs to be assessed. This assembly validation can be performed by mapping all reads back to the generated sequence, e.g. via BWA-MEM (Li 2013), and analyzing the distances of paired reads in this mapping, e.g. via REAPR (Hunt et al. 2013). Alternative approaches like implemented in KAT (Mapleson et al. 2017) inspect the assembly based on included *k*-mers. Most genome sequencing projects involve the generation of multiple assemblies with different tools and parameter settings. Selection of the best assembly can be challenging and criteria depend on the proposed research questions. The largest reasonable assembly, the assembly with the highest continuity, or the assembly resolving the highest number of genes might be of interest. Benchmarking Universal Single-Copy Orthologs (BUSCO) (Simão et al. 2015) is a frequently applied method to assess the assembly completeness and correctness. The underlying assumption is that all benchmarking genes should appear exactly once in the assembly. Different benchmarking sets exist for different taxonomic groups (Kriventseva et al. 2019). Due to a large phylogenetic distance to other sequenced species, this might not be perfectly accurate for the species of interest. However, the detection of single copy and complete genes is a good indicator for a high quality assembly. High numbers of duplicated BUSCOs can indicate separated haplophases. Recently, DOGMA (Dohmen et al. 2016) was released as an alternative tool for the analysis of sequence set completeness which also comes with an online version (https://domainworld-services.uni-muenster.de/dogma).

19.3.2 Gene Prediction

After generation and polishing of an assembly, the prediction of genes is often the next step. Besides protein encoding genes, there are also various RNA genes, transposable element genes, and numerous repeats which should be annotated as part of a genome project. In general, predictions are distinguished into (I) intrinsic approaches, which rely only on sequence properties, and (II) extrinsic approaches, which harness sequence similarity to previously annotated sequences to transfer annotation. However, frequently applied tools are designed to harness the power of both approaches (Table 19.3). AUGUSTUS (Stanke et al. 2006; Hoff and Stanke 2019) and GeneMark derivatives (Lomsadze et al. 2005, 2014; Ter-Hovhannisyan et al. 2008; Borodovsky and Lomsadze 2011) can predict genes ab initio without any external information. BUSCO can be applied to generate parameter files for this gene prediction process by assessing the gene structure of BUSCO genes (Waterhouse et al. 2018). In contrast to these ab initio approaches, GeMoMa (Keilwagen et al. 2016, 2018) combines external hints to construct a gene annotation based on sequence alignments. The exon intron structure of plant genes is posing a challenge to the gene prediction process, because tools need to account for interruptions of an open reading frame by on average four to five introns per gene (Pucker and Brockington 2018). Intron borders are often detected based on their conserved sequences: GT at the 5′ end and AG at the 3′ end. However, an average of at least 5% of all plant genes contains non-canonical splice sites, i.e. deviations from the GT-AG combination (Pucker and Brockington 2018; Pucker et al. 2017). Most gene prediction tools exclude non-canonical splice sites at least in the ab initio mode, because the number of possible gene models increases substantially when permitting many more possible intron positions. Therefore, external hints for intron positions are crucial to achieve an accurate prediction. If the identification of all isoforms of a gene is of interest, the accurate annotation of all exon intron borders is especially important. Expressed sequence tags (ESTs), contigs of a transcriptome assembly, or unassembled RNA-Seq reads can be aligned to the genomic sequence to generate hints. These sequences should originate from a broad range of different samples, e.g. collected under different environmental conditions,

Table 19.3 Plant gene prediction tools

Name	Availability	Link	References
AUGUSTUS	Installation required	https://github.com/Gaius-Augustus/Augustus	Stanke et al. (2006)
BRAKER1	Installation required	https://github.com/Gaius-Augustus/BRAKER	Hoff et al. (2016)
GeneMark	Installation required	http://exon.gatech.edu/GeneMark/license_download.cgi	Ter-Hovhannisyan et al. (2008), Borodovsky and Lomsadze (2011), Lomsadze et al. (2014)
GeMoMa	Jar file	http://www.jstacs.de/index.php/GeMoMa	Keilwagen et al. (2016, 2018)
Gnomon	Installation required	ftp://ftp.ncbi.nih.gov/toolbox/ncbi_tools ++/CURRENT	Souvorov et al. (2010)
MAKER2	Registration required	https://www.yandell-lab.org/software/maker.html	Holt and Yandell (2011)
SNAP	Installation required	https://github.com/KorfLab/SNAP	Korf (2004)

This table is an incomplete list of tools that can be applied for gene prediction on plant genome assemblies

from different tissues, and different developmental stages. The accurate alignment of transcript sequences to an assembly requires dedicated tools to account for introns. While BLAT (Kent 2002) can align long sequences, STAR (Dobin et al. 2013; Dobin and Gingeras 2015) is well suited for the split alignment of RNA-Seq reads. Dedicated tools like exonerate (Slater and Birney 2005) allow the alignment of previously annotated peptide sequences from other species. Resulting alignments can be converted into gene prediction hints. Annotation pipelines like MAKER2 (Holt and Yandell 2011), BRAKER1 (Hoff et al. 2016), and Gnomon (Souvorov et al. 2010) can integrate the information from different hint sources with ab initio prediction. While the prediction of protein encoding parts of a gene works relatively well, the annotation of untranslated regions (UTRs) and other non-coding sequences is still associated with a higher insecurity (Pucker et al. 2017; Haas et al. 2002; Fickett and Hatzigeorgiou 1997). Quality of the gene prediction process is in general not keeping pace with the rapid improvement of sequencing capacities and the frequent generation of highly contiguous assemblies (Salzberg 2019).

Technological progress allows the systematic investigation of non-protein encoding genes; e.g. through RNA-Seq experiments committed to the analysis of short RNAs. INFERNAL (Nawrocki and Eddy 2013) and tRNAscan-SE2 (Chan et al. 2019) are tools for the prediction of pure RNA genes.

Masking of repeats, e.g. via RepeatMasker (Smit et al. 2015), is frequently performed prior to the prediction of protein encoding genes, but this can actually have almost no or even detrimental effects on the prediction accuracy of certain gene families (Bayer et al. 2018). Although transposable elements and other repeats account for the major proportion of many plant genomes (Michael 2014; Vicient and Casacuberta 2017), the annotation of repeats is often performed poorly or omitted completely (Flutre et al. 2011; El Baidouri et al. 2015; Hoen et al. 2015). There is a plethora of annotation tools like RepeatScout (Price et al. 2005) and RepeatMasker (Smit et al. 2015). Bioinformatic pipelines were developed to account for weaknesses of single tools and to combine the strengths of many individual tools (Estill and Bennetzen 2009; Saha et al. 2008; Bergman and Quesneville 2007). One major issue with the TE and repeat annotation is the lack of a universal benchmarking study which could hint to the best tool for certain purposes (Hoen et al. 2015; Lerat 2010). While the annotation of protein encoding genes can be checked for completeness based on BUSCO (Simão et al. 2015) and DOGMA (Dohmen et al. 2016; Kemena et al. 2019), there is no such benchmarking data-set available for TEs.

Application examples Sequencing the genome of a plant species can provide insights into specific adaptations to local environmental conditions. *Crucihimalaya himalaica* is distributed at high altitudes at the Himalaya and the genome sequence reveals a reduced number of pathogen response genes as well as an increased number of DNA repair genes as response to a reduced amount of pathogens and an increased UV exposure, respectively (Kemena et al. 2019).

19.3.3 Re-sequencing and Variant Calling

Once a suitable reference genome sequence is available, re-sequencing projects can by-pass the laborious and expensive assembly step. Reads can be mapped to a reference sequence to identify differences between individuals of the same species or even between closely related species. Since the re-sequencing dataset does not need to provide sufficient data for a de novo assembly, the costs for re-sequencing are low compared to the initial genome project. Re-sequencing of over 1135 *A. thaliana* accessions revealed insights into the genomic diversity of this species (Alonso-Blanco et al. 2016). Since accessions are adapted to local environmental conditions, this project can reveal insights into adaptation mechanisms. Sequencing data also advances the understanding of population structures, genomic diversity between accessions, and genome evolution.

BWA-MEM (Li 2013) and bowtie2 (Langmead and Salzberg 2012) are frequently applied tools for the mapping of reads to a reference sequence (Table 19.4). The removal of PCR duplicates is necessary to avoid introducing a bias into following coverage analyses or variant callings. PCR duplicates are reads originating from a DNA fragment, which was amplified by PCR during the sequencing library preparation step. Functions like MarkDuplicates of Picard tools (Broad Institute 2019) allow the identification and removal of reads or read pairs originating from identical PCR products. This removal can be based on identical read sequences or identical positions in the mapping to a reference sequence. The detection of copy number variations depends on the equal representation of all genomic parts in the reads. PCR

Table 19.4 Read mapping tools

Name	Availability	Link	DNA/RNA	References
BWA-MEM	Installation required	https://github.com/lh3/bwa	DNA	Li (2013)
Bowtie 2	Installation required	https://github.com/BenLangmead/bowtie2	DNA	Langmead and Salzberg (2012)
GEM 3	Installation required	https://github.com/smarco/gem3-mapper	DNA	Marco-Sola et al. (2012)
bbmap	Jar file available	https://sourceforge.net/projects/bbmap	DNA	Bushnell (2019)
Novoalign	Trial available	http://www.novocraft.com/products/novoalign	DNA	NovoCraft (2010)
NextGenMap	Installation required	https://github.com/Cibiv/NextGenMap	DNA	Sedlazeck et al. (2013)
MAQ	Installation required	http://maq.sourceforge.net/maq-man.shtml	DNA	Li (2019)
RMAP	Installation required	https://github.com/smithlabcode/rmap	DNA	Smith et al. (2009)
MOSAIK	Installation required	https://github.com/wanpinglee/MOSAIK	DNA	Lee et al. (2014)
segemehl	Installation required	https://www.bioinf.uni-leipzig.de/Software/segemehl	RNA	Hoffmann et al. (2009)
STAR	Installation required	https://github.com/alexdobin/STAR	RNA	Dobin et al. (2013)
HISAT2	Binary available	https://ccb.jhu.edu/software/hisat2/manual.shtml	RNA	Kim et al. (2015)

This table is an incomplete list of tools which can be applied to map reads from second generation sequencing technologies against a reference sequence. While some tools are suitable for the continuous alignment of DNA reads, others can generate split alignments for RNA-Seq reads

duplicates could cause the identification of false positive duplications by producing a high numbers of identical reads which could display an apparent variant caused by a PCR error in an early amplification step. The identification of sequence variants is sensitive to PCR duplicates, because a certain number of reads displaying a variant is frequently used as filter criteria to remove false positive variant calls.

There are numerous tools for the detection of genomic differences based on a short read mapping (Table 19.5). Genome Analysis Tool Kit (GATK) (McKenna et al. 2010; der Auwera et al. 2013), samtools/bcftools (Li et al. 2009a), and VarDict

Table 19.5 Variant callers

Name	Availability	Link	Variants	References
DeepVariant	Installation required	https://github.com/google/deepvariant	Small	Poplin et al. (2018)
GATK	Jar file	https://software.broadinstitute.org/gatk/download	Small	McKenna et al. (2010), der Auwera et al. (2013)
SNVer	Installation required	http://snver.sourceforge.net	Small	Wei et al. (2011)
SAMtools	Jar file	http://samtools.sourceforge.net	Small	Li et al. (2009a)
VarDict	Installation required	https://github.com/AstraZeneca-NGS/VarDict	Small	Lai et al. (2016)
VarScan 2	Jar file	http://varscan.sourceforge.net	Small	Koboldt et al. (2012)
LoFreq	Binary available	https://csb5.github.io/lofreq/installation	Small	Wilm et al. (2012)
Platypus	Installation required	https://github.com/andyrimmer/Platypus	Small	Rimmer et al. (2014)
SOAPsnp	Installation required	https://sourceforge.net/projects/soapsnp	Small	Li et al. (2009b)
Atlas-SNP2	Installation required	https://sourceforge.net/projects/atlas2	Small	Shen et al. (2010)
FreeBayes	Installation required	https://github.com/ekg/freebayes	Small	Garrison and Marth (2012)
SVIM	Installation required	https://github.com/eldariont/svim	Large	Heller and Vingron (2018)
marginAlign	Installation required	https://github.com/benedictpaten/marginAlign	Large	Jain et al. (2015)
GraphMap	Installation required	https://github.com/isovic/graphmap	Large	Sović et al. (2016)
PoreSeq	Installation required	https://github.com/tszalay/poreseq	Large	Szalay and Golovchenko (2015)

This table is an incomplete list of tools which can be applied to identify sequence variants based on reads mapped against a reference sequence. While some tools are restricted to the identification of small variants, other can detect large structural variants

(Lai et al. 2016) can detect single nucleotide variations (SNVs) and small insertions/deletions (InDels). The rise of long read sequencing technologies added substantially to the sensitivity of the insertion/deletion detection. Moreover, it allows the identification of large scale structural rearrangements. GraphMap (Sović et al. 2016), marginAlign (Jain et al. 2015), and PoreSeq (Szalay and Golovchenko 2015) can align long reads to a reference sequence to call variants. Other tools like SVIM (Heller and Vingron 2018) rely on alignments generated by dedicated long read aligners like minimap2 (Li 2018) or BLASR (Chaisson and Tesler 2012). Identified variants can be subjected to downstream filtering; e.g. based on the number of supporting and contradicting reads.

Once the variants are identified, it is possible to assign functional annotations. Established tools for this purpose are SnpEff (Cingolani et al. 2012) and ANNOVAR (Wang et al. 2010). Based on the structural annotation of the reference sequence, SnpEff and ANNOVAR assign functional implications like "premature stop codon" or "frameshift" to single variants. Since these tools are predicting the effect for a single variant at a time, NAVIP (Baasner et al. 2019) was developed for the integrated annotation of all variants within one coding sequence. NAVIP accounts for combined effects of neighboring variants, e.g. two short InDels which are both causing a frameshift on their own, but result in a few substituted amino acids when considered together.

19.3.4 Mapping by Sequencing

Forward genetics Forward genetics describes the genetic screening of mutants which have been isolated based on an outstanding phenotype (Schneeberger 2014). Crossing a mutant with a wild type plant and selfing of the F1 offspring leads to a segregating F2 population. A large segregating population forms the basis for a forward genetics screen. Such a population contains members with the wild-type and mutant phenotypes, respectively. Except for the causal locus, the genotypes of this population should display a random distribution of alleles. Since this population is used for genetic mapping, it is called a mapping population. Genetic markers located near the causal mutation will co-segregate with this mutation. As a result of this linkage between the causal locus and flanking markers, one allele of the flanking markers should be over-represented in the mutant plants. Due to a gradually decreasing linkage, the frequency of the coupled marker allele should drop when moving away from the causal locus. Therefore, the allele frequency can be used to pinpoint loci of interest. Originally, the identification of the location of the causal mutation in the genome of a mutant has been a long-lasting procedure requiring a high number of genetic markers. Once a target region has been identified, this region was screened for candidate genes. In order to validate the link between the assumed candidate gene and the expected phenotype, complementation experiments were frequently conducted. In following studies, the molecular function of the mutated gene was often elucidated.

Next generation forward genetics Technological advances in next generation sequencing enable the use of small sequence variants as genetic markers. Since these small sequence variants occur in large numbers, the resolution of the resulting genetic map is extremely high. Allele frequencies at all sequence variants are calculated for identification of genomic regions associated with the phenotype of interest (Garcia et al. 2016). First approaches used bulk segregant analysis (BSA), where DNA from the mapping population is pooled based on the phenotypes of individuals and then sequenced, i.e. one pool comprises the wild type allele of a certain locus and the other pool the mutant allele of the respective locus. Next, reads are mapped against a reference genome sequence to detect sequence variants. In the next step, allele frequencies for all small sequence variants are calculated. High allele frequencies can indicate linkage with the causal locus. This approach is also known as mapping-by-sequencing (MBS) and allows the fast and simple identification of causal mutations through allele frequency deviations (Schneeberger 2014).

Mutagenesis Natural variation can provide mutants, but it is also possible to generate mutant plants via mutagenesis. DNA damaging agents deployed in these mutagenesis experiments can be classified as physical mutagens (e.g. gamma radiation and fast neutron bombardment) or chemical mutagens (e.g. ethyl methanesulfonate, diepoxybutane, sodium azide) (Sikora et al. 2011). In order to achieve maximal genetic variation with a minimum decrease in viability, mutagenic dosage and specific properties of the mutagen need to be considered (Sikora et al. 2011). High mutagenic dosages likely result in a high number of mutations in the individual genome, thus the high diversity around a causal mutation might impede the identification (Schneeberger 2014). If a mutagen introduces large genomic rearrangements (e.g. deletions or translocation of large regions), the resulting mutation density is typically low compared to a mutagen, which causes predominantly single nucleotide variations. Furthermore, large genomic rearrangements might impede or even prevent the identification of the causal mutation by breaking apart a set of linked genes.

Biological material Mapping-by-sequencing (MBS) can be based on four different sets of biological material. A classical mapping population scheme was frequently used during the first MBS experiments. This involved outcrossing of mutagenized plants with diverged strains followed by one round of selfing to generate the mapping population (Schneeberger et al. 2009; Cuperus et al. 2010). Sequencing was performed on two genomic F2 pools of mutant and wildtype plants, respectively. Starting with *A. thaliana*, this method was rapidly applied to other model organisms (Wenger et al. 2010; Leshchiner et al. 2012). An isogenic population is generated by crossing homozygous mutants with the non-mutagenized progenitor, resulting in segregation of subtle phenotypic differences in the F2 population (Abe et al. 2012). Therefore, the only segregating genetic variation is that induced by mutagens. MBS is performed as described above. Homozygosity mapping uses only the genomes of affected individuals, originally in the context of recessive disease alleles in inbred humans (Lander and Botstein 1987). In order to identify the causal homozygous mutation, the genomes are screened for regions with low heterozygosity. This approach enables MBS for species where a generation of a mapping population is not feasible (Lander and Botstein 1987; Singh et al. 2013) and no prior knowledge about the parental alleles (Voz et al. 2012) or crossing history is needed (Bowen et al. 2012). Sequencing of individual mutant genomes (Schneeberger 2014) is an expensive, but even more powerful approach. Phenotyping errors can contaminate pools in MBS, but this approach allows an in silico pooling.

Resolution and accuracy In general, correct phenotyping of each individual of the mapping population is essential for the accuracy of MBS approaches. Contamination of the mapping population with incorrectly phenotyped individuals results in a larger mapping interval, thus complicating the identification of the causal mutation (Greenberg et al. 2011). Therefore, the resolution of MBS depends on the sampling size of correctly phenotyped and genotyped individuals in the mapping population (Schneeberger 2014). However, the resolution is only slightly affected by the number of backcrossed generations (James et al. 2013). As with conventional methods (e.g. classic genetic markers), re-sequencing data can be used to fine map the trait(s) of interest in a crossing population (Schneeberger and Weigel 2011). The higher the number of recombinants analyzed, the narrower the final mapping interval. All variants can be considered as markers and thus the variant with the closest link to the trait hints towards the genomic position of the underlying locus. Due to the high marker density derived from natural polymorphisms in the recombinant mapping population, a stringent marker selection decreases the number of false-positive markers. However, at the same time the risk of excluding causal mutations increases, leading to a critical trade-off.

Mapping-by-sequencing applications SHOREmap demonstrated the applicability of MBS in *A. thaliana* (Schneeberger 2014; Schneeberger et al. 2009). Following projects applied MBS to various crop species including sugar beet (Ries et al. 2016), rice (Abe et al. 2012), maize (Liu et al. 2012), barley (Mascher et al. 2014), and cotton (Chen et al. 2015). Liu et al. applied a modification of MBS to maize for the identification of a drought tolerance locus: BSR-Seq (Liu et al. 2012). BSR-Seq uses RNA-Seq reads for the identification of causal mutations without any prior knowledge about polymorphic markers. As a proof of concept, RNA-Seq was performed for the recessive *glossy3* (*gl3*) mutation in a segregating F2 population. The *gl3* gene encodes a putative R2R3 type *myb* transcription factor, which regulates the biosynthesis of very-long-chain fatty acids, which are precursors of epicuticular waxes. Rice seedlings lacking *glossy3* show an extremely thick epicuticular wax on juvenile leaves. By using this alternative MBS approach the *gl3* locus was mapped to an interval of approximately 2 Mb. In summary, mapping-by-sequencing is a powerful technique, which will lead to (crop) plants that are well adapted to biotic and abiotic stresses in the future.

19.4 Transcriptomics

19.4.1 RNA-Seq

RNA-Seq, the sequencing of cDNAs, emerged as a valuable method for (1) gene expression analysis, (2) de novo transcriptome assembly, and (3) the generation of hints for the gene annotation. The Illumina sequencing workflow of cDNA is very similar to the sequencing of genomic DNA. Besides RNA-Seq, the direct sequencing of RNA became broadly available with ONT sequencing (Garalde et al. 2018). In addition, PacBio provides Iso-Seq to reveal the sequence of full length transcripts, which can facilitate gene annotation in plants (Minoche et al. 2015).

Gene expression analysis Short RNA-Seq reads replaced previous methods for systematic gene expression analyses like microarrays almost completely (Wang et al. 2009; Nagalakshmi et al. 2008; Mortazavi et al. 2008). Without any prior knowledge about the sequence, the abundance of transcripts can be quantified (Wang et al. 2009; Cheng et al. 2017a, b), e.g. by generating a de novo transcriptome assembly based on the RNA-Seq reads (see below) (Haak et al. 2018; Müller et al. 2017). RNA-Seq even allows to distinguish between different transcript isoforms of the same gene (Wang et al. 2009; Cheng et al. 2017a, b). Saturation of the signal as observed for microarrays is no longer an issue as the number of reads is proportional to the transcript abundance (Wang et al. 2009; Mortazavi et al. 2008). Low amounts of samples can be analyzed and transcripts with low abundance can be detected, because a single read would be sufficient to reveal the presence of a certain transcript (Wang et al. 2009; Hayashi et al. 2018). Transcript quantification can be performed based on alignments against a reference sequence, e.g. using STAR (Dobin et al. 2013), or alignment-free, e.g. via Kallisto (Bray et al. 2016) (Table 19.6). Information about the transcript abundance can be subjected to downstream analysis like the identification of differentially expressed genes between samples e.g. via DESeq2 (Love et al. 2014). An alternative approach is the identification of co-expressed genes or the construction of co-expression networks as described in (van Dam et al. 2017) and references therein.

De novo transcriptome assembly RNA-Seq reads contain comprehensive information about the transcript sequences. Therefore, a de novo assembly can be generated to reveal the sequences of transcripts present in the analyzed sample (Schliesky et al. 2012). De novo transcriptome assemblies were frequently applied to discover candidate genes which are responsible for a certain trait of interest (Han et al. 2017, 2018; Wu et al. 2017). One of the most popular transcriptome assemblers is Trinity (Haas et al. 2013) which comprises three sequentially applied modules. Trinity performs an in silico normalization of the provided reads, i.e. identical reads are filtered out to achieve a similar coverage depth for all transcripts. Supplying stranded RNA-Seq reads, i.e. reads originating from a specified strand, enables to distinguish between reads originating from mRNAs and reads originating from regulatory antisense transcripts. Trinity performed well in benchmarking studies (Hölzer and Marz 2019;

Table 19.6 RNA-Seq gene expression tools

Name	Availability	Link	Function	References
featureCounts	Binary available	http://bioinf.wehi.edu.au/featureCounts/	Read counting	Liao et al. (2014)
HTSeq	Installation required	https://htseq.readthedocs.io/en/release_0.11.1/	Read counting	Anders et al. (2015)
Kallisto	Installation required	https://pachterlab.github.io/kallisto/about		Bray et al. (2016)
DESeq 2	R package	https://www.bioconductor.org/packages//2.12/bioc/html/DESeq2.html	Differential gene expression analysis	Love et al. (2014)
Limma	R package	https://bioconductor.org/packages/release/bioc/html/limma.html	Differential gene expression analysis	Ritchie et al. (2015)
PIANO	R package	https://bioconductor.org/packages/release/bioc/html/piano.html	GO/pathway enrichment analysis	Väremo et al. (2013)
WEGO	Online	http://wego.genomics.org.cn/	GO enrichment analysis	Ye et al. (2018)
gProfiler	Online	https://biit.cs.ut.ee/gprofiler/gost	GO enrichment analysis	Reimand et al. (2007)
Mercator	Online	https://www.plabipd.de/portal/web/guest/mercator4	Pathway analysis	Schwacke et al. (2019)
MapMan	Online	https://plabipd.de/portal/mapman	Pathway analysis	Schwacke et al. (2019)
BioMart	Online	http://plants.ensembl.org/biomart/martview	Pathway analysis	Smedley et al. (2015)
Plant Reactome	Online	https://plantreactome.gramene.org	Pathway analysis	Naithani et al. (2017)

This table is an incomplete list of tools related to RNA-Seq analyses. Some tools allow the quantification of transcript abundances, while others are involved in the statistical analysis of the resulting abundance values

Behera et al. 2017), but there are more tools that can be evaluated on a given data set (Table 19.7). Several transcriptome assemblers including Cufflinks (Trapnell et al. 2010), Trinity (Haas et al. 2013), and StringTie (Pertea et al. 2015) allow the integration of a genome sequence for reference-based or genome-guided assembly.

After generation of an initial assembly, very short sequences as well as bacterial and fungal contamination sequences are usually filtered out based on sequence similarity to databases. Since no introns are included in assembled transcript sequences, the identification of protein coding regions can be performed by searching for open reading frames of sufficient length. ORFfinder (Wheeler et al. 2003), OrfPredictor (Min et al. 2005), and Transdecoder (Haas et al. 2013) can perform this task. Collapsing very similar sequences is sometimes required and can be performed by CD-HIT (Li and Godzik 2006; Fu et al. 2012). Once a final set of sequences is identified, the assignment of a functional annotation is usually the next step. Sequence similarity to functionally annotated databases like swissprot (Bairoch and Apweiler 2000; The

Table 19.7 De novo transcriptome assembly tools

Name	Availability	Link	References
Trinity	Installation required	https://github.com/trinityrnaseq/trinityrnaseq	Haas et al. (2013)
rnaSPAdes	Binary available	http://cab.spbu.ru/software/rnaspades	Bushmanova et al. (2018)
SPAdes	Binary available	http://cab.spbu.ru/software/spades	Bankevich et al. (2012)
Trans-ABySS	Installation required	https://github.com/bcgsc/transabyss	Robertson et al. (2010)
Bridger	Installation required	https://github.com/fmaguire/Bridger_Assembler	Chang et al. (2015)
SOAPdenovo-Trans	Installation required	https://github.com/aquaskyline/SOAPdenovo-Trans	Xie et al. (2014)
Oases	Installation required	https://github.com/dzerbino/oases	Schulz et al. (2012)
IDBA-Tran	Installation required	https://github.com/loneknightpy/idba	Peng et al. (2013)
BinPacker	Installation required	https://github.com/macmanes-lab/BinPacker	Liu et al. (2016)
Shannon	Installation required	https://github.com/sreeramkannan/Shannon	Kannan et al. (2016)

This table is an incomplete list of tools which can be applied to generate plant transcriptome assemblies based on RNA-Seq data

UniProt Consortium 2017) can be harnessed to transfer the functional annotation to the newly assembled sequences. InterProScan5 (Finn et al. 2017) assigns functional annotations including gene ontology (GO) terms and identifies Pfam domains.

Gene prediction hints Since RNA-Seq reads reveal transcript sequences, they can be incorporated in the prediction of genes. The alignment of RNA-Seq reads to a genome assembly indicates the positions of introns through gaps in the alignment. In addition, continuously aligned parts of RNA-Seq reads reveal exon positions. STAR (Dobin et al. 2013) and HISAT2 (Kim et al. 2015) are suitable tools for the mapping of RNA-Seq reads. If reads are already assembled into contigs, exonerate (Slater and Birney 2005) could be utilized to align transcript sequences to an assembly. Dedicated alignment tools also allow the incorporation of peptide sequences as hints by aligning the sequences of well annotated species against the new assembly. Examples for such peptide alignment tools are exonerate (Slater and Birney 2005) and BLAT (Kent 2002).

19.5 Future Directions

Recent developments in sequencing technologies enabled the cost-efficient generation of genome and transcriptome sequences for numerous plant species of interest (Bolger et al. 2014a; Jiao and Schneeberger 2017). Most of the traditional plant research already benefits from the availability of sequence information for the respective species of interest. This technological progress enables completely new research projects like comparative genomics of large taxonomic groups. Re-sequencing projects, which rely on a reference sequence for comparison, might be replaced by independent de novo genome assemblies for all samples of interest (Jiao and Schneeberger 2017).

The availability of large sequence data-sets will also lead to more data-based studies which just re-use the existing sequence data-sets. These publicly available data-sets can be harnessed to answer novel questions which could not have been addressed before (Pucker and Brockington 2018).

Availability of plant genome sequences can foster the research on and usage of orphan crops (Chang et al. 2018) and help during de novo domestication of crops (Fernie and Yan 2019). Intensifying research activity in this field is especially important to cope with global warming and climatic changes.

Acknowledgements We thank Jens Theine and Andreas Rempel for helpful comments on the manuscript.

References

Abe A, Kosugi S, Yoshida K, Natsume S, Takagi H, Kanzaki H et al (2012) Genome sequencing reveals agronomically important loci in rice using MutMap. Nat Biotechnol 30:174–178. https://doi.org/10.1038/nbt.2095

Alonso-Blanco C, Andrade J, Becker C, Bemm F, Bergelson J, Borgwardt KM et al (2016) 1135 Genomes reveal the global pattern of polymorphism in *Arabidopsis thaliana*. Cell 166:481–491. https://doi.org/10.1016/j.cell.2016.05.063

Andrews S (2010) FastQC a quality control tool for high throughput sequence data [internet]. https://www.bioinformatics.babraham.ac.uk/projects/fastqc/. Accessed on 14 Dec 2017

Anders S, Pyl PT, Huber W (2015) HTSeq—a python framework to work with high-throughput sequencing data. Bioinformatics 31:166–169. https://doi.org/10.1093/bioinformatics/btu638

Baasner J-S, Howard D, Pucker B (2019) Influence of neighboring small sequence variants on functional impact prediction. https://doi.org/10.1101/596718

Bairoch A, Apweiler R (2000) The SWISS-PROT protein sequence database and its supplement TrEMBL in 2000. Nucleic Acids Res 28:45–48

Bankevich A, Nurk S, Antipov D, Gurevich AA, Dvorkin M, Kulikov AS et al (2012) SPAdes: a new genome assembly algorithm and its applications to single-cell sequencing. J Comput Biol 19:455–477. https://doi.org/10.1089/cmb.2012.0021

Bayer PE, Edwards D, Batley J (2018) Bias in resistance gene prediction due to repeat masking. Nat Plants 4:762. https://doi.org/10.1038/s41477-018-0264-0

Behera S, Voshall A, Deogun JS, Moriyama EN (2017) Performance comparison and an ensemble approach of transcriptome assembly. In: 2017 IEEE international conference on bioinformatics and biomedicine (BIBM), pp 2226–2228. https://doi.org/10.1109/bibm.2017.8218005

Bergman CM, Quesneville H (2007) Discovering and detecting transposable elements in genome sequences. Brief Bioinform 8:382–392. https://doi.org/10.1093/bib/bbm048

Boetzer M, Pirovano W (2012) Toward almost closed genomes with GapFiller. Genome Biol 13:R56. https://doi.org/10.1186/gb-2012-13-6-r56

Boetzer M, Henkel CV, Jansen HJ, Butler D, Pirovano W (2011) Scaffolding pre-assembled contigs using SSPACE. Bioinformatics 27:578–579. https://doi.org/10.1093/bioinformatics/btq683

Boisvert S, Laviolette F, Corbeil J (2010) Ray: Simultaneous assembly of reads from a mix of high-throughput sequencing technologies. J Comput Biol 17:1519–1533. https://doi.org/10.1089/cmb.2009.0238

Bolger ME, Weisshaar B, Scholz U, Stein N, Usadel B, Mayer KF (2014a) Plant genome sequencing—applications for crop improvement. Curr Opin Biotechnol 26:31–37. https://doi.org/10.1016/j.copbio.2013.08.019

Bolger AM, Lohse M, Usadel B (2014b) Trimmomatic: a flexible trimmer for Illumina sequence data. Bioinforma Oxford Engl. 30:2114–2120. https://doi.org/10.1093/bioinformatics/btu170

Borodovsky M, Lomsadze A (2011) Eukaryotic gene prediction using GeneMark.hmm-E and GeneMark-ES. Curr Protoc Bioinforma Ed Board Andreas Baxevanis Al. CHAPTER: Unit-4.610. https://doi.org/10.1002/0471250953.bi0406s35

Bowen ME, Henke K, Siegfried KR, Warman ML, Harris MP (2012) Efficient mapping and cloning of mutations in Zebrafish by low-coverage whole-genome sequencing. Genetics 190:1017–1024. https://doi.org/10.1534/genetics.111.136069

Bradnam KR, Fass JN, Alexandrov A, Baranay P, Bechner M, Birol I et al (2013) Assemblathon 2: evaluating de novo methods of genome assembly in three vertebrate species. GigaScience 2:10. https://doi.org/10.1186/2047-217X-2-10

Bray NL, Pimentel H, Melsted P, Pachter L (2016) Near-optimal probabilistic RNA-seq quantification. Nat Biotechnol 34:525–527. https://doi.org/10.1038/nbt.3519

Broad Institute (2019) Picard toolkit [internet]. Broad Institute. https://github.com/broadinstitute/picard

Burton JN, Adey A, Patwardhan RP, Qiu R, Kitzman JO, Shendure J (2013) Chromosome-scale scaffolding of de novo genome assemblies based on chromatin interactions. Nat Biotechnol 31:1119–1125. https://doi.org/10.1038/nbt.2727

Bushmanova E, Antipov D, Lapidus A, Przhibelskiy AD (2018) rnaSPAdes: a de novo transcriptome assembler and its application to RNA-Seq data. https://doi.org/10.1101/420208

Bushnell B (2019) BBMap—Browse files at SourceForge.net. https://sourceforge.net/projects/bbmap/files/. Accessed on 28 May 2019

Chaisson MJ, Tesler G (2012) Mapping single molecule sequencing reads using basic local alignment with successive refinement (BLASR): application and theory. BMC Bioinform 13:238. https://doi.org/10.1186/1471-2105-13-238

Chaisson MJ, Brinza D, Pevzner PA (2009) De novo fragment assembly with short mate-paired reads: does the read length matter? Genome Res 19:336–346. https://doi.org/10.1101/gr.079053.108

Chan PP, Lin BY, Mak AJ, Lowe TM (2019) tRNAscan-SE 2.0: improved detection and functional classification of transfer RNA genes. https://doi.org/10.1101/614032

Chang Z, Li G, Liu J, Zhang Y, Ashby C, Liu D et al (2015) Bridger: a new framework for de novo transcriptome assembly using RNA-seq data. Genome Biol 16. https://doi.org/10.1186/s13059-015-0596-2

Chang Y, Liu H, Liu M, Liao X, Sahu SK, Fu Y et al (2018) The draft genomes of five agriculturally important African orphan crops. GigaScience. https://doi.org/10.1093/gigascience/giy152

Chen W, Yao J, Chu L, Yuan Z, Li Y, Zhang Y (2015) Genetic mapping of the nulliplex-branch gene (gb_nb1) in cotton using next-generation sequencing. Theor Appl Genet 128:539–547. https://doi.org/10.1007/s00122-014-2452-2

Chen F, Dong W, Zhang J, Guo X, Chen J, Wang Z et al (2018) The sequenced angiosperm genomes and genome databases. Front Plant Sci 9. https://doi.org/10.3389/fpls.2018.00418

Cheng C-Y, Krishnakumar V, Chan AP, Thibaud-Nissen F, Schobel S, Town CD (2017a) Araport11: a complete reannotation of the *Arabidopsis thaliana* reference genome. Plant J 89:789–804. https://doi.org/10.1111/tpj.13415

Cheng B, Furtado A, Henry RJ (2017b) Long-read sequencing of the coffee bean transcriptome reveals the diversity of full-length transcripts. GigaScience 6. https://doi.org/10.1093/gigascience/gix086

Chikhi R, Medvedev P (2014) Informed and automated k-mer size selection for genome assembly. Bioinformatics 30:31–37. https://doi.org/10.1093/bioinformatics/btt310

Chin C-S, Peluso P, Sedlazeck FJ, Nattestad M, Concepcion GT, Clum A et al (2016) Phased diploid genome assembly with single-molecule real-time sequencing. Nat Methods 13:1050–1054. https://doi.org/10.1038/nmeth.4035

Cingolani P, Platts A, Wang LL, Coon M, Nguyen T, Wang L et al (2012) A program for annotating and predicting the effects of single nucleotide polymorphisms, SnpEff. Fly (Austin) 6:80–92. https://doi.org/10.4161/fly.19695

Copetti D, Búrquez A, Bustamante E, Charboneau JLM, Childs KL, Eguiarte LE et al (2017) Extensive gene tree discordance and hemiplasy shaped the genomes of North American columnar cacti. Proc Natl Acad Sci 114:12003–12008. https://doi.org/10.1073/pnas.1706367114

Cuperus JT, Montgomery TA, Fahlgren N, Burke RT, Townsend T, Sullivan CM et al (2010) Identification of MIR390a precursor processing-defective mutants in Arabidopsis by direct genome sequencing. Proc Natl Acad Sci U S A 107:466–471. https://doi.org/10.1073/pnas.0913203107

der Auwera GAV, Carneiro MO, Hartl C, Poplin R, Angel G del, Levy-Moonshine A et al (2013) From FastQ data to high-confidence variant calls: the genome analysis toolkit best practices pipeline. Curr Protoc Bioinforma. 43:11.10.1–11.10.33. https://doi.org/10.1002/0471250953.bi1110s43

Dobin A, Gingeras TR (2015) Mapping RNA-seq reads with STAR. Curr Protoc Bioinforma 51:11.14.1–11.14.19. https://doi.org/10.1002/0471250953.bi1114s51

Dobin A, Davis CA, Schlesinger F, Drenkow J, Zaleski C, Jha S et al (2013) STAR: ultrafast universal RNA-seq aligner. Bioinforma Oxf Engl 29:15–21. https://doi.org/10.1093/bioinformatics/bts635

Dohmen E, Kremer LPM, Bornberg-Bauer E, Kemena C (2016) DOGMA: domain-based transcriptome and proteome quality assessment. Bioinformatics 32:2577–2581. https://doi.org/10.1093/bioinformatics/btw231

El Baidouri M, Kim KD, Abernathy B, Arikit S, Maumus F, Panaud O et al (2015) A new approach for annotation of transposable elements using small RNA mapping. Nucleic Acids Res 43:e84–e84. https://doi.org/10.1093/nar/gkv257

Estill JC, Bennetzen JL (2009) The DAWGPAWS pipeline for the annotation of genes and transposable elements in plant genomes. Plant Methods 5:8. https://doi.org/10.1186/1746-4811-5-8

Ewels P, Magnusson M, Lundin S, Käller M (2016) MultiQC: summarize analysis results for multiple tools and samples in a single report. Bioinformatics 32:3047–3048. https://doi.org/10.1093/bioinformatics/btw354

Fernie AR, Yan J (2019) De Novo domestication: an alternative route toward new crops for the future. Mol Plant 12:615–631. https://doi.org/10.1016/j.molp.2019.03.016

Fickett JW, Hatzigeorgiou AG (1997) Eukaryotic promoter recognition. Genome Res 7:861–878. https://doi.org/10.1101/gr.7.9.861

Finn RD, Attwood TK, Babbitt PC, Bateman A, Bork P, Bridge AJ et al (2017) InterPro in 2017—beyond protein family and domain annotations. Nucleic Acids Res 45:D190–D199. https://doi.org/10.1093/nar/gkw1107

Flutre T, Duprat E, Feuillet C, Quesneville H (2011) Considering transposable element diversification in De Novo annotation approaches. PLoS ONE 6. https://doi.org/10.1371/journal.pone.0016526

Fu L, Niu B, Zhu Z, Wu S, Li W (2012) CD-HIT: accelerated for clustering the next-generation sequencing data. Bioinformatics 28:3150–3152. https://doi.org/10.1093/bioinformatics/bts565

Gan X, Hay A, Kwantes M, Haberer G, Hallab A, Ioio RD et al (2016) The *Cardamine hirsuta* genome offers insight into the evolution of morphological diversity. Nat Plants 2:16167. https://doi.org/10.1038/nplants.2016.167

Garalde DR, Snell EA, Jachimowicz D, Sipos B, Lloyd JH, Bruce M et al (2018) Highly parallel direct RNA sequencing on an array of nanopores. Nat Methods 15:201–206. https://doi.org/10.1038/nmeth.4577

Garcia V, Bres C, Just D, Fernandez L, Tai FWJ, Mauxion J-P et al (2016) Rapid identification of causal mutations in tomato EMS populations via mapping-by-sequencing. Nat Protoc 11:2401–2418. https://doi.org/10.1038/nprot.2016.143

Garrison E, Marth G (2012) Haplotype-based variant detection from short-read sequencing. Q-Bio. http://arxiv.org/abs/1207.3907

Glenn TC (2011) Field guide to next-generation DNA sequencers. Mol Ecol Resour 11:759–769. https://doi.org/10.1111/j.1755-0998.2011.03024.x

Gnerre S, MacCallum I, Przybylski D, Ribeiro FJ, Burton JN, Walker BJ et al (2011) High-quality draft assemblies of mammalian genomes from massively parallel sequence data. Proc Natl Acad Sci 108:1513–1518. https://doi.org/10.1073/pnas.1017351108

Goodwin S, McPherson JD, McCombie WR (2016) Coming of age: ten years of next-generation sequencing technologies. Nat Rev Genet 17:333–351. https://doi.org/10.1038/nrg.2016.49

Greenberg MV, Ausin I, Chan SW, Cokus SJ, Cuperus JT, Feng S et al (2011) Identification of genes required for de novo DNA methylation in Arabidopsis. Epigenetics 6:344–354. https://doi.org/10.4161/epi.6.3.14242

Haak M, Vinke S, Keller W, Droste J, Rückert C, Kalinowski J et al (2018) High quality de Novo transcriptome assembly of *Croton tiglium*. Front Mol Biosci 5. https://doi.org/10.3389/fmolb.2018.00062

Haas BJ, Volfovsky N, Town CD, Troukhan M, Alexandrov N, Feldmann KA et al (2002) Full-length messenger RNA sequences greatly improve genome annotation. Genome Biol 3(research0029):1. https://doi.org/10.1186/gb-2002-3-6-research0029

Haas BJ, Papanicolaou A, Yassour M, Grabherr M, Blood PD, Bowden J et al (2013) De novo transcript sequence reconstruction from RNA-seq using the Trinity platform for reference generation and analysis. Nat Protoc 8:1494–1512. https://doi.org/10.1038/nprot.2013.084

Han Y, Wan H, Cheng T, Wang J, Yang W, Pan H et al (2017) Comparative RNA-seq analysis of transcriptome dynamics during petal development in *Rosa chinensis*. Sci Rep 7:43382. https://doi.org/10.1038/srep43382

Hayashi T, Ozaki H, Sasagawa Y, Umeda M, Danno H, Nikaido I (2018) Single-cell full-length total RNA sequencing uncovers dynamics of recursive splicing and enhancer RNAs. Nat Commun 9:619. https://doi.org/10.1038/s41467-018-02866-0

Heller D, Vingron M (2018) SVIM: structural variant Identification using mapped long reads. https://doi.org/10.1101/494096

Hoen DR, Hickey G, Bourque G, Casacuberta J, Cordaux R, Feschotte C et al (2015) A call for benchmarking transposable element annotation methods. Mob DNA 6. https://doi.org/10.1186/s13100-015-0044-6

Hoff KJ, Stanke M (2019) Predicting genes in single genomes with AUGUSTUS. Curr Protoc Bioinforma 65:e57. https://doi.org/10.1002/cpbi.57

Hoff KJ, Lange S, Lomsadze A, Borodovsky M, Stanke M (2016) BRAKER1: unsupervised RNA-seq-based genome annotation with GeneMark-ET and AUGUSTUS. Bioinformatics 32:767–769. https://doi.org/10.1093/bioinformatics/btv661

Hoffmann S, Otto C, Kurtz S, Sharma CM, Khaitovich P, Vogel J et al (2009) Fast mapping of short sequences with mismatches, insertions and deletions using index structures. PLoS Comput Biol 5:e1000502. https://doi.org/10.1371/journal.pcbi.1000502

Holt C, Yandell M (2011) MAKER2: an annotation pipeline and genome-database management tool for second-generation genome projects. BMC Bioinform 12:491. https://doi.org/10.1186/1471-2105-12-491

Hölzer M, Marz M (2019) De Novo transcriptome assembly: a comprehensive cross-species comparison of short-read RNA-Seq assemblers. GigaScience 8. https://doi.org/10.1093/gigascience/giz039

Hu Y, Green GS, Milgate AW, Stone EA, Rathjen JP, Schwessinger B (2019) Pathogen detection and microbiome analysis of infected wheat using a portable DNA sequencer. Phytobiomes J. https://doi.org/10.1094/pbiomes-01-19-0004-r

Hunt M, Kikuchi T, Sanders M, Newbold C, Berriman M, Otto TD (2013) REAPR: a universal tool for genome assembly evaluation. Genome Biol 14:R47. https://doi.org/10.1186/gb-2013-14-5-r47

Jain M, Fiddes I, Miga KH, Olsen HE, Paten B, Akeson M (2015) Improved data analysis for the MinION nanopore sequencer. Nat Methods 12:351–356. https://doi.org/10.1038/nmeth.3290

Jain M, Koren S, Miga KH, Quick J, Rand AC, Sasani TA et al (2018) Nanopore sequencing and assembly of a human genome with ultra-long reads. Nat Biotechnol 36:338–345. https://doi.org/10.1038/nbt.4060

James GV, Patel V, Nordström KJ, Klasen JR, Salomé PA, Weigel D et al (2013) User guide for mapping-by-sequencing in Arabidopsis. Genome Biol 14:R61. https://doi.org/10.1186/gb-2013-14-6-r61

Jiao W-B, Schneeberger K (2017) The impact of third generation genomic technologies on plant genome assembly. Curr Opin Plant Biol 36:64–70. https://doi.org/10.1016/j.pbi.2017.02.002

Jiao W-B, Accinelli GG, Hartwig B, Kiefer C, Baker D, Severing E et al (2017) Improving and correcting the contiguity of long-read genome assemblies of three plant species using optical mapping and chromosome conformation capture data. Genome Res https://doi.org/10.1101/gr.213652.116

Kajitani R, Toshimoto K, Noguchi H, Toyoda A, Ogura Y, Okuno M et al (2014) Efficient de novo assembly of highly heterozygous genomes from whole-genome shotgun short reads. Genome Res 24:1384–1395. https://doi.org/10.1101/gr.170720.113

Kannan S, Hui J, Mazooji K, Pachter L, Tse D (2016) Shannon: an information-optimal de Novo RNA-Seq assembler. https://doi.org/10.1101/039230

Keilwagen J, Wenk M, Erickson JL, Schattat MH, Grau J, Hartung F (2016) Using intron position conservation for homology-based gene prediction. Nucleic Acids Res 44:e89. https://doi.org/10.1093/nar/gkw092

Keilwagen J, Hartung F, Paulini M, Twardziok SO, Grau J (2018) Combining RNA-seq data and homology-based gene prediction for plants, animals and fungi. BMC Bioinform 19:189. https://doi.org/10.1186/s12859-018-2203-5

Kemena C, Dohmen E, Bornberg-Bauer E (2019) DOGMA: a web server for proteome and transcriptome quality assessment. Nucleic Acids Res. https://doi.org/10.1093/nar/gkz366

Kent WJ (2002) BLAT—The BLAST-like alignment tool. Genome Res 12:656–664. https://doi.org/10.1101/gr.229202

Kim D, Langmead B, Salzberg SL (2015) HISAT: a fast spliced aligner with low memory requirements. Nat Methods 12:357–360. https://doi.org/10.1038/nmeth.3317

Koboldt DC, Zhang Q, Larson DE, Shen D, McLellan MD, Lin L et al (2012) VarScan 2: somatic mutation and copy number alteration discovery in cancer by exome sequencing. Genome Res 22:568–576. https://doi.org/10.1101/gr.129684.111

Kolmogorov M, Yuan J, Lin Y, Pevzner PA (2019) Assembly of long, error-prone reads using repeat graphs. Nat Biotechnol 37:540. https://doi.org/10.1038/s41587-019-0072-8

Koren S, Walenz BP, Berlin K, Miller JR, Bergman NH, Phillippy AM (2017) Canu: scalable and accurate long-read assembly via adaptive k-mer weighting and repeat separation. Genome Res. https://doi.org/10.1101/gr.215087.116

Korf I (2004) Gene finding in novel genomes. BMC Bioinform 5:59. https://doi.org/10.1186/1471-2105-5-59

Kriventseva EV, Kuznetsov D, Tegenfeldt F, Manni M, Dias R, Simão FA et al (2019) OrthoDB v10: sampling the diversity of animal, plant, fungal, protist, bacterial and viral genomes for evolutionary and functional annotations of orthologs. Nucleic Acids Res 47:D807–D811. https://doi.org/10.1093/nar/gky1053

Lai Z, Markovets A, Ahdesmaki M, Chapman B, Hofmann O, McEwen R et al (2016) VarDict: a novel and versatile variant caller for next-generation sequencing in cancer research. Nucleic Acids Res 44:e108. https://doi.org/10.1093/nar/gkw227

Lander ES, Botstein D (1987) Homozygosity mapping: a way to map human recessive traits with the DNA of inbred children. Science 236:1567–1570

Langmead B, Salzberg SL (2012) Fast gapped-read alignment with Bowtie 2. Nat Methods 9:357–359. https://doi.org/10.1038/nmeth.1923

Lee W-P, Stromberg MP, Ward A, Stewart C, Garrison EP, Marth GT (2014) MOSAIK: a hash-based algorithm for accurate next-generation sequencing short-read mapping. PLoS ONE 9:e90581. https://doi.org/10.1371/journal.pone.0090581

Lerat E (2010) Identifying repeats and transposable elements in sequenced genomes: how to find your way through the dense forest of programs. Heredity 104:520–533. https://doi.org/10.1038/hdy.2009.165

Leshchiner I, Alexa K, Kelsey P, Adzhubei I, Austin-Tse CA, Cooney JD et al (2012) Mutation mapping and identification by whole-genome sequencing. Genome Res 22:1541–1548. https://doi.org/10.1101/gr.135541.111

Li H (2013) Aligning sequence reads, clone sequences and assembly contigs with BWA-MEM. http://arxiv.org/abs/1303.3997

Li H (2016) Minimap and miniasm: fast mapping and de novo assembly for noisy long sequences. Bioinforma Oxf Engl 32:2103–2110. https://doi.org/10.1093/bioinformatics/btw152

Li H (2018) Minimap2: pairwise alignment for nucleotide sequences. Bioinforma Oxf Engl 34:3094–3100. https://doi.org/10.1093/bioinformatics/bty191

Li H (2019) MAQ [internet]. Available: http://maq.sourceforge.net/maq-manpage.shtml#12. Accessed on 28 May 2019

Li W, Godzik A (2006) Cd-hit: a fast program for clustering and comparing large sets of protein or nucleotide sequences. Bioinformatics 22:1658–1659. https://doi.org/10.1093/bioinformatics/btl158

Li H, Handsaker B, Wysoker A, Fennell T, Ruan J, Homer N et al (2009a) The sequence alignment/map format and SAMtools. Bioinformatics 25:2078–2079. https://doi.org/10.1093/bioinformatics/btp352

Li R, Li Y, Fang X, Yang H, Wang J, Kristiansen K et al (2009b) SNP detection for massively parallel whole-genome resequencing. Genome Res 19:1124–1132. https://doi.org/10.1101/gr.088013.108

Li C, Lin F, An D, Wang W, Huang R (2017) Genome sequencing and assembly by long reads in plants. Genes 9. https://doi.org/10.3390/genes9010006

Liao Y, Smyth GK, Shi W (2014) featureCounts: an efficient general purpose program for assigning sequence reads to genomic features. Bioinforma Oxf Engl 30:923–930. https://doi.org/10.1093/bioinformatics/btt656

Lightfoot DJ, Jarvis DE, Ramaraj T, Lee R, Jellen EN, Maughan PJ (2017) Single-molecule sequencing and Hi-C-based proximity-guided assembly of amaranth (*Amaranthus hypochondriacus*) chromosomes provide insights into genome evolution. BMC Biol 15:74. https://doi.org/10.1186/s12915-017-0412-4

Lin HC, Goldstein S, Mendelowitz L, Zhou S, Wetzel J, Schwartz DC et al (2012) AGORA: assembly guided by optical restriction alignment. BMC Bioinform 13:189. https://doi.org/10.1186/1471-2105-13-189

Liu S, Yeh C-T, Tang HM, Nettleton D, Schnable PS (2012) Gene mapping via bulked segregant RNA-Seq (BSR-Seq). PLoS ONE 7. https://doi.org/10.1371/journal.pone.0036406

Liu J, Li G, Chang Z, Yu T, Liu B, McMullen R et al (2016) BinPacker: packing-based De Novo transcriptome assembly from RNA-seq data. PLoS Comput Biol 12. https://doi.org/10.1371/journal.pcbi.1004772

Loman NJ, Quick J, Simpson JT (2015) A complete bacterial genome assembled de novo using only nanopore sequencing data. Nat Methods 12:733–735. https://doi.org/10.1038/nmeth.3444

Lomsadze A, Ter-Hovhannisyan V, Chernoff YO, Borodovsky M (2005) Gene identification in novel eukaryotic genomes by self-training algorithm. Nucleic Acids Res 33:6494–6506. https://doi.org/10.1093/nar/gki937

Lomsadze A, Burns PD, Borodovsky M (2014) Integration of mapped RNA-Seq reads into automatic training of eukaryotic gene finding algorithm. Nucleic Acids Res 42:e119. https://doi.org/10.1093/nar/gku557

Love MI, Huber W, Anders S (2014) Moderated estimation of fold change and dispersion for RNA-seq data with DESeq2. Genome Biol 15. https://doi.org/10.1186/s13059-014-0550-8

Luo R, Liu B, Xie Y, Li Z, Huang W, Yuan J et al (2012) SOAPdenovo2: an empirically improved memory-efficient short-read de novo assembler. GigaScience 1:18. https://doi.org/10.1186/2047-217X-1-18

Mapleson D, Garcia Accinelli G, Kettleborough G, Wright J, Clavijo BJ (2017) KAT: a K-mer analysis toolkit to quality control NGS datasets and genome assemblies. Bioinformatics 33:574–576. https://doi.org/10.1093/bioinformatics/btw663

Marco-Sola S, Sammeth M, Guigó R, Ribeca P (2012) The GEM mapper: fast, accurate and versatile alignment by filtration. Nat Methods 9:1185–1188. https://doi.org/10.1038/nmeth.2221

Mardis ER (2008a) Next-generation DNA sequencing methods. Annu Rev Genomics Hum Genet 9:387–402. https://doi.org/10.1146/annurev.genom.9.081307.164359

Mardis ER (2008b) The impact of next-generation sequencing technology on genetics. Trends Genet TIG. 24:133–141. https://doi.org/10.1016/j.tig.2007.12.007

Margulies M, Egholm M, Altman WE, Attiya S, Bader JS, Bemben LA et al (2005) Genome sequencing in microfabricated high-density picolitre reactors. Nature 437:376–380. https://doi.org/10.1038/nature03959

Mascher M, Jost M, Kuon J-E, Himmelbach A, Aßfalg A, Beier S et al (2014) Mapping-by-sequencing accelerates forward genetics in barley. Genome Biol 15:R78. https://doi.org/10.1186/gb-2014-15-6-r78

McCoy RC, Taylor RW, Blauwkamp TA, Kelley JL, Kertesz M, Pushkarev D et al (2014) Illumina TruSeq synthetic long-reads empower De Novo assembly and resolve complex, highly-repetitive transposable elements. PLoS ONE 9:e106689. https://doi.org/10.1371/journal.pone.0106689

McKenna A, Hanna M, Banks E, Sivachenko A, Cibulskis K, Kernytsky A et al (2010) The genome analysis toolkit: a MapReduce framework for analyzing next-generation DNA sequencing data. Genome Res 20:1297–1303. https://doi.org/10.1101/gr.107524.110

Metzker ML (2009) Sequencing in real time. Nat Biotechnol 27:150–151. https://doi.org/10.1038/nbt0209-150

Metzker ML (2010) Sequencing technologies—the next generation. Nat Rev Genet 11:31–46. https://doi.org/10.1038/nrg2626

Michael TP (2014) Plant genome size variation: bloating and purging DNA. Brief Funct Genomics 13:308–317. https://doi.org/10.1093/bfgp/elu005

Michael TP, Jupe F, Bemm F, Motley ST, Sandoval JP, Lanz C et al (2018) High contiguity *Arabidopsis thaliana* genome assembly with a single nanopore flow cell. Nat Commun 9:541. https://doi.org/10.1038/s41467-018-03016-2

Min XJ, Butler G, Storms R, Tsang A (2005) OrfPredictor: predicting protein-coding regions in EST-derived sequences. Nucleic Acids Res 33:W677–W680. https://doi.org/10.1093/nar/gki394

Minoche AE, Dohm JC, Schneider J, Holtgräwe D, Viehöver P, Montfort M et al (2015) Exploiting single-molecule transcript sequencing for eukaryotic gene prediction. Genome Biol 16:184. https://doi.org/10.1186/s13059-015-0729-7

Monat C, Padmarasu S, Lux T, Wicker T, Gundlach H, Himmelbach A et al (2019) TRITEX: chromosome-scale sequence assembly of Triticeae genomes with open-source tools. https://doi.org/10.1101/631648

Mortazavi A, Williams BA, McCue K, Schaeffer L, Wold B (2008) Mapping and quantifying mammalian transcriptomes by RNA-Seq. Nat Methods 5:621–628. https://doi.org/10.1038/nmeth.1226

Müller M, Seifert S, Lübbe T, Leuschner C, Finkeldey R (2017) De novo transcriptome assembly and analysis of differential gene expression in response to drought in European beech. PLoS ONE 12:e0184167. https://doi.org/10.1371/journal.pone.0184167

Myers JEW (2016) A history of DNA sequence assembly. It-Inf Technol 58:126–132. https://doi.org/10.1515/itit-2015-0047

Nagalakshmi U, Wang Z, Waern K, Shou C, Raha D, Gerstein M et al (2008) The transcriptional landscape of the yeast genome defined by RNA sequencing. Science 320:1344–1349. https://doi.org/10.1126/science.1158441

Naithani S, Preece J, D'Eustachio P, Gupta P, Amarasinghe V, Dharmawardhana PD et al (2017) Plant Reactome: a resource for plant pathways and comparative analysis. Nucleic Acids Res 45:D1029–D1039. https://doi.org/10.1093/nar/gkw932

Nawrocki EP, Eddy SR (2013) Infernal 1.1: 100-fold faster RNA homology searches. Bioinformatics. 29:2933–2935. https://doi.org/10.1093/bioinformatics/btt509

Nguyen SH, Duarte TPS, Coin LJM, Cao MD (2017) Real-time demultiplexing Nanopore barcoded sequencing data with npBarcode. Bioinformatics 33:3988–3990. https://doi.org/10.1093/bioinformatics/btx537

NovoCraft (2010) NovoAlign [internet]. http://www.novocraft.com/products/novoalign/. Accessed on 27 May 2019

Paulino D, Warren RL, Vandervalk BP, Raymond A, Jackman SD, Birol I (2015) Sealer: a scalable gap-closing application for finishing draft genomes. BMC Bioinform 16. https://doi.org/10.1186/s12859-015-0663-4

Payne A, Holmes N, Rakyan V, Loose M (2018) Whale watching with BulkVis: a graphical viewer for Oxford Nanopore bulk fast5 files. https://doi.org/10.1101/312256

Peng Y, Leung HCM, Yiu S-M, Lv M-J, Zhu X-G, Chin FYL (2013) IDBA-tran: a more robust de novo de Bruijn graph assembler for transcriptomes with uneven expression levels. Bioinformatics 29:i326–i334. https://doi.org/10.1093/bioinformatics/btt219

Pertea M, Pertea GM, Antonescu CM, Chang T-C, Mendell JT, Salzberg SL (2015) StringTie enables improved reconstruction of a transcriptome from RNA-seq reads. Nat Biotechnol 33:290–295. https://doi.org/10.1038/nbt.3122

Peterson DG, Arick M (2018) Sequencing plant genomes, pp 1–85. https://doi.org/10.1007/124_2018_18

Phillippy AM (2017) New advances in sequence assembly. Genome Res 27:xi–xiii. https://doi.org/10.1101/gr.223057.117

Pomerantz A, Peñafiel N, Arteaga A, Bustamante L, Pichardo F, Coloma LA et al (2018) Real-time DNA barcoding in a rainforest using nanopore sequencing: opportunities for rapid biodiversity assessments and local capacity building. GigaScience 7. https://doi.org/10.1093/gigascience/giy033

Poplin R, Chang P-C, Alexander D, Schwartz S, Colthurst T, Ku A et al (2018) A universal SNP and small-indel variant caller using deep neural networks. Nat Biotechnol 36:983–987. https://doi.org/10.1038/nbt.4235

Price AL, Jones NC, Pevzner PA (2005) De novo identification of repeat families in large genomes. Bioinforma Oxf Engl 21(Suppl 1):i351–i358. https://doi.org/10.1093/bioinformatics/bti1018

Pucker B, Brockington SF (2018) Genome-wide analyses supported by RNA-Seq reveal non-canonical splice sites in plant genomes. BMC Genom 19:980. https://doi.org/10.1186/s12864-018-5360-z

Pucker B, Holtgräwe D, Sörensen TR, Stracke R, Viehöver P, Weisshaar B (2016) A De Novo genome sequence assembly of the *Arabidopsis thaliana* accession Niederzenz-1 displays presence/absence variation and strong synteny. PLoS ONE 11:e0164321. https://doi.org/10.1371/journal.pone.0164321

Pucker B, Holtgräwe D, Weisshaar B (2017) Consideration of non-canonical splice sites improves gene prediction on the *Arabidopsis thaliana* Niederzenz-1 genome sequence. BMC Res Notes 10. https://doi.org/10.1186/s13104-017-2985-y

Pucker B, Holtgräwe D, Stadermann KB, Frey K, Huettel B, Reinhardt R et al (2019) A chromosome-level sequence assembly reveals the structure of the Arabidopsis thaliana Nd-1 genome and its gene set. PLoS ONE 14:e0216233. https://doi.org/10.1371/journal.pone.0216233

QIAGEN (2016) QIAGEN bioinformatics—sample to Insight. In: QIAGEN bioinformatics [internet]. https://www.qiagenbioinformatics.com/. Accessed 16 Dec 2018

Quail MA, Smith M, Coupland P, Otto TD, Harris SR, Connor TR et al (2012) A tale of three next generation sequencing platforms: comparison of Ion Torrent, Pacific Biosciences and Illumina MiSeq sequencers. BMC Genom 13:341. https://doi.org/10.1186/1471-2164-13-341

Reimand J, Kull M, Peterson H, Hansen J, Vilo J (2007) g:profiler—a web-based toolset for functional profiling of gene lists from large-scale experiments. Nucleic Acids Res 35:W193–W200. https://doi.org/10.1093/nar/gkm226

Ries D, Holtgräwe D, Viehöver P, Weisshaar B (2016) Rapid gene identification in sugar beet using deep sequencing of DNA from phenotypic pools selected from breeding panels. BMC Genomics 17. https://doi.org/10.1186/s12864-016-2566-9

Rimmer A, Phan H, Mathieson I, Iqbal Z, Twigg SRF, Wgs500 Consortium, et al (2014) Integrating mapping-, assembly- and haplotype-based approaches for calling variants in clinical sequencing applications. Nat Genet. 46:912–918. https://doi.org/10.1038/ng.3036

Ritchie ME, Phipson B, Wu D, Hu Y, Law CW, Shi W et al (2015) limma powers differential expression analyses for RNA-sequencing and microarray studies. Nucleic Acids Res 43:e47. https://doi.org/10.1093/nar/gkv007

Robertson G, Schein J, Chiu R, Corbett R, Field M, Jackman SD et al (2010) De novo assembly and analysis of RNA-seq data. Nat Methods 7:909–912. https://doi.org/10.1038/nmeth.1517

Ruan J, Li H (2019) Fast and accurate long-read assembly with wtdbg2. https://doi.org/10.1101/530972

Saha S, Bridges S, Magbanua ZV, Peterson DG (2008) Computational approaches and tools used in identification of dispersed repetitive DNA sequences. Trop Plant Biol 1:85–96. https://doi.org/10.1007/s12042-007-9007-5

Salzberg SL (2019) Next-generation genome annotation: we still struggle to get it right. Genome Biol 20:92. https://doi.org/10.1186/s13059-019-1715-2

Salzberg SL, Phillippy AM, Zimin A, Puiu D, Magoc T, Koren S et al (2012) GAGE: a critical evaluation of genome assemblies and assembly algorithms. Genome Res 22:557–567. https://doi.org/10.1101/gr.131383.111

Sanger F, Coulson AR (1975) A rapid method for determining sequences in DNA by primed synthesis with DNA polymerase. J Mol Biol 94:441–448

Sanger F, Nicklen S, Coulson AR (1977) DNA sequencing with chain-terminating inhibitors. Proc Natl Acad Sci U S A 74:5463–5467

Schadt EE, Turner S, Kasarskis A (2010) A window into third-generation sequencing. Hum Mol Genet 19:R227–R240. https://doi.org/10.1093/hmg/ddq416

Schliesky S, Gowik U, Weber APM, Bräutigam A (2012) RNA-Seq assembly—are we there yet? Front Plant Sci 3:220. https://doi.org/10.3389/fpls.2012.00220

Schneeberger K (2014) Using next-generation sequencing to isolate mutant genes from forward genetic screens. Nat Rev Genet 15:662–676. https://doi.org/10.1038/nrg3745

Schneeberger K, Weigel D (2011) Fast-forward genetics enabled by new sequencing technologies. Trends Plant Sci 16:282–288. https://doi.org/10.1016/j.tplants.2011.02.006

Schneeberger K, Ossowski S, Lanz C, Juul T, Petersen AH, Nielsen KL et al (2009) SHOREmap: simultaneous mapping and mutation identification by deep sequencing. Nat Methods 6:550–551. https://doi.org/10.1038/nmeth0809-550

Schulz MH, Zerbino DR, Vingron M, Birney E (2012) Oases: robust de novo RNA-seq assembly across the dynamic range of expression levels. Bioinformatics 28:1086–1092. https://doi.org/10.1093/bioinformatics/bts094

Schwacke R, Ponce-Soto GY, Krause K, Bolger AM, Arsova B, Hallab A et al (2019) MapMan4: a refined protein classification and annotation framework applicable to multi-omics data analysis. Mol Plant. https://doi.org/10.1016/j.molp.2019.01.003

Sedlazeck FJ, Rescheneder P, von Haeseler A (2013) NextGenMap: fast and accurate read mapping in highly polymorphic genomes. Bioinformatics 29:2790–2791. https://doi.org/10.1093/bioinformatics/btt468

Shariat B, Movahedi NS, Chitsaz H, Boucher C (2014) HyDA-Vista: towards optimal guided selection of k-mer size for sequence assembly. BMC Genom 15:S9. https://doi.org/10.1186/1471-2164-15-S10-S9

Shen Y, Wan Z, Coarfa C, Drabek R, Chen L, Ostrowski EA et al (2010) A SNP discovery method to assess variant allele probability from next-generation resequencing data. Genome Res 20:273–280. https://doi.org/10.1101/gr.096388.109

Shendure J, Ji H (2008) Next-generation DNA sequencing. Nat Biotechnol 26:1135–1145. https://doi.org/10.1038/nbt1486

Shendure J, Mitra RD, Varma C, Church GM (2004) Advanced sequencing technologies: methods and goals. Nat Rev Genet 5:335–344. https://doi.org/10.1038/nrg1325

Sikora P, Chawade A, Larsson M, Olsson J, Olsson O (2011) Mutagenesis as a tool in plant genetics, functional genomics, and breeding. Int J Plant Genomics https://doi.org/10.1155/2011/314829

Simão FA, Waterhouse RM, Ioannidis P, Kriventseva EV, Zdobnov EM (2015) BUSCO: assessing genome assembly and annotation completeness with single-copy orthologs. Bioinforma Oxf Engl 31:3210–3212. https://doi.org/10.1093/bioinformatics/btv351

Simpson JT, Durbin R (2012) Efficient de novo assembly of large genomes using compressed data structures. Genome Res 22:549–556. https://doi.org/10.1101/gr.126953.111

Singh R, Leslie Low E-T, Ooi LC-L, Ong-Abdullah M, Chin TN, Nagappan J et al (2013) The oil palm Shell gene controls oil yield and encodes a homologue of SEEDSTICK. Nature 500:340–344. https://doi.org/10.1038/nature12356

Slater GS, Birney E (2005) Automated generation of heuristics for biological sequence comparison. BMC Bioinform 6:31. https://doi.org/10.1186/1471-2105-6-31

Smedley D, Haider S, Durinck S, Pandini L, Provero P, Allen J et al (2015) The BioMart community portal: an innovative alternative to large, centralized data repositories. Nucleic Acids Res 43:W589–W598. https://doi.org/10.1093/nar/gkv350

Smit A, Hubley R, Green P (2015) RepeatMasker frequently open-4.0 [internet]. http://www.repeatmasker.org/
Smith AD, Chung W-Y, Hodges E, Kendall J, Hannon G, Hicks J et al (2009) Updates to the RMAP short-read mapping software. Bioinformatics 25:2841–2842. https://doi.org/10.1093/bioinformatics/btp533
Somssich M (2018) A short history of *Arabidopsis thaliana* (L.) Heynh. Columbia-0 [internet]. PeerJ Inc. Report No. e26931v4. https://doi.org/10.7287/peerj.preprints.26931v4
Souvorov A, Kapustin Y, Kiryutin B, Chetvernin V, Tatusova T, Lipman D (2010) Gnomon—NCBI eukaryotic gene prediction tool. http://www.ncbi.nlm.nih.gov/core/assets/genome/files/Gnomon-description.pdf
Sović I, Šikić M, Wilm A, Fenlon SN, Chen S, Nagarajan N (2016) Fast and sensitive mapping of nanopore sequencing reads with GraphMap. Nat Commun 7:11307. https://doi.org/10.1038/ncomms11307
Stanke M, Keller O, Gunduz I, Hayes A, Waack S, Morgenstern B (2006) AUGUSTUS: *ab initio* prediction of alternative transcripts. Nucleic Acids Res 34:W435–W439. https://doi.org/10.1093/nar/gkl200
Stoiber M, Brown J (2017) BasecRAWller: streaming nanopore basecalling directly from raw signal. https://doi.org/10.1101/133058
Szalay T, Golovchenko JA (2015) De novo sequencing and variant calling with nanopores using PoreSeq. Nat Biotechnol 33:1087–1091. https://doi.org/10.1038/nbt.3360
Tang H, Lyons E, Town CD (2015) Optical mapping in plant comparative genomics. GigaScience 4. https://doi.org/10.1186/s13742-015-0044-y
Ter-Hovhannisyan V, Lomsadze A, Chernoff YO, Borodovsky M (2008) Gene prediction in novel fungal genomes using an *ab initio* algorithm with unsupervised training. Genome Res 18:1979–1990. https://doi.org/10.1101/gr.081612.108
The UniProt Consortium (2017) UniProt: the universal protein knowledgebase. Nucleic Acids Res 45:D158–D169. https://doi.org/10.1093/nar/gkw1099
Trapnell C, Williams BA, Pertea G, Mortazavi A, Kwan G, Van MB et al (2010) Transcript assembly and quantification by RNA-Seq reveals unannotated transcripts and isoform switching during cell differentiation. Nat Biotechnol 28:511–515. https://doi.org/10.1038/nbt.1621
Tyler AD, Mataseje L, Urfano CJ, Schmidt L, Antonation KS, Mulvey MR et al (2018) Evaluation of Oxford Nanopore's MinION sequencing device for microbial whole genome sequencing applications. Sci Rep 8:10931. https://doi.org/10.1038/s41598-018-29334-5
van Dam S, Võsa U, van der Graaf A, Franke L, de Magalhães JP (2017) Gene co-expression analysis for functional classification and gene–disease predictions. Brief Bioinform 19:575–592. https://doi.org/10.1093/bib/bbw139
Väremo L, Nielsen J, Nookaew I (2013) Enriching the gene set analysis of genome-wide data by incorporating directionality of gene expression and combining statistical hypotheses and methods. Nucleic Acids Res 41:4378–4391. https://doi.org/10.1093/nar/gkt111
Vaser R, Sović I, Nagarajan N, Šikić M (2017) Fast and accurate de novo genome assembly from long uncorrected reads. Genome Res 27:737–746. https://doi.org/10.1101/gr.214270.116
Vicient CM, Casacuberta JM (2017) Impact of transposable elements on polyploid plant genomes. Ann Bot 120:195–207. https://doi.org/10.1093/aob/mcx078
Voz ML, Coppieters W, Manfroid I, Baudhuin A, Von Berg V, Charlier C et al (2012) Fast homozygosity mapping and identification of a Zebrafish ENU-induced mutation by whole-genome sequencing. PLoS ONE 7. https://doi.org/10.1371/journal.pone.0034671
Walker BJ, Abeel T, Shea T, Priest M, Abouelliel A, Sakthikumar S et al (2014) Pilon: an Integrated Tool for Comprehensive microbial variant detection and genome assembly improvement. PLoS ONE 9:e112963. https://doi.org/10.1371/journal.pone.0112963
Wang Z, Gerstein M, Snyder M (2009) RNA-Seq: a revolutionary tool for transcriptomics. Nat Rev Genet 10:57–63. https://doi.org/10.1038/nrg2484

Wang K, Li M, Hakonarson H (2010) ANNOVAR: functional annotation of genetic variants from high-throughput sequencing data. Nucleic Acids Res 38:e164. https://doi.org/10.1093/nar/gkq603

Waterhouse RM, Seppey M, Simão FA, Manni M, Ioannidis P, Klioutchnikov G et al (2018) BUSCO applications from quality assessments to gene prediction and phylogenomics. Mol Biol Evol 35:543–548. https://doi.org/10.1093/molbev/msx319

Wei Z, Wang W, Hu P, Lyon GJ, Hakonarson H (2011) SNVer: a statistical tool for variant calling in analysis of pooled or individual next-generation sequencing data. Nucleic Acids Res 39:e132–e132. https://doi.org/10.1093/nar/gkr599

Wenger JW, Schwartz K, Sherlock G (2010) Bulk segregant analysis by high-throughput sequencing reveals a novel xylose utilization gene from Saccharomyces cerevisiae. PLoS Genet 6. https://doi.org/10.1371/journal.pgen.1000942

Wheeler DL, Church DM, Federhen S, Lash AE, Madden TL, Pontius JU et al (2003) Database resources of the National Center for Biotechnology. Nucleic Acids Res 31:28–33

Wilm A, Aw PPK, Bertrand D, Yeo GHT, Ong SH, Wong CH et al (2012) LoFreq: a sequence-quality aware, ultra-sensitive variant caller for uncovering cell-population heterogeneity from high-throughput sequencing datasets. Nucleic Acids Res 40:11189–11201. https://doi.org/10.1093/nar/gks918

Wu S, Lei J, Chen G, Chen H, Cao B, Chen C (2017) De novo transcriptome assembly of chinese kale and global expression analysis of genes involved in glucosinolate metabolism in multiple tissues. Front Plant Sci 8. https://doi.org/10.3389/fpls.2017.00092

Xie Y, Wu G, Tang J, Luo R, Patterson J, Liu S et al (2014) SOAPdenovo-Trans: de novo transcriptome assembly with short RNA-Seq reads. Bioinformatics 30:1660–1666. https://doi.org/10.1093/bioinformatics/btu077

Ye J, Zhang Y, Cui H, Liu J, Wu Y, Cheng Y et al (2018) WEGO 2.0: a web tool for analyzing and plotting GO annotations, update. Nucleic Acids Res 2018(46):W71–W75. https://doi.org/10.1093/nar/gky400

Zerbino DR, Birney E (2008) Velvet: algorithms for de novo short read assembly using de Bruijn graphs. Genome Res 18:821–829. https://doi.org/10.1101/gr.074492.107

Zimin AV, Marçais G, Puiu D, Roberts M, Salzberg SL, Yorke JA (2013) The MaSuRCA genome assembler. Bioinforma Oxf Engl. 29:2669–2677. https://doi.org/10.1093/bioinformatics/btt476

Chapter 20
Molecular Interventions to Ameliorate Environmental Stresses in Orchids

Jaspreet K. Sembi, Devina Ghai and Jagdeep Verma

Abstract Orchids constitute one of the largest, diverse, cosmopolitan and highly evolved plant families, the Orchidaceae. Due to their poorly developed reproductive barriers, they are in a constant state of evolutionary flux and active speciation. Epiphytic habit in orchids, expose these plants to light and water stress which in turn encourage modifications and adaptations. The flowers have a complex structure with one of the petal being variously modified into a labellum, which forms the basis of their immense importance as ornamentals. As a result, they are in huge demand in floriculture industry. Additionally, orchids have been significantly reported in ancient therapeutic scriptures. There have been a large number of reports of their curative and restorative role potential. They are considered to be highly habitat specific and show poor propagation in nature owing to their microscopic seeds with undifferentiated embryos and no endosperm, and dependence on a symbiotic association with mycorrhizal fungi for continued germination. They are also considered to be good indicators of environmental degradation. Owing to their immense demand in floriculture industry as well as therapeutics, these plants have been under constant stress from various environmental sources as well as anthropogenic pressures. With the advent of molecular technology, there have been constant efforts to ameliorate the deleterious effects of environment stresses in orchids. The present review deals with the various molecular interventions involved in dealing with these environmental stresses.

Keywords Orchid · Stress · Oxidative · Osmotic · Antioxidants · Osmoprotectants

---

J. K. Sembi (✉) · D. Ghai
Department of Botany, Panjab University, Chandigarh 160014, India
e-mail: jaspreet.sembi@pu.ac.in

J. Verma
Department of Botany, Government College, Rajgarh, Sirmaur,
Himachal Pradesh 173101, India

S. P. Singh et al. (eds.), *Molecular Approaches in Plant Biology and Environmental Challenges*, Energy, Environment, and Sustainability,
https://doi.org/10.1007/978-981-15-0690-1_20

20.1 Plant and Environmental Challenges

Orchids have a place of pride among the flowering plants. These plants are valued for their attractive floral morphology as well as curative properties. Increase in their utilization for commercial purposes, habitat destruction and anthropogenic pressures have pushed these to being endangered of survival. Their low rates of germination in nature, recalcitrant nature and slow growing habit, dependency on mycorrhizal association and prolonged life cycles are some of the inherent factors which contribute to a threat for their survival.

Plants being fixed to a particular place are exposed to extreme environmental conditions which causes stress to their existence. As a result they have developed two major strategies to cope up, namely stress avoidance and stress tolerance. Stress avoiding plants adapt themselves mainly morphologically by changing water uptake, developing thick leaves and cuticle, regulating stomatal movement, developing succulent stems with spinose leaves etc. Stress tolerant plants perceive stress and initiate physiological, molecular and biochemical changes which are in turn determined by the sensitivity of plant and severity of stress. The plant sensitivity, on the other hand, is controlled by the genotype, developmental stage and its prior exposure to stress. Exposure to a type of stress often leads to a cascade of events leading to other type of stresses as well. These harsh conditions rapidly release signal transduction cues like antioxidants, proteins, osmoregulators, and secondary metabolites which are manifested in various plant metabolic pathways. Perception by the plant is the first step to a stress stimulus which leads to secondary signals like non-protein molecules such as calcium ions, reactive oxygen species (ROS) and inositol phosphates (IPs) and changes in their cystosolic concentrations initiates further signal transduction pathways involved in growth and development processes (Xiong et al. 2002).

Environmental stress can be categorized broadly into two types, abiotic and biotic. Abiotic includes stress generated by non-living components of the environment such as water, salt, temperature, heavy metal, sound and light whereas biotic stress is a result of attack or association in the form of various micro or macro organisms such as bacteria, fungi or herbivores. A number of molecular and cellular factors play a role in ameliorating these stresses, the details of which are discussed below and are summarized in Tables 20.1 and 20.2.

20.2 Plant Perception and Response to Stress

Gene regulation is the fundamental aspect of any cellular change occurring within the plants. The stress acts as a stimulant for the activation of signal transduction pathway mainly CDPK and MAPK. These pathways activate certain transcription factors such as *MYB*, *MYC*, *NAC*, *WRKY*, *ERF*, *GRAS* etc. which regulate the expression of specific genes leading to stress tolerance (Fig. 20.1). At times, the expression of genes is also specific to a certain type of stress. The summarized gene expression in

Table 20.1 Stress related genes/proteins reported in orchids

Stress	Plant	Gene/proteins	Functions	References
Salt	*Dendrobium catenatum*	*GRAS*	Upregulated in salt stress	Zeng et al. (2019)
	Dendrobium nobile	*DnWRKY11*	Upregulated in salt stress Increased tolerance to salt in transgenic tobacco	Xu et al. (2014)
	Dendrobium officinale	*DnWRYK29*	Upregulated in salt stress Increased sensitivity to salt in transgenic tobacco	Xu et al. (2015)
		DoPMM	Upregulated in salt stress	He et al. (2017a)
		DoMYC2	Increased sensitivity to salt in transgenic *Arabidopsis*	Zhu et al. (2017)
		DofLEA	Upregulated in salt stress	Ling et al. (2016)
		DoUGE	Increased tolerance to salt	Yu et al. (2017)
	Vanda coerulea	*VcPAL*	Upregulated in salt stress	Nag and Kumaria (2018)
Light	*Phalaenopsis hybrid* Fortune Saltzman	*NAC* *MYB* *ARF*	Protects photosynthetic machinery from light damage	Li et al. (2014)
	Vanda coerulea	*VcPAL*	Upregulated in UV-B Downregulated in dark	Nag and Kumaria (2018)
Heat	*Dendrobium catenatum*	*GRAS*	Upregulated in heat stress	Zeng et al. (2019)
	Dendrobiun nobile	*DnSIZ1*	Increased tolerance to heat	Liu et al. (2015)
	Dendrobium officinale	*DofLEAs*	Upregulated in heat stress	Ling et al. (2016)
Cold	*Dendrobium nobile*	*DnSIZ1*	Upregulated in cold stress	Liu et al. (2015)
	Dendrobium officinale	*PMM*	Upregulated in cold stress	He et al. (2017a)

(continued)

Table 20.1 (continued)

Stress	Plant	Gene/proteins	Functions	References
		WRKY	Upregulated in cold stress	He et al. (2017c)
	Oncidium Gower Ramsey	*OncPSY*	Upregulated in cold stress	Lee et al. (2012)
	Phalaenopsis amabilis	LTP	Improved adaptive responses to cold stress	Qin et al. (2011)
	Phlaenopsis aphrodite	*PaCBF1*	Upregulated in cold stress	Peng et al. (2014)
		PaDHN1	Upregulated in cold stress	
		PaCDPK1	Upregulated in cold stress	Tsai et al. (2007)
	Vanda coerulea	*VcPAL*	Upregulated in cold stress	Nag and Kumaria (2018)
Water	*Dendrobium nobile*	*DnWRKY11*	Increased tolerance to drought in transgenic tobacco	Xu et al. (2014)
	Dendrobium officinale	*DnWRKY29*	Increased sensitivity to drought in transgenic tobacco	Xu et al. (2015)
		DoUGE	Upregulated in drought stress	Yu et al. (2017)
	Dendrobium wangliangii	R2R2-type MYB	Inducing lateral root development in drought leading to stress avoidance	Zhao et al. (2018)
	Phaius tankervilliae	*PtNCED1*	Upregulated in desiccated protocorms	Lee et al. (2018)
	Vanda coerulea	*VcPAL*	Upregulated in drought stress	Nag and Kumaria (2018)
Wounding	*Dendrobium nobile*	*DnSIZ1*	Upregulated in wound stress	Liu et al. (2015)
	Phalaenopsis amabilis	*PaPTP1*	Upregulated in wound stress	Fu et al. (2011)
	Phalaenopsis aphrodite	*PaCDPK1*	Upregulated in wound stress	Tsai et al. (2007)

(continued)

Table 20.1 (continued)

Stress	Plant	Gene/proteins	Functions	References
	Vanda coerulea	*VcPAL*	Upregulated by wound stress	Nag and Kumaria (2018)
Biotic	*Cymbidium hybridum*	*WRKY*	Induced on symbiotic association with *Tulasnella*-like *Rhizoctonia* and *Umbelopsis*	Zhao et al. (2013)
	Dendrobium Hickam Deb	CymMV coat protein	Induced resistance to CymMV	Chang et al. (2005)
	Dendrobium officinale	*WRKY*	Induced on association with *Tulasnella*-like mycorrhizal fungus	Wang et al. (2018)
	Phalaenopsis amabilis	*PaECR1*	Increased susceptibility to *Erwinia chrysanthemi*	Fu et al. (2012)
	Phalaenopsis aphrodite	*PaCDPK1*	Induced on infection by *Erwinia chrysanthemi*	Tsai et al. (2007)
	Phalaenopsis orchid 'TS340'	CymMV coat protein	Induced resistance to CymMV	Liao et al. (2004)
	Phalaenopsis Sogo Yukidian cultivar V3	Hairpin protein	Induced PAMP mediated immunity	Chuang et al. (2014)
	Orchis morio	*PAL*	Upregulated in biotic stress	Beyrle et al. (1995)
	Oncidium 'Sherry Baby cultivar OM8'	Sweet pepper ferredoxin-like protein	Induced resistance Erwinia carotovora	Liau et al. (2003)
	Oncidium sphacelatum	*PAL*	Induced in green protocorm associated with *Ceratobasidium* sp.	Valadares et al. (2014)
		CHS		
		naringenin-3-dioxygenase		

Table 20.2 List of genes/proteins related to osmotic and oxidative stress in orchids

Stress	Plant name	Gene/protein	References
Osmotic	*Dendrobium candidum*	LEA	Yang et al. (2010)
	Dendrobium officinale	*DoUGE*	Yu et al. (2017)
		DoNI	Gao et al. (2016)
		DoUGP	Wan et al. (2017)
		DoPMM	He et al. (2017a)
		DoCSLA1-8	He et al. (2015)
		DoLEA	Ling et al. (2016)
	Dendrobium wangliangii	LEA	Zhao et al. (2018)
	Phalaenopsis cv. 'Red Sky'	*SAMS*	Yuan et al. (2015)
Oxidative	*Dendrobium nobile*	*DnWRKY11*	Xu et al. (2014)
	Oncidium Gower Ramsey	*OncPSY*	Lee et al. (2012)
	Oncidium sphacelatum	*NCED1*	Valadares et al. (2014)
		PAL	
		CHS	
		naringenin-3-dioxygenase	
	Phaius tankervilliae	*PtNCED1*	Lee et al. (2018)
	Phalaenopsis bellina	*PbbHLH4*	Chuang et al. (2018)
	Phalaenopsis white hybrid	Xanthine oxidase	Tewari et al. (2009)

various stresses is given in Table 20.1. However, there are also reports of crosstalk between various stress related pathways such as methylerythritol-4-phosphate (MEP) pathway, phenylpropanoid pathway, carotenoid synthesis, carbohydrate metabolism, fatty acid biosynthesis, which help the plant to endure different types of stresses occurring together.

The biotic or abiotic stress generally results in either osmotic stress or oxidative stress. In osmotic stress, there is decrease in the turgor pressure of the plant cell due to reduction in water content. This osmotic stress can occur through heat, salt or drought stress. There are two pathways by which plant endures osmotic stress, ABA dependent and ABA independent pathways. Aside from this, plant under osmotic stress also produces different type of osmoprotectants, like chaperons (HSPs), polyamine, sugars, proline, alkaloids, carotenoids and flavonoids, which can increase the water potential of the cell. In oxidative stress, there is formation of reactive oxygen species (ROS) due to presence of unbound electrons. The ROS are quite detrimental for plant growth as these molecules disturbs the homeostasis and causes ion toxicity. The redox balance can be restored when such active species are sequestered by antioxidants such as APX, GPX, CAT, Tocopherols etc.

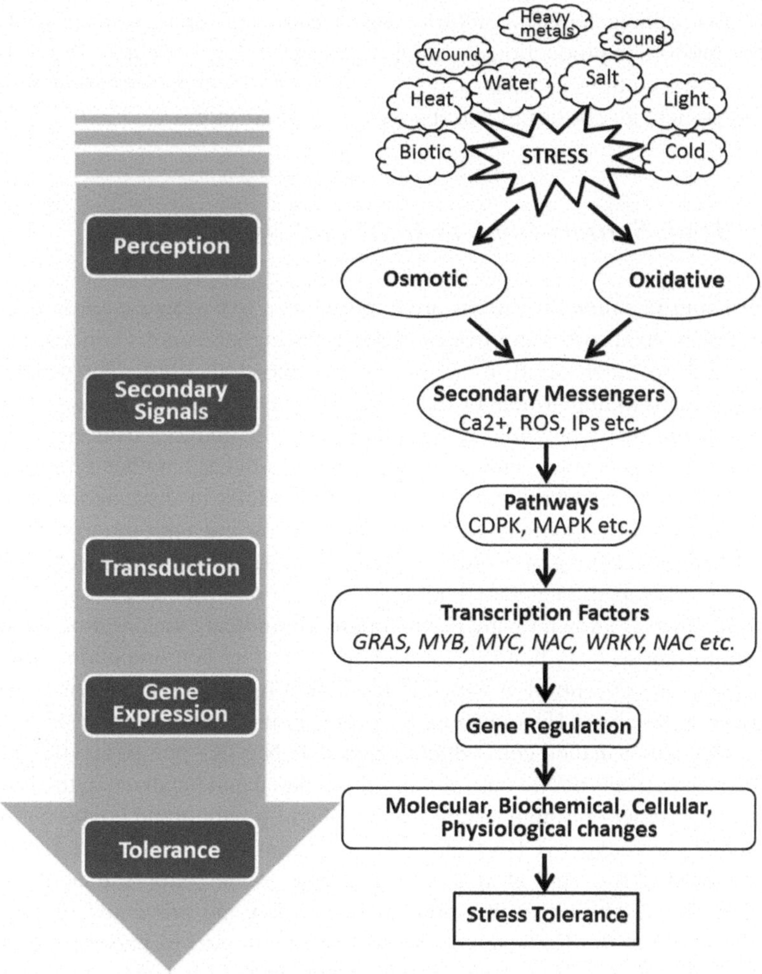

Fig. 20.1 General outline of mechanisms of stress tolerance in orchids

20.3 Abiotic Stress Activates Differential Gene Expression

Abiotic stress is caused by the non-living components of the environment like salt, light, temperature, water, sound, heavy metal etc. Abiotic stress elicits a cascade of intertwined processes at physiological, cellular, biochemical and molecular levels (Grover et al. 2001). All these various types of stresses rarely act individually and their action is almost simultaneous and combined (Lichtenthaler 1996). The plant develops certain adaptations which can be morphological or physiological such as increase in lignin content, cuticle thickness of the leaves, activation of certain signal transduction pathways involved in production of osmoprotectants and antioxidants. The following sections enumerate the various changes involved at molecular, biochemical, cellular and physiological levels during various types of abiotic stress.

20.3.1 High Salinity Induces Ionic and Osmotic Stress

Plants growing in saline conditions are exposed to a hypertonic environment. The roots are unable to absorb water from soil due to its higher osmotic concentration. In turn, this leads to water loss from the plants (Serrano et al. 1999). The water deficit in the plants leads to impairment or reduction of many vital physiological processes. This decrease in water potential in the plant cells also induces osmotic stress and ion toxicity. In a way, one type of stress is manifested in another type of stress. The disruption of the balance in water potential results in disrupted intracellular homeostasis normally maintained by Na+/K+ pumps and inhibition of enzymatic activity. During the course of evolution, plants have devised several mechanisms for instilling salt tolerance. The plant reaction to stress is controlled by the specific gene expression which is known to ameliorate or simply indicate the onset of stress.

One important strategy that salt tolerant species adapt is to maintain a low cytoplasmic Na+ ion concentration which is facilitated by the regulation of expression and activity of K+, Na+, H+ transporters. Under saline conditions, the salt accumulates organic solutes in their cells which manipulate Na+/K+ pumps and balances the ion homeostasis. The salt/Na ions remain accumulated and localized in the roots and its upward movement through xylem is controlled by membrane bound antiporters like cation/H+ exchanger (CHX) and salt overly sensitive (SOS) as in *Arabidopsis* (Maathuis et al. 2014; Hall et al. 2006). In *Dendrobium* sonia 'Earsakul', similar accumulations of salts were observed in root region and prevented its uptake to leaves. In addition, the K+ levels in leaves are maintained to prevent preferential binding of Na+ with the membrane receptors which can hinder the normal growth and development of plant (Abdullakasim et al. 2018).

Another approach would be activation of transcription factors and other signalling kinases. One major plant specific transcription factor gene family is *GRAS*, which play an important role in major physiological pathways and regulatory networks. The *GRAS* genes were upregulated during salt stress in *Dendrobium catenatum* (Zeng et al. 2019). Similarly, *WRKY* transcription factor constitute one of the largest gene family of positive and negative regulators involved in stress responses and other physiological and development processes. In *D. nobile*, *WRKY* was found to be upregulated in salt stress (Xu et al. 2014). However, constitutive expression of *DnWRYK29* from *D. officinale* in tobacco decreased the tolerance of plant to salt stress (Xu et al. 2015). Most of the members of the *WRKY* gene family favoured enhanced tolerance to abiotic stress. A *GmWRKY27* gene was reported to interact with a *GmMYB174* gene to reduce the levels of the negative stress tolerance transcription factor, *GmNAC29*, which manifests in improved salinity tolerance (Wang et al. 2015). A *NAC* gene was isolated and characterized from *D. candidum* (Zhao et al. 2011). A *phosphomannomutase* (*DoPMM*) gene was expressed on NaCl treatment (He et al. 2017a). Ectopic expression in *Arabidopsis* of a bHLH family *DoMYC2* transcription factor, which functions in terpenoid indol alkaloid pathway and involved in stress responses makes it to be salt tolerant (Zhu et al. 2017). A number of other reports on genes affecting salt tolerance to the plants are involved in Phenylpropanoid pathway, carotenoid synthesis and carbohydrate metabolism (Nag and Kumaria 2018; Lee et al. 2012; He et al. 2015; Yu et al. 2017).

20.3.2 Plant Adjustments to Variable Light Intensities

Over the last few decades, depletion of ozone has resulted in high incidence of UV-B radiations. The extreme intensities of light lead to subcellular injuries like DNA damage, chlorophyll degradation and disruption of photosynthetic pathways specially PSII due to the release of manganese ions from oxygen evolving complex (Szymańska et al. 2017). Light stress also leads to increased levels of photo-oxidation which causes accumulation of ROS. Upregulation of ferrodoxin-NADP reductase in *Cymbidium tracyanum* links it to photoprotection by activating the electron cycles (Li et al. 2018). Carotenoids protect the chloroplast against oxidative damage occurring mainly due to extreme light conditions. These are synthesized in MEP pathway in chloroplast. A number of other metabolites of this pathway cross talk with ABA and GA biosynthetic pathways leading to a significant role in plant stress. Flavonoids and other phenolic compounds help in UV-B photo-protection in leaves (Caldwell et al. 1983; Li et al. 1993). Light induces the localization of flavonoids in velamen of the roots which is associated with epiphytic habit (Chomicki et al. 2015). A *phenylalanine ammonia lyase* (*PAL*) transcript, an important member of the phenylpropanoid pathway, in *Vanda coerulea* was found to be upregulated in UV-B and downregulated in dark (Nag and Kumaria 2018). Another *PAL* gene and a *chalcone synthase* (*CHS*) gene expressing in green protocorms may be due to effect of light on regulation of genes involved in flavonoid biosynthesis (Kubasek et al. 1992). In *Phalaenopsis* hybrid, transcription factors like *NAC*, *MYB* and *ARF* were expressed in warm day cool night conditions (Li et al. 2014). Hairpin proteins which are secreted by pathogenic bacteria might be indirectly related to protecting the photosynthetic machinery from light damage by evoking responses like enhanced expression of light stress regulated genes in *Phalaenopsis* (Chuang et al. 2014).

20.3.3 Plants 'Feel the Heat' by Changing Cellular and Biomolecular Environment

With ever changing climate, there is increase of temperature across globe. Plant responds to heat by a number of processes but a comprehensive understanding to these thermotolerance mechanisms is lacking. The consequences of temperature stress are seen in growth and developmental processes like abnormal plant morphology and stunted growth, leaf chlorosis, necrosis and senescence. The reproductive parts of the plant are worst affected with reduced number of floral buds, abnormal growth and fertility of pollen and ovules affecting the overall fruit quality (Cao et al. 2008). Extreme heat stress causes complete impairment of vital processes. All these phenotypic consequences are manifested due to several physiological changes occurring inside the plant like altered cellular water potential and membrane instability leading to loss of cell turgidity, lipid peroxidation, reduction in soluble proteins, decreased photosynthetic rate and carbon assimilation. Accumulation of reactive oxygen species (ROS) causes degradation of various macromolecules present in the cells and ultimately cell death.

Plant avoids stress by producing thick cuticle and sunken stomata, and stomatal closure on the onset of extreme conditions. It has another strategy to tolerate excessive heat by producing various antioxidants (for redox homeostasis and scavenging of free ROS), osmoprotectants (proline, glycine betaine, treahalose), Late embryogenesis abundant (LEA) protein, heat shock proteins (HSPs) etc. These molecules are produced as by-products of various signalling pathways like MAPK (mitogen activating protein kinases) and CDPK (cyclin-dependant protein kinases), which are triggered on the onset of heat stress. A number of genes are upregulated ensuring heat stress indicating their positive role in combating stress. *GRAS* gene family shows higher expression during heat stress in *D. officinale* (Zeng et al. 2019). Similarly, ectopic expression of *DofLEAs* genes from *D. officinale*, increases the tolerance to heat stress in *E. coli* (Ling et al. 2016). The *DoGMP1* and *DoGMP2* genes, expressing during heat stress, contains a heat responsive element, HSE (He et al. 2017b). Production of calcium, methyl jasmonate, and salicylic acid has been reported in heat stress conditions in *Phalaenopsis* seedlings (Huageng et al. 2011). A *Dendrobium nobile DnSIZ1* involved in SUMOlyation, an important post translational modification, increased its expression during high temperature (Liu et al. 2015). Lipid peroxidation has been reported to increase in *Phalaenopsis* under temperature-stress. Temperature stress induces production of active oxygen species scavenging protective enzymes in *Phalaenopsis* (Ali et al. 2005). Researchers have used promoters inducible by cytokinins, desiccation, and heat shock to design transgenics. A *Dendrobium* Sonia *cytokinin oxidase* gene (*DSCKX1*) gene was transferred for combating heat stress (Yang et al. 2003).

20.3.4 Plant Acclimates to Cold Stress by Producing Osmoprotectants

Orchid plants of tropical and sub-tropical origin are more prone to cold temperature. There is impairment in their growth and development processes due to extreme low temperature. Physiological processes like photosynthesis, translocation and respiration are retarded along with production of ROS and membrane instability. The changing temperature regime disturbs the electron transport chain and produces ROS which injures the intracellular structures (Moller 2001). Under prolonged conditions, plants are unable to recover from cold stress. Yuan et al. (2015) reported that low temperatures affect the photosynthetic ability due to degradation of large subunit of RuBisCO in *Phalaenopsis*. This cellular dysfunction was repaired by the production of heat shock proteins (HSPs). Presence of a higher number of HSPs and Resistance (*R*) genes in *Dendrobium catenatum* than *Phalaenopsis equestris* in a comparative study (Zhang et al. 2016a) indicates towards a wider and efficient tolerance to a variety of adverse environments in *D. catenatum*. Similarly, in *P. amabilis*, the leaves became soft and were shed off (Qin et al. 2011). The tolerance to cold stress increases by accumulation of high levels of sugars and energy carriers like ATP which can be immediately utilized to repair the damage to cellular membranes to ensure stress free membrane transport (Dobrota 2007; Fristedt et al. 2015). In *P. aphrodite* plants, ion leakage was directly proportional to the duration of cold conditions (Peng et al. 2014) while this was not the case in flowers as there was accumulation of higher levels of starch and soluble sugars in floral tissues (Chang et al. 2003). A cis-regulatory element, ABRE which was related to responsiveness to ABA and MYC recognition sites were associated with *PaCBF1* gene in *P. aphrodite* (Peng et al. 2014). Qin et al. (2011) reported that a Lipid Transfer Protein (LTP) in *P. amabilis* helped by accumulating several solutes and maintained the optimum electrolyte level within the cells as compared with controls. An ethylene responsive transcription factor 1B showed reduced expression in *Phalaenopsis* cv. 'Red Sky' (Yuan et al. 2015) but previous reports of its upregulation in rice are also there (Cao et al. 2006). A study in *Phalaenopsis*, which is a cold sensitive species due to its sub-tropical origin, shows that these plants counter the stress conditions by restructuring their cytoskeleton and changing carbon metabolism pathways. A proteomic analysis shows the upregulation of cytoskeleton proteins and enzymes in photosynthesis, respiration, protein synthesis, signal transduction like S-adenosylmethionine synthetase (SAMS), phosphoribulokinase, sedoheptulose-1,7-bisphosphate, rubisco activase, pyruvate dehydrogenase E1 beta, triosephosphate isomerise, enolase, SAL1 phosphatase, V-type ATPase etc. The enhanced level of these enzymes help the plant to adapt to low temperature by optimising these processes to maintain a balance of energy supply and carbon sequestration (Yuan et al. 2015). Upregulation of legumin, which is a seed-storage protein which immediately provides carbon skeletons and nitrogen by hydrolysis, to the germinating entities (Wind and Häger 1996), under cold stress was also reported in this study (Yuan et al. 2015). Cold adaptation and cold tolerance are quantitative traits which are strictly regulated by particular gene expression leading to specific enzyme activities and accumulation of specific metabolites (Theocharis et al. 2012; Zhu et al. 2007). A higher expression of genes encoding DREB, zinc finger proteins, LEA, HSPs, adenylate kinase, calcium-transporting ATPase, glycosyl hydrolase family proteins, protein ligases, and membrane transporters during cold adaptation was reported (Wu et al. 2016). *WRKY* genes in *D. officinale* are induced by low temperature stress due to the presence of W-box, ABA responsive and other low temperature responsive elements (Wu et al. 2016; He et al. 2017b). An up-regulation of a number of other genes were observed as a response to cold stress such as *C-repeat binding factor1* (*PaCBF1*) and *dehydrin1* (*PaDHN1*) which have C-repeat (CRT) or dehydration responsive element (DRE) in their promoter regions *in P. aphrodite* (Peng et al. 2014), *VcPAL* transcript in *Vanda coerulea* (Nag and Kumaria 2018), *OncPSY* in *Oncidium* Gower Ramsey (Lee et al. 2012), *DoPMM* gene (He et al. 2017a), *DoGMP1* and *DoGMP2* genes having low-temperature responsive element in *D. officinale* (He et al. 2017b), *DnSIZ1* in *D. nobile* (Liu et al. 2015) and *PaCDPK1* gene in *P. amabilis* (Tsai et al. 2007).

20.3.5 Plant Water Relations and Adaptations by Subcellular Responses

Water stress is basically unavailability of water which can be due to low absorption by the roots or rapid loss by transpiration. The immediate response of the plant is closure of stomata due to ABA accumulation, and which triggers photorespiration process in which utilization of light and carbon is compromised and leads to a decreased chlorophyll content. Water deficit hampers the activity of various photosynthetic enzymes and leads to production of ROS and resultant chloroplast damage and activation of lipid peroxidation Vascular epiphytes are more prone to drought stress than terrestrial ones (Zhang et al. 2015) and therefore have developed specific strategies for drought tolerance (Zhang et al. 2015; Rada and Jaimez 1992). Similarly, orchids have also devised certain morphological and anatomical adaptation to tolerate drought stress. These plants have developed water storage organs like pseudobulbs and thick fleshy leaves to ensure rapid supply of water at minimal loss. The presence of velamen on the roots for easy absorption of atmospheric moisture, thick cuticle with suberized exodermis, dense leaf veins with less and sunken stomata, and extensive root system are other specialized features (Zhang et al. 2016b). On these lines, the drought tolerant *D. wangliangii* has developed water storage organs, succulent deciduous leaves with thick cuticles and reduced stomatal conductance (Zhao et al. 2018). Orchids also utilise facultative CAM pathway to endure drought (Zhang et al. 2014).

Accumulation of ABA and osmoprotectants such as HSP, LEA proteins, proline and mannitol etc. occurs to maintain the osmotic homeostasis and cell turgor. In a terrestrial orchid, *Phaius tankervilliae*, a 9-cis-epoxycarotenoid dioxygenase (*PtNCED1*) gene, was expressed in developing seeds and protocorms under water stress along with accumulation of ABA (Lee et al. 2018). A number of proteins are upregulated to help the plant in combating stress. Upregulation of RuBisCO activase and phosphoglycolate phosphatase in *Cymbidium tracyanum*, catalase in *Cymbidium sinense* and Superoxide dismutase (SOD) and ascorbate peroxidase (APX) in *C. tracyanum* (Li et al. 2018) has been reported. Favouring of oxygenation of RuBP by RuBisCO due to reduced level of intracellular CO_2 by stomatal closure resulting from extreme drought conditions lead to photorespiration (Li et al. 2018). Water deficit stress can also induce the plant to flower as reported in some *Cattleya* species (Cardoso et al. 2010). The DNA methylation levels also decreased in *Dendrobium huoshanense* (Fan et al. 2012). The same study confirms that endogenous NO (Nitric oxide) protects the plant from the oxidative damage by maintaining high relative water content but exogenous supply of NO may inhibit or disrupt the normal physiological metabolism.

Drought tolerance involves interaction of various physiological responses involving stress perception and signal transduction leading to stunted growth, reduced transpiration, osmolyte accumulation, development of lateral root meristem, activation of photosynthetic adaptation and antioxidant pathways (Zhao et al. 2018; Baldoni et al. 2015; Basu et al. 2016; Joshi et al. 2016), all of which are strictly controlled by specific gene expression. *WRKY* gene family is one family which has contrasting roles to play in abiotic stress. In *D. officinale*, the expression of *DnWRKY29* enhances sensitivity to drought (Xu et al. 2015) while another *WRKY* gene, *DnWRKY11* gene helps in enhancing seed tolerance to drought (Xu et al. 2014). A number of WRKYs, NACs, bZIPs, and DREBs were found to be up-regulated in *D. wangliangii* facilitating an improved tolerance to drought (Zhao et al. 2018). A TCA element responsive to salicylic acid, MBS element and CCAAT-box responsive to drought stress were found to be associated with expression of *DoUGE* gene in *D. officinale* (Yu et al. 2017). A *Vanda coerulea PAL* gene showed higher expression in drought stress (Nag and Kumaria 2018).

Various pathways are activated on incidence of drought stress. MAPK6, a key player in ethylene biosynthesis, was reported in *D. wangliangii* indicating the activation of ethylene signalling in drought (Zhao et al. 2018). Lipid metabolism was also triggered to produce lipid transfer proteins and lipases as a drought adaption strategy along with production of R2R3-type MYBs which have a role in development of lateral root meristem (Zhao et al. 2018).

20.3.6 Stress Management by Production of Stress Specific Protein

Sound vibrations (SV) causes mechanical stresses which alter the growth and development patterns of plants by activating certain specialised cellular processes specific signal transduction pathways. The levels of salicylic acid have also been reported to increase during sound stress (Ghosh et al. 2016) in addition to increase in levels of ATP, protective enzymes and hormones, certain osmolytes like soluble proteins, sugar content etc. (Hassanien et al. 2014; da Teixeira Silva and Dobránszki 2010; Mishra et al. 2016). Sound stress has also been reported in *D. candidum*, where the SOD, CAT, POD and APX activities increased upon incidence of stressful sound vibrations. The resultant oxidative stress also led to activation of lipid peroxidation (Li et al. 2008). On the other hand, there have been favourable reports in *D. officinale* where sound at ultrasonic levels have promoted shoot production in micropropagation (Wei et al. 2012).

Wounding attracts pathogens infestation at the wound site facilitating their entry into the plant. As a result, plants have evolved constitutive (presence of physical barriers like cuticle/lignin, toxins to deter the wound causing organism like herbivores etc.), rapid induced (oxidative reaction and defense-related gene expression) and late induced mechanisms (callose deposition, accumulation of proteinase inhibitors and hydrolytic enzymes) of defence to combat this stress and maintain a healthy microenvironment of the plant (Savatin et al. 2014). A *PaPTP1* (protein tyrosine phosphorylation) transcript, involved in tyrosine phosphorylation and dephosphorylation which play a role in ABA signalling, increased in wound treated leaves in *P. amabilis* (Fu et al. 2011). The activation of specific MAP kinases on wounding also suggests the involvement of tyrosine phosphorylation in intracellular signal transduction (Fu et al. 2011). A number of proteins are also reported to be expressed in wounding like PAL in *Vanda coerulea* (Nag and Kumaria 2018), PaCDPK1 in *P. amabilis* (Tsai et al. 2007), DnSIZ1 in *D. nobile* (Liu et al. 2015) which indicate stress management by the plants.

Accumulation of heavy metals (HMs) above the optimum levels can lead to toxicity due to abnormal binding with several vital macromolecules like DNA and nuclear proteins. This leads to production of ROS (Emamverdian et al. 2015). In *Bipinnula fimbriata* plantlets growing in HMs soils, a higher root growth was observed (Herrera et al. 2018). In non mycorrhizal roots, epoxycarotenoid dioxygenase and ABA was detected which may function as a response controller against HMs stress and contribute to decrease in plant toxicity (Bücker-Neto et al. 2017). HMs stress induces complete change in the proteome of mycorrhizal and non-mycorrhizal roots segments in *B. fimbriata* which facilitate the establishment of plant in polluted habitat (Herrera et al. 2018).

20.4 Plant-Microbe Interactions Evokes Specialised Reactions During Biotic Stress

Viruses are one of the major pathogen affecting orchids. The threat to these are high specially in commercial production of orchids for floricultural purposes. Out of more than 50 viruses affecting orchids, Cymbidium mosaic virus (CymMV) and Odontoglossum ringspot virus (ORSV) are the most common ones worldwide. Plants have evolved specialized pathways against biotic stress and these include ROS production, defence gene activation and phytohormone responses (Hou et al. 2009; Pieterse et al. 2009). Systematic acquired resistance (SAR) is a potent broad-spectrum defence mechanism that gives protection to the plant by activating salicylic acid (SA) signalling. *PhaPR1* is a marker for efficient SA signalling (Chen et al. 2013). Resistance to CymMV was successfully achieved by transformation using the coat protein (CP) gene in *Phalaenopsis* (Liao et al. 2004) and *Dendrobium* (Chang et al. 2005). In *P. amabilis*, upregulation of certain functional proteins like ATP sulfurylase, thioredoxin H-type, RuBisCO larger subunit, NADP specific isocitrate dehydrogenase, cinnamoyl CoA reductase etc. leads to better stress tolerance (Lai et al. 2013).

Very long chain fatty acids (VLCFAs) help in biosynthesis of cuticle and sphingolipids which are thought to provide a barrier against pathogens which colonize the surface (Reina-Pinto and Yephremov 2009) and contribute to plant innate immunity. This includes cues for elicitor in defence mechanisms and programmed cell death (Takahashi et al. 2009). These signals may affect activity of transcription factors like *MYB*, *bZIP* and *WRKY*. The VLCFA biosynthesis pathway is also activated in orchid-*Erwinia* interaction, where *PaECR1*-silenced plants were more susceptible (Fu et al. 2012). *WRKY* genes are associated with biotic stress in *Phalaenopsis amabilis*. Here, ROS accumulation occurs at the site of pathogenesis mainly due to decrease in peroxidase activity (Fu et al. 2012). Induction of calcium-dependent protein kinases (CDPK) was triggered after infection with *Erwinia chrysanthemi* (Tsai et al. 2007). Another gene, sweet pepper ferredoxin-like protein (*pflp*) is also a potential gene to confer resistance against soft rot disease caused by *Erwinia carotovora* in orchids (Liau et al. 2003). An infection by *Pectobacterium* spp. causing bacterial soft rot in *Oncidium*, causes calcium ion influx, hydrogen peroxide accumulation, and NADPH oxidase activation (Lin et al. 2015). Gene stacking in *Phalaenopsis* with transforming a CymMV CP gene and an antimicrobial *pflp* gene confers resistant to both CymMV and soft-rot disease produced by *Pectobacterium carotovora* (Chan et al. 2005). The pathogen interacts with the plant through special molecules called pathogen associated molecular patterns (PAMPs) such as flagellin and lipopolysaccharides from bacteria, and chitin and ergosterol from fungi (Fu et al. 2012). Hairpin proteins, secreted by the pathogenic bacteria, increases the resistance to plant diseases by activating PAMP-induced immunity (Chuang et al. 2014).

Fungus-induced calcium-dependent protein kinases (CDPKs) have an important role in regulation of *D. officinale* symbiosis (Zhao et al. 2013). Accumulation of *PAL* and *chalcone synthese* (*CHS*) may be associated with production of defence molecules (phytoalexins) and signal molecules (Schenkluhn et al. 2010). A higher *PAL* expression was noticed in symbiotic protocorms of *Orchis morio* (Beyrle et al. 1995). *PAL*, *CHS* and *naringenin-3-dioxygenase* were detected in green protocorms of *Oncidium sphacelatum* (Valadares et al. 2014). Phytoalexins are known restrict fungal growth in photosynthetic tissues in orchids (Shimura et al. 2007). Fungus-responsive elements like EIRE and ELI-box3 were predicted from the promoter region of *WRKY* genes in *D. officinale* justified its receptivity towards favoured seed germination in the presence of *Tulasnella*-like mycorrhizal fungus (Wang et al. 2018). Similarly, *WRKY* genes were induced in roots of *Cymbidium hybridum* upon symbiotic association with *Tulasnella*-like *Rhizoctonia* and *Umbelopsis* (Zhao et al. 2014). *DoWRKY* genes are also important for establishing a symbiotic relationship with *Sebacina* sp. (Zhao et al. 2013). An antagonistic relationship of orchid mycorrhiza is given in a study on *Serapias vomeracea* and a mycorrhizal fungus, *Tulasnella calospora* which provides no evidence of any strong defence activation due to this association, however, some nodulin-like *ENODL* genes, were significantly up-regulated (Perotto et al. 2014). A number of *ENODL* genes also get induced in response to arbuscular mycorrhizal fungi inoculation (Harrison 1998). A high expression of β-tubulin, xanthine dehydrogenase, and DEAD-box ATP RNA helicase fortifies *Oncidium sphacelatum* against *Thanatephorus* sp. (RG26 strain) (López-Chávez et al. 2016).

20.5 Fluctuations of Water Potential Causes Osmotic Stress

Altering environmental water potential leads to disrupting of normal cellular functioning causes osmotic stress to plants. As a result, a number of changes at phenotypic, metabolic, cellular and molecular levels are established (Hasegawa et al. 2000). These responses are activated by primary or secondary osmotic stress signals. The secondary signals can be phytohormones (ABA, ethylene), reactive oxygen species (ROS) and intracellular second messengers (phospholipids) (Xiong and Zhu 2002). Osmolytes are the soluble molecules [carbohydrates (Da Silva and Arrabaça 2004), betaine (Wyn Jones and Storey 1981) and proline (Aspinall and Paleg 1981)] which function to optimize the cellular osmotic potential by stabilizing the intracellular salt concentrations. Their production is obligatory to all living cells under the influence of hyperosmotic stress (Yancey et al. 1982). Dehydration also triggers ABA biosynthesis which activates a specific set of genes which are induced by various stresses like drought, salt, and cold stresses, and also by ABA (Boudsocq 2005). Water loss causes decrease in cell volume leading to loss of cell turgor which causes decrease in leaf expansion and plant growth rate. The osmotic potential of cell sap decreases to compensate for cell hydration. Osmolytes come into picture at this stage and are produced to counter the loss of cell turgor (Beck et al. 2007). Drought stress causes mainly osmotic stress whereas salt stress manifests in ion toxicity. Cold stress leads to drought condition but also impacts the enzyme activities which are highly temperature specific.

Osmoprotectants, the small molecules acting as osmolytes, maintain the cell turgor during osmotic stress. Soluble sugars such as glucose, fructose, and oligosaccharides (raffinose and stachyose) accumulate in cells during these conditions (Qin et al. 2011) and protect cellular membranes from injury by dehydration and freezing, and ensure membrane stability (Anchordoguy et al. 1987). Proline also protects the membranes and proteins against the adverse effects of temperature extremes (Santarius 1992; Santoro et al. 1992). In *D. officinale*, metabolites like proline, alanine, isoleucine, aspartic acid, sucrose etc. accumulated during cold stress (Wu et al. 2016). However, in *D. wangliangii*, proline biosynthesis genes were not activated during drought stress but trehalose biosynthesis gene was upregulated (Zhao et al. 2018). An enhanced expression of *SAMS* which is a precursor for ethylene and polyamine biosynthesis was observed in cold treated *Phalaenopsis* (Yuan et al. 2015).

Water soluble polysaccharides also serve as osmolytes to maintain the osmotic balance of plants (Iraki et al. 1989). A transgenic with *DoUGE* (UDP glucose 4-epimerase) gene which affects sugar metabolism, enhanced the tolerance to salt and drought stress by bioaccumulation of these polysaccharide (Yu et al. 2017). Cell wall mannan polysaccharides contribute in maintaining cell wall integrity by minimizing cellular damage during stress (He et al. 2015). In *D. officinale*, a number of genes of the mannan biosynthetic pathway like *DoNI* (*alkaline/neutral invertase*; Gao et al. 2016), *DoUGP* (*UDP-glucose pyrophosphorylase*; Wan et al. 2017), *DoPMM* (*phosphomannomutase*; He et al. 2017a), *DoCSLA1-8 (cellulose synthase like A*; He et al. 2015) have been isolated and characterized which improved the osmotic stress tolerance. Accumulation of soluble sugars gave higher freezing tolerance to *D. huoshanense* (Jin et al. 2016).

LEA proteins also have osmoprotective function to counter against various environmental stresses. These offer protection against desiccation by showing antioxidant activity, hydration buffering, ion binding, stabilization of membranes and protein, and nucleic acid interactions (Tunnacliffe and Wise 2007; Hong-Bo et al. 2005). Expression of LEA proteins in *D. candidum* gave the plants increased tolerance to salinity stress (Yang et al. 2010). Similarly, *DoLEA*s expression in *D. officinale* promotes stress tolerance (Ling et al. 2016). LEA proteins were also observed in *D. wangliangii* during drought stress (Zhao et al. 2018). Transposable elements lead to genome restructuring as a reaction to environmental stresses. In *D. officinale*, a number of Tyl-copia retrotransposons were activated under osmotic stress treatments (Gao 2016).

20.6 Elucidation of Key Players During Oxidative Stress

The major consequence of various biotic and abiotic stresses is the reactive oxygen species (ROS) which are produced mainly in chloroplasts as a result of reduction of oxygen during erroneous metabolism. The accumulation of ROS creates a toxic microenvironment in the cells leading to degradation of macromolecules like proteins, lipids, nucleic acids, damaging to photosynthetic pigments and cellular membranes (Alscher et al. 1997). The plant defence to overcome this is to counter these free radicals by activating the production of antioxidants which remain the key regulators of stress control. Antioxidants can be divided into two broad categories: enzymatic and non-enzymatic. Enzymatic antioxidants are ascorbate peroxidase (APX), catalase (CAT), dehydroascorbate reductase (DHAR), glutathione peroxidase (GPX), glutathione reductase (GR), guaiacol peroxidase (GP), glutathione-S-transferase (GST), monodehydroascorbate reductase (MDHAR) and superoxide dismutase (SOD). The non-enzymatic ones include ascorbic acid (AA), carotenoids, flavonoids, osmolyte proline, phenols, reduced glutathione and α-tocopherol (Smirnoff 1995; Meyer et al. 2012; Miller et al. 2010).

In a study on *Oncidium sphacelatum*, accumulation of SOD and APX was observed in green protocorms (Valadares et al. 2014). In *Cyperipedium* sp., an *SOD* gene was involved stress tolerance related signalling pathway (Guo et al. 2018). Greater expression of *APX* was also recorded in *Cymbidium tracyanum* in a similar study (Li et al. 2018). Enhanced activities of enzymatic antioxidants like CAT, SOD etc. was observed in *DnWRKY11* transgenic tobacco showing better phenotypic characters than the wild type under stress conditions; lower content of malondialdehyde (MDA) was also observed (Xu et al. 2014). MDA levels are proposed to be heightened leading to damage due to stress in *Dendrobium huoshanense* (Fan et al. 2012). Similar increased MDA levels are reported in *Phalaenopsis* during temperature-induced stress (Ali et al. 2005).

Carotenoids are synthesized in MEP pathway occurring in plastids and help in fortification of these organelles during oxidative damage especially by extreme light conditions, by quenching of singlet and triplet chlorophyll states (Moise et al. 2014). Carotenoid derived ABA and GA, are also produced in this pathway and contribute significantly to plant stress responses (Nambara and Marion-Poll 2005). In *Oncidium*, a *phytoene synthase* (*PSY*) gene was characterized, which is a pivotal enzyme in biosynthesis of carotenoids (Lee et al. 2012). Interestingly, the expression of this gene was also observed in roots which can be corroborated the fact that orchid roots are photosynthetic and carotenoids are vital for guarding chlorophylls against oxidative damage by photo-oxidation (Britton et al. 1995; Huang et al. 2009). In a study on *Oncidium sphacelatum*, 9-cis-epoxycarotenoid dioxgenase (NCED) and carotenoid cleavage dioxygenase (CCD), key enzymes in ABA biosynthesis were observed in green protocorm than achlorophyllous protocorm (Valadares et al. 2014). A *PtNCED* gene was also found to be upregulated in water stressed protocorms of *Phaius tankervilliae* (Lee et al. 2018). Flavonoids are also known to protect plants against stress especially light. In epiphytic orchids, root velamen is site of flavonoid accumulation as this prevents the light damage to chlorophyllous roots (Chomicki et al. 2015). PAL, CHS and naringenin-3-dioxygenase, important enzymes of flavonoid pathways were observed in *Oncidium sphacelatum* under biotic stress (Valadares et al. 2014). Jasmonates are a significant part of the hormonal network of oxidative stress responses and can help in ameliorating a number of stresses specially drought and salinity by enhancing the activity of antioxidant enzymes (Riemann et al. 2015). Increase in jasmonic acid (JA) levels was observed in *Cymbidium sinense* in water deficit conditions (Li et al. 2018). JA was also reported to regulate the proliferation of orchid mycorrhiza (Valadares et al. 2014). *PbbHLH4*, a transcription factor regulates biosynthesis of floral monoterpenes in *Phalaenopsis bellina* mainly to attract pollinators but it may also have a role in stress responses (Chuang et al. 2018). A bibenzyl synthase, which has an established role in phytoalexin biosynthesis in orchids (Reinecke and Kindl 1994), was produced in green protocorms, suggesting its role in regulation of orchid mycorrhizae at the cellular level and a probable role in non-mycorrhizal interactions also (Valadares et al. 2014). Both excess light and drought stress caused oxidative stress for *D. wangliangii* (Zhao et al. 2018). Polysaccharides also act as ROS scavengers (Duan and Kasper 2011; Fry et al. 2015) as has also been reported in studies on *Dendrobium*, where the polysaccharides scavenge ROS and show antioxidant activity leading to stress resistance (Fan et al. 2009). It has been reported that NO retards xanthine oxide mediated responses in *Phalaenopsis* (Tewari et al. 2009).

20.7 Conclusions

Stress is one of the global factors responsible for delimiting plant growth and yield. There have been a lot of efforts for devising strategies for ameliorating stress tolerance amongst plants. Molecular approaches are the most important tools for identification and management of key players involved in stress responses. Stress is a quantitative trait and involves multitude of genes influencing cascade of subcellular processes leading to stress perception, response and tolerance. Although a quantum of work has been done on stress in different crop plants, initiatives on stress management in orchids is not commensurate with the potential of this unique plant group. This chapter provides insights to various molecular interventions reported in orchids to ameliorate different types of stresses and will pave way for planning of future initiatives for stress engineering in orchids.

Acknowledgements JKS is grateful to CSIR for research support No. 38(1443)/17/EMR-II dated 25/5/17. DG is grateful for Junior Research Fellowship (File No. 09/135(0809)/2018-EMR-I) by CSIR. We are also grateful to Department of Science and Technology, Government of India for partial financial support under Promotion of University Research and Scientific Excellence (PURSE) grant scheme.

References

Abdullakasim S, Kongpaisan P, Thongjang P, Saradhuldhat P (2018) Physiological responses of potted *Dendrobium* orchid to salinity stress. Hortic Environ Biotechnol 59:491–498. https://doi.org/10.1007/s13580-018-0057-4

Ali MB, Hahn EJ, Paek KY (2005) Effects of temperature on oxidative stress defense systems, lipid peroxidation and lipoxygenase activity in *Phalaenopsis*. Plant Physiol Biochem 43:213–223. https://doi.org/10.1016/j.plaphy.2005.01.007

Alscher RG, Donahue JL, Cramer CL (1997) Reactive oxygen species and antioxidants: relationships in green cells. Physiol Plant 100:224–233. https://doi.org/10.1034/j.1399-3054.1997.1000203.x

Anchordoguy TJ, Rudolph AS, Carpenter JF, Crowe JH (1987) Modes of interaction of cryoprotectants with membrane phospholipids during freezing. Cryobiology 24:324–331. https://doi.org/10.1016/0011-2240(87)90036-8

Aspinall D, Paleg LG (1981) Proline accumulation: physiological aspects. In: Paleg LG, Aspinall D (eds) The physiology and biochemistry of drought resistance in plants. Academic Press, Sidney, pp 205–241

Baldoni E, Genga A, Cominelli E (2015) Plant MYB transcription factors: their role in drought response mechanisms. Int J Mol Sci 16:15811–15851. https://doi.org/10.3390/ijms160715811

Basu S, Ramegowda V, Kumar A, Pereira A (2016) Plant adaptation to drought stress. F1000Research 5:1554. https://doi.org/10.12688/f1000research.7678.1

Beck EH, Fettig S, Knake C, Hartig K, Bhattarai T (2007) Specific and unspecific responses of plants to cold and drought stress. J Biosci 32:501–510. https://doi.org/10.1007/s12038-007-0049-5

Beyrle HF, Smith SE, Franco C, Peterson RL (1995) Colonization of *Orchis morio* protocorms by a mycorrhizal fungus: effects of nitrogen nutrition and glyphosate in modifying the responses. Can J Bot 73:1128–1140. https://doi.org/10.1139/b95-123

Boudsocq M (2005) Osmotic signaling in plants. Multiple pathways mediated by emerging kinase families. Plant Physiol 138:1185–1194. https://doi.org/10.1104/pp.105.061275

Britton G, Liaaen-Jensen S, Pfander H (1995) Carotenoids. In: Biosynthesis and metabolism, vol 3. Birkhäuser Verlag, Switzerland, p 414

Bücker-Neto L, Paiva ALS, Machado RD, Arenhart RA, Margis-Pinheiro M (2017) Interactions between plant hormones and heavy metals responses. Genet Mol Biol 40:373–386. https://doi.org/10.1590/1678-4685-GMB-2016-0087

Caldwell MM, Robberecht R, Flint SD (1983) Internal filters: prospects for UV-acclimation in higher plants. Physiol Plant 58:445–450. https://doi.org/10.1111/j.1399-3054.1983.tb04206.x

Cao Y, Song F, Goodman RM, Zheng Z (2006) Molecular characterization of four rice genes encoding ethylene-responsive transcriptional factors and their expressions in response to biotic and abiotic stress. J Plant Physiol 163:1167–1178. https://doi.org/10.1016/j.jplph.2005.11.004

Cao YY, Duan H, Yang LN, Wang ZQ, Zhou SC, Yang JC (2008) Effect of heat stress during meiosis on grain yield of rice cultivars differing in heat tolerance and its physiological mechanism. Acta Agron Sin 34:2134–2142. https://doi.org/10.1016/S1875-2780(09)60022-5

Cardoso JC, Ono EO, Rodrigues JD (2010) Gibberellic acid and water regime in the flowering induction of *Brassocattleya* and *Cattleya* hybrid orchids. Hortic Bras 28:395–398. https://doi.org/10.1590/S0102-05362010000400003

Chan YL, Lin KH, Sanjaya Liao LJ, Chen WH, Chan MT (2005) Gene stacking in *Phalaenopsis* orchid enhances dual tolerance to pathogen attack. Transgenic Res 14:279–288. https://doi.org/10.1007/s11248-005-0106-5

Chang CH, Lee N, Chang TH (2003) Flower and bud wilting of potted *Phalaenopsis* caused by ethylene and darkness. J Chin Soc Hort Sci 49:173–182

Chang C, Chen YC, Hsu YH, Wu JT, Hu CC, Chang WC, Lin NS (2005) Transgenic resistance to Cymbidium mosaic virus in *Dendrobium* expressing the viral capsid protein gene. Transgenic Res 14:41–46. https://doi.org/10.1007/s11248-004-2373-y

Chen JC, Lu HC, Chen CE, Hsu HF, Chen HH, Yeh HH (2013) The NPR1 ortholog PhaNPR1 is required for the induction of PhaPR1 in *Phalaenopsis aphrodite*. Bot Stud 54:31. https://doi.org/10.1186/1999-3110-54-31

Chomicki G, Bidel LPR, Ming F, Coiro M, Zhang X, Wang Y, Baissac Y, Jay-Allemand C, Renner SS (2015) The velamen protects photosynthetic orchid roots against UV-B damage, and a large dated phylogeny implies multiple gains and losses of this function during the Cenozoic. New Phytol 205:1330–1341. https://doi.org/10.1111/nph.13106

Chuang HW, Chang PY, Syu YY (2014) Harpin protein, an elicitor of disease resistance, acts as a growth promoter in *Phalaenopsis* orchids. J Plant Growth Regul 33:788–797. https://doi.org/10.1007/s00344-014-9425-1

Chuang YC, Hung YC, Tsai WC, Chen WH, Chen HH (2018) PbbHLH4 regulates floral monoterpene biosynthesis in *Phalaenopsis* orchids. J Exp Bot 69:4363–4377. https://doi.org/10.1093/jxb/ery246

Da Silva JM, Arrabaça MC (2004) Photosynthesis in the water-stressed C4 grass *Setaria sphacelata* is mainly limited by stomata with both rapidly and slowly imposed water deficits. Physiol Plant 121:409–420. https://doi.org/10.1111/j.1399-3054.2004.00328.x

da Teixeira Silva J, Dobránszki J (2010) Sonication and ultrasound: impact on plant growth and development. Plant Cell Tissue Organ Cult 117:131–143. https://doi.org/10.1007/s11240-014-0429-0

Dobrota C (2007) Energy dependant plant stress acclimation. In: Amils R, Ellis-Evans C, Hinghofer-Szalkay H (eds) Life in extreme environments. Springer, Dordrecht, Netherlands, pp 277–285. https://doi.org/10.1007/978-1-4020-6285-8_17

Duan J, Kasper DL (2011) Oxidative depolymerization of polysaccharides by reactive oxygen/nitrogen species. Glycobiology 21:401–409. https://doi.org/10.1093/glycob/cwq171

Emamverdian A, Ding Y, Mokhberdoran F, Xie Y (2015) Heavy metal stress and some mechanisms of plant defense response. Sci World J 2015:756120. https://doi.org/10.1155/2015/756120

Fan Y, He XJ, Zhou S, Luo A, He T, Chun Z (2009) Composition analysis and antioxidant activity of polysaccharide from *Dendrobium denneanum*. Int J Biol Macromol 45:169–173. https://doi.org/10.1016/j.ijbiomac.2009.04.019

Fan H, Li T, Guan L, Li Z, Guo N, Cai Y, Lin Y (2012) Effects of exogenous nitric oxide on antioxidation and DNA methylation of *Dendrobium huoshanense* grown under drought stress. Plant Cell Tissue Organ Cult 109:307–314. https://doi.org/10.1007/s11240-011-0096-3

Fristedt R, Martins NF, Strenkert D, Clarke CA, Suchoszek M, Thiele W, Schöttler MA, Merchant SS (2015) The thylakoid membrane protein CGL160 supports CF1CF0 ATP synthase accumulation in *Arabidopsis thaliana*. PLoS ONE 10:e0121658. https://doi.org/10.1371/journal.pone.0121658

Fry SC, Dumville JC, Miller JG (2015) Fingerprinting of polysaccharides attacked by hydroxyl radicals in vitro and in the cell walls of ripening pear fruit. Biochem J 357:729–737. https://doi.org/10.1042/bj3570729

Fu SF, Lin CW, Kao TW, Huang DD, Huang HJ (2011) *PaPTP1*, a gene encoding protein tyrosine phosphatase from orchid, *Phalaenopsis amabilis*, is regulated during floral development and induced by wounding. Plant Mol Biol Rep 29:106–116. https://doi.org/10.1007/s11105-010-0216-y

Fu SF, Tsai TM, Chen YR, Liu CP, Haiso LJ, Syue LH, Yeh HH, Huang HJ (2012) Characterization of the early response of the orchid, *Phalaenopsis amabilis*, to *Erwinia chrysanthemi* infection using expression profiling. Physiol Plant 145:406–425. https://doi.org/10.1111/j.1399-3054.2012.01582.x

Gao Y (2016) Activation of Ty1-copia group retrotransposons of *Dendrobium officinale* under abiotic stress conditions. Hereditary Genet 5:2161–1041. https://doi.org/10.4172/2161-1041.1000161

Gao F, Cao XF, Si JP, Chen ZY, Duan CL (2016) Characterization of the alkaline/neutral invertase gene in *Dendrobium officinale* and its relationship with polysaccharide accumulation. Genet Mol Res 15:1–8. https://doi.org/10.4238/gmr.15027647

Ghosh R, Mishra RC, Choi B, Kwon YS, Bae DW, Park S-C, Jeong M-J, Bae H (2016) Exposure to sound vibrations lead to transcriptomic, proteomic and hormonal changes in *Arabidopsis*. Sci. Rep. 6:33370. https://doi.org/10.1038/srep37484

Grover A, Kapoor A, Lakshmi OS, Agarwal S, Sahi C, Katiyar-Agarwal S, Agarwal M, Dubey H (2001) Understanding molecular alphabets of the plant abiotic stress responses. Curr Sci 80:206–216. https://doi.org/10.2307/24104280

Guo YY, Zhang YQ, Zhang GQ, Huang LQ, Liu ZJ (2018) Comparative transcriptomics provides insight into the molecular basis of species diversification of section Trigonopedia (*Cypripedium*) on the Qinghai-Tibetan Plateau. Sci. Rep. 8:11640. https://doi.org/10.1038/s41598-018-30147-9

Hall D, Evans AR, Newbury HJ, Pritchard J (2006) Functional analysis of CHX21: a putative sodium transporter in *Arabidopsis*. J Exp Bot 57:1201–1210. https://doi.org/10.1093/jxb/erj092

Harrison MJ (1998) Development of the arbuscular mycorrhizal symbiosis. Curr Opin Plant Biol 1:360–365. https://doi.org/10.1016/1369-5266(88)80060-8

Hasegawa PM, Bressan RA, Zhu JK, Bohnert HJ (2000) Plant cellular and molecular responses to high salinity. Annu Rev Plant Physiol Plant Mol Biol 51:463–499. https://doi.org/10.1146/annurev.arplant.51.1.463

Hassanien RHE, Hou T, Li Y, Li B (2014) Advances in effects of sound waves on plants. J Integr Agric 13:335–348. https://doi.org/10.1016/S2095-3119(13)60492-X

He C, Zhang J, Liu X, Zeng S, Wu K, Yu Z, Wang X, da Teixeira Silva JA, Lin Z, Duan J (2015) Identification of genes involved in biosynthesis of mannan polysaccharides in *Dendrobium officinale* by RNA-seq analysis. Plant Mol Biol 88:219–231. https://doi.org/10.1007/s11103-015-0316-z

He C, Zeng S, da Teixeira Silva JA, Yu Z, Tan J, Duan J (2017a) Molecular cloning and functional analysis of the *phosphomannomutase* (*PMM*) gene from *Dendrobium officinale* and evidence for the involvement of an abiotic stress response during germination. Protoplasma 254:1693–1704. https://doi.org/10.1007/s00709-016-1044-1

He C, Yu Z, da Teixeira Silva JA, Zhang J, Liu X, Wang X, Zhang X, Zeng S, Wu K, Tan J, Ma G, Luo J, Duan J (2017b) DoGMP1 from *Dendrobium officinale* contributes to mannose content of water-soluble polysaccharides and plays a role in salt stress response. Sci Rep 7:41010. https://doi.org/10.1038/srep41010

He C, da Teixeira Silva JA, Tan J, Zhang J, Pan X, Li M, Luo J, Duan J (2017c) A genome-wide identification of the *WRKY* family genes and a survey of potential *WRKY* target genes in *Dendrobium officinale*. Sci Rep 7:9200. https://doi.org/10.1038/s41598-017-07872-8

Herrera H, Valadares R, Oliveira G, Quiroz A, Almonacid Muñoz L, Nascimento S, Bashan Y, Arriagada C (2018) Adaptation and tolerance mechanisms developed by mycorrhizal *Bipinnula fimbriata* plantlets (Orchidaceae) in a heavy metal-polluted ecosystem. Mycorrhiza 28:651–663. https://doi.org/10.1007/s00572-018-0858-4

Hong-Bo S, Zong-Suo L, Ming-An S (2005) LEA proteins in higher plants: structure, function, gene expression and regulation. Colloids Surfaces B Biointerfaces 45:131–135. https://doi.org/10.1016/j.colsurfb.2005.07.017

Hou S, Yang Y, Zhou JM (2009) The multilevel and dynamic interplay between plant and pathogen. Plant Signal Behav 4:283–293. https://doi.org/10.4161/psb.4.4.8155

Huageng Y, Suliang Y, Huijuan C, Chongfa Y, Fusun Y, Zifan L (2011) Effect of exogenous methyl jasmonate, calcium and salicylic acid on the heat tolerance in *Phalaenopsis* seedlings under high temperature stress. Chin Agric Sci Bull 28:029

Huang JZ, Cheng TC, Wen PJ, Hsieh MH, Chen FC (2009) Molecular characterization of the *Oncidium* orchid *HDR* gene encoding 1-hydroxy-2-methyl-2-(E)-butenyl 4-diphosphate reductase, the last step of the methylerythritol phosphate pathway. Plant Cell Rep 28:1475–1486. https://doi.org/10.1007/s00299-009-0747-6

Iraki NM, Bressan RA, Carpita NC (1989) Extracellular polysaccharides and proteins of tobacco cell cultures and changes in composition associated with growth-limiting adaptation to water and saline stress. Plant Physiol 91:54–61. https://doi.org/10.1104/pp.91.1.54

Jin Q, Jiao C, Sun S, Song C, Cai Y, Lin Y, Fan H, Zhu Y (2016) Metabolic analysis of medicinal *Dendrobium officinale* and *Dendrobium huoshanense* during different growth years. PLoS ONE 11:1–17. https://doi.org/10.1371/journal.pone.0146607

Joshi R, Wani SH, Singh B, Bohra A, Dar ZA, Lone AA, Pareek A, Singla-Pareek SL (2016) Transcription factors and plants response to drought stress: current understanding and future directions. Front Plant Sci 7:1029. https://doi.org/10.3389/fpls.2016.01029

Kubasek WL, Shirley BW, McKillop A, Goodman HM, Briggs W, Ausubel FM (1992) Regulation of flavonoid biosynthetic genes in germinating *Arabidopsis* seedlings. Plant Cell 4:1229–1236. https://doi.org/10.1105/tpc.4.10.1229

Lai T, Deng Y, Zhang P, Chen Z, Hu F, Zhang Q, Hu Y, Shi N (2013) Proteomics-based analysis of *Phalaenopsis amabilis* in response toward Cymbidium mosaic virus and/or Odontoglossum ringspot virus infection. Am. J. Plant Sci. 04:1853–1862. https://doi.org/10.4236/ajps.2013.49228

Lee W, Huang J, Chen F (2012) Ectopic expression of *Oncidium* gower ramsey phytoene synthase gene caused pale flower and reduced leaf size in transgenic tobacco. J Int Coop 7:133–150

Lee Y-I, Chen M-C, Lin L, Chung M-C, Leu W-M (2018) Increased expression of 9-Cis-epoxycarotenoid dioxygenase, *PtNCED1*, associated with inhibited seed germination in a terrestrial orchid, *Phaius tankervilliae*. Front Plant Sci 9:1043. https://doi.org/10.3389/fpls.2018.01043

Li J, Ou-Lee TM, Raba R, Amundson RG, Last RL (1993) *Arabidopsis* flavonoid mutants are hypersensitive to UV-B irradiation. Plant Cell 5:171–179. https://doi.org/10.1105/tpc.5.2.171

Li B, Wei J, Wei X, Tang K, Liang Y, Shu K, Wang B (2008) Effect of sound wave stress on antioxidant enzyme activities and lipid peroxidation of *Dendrobium candidum*. Colloids Surfaces B Biointerfaces 63:269–275. https://doi.org/10.1016/j.colsurfb.2007.12.012

Li DM, Lǚ FB, Zhu GF, Sun YB, Xu YC, Jiang MD, Liu JW, Wang Z (2014) Identification of warm day and cool night conditions induced flowering-related genes in a *Phalaenopsis* orchid hybrid by suppression subtractive hybridization. Genet Mol Res 13:7037–7051. https://doi.org/10.4238/2014.February.14.7

Li JW, Chen XD, Hu XY, Ma L, Zhang SB (2018) Comparative physiological and proteomic analyses reveal different adaptive strategies by *Cymbidium sinense* and *C. tracyanum* to drought. Planta 247:69–97. https://doi.org/10.1007/s00425-017-2768-7

Liao LJ, Pan IC, Chan YL, Hsu YH, Chen WH, Chan MT (2004) Transgene silencing in *Phalaenopsis* expressing the coat protein of Cymbidium Mosaic Virus is a manifestation of RNA-mediated resistance. Mol Breed 13:229–242. https://doi.org/10.1023/B:MOLB.0000022527.68551.30

Liau C-H, Lu J-C, Prasad V, Hsiao H, You S-J, Lee J, Yang N-S, Huang H-E, Feng T-Y, Chen W, Chan M-T (2003) The Sweet pepper ferredoxin-like protein (pflp) conferred resistance against soft rot disease in *Oncidium* orchid. Transgenic Res 12:329–336. https://doi.org/10.1023/A:1023343620729

Lichtenthaler HK (1996) Vegetation stress: an introduction to the stress concept in plants. J Plant Physiol 148:4–14. https://doi.org/10.1016/s0176-1617(96)80287-2

Lin YH, Lee PJ, Shie WT, Chern LL, Chao YC (2015) Pectobacterium chrysanthemi as the dominant causal agent of bacterial soft rot in Oncidium "Grower Ramsey". Eur J Plant Pathol 142:331–343. https://doi.org/10.1007/s10658-015-0618-7

Ling H, Zeng X, Guo S (2016) Functional insights into the late embryogenesis abundant (LEA) protein family from *Dendrobium officinale* (Orchidaceae) using an *Escherichia coli* system. Sci Rep 6:39693. https://doi.org/10.1038/srep39693

Liu F, Wang X, Su M, Yu M, Zhang S, Lai J, Yang C, Wang Y (2015) Functional characterization of DnSIZ1, a SIZ/PIAS-type SUMO E3 ligase from *Dendrobium*. BMC Plant Biol 15:225. https://doi.org/10.1186/s12870-015-0613-3

López-Chávez MY, Guillén-Navarro K, Bertolini V, Encarnación S, Hernández-Ortiz M, Sánchez-Moreno I, Damon A (2016) Proteomic and morphometric study of the in vitro interaction between *Oncidium sphacelatum* Lindl. (Orchidaceae) and *Thanatephorus* sp. RG26 (Ceratobasidiaceae). Mycorrhiza 26:353–365. https://doi.org/10.1007/s00572-015-0676-x

Maathuis FJM, Ahmad I, Patishtan J (2014) Regulation of Na+ fluxes in plants. Front Plant Sci 5:467. https://doi.org/10.3389/fpls.2014.00467

Meyer Y, Belin C, Delorme-Hinoux V, Reichheld J-P, Riondet C (2012) Thioredoxin and glutaredoxin systems in plants: molecular mechanisms, crosstalks, and functional significance. Antioxid Redox Signal 17:1124–1160. https://doi.org/10.1089/ars.2011.4327

Miller G, Suzuki N, Ciftci-Yilmaz S, Mittler R (2010) Reactive oxygen species homeostasis and signalling during drought and salinity stresses. Plant Cell Environ 33:453–467. https://doi.org/10.1111/j.1365-3040.2009.02041.x

Mishra RC, Ghosh R, Bae H (2016) Plant acoustics: in the search of a sound mechanism for sound signaling in plants. J Exp Bot 67:4483–4494. https://doi.org/10.1093/jxb/erw235

Moise AR, Al-Babili S, Wurtzel ET (2014) Mechanistic aspects of carotenoid biosynthesis. Chem Rev 114:164–193. https://doi.org/10.1021/cr400106y

Moller IM (2001) Plant Mitochondria and oxidative stress : electron transport, NADPH turnover, and metabolism of reactive oxygen species. Annu Rev Plant Physiol Plant Mol Biol 52:561–591. https://doi.org/10.1146/annurev.arplant.52.1.561

Nag S, Kumaria S (2018) *In silico* characterization and transcriptional modulation of phenylalanine ammonia lyase (PAL) by abiotic stresses in the medicinal orchid *Vanda coerulea* Griff. ex Lindl. Phytochemistry 156:176–183. https://doi.org/10.1016/j.phytochem.2018.09.012

Nambara E, Marion-Poll A (2005) Abscisic acid biosynthesis and catabolism. Annu Rev Plant Biol 56:165–185. https://doi.org/10.1146/annurev.arplant.56.032604.144046

Peng PH, Lin CH, Tsai HW, Lin TY (2014) Cold response in *Phalaenopsis aphrodite* and characterization of PaCBF1 and PaICE1. Plant Cell Physiol 55:1623–1635. https://doi.org/10.1093/pcp/pcu093

Perotto S, Rodda M, Benetti A, Sillo F, Ercole E, Rodda M, Girlanda M, Murat C, Balestrini R (2014) Gene expression in mycorrhizal orchid protocorms suggests a friendly plant-fungus relationship. Planta 239:1337–1349. https://doi.org/10.1007/s00425-014-2062-x

Pieterse CMJ, Leon-Reyes A, Van der Ent S, Van Wees SCM (2009) Networking by small-molecule hormones in plant immunity. Nat Chem Biol 5:308. https://doi.org/10.1038/nchembio.164

Qin X, Liu Y, Mao S, Li T, Wu H, Chu C, Wang Y (2011) Genetic transformation of lipid transfer protein encoding gene in *Phalaenopsis amabilis* to enhance cold resistance. Euphytica 177:33–43. https://doi.org/10.1007/s10681-010-0246-4

Rada F, Jaimez R (1992) Comparative ecophysiology and anatomy of terrestrial and epiphytic *Anthurium bredemeyeri* schott in a tropical andean cloud forest. J Exp Bot 43:723–727. https://doi.org/10.1093/jxb/43.5.723

Reina-Pinto JJ, Yephremov A (2009) Surface lipids and plant defenses. Plant Physiol Biochem 47:540–549. https://doi.org/10.1016/j.plaphy.2009.01.004

Reinecke T, Kindl H (1994) Inducible enzymes of the 9,10-dihydro-phenanthrene pathway. Sterile orchid plants responding to fungal infection. AGRIS 7:449–454. https://doi.org/10.1094/mpmi-7-0449

Riemann M, Dhakarey R, Hazman M, Miro B, Kohli A, Nick P (2015) Exploring jasmonates in the hormonal network of drought and salinity responses. Front Plant Sci 6:1077. https://doi.org/10.3389/fpls.2015.01077

Santarius KA (1992) Freezing of isolated thylakoid membranes in complex media. VIII. Differential cryoprotection by sucrose, proline and glycerol. Physiol Plant 84:87–93. https://doi.org/10.1111/j.1399-3054.1992.tb08769.x

Santoro MM, Liu Y, Khan SMA, Hou LX, Bolen DW (1992) Increased thermal stability of proteins in the presence of naturally occurring osmolytes. Biochemistry 31:5278–5283. https://doi.org/10.1021/bi00138a006

Savatin DV, Gramegna G, Modesti V, Cervone F (2014) Wounding in the plant tissue: the defense of a dangerous passage. Front Plant Sci 5:470. https://doi.org/10.3389/fpls.2014.00470

Schenkluhn L, Hohnjec N, Niehaus K, Schmitz U, Colditz F (2010) Differential gel electrophoresis (DIGE) to quantitatively monitor early symbiosis- and pathogenesis-induced changes of the *Medicago truncatula* root proteome. J Proteomics 73:753–768. https://doi.org/10.1016/j.jprot.2009.10.009

Serrano R, Mulet J, Rios G, Marquez JA, de Larrinoa IF, Leube MP, Mendizabal I, Pascual-Ahuir A, Proft M, Ros R (1999) Montesinos C (1999) A glimpse of the mechanisms of ion homeostasis during salt stress. J Exp Bot 50:1023–1036. https://doi.org/10.1093/jxb/50.special_issue.1023

Shimura H, Matsuura M, Takada N, Koda Y (2007) An antifungal compound involved in symbiotic germination of *Cypripedium macranthos* var. rebunense (Orchidaceae). Phytochemistry 68:1442–1447. https://doi.org/10.1016/j.phytochem.2007.03.006

Smirnoff N (1995) Antioxidant systems and plant response to the environment. In: Smirnoff V (ed) Environment and plant metabolism: flexibility and acclimation. BIOS Scientific Publishers, Oxford, pp 217–243

Szymańska R, Ślesak I, Orzechowska A, Kruk J (2017) Physiological and biochemical responses to high light and temperature stress in plants. Environ Exp Bot 139:165–177. https://doi.org/10.1016/j.envexpbot.2017.05.002

Takahashi Y, Berberich T, Kanzaki H, Matsumura H, Saitoh H, Kusano T, Terauchi R (2009) Unraveling the roles of sphingolipids in plant innate immunity. Plant Signal Behav 4:536–538. https://doi.org/10.4161/psb.4.6.8583

Tewari RK, Kumar P, Kim S, Hahn EJ, Paek KY (2009) Nitric oxide retards xanthine oxidase-mediated superoxide anion generation in *Phalaenopsis* flower: an implication of NO in the senescence and oxidative stress regulation. Plant Cell Rep 28:267–279. https://doi.org/10.1007/s00299-008-0632-8

Theocharis A, Clément C, Barka EA (2012) Physiological and molecular changes in plants grown at low temperatures. Planta 235:1091–1105. https://doi.org/10.1007/s00425-012-1641-y

Tsai TM, Chen YR, Kao TW, Tsay WS, Wu CP, Huang DD, Chen WH, Chang CC, Huang HJ (2007) *PaCDPK1*, a gene encoding calcium-dependent protein kinase from orchid, *Phalaenopsis amabilis*, is induced by cold, wounding, and pathogen challenge. Plant Cell Rep 26:1899–1908. https://doi.org/10.1007/s00299-007-0389-5

Tunnacliffe A, Wise MJ (2007) The continuing conundrum of the LEA proteins. Naturwissenschaften 94:791–812. https://doi.org/10.1007/s00114-007-0254-y

Valadares RBS, Perotto S, Santos EC, Lambais MR (2014) Proteome changes in *Oncidium sphacelatum* (Orchidaceae) at different trophic stages of symbiotic germination. Mycorrhiza 24:349–360. https://doi.org/10.1007/s00572-013-0547-2

Wan RL, Sun J, He T, Hu YD, Zhao Y, Wu Y, Chun Z (2017) Cloning cDNA and functional characterization of UDP-glucose pyrophosphorylase in *Dendrobium officinale*. Biol Plant 61:147–154. https://doi.org/10.1007/s10535-016-0645-z

Wang F, Chen HW, Li QT, Wei W, Li W, Zhang WK, Ma B, Bi YD, Lai YC, Liu XL, Man WQ, Zhang JS, Chen SY (2015) *GmWRKY27* interacts with *GmMYB174* to reduce expression of *GmNAC29* for stress tolerance in soybean plants. Plant J 83:224–236. https://doi.org/10.1111/tpj.12879

Wang T, Song Z, Wei L, Li L (2018) Molecular characterization and expression analysis of *WRKY* family genes in *Dendrobium officinale*. Genes Genomics 40:265–279. https://doi.org/10.1007/s13258-017-0602-z

Wei M, Yang C, Wei S (2012) Enhancement of the differentiation of protocorm-like bodies of *Dendrobium officinale* to shoots by ultrasound treatment. J Plant Physiol 169:770–774. https://doi.org/10.1016/j.jplph.2012.01.018

Wind C, Häger KP (1996) Legumin encoding sequences from the Redwood family (Taxodiaceae) reveal precursors lacking the conserved Asn-Gly processing site. FEBS Lett 383:46–50. https://doi.org/10.1016/0014-5793(96)00216-5

Wu Z-G, Jiang W, Chen S-L, Mantri N, Tao Z-M, Jiang C-X (2016) Insights from the cold transcriptome and metabolome of *Dendrobium officinale*: global reprogramming of metabolic and gene regulation networks during cold acclimation. Front Plant Sci 7:1653. https://doi.org/10.3389/fpls.2016.01653

Wyn Jones RG, Storey R (1981) Betaines. In: Paleg LG, Aspinall D (eds) The physiology and biochemistry of drought resistance in plants. Academic Press, Sidney, pp 171–204

Xiong L, Zhu JK (2002) Molecular and genetic aspects of plant responses to osmotic stress. Plant Cell Environ 25:131–139. https://doi.org/10.1046/j.1365-3040.2002.00782.x

Xiong L, Schumaker KS, Zhu J-K (2002) Cell Signaling during cold, drought, and salt stress. Plant Cell 14:S165–S183. https://doi.org/10.1105/tpc.000596

Xu XB, Pan YY, Wang CL, Ying QC, Song HM, Wang HZ (2014) Overexpression of *DnWRKY11* enhanced salt and drought stress tolerance of transgenic tobacco. Biol. 69:994–1000. https://doi.org/10.2478/s11756-014-0398-0

Xu X, Wang C, Ma X, Pan Y, Ying Q, Song H, Wang H (2015) Overexpression of *DnWRKY29* in tobacco impaired plants tolerance to salt and drought stresses. Russ J Plant Physiol 62:262–269. https://doi.org/10.1134/S1021443715010173

Yancey PH, Clark ME, Hand SC, Bowlus RD, Somero GN (1982) Living with water stress: evolution of osmolyte systems. Science 217:1214–1222. https://doi.org/10.1126/science.7112124

Yang SH, Yu H, Goh CJ (2003) Functional characterisation of a *cytokinin oxidase* gene *DSCKX1* in *Dendrobium* orchid. Plant Mol Biol 51:237–248. https://doi.org/10.1023/A:1021115816540

Yang X, Wang Y, Luo J (2010) Transformation of lea3 gene into *Dendrobium candidum* Wall. ex Lindl. for enhancing its salt tolerance. Chin. J. Appl. Environ. Biol. 16:622–626 (in Chinese with English abstract)

Yu Z, He C, da Teixeira Silva JA, Zhang G, Dong W, Luo J, Duan J (2017) Molecular cloning and functional analysis of DoUGE related to water-soluble polysaccharides from *Dendrobium officinale* with enhanced abiotic stress tolerance. Plant Cell Tissue Organ Cult 131:579–599. https://doi.org/10.1007/s11240-017-1308-2

Yuan XY, Liang F, Jiang SH, Wan MF, Ma J, Zhang XY, Cui B (2015) Differential protein expression in *Phalaenopsis* under low temperature. Appl Biochem Biotechnol 175:909–924. https://doi.org/10.1007/s12010-014-1345-9

Zeng X, Ling H, Chen X, Guo S (2019) Genome-wide identification, phylogeny and function analysis of GRAS gene family in *Dendrobium catenatum* (Orchidaceae). Gene 705:5–15. https://doi.org/10.1016/j.gene.2019.04.038

Zhang Z, He D, Niu G, Gao R (2014) Concomitant CAM and C3 photosynthetic pathways in *Dendrobium officinale* plants. J Am Soc Hortic Sci 139:290–298. https://doi.org/10.21273/jashs.139.3.290

Zhang S-B, Dai Y, Hao G-Y, Li J-W, Fu X-W, Zhang J-L (2015) Differentiation of water-related traits in terrestrial and epiphytic *Cymbidium* species. Front Plant Sci Front Plant Sci 6:260. https://doi.org/10.3389/fpls.2015.00260

Zhang GQ, Xu Q, Bian C, Tsai WC, Yeh CM, Liu KW, Yoshida K, Zhang LS, Chang SB, Chen F, Shi Y, Su YY, Zhang YQ, Chen LJ, Yin Y, Lin M, Huang H, Deng H, Wang ZW, Zhu SL, Zhao X, Deng C, Niu SC, Huang J, Wang M, Liu GH, Yang HJ, Xiao XJ, Hsiao YY, Wu WL, Chen YY, Mitsuda N, Ohme-Takagi M, Luo YB, Van De Peer Y, Liu ZJ (2016a) The *Dendrobium catenatum* Lindl genome sequence provides insights into polysaccharide synthase, floral development and adaptive evolution. Sci Rep 6:19029. https://doi.org/10.1038/srep19029

Zhang W, Hu H, Zhang S-B (2016b) Divergent adaptive strategies by two co-occurring epiphytic orchids to water stress: escape or avoidance? Front Plant Sci 7:588. https://doi.org/10.3389/fpls.2016.00588

Zhao P, Wang W, Sun M (2011) Characterization and expression pattern analysis of *DcNAC* gene in somatic embryos of *Dendrobium candidum* Wall Ex Lindl. Plant Cell Tissue Organ Cult 107:151–159. https://doi.org/10.1007/s11240-011-9968-9

Zhao MM, Zhang G, Zhang DW, Hsiao YY, Guo SX (2013) ESTs analysis reveals putative genes involved in symbiotic seed germination in *Dendrobium officinale*. PLoS ONE 8:e72705. https://doi.org/10.1371/journal.pone.0072705

Zhao X, Zhang J, Chen C, Yang J, Zhu H, Liu M, Lv F (2014) Deep sequencing-based comparative transcriptional profiles of *Cymbidium hybridum* roots in response to mycorrhizal and non-mycorrhizal beneficial fungi. BMC Genom 15:747. https://doi.org/10.1186/1471-2164-15-747

Zhao D, Shi Y, Senthilkumar HA, Qiao Q, Wang Q, Shen Y, Hu G (2018) Enriched networks 'nucleoside/nucleotide and ribonucleoside/ribonucleotide metabolic processes' and 'response to stimulus' potentially conferred to drought adaptation of the epiphytic orchid *Dendrobium wangliangii*. Physiol Mol Biol Plants 25:31–45. https://doi.org/10.1007/s12298-018-0607-3

Zhu J, Dong CH, Zhu JK (2007) Interplay between cold-responsive gene regulation, metabolism and RNA processing during plant cold acclimation. Curr Opin Plant Biol 10:290–295. https://doi.org/10.1016/j.pbi.2007.04.010

Zhu Y, Meng C, Zhu L, Li D, Jin Q, Song C, Cai Y, Fan H, Lin Y (2017) Cloning and characterization of DoMYC2 from *Dendrobium officinale*. Plant Cell Tissue Organ Cult 129:533–541. https://doi.org/10.1007/s11240-017-1198-3

Chapter 21
Development of Bryophytes as a New Model System to Understand the Phenomenon of Terrestrialization with Environmental Changes

Sandhya Yadav, Subhankar Biswas, Akanksha Srivastava and Yogesh Mishra

Abstract Bryophytes are earliest diverging lineages of the extant land plants with around 25,000 species distributed all over the world. Bryophytes can be further classified into three main classes viz. Liverworts, Hornworts and Mosses that grow on a wide range of habitats. Bryophytes, with high ecological and economic values, occupy a very important position in the evolution of terrestrial plants. During the transition of aquatic to terrestrial habitat (terrestrialization), bryophytes got exposed to global climate changes as well as dehydrating atmosphere of terrestrial habitats that led to the desiccation of plant tissues. In order to tolerate the environmental alterations and to protect themselves from abiotic stresses, bryophytes must have enabled themselves to develop certain adaptive strategies. In order to understand these adaptive strategies at molecular level, attempts have been made to develop certain bryophytes as new model system such as *Physcomitrella patens* and *Marchantia polymorpha*. In the current chapter we will addresses how does these model systems have been used to address the uniqueness of bryophytes in terms of their capabilities behind the conquering the land i.e. terrestrialization.

Keywords Alternation of generation · Bryophytes · *Marchantia polymorpha* · *Physcomitrella patens* · Terrestrialization

21.1 Introduction

Life was originated for the first time in the water and it was a big challenge for the scientist to find out the process of migrations of plants from water to land. Aquatic plants had shifted to the land because of certain reasons such as low availability of CO_2 in water, low pH of water bodies and might be due to frequently wave splashed.

S. Yadav · S. Biswas · A. Srivastava · Y. Mishra (✉)
Department of Botany, Centre of Advanced Study in Botany, Institute of Science, Banaras Hindu University, Varanasi 221005, India
e-mail: ymishra@bhu.ac.in

S. P. Singh et al. (eds.), *Molecular Approaches in Plant Biology and Environmental Challenges*, Energy, Environment, and Sustainability,
https://doi.org/10.1007/978-981-15-0690-1_21

This invasion of land by plants from the water is broadly considered as "terrestrialization". The emergence and evolution of sporophytic land plants which was one of the most important events in the history of Earth is the matter of controversy. In literature two conflicting theories are available concerning the origin of sporophyte in land plants: the "transformation" theory and the "interpolation" theory (Blackwell 2003). According to the "transformation" theory, sporophytes are the direct modification of certain algal gametophyte (e.g. *Ulva*) with the specific functions of spore productions. However, the "interpolation" theory suggests that sporophyte generation progenated from a haploid, green algal thallus where repeated mitotic cell division of a zygote resulted in the formation of dependent sporophyte. Life cycle of the bryophytes (where gametophyte is dominant and sporophyte is dependent on it) is the best evidence which support this theory. Numerous morphological, cytological, biochemical and molecular evidences has also lent support to interpolation theory which suggests the evolution of bryophytes has been taken place from algae and later these first land plants evolved into the pteridophytes (first vascular plants).

Bryophytes are considered as earliest diverging lineages of the extant land plants that crawl to land. Bryophytes comprise almost 17,000 species and are categorized into Liverworts, Hornworts and Mosses which comprises ~7000, ~100 and ~1000 species, respectively (Qui and Palmer 1999). Instead of having vast diversity of plant kingdom, bryophytes are unique in being the archaic and in the evolutionary point of view because they serve as connecting link between algae and pteridophytes. Therefore, to understand this adventures voyage of terrestrialization by early life forms, the bryophytes are indispensable to study. They are considered as 'amphibious group' of plant kingdom because of the ability to grow at the interface of water and land. During the process of terrestrialization, bryophytes becomes greatly adapted for the differential harsh stresses, which makes them capable enough to grow on diverse habitats ranging from barren arctic and alpine ground to hot deserts, bottom of lakes to the canopy of tropical rain forests etc. Ecologically, the bryophytes also possess a significant position by playing the role of pioneers. The arrival of bryophytes on the land, made the earth surface more likely to be colonized by their descendants. After the evolution of early land plants, they underwent a set of changes i.e. morphological, physiological, anatomical, biochemical and genetic changes that led to the foundation of modern plants. This definitely raised certain key questions

(1) How did this green branch of life plunged from water-to-land?
(2) How did they enable to balance the water level and how they dealt with the consequences of desiccation, photo damage and UV radiations?
(3) How did they evolve the process of acquisition of CO_2 and the biochemistry of photosynthesis?
(4) What are those morphological, physiological, anatomical, biochemical and genetic changes which helped them in conquering the land successfully?

Keeping in the view of above questions, in current chapter we are aiming to give a comprehensive account of the land plants evolution via bryophytes.

21.2 Alternation of Generations in Land Plants

Every organism, except prokaryotes, goes through the sexual reproduction which involves a regular alteration between meiosis and fertilization. The closest extant algal relatives of land plants, the charophytes have a haplontic life cycle with dominant haploid stage, where gametes are produced through the mitosis which after fertilization forms the zygote, only diploid cell in the entire life cycle. Zygotic cells further undergo meiosis and eventually form haploid spores. In contrast, the land plants usually express two alternating generations one is the multicellular diploid sporophyte and another is multicellular haploid gametophyte which is termed as haplodiplontic life cycle. There are two controversial theories which explain the origins of alternation in generation (and sporophyte origin) in land plants: the "transformation" theory and the "interpolation" theory.

21.2.1 Two Theories Behind the Origin of Alternating Generation in Land Plants

21.2.1.1 Transformation Theory (Homologous or Modification Theory)

In 1878, Pringsheim initially proposed this theory to explain the origins of alternating generations in land plants. According to this theory, sporophytes have directly been modified from algal gametophytes with addition to spore forming capabilities. This theory is based on the life cycle of certain algae (*Ulva, Ectocarpous* and *Cladophora*) which are considered as ancestors of modern plants and undergo an alternation of isomorphic generation where gametophytes and sporophytes are similar or homologous. Prior to land invasion these algal ancestors have been hypothesized to give rise the land plants with their sporophytes by directly modifying their gametophytes. Algal based origin of land plants theory suggested that they must have acquired/evolved certain adaptive strategies to cope up on harsh habitat. To support this theory there are many evidences as *Chlorokybus,* an algae grow successfully in sub aerial and terrestrial environment (Graham et al. 2012) by probably coping up the long way of water scarcity. *Klebsormidiales* having traits of desiccation tolerance (Elster et al. 2008), shows full photosynthetic recovery after a long period of water deficient condition (Karsten et al. 2010). Another trait of algae that supposed to assist them in successful landing is the occurrences of beta type carbonic anhydrase i.e. enzyme that enable them to use bicarbonates as an alternative source of carbon onto the land, found in many streptophytes (Arancibia-Avila et al. 2001), that give a hand to remodeling the additional carbon acquisition strategies along with, provide the flexibility to obtaining carbon for photosynthesis. Another important feature of algae that hold onto the theory is occurrence of sporopollenin and lignin like phenolic cell wall (Delwiche et al. 1989). In some cyanobacteria, mycosporine like amino acids (MAA) are found, that provided them endurance during UV radiations and diverse oxidative stress (Aharon and Gunde-Cimerman 2007). Therefore, on the behalf of such evidences a group of scientists support the thought that algae bequeathed these terrestrial traits to embryophytes descendants.

However, the main problem regards to transformation theory was that it does not explains, how the development of the dependent sporophyte came into the existence in the life cycle. Isomorphic alternation of generation was the further problem for this theory as modern land plants have alternative heteromorphic generations in their life cycle.

21.2.1.2 Interpolation Theory (Antithetic or Intercalation Theory)

Interpolation theory (Bower 1908) is an alternative hypothesis for the origins of alternating generations in land plants. According to this theory, land plants evolved from the algal ancestors as in transformation theory but only the gametophyte was thought to be present initially in algal life cycle and sporophyte arising subsequently by a delay of zygotic meiosis with zygotic mitoses on gametophyte, when it started occupying the land. Further gradual evolution of an independent sporophyte in terms of capacity to increase in body size, developing greater number of dispersal structures such as spores later evolved as seeds etc. led to the transition from aquatic to terrestrial plant life on Earth. Bryophytes with distinct and independent gametophyte and dependent sporophyte have strengthened this theory. In light of this theory we can assume that bryophytes are the first land plants derived directly from a green algal progenitor and which in due course of evolution led to the development of pteridophytes. Although this theory is more plausible than the transformation theory, still there is phylogenetic dilemma concerning to the transition of pteridophytes from bryophytes because of having large evolutionary gaps between these two groups not only in life cycle but also at the structural/developmental level. In spite of all dilemmas, this theory has got support from the variety of morphological, cytological, biochemical and molecular evidences which have been discussed in current chapter. Beside these evidences, interpolation theory has also been strengthened by the life cycle exhibited by the charophytes which has been considered to the closest relative of land plants. Charophytes such as *Chara, Nitella* and *Coleochaete* exhibits halontic life cycle with zygotic meiosis and no sporophyte which suggests that only interpolation theory is only feasible to explain the origin of sporophytes in land plants.

21.2.1.3 Transition of Dominant Gametophytes to Dominant Sporophytes

In due course of evolution, the transition of dominant gametophytic to dominant sporophytic generation clearly indicates that diploidization of plant body was obligatory for conquering the land by flowering plants. As the Charophycean algae follow the haplontic life cycle i.e. gametophyte is the major dominating multicellular haploid phase and a unicellular diploid zygote. In lower thalloid bryophytes, the multicellular haploid phase is independent, but diploid sporophytic phase is initially completely dependent for nutrition on its parents i.e., gametophytes. Later on the sporophytic phase of bryophytes became small stalked capsule and partially gametophytic dependent like mosses. Subsequently, this multicellular diploid phase evolved into a highly successful spore producing dispersal unit that emerged into recent diversified and astonishing forms of green branch of life. In vascular plants like ferns, gymnosperms and flowering plants, sporophytes are large, leafy, free-living plants. This shift involved the liberation of the sporophyte from complete physiological dependence on its gametophyte to achieve a free living status.

21.3 Process of Terrestrialization via Bryophytes

On the basis of recent evolutionary and phylogenetic evidences it is believed that bryophytes are the earliest land plants which not only succeeded in perfect landing in Ordovician period but also dominated the terrestrial habitat by evolving exceptionally well stress tolerant strategies. But what are those adaptive features that assisted them to succeed terrestrial habitat with high solar radiation, a drying atmosphere, high temperature alterations and limited access of nutrients is still partially explained. Bryophytes placed at the base of the lineages of terrestrial species, therefore, it is important to addresses, what uniqueness bryophyte possess, that made them capable to conquer the land successfully.

21.3.1 Unique Features of Bryophytes Towards Being Pioneer of Terrestrialization

Bryophytes are well known as "amphibian of plant kingdom" because they usually live in soil but they need water for their sexual reproduction. They are diverse in morphology and anatomy. Morphologically they ranges from thalloid (liverworts and hornworts) to leafy-forms (mosses). The leaf like structures of mosses mimics the structure of higher plants. For the first time cellular differentiation was noticed in the gametophytes of thalloid bryophytes, which is differentiated in upper photosynthetic zone and lower storage region. The functional diversification of cells shows an evolutionary advancement. Other traits are also found in bryophytes such as presence of scales and rhizoids which play the role of roots, air-pores (thalloid members) and pseudo-stomata (sporophytes of hornworts and capsules of mosses) for gaseous exchange. Modified absorbing and anchoring cell-types were found in several green algae but it is more prominent and functionally active in bryophytes. Although the rhizoids and scales were not as strong as root-system of flowering plants but played the essential role in anchoring and absorbing the nutritions on land. These characters are still present in higher plant which suggests that bryophytes were pioneer of terrestrialization.

21.3.2 Adaptive Strategies of Bryophytes Towards Terrestrialization

Morphological, anatomical, physiochemical and molecular phylogenetic study of plant kingdom strongly suggested that bryophytes stand in between green algae and seed plants in the course of evolution (Fig. 21.1). With the course of time and Darwinian selection the modern day angiosperms were arrived and dominated the earth crust. Bryophytes are diverse in life forms, habitat, and chemical composition because of having characteristics of both aquatic algae and higher plants. To survive the land for very first time, bryophytes have evolved certain adaptive strategies that led these organisms to be so sturdy to thrive on harsh environment. Molecular studies revealed that to conquer the land, the first requirement of bryophytes was three-dimensional (3D) growth and cellular differentiation with specialized functions. The key player of the 3D growth inductions has been identified in *P. patens* and homolog of this gene is found in *Arabidopsis thaliana* where they play diverse functions (Moody et al. 2018). The development of a sporophytic apical meristem is one of the most critical events of terrestrialization, which is regulated by class I KNOX

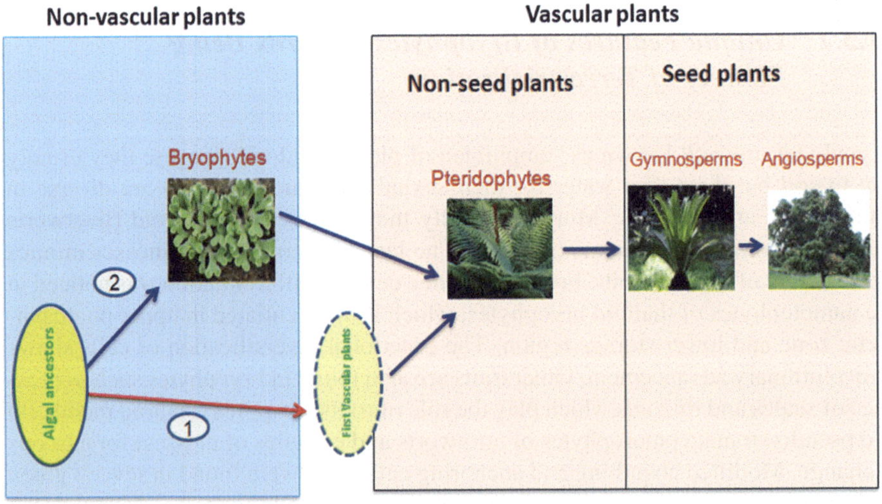

Fig. 21.1 Schematic representation of two theories of terrestrialization (1—transformation; 2—interpolation). The solid lines represents where the clear evidences are available however, the dashed lines represents where exact evidences are missing

(KNOX1) gene family in modern land plants. The orthologs of this gene family have been reported in *P. patens* that reveal that all KNOX1 genes in this species are expressed in the apical cell and meristematic region of the sporophyte. The presence of KNOX1 mediated gene regulatory network in *P. patens* for the sporophytic apical meristem suggest that it was conserved among land plants, and was might be required early in the process of terrestrialization (Hay and Tsiantis 2010).

In the process of movement from water to land, the development of roots has played a very important role. In angiosperms the development of root hairs in multicellular roots are regulated by ROOT HAIR DEFECTIVE SIX-LIKE (RSL) class I genes, which are classified as members of the bHLH VIIIc subfamily. The genome search of *P. patens* and *M. polymorpha* suggest that the presence of bHLH family genes in these two bryophytes. In these two organisms, bHLH members are known to be involved in the development of rhizoids in the gametophyte generation (Jang et al. 2011; Menand et al. 2007; Pires and Dolan 2010).

Stomata are another vital element of seed plants for uptake of CO_2 from atmosphere and many other important physiological functions. Development and patterning of stomata are determined by several genes, some of them also found in basal land plants, which suggests a tightly regulated genetic evolution of stomata (Chater et al. 2017). Stomata located on the sporophyte of *P. patens* found to respond environmental cues (Raven 2002). ABA is one of the key regulators of stomata and found to regulate moss stomata like higher plants (Stevenson et al. 2016). From above observation it is concluded that ABA-dependent stomata regulatory components were developed and functional before the origin of higher plants.

As bryophytes were the first land plants so desiccation was the most dominant among all the stresses. Cuticle like structure is present in *P. patens* which are structurally and chemically similar to *A. thaliana* and involved in desiccation tolerance of this moss (Buda et al. 2013). Regulatory pathway of cuticle formation in *P. patens*is similar to higher plants but presented in a simplest form while the mechanism is more sophisticated in case of higher plants (Cui et al. 2016). pPORS is an ancient type of polyketide synthase, required for proper leaf-cuticle development and reduction of dehydration found in *P. patens* (Li et al. 2018). These observations suggested as the cuticle mediated reduction of water loss is an ancient type of mechanism that come along with basal plants and furthermore help them to conquer the land.

Aquaporins are water channels that involves in several physiological processes. It delays the desiccation of leaves in *P. patens* during drying condition and increased the absorption of water which eventually plays an important role in response to hydration and dehydration cycle (Lienard et al. 2008).

Beside above mentioned strategies, early light-induced proteins (ELIP) also play a significant protective role in desiccation tolerance by securing the photosynthetic machinery of desiccation tolerant plants because photosynthesis is very sensitive to drought or dehydration (Van Buren et al. 2019). The domain wise studies of ELIPS denoted that the evolution of ELIP mediated drought tolerance was evolved convergently in plants from bryophytes.

Photooxidative damage is the most dangerous threat for all the land plants and each group of plants harbors few mechanisms to avoid or mitigate photoxidation by high light intensity. But, interestingly Bryophytes like *M. polymorpha* possess genes encoding flavodiiron proteins (FLV) which are involved in alternative electron flow (AEF) like cyanobacteria (Shimakawa et al. 2017). This study showed that AEF by FLV induces non-photochemical quenching (NPQ) and reduces the electron transport rate through electron transport chain (ETS) of chloroplast. Further studies has reported that *P. patens* possess two distinct mechanisms for non-photochemical quenching—algal type light-harvesting complex stress-related (LHCSR) and plant-type S subunit of Photosystem II (PSBS)-dependent mechanisms (Pinnola et al. 2013).

Ultraviolet radiation (UV) is another destructive stress at terrestrial habitats therefore bryophytes must have evolved protective measure to face it. Although bryophytes grow dominantly on shaded habitats but a large number of species grows in open sunlight. Bryophytes possess a lot of secondary metabolites like flavonoids, phenylpropanoids, sporopollenin, lignin, which absorb the UV-B radiation and in this way played a major protective role in bryophytes (Rozema et al. 2009). This assumption get strengthen by recent study shows that *P. patens is* more UV-B resistant than *A. thaliana* (Wolf et al. 2010). Apart from it, UV-B mediated DNA damage in *P. patens* found to use a different homologous recombination based damaged DNA repair machinery using RAD51 which is non-functional in *A. thaliana* (Markmann-Mulisch et al. 2007).

In the process terrestrialization phytohormones are supposed to play an important role. Auxin is one of the most important phytohormone that controls a vast array of developmental processes in higher plants and is found throughout the plant kingdom. Recent studies showed that the role of auxin is bit different in *P. patens* but at the perception level it is quite similar to higher plants (Prigge et al. 2010). Gibbrellic acid mediated signaling is well known for playing crucial role in growth and developmental pathways in flowering plants. In the moss *P. patens*, it has been shown that GA is not produced here rather ent-kaurene metabolites may play the role of GA (Miyazaki et al. 2018). GA-DELLA mediated signaling is not reported in bryophytes they may be evolved after the divergence of flowering plants from this basal group (Yasumura et al. 2007).

Small RNAs are key regulator of eukaryotic gene expression. In *P. patens*, heterochromatic siRNAs and their biogenesis pathway are identical to flowering plants (Coruh et al. 2015). So, the siRNA mediated post-transcriptional regulation of gene-expression may be conserved throughout the plant kingdom such evidences further validate this theory of terrestrialization.

Another example which further attests the bryophytes based theory of terrestrialization is the common mechanism of post-translational modifications of proteins between bryophytes and higher plants. For example N-glycosylation of proteins in *P. patens* and higher plants (*A. thaliana*) share common mechanism (Vietor et al. 2003). So, this process of post translational modifications was supposed to evolve before the evolution of terrestrial plants.

Apart from above examples certain evidences are also available on metabolic events such as sulphur assimilation. Sulphur assimilation mechanism of *P. patens* is much more robust than higher plants (Hermsen et al. 2010). This mechanism also indicated that after separation from seed plants evolution of bryophytes did not halt but they became more likely to grow on the changing environment.

21.4 Emergence of New Bryophyte Model System to Understand the Secrets of Terrestrialization

In the last two decades, a shift in the study of bryophytes were taken place from taxonomy to molecular biology throughout the world. In this scenario two land mark study has taken place i.e. the complete genome sequencing projects of two bryophytes such as *P. patens* (Lang et al. 2008) and *M. polymorpha* L (Bowman et al. 2017) came into picture. The results from these sequencing projects further attested the bryophyte based theory of terrestrialization. The complete genome sequence information of these two bryophytes have given deeper insights of the molecular events that has taken place during the process of terrestrialization, which have already been discussed in above subsection of this chapter.

21.5 Conclusions and Future Perspectives

With the advancement in the DNA sequencing technologies especially concerning to the next-generation sequencing (NGS) has decreased the costs per megabase that led to the increase in the number and diversity of sequenced genomes including bryophytes. Informations gathered from the genome sequencing have accelerated the identification of both gene losses and additions during the course of evolution. Progress in genome sequencing of bryophytes has assisted in the exploration of the evolutionary path of land plants. Recently, genomes of two bryophytes namely *P. patens* and *M. polymorpha* has been sequenced and they have been established as model organisms to study the molecular level adaptation acquired by bryophytes when they started invading the land for the first time in history of Earth. Functional studies in these two model organisms have started providing the valuable Informations regarding evolution of underlying genetic regulatory mechnisms and if this progress will remain in the same pace then bryophytes may become the *E. coli* of plant molecular biology in near future. Furthermore, study of the phenomenon of terrestrialization will help us to know how the plant responds to future climate changes and perturbations. Comprehensive understanding of the phenomenon of terrestrialization at molecular level will not only strengthen the basic concept of stress adaptation mechanisms in multi-stress environment, but it will also open the possibilities for the mining of novel stress tolerance genes which can be used for development of stress tolerant crop plants in near future. In the end, only tracking changes during the course of evolution will truly show how plants evolve and adapt to life on land.

Acknowledgements Sandhya Yadav is thankful to UGC for Junior Research fellowship (JRF), Subhankar Biswas is grateful to CSIR for JRF and Akansha Srivastava is gratified INSPIRE for JRF. Dr. Yogesh Mishra is thankful to UGC- Start up grant.

References

Aharon O, Gunde-Cimerman N (2007) Mycosporines and mycosporine-like amino acids: UV protectants or multipurpose secondary metabolites? FEMS Microbiol Lett 269:1–10. https://doi.org/10.1111/j.1574-6968.2007.00650.x

Arancibia-Avila P, Coleman JR, Russin WA, Graham JM, Graham LE (2001) Carbonic anhydrase localization in charophycean green algae: ecological and evolutionary significance. Int J Plant Sci 162:127–135. https://doi.org/10.1086/317908

Blackwell WH (2003) Two theories of origin of the land-plant sporophyte: which is left standing? Bot Rev 69:125. https://doi.org/10.1663/0006-8101

Bower FO (1908) The origin of a land flora. A theory based upon the facts of alternation. MacMillan and Co., Limited, New York. https://doi.org/10.5962/bhl.title.1698

Bowman JL, Kohchi T, Yamato KT, Jenkins J, Shu S, Ishizaki K, Yamaoka S, Nishihama R, Nakamura Y, Berger F, Adam C, Aki SS, Althoff F, Araki T, Arteaga-Vazquez MA, Balasubrmanian SK, Barry K, Bauer D, Boehm CR, Briginshaw L, Caballero-Perez J, Catarino B, Chen F, Chiyoda S, Chovatia M, Davies KM, DelmansM Demura T, Dierschke T, Dolan L, Dorantes-Acosta AE, Eklund DM, Florent SN, Flores-Sandoval E, Fujiyama A, Fukuzawa H, Galik B, Grimanelli D, Grimwood J, Grossniklaus U, Hamada T, Haseloff J, Hetherington AJ, Higo A, Hirakawa Y, Hundley HN, Ikeda Y, Inoue K, Inoue S, Ishida S, Jia Q, Kakita M, Kanazawa T, Kawai Y, Kawashima T, Kennedy M, Kinose K, Kinoshita T, Kohara Y, Koide E, Komatsu K, Kopischke S, Kubo M, Kyozuka J, Lagercrantz U, Lin SS, Lindquist E, Lipzen AM, Lu CW, Luna ED, Martienssen RA, Minamino N, Mizutani M, Mizutani M, Mochizuki N, Monte I, Mosher R, Nagasaki H, Nakagami H, Naramoto S, Nishitani K, Ohtani M, Okamoto T, Okumura M, Phillips J, Pollak B, Reinders A, Ro vekamp M, Sano R, Sawa S, Schmid MW, Shirakawa M, Solano R, Spunde A, Suetsugu N, Sugano S, Sugiyama A, Sun R, Suzuki Y, Takenaka M, Takezawa D, Tomogane H, Tsuzuki M, Ueda T, Umeda M, Ward JM, Watanabe Y, Yazaki K, Yokoyama R, Yoshitake Y, Yotsui Y, Zachgo, Schmutz J (2017) Insights into land plant evolution garnered from the *Marchantia polymorpha* Genome. Cell 171:287–304. https://doi.org/10.1016/j.cell.2017.09.030

Buda GJ, Barnes WJ, Fich EA, Park S, Yeats TH, Zhao L, Domozych DS, Rosea JKC (2013) An ATP binding cassette transporter is required for cuticular wax deposition and desiccation tolerance in the moss *Physcomitrella patens*. Plant Cell 25:4000–4013. https://doi.org/10.1105/tpc.113.117648

Chater CCC, Caine RS, Fleming AJ, Gray JE (2017) Origins and evolution of stomatal development. Plant Physiol 174:624–638. https://doi.org/10.1104/pp.17.00183

Coruh C, Cho SH, Shahid S, Liu Q, Wierzbicki A, Axtella MJ (2015) Comprehensive annotation of *Physcomitrella patens* small RNA loci reveals that the heterochromatic short interfering RNA pathway is largely conserved in land plants. Plant Cell 27:2148–2162. https://doi.org/10.1105/tpc.15.00228

Cui F, Brosche M, Lehtonen MT, Amiryousefi A, Xu E, Punkkinen M, Valkonen JPT, Fujii H, Overmyer K (2016) Dissecting abscisic acid signaling pathways involved in cuticle formation. Mol Plant 9:926–938. https://doi.org/10.1016/j.molp.2016.04.001

Delwiche CF, Graham LE, Thomson N (1989) Lignin-like compounds and sporopollenin coleochaete, an algal model for land plant ancestry. Science 245:399–401. https://doi.org/10.1126/science.245.4916.399

Elster JD, Peter D, L'ubomir K, Lucia V, Katarina Š, Antonio BP (2008) Freezing and desiccation injury resistance in the filamentous green alga *Klebsormidium* from the Antarctic, Arctic and Slovakia. Biologia 63:843–851. https://doi.org/10.2478/s11756-008-0111-2

Graham LE, Arancibia-Avila P, Taylor WA, Strother PK, Cook ME (2012) Aeroterrestrial Coleochaete (Streptophyta, Coleochaetales) models early plant adaptation to land. Am J Bot 99:130–144. https://doi.org/10.3732/ajb.1100245

Hay A, Tsiantis M (2010) KNOX genes: versatile regulators of plant development and diversity. Development 137:3153–3165. https://doi.org/10.1242/dev.030049

Hermsen C, Koprivova A, Matthewman C, Wesenberg D, Krauss GJ, Kopriva S (2010) Regulation of sulfate assimilation in *Physcomitrella* patens: mosses are different! Planta 232:461–470. https://doi.org/10.1007/s00425-010-1190-1

Jang G, Yi K, Pires ND, Menand B, Dolan L (2011) RSL genes are sufficient for rhizoid system development in early diverging land plants. Development 138:2273–2281. https://doi.org/10.1242/dev.060582

Karsten U, Lütz C, Holzinger A (2010) Ecophysiological performance of the aeroterrestrial green alga *Klebsormidium crenulatum* (charophyceae, streptophyta) isolated from an alpine soil crust with an emphasis on desiccation stress. J Phycol 46:1187–1197. https://doi.org/10.1111/j.1529-8817.2010.00921.x

Lang D, Zimmer AD, Rensing SA, Reski R (2008) Exploring plant biodiversity: the *Physcomitrella* genome and beyond. Trends Plant Sci 13:542–549

Li L, Aslam M, Rabbi F, Vanderwel MC, Ashton NW, Suh D (2018) PpORS, an ancient type III polyketide synthase, is required for integrity of leaf cuticle and resistance to dehydration in the moss, *Physcomitrella patens*. Planta 247:527–541. https://doi.org/10.1007/s00425-017-2806-5

Lienard D, Durambur G, Kiefer-Meyer MC, Nogue F, Menu-Bouaouiche L, Charlot F, Gomord V, Lassalles J (2008) Water transport by aquaporins in the extant plant Physcomitrella patens. Plant Physiol 146:1207–1218. https://doi.org/10.1104/pp.107.111351

Markmann-Mulisch U, Wendeler E, Zobell O, Schween G, Steinbiss H, Reissa B (2007) Differential requirements for RAD51 in *Physcomitrella patens* and *Arabidopsis thaliana* development and DNA damage repair. Plant Cell 19:3080–3089. https://doi.org/10.1105/tpc.107.054049

Menand B, Yi K, Jouannic S, Hoffmann L, Ryan E, Linstead P, Schaefer DG, Dolan L (2007) An ancient mechanism controls the development of cells with a rooting function in land plants. Science 316:1477–1480. https://doi.org/10.1126/science.1142618

Miyazaki S, Hara M, Ito S, Tanaka K, Asami T, Hayashi K, Kawaide H, Nakajima M (2018) An ancestral gibberellin in a moss *Physcomitrella patens*. Mol Plant 11:1097–1100. https://doi.org/10.1016/j.molp.2018.03.010

Moody LA, Kelly S, Rabbinowitsch E, Langdale JA (2018) Genetic regulation of the 2D to 3D growth transition in the moss Physcomitrella patens. Curr Biol 28:473–478. https://doi.org/10.1016/j.pbi.2018.10.001

Pinnola A, Dall'Osto L, Gerotto C, Morosinotto T, Bassi R, Alboresia A (2013) Zeaxanthin binds to light-harvesting complex stress-related protein to enhance nonphotochemical quenching in *Physcomitrella patens*. Plant Cell 25:3519–3534. https://doi.org/10.1105/tpc.113.114538

Pires N, Dolan L (2010) Origin and diversification of basic-helix-loophelix proteins in plants. Mol Biol Evol 27:862–874. https://doi.org/10.1093/molbev/msp288

Prigge MJ, Lavy M, Ashton NW, Estelle M (2010) *Physcomitrella patens* auxin-resistant mutants affect conserved elements of an auxin-signaling pathway. Curr Biol 20:1907–1912. https://doi.org/10.1016/j.cub.2010.08.050

Qui YL, Palmer JD (1999) Phylogeny of early land plants: insights from genes and genomes. Trends Plant Sci 4:26–30. https://doi.org/10.1016/s1360-1385(98)01361-2

Raven JA (2002) Selection pressures on stomatal evolution. New Phytol 153:371–386. https://doi.org/10.1046/j.0028-646x.2001.00334.x

Rozema J, Blokker P, Mayoral Fuertes MA, Broekman R (2009) UV-B absorbing compounds in present-day and fossil pollen, spores, cuticles, seed coats and wood: evaluation of a proxy for solar UV radiation. Photochem Photobiol Sci 8:1233–1243. https://doi.org/10.1039/b904515e

Shimakawa G, Ishizaki K, Tsukamoto S, Tanaka M, Sejima T, Miyake (2017) The liverwort, Marchantia, drives alternative electron flow using a flavodiiron protein to protect PSI. Plant Physiol 173:1636–1647. https://doi.org/10.1104/pp.16.01038

Stevenson SR, Kamisugi Y, Trinh CH, Schmutz J, Jenkins JW, Grimwood J, Muchero W, Tuskan GA, Rensing SA, Lang D, Reski R, Melkonian M, Rothfels CJ, Li FW, Larsson A, Wong GK, Edwards TA, Cuming AC (2016) Genetic analysis of *Physcomitrella patens* identifies ABSCISIC ACID NON-RESPONSIVE, a regulator of ABA responses unique to basal land plants and required for desiccation tolerance. Plant Cell 28:1310–1327. https://doi.org/10.1105/tpc.16.00091

Van Buren R, Pardo J, Wai CM, Evans S, Bartelsd D (2019) Massive tandem proliferation of ELIPs supports convergent evolution of desiccation tolerance across land plants. Plant Physiol 179:1040–1049

Vietor R, Loutelier-Bourhis C, Fitchette A, Margerie P, Gonneau M, Faye L, Lerouge P (2003) Protein N-glycosylation is similar in the moss *Physcomitrella patens* and in higher plants. Planta 218:269–275

Wolf L, Rizzini L, Stracke R, Ulm R, Rensing SA (2010) The molecular and physiological responses of *Physcomitrella patens* to ultraviolet-B radiation. Plant Physiol 153:1123–1134

Yasumura Y, Crumpton-Taylor M, Fuentes S, Harberd NP (2007) Step-by-Step Acquisition of the Gibberellin-DELLA Growth-regulatory Mechanism during Land-Plant Evolution. Curr Biol 17:1225–1230

Chapter 22
Role of Endosymbionts in Nutritional Uptake of Sap Sucking Insects

Sanchita Singh, Surjeet Kumar Arya, Gurminder Kaur, Gauri Saxena and Praveen Chandra Verma

Abstract Insects are the most diverse and abundant group, comprising 95% of the animal clade. The bacteria domesticate the host and can potentially serve numerous purposes of the host. Insects of the suborder Sternorrhyncha of order Hemiptera feed on plant sap. These exhibit types of endosymbiosis, from obligate mutualism to facultative parasitism. Many beneficial services are given to the sap sucking host insect by bacterial endosymbionts, supply of nutrition, mainly the essential amino acids is one of them. Here we are discussing nutritional function of endosymbionts of hemipteran host.

Keywords Endosymbionst · Sap sucking insects · Hemiptera and nutrition

22.1 Introduction

All animals are colonized by microorganisms. This microbial colonization can have multiple habitats ranging from, external body tissue to blood to gut. A considerable population of microbes is harbored inside the animal body. These internally present microbes are called as endosymbionts. These symbionts can have mutualistic, commensalistic or parasitic associations depending upon the effect of their interactions

S. Singh · G. Saxena
Department of Botany, University of Lucknow, Lucknow, Uttar Pradesh 226001, India

S. Singh · S. K. Arya · P. C. Verma (✉)
Council of Scientific and Industrial Research—National Botanical Research Institute (CSIR-NBRI), Rana Pratap Marg, Lucknow, Uttar Pradesh 226001, India
e-mail: praveencverma@nbri.res.in

S. K. Arya · P. C. Verma
Academy of Scientific and Innovative Research (AcSIR), Ghaziabad 201002, India

G. Kaur
Faculty of Bioscience, Institute of Bioscience and Technology, Shri Ramswaroop Memorial University, Lucknow-Deva Road, Barabanki, Uttar Pradesh 225003, India

S. P. Singh et al. (eds.), *Molecular Approaches in Plant Biology and Environmental Challenges*, Energy, Environment, and Sustainability,
https://doi.org/10.1007/978-981-15-0690-1_22

with the host (O'Neill et al. 1997). A majority of animals show association with gut microbiome having different composition (Brooks et al. 2016; McFall-Ngai et al. 2013).

The class insecta is one of the most successful groups of the kingdom animalia. It contains approximately 95% of all the known animal species. In order to attain this success, insects adapted many characteristics like extremely efficient immune system, the obligatory association of each insect with a different prokaryotic endosymbionts and others. Insects have evolved obligate, mutualistic interactions with bacteria. This interaction is widely present in insects.

The order hemiptera commonly called "true bugs" comprises of numerous phytophagous sap sucking insect. The Sap sucking insects form an important group of phytophagous insects, they are also called as sap feeders as they feed on the sugary plant phloem sap. These insects have sucking mouthparts in which the maxillae and mandibles are slender bristle-like organs closed in a sheath formed by the labium. Thus the mouthpart forms a beak like structure which is inserted in the tissues and fluid is sucked.

In recent studies the insect order hemiptera has served as a very efficient system for the study of "insect-endosymbiont" interaction. The traits of endosymbionts that are useful to the host are defined as their functional traits. The two main functions of the endosymbionts of sap sucking insects are; those which are beneficial to the insect under specific ecological conditions and those which play a role in metabolic activities of the insect. The metabolic evolution of endosymbionts has taken place in such a way that they complement the natural diet of their host (Gosalbes et al. 2010). The endosymbionts of insects are classified into Primary and Secondary endosymbionts. The primary endosymbionts (PS) provide essential amino acids and have an obligatory relation with the host. These display a phylogenetic congruence with their host (Buchner 1965; Baumann 2005). On the other hand the secondary endosymbionts (SS) show a facultative relationship and possess a short evolutionary history with their host (Baumann 2005; Dale and Moran 2006). The various functional roles served by the secondary endosymbionts on their hosts are increasing tolerance to heat stress (Montllor et al. 2002), providing fitness benefits (Kaiser et al. 2010), causing host plant specialization (Tsuchida et al. 2004), increasing resistance to parasitic wasps (Oliver et al. 2003), conferring invasiveness (Feldhaar 2011). The secondary endosymbionts have also been reported to influence the host's ability to be a pest. There is ever-increasing evidence that variations in microbial association can alter host phenotype (Hamilton et al. 2016; Kohl et al. 2014; Sampson et al. 2016; Surana and Kasper 2017). Any sort of imbalance in this symbiotic association can directly affect the fitness of the host insect.

Most of the research focuses on insect-endosymbiont interactions, localization and identification of endosymbionts. In this chapter we highlight the reported nutritional roles of mutualistic endosymbionts exclusively on plant sap-feeding hemipterian host with a special emphasis on multi-partner endosymbiosis in hemipterans insects.

22.1.1 Effects of Diet on Insects Endosymbionts

The microbial community of insect gut is majorly structured by the diet of the insect. Many studies have been carried out comparing the effects of gut micro-biome of insects feeding on natural diet and artificial diet, and in some studies comparison have been done between the diets having varied composition of chief nutritional classes like sugar, lipid, fiber, protein (Dillon and Charnley 2002; Koch et al. 2012; Kuechler et al. 2011; Santo Domingo et al. 1998; Tang et al. 2012). The diet of an insect can affect its gut microbiota in two ways—directly or indirectly.

Direct effects—Any change in the composition of the food ingested by the insect can efficiently alter the population of microbiota associated with the food. Also, the environment of gut lumen favors only those microbial taxa which are most efficient in utilizing the food derived nutrients.
Indirect effects—In the insects where the gut microbiota is weakly overlapped by the food-associated microbiota, food has a significant impact on the digestion, immunity and anatomy of gut (Tang et al. 2012; Andert et al. 2010; Sudakaran et al. 2012).

These reports are important in paving a path towards the better understanding of the processes involved in the determination of direction and the scale of microbial response to diet.

22.1.2 Co-evolution and Genome Reduction—Evidences for Nutritional Function of Endosymbionts

The Insect host has a hold over nutrient availability and offers a thorough immunological examination, this creates unfriendly atmosphere for the colonizing microbes. In order to establish a successful symbiotic relation with the host the microbes show specific adaptations involving co-evolution with the host, genome reduction and a decrease in its capability to survive independent of the host (Moulder 1979; Peterson and Artis 2014; Ray et al. 2009; McCutcheon and Moran 2012). Before 2006, the cellular genomes reaching to 500 kb were considered as the smallest known genomes. In recent studies, genome size of symbiotic bacteria have been reported to have size ranging from 139 to 250 kb (Glass et al. 2006). In spite of this high reduction in the genome size, genes coding for enzymes which are responsible for biosynthesis of nutrients required by insect host are retained. The genome of typical PS of *B. aphidicola* is extremely reduced, but the genes responsible for provision of nutrients to the host are preserved (Shigenobu et al. 2000; Hansen and Moran 2011; Poliakov et al. 2011). Similarly *I. capsulate* despite of showing a high reduction in its genome has maintained all genes involved in biosynthetic pathways of EAA (Hosokawa et al. 2006; Nikoh et al. 2011).

Hansen and Moran (2011) their studied gene expression comparison of bacteriocytes and other body tissues of the aphid and revealed that bacteriocytes show an up-regulation of glutamate synthase and glutamine synthetase gene. This suggested an integration between gene expression of host and abilities of symbionts in bacteriocytes. In 2016 a study by Calle-Espinosa et al. (2016) has suggested that in a symbiotic association between *Candidatus* Portiera aleyrodidarum and whitefly, the production of key metabolites of the bacteria like, carotenoids and EAAs synthesis is coupled with energy generation. So, to say the host domesticate the symbiont and which in turn assigns a part of its gene stock to the processes which are of more advantageous to the host than to the symbiont itself.

22.2 Nutritional Function

Insects cannot synthesize many nutrients like essential amino acids (EAA), sterols, vitamin B and polysaccharides (in case insect is feeding on fiber rich diet) therefore these insects are associated with microbes which supply them with these nutrients and enhance the insects' capability to use unbalanced diet or diet with low nutritional value. This can be achieved by two ways namely, **Nutrient Provisioning and Nutritional Symbiosis**. Insects accommodate endosymbionts within bacteriocytes which directly provide amino acids and cofactors to the insect host. This is called nutrient provisioning. In nutrient symbiosis gut bacteria is digested and utilized as source of nutrients. These are also known as nutritional bacteria. The high throughput sequencing technologies has transformed our understanding of nutritional communications in insect–microbial symbioses. This enables us to deduce the metabolic ability of the partners from the transcriptome or genome (Hansen and Moran 2011).

22.2.1 Essential Amino Acids (EAAs)

The universal prevalence of symbioses in plant sap-feeding insects and the preservation of the genes involved in biosynthesis of EAAs in all the tested symbionts indicated towards the role of microbes in the EAA provisioning in these insects (McCutcheon and Moran 2012). In plant phloem sap the free amino acids are the primary source of nitrogen. Phloem sap has an unbalanced composition (less than 20%) of EAAs (Douglas 2006). There are 10 amino acids having contribution in biosynthesis of insect proteins which are not synthesized by the insects de novo. The plant phloem sap diet of insects lacks these EAA. These 10 amino acids are provided to the insect host by their endosymbionts. The significance of symbionts in providing EAA has been studied intensively in sap sucking hemipterans. Few examples of endosymbionts involved in EAAs provision to hemipteran insects are as follows: *Ishikawaella capsulatus* in plataspid stinkbugs, *Buchnera* in *Acyrthosiphon pisum* and in *Myzus persicae* (Sulzer) and *Candidatus Tremblaya* in mealybugs. There are symbiotic associations where the primary symbiont (PS) is the only endosymbiont and in such cases the all the 10 EAAs are provided by the PS, whereas in case of multiple endosymbiotic partners there is a clear division in the production of EAAs between the PS and the secondary symbionts (SS). The long term co-evolution interaction between the co-existing microbes and host, moulds the phenotypic traits of the endosymbionts. In case of multi-partner endosymbiosis nutritional function is sustained by showing partitioning in the genetic capability between different endosymbionts for metabolic function. Below mentioned are examples of three EAAs whose synthesis is shared by the endosymbionts.

Methionine biosynthetic pathways in the auchenorrhynchan symbioses is shared by different endosymbionts. The initial reactions of methionine biosynthesis are mediated by *Ca.* Sulcia muelleri (PS). These reactions are common to lysine and threonine biosynthesis also. The later reactions are mediated by companion symbionts also known as secondary symbionts (SS), these reactions are specifically a part of methionine biosynthesis (Bennett and Moran 2013; McCutcheon et al. 2009; McCutcheon and Moran 2010) (Table 22.1).

Whereas the biosynthesis of **arginine** in multi-partner endosymbiosis can take place in diverse ways such as

(a) exclusively in the PS for example in some sternorrhyncha and all auchenorrhyncha
(b) divided between the symbionts for example, in the nested '*Ca.* Tremblaya princeps'–'*Ca.* Moranella endobia' or
(c) replicated between the two symbionts for example, in the psyllid *Ctenarytaina eucalypti* (McCutcheon and Von Dohlen 2011; Sloan and Moran 2012) (Table 22.1).

There are various ways are marked for the synthesis of **tryptophan** also (McCutcheon et al. 2009; McCutcheon and Von Dohlen 2011; Sloan and Moran 2012; Lamelas et al. 2011) (Table 22.1). For example;

(a) In Spittlebug—the PS *Ca.* Sulcia muelleri is accountable for all the reactions used for phenylalanine and tryptophan synthesis. Here the SS mediates the reactions responsible only for tryptophan synthesis.
(b) In the aphid *Cinara cedri*—in addition to the shikimate pathway the *B. aphidicola* (PS) is dedicated to the initial reaction of tryptophan synthesis.
(c) In mealybug *P. citri*, the shikimate pathway is partitioned between the PS '*Ca.* Tremblaya princeps' and the SS '*Ca.* Moranella endobia'.

22.2.2 Sterols

Sterols are important component of membranes of eukaryotes and also play a role in animal hormone biosynthesis (Behmer and Nes 2003; Espenshade and Hughes 2007). Different plant species have different composition and the amount of phytosterols. This character is of specific importance to phytophagous feeding because of two main reasons first, insects cannot synthesize sterols on their own and thus rely on

Table 22.1 Essential amino acid biosynthesis pathways shared between endosymbionts

Host insect	EAA supplied to host	Primary symbiont		Secondary symbiont		References
		Name	Function	Name	Function	
Sharpshooters	Methionine	*Candidatus* Sulcia muelleri	Convert aspartate to homoserine	*Candidatus* Baumannia cicadellinicola	Convert homoserine to methionine	(Bennett and Moran 2013; McCutcheon et al. 2009; McCutcheon and Moran 2010)
Leafhoppers	Methionine	*Candidatus* Sulcia muelleri	Convert aspartate to homoserine	*Candidatus* Nasuia deltocephalinicola	Convert homoserine to methionine	(Bennett and Moran 2013; McCutcheon et al. 2009; McCutcheon and Moran 2010)
Cicadas	Methionine	*Candidatus* Sulcia muelleri	Convert aspartate to homoserine	*Candidatus* Hodgkinia cicadicola	Convert homoserine to methionine	(Bennett and Moran 2013; McCutcheon et al. 2009; McCutcheon and Moran 2010)
Spittlebugs	Methionine	*Candidatus* Sulcia muelleri	Convert aspartate to homoserine	*Candidatus* Zinderia insecticola	Convert homoserine to methionine	(Bennett and Moran 2013; McCutcheon et al. 2009; McCutcheon and Moran 2010)
Mealybug	Arginine	*Candidatus* Tremblaya princeps	Convert citrulline to arginosuccinate	*Candidatus* Moranella endobia	Convert ornithine to citrulline and convert arginosuccinate to arginine	(McCutcheon and Von Dohlen 2011; Sloan and Moran 2012)

(continued)

Table 22.1 (continued)

Host insect	EAA supplied to host	Primary symbiont		Secondary symbiont		References
		Name	Function	Name	Function	
Psyllid	Arginine	*Candidatus* Carsonella ruddii	Convert citrulline to arginine	*Candidatus* Moranella-like	Converts ornithine to arginine	(McCutcheon and Von Dohlen 2011; Sloan and Moran 2012)
Spittlebug	Tryptophan	*Candidatus* Sulcia muelleri	Convert phosphoenol-pyruvate (PEP) and erythrose 4-phosphate (E4P) to chorismate	*Candidatus* Zinderia insecticola	Convert chorismate to tryptophan	(McCutcheon et al. 2009; McCutcheon and Von Dohlen 2011; Sloan and Moran 2012; Lamelas et al. 2011)
Aphid	Tryptophan	*Buchnera aphidicola*	Convert PEP and E4P to anthranilate	*Serratia symbiotica*	Convert anthranilate to tryptophan	(McCutcheon et al. 2009; McCutcheon and Von Dohlen 2011; Sloan and Moran 2012; Lamelas et al. 2011)
Mealybug	Tryptophan	*Candidatus* Tremblaya princeps	Convert PEP and E4P to dehydroquinate and 3-convert dehydroshikimate to chorismate	*Candidatus* Tremblaya princeps	Convert dehydroquinate to shikimate and convert chorismate to tryptophan	(McCutcheon et al. 2009; McCutcheon and Von Dohlen 2011; Sloan and Moran 2012; Lamelas et al. 2011)

exogenous source for the same; second different insects use phytosterols differently (Behmer and Nes 2003; Behmer and Elias 1999). Certain beetles and planthoppers obtain sterols from their fungal endosymbionts (Janson 2009); sterol synthesis is not done by bacterial symbionts (Douglas 2009). The sterol profile of planthoppers (Hemiptera) and *Anobiidae* (Coleoptera) possessing yeast symbionts showed the presence of ergosterol (fungal sterol) and similar compounds. This indicated that insect sterol has a fungal source (Nasir and Noda 2003; Noda and Koizumi 2003).

22.2.3 B Vitamins

Insects are incapable of synthesizing eight B vitamins which act as co-enzymes in numerous enzymatic reactions. Majority of insects fulfill their requirements for B vitamin through exogenous sources like diet, symbiotic microorganisms or by a mixture of both. Studies involving B vitamin deficient biomarkers provide a new insight for investigation of effect of B vitamin deficiency on insect fitness and physiology (Douglas 2017). The central idea of endosymbiotic association between *Wolbachia* and the *Cimex lectularius* (Bed bug) is obligate mutualism. Studies have suggested *C. lectularius* is provided with B vitamins in exchange of accommodation and transmission as an inherited component of oocyte (Hosokawa et al. 2010; Nikoh et al. 2014).

22.3 Biogeochemical Cycles and Endosymbionts

The ecosystems depend vitally on the insects and their gut microbial colonies because they act as mediators in the biogeochemical carbon cycle, in plant biomass decomposition (Bignell et al. 1997; Fierer et al. 2009), nitrogen cycle and nitrogen fixation (Fox-Dobbs et al. 2010). Raven (1983) reported that plant phloem and xylem sap are poor in nutrient and this leads to unbalanced diets of sap feeders. It is widely accepted that the insect symbionts posses the biosynthetic abilities to complement the diet of their host and this plays a critical role in flourishing of sap feeding insects (Buchner 1965; Douglas 1995).

22.3.1 Tripartite Coalition of Hemipterians, Symbionts and Nitrogen

Nitrogen majorly determines the increase in population and the abundance of plant sap-sucking insects (McNeill 1977). The phytophagous insects depend on mutualistic bacteria having a devoted nitrogen metabolism which balances the host's requirement for nitrogen (Cook and Davidson 2006). Nitrogen recycling is a process in which the symbiotic microbes transform the nitrogenous waste of the animals into nitrogenous compounds like essential amino acids which have nutritional value to the animals. In 2014, Wilkinson and Douglas (Whitehead et al. 1992) suggested two evidences that the bacteria in the intact symbiotic association with the pea aphid (*Acyrthosiphon pisum*) posses a transport system for the externally provided ammonium, also these bacteria use the ammonia derived from aphids. To investigate the same two strategies were followed, one was to explore the ability of the bacteria for the exogenous up take of the ammonia and second one involved the comparison of the concentration of ammonia in the honey dew of the aphids with and without symbiotic bacteria. Further, studies showed on altering the concentrations of the dietary nitrogen (45–270 mM amino acids) of *A. pisum* the responses were noted and the influence of antibiotic chlortetracycline on these responses was observed. This study pointed out that the antibiotic chlortetracycline as well as the altered levels of nitrogen in the diet causes a selective disruption in the aphid-bacteria symbiosis (Prosser et al. 1992). Reports have shown the shield bug, *Parastrachia japonensis* (Hemiptera: Parastrachiidae) colonizing Erwinia-like bacteria which recycles the uric acid using enzymes like uricase, allantoinase and allantoicase. This contribute to long term survival of the insects on water (Kashima et al. 2006). *N. lugens* does not excrete uric acid, here all the uric acid is re-utilised by yeast-like symbionts harboring in the fat body (Sasaki et al. 1996; Hongoh and Ishikawa 1997; Hongoh et al. 2000).

22.4 Concluding Remarks

The factors affecting the host-symbiont relationship have an influence on the roles played by the symbionts and the traits of the insect. Microbial interactions play a significant role in insect physiology and thus there is a critical need of expanding our knowledge on the subject. The partnership between Hemiptera and endosymbionts serves as rich biological system in studying the inter-relationship between the two. Advancement in functional genomics and molecular biology has revolutionized our knowledge regarding the nutritional role of insect endosymbionts. Owing to the enormous diversity in ecology and taxonomy of phloem sap feeding insects it is hard to generalize the microbes harbored by them. A very common trait which is observed in majority of the hemipterans is the nutritional dependency of the insect on their intracellular symbionts. An integration of the gene products of both the host and that of the symbionts is responsible for this nutritional reliance. For better admiration of this network a detailed understanding of the interaction is required. Further, the knowledge attained from this chapter may be exploited from pest management strategies and the developing knowledge can also be integrated with immune function of symbionts for a wider insect-biology study. This chapter provides an affluent source to understand the nutritional function of endosymbioses in Hemipteran insects.

References

Andert J, Marten A, Brandl R, Brune A (2010) Inter- and intraspecific comparison of the bacterial assemblages in the hindgut of humivorous scarab beetle larvae (Pachnoda spp.). FEMS Microbiol Ecol 74(2):439–449. https://doi.org/10.1111/j.1574-6941.2010.00950.x

Baumann P (2005) Biology of bacteriocyte-associated endosymbionts of plant sap-sucking insects. Annu Rev Microbiol 59:155–189. https://doi.org/10.1146/annurev.micro.59.030804.121041

Behmer ST, Elias DO (1999) The nutritional significance of sterol metabolic constraints in the generalist grasshopper *Schistocerca americana*. J Insect Physiol 45(4):339–348. https://doi.org/10.1016/s0022-1910(98)00131-0

Behmer ST, Nes WD (2003) Insect sterol nutrition and physiology: a global overview. Adv Insect Physiol 31:1–72. https://doi.org/10.1016/s0065-2806(03)31001-x

Bennett GM, Moran NA (2013) Small, smaller, smallest: the origins and evolution of ancient dual symbioses in a phloem-feeding insect. Genome Biol Evol 5(9):1675–1688. https://doi.org/10.1093/gbe/evt118

Bignell DE, Eggleton P, Nunes L, Thomas KL (1997) Termites as mediators of carbon fluxes in tropical forest: budgets for carbon dioxide and methane emissions. Forests Insects 109–134

Brooks AW, Kohl KD, Brucker RM, van Opstal EJ, Bordenstein SR (2016) Phylosymbiosis: relationships and functional effects of microbial communities across host evolutionary history. PLoS Biol 14(11):e2000225. https://doi.org/10.1371/journal.pbio.2000225

Buchner P (1965) Endosymbiosis of animals and plant microorganisms. Wiley, New York, USA

Calle-Espinosa J, Ponce-de-Leon M, Santos-Garcia D, Silva FJ, Montero F, Peretó J (2016) Nature lessons: the whitefly bacterial endosymbiont is a minimal amino acid factory with unusual energetics. J Theor Biol 407:303–317. https://doi.org/10.1016/j.jtbi.2016.07.024

Cook SC, Davidson DW (2006) Nutritional and functional biology of exudate-feeding ants. Entomol Exp Appl 118(1):1–10. https://doi.org/10.1111/j.1570-7458.2006.00374.x

Dale C, Moran NA (2006) Molecular interactions between bacterial symbionts and their hosts. Cell 126(3):453–465. https://doi.org/10.1016/j.cell.2006.07.014

Dillon R, Charnley K (2002) Mutualism between the desert locust *Schistocerca gregaria* and its gut microbiota. Res Microbiol 153(8):503–509

Douglas AE (1995) The ecology of symbiotic micro-organisms. In: Advances in ecological research, vol 26. Academic Press, pp 69–103

Douglas AE (2006) Phloem-sap feeding by animals: problems and solutions. J Exp Bot 57(4):747–754

Douglas AE (2009) The microbial dimension in insect nutritional ecology. Funct Ecol 23(1):38–47

Douglas AE (2017) The B vitamin nutrition of insects: the contributions of diet, microbiome and horizontally acquired genes. Curr Opin Insect Sci 23:65–69. https://doi.org/10.1016/j.cois.2017.07.012

Espenshade PJ, Hughes AL (2007) Regulation of sterol synthesis in eukaryotes. Annu Rev Genet 41:401–427. https://doi.org/10.1146/annurev.genet.41.110306.130

Feldhaar H (2011) Bacterial symbionts as mediators of ecologically important traits of insect hosts. Ecol Entomol 36(5):533–543. https://doi.org/10.1007/s00248-016-0829-2

Fierer N, Strickland MS, Liptzin D, Bradford MA, Cleveland CC (2009) Global patterns in belowground communities. Ecol Lett 12(11):1238–1249. https://doi.org/10.1111/j.1461-0248.2009.01360.x

Fox-Dobbs K, Doak DF, Brody AK, Palmer TM (2010) Termites create spatial structure and govern ecosystem function by affecting N2 fixation in an East African savanna. Ecology 91(5):1296–1307. https://doi.org/10.1890/09-0653.1

Glass JI, Assad-Garcia N, Alperovich N, Yooseph S, Lewis MR, Maruf M, Hutchison CA, Smith HO, Venter JC (2006) Essential genes of a minimal bacterium. Proc Natl Acad Sci 103(2):425–430. https://doi.org/10.1073/pnas.0510013103

Gosalbes MJ, Latorre A, Lamelas A, Moya A (2010) Genomics of intracellular symbionts in insects. Int J Med Microbiol 300(5):271–278. https://doi.org/10.1016/j.ijmm.2009.12.001

Hamilton PT, Peng F, Boulanger MJ, Perlman SJ (2016) A ribosome-inactivating protein in a Drosophila defensive symbiont. Proc Natl Acad Sci 113(2):350–355. https://doi.org/10.1073/pnas.1518648113

Hansen AK, Moran NA (2011) Aphid genome expression reveals host–symbiont cooperation in the production of amino acids. Proc Natl Acad Sci 108(7):2849–2854. https://doi.org/10.1073/pnas.1013465108

Hongoh Y, Ishikawa H (1997) Uric acid as a nitrogen resource for the brown planthopper, *Nilaparvata lugens*: studies with synthetic diets and aposymbiotic insects. Zool Sci 14(4):581–587

Hongoh Y, Sasaki T, Ishikawa H (2000) Cloning, sequence analysis and expression in *Escherichia coli* of the gene encoding a uricase from the yeast-like symbiont of the brown planthopper, *Nilaparvata lugens*. Insect Biochem Mol Biol 30(2):173–182

Hosokawa T, Kikuchi Y, Nikoh N, Shimada M, Fukatsu T (2006) Strict host-symbiont cospeciation and reductive genome evolution in insect gut bacteria. PLoS Biol 4(10):e337. https://doi.org/10.1371/journal.pbio.0040337

Hosokawa T, Koga R, Kikuchi Y, Meng XY, Fukatsu T (2010) Wolbachia as a bacteriocyte-associated nutritional mutualist. Proc Natl Acad Sci 107(2):769–774. https://doi.org/10.1073/pnas.0911476107

Janson EM, Grebenok RJ, Behmer ST, Abbot P (2009) Same host-plant, different sterols: variation in sterol metabolism in an insect herbivore community. J Chem Ecol 35(11):1309. https://doi.org/10.1007/s10886-009-9713-6

Kaiser W, Huguet E, Casas J, Commin C, Giron D (2010) Plant green-island phenotype induced by leaf-miners is mediated by bacterial symbionts. Proc Roy Soc B: Biol Sci 277(1692):2311–2319. https://doi.org/10.1098/rspb.2010.0214

Kashima T, Nakamura T, Tojo S (2006) Uric acid recycling in the shield bug, *Parastrachia japonensis* (Hemiptera: Parastrachiidae), during diapause. J Insect Physiol 52(8):816–825. https://doi.org/10.1016/j.jinsphys.2006.05.003

Koch H, Cisarovsky G, Schmid-Hempel P (2012) Ecological effects on gut bacterial communities in wild bumblebee colonies. J Anim Ecol 81(6):1202–1210. https://doi.org/10.1111/j.1365-2656.2012.02004.x

Kohl KD, Weiss RB, Cox J, Dale C, Denise Dearing M (2014) Gut microbes of mammalian herbivores facilitate intake of plant toxins. Ecol Lett 17(10):1238–1246. https://doi.org/10.1111/ele.12329

Kuechler SM, Dettner K, Kehl S (2011) Characterization of an obligate intracellular bacterium in the midgut epithelium of the bulrush bug *Chilacis typhae* (Heteroptera, Lygaeidae, Artheneinae). Appl Environ Microbiol 77(9):2869–2876. https://doi.org/10.1128/AEM.02983-10

Lamelas A, Gosalbes MJ, Moya A, Latorre A (2011) New clues about the evolutionary history of metabolic losses in bacterial endosymbionts, provided by the genome of *Buchnera aphidicola* from the aphid *Cinara tujafilina*. Appl Environ Microbiol 77(13):4446–4454. https://doi.org/10.1128/AEM.00141-11

McCutcheon JP, Moran NA (2010) Functional convergence in reduced genomes of bacterial symbionts spanning 200 My of evolution. Genome Biol Evol 2:708–718. https://doi.org/10.1093/gbe/evq055

McCutcheon JP, Moran NA (2012) Extreme genome reduction in symbiotic bacteria. Nat Rev Microbiol 10(1):13. https://doi.org/10.1038/nrmicro2670

McCutcheon JP, Von Dohlen CD (2011) An interdependent metabolic patchwork in the nested symbiosis of mealybugs. Curr Biol 21(16):1366–1372. https://doi.org/10.1016/j.cub.2011.06.051

McCutcheon JP, McDonald BR, Moran NA (2009) Convergent evolution of metabolic roles in bacterial co-symbionts of insects. Proc Natl Acad Sci 106(36):15394–15399. https://doi.org/10.1073/pnas.0906424106

McFall-Ngai M, Hadfield MG, Bosch TC, Carey HV, Domazet-Lošo T, Douglas AE, Dubilier N, Eberl G, Fukami T, Gilbert SF, Hentschel U (2013) Animals in a bacterial world, a new imperative for the life sciences. Proc Natl Acad Sci 110(9):3229–3236. https://doi.org/10.1073/pnas.1218525110

McNeill S (1977) The role of nitrogen in the development of insect-plant relationships. Biochemical aspects of plant and animal coevolution, pp 77–98

Montllor CB, Maxmen A, Purcell AH (2002) Facultative bacterial endosymbionts benefit pea aphids *Acyrthosiphon pisum* under heat stress. Ecol Entomol 27(2):189–195

Moulder JW (1979) The cell as an extreme environment. Proc Roy Soc Lond Ser B: Biol Sci 204(1155):199–210

Nasir H, Noda H (2003) Yeast-like symbiotes as a sterol source in anobiid beetles (Coleoptera, Anobiidae): possible metabolic pathways from fungal sterols to 7-dehydrocholesterol. Arch Insect Biochem Physiol 52(4):175–182. https://doi.org/10.1002/arch.10079

Nikoh N, Hosokawa T, Oshima K, Hattori M, Fukatsu T (2011) Reductive evolution of bacterial genome in insect gut environment. Genome Biol Evol 3:702–714. https://doi.org/10.1093/gbe/evr064

Nikoh N, Hosokawa T, Moriyama M, Oshima K, Hattori M, Fukatsu T (2014) Evolutionary origin of insect–Wolbachia nutritional mutualism. Proc Natl Acad Sci 111(28):10257–10262. https://doi.org/10.1073/pnas.1409284111

Noda H, Koizumi Y (2003) Sterol biosynthesis by symbiotes: cytochrome P450 sterol C-22 desaturase genes from yeastlike symbiotes of rice planthoppers and anobiid beetles. Insect Biochem Mol Biol 33(6):649–658. https://doi.org/10.1016/S0965-1748(03)00056-0

O'Neill SL, Werren JH, Hoffmann AA (1997) Influential passengers: inherited microorganisms and arthropod reproduction

Oliver KM, Russell JA, Moran NA, Hunter MS (2003) Facultative bacterial symbionts in aphids confer resistance to parasitic wasps. Proc Natl Acad Sci 100(4):1803–1807. https://doi.org/10.1073/pnas.0335320100

Peterson LW, Artis D (2014) Intestinal epithelial cells: regulators of barrier function and immune homeostasis. Nat Rev Immunol 14(3):141. https://doi.org/10.1038/nri3608

Poliakov A, Russell CW, Ponnala L, Hoops HJ, Sun Q, Douglas AE, van Wijk KJ (2011) Large-scale label-free quantitative proteomics of the pea aphid-Buchnera symbiosis. Mol Cell Proteomics 10(6):M110–007039. https://doi.org/10.1074/mcp.m110.007039

Prosser WA, Simpson SJ, Douglas AE (1992) How an aphid (*Acyrthosiphon pisum*) symbiosis responds to variation in dietary nitrogen. J Insect Physiol 38(4):301–307. https://doi.org/10.1016/0022-1910(92)90130-6

Raven JA (1983) Phytophages of xylem and phloem: a comparison of animal and plant sap-feeders. In: Advances in ecological research, vol 13. Academic Press, pp 135–234. https://doi.org/10.1016/s0065-2504(08)60109-9

Ray K, Marteyn B, Sansonetti PJ, Tang CM (2009) Life on the inside: the intracellular lifestyle of cytosolic bacteria. Nat Rev Microbiol 7(5):333. https://doi.org/10.1038/nrmicro2112

Sampson TR, Debelius JW, Thron T, Janssen S, Shastri GG, Ilhan ZE, Challis C, Schretter CE, Rocha S, Gradinaru V, Chesselet MF (2016) Gut microbiota regulate motor deficits and neuroinflammation in a model of Parkinson's disease. Cell 167(6):1469–1480. https://doi.org/10.1016/j.cell.2016.11.018

Santo Domingo JW, Kaufman MG, Klug MJ, Holben WE, Harris D, Tiedje JM (1998) Influence of diet on the structure and function of the bacterial hindgut community of crickets. Mol Ecol 7(6):761–767. https://doi.org/10.1007/s00253-005-0215-3

Sasaki T, Kawamura M, Ishikawa H (1996) Nitrogen recycling in the brown planthopper, *Nilaparvata lugens*: involvement of yeast-like endosymbionts in uric acid metabolism. J Insect Physiol 42(2):125–129. https://doi.org/10.1016/0022-1910(95)00086-0

Shigenobu S, Watanabe H, Hattori M, Sakaki Y, Ishikawa H (2000) Genome sequence of the endocellular bacterial symbiont of aphids Buchnera sp. APS. Nature 407(6800):81. https://doi.org/10.1038/35024074

Sloan DB, Moran NA (2012) Genome reduction and co-evolution between the primary and secondary bacterial symbionts of psyllids. Mol Biol Evol 29(12):3781–3792. https://doi.org/10.1093/molbev/mss180

Sudakaran S, Salem H, Kost C, Kaltenpoth M (2012) Geographical and ecological stability of the symbiotic mid-gut microbiota in European firebugs, *Pyrrhocoris apterus* (Hemiptera, Pyrrhocoridae). Mol Ecol 21(24):6134–6151. https://doi.org/10.1111/mec.12027
Surana NK, Kasper DL (2017) Moving beyond microbiome-wide associations to causal microbe identification. Nature 552(7684):244. https://doi.org/10.1038/nature25019
Tang X, Freitak D, Vogel H, Ping L, Shao Y, Cordero EA, Andersen G, Westermann M, Heckel DG, Boland W (2012) Complexity and variability of gut commensal microbiota in polyphagous lepidopteran larvae. PLoS ONE 7(7):e36978. https://doi.org/10.1371/journal.pone.0036978
Tsuchida T, Koga R, Fukatsu T (2004) Host plant specialization governed by facultative symbiont. Science 303(5666):1989. https://doi.org/10.1126/science.1094611
Whitehead LF, Wilkinson TL, Douglas AE (1992) Nitrogen recycling in the pea aphid (*Acyrthosiphon pisum*) symbiosis. Proc Roy Soc Lond Ser B: Biol Sci 250(1328):115–117. https://doi.org/10.1098/rspb.1992.0138

The manufacturer's authorised representative in the EU is Springer Nature Customer Service Centre GmbH, Europaplatz 3, 69115 Heidelberg, Germany. If you have any concerns regarding our products, please contact ProductSafety@springernature.com

Printed and bound by CPI Group (UK) Ltd, Croydon, CR0 4YY
30/09/2026
02225452-0019